T0226522

ENGINEERING TRIBOLOGY

ENGINEERING TRIBOLOGY

THIRD EDITION

Gwidon W. Stachowiak

*School of Mechanical Engineering,
University of Western Australia,
Perth, Australia*

Andrew W. Batchelor

*School of Engineering,
Monash University Malaysia,
Bandar Sunway, Selangor, Malaysia*

ELSEVIER
BUTTERWORTH
HEINEMANN

Amsterdam • Boston • Heidelberg • London
New York • Oxford • Paris • San Diego
San Francisco • Singapore • Sydney • Tokyo

Elsevier Butterworth–Heinemann

30 Corporate Drive, Suite 400, Burlington, MA 01803, USA

Linacre House, Jordan Hill, Oxford OX2 8DP, UK

Copyright © 2005, Elsevier Inc. All rights reserved.

No part of this publication may be reproduced, stored in a retrieval system, or transmitted in any form or by any means, electronic, mechanical, photocopying, recording, or otherwise, without the prior written permission of the publisher.

Permissions may be sought directly from Elsevier's Science & Technology Rights Department in Oxford, UK: phone: (+44) 1865 843830, fax: (+44) 1865 853333, e-mail: permissions@elsevier.co.uk. You may also complete your request on-line via the Elsevier homepage (http://elsevier.com), by selecting "Customer Support" and then "Obtaining Permissions."

Recognizing the importance of preserving what has been written, Elsevier prints its books on acid-free paper whenever possible.

Library of Congress Cataloging-in-Publication Data

Stachowiak, G. W. (Gwidon W.)

Engineering tribology / Gwidon W. Stachowiak, Andrew W. Batchelor.-- 3rd ed.

   p. cm.

Includes bibliographical references and index.

ISBN 0-7506-7836-4 (casebound : alk. paper) 1. Tribology. I. Batchelor, A. W. (Andrew W.) II. Title.

TJ1075.S78 2005

621.8'9--dc22

                      2005014320

British Library Cataloguing-in-Publication Data

A catalogue record for this book is available from the British Library.

ISBN-13: 978-0-7506-7836-0

ISBN-10: 0-7506-7836-4

For information on all Elsevier Butterworth–Heinemann publications visit our Web site at www.books.elsevier.com

Printed and bound by CPI Group (UK) Ltd, Croydon, CR0 4YY

Transferred to Digital Print 2011

**Working together to grow
libraries in developing countries**

www.elsevier.com | www.bookaid.org | www.sabre.org

ELSEVIER   BOOK AID
             International   Sabre Foundation

To the most important persons in our lives
Grazyna Stachowiak
Gwidon (Jr.) Stachowiak
and
Valli M. Batchelor
Diana, Vicky & Vincent Batchelor

CONTENTS

PREFACE

Several years ago, the idea arose to write a general book on tribology. Students often requested a suitable book for the study of tribology and there were problems in recommending any one textbook. Existing textbooks were either too specialized or too literal. Many books provided exhaustive reviews of friction and wear data while others provided detailed description of the lubrication and wear problems occurring in machinery. A book which explains the concepts of tribology in terms useful to engineering students and engineers was, however, lacking. In many cases the basic models of friction and wear were not explained adequately. As a result more sophisticated concepts could not be understood. The interdisciplinary nature of tribology with knowledge drawn from different disciplines such as mechanical engineering, materials science, chemistry and physics leads to a general tendency for the chemist to describe in detail, for example, lubricant additives, the mechanical engineer to discuss, for example, pad bearings and so on, with no overall guide to the subject. In this book, the interaction between these different fields of knowledge to achieve the final result, the control of friction and wear, is emphasized. The interdisciplinary view of tribology was largely developed by Professor Alastair Cameron about three decades ago and has proved to be the most successful way of analysing friction and wear problems.

In many cases tribology is viewed as an inaccessible subject which does not produce useful answers. In this book we try to redress this problem. Rutherford's maxim, that 'any good scientific theory is explainable to the average barmaid', is applied in this book with various concepts explained in the simplest possible terms with supporting illustrations.

In this third edition of 'Engineering Tribology' we aim to update the contents of the second edition while maintaining its style. In this edition a number of extra topics have been included to make the book more comprehensive. The listings of literature citations have been extended to include recent findings from tribology research. Extra diagrams have also been included where it was found that the readability of the original text could be improved. Computer programs used in the numerical analysis have been upgraded to allow for the friction stress and friction coefficient calculations. A new chapter on 'Current Trends in Tribology' has been written to conclude the book. To provide opportunities for active learning, a series of revision questions are provided at the end of each chapter. This should be helpful in assessing the understanding for both students and lecturers. Despite all these changes, the purpose of writing 'Engineering Tribology' remains the same, i.e., to provide a reader-friendly and comprehensive introduction to the subject of tribology and its implications for engineering. This edition, like the previous editions, is intended for final year under-graduate and post-graduate students and professional engineers. The subject matter of the book is also relevant to mechanical and materials engineering, applied chemistry, physics and biomedical courses.

Gwidon W. Stachowiak

Andrew W. Batchelor

ACKNOWLEDGEMENTS

Any book depends on the efforts of many different people and this book is no exception. Firstly, we would like to thank Professor Duncan Dowson for his personal input, enthusiasm, encouragement and meticulous checking of the manuscript and very many constructive comments and remarks. We would also like to thank Ms Grazyna Stachowiak for very detailed research, review of technical material, proof-reading, many constructive discussions, SEM micrographs and preparation of index; Dr Pawel Podsiadlo for his help in converting the computer programs into Matlab, useful discussions on wavelets; Dr Karl Stoffel for useful discussions on orthopaedic implants; Professor Nic Spencer for helpful comments on boundary lubrication; Professor Hugh Spikes for help and ideas with the revision questions; Professor Stephen Hsu for useful discussions on future trends in tribology; Professor Koji Kato for helpful discussions and providing a friendly environment at Tohoku University during the final stages of manuscript preparation; Gosia Wlodarczak-Sarnecka for the design of the book cover; Longin Sarnecki for the cover photo; Dr Philippa O'Neill for thorough checking of some of the chapters; and Dr Nathan Scott for the preparation of the illustrations. Without Nathan's illustrations the book would be diminished in terms of readability and aesthetics. We also would like to thank the Library of the University of Western Australia for their help in finding all those references and the School of Mechanical Engineering, University of Western Australia, for its help during the preparation of the manuscript. The support of the School of Engineering, Monash University Malaysia, is also gratefully acknowledged.

Finally, we would like to thank the following publishers for granting us permission to reproduce the figures listed below:

Figure 9.7: Society of Tribologists and Lubrication Engineers. From Tribology Transactions, Vol. 31, 1988, pp. 214-227.

Figures 13.5 and 13.11: Japanese Society of Tribologists. From Journal of Japan Society of Lubrication Engineers, Vol. 31, 1986, pp. 883-888 and Vol. 28, 1983, pp. 53-56, respectively.

Figures 14.2 and 15.2: Royal Society of London. From Proceedings of the Royal Society of London, Vol. 394, 1984, pp. 161-181 and Vol. 230, 1955, pp. 531-548, respectively.

Figure 16.6: The American Society of Mechanical Engineers. From Transactions of the ASME, Journal of Lubrication Technology, Vol. 101, 1979, pp. 212-219.

Figures 11.41 and 16.22 were previously published in Wear, Vol. 113, 1986, pp. 305-322 and Vol. 17, 1971, pp. 301-312, respectively.

1   I N T R O D U C T I O N

1.1   BACKGROUND

Tribology in a traditional form has been in existence since the beginning of recorded history. There are many well documented examples of how early civilizations developed bearings and low friction surfaces [1]. The scientific study of tribology also has a long history, and many of the basic laws of friction, such as the proportionality between normal force and limiting friction force, are thought to have been developed by Leonardo da Vinci in the late 15th century. However, the understanding of friction and wear languished in the doldrums for several centuries with only fanciful concepts to explain the underlying mechanisms. For example, it was proposed by Amonton in 1699 that surfaces were covered by small spheres and that the friction coefficient was a result of the angle of contact between spheres of contacting surfaces. A reasonable value of friction coefficient close to 0.3 was therefore found by assuming that motion was always to the top of the spheres. The relatively low priority of tribology at that time meant that nobody really bothered to question what would happen when motion between the spheres was in a downwards direction. Unlike thermodynamics, where fallacious concepts like 'phlogiston' were rapidly disproved by energetic researchers such as Lavoisier in the late 18th century, relatively little understanding of tribology was gained until 1886 with the publication of Osborne Reynolds' classical paper on hydrodynamic lubrication. Reynolds proved that hydrodynamic pressure of liquid entrained between sliding surfaces was sufficient to prevent contact between surfaces even at very low sliding speeds. His research had immediate practical application and led to the removal of an oil hole from the load line of railway axle bearings. The oil, instead of being drained away by the hole, was now able to generate a hydrodynamic film and much lower friction resulted. The work of Reynolds initiated countless other research efforts aimed at improving the interaction between two contacting surfaces, and which continue to this day. As a result journal bearings are now designed to high levels of sophistication. Wear and the fundamentals of friction are far more complex problems, the experimental investigation of which is dependent on advanced instrumentation such as scanning electron microscopy and atomic force microscopy. Therefore, it has only recently been possible to study these processes on a microscopic scale where a true understanding of their nature can be found.

Tribology is therefore a very new field of science, most of the knowledge being gained after the Second World War. In comparison many basic engineering subjects, e.g., thermodynamics, mechanics and plasticity, are relatively old and well established. Tribology is still in an imperfect state and subject to some controversy which has impeded the diffusion of information to technologists in general. The need for information is nevertheless critical; even simple facts such as the type of lubricant that can be used in a particular application, or preventing the contamination of oil by water must be fully understood by an engineer. Therefore this book is devoted to these fundamental engineering tribology principles.

1.2 MEANING OF TRIBOLOGY

Tribology, which focuses on friction, wear and lubrication of interacting surfaces in relative motion, is a new field of science defined in 1967 by a committee of the Organization for Economic Cooperation and Development. '**Tribology**' is derived from the Greek word '**tribos**' meaning rubbing or sliding. After an initial period of scepticism, as is inevitable for any newly introduced word or concept, the word 'tribology' has gained gradual acceptance. As the word tribology is relatively new, its meaning is still unclear to the wider community and humorous comparisons with tribes or tribolites tend to persist as soon as the word 'tribology' is mentioned.

Wear is the major cause of material wastage and loss of mechanical performance and any reduction in wear can result in considerable savings. Friction is a principal cause of wear and energy dissipation. Considerable savings can be made by improved friction control. It is estimated that one-third of the world's energy resources in present use is needed to overcome friction in one form or another. Lubrication is an effective means of controlling wear and reducing friction. Tribology is a field of science which applies an operational analysis to problems of great economic significance such as reliability, maintenance and wear of technical equipment ranging from household appliances to spacecraft.

The question is why 'the interacting surfaces in relative motion' (which essentially means rolling, sliding, normal approach or separation of surfaces) are so important to our economy and why they affect our standard of living. The answer is that surface interaction dictates or controls the functioning of practically every device developed by man. Everything that man makes wears out, almost always as a result of relative motion between surfaces. An analysis of machine break-downs shows that in the majority of cases failures and stoppages are associated with interacting moving parts such as gears, bearings, couplings, sealings, cams, clutches, etc. The majority of problems accounted for are tribological. Our human body also contains interacting surfaces, e.g., human joints, which are subjected to lubrication and wear. Despite our detailed knowledge covering many disciplines, the lubrication of human joints is still far from fully understood.

Tribology affects our lives to a much greater degree than is commonly realized. For example, long before the deliberate control of friction and wear was first promoted, human beings and animals were instinctively modifying friction and wear as it affected their own bodies. It is common knowledge that the human skin becomes sweaty as a response to stress or fear. It has only recently been discovered that sweating on the palms of hands or soles of feet of humans and dogs, but not rabbits, has the ability to raise friction between the palms or feet and a solid surface [2]. In other words, when an animal or human senses danger, sweating occurs to promote either rapid flight from the scene of danger, or else the ability to firmly hold a weapon or climb the nearest tree.

A general result or observation derived from innumerable experiments and theories is that tribology comprises the study of:

- · the characteristics of films of intervening material between contacting bodies, and
- · the consequences of either film failure or absence of a film which are usually manifested by severe friction and wear.

Film formation between any pair of sliding objects is a natural phenomenon which can occur without human intervention. Film formation might be the fundamental mechanism preventing the extremely high shear rates at the interface between two rigid sliding objects. Non-mechanical sliding systems provide many examples of this film formation. For example, studies of the movement between adjacent geological plates on the surface of the earth reveal that a thin layer of fragmented rock and water forms between opposing rock masses. Chemical reactions between rock and water initiated by prevailing high temperatures (about 600°C) and pressures (about 100 [MPa]) are believed to improve the lubricating function of the material in this layer [3]. Laboratory tests of model faults reveal that sliding initiates the formation of a self-sliding layer of fragmented rock at the interface with solid rock. A pair of self-sealing layers attached to both rock masses prevent the leakage of water necessary for the lubricating action of the inner layer of fragmented rock and water [3]. Although the thickness of the intervening layer of fragmented rock is believed to be between 1 and 100 [m] [3], this thickness is insignificant when compared to the extent of geological plates and these layers can be classified as 'films'. Sliding on a geological scale is therefore controlled by the properties of these 'lubricating films', and this suggests a fundamental similarity between all forms of sliding whether on the massive geological scale or on the microscopic scale of sliding between erythrocytes and capillaries. The question is, why do such films form and persist? A possible reason is that a thin film is mechanically stable, i.e., it is very difficult to completely expel such a film by squeezing between two objects. It is not difficult to squeeze out some of the film but its complete removal is virtually impossible. Although sliding is destructive to these films, i.e., wear occurs, it also facilitates their replenishment by entrainment of a 'lubricant' or else by the formation of fresh film material from wear particles.

Film formation between solid objects is intrinsic to sliding and other forms of relative motion, and the study and application of these films for human benefits are the *raison d'etre* of tribology.

In simple terms it appears that the practical objective of tribology is to minimize the two main disadvantages of solid to solid contact: friction and wear, but this is not always the case. In some situations, as illustrated in Figure 1.1, minimizing friction and maximizing wear or minimizing wear and maximizing friction or maximizing both friction and wear is desirable. For example, reduction of wear but not friction is desirable in brakes and lubricated clutches, reduction of friction but not wear is desirable in pencils, increase in both friction and wear is desirable in erasers.

Lubrication

Thin low shear strength layers of gas, liquid and solid are interposed between two surfaces in order to improve the smoothness of movement of one surface over another and to prevent damage. These layers of material separate contacting solid bodies and are usually very thin and often difficult to observe. In general, the thicknesses of these films range from **1** to **100** [μm], although thinner and thicker films can also be found. Knowledge that is related to enhancing or diagnosing the effectiveness of these films in preventing damage in solid contacts is commonly known as '**lubrication**'. Although there are no restrictions on the type of material required to form a lubricating film, as gas, liquid and certain solids are all effective, the material type does influence the limits of film effectiveness. For example, a gaseous film is suitable for low contact stress while solid films are usually applied to slow sliding speed contacts. Detailed analysis of gaseous or liquid films is usually termed '**hydrodynamic lubrication**' while lubrication by solids is termed '**solid lubrication**'. A specialized form of hydrodynamic lubrication involving physical interaction between the contacting bodies and the liquid lubricant is termed '**elastohydrodynamic lubrication**' and is of considerable practical significance. Another form of lubrication involves the chemical interactions between contacting bodies and the liquid lubricant and is termed '**boundary and extreme pressure lubrication**'. In the absence of any films, the only reliable means of ensuring relative movement is to maintain, by external force fields, a small distance of separation between the opposing surfaces. This, for example, can be achieved by the application of magnetic forces, which is the operating principle of magnetic levitation or 'maglev'. Magnetic levitation is, however, a highly specialized technology that is still at the experimental stage. A form of lubrication that operates by the same principle, i.e., forcible separation of the contacting bodies involving an external energy source, is '**hydrostatic lubrication**' where liquid or gaseous lubricant is forced into the space between contacting bodies.

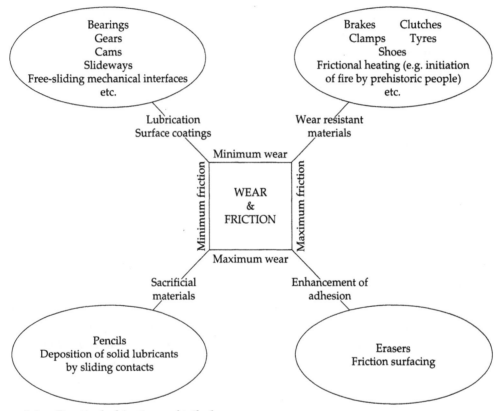

FIGURE 1.1   Practical objectives of tribology.

Liquid lubrication is a technological nuisance since filters, pumps and cooling systems are required to maintain the performance of the lubricant over a period of time. There are also environmental issues associated with the disposal of the used lubricants. Therefore '**solid lubrication**' and '**surface coatings**' are the subject of intense research.

The principal limitations of, in particular, liquid lubricants are the loss of load carrying capacity at high temperature and degradation in service. The performance of the lubricant depends on its composition and its physical and chemical characteristics.

From the practical engineering viewpoint, prediction of lubricating film characteristics is extremely important. Although such predictions are possible there always remains a certain degree of empiricism in the analysis of film characteristics. Prediction methods for liquid or gaseous films involve at the elementary level hydrodynamic, hydrostatic and elastohydrodynamic lubrication. For more sophisticated analyses '**computational methods**' must be used. There is still, however, no analytical method for determining the limits of solid films.

Wear

Film failure impairs the relative movement between solid bodies and inevitably causes severe damage to the contacting surfaces. The consequence of film failure is severe wear. Wear in these circumstances is the result of adhesion between contacting bodies and is termed '**adhesive wear**'. When the intervening films are partially effective then milder forms of wear occur and these are often initiated by fatigue processes due to repetitive stresses under either sliding or rolling. These milder forms of wear can therefore be termed '**fatigue wear**'. On the other hand, if the film material consists of hard particles or merely flows against one body without providing support against another body then a form of wear, which sometimes can be very rapid, known as '**abrasive wear**' occurs. Two other associated forms of wear are '**erosive wear**' (due to impacting particles) and '**cavitation wear**' which is caused by fast flowing liquids. In some practical situations the film material is formed by chemical attack of either contacting body and while this may provide some lubrication, significant wear is virtually inevitable. This form of wear is known as '**corrosive wear**' and when atmospheric oxygen is the corroding agent, then '**oxidative wear**' is said to occur. When the amplitude of movement between contacting bodies is restricted to, for example, a few micrometres, the film material is trapped within the contact and may eventually become destructive. Under these conditions '**fretting wear**' may result. There are also many other forms or mechanisms of wear. Almost any interaction between solid bodies will cause wear. Typical examples are '**impact wear**' caused by impact between two solids, '**melting wear**' occurring when the contact loads and speeds are sufficiently high to allow for the surface layers of the solid to melt and '**diffusive wear**' occurring at high interface temperatures. This dependence of wear on various operating conditions can be summarized in a flowchart shown in Figure 1.2.

1.3 COST OF FRICTION AND WEAR

The enormous cost of tribological deficiencies to any national economy is mostly caused by the large amount of energy and material losses occurring simultaneously on virtually every mechanical device in operation. When reviewed on the basis of a single machine, the losses are small. However, when the same loss is repeated on perhaps a million machines of a similar type, then the costs become very large.

For example, about two hundred years ago, it was suggested by Jacobs Rowe that by the application of the rolling element bearing to the carriages the number of horses required for all the carriages and carts in the United Kingdom could be halved. Since the estimated national total number of horses involved in this form of transportation was at that time about 40,000, the potential saving in horse-care costs was about one million pounds per annum at early 18th century prices [1,4].

In more contemporary times the simple analysis reveals that supplying all the worm gear drives in the United States with a lubricant that allows a relative increase of 5% in the mechanical efficiency compared to a conventional mineral oil would result in savings of about US$0.6 billion per annum [5]. The reasoning is that there are 3 million worm gears operating in the U.S.A. with an average power rating of about 7.5 [KW]. The annual national savings of energy would be 9.8 billion kilowatt-hours and the corresponding value of this energy is 0.6 billion US$ at an electricity cost of 0.06 US$ per kilowatt-hour.

These examples suggest that a form of 'tribology equation' can be used to obtain a simple estimate of either costs or benefits from existing or improved tribological practice. Such equation can be summarized as:

> **Total Tribological Cost/Saving = Sum of Individual Machine Cost/Saving ×
> Number of Machines**

This equation can be applied to any other problem in order to roughly estimate the relevance of tribology to a particular situation.

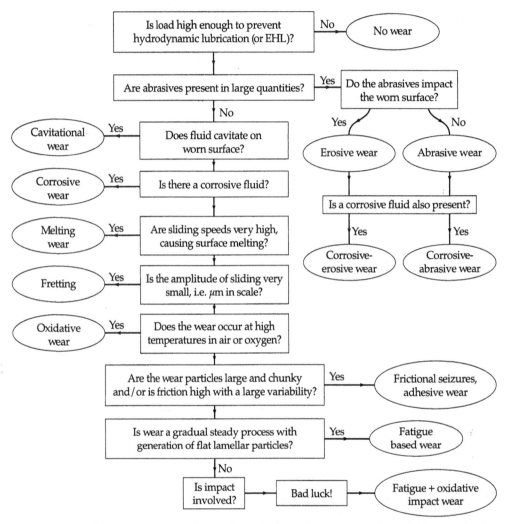

FIGURE 1.2   Flowchart illustrating the relationship between operating conditions and type of wear.

It was estimated by Peter Jost in 1966 that by the application of the basic principles of tribology, the economy of U.K. could save approximately £515 million per annum at 1965 values [6]. A similar report published in West Germany in 1976 revealed that the economic losses caused by friction and wear cost about 10 billion DM per annum, at 1975 values, which is equivalent to 1% of the Gross National Product [7]. About 50% of these losses were due to abrasive wear. In the U.S.A. it has been estimated that about 11% of total annual energy can be saved in the four major areas of transportation, turbo machinery, power generation and industrial processes through progress in tribology [8]. For example, tribological improvements in cars alone can save about 18.6% of total annual energy consumed by cars in the U.S.A., which is equivalent to about 14.3 billion US$ per annum [9]. In the U.K. the possible national energy savings achieved by the application of tribological principles and practices have been estimated to be between £468 to £700 million per annum [10]. The economics of tribology are of such gigantic proportions that tribological programmes have been established by industry and governments in many countries throughout the world.

The problems of tribology economics are of extreme importance to an engineer. For example, in pneumatic transportation of material through pipes, the erosive wear at bends can be up to 50 times more than in straight sections [11]. Apparently non-abrasive materials such as sugar cane [12] and wood chips can actually cause abrasive wear. Many tribological failures are associated with bearings. Simple bearing failures on modern generator sets in the U.S.A. cost about US$25,000 per day while to replace a £200,000 bearing in a single point mooring on a North Sea Oil Rig a contingency budget of about £1 million is necessary [13]. In addition there are some production losses which are very costly. The total cost of wear for a single US naval aircraft has been estimated to be US$243 per flight hour [14]. About 1000 megatonnes of material is excavated in Australia. Much of this is material waste which must be handled in order to retrieve metalliferous ores or coal. The cost of wear is around 2% of the saleable product. The annual production by a large iron ore mining company might be as high as 40 megatonnes involving a direct cost through the replacement of wearing parts of A$6 million per annum at 1977 values [15,16].

The analysis of the causes of friction and wear can have direct commercial implications, even in terms of who bears the cost of excessive wear or friction. For example, in one instance of a gas turbine that suffered excessive damage to its first-stage blades, detailed analysis of the cause of wear helped determine whether the owner or the insurance company would pay for the damage [17].

As soon as the extent of economic losses due to friction and wear became clear, researchers and engineers rejected many of the traditional limitations to mechanical performance and have found or are looking for new materials and lubricants to overcome these limits. Some of these improvements are so radical that the whole technology and economics of the product may change. A classic example is the adiabatic engine. The principle behind this development is to remove the oil and the lubricating system and use a dry, high temperature self-lubricating material. If the engine can operate adiabatically at high temperatures, heat previously removed by the now obsolete radiator can be turned to mechanical work. As a result, a fuel efficient, light weight engine might be built which will lead to considerable savings in fuels, oils and vehicle production costs. A fuel efficient engine is vital in reducing transportation and agricultural costs and therefore is a very important research and development task.

New ventures, even if they involve mostly conventional technologies, such as mining and processing of oil sands, can impose arduous conditions on equipment and necessitate new wear-resistant materials. Oil sands slurries are capable of wearing out a high chromium white cast iron pump impeller after only 3 months of service [18]. It is thus necessary to find new, hard yet tough, materials for better wear resistance and extended service life for the equipment.

Other examples of such innovations include surface treated cutters for sheep shearing, surface hardened soil engaging tools, polyethylene pipes for coal slurries and ion implanted titanium alloys for orthopaedic endoprostheses. Whenever wear and friction limit the function or durability of a device or appliance, there is a scope for tribology to offer some improvement.

In general terms, wear can effectively be controlled by selecting materials with a specific properties as illustrated in Figure 1.3. However, more detailed information on wear mechanisms and wear control is given in Chapters 11-16.

1.4   SUMMARY

Although the study of friction and wear caught the attention of many eminent scientists during the course of the past few centuries, consistent and sustained scientific investigation into friction and wear is a relatively recent phenomenon. Tribology is therefore a comparatively young science where rigorous analytical concepts have not yet been established to provide a clear guide to the complex characteristics of wear and friction. Much of the tribological research is applied or commercially orientated and already a wide range of wear resistant or friction reducing materials have been developed. The concept of developing special materials and coatings to overcome friction and wear problems is becoming a reality. Most analytical models and experimental knowledge of tribology have been completed in the past few decades, and some time in the future our understanding of the mechanisms of friction and wear may be radically changed and improved.

Critical materials property	Wear mechanism							
	Abrasive	Erosive	Cavitation	Corrosive	Fretting	Adhesive	Melting	Fatigue
Hardness	✓	✓	○	○	○	✓	○	○
Toughness	○	✓	✓	○	○	○	○	✓
Fatigue resistance	✓	✓	✓	○	✓	○	○	✓
Inertness	○	○	○	✓	✓①	○	○	○
High melting point	○	○	○	○	○	✓	✓	○
Heterogeneous microstructure	✓	○	○	✗②	○	✓	○	○
Non-metallic character	○	○	○	✓	○	✓	○	○

✓ Important   ① Fretting in air for metals
○ Marginal   ② Homogeneous microstructure inhibits electrochemical corrosion and, with it,
✗ Unfavourable   most forms of corrosive wear

FIGURE 1.3   General materials selection guide for wear control.

The bewildering range of experimental data and theories compiled so far has helped to create an impression that tribology, although undoubtedly important, is somehow mysterious and not readily applicable to engineering problems. Tribology cannot, however, be ignored as many governments and private studies have consistently concluded that the cost of friction and wear imposes a severe burden on industrialized countries. Part of the difficulty in controlling friction and wear is that the total cost in terms of energy and material wastage is spread over every type of industry. Although to the average engineer the cost of friction and wear may appear small, when the same costs are totalled for an entire country a very large loss of resources becomes apparent. The widely distributed incidence of tribological problems means that tribology cannot be applied solely by specialists but instead many engineers or technologists should have working knowledge of this subject.

The basic concept of tribology is that friction and wear are best controlled with a thin layer or intervening film of material separating sliding, rolling and impacting bodies. There is almost no restriction on the type of material that can form such a film and some solids, liquids and gases are equally effective. If no film material is supplied then the process of wear itself may generate a substitute film. The aim of tribology is either to find the optimum film material for a given application or to predict the sequence of events when a sliding/rolling/impacting contact is left to generate its own intervening film. The purpose of this book is to present the scientific principles of tribology as currently understood and to illustrate their applications to practical problems.

REVISION QUESTIONS

1.1   Is it always desirable to minimize friction?

1.2   Is it always desirable to minimize wear?

1.3   Name a few tribological interfaces and surfaces in and on the human body. Suggest consequences if there is a tribofailure, e.g., lubrication failure or wear damage, at these sites.

1.4   The diameter of a large rotating journal bearing is 400 [mm] and the film thickness of oil between the shaft and the journal is 20 [μm]. Calculate the ratio of bearing radius to film thickness.

1.5   The thickness of the fault zone between two sliding continental plates is approximately 1 m. Calculate the ratio of the width, i.e., the distance across the fault, of sliding rock to the thickness of the fault zone (i.e., the 'film thickness'). Hint: the range of violent sliding (the width of sliding rock) during a tremor or earthquake is approximately 100 [km]. It may also be assumed that the continental plates are sliding past each other, not subducting (one plate going below the other). Subducting plates give rise to very widespread tremors, greater than, for example, 100 [km] in range. Subducting plates can also cause tsunamis.

1.6   A nanotechnological device involves a bearing 20 [nm] in diameter. Using the same film thickness ratios calculated in questions 1.4 and 1.5, calculate a likely film thickness. Comment on the physical reality or type of material, e.g., liquid or gas, in this film.

REFERENCES

1   D. Dowson, History of Tribology, Longman Group Limited, 1979.

2   S. Adelman, C.R. Taylor and N.C. Heglund, Sweating on Paws and Palms: What Is Its Function, *American Journal of Physiology*, Vol. 29, 1975, pp. 1400-1402.

3   N.H. Sleep and M.L. Blanpied, Creep, Compaction and the Weak Rheology of Major Faults, *Nature*, Vol. 359, 1992, pp. 687-692.

4   B.W. Kelley, Lubrication of Concentrated Contacts, Interdisciplinary Approach to the Lubrication of Concentrated Contacts, Troy, New York, NASA SP-237, 1969, pp. 1-26.

5   P.A. Pacholke and K.M. Marshek, Improved Worm Gear Performance With Colloidal Molybdenum Disulfide Containing Lubricants, *Lubrication Engineering*, Vol. 43, 1986, pp. 623-628.

6   Lubrication (Tribology) - Education and Research. A Report on the Present Position and Industry Needs, (Jost Report), Department of Education and Science, HM Stationary Office, London, 1966.

7   Research Report (T76-38) Tribologie (Code BMFT-FBT76-38), Bundesministerium Fur Forschung und Technologie (Federal Ministry for Research and Technology), West Germany, 1976.

8   Strategy for Energy Conservation Through Tribology, ASME, New York, November, 1977.

9   L.S. Dake, J.A. Russell and D.C. Debrodt, A Review of DOE ECT Tribology Surveys, *Transactions ASME, Journal of Tribology*, Vol. 108, 1986, pp. 497-501.

10   H.P. Jost and J. Schofield, Energy Savings Through Tribology: A Techno-Economic Study, *Proc. Inst. Mech. Engrs., London*, Vol. 195, No. 16, 1981, pp. 151-173.

11   M.H. Jones and D. Scott (editors), Industrial Tribology, The Practical Aspects of Friction, Lubrication and Wear, Elsevier, Amsterdam, 1983.

12   K.F. Dolman, Alloy Development: Shredder Hammer Tips, Proc. 5th Conference of Australian Society of Sugar Cane Technologists, 1983, pp. 281-287.

13   E.W. Hemingway, Preface, Proc. Int. Tribology Conference, Melbourne, The Institution of Engineers, Australia, National Conference Publication No. 87/18, December, 1987.

14   M.J. Devine (editor), Proceedings of a Workshop on Wear Control to Achieve Product Durability, sponsored by the Office of Technology Assessment, United States Congress, Naval Air Development Centre, Warminster, 1977.

15   C.M. Perrott, Ten Years of Tribology in Australia, *Tribology International*, Vol. 11, 1978, pp. 35-36.

16   P.F. Booth, Metals in Mining-Wear in the Mining Industry, *Metals Austr.*, Vol. 9, 1977, pp. 7-9.

17   J.M. Gallardo, J.A. Rodriguez and E.J. Herrera, Failure of Gas Turbine Blades, *Wear*, Vol. 252, 2002, pp. 264-268.

18   R.J. Llewellyn, S.K. Yick and K.F. Dolman, Scouring Erosion Resistance of Metallic Materials Used in Slurry Pump Service, *Wear*, Vol. 256, 2004, pp. 592-599.

PHYSICAL PROPERTIES
O F
LUBRICANTS

2.1   INTRODUCTION

Before discussing lubrication and wear mechanisms, some information about lubricants is necessary. What are lubricants made of, and what are their properties? Are oils different from greases? Can mineral oils be used in high performance engines? Which oils are the most suitable for application to gears, bearings, etc.? What criteria should they meet? What is the oil viscosity, viscosity index, pressure-viscosity coefficient? How can these parameters be determined? What are the thermal properties and temperature characteristics of lubricants? An engineer should know the answers to all these questions.

In simple terms, the function of a lubricant is to control friction and wear in a given system. The basic requirements therefore relate to the performance of the lubricant, i.e., its influence upon friction and wear characteristics of a system. Another important aspect is the lubricant quality, which reflects its resistance to degradation in service. Most of the present day lubricant research is dedicated to the study, prevention and monitoring of oil degradation, since the lifetime of an oil is as important as its initial level of performance. Apart from suffering degradation in service, which may cause damage to the operating machinery, an oil may cause corrosion of contacting surfaces. The oil quality, however, is not the only consideration. Economic considerations are also important. For example, in large machinery holding several thousand litres of lubricating oil, the cost of the oil can be very high.

In this chapter the fundamental physical properties of lubricants such as viscosity, viscosity temperature dependence, viscosity index, pour point, flash point, volatility, oxidation stability, thermal stability, etc., together with the appropriate units and the ways of measuring these values will be outlined. The basic composition of oils and greases will be discussed in the next chapter.

2.2   OIL VISCOSITY

The parameter that plays a fundamental role in lubrication is oil viscosity. Different oils exhibit different viscosities. In addition, oil viscosity changes with temperature, shear rate and pressure and the thickness of the generated oil film is usually proportional to it. So, at first glance, it appears that the more viscous oils would give better performance, since the generated films would be thicker and a better separation of the two surfaces in contact would be achieved. This unfortunately is not always the case since more viscous oils require more power to be sheared. Consequently the power losses are higher and more heat is generated resulting in a substantial increase in the temperature of the contacting surfaces, which may lead to the failure of the component. For engineering applications the oil viscosity is usually chosen to give optimum performance at the required temperature. Knowing the temperature at which the oil is expected to operate is critical as oil viscosity is extremely temperature dependent. The viscosity of different oils varies at different rates with temperature. It can also be affected by the velocities of the operating surfaces (shear rates). The knowledge of the viscosity characteristics of a lubricant is therefore very important in the design and in the prediction of the behaviour of a lubricated mechanical system.

In this chapter a simplified concept of viscosity, sufficient for most engineering applications, is considered. Refinements to this model incorporating, for example, transfer of momentum between the adjacent layers of lubricant and transient visco-elastic effects, can be found in more specialized literature.

Dynamic Viscosity

Consider two flat surfaces separated by a film of fluid of thickness 'h' as shown in Figure 2.1. The force required to move the upper surface is proportional to the wetted area 'A' and the velocity gradient 'u/h', as the individual fluid layer in a thicker film will be subjected to less shear than in a thin film, i.e.:

$$\mathbf{F} \; \alpha \; \mathbf{A} \times \mathbf{u/h} \tag{2.1}$$

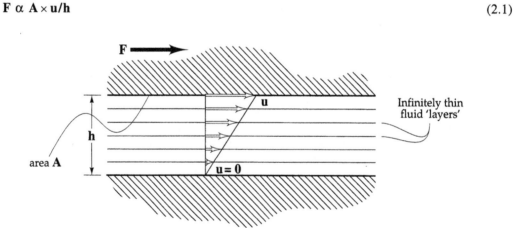

FIGURE 2.1   Schematic representation of the fluid separating two surfaces.

This relationship is maintained for most fluids. Different fluids will exhibit a different proportionality constant 'η', called the '**dynamic viscosity**'. The relationship (2.1) can be written as:

$$\mathbf{F} = \eta \times \mathbf{A} \times \mathbf{u/h} \tag{2.2}$$

Rearranging gives:

$$\eta = (\mathbf{F/A}) \; / \; (\mathbf{u/h})$$

or

$$\eta = \tau \; / \; (\mathbf{u/h}) \tag{2.3}$$

where:

η   is the dynamic viscosity [Pas];

τ   is the shear stress acting on the fluid [Pa];

u/h   is the shear rate, i.e., velocity gradient normal to the shear stress [s^{-1}].

Before the introduction of the SI system the most commonly used dynamic viscosity unit was the Poise. Incidentally this name originated not from an engineer but from a French medical doctor, Poiseuille, who studied the flow of blood. For practical applications the Poise [P] was far too large, thus a smaller unit, the centipoise [cP], was more commonly used. The SI unit for dynamic viscosity is Pascal-second [Pas]. The relationship between Poise and Pascal-second is as follows:

1 [P] = 100 [cP] ≈ 0.1 [Pas]

Kinematic Viscosity

Kinematic viscosity is defined as the ratio of dynamic viscosity to fluid density:

$$\upsilon = \eta/\rho \qquad\qquad (2.4)$$

where:

υ   is the kinematic viscosity [m^2/s];

η   is the dynamic viscosity [Pas];

ρ   is the fluid density [kg/m^3].

Before the introduction of the SI system the most commonly used kinematic viscosity unit was the Stoke [S]. This unit, however, was often too large for practical applications, thus a smaller unit, the centistoke [cS], was used. The SI unit for kinematic viscosity is [m^2/s], i.e.:

1 [S] = 100 [cS] = 0.0001 [m^2/s]

The densities of lubricating oils are usually in the range between 700 and 1200 [kg/m^3] (0.7 - 1.2 [g/cm^3]). The typical density of mineral oil is **850** [kg/m^3] (0.85 [g/cm^3]). To find the dynamic viscosity of any oil in [cP] or [Pas] the viscosity of this oil in [cS] is multiplied by its density in [g/cm^3], hence for a typical mineral oil:

viscosity in [cP] = viscosity in [cS] × 0.85 [g/cm^3] or

viscosity in [Pas] = viscosity in [cS] × 0.85 [g/cm^3] × 10^{-3}

2.3   VISCOSITY TEMPERATURE RELATIONSHIP

The viscosity of lubricating oils is extremely sensitive to the operating temperature. With increasing temperature the viscosity of oils falls quite rapidly. In some cases the viscosity of oil can fall by about 80% with a temperature increase of 25°C. From the engineering viewpoint it is important to know the viscosity at the operating temperature since it influences the lubricant film thickness separating two surfaces. The oil viscosity at a specific temperature can be either calculated from the viscosity-temperature equation or obtained from the viscosity-temperature ASTM chart.

Viscosity-Temperature Equations

There are several viscosity-temperature equations available. Some of them are purely empirical while others are derived from theoretical models. The most commonly used equations are summarized in Table 2.1 [33]. The most accurate of these is the Vogel equation. Three viscosity measurements at different temperatures for a specific oil are needed in order to determine the three constants in this equation. The oil viscosity can then be calculated at the required temperature, or the operating temperature can be calculated if the viscosity is known. Apart from being very accurate the Vogel equation is useful in numerical analysis. A computer program '**VISCOSITY**' for the Vogel equation is listed in the Appendix. Based on the three temperature-viscosity measurements the program calculates the viscosity at the required temperature. It also includes the option for the calculation of temperature corresponding to a given viscosity.

TABLE 2.1   Viscosity-temperature equations (adapted from [33]).

Name	Equation	Comments
Reynolds	$\eta = be^{-aT}$	Early equation; accurate only for a very limited temperature range
Slotte	$\eta = a/(b + T)^c$	Reasonable; useful in numerical analysis
Walther	$(\upsilon + a) = bd^{1/T^c}$	Forms the basis of the ASTM viscosity-temperature chart
Vogel	$\eta = ae^{b/(T - c)}$	Most accurate; very useful in engineering calculations

where:

$\mathbf{a, b, c, d}$ are constants;

υ is the kinematic viscosity [m^2/s];

$\mathbf{T}$ is the absolute temperature [K].

Viscosity-Temperature Chart

The most widely used chart is the ASTM (American Society for Testing Materials) Viscosity-Temperature chart (ASTM D341), which is entirely empirical and is based on Walther's equation Table 2.1.

$$(\upsilon + a) = bd^{1/T^c} \tag{2.5}$$

In deriving the bases for the ASTM chart, logs were taken from Walther's equation and '**d**' was assumed to equal **10**. The equation was then written in the form:

$$\log_{10}(\upsilon + a) = \log_{10}b + 1/T^c \tag{2.6}$$

It has been found that if 'v' is in [cS] then '**a**' is approximately equal to **0.6**. After substituting this into the equation, the logs were taken again in the manner shown below:

$$\log_{10}\log_{10}(v_{cS} + 0.6) = a' - c\log_{10}T \qquad (2.7)$$

where:

   a', **c**   are constants.

Although equation (2.7) forms successful bases for the ASTM viscosity-temperature chart, where the ordinate is $\log_{10}\log_{10}(v_{cS} + 0.6)$ and the abscissa is $\log_{10}T$, from the mathematical viewpoint the above derivation is incorrect. This is because when taking logs, equation (2.6) should be in the form:

$$\log_{10}\log_{10}(v_{cS} + 0.6) = \log_{10}(\log_{10}b + 1/T^c)$$

but,

$$a' - c\log_{10}T \neq \log_{10}(\log_{10}b + 1/T^c)$$

Despite this the ASTM chart is quite successful and works very well for mineral and synthetic oils under normal conditions. It is so well standardized that the viscosity-temperature characteristics are sometimes specified as 'ASTM slope'.

2.4   VISCOSITY INDEX

Different oils may have different ASTM slopes as shown in Figure 2.2. As early as 1920 it was known that Pennsylvania crude oils were better than the Gulf Coast (Texan) crude oils. Pennsylvania crude had the best viscosity temperature characteristics while the Gulf Coast crude had the worst since its viscosity varied much more with temperature. From the engineering viewpoint there was a need for a parameter which would accurately describe the viscosity-temperature characteristics of the oils. In 1929 a '**Viscosity Index**' was developed by Dean and Davis [1,2]. The viscosity index is an entirely empirical parameter that compares the kinematic viscosity of the oil of interest to the viscosities of two reference oils that have a considerable difference in sensitivity of viscosity to temperature. The reference oils have been selected in such a way that one has a viscosity index equal to zero (VI=0) and the other has a viscosity index equal to one hundred (VI=100) at 100°F (37.8°C), but they both have the same viscosity as the oil of interest at 210°F (98.89°C), as illustrated in Figure 2.3.

Since Pennsylvania and Gulf Coast oils have the same viscosity at 210°F (98.9°C) they were initially selected as reference oils. Oils made from Pennsylvania crude were assigned the viscosity index of **100** whereas oils made from the Gulf Coast crude the viscosity index of **0**.

The viscosity index can be calculated from the following formula:

$$\textbf{VI} = \textbf{(L} - \textbf{U)} \, / \, \textbf{(L} - \textbf{H)} \times \textbf{100} \qquad (2.8)$$

Firstly the kinematic viscosity of the oil of interest is measured at 40°C ('**U**') and at 100°C. Then the values of '**L**' and '**H**' that correspond to the viscosity at 100°C of the oil of interest are read from Table 2.2 [3] (ASTM D2270). Substituting the obtained values of '**U**', '**L**' and '**H**' into the above equation yields the viscosity index.

Note that the viscosity index is an inverse measure of the decline in oil viscosity with temperature. High values indicate that the oil shows less relative decline in viscosity with temperature. The viscosity index of most of the refined mineral oils available on the market is about **100**, whereas multigrade and synthetic oils have higher viscosity indices of about **150**.

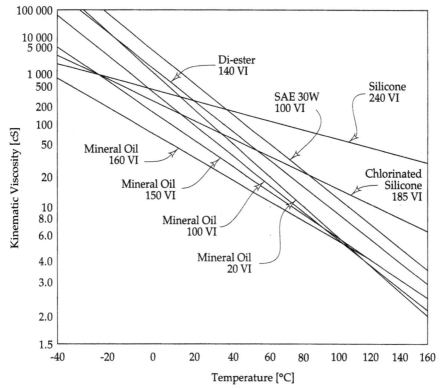

FIGURE 2.2   Viscosity-temperature characteristics of selected oils (adapted from [29 and 22]).

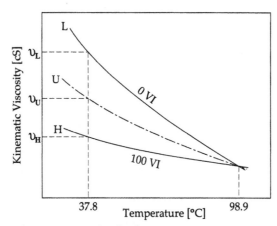

FIGURE 2.3   Evaluation of viscosity index [23].

EXAMPLE

Find the viscosity index of an oil that has a kinematic viscosity at 40°C of $\upsilon_{40°C}$ = 135 [cS] and at 100°C of $\upsilon_{100°C}$ = 17 [cS].

From Table 2.2 for $\upsilon_{100°C}$ = 17 [cS], **L** = 369.4 and **H** = 180.2 can be found. Substituting into the viscosity index equation yields:

VI = (369.4 − 135) / (369.4 − 180.2) × 100 = 123.9

2.5 VISCOSITY PRESSURE RELATIONSHIP

Lubricant viscosity increases with pressure. For most lubricants this effect is considerably larger than the effect of temperature or shear when the pressure is significantly above atmospheric. This is of particular importance in the lubrication of heavily loaded concentrated contacts, which can be found, for example, in rolling contact bearings and gears. The pressures encountered in these contacts can be so high and the rate of pressure rise so rapid that the lubricant behaves like a solid rather than a liquid. The phenomenon of viscosity increasing with pressure and the possibility of lubricant failure by fracture rather than viscous shear is often observed but not always recognized. For example, when asphalt or pitch is hit with a hammer it may shatter; however, when placed on an incline it might slowly flow downhill.

A number of attempts have been made to develop a formula describing the relationship between pressure and viscosity of lubricants. Some have been quite satisfactory, especially at low pressures, while others have been quite complex and not easily applicable in practice. The best known equation to calculate the viscosity of a lubricant at moderate pressures (close to atmospheric) is the Barus equation [4,5]. The application of this equation to pressures above 0.5 [GPa] can, however, lead to serious errors [6]. The equation becomes even more unreliable if the ambient temperature is high. The Barus equation is of the form:

$$\eta_p = \eta_0 e^{\alpha p} \qquad\qquad (2.9)$$

where:

η_p is the viscosity at pressure 'p' [Pas];

η_0 is the atmospheric viscosity [Pas];

α is the pressure-viscosity coefficient [m^2/N], which can be obtained by plotting the natural logarithm of dynamic viscosity 'η' versus pressure 'p'. The slope of the graph is 'α';

p is the pressure of concern [Pa].

For higher pressures Chu and Cameron [7] suggested that the following formula can be used:

$$\eta_p = \eta_0(1 + C \times p)^n \qquad\qquad (2.10)$$

where:

C, n are constants, 'n' is approximately **16** for most cases and 'C' can be obtained from the diagram shown in Figure 2.4 [7,8].

The pressure-viscosity coefficient is a function of the molecular structure of the lubricant and its physical characteristics such as molecular interlocking, molecular packing and rigidity and viscosity-temperature characteristics. There are various formulae available to calculate it. One of the early equations was derived by Wooster [5]:

$$\alpha = (0.6 + 0.965\log_{10}\eta_0) \times 10^{-8} \qquad\qquad (2.11)$$

where:

α is the pressure-viscosity coefficient [m^2/N];

η_0 is the atmospheric viscosity [cP], i.e., 1[cP] = 10^{-3}[Pas].

TABLE 2.2   Data for the evaluation of viscosity index [3].

v_{100}	L	H	v_{100}	L	H	v_{100}	L	H	v_{100}	L	H	v_{100}	L	H
2.00	7.994	6.394	8.30	106.9	63.05	14.6	283.0	143.9	21.8	575.6	261.5	41.0	1810	676.6
2.10	8.640	6.894	8.40	109.2	64.18	14.7	286.4	145.3	22.0	585.2	264.9	41.5	1851	689.1
2.20	9.309	7.410	8.50	111.5	65.32	14.8	289.7	146.8	22.2	595.0	268.6	42.0	1892	701.9
2.30	10.00	7.944	8.60	113.9	66.48	14.9	293.0	148.2	22.4	604.3	272.3	42.5	1935	714.9
2.40	10.71	8.496	8.70	116.2	67.64	15.0	296.5	149.7	22.6	614.2	275.8	43.0	1978	728.2
2.50	11.45	9.063	8.80	118.5	68.79	15.1	300.0	151.2	22.8	624.1	279.6	43.5	2021	741.3
2.60	12.21	9.647	8.90	120.9	69.94	15.2	303.4	152.6	23.0	633.6	283.3	44.0	2064	754.4
2.70	13.00	10.25	9.00	123.3	71.10	15.3	306.9	154.1	23.2	643.4	286.8	44.5	2108	767.6
2.80	13.80	10.87	9.10	125.7	72.27	15.4	310.3	155.6	23.4	653.8	290.5	45.0	2152	780.9
2.90	14.63	11.50	9.20	128.0	73.42	15.5	313.9	157.0	23.6	663.3	294.4	45.5	2197	794.5
3.00	15.49	12.15	9.30	130.4	74.57	15.6	317.5	158.6	23.8	673.7	297.9	46.0	2243	808.2
3.10	16.36	12.82	9.40	132.8	75.73	15.7	321.1	160.1	24.0	683.9	301.8	46.5	2288	821.9
3.20	17.26	13.51	9.50	135.3	76.91	15.8	324.6	161.6	24.2	694.5	305.6	47.0	2333	835.5
3.30	18.18	14.21	9.60	137.7	78.08	15.9	328.3	163.1	24.4	704.2	309.4	47.5	2380	849.2
3.40	19.12	14.93	9.70	140.1	79.27	16.0	331.9	164.6	24.6	714.9	313.0	48.0	2426	863.0
3.50	20.09	15.66	9.80	142.7	80.46	16.1	335.5	166.1	24.8	725.7	317.0	48.5	2473	876.9
3.60	21.08	16.42	9.90	145.2	81.67	16.2	339.2	167.7	25.0	736.5	320.9	49.0	2521	890.9
3.70	22.09	17.19	10.0	147.7	82.87	16.3	342.9	169.2	25.2	747.2	324.9	49.5	2570	905.3
3.80	23.13	17.97	10.1	150.3	84.08	16.4	346.6	170.7	25.4	758.2	328.8	50.0	2618	919.6
3.90	24.19	18.77	10.2	152.9	85.30	16.5	350.3	172.3	25.6	769.3	332.7	50.5	2667	933.6
4.00	25.32	19.56	10.3	155.4	86.51	16.6	354.1	173.8	25.8	779.7	336.7	51.0	2717	948.2
4.10	26.50	20.37	10.4	158.0	87.72	16.7	358.0	175.4	26.0	790.4	340.5	51.5	2767	962.9
4.20	27.75	21.21	10.5	160.6	88.95	16.8	361.7	177.0	26.2	801.6	344.4	52.0	2817	977.5
4.30	29.07	22.05	10.6	163.2	90.19	16.9	365.6	178.6	26.4	812.8	348.4	52.5	2867	992.1
4.40	30.48	22.92	10.7	165.8	91.40	17.0	369.4	180.2	26.6	824.1	352.3	53.0	2918	1007
4.50	31.96	23.81	10.8	168.5	92.65	17.1	373.3	181.7	26.8	835.5	356.4	53.5	2969	1021
4.60	33.52	24.71	10.9	171.2	93.92	17.2	377.1	183.3	27.0	847.0	360.5	54.0	3020	1036
4.70	35.13	25.63	11.0	173.9	95.19	17.3	381.0	184.9	27.2	857.5	364.6	54.5	3073	1051
4.80	36.79	26.57	11.1	176.6	96.45	17.4	384.7	186.5	27.4	869.0	368.3	55.0	3126	1066
4.90	38.50	27.53	11.2	179.4	97.71	17.5	388.9	188.1	27.6	880.6	372.3	55.5	3180	1082
5.00	40.23	28.49	11.3	182.1	98.97	17.6	392.7	189.7	27.8	892.3	376.4	56.0	3233	1097
5.10	41.99	29.46	11.4	184.9	100.2	17.7	396.7	191.3	28.0	904.1	380.6	56.5	3286	1112
5.20	43.76	30.43	11.5	187.6	101.5	17.8	400.7	192.9	28.2	915.8	384.6	57.0	3340	1127
5.30	45.53	31.40	11.6	190.4	102.8	17.9	404.6	194.6	28.4	927.6	388.8	57.5	3396	1143
5.40	47.31	32.37	11.7	193.3	104.1	18.0	408.6	196.2	28.6	938.6	393.0	58.0	3452	1159
5.50	49.09	33.34	11.8	196.2	105.4	18.1	412.6	197.8	28.8	951.2	396.6	58.5	3507	1175
5.60	50.87	34.32	11.9	199.0	106.7	18.2	416.7	199.4	23.0	963.4	401.1	59.0	3563	1190
5.70	52.64	35.29	12.0	201.9	108.0	18.3	420.7	201.0	29.2	975.4	405.3	59.5	3619	1206
5.80	54.42	36.26	12.1	204.8	109.4	18.4	424.9	202.6	29.4	987.1	409.5	60.0	3676	1222
5.90	56.20	37.23	12.2	207.8	110.7	18.5	429.0	204.3	29.6	998.9	413.5	60.5	3734	1238
6.00	57.97	38.19	12.3	210.7	112.0	18.6	433.2	205.9	29.8	1011	417.6	61.0	3792	1254
6.10	59.74	39.17	12.4	213.6	113.3	18.7	437.3	207.6	30.0	1023	421.7	61.5	3850	1270
6.20	61.52	40.15	12.5	216.6	114.7	18.8	441.5	209.3	30.5	1055	432.4	62.0	3908	1286
6.30	63.32	41.13	12.6	219.6	116.0	18.9	445.7	211.0	31.0	1086	443.2	62.5	3966	1303
6.40	65.18	42.14	12.7	222.6	117.4	19.0	449.9	212.7	31.5	1119	454.0	63.0	4026	1319
6.50	67.12	43.18	12.8	225.7	118.7	19.1	454.2	214.4	32.0	1151	464.9	63.5	4087	1336
6.60	69.16	44.24	12.9	228.8	120.1	19.2	458.4	216.1	32.5	1184	475.9	64.0	4147	1352
6.70	71.29	45.33	13.0	231.9	121.5	19.3	462.7	217.7	33.0	1217	487.0	64.5	4207	1369
6.80	73.48	46.44	13.1	235.0	122.9	19.4	467.0	219.4	33.5	1251	498.1	65.0	4268	1386
6.90	75.72	47.51	13.2	238.1	124.2	19.5	471.3	221.1	34.0	1286	509.6	65.5	4329	1402
7.00	78.00	48.57	13.3	241.2	125.6	19.6	475.7	222.8	34.5	1321	521.1	66.0	4392	1419
7.10	80.25	49.61	13.4	244.3	127.0	19.7	479.7	224.5	35.0	1356	532.5	66.5	4455	1436
7.20	82.39	50.69	13.5	247.4	128.4	19.8	483.9	226.2	35.5	1391	544.0	67.0	4517	1454
7.30	84.53	51.78	13.6	250.6	129.8	19.9	488.6	227.7	36.0	1427	555.6	67.5	4580	1471
7.40	86.66	52.88	13.7	253.8	131.2	20.0	493.2	229.5	36.5	1464	567.1	68.0	4645	1488
7.50	88.85	53.98	13.8	257.0	132.6	20.2	501.5	233.0	37.0	1501	579.3	68.5	4709	1506
7.60	91.04	55.09	13.9	260.1	134.0	20.4	510.8	236.4	37.5	1538	591.3	69.0	4773	1523
7.70	93.20	56.20	14.0	263.3	135.4	20.6	519.9	240.1	38.0	1575	603.1	69.5	4839	1541
7.80	95.43	57.31	14.1	266.6	136.8	20.8	528.8	243.5	38.5	1613	615.0	70.0	4905	1558
7.90	97.72	58.45	14.2	269.8	138.2	21.0	538.4	247.1	39.0	1651	627.1			
8.00	100.0	59.60	14.3	273.0	139.6	21.2	547.5	250.7	39.5	1691	639.2			
8.10	102.3	60.74	14.4	276.3	141.0	21.4	556.7	254.2	40.0	1730	651.8			
8.20	104.6	61.89	14.5	279.6	142.4	21.6	566.4	257.8	40.5	1770	664.2			

v_{100} is the kinematic viscosity of the oil of interest at 100°C in [cS].

Although this exponential law fits most lubricants it is not particularly accurate. There are other equations for the calculation of the pressure-viscosity coefficient available in the literature. It is often reported that some of these equations are accurate for certain fluids but unsuitable for others. One of the best formulae for the analytical determination of the pressure-viscosity coefficient is the empirical expression developed by So and Klaus [9]. A combination of linear and nonlinear regression analyses with atmospheric viscosity, density and the viscosity temperature property (modified ASTM slope) was applied to obtain the following expression:

$$\alpha = 1.216 + 4.143 \times (\log_{10} \upsilon_0)^{3.0627} + 2.848 \times 10^{-4} \times b^{5.1903} (\log_{10} \upsilon_0)^{1.5976} - 3.999 \times (\log_{10} \upsilon_0)^{3.0975} \rho^{0.1162}$$

$$(2.12)$$

where:

α is the pressure-viscosity coefficient [$\times 10^{-8} \, m^2/N$];

υ_0 is the kinematic viscosity at the temperature of interest [cS];

b is the ASTM slope of a lubricant divided by **0.2**;

ρ is the atmospheric density at the temperature of interest in [g/cm^3].

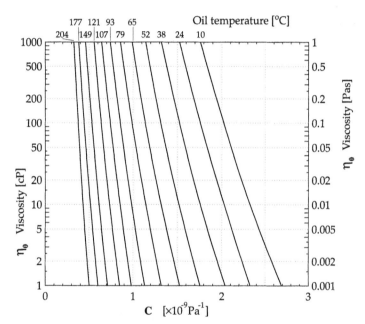

FIGURE 2.4   Graph for the determination of the constant 'C' (adapted from [8]).

One of the problems associated with available formulae is that they are most accurate when calculating the pressure-viscosity coefficients at low shear rates. In many engineering applications, especially in heavily loaded concentrated contacts, the lubricant operates under very high shear rates and precise values of the pressure-viscosity coefficient are needed for the evaluation of the minimum film thickness. Fortunately, an accurate value of this coefficient can be determined experimentally and this problem is discussed later.

If an accurate analytical formula, for the relationship between viscosity, pressure and shear rate, could be developed it would certainly be useful. Such a formula could provide a relationship between the fundamental parameters of the lubricant and the pressure-viscosity coefficient as opposed to being a strictly empirical equation. This would open up the possibilities of modifying the chemical makeup of the lubricant in order to achieve the desired pressure-viscosity coefficient for specific applications. A limited attempt to find such a relationship was reported by Johnston [15].

The rise in viscosity with pressure varies between oils, and there is a considerable difference between paraffinic and naphthenic oils. According to Klamann [16] the pressure-viscosity coefficient of the Barus equation or **'alpha value'** is between 1.5 and $2.4 \times 10^{-8} [m^2/N]$ for paraffinic oils, and between 2.5 and $3.5 \times 10^{-8} [m^2/N]$ for aromatic oils. For aromatic extracts of oil, the pressure-viscosity coefficient is much higher, but this is of limited practical significance. The value of the pressure-viscosity coefficient is in general reduced at higher temperatures, with naphthenic oils being the most severely affected. In some cases even at 80°C there is a substantial reduction in the pressure-viscosity coefficient. This effect is not so pronounced in paraffinic oils, thus they usually generate more stable lubricating films over a wider temperature range, from ambient to typical bearing and gear temperatures. Water, by contrast, shows only a small rise, almost negligible, in viscosity with pressure. More interestingly, at temperatures close to its freezing point it shows a decline in viscosity with pressure [17,18]. Obviously, water is not a particularly good lubricant, although it can function surprisingly well in some situations.

There are many other formulae for viscosity-pressure relationships. A short review of some of the empirical formulae for the viscosity-pressure relationships is given in [9,10]. These formulae allow for the calculations of viscosity changes with pressure under various conditions and to various degrees of accuracy.

An expression which is suitable for computational applications was initially proposed by Roelands [11,12] and developed further by Houpert [12,13]. To calculate lubricant viscosity at a specific pressure a particular form of the Barus equation was proposed:

$$\eta_R = \eta_0 e^{\alpha^* p} \tag{2.13}$$

where:

η_R is the viscosity at pressure 'p' and temperature 'θ' [Pas];

η_0 is the atmospheric viscosity [Pas];

α^* is the Roelands pressure-viscosity coefficient, which is a function of both 'p' and 'θ' [m²/N];

p   is the pressure of interest [Pa].

The Roelands pressure-viscosity coefficient 'α*' can be calculated from the formula:

$$\alpha^* p = [\ln \eta_0 + 9.67] \left\{ \left(\frac{\theta - 138}{\theta_0 - 138} \right)^{-S_0} (1 + 5.1 \times 10^{-9} p)^Z - 1 \right\} \tag{2.14}$$

where:

θ_0 is a reference or ambient temperature [K];

η_0 is the atmospheric viscosity [Pas];

Z, S_0 are constants, characteristic for a specific oil, and independent of temperature and pressure. These constants can be calculated from the following formulae [12]:

$$Z = \frac{\alpha}{5.1 \times 10^{-9} [\ln \eta_0 + 9.67]}$$

$$S_0 = \frac{\beta(\theta_0 - 138)}{\ln \eta_0 + 9.67}$$

where:

α   is the pressure-viscosity coefficient $[m^2/N]$;

β   is given by the following expression [12,13]:

$$\beta = [\ln \eta_0 + 9.67]\,[1 + 5.1 \times 10^{-9}\,p]^Z \left[\frac{S_0}{(\theta_0 - 138)}\right]$$

The above formula appears to be more comprehensive than those previously described since it takes into account the simultaneous effects of temperature and pressure. The values of 'α' and dynamic viscosity 'η_0' for some commonly used lubricants are given in Table 2.3 [12,14].

TABLE 2.3   Dynamic viscosity and pressure-viscosity coefficients of some commonly used lubricants (adapted from [12]).

Lubricants	Dynamic viscosity η_0 measured at atmospheric pressure $[\times 10^{-3}\ Pas]$			Pressure-viscosity coefficient α $[\times 10^{-9}\ m^2/N]$		
	30°C	60°C	100°C	30°C	60°C	100°C
High VI oils						
Light machine oil	38	12.1	5.3	-	18.4	13.4
Heavy machine oil	153	34	9.1	23.7	20.5	15.8
Heavy machine oil	250	50.5	12.6	25.0	21.3	17.6
Cylinder oil	810	135	26.8	34	28	22
Medium VI oils						
Spindle oil	18.6	6.3	2.4	20	16	13
Light machine oil	45	12	3.9	28	20	16
Medicinal white oil	107	23.3	6.4	29.6	22.8	17.8
Heavy machine oil	122	26.3	7.3	27.0	21.6	17.5
Heavy machine oil	171	31	7.5	28	23	18
Low VI oils						
Spindle oil	30.7	8.6	3.1	25.7	20.3	15.4
Heavy machine oil	165	30.0	6.8	33.0	23.8	16.0
Heavy machine oil	310	44.2	9.4	34.6	26.3	19.5
Cylinder oil	2000	180	24	41.5	29.4	25.0
Other fluids and lubricants						
Water	0.80	0.47	0.28	0	0	0
Ethylene oxide- propylene oxide copolymer	204	62.5	22.5	17.6	14.3	12.2
Castor oil	360	80	18.0	15.9	14.4	12.3
Di(2-ethylhexyl) phthalate	43.5	11.6	4.05	20.8	16.6	13.5
Glycerol (glycerine)	535	73	13.9	5.9	5.5	3.6
Polypropylene glycol 750	82.3	-	-	17.8	-	-
Polypropylene glycol 1500	177	-	-	17.4	-	-
Tri-arylphosphate ester	25.5	-	-	31.6	-	-

2.6 VISCOSITY-SHEAR RATE RELATIONSHIP

From the engineering viewpoint, it is essential to know the value of the lubricant viscosity at a specific shear rate. For simplicity it is usually assumed that the fluids are Newtonian, i.e., their viscosity is proportional to shear rate as shown in Figure 2.5.

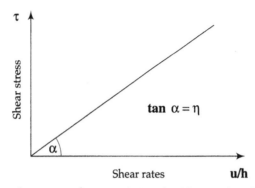

FIGURE 2.5   Shear stress - shear rate characteristic of a Newtonian fluid.

For pure mineral oils this is usually true up to relatively large shear rates of 10^5 - 10^6 [s^{-1}] [31], but at the higher shear rates frequently encountered in engineering applications this proportionality is lost and the lubricant begins to behave as a non-Newtonian fluid. In these fluids the viscosity depends on shear rate, i.e., these fluids do not exhibit a single value of viscosity over the range of shear rates. Non-Newtonian behaviour is, in general, a function of the structural complexity of a fluid. For example, liquids like water, benzene and light oils are Newtonian. These fluids have a loose molecular structure that is not affected by shearing action. On the other hand, fluids in which the suspended molecules form a structure that interferes with the shearing of the suspension medium are considered to be non-Newtonian. Typical examples of such fluids are water-oil emulsions, polymer thickened oils and, in extreme cases, greases. The non-Newtonian behaviour of some selected fluids is shown in Figure 2.6.

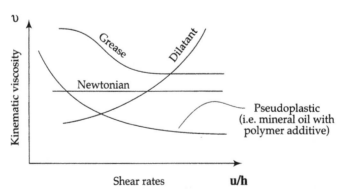

FIGURE 2.6   Viscosity - shear rate characteristics for some non-Newtonian fluids.

There are two types of non-Newtonian behaviour that are important from the engineering viewpoint: pseudoplastic and thixotropic behaviour.

Pseudoplastic Behaviour

Pseudoplastic behaviour is also known in the literature as shear thinning and is associated with the thinning of the fluid as the shear rate increases. This is illustrated in Figure 2.7.

During the process of shearing in polymer fluids, long molecules that are randomly orientated and with no connected structure tend to align, giving a reduction in apparent viscosity. In emulsions a drop in viscosity is due to orientation and deformation of the emulsion particles. The process is usually reversible. Multigrade oils are particularly susceptible to this type of behaviour; they shear thin with increased shear rates, as shown in Figure 2.8 [38].

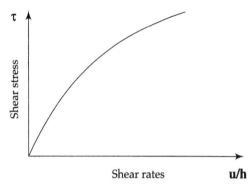

FIGURE 2.7   Pseudoplastic behaviour.

The contrary phenomenon to shear thinning is shear thickening, also referred to as dilatancy. Dilatant fluids are usually suspensions with a high solid content. The increase in viscosity with shear rate is attributed to the rearranging of the particles suspended in the fluid, resulting in the dilation of voids between the particles. This behaviour can be related to the arrangement of the fluid molecules. The theory is that in the non-shear condition molecules adopt a close packed formation, which gives the minimum volume of voids. When shear is applied the molecules move to an open pack formation, dilating the voids. As a result, there is an insufficient amount of fluid to fill the voids, giving an increased resistance to flow. An analogy to such fluids is seen when walking on wet sand where footprints are always dry.

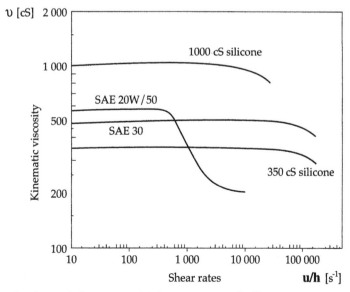

FIGURE 2.8   Pseudoplastic behaviour of lubricating oils [38].

Thixotropic Behaviour

Thixotropic behaviour, also known in the literature as shear duration thinning, is shown in Figure 2.9. It is associated with a loss of consistency of the fluid as the duration of shear increases. During the process of shearing, it is believed that the structure of the thixotropic fluid is being broken down. The destruction of the fluid structure progresses with time, giving a reduction in apparent viscosity, until a certain balance is reached where the structure rebuilds itself at the same rate as it is destroyed. At this stage the apparent viscosity attains a steady value. In some cases the process is reversible, i.e., viscosity returns to its original value when shear is removed, but permanent viscosity loss is also possible.

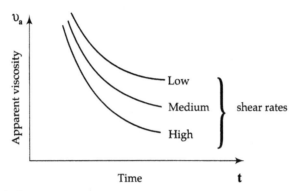

FIGURE 2.9   Thixotropic behaviour.

A converse effect to thixotropic behaviour, i.e., thickening of the fluid with the duration of shearing, can also occur with some fluids. This phenomenon is known in the literature as inverse thixotropy or rheopectic behaviour [19]. An example of a fluid with such properties is synovial fluid, a natural lubricant found in human and animal joints. It was found that the viscosity of synovial fluid increases with the duration of shearing [20,39]. It seems that the longer the duration of shearing the better the lubricating film which is generated by the body.

2.7   VISCOSITY MEASUREMENTS

Various viscosity measurement techniques and instruments have been developed over the years. The most commonly used in engineering applications are capillary and rotational viscometers. In general, capillary viscometers are suitable for fluids with negligible non-Newtonian effects and rotational viscometers are suitable for fluids with significant non-Newtonian effects. Some viscometers have a special heating bath built in, in order to control and measure the temperature, so that the viscosity-temperature characteristics can be obtained. In most cases water is used in the heating bath. Water is suitable for the temperature range between 0° and 99°C. For higher temperatures mineral oils are used and for low temperatures down to -54°C, ethyl alcohol or acetone is used.

Capillary Viscometers

Capillary viscometers are based on the principle that a specific volume of fluid will flow through the capillary (ASTM D445, ASTM D2161). The time necessary for this volume of fluid to flow gives the '**kinematic viscosity**'. Flow through the capillary must be laminar and the deductions are based on Poiseuille's law for steady viscous flow in a pipe. There is a number of such viscometers available and some of them are shown in Figure 2.10.

Assuming that the fluids are Newtonian, and neglecting end effects, the kinematic viscosity can be calculated from the formula:

$$\upsilon = \pi r^4 glt \,/\, 8LV = k(t_2 - t_1) \qquad\qquad (2.15)$$

where:

υ	is the kinematic viscosity [m^2/s];
r	is the capillary radius [m];
l	is the mean hydrostatic head [m];
g	is the earth acceleration [m/s^2];
L	is the capillary length [m];
V	is the flow volume of the fluid [m^3];
t	is the flow time through the capillary, $t = (t_2 - t_1)$, [s];
k	is the capillary constant, which has to be determined experimentally by applying a reference fluid with known viscosity, e.g., by applying freshly distilled water. The capillary constant is usually given by the manufacturer of the viscometer.

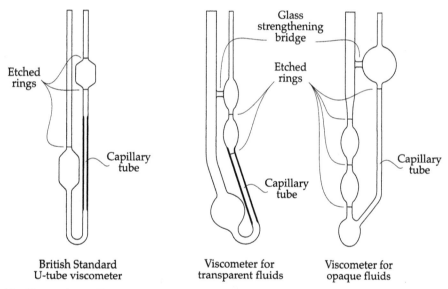

FIGURE 2.10   Typical capillary viscometers (adapted from [23]).

In order to measure the viscosity of the fluid by one of the viscometers shown in Figure 2.10, the container is filled with oil between the etched lines. The measurement is then made by timing the period required for the oil meniscus to flow from the first to the second timing mark. This is measured with an accuracy to within 0.1 [s].

Kinematic viscosity can also be measured by so-called 'short tube' viscometers. In the literature they are also known as efflux viscometers. As in capillary viscometers, viscosity is determined by measuring the time necessary for a given volume of fluid to discharge under gravity through a short tube orifice in the base of the instrument. The most commonly used viscometers are Redwood, Saybolt and Engler. The operation principle of these viscometers is the same, and they only differ by the orifice dimensions and the volume of fluid discharged. Redwood viscometers are used in the United Kingdom, Saybolt in Europe and Engler mainly in former Eastern Europe. The viscosities measured by these viscometers are quoted in terms of the time necessary for the discharge of a certain volume of fluid. Hence the viscosity is sometimes found as being quoted in Redwood and Saybolt seconds. The viscosity measured on Engler viscometers is quoted in Engler degrees, which is the time for the fluid to discharge divided by the discharge time of the same volume of water at the same temperature. Redwood and Saybolt seconds and Engler degrees can be converted into centistokes as shown in Figure 2.11. These particular types of viscometers are gradually becoming obsolete. A typical short tube viscometer is shown in Figure 2.12.

In order to extend the range of kinematic, Saybolt Universal, Redwood No. 1 and Engler viscosity scales only (Figure 2.11), a simple operation is performed. The viscosities on these scales that correspond to the viscosity between 100 and 1000 [cS] on the kinematic scale are multiplied by a factor of 10, which gives the required extension. For example:

$$4000\,[cS] = 400\,[cS] \times 10 \approx 1850\,[SUS] \times 10 = 18500\,[SUS] \approx 51\,[Engler] \times 10 = 510\,[Engler]$$

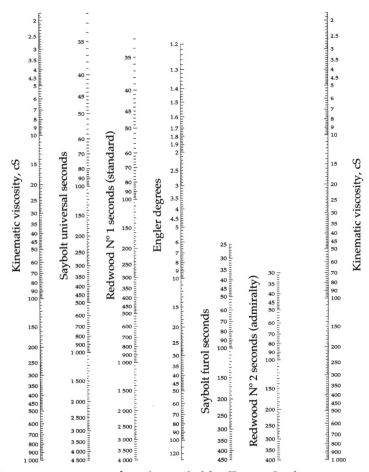

FIGURE 2.11   Viscosity conversion chart (compiled by Texaco Inc.).

Rotational Viscometers

Rotational viscometers are based on the principle that the fluid viscosity is related to the force required to generate shear between two surfaces separated by a film of fluid (ASTM D2983). In these viscometers one of the surfaces is stationary and the other is rotated by an external drive and the fluid fills the space in between. The measurements are conducted by applying either a constant torque and measuring the changes in the speed of rotation or applying a constant speed and measuring the changes in the torque. These viscometers give the '**dynamic viscosity**'. There are two main types of these viscometers: rotating cylinder and cone-on-plate viscometers.

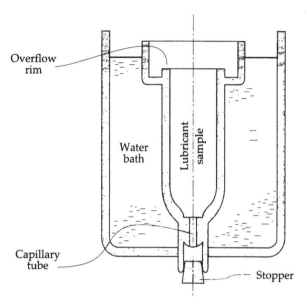

FIGURE 2.12 Schematic diagram of a short tube viscometer.

· *Rotating Cylinder Viscometer*

The rotating cylinder viscometer, also known as a 'Couette viscometer', consists of two concentric cylinders with an annular clearance filled with fluid as shown in Figure 2.13. The inside cylinder is stationary and the outside cylinder rotates at constant velocity. The force necessary to shear the fluid between the cylinders is measured. The velocity of the cylinder can be varied so that the changes in viscosity of the fluid with shear rate can be assessed. Care needs to be taken with non-Newtonian fluids as these viscometers are calibrated for Newtonian fluids. Different cylinders with a range of radial clearances are used for different fluids. For Newtonian fluids the dynamic viscosity can be estimated from the formula:

$$\eta = M(1/r_b^2 - 1/r_c^2) \; / \; 4\pi d\omega = kM \; / \; \omega \qquad\qquad (2.16)$$

where:

η is the dynamic viscosity [Pas];

r_b, r_c are the radii of the inner and outer cylinders, respectively [m];

M is the shear torque on the inner cylinder [Nm];

ω is the angular velocity [rad/s];

d is the immersion depth of the inner cylinder [m];

k is the viscometer constant, usually supplied by the manufacturer for each pair of cylinders [m^{-3}].

When motor oils are used in European and North American conditions, the oil viscosity data at -18°C is required in order to assess the ease with which the engine starts. A specially adapted rotating cylinder viscometer, known in the literature as the 'Cold Cranking Simulator' (CCS), is used to obtain these measurements (ASTM D2602). A schematic diagram of this viscometer is shown in Figure 2.14.

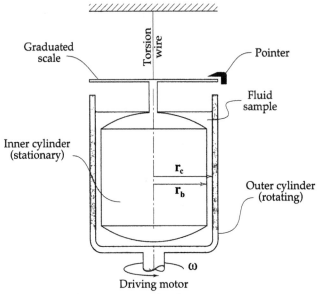

FIGURE 2.13 Schematic diagram of a rotating cylinder viscometer.

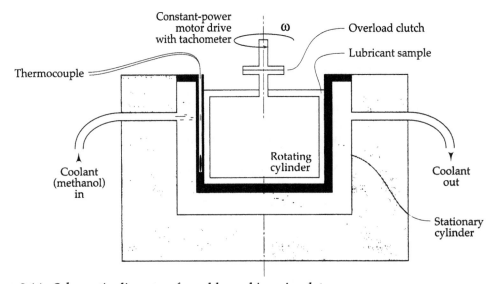

FIGURE 2.14 Schematic diagram of a cold cranking simulator.

The inner cylinder is rotated at constant power in the cooled lubricant sample of volume about 5 [ml]. The viscosity of the oil sample tested is assessed by comparing the rotational speed of the test oil with the rotational speed of the reference oil under the same conditions. The measurements provide an indication of the ease with which the engine will turn at low temperatures and with limited available starting power. In the case of very viscous fluids, two cylinder arrangements with a small clearance might be impractical because of the very high viscous resistance; thus a single cylinder is rotated in fluid held in a cylindrical container with a much larger clearance and measurements are calibrated against measurements obtained with reference fluids.

· *Cone on Plate Viscometer*

The cone on plate viscometer consists of a conical surface and a flat plate. Either of these surfaces can be rotated. The clearance between the cone and the plate is filled with the fluid and the cone angle ensures a constant shear rate in the clearance space. The advantage of this viscometer is that a very small sample volume of fluid is required for the test. In some of these viscometers, the temperature of the fluid sample is controlled during tests. This is achieved by circulating pre-heated or cooled external fluid through the plate of the viscometer. These viscometers can be used with both Newtonian and non-Newtonian fluids as the shear rate is approximately constant across the gap. The schematic diagram of this viscometer is shown in Figure 2.15.

The dynamic viscosity can be estimated from the formula:

$$\eta = 3M\alpha\cos^2\alpha(1 - \alpha^2/2) \, / \, 2\pi\omega r^3 = kM \, / \, \omega \qquad (2.17)$$

where:

η	is the dynamic viscosity [Pas];
r	is the radius of the cone [m];
M	is the shear torque on the cone [Nm];
ω	is the angular velocity [rad/s];
α	is the cone angle [rad];
k	is the viscometer constant, usually supplied by the manufacturer [m^{-3}].

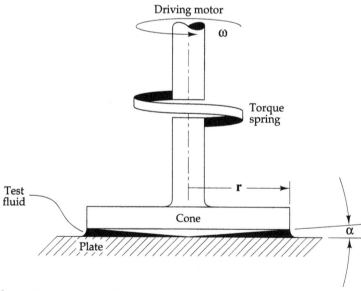

FIGURE 2.15 Schematic diagram of a cone on plate viscometer.

Other Viscometers

Many other types of viscometers, based on different principles of measurement, are also available. Most commonly used in many laboratories is the 'Falling Ball Viscometer'. A glass tube is filled with the fluid to be tested and a steel ball is dropped into the tube. The measurement is then made by timing the period required for the ball to fall from the first to the second timing mark, etched on the tube. The time is measured with an accuracy to within 0.1 [s]. This viscometer can also be used for the determination of viscosity changes under pressure. A schematic diagram of this viscometer is shown in Figure 2.16.

The dynamic viscosity is estimated from the formula:

$$\eta = 2r^2(\rho_b - \rho)gF / 9v \tag{2.18}$$

where:

η	is the dynamic viscosity [Pas];
r	is the radius of the ball [m];
ρ_b	is the density of the ball [kg/m^3];
ρ	is the density of the fluid [kg/m^3];
g	is the gravitational constant [m/s^2];
v	is the velocity of the ball [m/s];
F	is the correction factor.

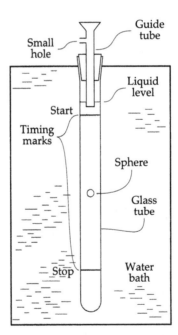

FIGURE 2.16   Schematic diagram of a 'Falling Ball Viscometer'.

The correction factor can be calculated from the formula given by Faxen [19]:

$$F = 1 - 2.104(d/D) + 2.09(d/D)^3 - 0.9(d/D)^5 \tag{2.19}$$

where:

d	is the diameter of the ball [m];
D	is the internal diameter of the tube [m].

There are also many other specialized viscometers designed to perform viscosity measurements under specific conditions, e.g., under high pressures, on very small volumes of fluid, etc. These viscometers are described in more specialized literature [e.g., 21].

2.8   VISCOSITY OF MIXTURES

In industrial practice it might be necessary to mix two similar fluids of different viscosities in order to achieve a mixture of a certain viscosity. The question is, how much of fluid '**A**' should be mixed with fluid '**B**'? This can simply be worked out by using ASTM viscosity paper with linear abscissa representing percentage quantities of each of the fluids. The viscosity of each of the fluids at the same temperature is marked on the ordinate on each side of the graph as shown in Figure 2.17. A straight line is drawn between these points and intersects a horizontal line which corresponds to the required viscosity. A vertical line drawn from the point of intersection crosses the abscissa, indicating the proportions needed of the two fluids. In the example of Figure 2.17, 20% of the less viscous component is mixed with 80% of the more viscous component to give the 'required viscosity'.

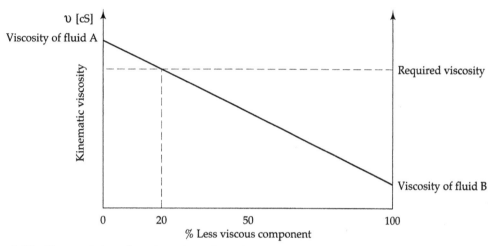

FIGURE 2.17   Determining the viscosity of a mixture.

2.9   OIL VISCOSITY CLASSIFICATION

There are several widely used oil viscosity classifications. The most commonly used are SAE (Society of Automotive Engineers), ISO (International Organization for Standardization) and military specifications.

SAE Viscosity Classification

Oils used in combustion engines and power transmissions are graded according to SAE J300 and SAE J306 classifications, respectively. A recent SAE classification establishes eleven engine oil and seven transmission oil grades [34,35]. Engine oil viscosities for different SAE grades are shown in Table 2.4.

Note that the viscosity in column 2 (Table 2.4) is the dynamic viscosity while column 3 shows the kinematic viscosity. The low temperature viscosity was measured by the 'cold-cranking simulator' and is an indicator of cold weather starting ability. The viscosity measurements at 100°C are related to the normal operating temperature of the engine. The oils without a '**W**' suffix are called '**monograde oils**' since they meet only one SAE grade. The oils with a '**W**' suffix, which stands for 'winter', have good cold starting capabilities. For climates where the temperature regularly drops below zero Celsius, engine and transmission oils are formulated in such a manner that they give low resistance at start, i.e., their viscosity is low at the starting temperature. Such oils have a higher viscosity index, achieved by adding viscosity improvers (polymeric additives) to the oil and are called '**multigrade oils**'. For example, SAE 20W/50 has a viscosity of SAE 20 at -18°C and viscosity of SAE 50 at 100°C as is illustrated in Figure 2.18. The problem associated with the use of multigrade oils is that they usually shear thin, i.e., their viscosity drops significantly with increased shear rates, due to the polymeric additives. This has to be taken into account when designing machine components lubricated by these oils. The drop in viscosity can be significant, and with some viscosity improvers even a permanent viscosity loss at high shear rates may occur due to the breaking up of molecules into smaller units. The viscosity loss affects the thickness of the lubricating film and subsequently affects the performance of the machine.

TABLE 2.4   SAE classification of engine oils [34].

SAE viscosity grade	Viscosity [cP] at temp [°C] max		Kinematic viscosity [cS] at 100°C	
	Cranking	Pumping	min	max
0W	3 250 at -30	30 000 at -35	3.8	-
5W	3 500 at -25	30 000 at -30	3.8	-
10W	3 500 at -20	30 000 at -25	4.1	-
15W	3 500 at -15	30 000 at -20	5.6	-
20W	4 500 at -10	30 000 at -15	5.6	-
25W	6 000 at -5	30 000 at -10	9.3	-
20	-	-	5.6	< 9.3
30	-	-	9.3	< 12.5
40	-	-	12.5	< 16.3
50	-	-	16.3	< 21.9
60	-	-	21.9	< 26.1

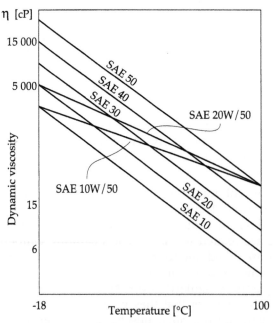

FIGURE 2.18 Viscosity-temperature graph for some monograde and multigrade oils (not to scale, adapted from [12]).

SAE classification of transmission oils is very similar to that of engine oils. The only difference is that the winter grade is defined by the temperature at which the oil reaches the viscosity of 150,000 [cP]. This is the maximum oil viscosity that can be used without causing damage to gears. The classification also permits multigrading. The transmission oil viscosities for different SAE grades are shown in Table 2.5 [35].

TABLE 2.5   SAE classification of transmission oils [35].

SAE viscosity grade	Max. temp. for viscosity of 150 000 cP [°C]	Kinematic viscosity [cS] at 100°C	
		min	max
70W	-55	4.1	-
75W	-40	4.1	-
80W	-26	7.0	-
85W	-12	11.0	-
90	-	13.5	< 24.0
140	-	24.0	< 41.0
250	-	41.0	-

It should also be noted that transmission oils have higher classification numbers than engine oils. As can be seen from Figure 2.19 this does not mean that they are more viscous than the engine oils. The higher numbers simply make it easier to differentiate between engine and transmission oils.

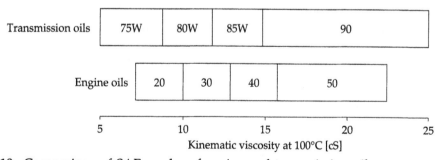

FIGURE 2.19   Comparison of SAE grades of engine and transmission oils.

ISO Viscosity Classification

The ISO (International Standards Organization) viscosity classification system was developed in the USA by the American Society of Lubrication Engineers (ASLE) and in the United Kingdom by The British Standards Institution (BSI) for all industrial lubrication fluids. It is now commonly used throughout industry. The industrial oil viscosities for different ISO viscosity grade numbers are shown in Table 2.6 [36] (ISO 3448).

2.10   LUBRICANT DENSITY AND SPECIFIC GRAVITY

Lubricant density is important in engineering calculations and sometimes offers a simple way of identifying lubricants. Density or specific gravity is often used to characterize crude oils. It gives a rough idea of the amount of gasoline and kerosene present in the crude. The oil density, however, is often confused with specific gravity.

Specific gravity is defined as the ratio of the mass of a given volume of oil at temperature 't_1' to the mass of an equal volume of pure water at temperature 't_2' (ASTM D941, D1217, D1298).

TABLE 2.6   ISO classification of industrial oils [36].

ISO viscosity grade	Kinematic viscosity limits [cSt] at 40°C		
	min.	midpoint	max.
ISO VG 2	1.98	2.2	2.42
ISO VG 3	2.88	3.2	3.52
ISO VG 5	4.14	4.6	5.06
ISO VG 7	6.12	6.8	7.48
ISO VG 10	9.00	10	11.0
ISO VG 15	13.5	15	16.5
ISO VG 22	19.8	22	24.2
ISO VG 32	28.8	32	35.2
ISO VG 46	41.4	46	50.6
ISO VG 68	61.2	68	74.8
ISO VG 100	90.0	100	110
ISO VG 150	135	150	165
ISO VG 220	198	220	242
ISO VG 320	288	320	352
ISO VG 460	414	460	506
ISO VG 680	612	680	748
ISO VG 1000	900	1000	1100
ISO VG 1500	1350	1500	1650

For petroleum products the specific gravity is usually quoted using the same temperature of 60°F (15.6°C). Density, on the other hand, is the mass per unit volume of oil [kg/m^3].

In the petroleum industry an API (American Petroleum Institute) unit is used which is a derivative of the conventional specific gravity. The API scale is expressed in degrees which in some cases are more convenient to use than specific gravity readings. The API specific gravity is defined as [23]:

$$\textbf{Degrees API} = (\textbf{141.5 / s}) - \textbf{131.5} \tag{2.20}$$

where:

s   is the specific gravity at 15.6°C (60°F).

As mentioned already the density of a typical mineral oil is about 850 [kg/m^3] and, since the density of water is about 1000 [kg/m^3], the specific gravity of mineral oils is typically 0.85.

2.11   THERMAL PROPERTIES OF LUBRICANTS

The most important thermal properties of lubricants are specific heat, thermal conductivity and thermal diffusivity. These three parameters are essential in assessing heating effects in lubrication, e.g., the cooling properties of the oil, the operating temperature of the surfaces, etc. They are also important in bearing design.

Specific Heat

Specific heat varies linearly with temperature and increases with increasing polarity or hydrogen bonding of the molecules. The specific heat of an oil is usually half that of water. For mineral and synthetic hydrocarbon based lubricants, specific heat is in the range from about 1800 [J/kgK] at 0°C to about 3300 [J/kgK] at 400°C. For a rough estimation of specific heat, the following formula can be used [5]:

$$\sigma = (1.63 + 0.0034\theta) / s^{0.5} \tag{2.21}$$

where:

σ is the specific heat [kJ/kgK];

θ is the temperature of interest [°C];

s   is the specific gravity at 15.6°C.

Thermal Conductivity

Thermal conductivity also varies linearly with temperature and is affected by polarity and hydrogen bonding of the molecules. The thermal conductivity of most of the mineral and synthetic hydrocarbon based lubricants is in the range between 0.14 [W/mK] at 0°C and 0.11 [W/mK] at 400°C. For a rough estimation of thermal conductivity the following formula can be used [5]:

$$K = (0.12 / s) \times (1 - 1.667 \times 10^{-4}\theta) \tag{2.22}$$

where:

K   is the thermal conductivity [W/mK];

θ is the temperature of interest [°C];

s   is the specific gravity at 15.6°C.

Thermal Diffusivity

Thermal diffusivity is the parameter describing the temperature propagation into the solids which is defined as:

$$\chi = K / \rho\sigma \tag{2.23}$$

where:

χ is the thermal diffusivity [m²/s];

K   is the thermal conductivity [W/mK];

ρ is the density [kg/m³];

σ is the specific heat [J/kgK].

The values of density, specific heat, thermal conductivity and thermal diffusivity for some typical materials are given in Table 2.7.

2.12 TEMPERATURE CHARACTERISTICS OF LUBRICANTS

The temperature characteristics are important in the selection of a lubricant for a specific application. In addition the temperature range over which the lubricant can be used is of extreme importance. At high temperatures, oils decompose or degrade by thermal decomposition or oxidation, while at low temperatures oils may become near solid or even freeze. During service, oils may release deposits and lacquers on contacting surfaces, form emulsions with water or produce a foam when vigorously churned. These effects are undesirable and have been the subject of intensive research. The degradation of oil does not just affect the oil, but more importantly can lead to damage or failure of the lubricated contacts. It may also result in detrimental secondary effects to the operating machinery. A prime example of secondary damage is corrosion caused by the acidity of oxidized oils. The most important temperature characteristics of a lubricant are its pour point, flash point, volatility, oxidation and thermal stability.

TABLE 2.7   Density, specific heat, thermal conductivity and thermal diffusivity values for some typical materials.

Material	Density at 20°C [kg/m^3]	Specific heat at 20°C [J/kgK]	Thermal conductivity at 100°C [W/mK]	Thermal diffusivity at 100°C [× 10^{-6} m^2/s]
Mineral oil	700 - 1 200	1 670	0.14	0.059 - 0.102
Water	1 000	4 184	0.58	0.16
Steel	7 800	460	46.7	13.02
Bronze	8 800	380	50 - 65	14.95 - 19.44
Brass	8 900	380	80 - 105	23.66 - 31.05
Aluminium (pure)	2 600	870	230	101.68
Aluminium (alloy)	2 700	870	120 - 170	51.09 - 72.37

Pour Point and Cloud Point

The pour point of an oil (ASTM D97, D2500) is the lowest temperature at which the oil will just flow when it is cooled. In order to determine the pour point the oil is first heated to ensure solution of all ingredients and elimination of any influence of past thermal treatment. It is then cooled at a specific rate and, at decrements of 3°C, the container is tilted to check for any movement. The temperature 3°C above the point at which the oil stops moving is recorded as the pour point. This oil property is important in the lubrication of any system exposed to low temperature, such as automotive engines, construction machines, military and space applications. When oil ceases to flow this indicates that sufficient wax crystallization has occurred or that the oil has reached a highly viscous state. At this stage waxes or high molecular weight paraffins precipitate from the oil. The waxes form the interlocking crystals that prevent the remaining oil from flowing. This is a critical point since the successful operation of a machine depends on the continuous supply of oil to the moving parts. The viscosity of the oil at the pour point is usually very large, i.e., several hundred [Pas] [24], but the exact value is of little practical significance since what is important is the minimum temperature at which the oil can be used.

The cloud point is the temperature at which paraffin wax and other materials begin to precipitate. The onset of wax precipitation causes a distinct cloudiness or haze visible in the bottom of a jar of oil. This phenomenon has some practical applications in capillary or wick fed systems in which the wax formed may obstruct the oil flow. It is limited only to transparent fluids since measurement is based purely on observation. If the cloud point of an oil is observed at a temperature higher than the pour point, the oil is said to have a 'Wax Pour Point'. If the pour point is reached without a cloud point the oil shows a simple 'Viscosity Pour Point'.

There is also another critical temperature known as the 'Flock Point', which is primarily limited to refrigerator oils. It is the temperature at which the oil separates from the mixture, which consists of 90% refrigerant and 10% oil. The Flock point provides an indication of how the oil reacts with a refrigerant, such as Freon, at low temperature.

Flash Point and Fire Point

The 'flash point' of the lubricant is the temperature at which its vapour will ignite. In order to determine the flash point the oil is heated at a standard pressure to a temperature which is just high enough to produce sufficient vapour to form an ignitable mixture with air. This is the flash point. The 'fire point' of an oil is the temperature at which enough vapour is produced to sustain burning after ignition. The schematic diagram of a flash and fire point apparatus is shown in Figure 2.20.

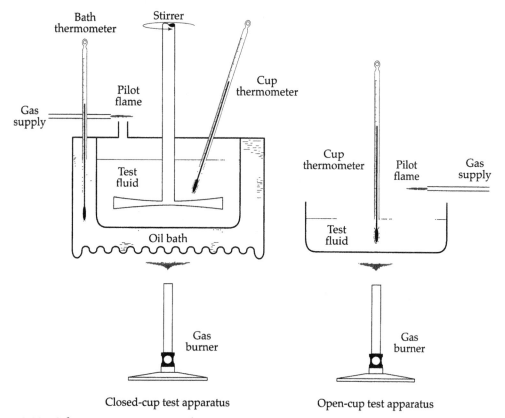

FIGURE 2.20   Schematic diagram of the flash and fire point apparatus.

Flash and fire points (ASTM D92, D93, D56, D1310) are very important from a safety view point since they constitute the only factors that define the fire hazard of a lubricant. In general, the flash point and fire point of oils increase with increasing molecular weight. For a typical lubricating oil, the flash point is about 210°C whereas the fire point is about 230°C.

Volatility and Evaporation

In many applications the loss of lubricant due to evaporation can be significant. The temperature has a controlling influence on evaporation. At elevated temperatures in particular, oils may become more viscous and greases tend to stiffen and eventually dry out because of evaporation. Volatile components of the lubricant may be lost through evaporation, resulting in a significant increase in viscosity and a further temperature rise due to higher friction, which in turn causes further oil losses due to evaporation. Volatility of lubricants is expressed as a direct measure of evaporation losses (ASTM D2715). In order to determine the lubricant volatility, a known quantity of lubricant is exposed in a vacuum thermal balance device. The evaporated material is collected on a condensing surface and the decreasing weight of the original material is expressed as a function of time. Depending on available equipment it is possible to obtain quantitative evaporation data together with some information on the identity of the volatile products. Frequently the evaporation rates are determined at various temperatures. A schematic diagram of an evaporation test apparatus is shown in Figure 2.21.

In this device a known quantity of oil is placed in a specially designed cup. The air enters the periphery of the cup and flows across the surface of the sample and exits through the centrally located tube. Prior to the test the cell is preheated to the required temperature in an oil bath. The flow rate of air is about 2 [litres/min]. The cup is aerated for 22 hours then cooled and weighed at the end of the test. The percentage of lost mass gives the evaporation rate.

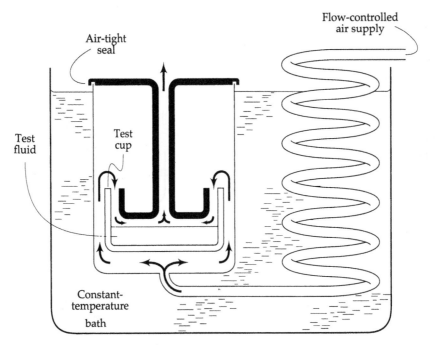

FIGURE 2.21 Schematic diagram of the evaporation test apparatus.

Oxidation Stability

Oxidation stability (ASTM D943, D2272, D2893, D1313, D2446) is the resistance of a lubricant to molecular breakdown or rearrangement at elevated temperatures in the ordinary air environment. Lubricating oils can oxidize when exposed to air, particularly at elevated temperatures, and this has a very strong influence on the life of the oil. The rate of oxidation depends on the degree of oil refinement, temperature, presence of metal catalysts and operating conditions [25,26]. It increases with temperature.

Oxidation of oils is a complex process as different compounds are generated at different temperatures. For example, at about 150°C organic acids are produced whereas at higher temperatures aldehydes are formed [24]. The oxidation rates vary between different compounds, as shown in the frame below.

One way of improving oxidation stability is to remove the hydrocarbon type aromatics and molecules containing sulphur, oxygen, nitrogen, etc. This is achieved through refining. More refined oil has better oxidation stability. It is also more expensive and has poorer boundary lubrication characteristics, so oil selection for a particular application is always a compromise, depending on the type of job the oil is expected to perform. Oxidation can also be controlled by additives that attack the hyperoxides formed in the initial stages of oxidation or break the chain reaction mechanism by scavenging free radicals. The products of oxidation usually consist of acidic compounds, sludge and lacquers. All of these compounds cause oil to become more corrosive, more viscous and also cause the deposition of insoluble products on working surfaces, restricting the flow of oil in operating machinery. This interferes with the performance of the machinery. Oxidation stability is a very important oil characteristic, especially where extended life is required, e.g., turbines, transformers, hydraulic and heat transfer units, etc. A lubricant with limited oxidation stability requires more frequent maintenance or replacement, resulting in higher operating costs. Under more severe conditions the oil changes may need to take place more frequently, hence the operating costs will be even higher. Many tests have been devised to assess the oxidation characteristics of oils and there is no clear rationale for selecting a particular test [32]. Some have been devised for specific applications, for example, the assessment of oxidation characteristics of railway diesel engine lubricants [27]. In most test apparatus the oil is in contact with selected catalysts and is exposed to air or oxygen, and the effects are measured in terms of acid or sludge formed, viscosity change, etc. A schematic diagram of a typical oxidation apparatus is shown in Figure 2.22.

In this apparatus oxygen is passed through the oil sample placed in the reaction vessel. The reaction vessel consists of a large test tube with a smaller central removable oxygen inlet tube which supports the steel-copper catalyst coil. At the end of the tube there is a water cooled condenser which returns the more volatile components to the reaction. About 300 [ml] of oil together with 60 [ml] of distilled water is placed in the test tube. The flow rate of oxygen is about 0.5 [litre/min] and the test is conducted at a temperature of 95°C. During the test acidic compounds are produced in the tube, and the neutralization number determined at the end of the test is a measure of the oxidation stability of the oil. The tests are usually run over a specific period of time. It has to be mentioned, however, that ASTM oxidation tests are still under revision [28] and new techniques are being developed. For example, Differential Scanning Calorimetry has been employed to assess oxidation stability of oils [e.g., 40-44].

Thermal Stability

When heated above a certain temperature oils will start to decompose, even if no oxygen is present. Thermal stability is the resistance of the lubricant to molecular breakdown or molecular rearrangement at elevated temperatures in the absence of oxygen. When heated, mineral oils break down to methane, ethane and ethylene. Thermal stability can be improved by the refining process, but not by additives. It can be measured by placing the oil in a closed vessel with a manometer monitoring the rate of pressure increase when the container is heated at a specific rate under nitrogen atmosphere. Mineral oils with a substantial percentage of **C-C** single bonds have a thermal stability limit of about 350°C. Synthetic oils, in general, exhibit better thermal stability than mineral oils. However, there can be exceptions. For example, synthetic hydrocarbons produced by the polymerization or oligomerization process, although possessing the same basic structures as mineral oils, have a thermal stability limit 28°C or more below that of mineral oils [22]. Lubricants with aromatic linkages or with aromatic linkages and methyl groups as side chains exhibit a thermal stability limit of about 460°C. The additives used for lubrication improvement usually have a thermal stability below that of base oils. In general, thermal degradation of the oil takes place at much higher temperatures than oxidation. Thus the maximum

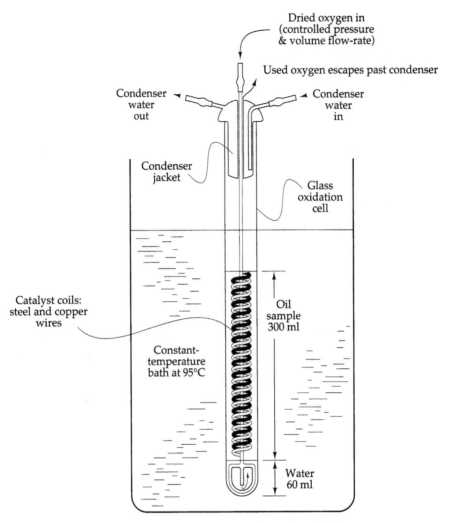

FIGURE 2.22 Schematic diagram of an oxidation test apparatus [23].

temperature at which an oil can be used is determined by its oxidation stability. In Figures 2.23 and 2.24 the relationships between lubricant life and temperature are shown for mineral and synthetic oils, respectively [29].

2.13 OTHER LUBRICANT CHARACTERISTICS

There are many other lubricant characteristics described in the literature and the most frequently used include surface tension, neutralization number and carbon residue.

Surface Tension

Various lubricants generally show some differences in the degree of wetting and spreading on surfaces. Furthermore, the same lubricant can show different wetting and spreading characteristics depending on the degree of oxidation or on the modification of the lubricant by additives. The phenomena of wetting and spreading are dependent on surface tension ·(ASTM D971, D2285), which is especially sensitive to additives, e.g., less than 0.1 wt% of silicone in mineral oil will reduce the surface tension of the oil to that of silicone [22]. Surface and interfacial tension are related to the free energy of the surface, and the attraction between

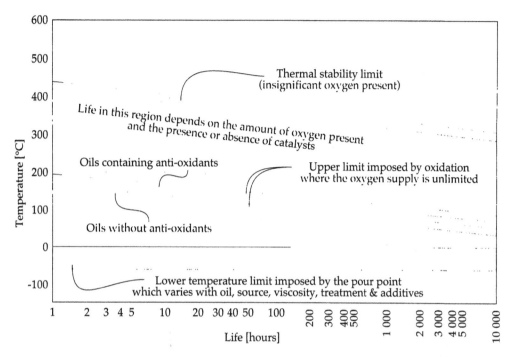

FIGURE 2.23 Temperature-life limits for mineral oils [29].

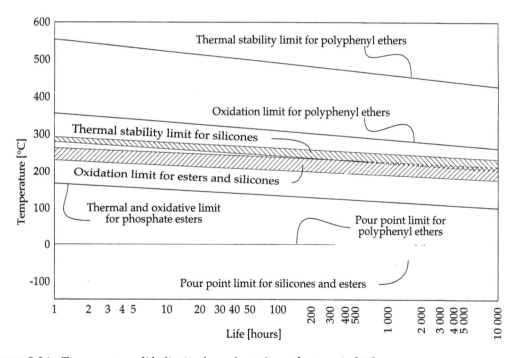

FIGURE 2.24 Temperature-life limits for selected synthetic oils [29].

the surface molecules is responsible for these phenomena. Surface tension refers to the free energy at a gas-liquid interface, while interfacial tension takes place at the interface between two immiscible liquids. Surface tension can be measured by the du Noy ring method (ASTM D971). A schematic diagram of surface tension measurement principles is shown in Figure

2.25. It involves the measurement of the force necessary to detach the platinum wire ring from the surface of the liquid. The surface tension is then calculated from the following formula [22]:

$$\sigma_s = F / 4\pi r \qquad (2.24)$$

where:

σ_s is the surface tension [N/m];

F   is the force [N];

r   is the radius of the platinum ring [m].

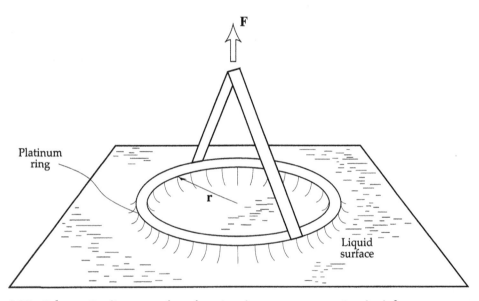

FIGURE 2.25   Schematic diagram of surface tension measurement principles.

Typical values of surface tension for some basic fluids are shown in Table 2.8 [22]. Surface tension is frequently used together with the neutralization number as a measure of oil deterioration in transformers, hydraulic systems and turbines. Interfacial tension between two immiscible liquids is approximately equal to the difference in the surface tension between the two liquids.

TABLE 2.8   Surface tension of some basic fluids [22].

Fluid	Surface tension [$\times 10^{-3}$ N/m]
Water	72
Mineral oils	30 - 35
Esters	30 - 35
Methylsilicone	20 - 22
Fluorochloro compounds	15 - 18
Perfluoropolyethers	19 - 21

Neutralization Number

The neutralization number of a lubricant (ASTM D974, D664) is the quantity in milligrams of potassium hydroxide (KOH) per gram of oil necessary to neutralize acidic or alkaline compounds present in the lubricant. The procedure described in D664 is the most popular method for determining the acidic condition of the oil. The results are reported as a **Total Acid Number** (TAN) for acidic oils and as a **Total Base Number** (TBN) for alkaline oils. TAN is expressed as the amount of potassium hydroxide in milligrams necessary to neutralize 1 gram of oil. TBN is the amount of potassium hydroxide in milligrams necessary to neutralize the hydrochloric acid (HCl) which would be required to remove the basicity in 1 gram of oil. So, the TAN is a measure of acidic matter remaining in the oil and the TBN is the measure of alkaline matter remaining in the oil. In general, TBN applies only to the oil supplied with alkaline additives to suppress sulphur based acid formation in the presence of low grade fuels such as diesel engine lubricants. Thus TBN is a negative measure of oil acidity and a minimum value should be maintained. On the other hand, the TAN number applies to most oils since they are normally weakly acidic. During the test, the neutralizing solution is added until all acid or alkaline ingredients are neutralized. The neutralization number is useful in assessing changes in the lubricant that occur during service under oxidizing conditions. It is frequently used in conjunction with the other parameters, such as interfacial tension, in lubricant condition monitoring. The best test results are achieved in systems which are relatively free of contaminants such as steam turbine generators, transformers, hydraulic systems, etc. It can also be used in the condition monitoring of oils operating in engines, compressors, gears and as cutting fluids. Usually a limiting neutralization number is established as a criterion indicating when oil needs to be changed or reclaimed.

Carbon Residue

At temperatures of 300°C or more in the absence of air, oils may decompose to produce low molecular weight fragments from the large molecular weight species typically found in mineral oils. The fragmented or 'cracked' hydrocarbon molecules either recombine to form tarry deposits (asphaltenes) or are released to the atmosphere as volatile components [30]. The deposits are undesirable in almost all cases and most lubricating oils are tested for deposit forming tendencies. The carbon residue (ASTM D189, D524) is determined by weighing the residue after the oil has been heated to a high temperature in the absence of air. The carbon residue parameter is of little importance in the case of synthetic oils because of their good thermal stability. It is also infrequently used in characterizing well refined lubricants.

2.14 OPTICAL PROPERTIES OF LUBRICANTS

Refractive Index

The refractive index (ASTM D1218, D1747) is defined as the ratio of the velocity of a specified wavelength of light in air to that in the oil under test and it can be measured by an Abbe refractometer. It is a function of temperature and pressure. The refractive index of very viscous lubricants is measured at temperatures between 80 and 100°C and of typical oils at 20°C. Refractive index is sensitive to oil composition and hence it is useful in characterizing base stocks. It is very important in calculations of minimum film thickness in experiments involving optical interferometry, as discussed later in Chapter 7. For most mineral oils, the value of the refractive index at atmospheric pressure is about **1.51** [12]. It can also be roughly estimated from the formula:

$$(n^2 - 1) / (n^2 + 2) = \rho c \qquad\qquad (2.25)$$

where:

   n   is the refractive index of the lubricant;

   ρ is the oil lubricant density [g/cm^3];

   c   is a constant. For example, for SAE 30, **c** = 0.33 [12].

2.15   ADDITIVE COMPATIBILITY AND SOLUBILITY

Additives used in lubricants should be compatible with each other and soluble in the lubricant. These additive features are defined as additive compatibility and additive solubility.

Additive Compatibility

Two or more additives in an oil are compatible if they do not react with each other and if their individual properties are beneficial to the functioning of the system. It is usually considered that additives are compatible if they do not give visible evidence of reacting together, such as a change in colour or smell. This also refers to the compatibility of two or more finished lubricants.

Lubricants should also be compatible with the component materials used in a specific application. For example, mineral oils are incompatible with natural rubber, and phosphate esters are incompatible with many different rubbers. Mineral oils give very poor performance with red hot steels because they produce carburization while rapeseed oil does not have this problem. In most industries these problems can be overcome by careful selection of lubricants. On the other hand, in some industries, such as pharmaceutical and food processing where oil leaks are unacceptable, process fluids might be used as lubricants. For example, in sugar refining high viscosity syrups and molasses can be used, if necessary, to lubricate the bearings, but they are in general poor lubricants and their use may lead to severe equipment or machinery problems.

Additive Solubility

The additive must dissolve well in petroleum products. It must remain dissolved over the entire operating temperature range. Separation of an additive in storage or in service is highly undesirable. For example, elemental sulphur could be used as an additive in extreme conditions of temperature and pressure but it is insoluble in oil under atmospheric conditions and would separate during storage and service.

2.16   LUBRICANT IMPURITIES AND CONTAMINANTS

Water Content

Water content (ASTM D95, D1744, D1533, D96) is the amount of water present in the lubricant. It can be expressed as parts per million, percent by volume or percent by weight. It can be measured by centrifuging, distillation and voltametry. The most popular, although least accurate, method of water content assessment is the centrifuge test. In this method a 50% mixture of oil and solvent is centrifuged at a specified speed until the volumes of water and sediment observed are stable. Apart from water, solids and other solubles are also separated and the results obtained do not correlate well with those obtained by the other two methods.

The distillation method is a little more accurate and involves distillation of oil mixed with xylene. Any water present in the sample condenses in a graduated receiver.

The voltametry method is the most accurate. It employs electrometric titration, giving the water concentration in parts per million.

Corrosion and oxidation behaviour of lubricants is critically related to water content. An oil mixed with water gives an emulsion. An emulsion has a much lower load carrying capacity than pure oil and lubricant failure followed by damage to the operating surfaces can result. In general, in applications such as turbine oil systems, the limit on water content is 0.2% and for hydraulic systems 0.1%. In dielectric systems excessive water content has a significant effect on dielectric breakdown. Usually the water content in such systems should be kept below 35 [ppm].

Sulphur Content

Sulphur content (ASTM D1266, D129, D1662) is the amount of sulphur present in an oil. It can have some beneficial, as well as some detrimental, effects on operating machinery. Sulphur is a very good boundary agent, which can effectively operate under extreme conditions of pressure and temperature. On the other hand, it is very corrosive. A commonly used technique for the determination of sulphur content is the bomb oxidation technique. It involves the ignition and combustion of a small oil sample under pressurised oxygen. The sulphur from the products of combustion is extracted and weighed.

Ash Content

There is some quantity of noncombustible material present in a lubricant which can be determined by measuring the amount of ash remaining after combustion of the oil (ASTM D482, D874). The contaminants may be wear products, solid decomposition products from a fuel or lubricant, atmospheric dust entering through a filter, etc. Some of these contaminants are removed by an oil filter but some settle into the oil. To determine the amount of contaminant, the oil sample is burned in a specially designed vessel. The residue that remains is then ashed in a high temperature muffle furnace and the result displayed as a percentage of the original sample. The ash content is used as a means of monitoring oils for undesirable impurities and sometimes additives. In used oils it can also indicate contaminants such as dirt, wear products, etc.

Chlorine Content

The amount of chlorine in a lubricant should be at an optimum level. Excess chlorine causes corrosion whereas an insufficient amount of chlorine may cause wear and frictional losses to increase. Chlorine content (ASTM D808, D1317) can be determined either by a bomb test which provides the gravimetric evaluation or by a volumetric test which gives chlorine content, after reacting with sodium metal to produce sodium chloride, then titrating with silver nitride [22].

2.17 SOLUBILITY OF GASES IN OILS

Almost all gases are soluble in oil to a certain extent. Oxygen dissolved in oil affects friction and wear of metal surfaces and this is discussed in the next chapters. Bubbles of gas (usually air) which are released in the oil of hydraulic systems due to the drop in pressure may cause a drastic increase in the compressibility of the hydraulic fluid, affecting the overall performance of the system.

The solubility of a gas in a liquid is calculated from the Ostwald coefficient, which is defined as the ratio of the volume of dissolved gas to the volume of solvent liquid at the test temperature and pressure. For example, if the Ostwald coefficient is equal to 0.2 then 5 litres of oil will contain $0.2 \times 5 = 1$ litre of dissolved gas. The solubility of a gas in a liquid is usually proportional to pressure, so that the Ostwald coefficient, defined in terms of the volume of gas, remains constant. On the other hand, to define this coefficient in terms of mass would require the introduction of a pressure proportionality parameter. Hence the coefficient defined in terms of volume of gas is commonly used. The formulae necessary for the evaluation of the Ostwald coefficient (ASTM D2779) are empirical and the procedure is carried out in two steps.

In the first step the Ostwald coefficient for a reference liquid which is only a function of temperature is calculated from the formula:

$$\mathbf{C_{o,r} = 0.3} \times \mathbf{e}^{[(0.639(700\text{-}T)/T) \times \ln(3.333C_{o,d})]}$$
(2.26)

where:

$\mathbf{C_{o,r}}$ is the Ostwald coefficient of the reference liquid at the specified temperature;

$\mathbf{T}$ is the absolute temperature [K];

$\mathbf{C_{o,d}}$ is the Ostwald coefficient for a specific gas dissolved in the reference liquid at standard temperature (273K).

Next, the Ostwald coefficient for the lubricant of interest is evaluated based on the lubricant density 'ρ' and the calculated reference fluid coefficient '$C_{o,r}$', i.e.:

$$\mathbf{C_o = 7.70} \times \mathbf{C_{o,r}} \times \mathbf{(980 - \rho)}$$
(2.27)

where:

ρ   is the density of the oil in [kg/m^3].

The Ostwald coefficients '$C_{o,d}$' for typical gases dissolved in hydrocarbons at the standard temperature of 273K are shown in Table 2.9 [37].

TABLE 2.9   Ostwald coefficients '$C_{o,d}$' for typical gases dissolved in hydrocarbons at 273K [37].

Gas	$\mathbf{C_{o,d}}$
Helium	0.010
Neon	0.021
Hydrogen	0.039
Nitrogen	0.075
Air	0.095
Carbon monoxide	0.10
Oxygen	0.15
Argon	0.23
Methane	0.31
Carbon dioxide	1.0
Krypton	1.3

One of the serious limitations of the method above is that it applies only to mineral oils. A more general formula based on a combination of linear regression of experimental results and detailed application of solvation theory has been developed by Beerbower [37]. Two new parameters were introduced in the formula: a measure of the solvation capacity of the lubricant '∂_1' and a gas solubility parameter '∂_2'. The previously used formulae for the determination of the Ostwald coefficient for a particular lubricant were replaced by the following, single expression:

$$\mathbf{ln C_0 = [0.0395(\partial_1 - \partial_2)^2 - 2.66] \times (1 - 273/T) - 0.303\partial_1 - 0.0241(17.6 - \partial_2)^2 + 5.731} \qquad (2.28)$$

Values of '∂_1' and '∂_2' parameters for some typical lubricants and gases are shown in Table 2.10 [37]. This formula gives good results for temperatures above 0°C (273K). At sub-zero temperatures, however, experimental confirmation of the calculated values would be necessary.

It has to be pointed out that some gases such as oxygen are reactive to most hydrocarbons and so are continuously absorbed by mineral oil instead of saturating to some equilibrium value. This phenomenon is related to oil oxidation and will be discussed in the next chapter. The solubility of gaseous oxygen, however, remains unaffected by the gradual oxidation process until most of the oil changes its composition.

TABLE 2.10   Values of '∂_1' and '∂_2' parameters for some typical lubricants and gases [37].

∂_1		∂_2	
Lubricant	∂_1 [MPa$^{0.5}$]	Gas	∂_2 [MPa$^{0.5}$]
Di-2-ethylhexyl adipate	18.05	He	3.35
Di-2-ethylhexyl sebacate	17.94	Ne	3.87
Trimetholylpropane pelargonate	18.18	H_2	5.52
Pentaerythritol caprylate	18.95	N_2	6.04
Di-2-ethylhexyl phthalate	18.97	Air	6.69
Diphenoxy diphenylene ether	23.21	CO	7.47
Polychlorotrifluoro ethylene	15.19	O_2	7.75
Polychlorotrifluoro ethylene	15.55	Ar	7.77
Polychlorotrifluoro ethylene	15.77	CH_4	9.10
Dimethyl silicone	15.14	CO_2	14.81
Methyl phenyl silicone	18.41	Kr	10.34
Perfluoropolyglycol	14.20		
Tri-2-ethylhexyl phosphate	18.29		
Tricresyl phosphate	18.82		

EXAMPLE

Find the quantity of air that could be dissolved in 1 litre of dimethyl silicone oil at 100°C.

From Table 2.10, for dimethyl silicone oil ∂_1 = 15.14 [MPa$^{0.5}$] and for air ∂_2 = 6.69 [MPa$^{0.5}$]. Absolute oil temperature is 373K.

Substituting these values into the above equation yields the Ostwald coefficient of air in dimethyl silicone at 373K, i.e.:

> **ln C$_0$** = [0.0395 × (15.14 − 6.69)^2 − 2.66] × (1 − 273/373) − 0.303 × 15.14 − 0.0241 ×
>   (17.60 − 6.69)^2 + 5.731
>
>   = (2.8204 − 2.66) × 0.2681 − 4.5874 − 2.8686 + 5.731
>
>   = −1.6820
>
>   **C$_0$** = **0.1860**
>
> Which means that in every litre of dimethyl silicone oil, approximately 186 [ml] of air can be dissolved at 100°C.

2.18   SUMMARY

The fundamental physical properties of a lubricant that determine its lubrication and performance characteristics have been discussed in this chapter. There are many other parameters that describe the different physical properties of an oil, which can be found in the literature. In most instances, however, the specified parameters are those mentioned in this chapter. The most frequently specified parameters are those that describe the oil's lubrication characteristics and some of its main performance characteristics. In some cases there might be little variation between oils for a given parameter, or sometimes the importance of a particular parameter is not sufficiently appreciated. With the rapid development of synthetic lubricants, there will most likely be more profound differences between lubricants, and hence a greater range of specifications required. Further investigations into the lubrication mechanisms of oils could reveal even more controlling parameters.

REVISION QUESTIONS

2.1.   What range of viscosity covers most oils?

2.2   In general, how many temperatures are needed to specify the viscosity of an oil and why?

2.3   How is the viscosity index defined and can it be accurately used to evaluate modern oils?

2.4   What is the disadvantage of the viscosity index?

2.5   Is the viscosity of low viscosity index (VI) oils less sensitive to temperature than high VI oils?

2.6   What other parameter, apart from viscosity, is significant when dealing with synthetic oils?

2.7   Many animal and plant oils and fats make excellent lubricants yet they are rarely used. Why should this be?

2.8   When choosing a lubricant for food and drug processing machinery what property of the oil must be considered?

2.9   An oil has a viscosity index of **VI = 92** and density of ρ = **895** [kg/m^3]. What is the dynamic viscosity η of this oil at **20°C**?

   The kinematic viscosity υ at **40°C** of a reference oil having zero viscosity index and the same kinematic viscosity at **100°C** as the oil investigated is $\upsilon_{40°C,ref}$ = **253.89** [cS]. (Ans. $\eta_{20°C}$ = **253.89** [cP])

2.10   A lubricant's viscosity at **-20°C** is **1000** [cS] and the ASTM slope for this lubricant is **0.6**. What is the viscosity index of this lubricant? (Ans. VI = 170)

2.11   What is the viscosity index of the mixture consisting of **400** litres of SAE 10W and **200** litres of SAE 20W?

   Oil density of both SAE 10W and SAE 20W is ρ = **850** [kg/m^3]. (Ans. VI = 48)

2.12   What is the SAE grade of the lubricant that has viscosity index of about **VI = 120** and dynamic viscosity at **20°C** of $\eta_{20°C}$ = **0.432** [Pas]. The lubricant's density is ρ = **900** [kg/m^3]. (Ans. SAE20W/50)

2.13   Three viscosities of a lubricating oil were determined at three different temperatures, i.e., **1000** [cS] at **0°C**, **75** [cS] at **40°C** and **10** [cS] at **100°C**. Using the Vogel viscosity-temperature equation determine the oil's viscosity at **-20°C**. What is the viscosity index of this oil? The lubricant's density is ρ = **900** [kg/m^3]. (Ans. $\eta_{-20°C}$ = 7.33 [cP])

REFERENCES

1   E.W. Dean and G.H.B. Davis, Viscosity Variations of Oils With Temperature, *Chem. and Met. Eng.*, Vol. 36, 1929, pp. 618-619.

2   G.H.B. Davis, G.M. Lapeyrouse and E.W. Dean, Applying Viscosity Index to Solution of Lubricating Problems, *Journal of Oil and Gas*, Vol. 30, 1932, pp. 92-93.

3   Standard Practice for Calculating Viscosity Index From Kinematic Viscosity at 40 and 100°C, ASTM D2270 - 86, 1986.

4   C. Barus, Isotherms, Isopiestics and Isometrics Relative to Viscosity, *American Journal of Science*, Vol. 45, 1893, pp. 87-96.

5   A. Cameron, Basic Lubrication Theory, Ellis Horwood Limited, 1981.

6   L.B. Sargent Jr, Pressure-Viscosity Coefficients of Liquid Lubricants, *ASLE Transactions*, Vol. 26, 1983, pp. 1-10.

7   P.S.Y. Chu and A. Cameron, Pressure Viscosity Characteristics of Lubricating Oils, *Journal of the Institute of Petroleum*, Vol. 48, 1962, pp. 147-155.

8   A. Cameron, The Principles of Lubrication, Longmans Green and Co. Ltd., 1966.

9   B.Y.C. So and E.E. Klaus, Viscosity-Pressure Correlation of Liquids, *ASLE Transactions*, Vol. 23, 1980, pp. 409-421.

10 H. van Leeuwen, Discussion to the paper by L.B. Sargent on Pressure-Viscosity Coefficients of Liquid Lubricants, *ASLE Transactions*, Vol. 26, 1983, pp. 1-10.

11 C.J.A. Roelands, Correlational Aspect of Viscosity-Temperature-Pressure Relationships of Lubricating Oils, PhD thesis, Delft University of Technology, The Netherlands, 1966.

12 R. Gohar, Elastohydrodynamics, Ellis Horwood Limited, 1988.

13 L. Houpert, New Results of Traction Force Calculation in EHD Contacts, *Transactions ASME, Journal of Lubrication Technology*, Vol. 107, 1985, pp. 241-248.

14 Engineering Science Data Unit International Limited, Film Thickness in Lubricated Hertzian Contacts (EHL), Part 1, No. 85027, October, 1985.

15 W.G. Johnston, A Method to Calculate the Pressure-Viscosity Coefficient from Bulk Properties of Lubricants, *ASLE Transactions*, Vol. 24, 1981, pp. 232-238.

16 D. Klamann, Lubricants and Related Products, Publ. Verlag Chemie, Weinheim, 1984, pp. 51-83.

17 K.E. Bett and J.B. Cappi, Effect of Pressure on the Viscosity of Water, *Nature*, Vol. 207, 1965, pp. 620-621.

18 J. Wonham, Effect of Pressure on the Viscosity of Water, *Nature*, Vol. 215, 1967, pp. 1053-1054.

19 J. Halling, Principles of Tribology, The MacMillan Press, 1975.

20 P.L. O'Neill and G.W. Stachowiak, The Lubricating Properties of Arthritic Synovial Fluid, 1st World Congress in Bioengineering, San Diego, Vol. II, 1990, pp. 269.

21 Z. Rymuza, Tribology of Miniature Systems, Elsevier, Tribology Series 13, 1989.

22 E.R. Booser, CRC Handbook of Lubrication, Volume I and II, CRC Press, Boca Raton, Florida, 1984.

23 J.J. O'Connor, J. Boyd and E.A. Avallone, Standard Handbook of Lubrication Engineering, McGraw-Hill Book Company, 1968.

24 H.H. Zuidema, The Performance of Lubricating Oils, Reinhold Publ., New York, 1952, pp. 26-28.

25 E.E. Klaus, D.I. Ugwuzor, S.K. Naidu and J.L. Duda, Lubricant-Metal Interaction Under Conditions Simulating Automotive Bearing Lubrication, Proc. JSLE International Tribology Conf., 8-10 July 1985, Tokyo, Japan, Elsevier, pp. 859-864.

26 T. Colclough, Role of Additives and Transition Metals in Lubricating Oil Oxidation, *Ind. Eng. Chem. Res.*, Vol. 26, 1987, pp. 1888-1895.

27 R.D. Stauffer and J.L. Thompson, Improved Bench Oxidation Tests for Railroad Diesel Engine Lubricants, *Lubrication Engineering*, Vol. 44, 1988, pp. 416-423.

28 T.M. Warne, Oxidation Test Method Development by ASTM Subcommittee D02. 09. ASTM Technical Publication 916, editors W.H. Stadtmiller and A.N. Smith, ASTM Committee D-2, Symposium on Petroleum Products and Lubricants, Miami, Florida USA, 5 December 1983.

29 M.H. Jones and D. Scott, Industrial Tribology, The Practical Aspects of Friction, Lubrication and Wear, Elsevier, 1983.

30 T.I. Fowle, Lubricants for Fluid Film and Hertzian Contact Conditions, *Proc. Inst. Mech. Engrs.*, Vol. 182, Pt. 3A, 1967-1968, pp. 508-584.

31 R.S. Porter and J.F. Johnson, Viscosity Performance of Lubricating Base Oils at Shears Developed in Machine Elements, *Wear*, Vol. 4, 1961, pp. 32-40.

32 A.N. Smith, Turbine Lubricant Oxidation: Testing, Experience and Prediction, ASTM Technical Publication 916, editors, W.H. Stadtmiller and A.N. Smith, ASTM Committee D-2, Symposium on Petroleum Products and Lubricants, Miami, Florida USA, 5 December 1983.

33 R.F. Crouch and A. Cameron, Viscosity-Temperature Equations for Lubricants, *Journal of the Institute of Petroleum*, Vol. 47, 1961, pp. 307-313.

34 SAE Standard, Engine Viscosity Classification, SAE J300, June 1989.

35 SAE Standard, Axle and Manual Transmission Lubricant Viscosity Classification, SAE J306, March 1985.

36 International Standard, Industrial Liquid Lubricants, ISO Viscosity Classification, ISO 3448, 1975.

37 A. Beerbower, Estimating the Solubility of Gases in Petroleum and Synthetic Lubricants, *ASLE Transactions*, Vol. 23, 1980, pp. 335-342.

38 H. Spikes, Lecture Notes, Imperial College of Science and Technology, London, 1982.

39 P.L. O'Neill and G.W. Stachowiak, The Thixotropic Behaviour of Synovial Fluid, Proc. XIV I.S.B. Congress in Biomechanics, Paris, 4-8 July, 1993, editors: S. Bouisset, S. Metral and H. Mond, Publ. Societe de Biomecanique, Vol. II, 1993, pp. 970-971.

40 R.E. Kauffman and W.E. Rhine, Development of a Remaining Useful Life of a Lubricant Evaluation Technique. Part I: Differential Scanning Calorimetric Techniques, *Lubrication Engineering*, Vol. 44, 1988, pp. 154-161.

41 S.M. Hsu, A.L. Cummings and D.B. Clark, Differential Scanning Calorimetry Test for Oxidation Stability of Engine Oil, Proc. of the Conf. on Measurements and Standards for Recycled Oil, National Bureau of Standards, Washington, DC, 1982, pp. 195-207.

42 W.F. Bowman and G.W. Stachowiak, Determining the Oxidation Stability of Lubricating Oils Using Sealed Pan Differential Scanning Calorimetry (SPDSC), *Tribology International*, Vol. 29, No. 1, 1996, pp. 27-34.

43 W.F. Bowman and G.W. Stachowiak, New Criteria to Assess the Remaining Useful Life of Industrial Turbine Oils, *Lubrication Engineering*, Vol. 52, No. 10, 1996, pp. 745-750.

44 W.F. Bowman and G.W. Stachowiak, Application of Sealed Capsule Differential Scanning Calorimetry, Part I: Predicting the Remaining Useful Life of Industry Used Turbine Oils, *Lubrication Engineering*, Vol. 54, No. 10, 1998, pp. 19-24.

3

L U B R I C A N T S
A N D
T H E I R   C O M P O S I T I O N

3.1   INTRODUCTION

In the previous chapter physical properties of lubricants, specifically oils and greases, have been outlined. The questions which remain, however, are what is the chemical composition of the lubricant? Will the lubricant's composition differ depending on its application? What is the basic makeup of the oil used to lubricate machinery? It is readily appreciated that oil manufacturers are so circumspect about the formulation of their products. The oil ingredients are not listed on the lubricant containers as they are listed on the packaging of most other products. How then, can one differentiate between various oils? What are the differences between mineral and synthetic oils? What are the typical additives used in oils? What is their purpose and mechanism of action? What is the composition and properties of grease lubricants? In what applications are greases usually used? An engineer, as a potential user of lubricants, should know the answers to these questions.

Oils can be of two different origins, biological and non-biological, and this provides a vast array of hydrocarbon compounds. These substances are usually present as complex mixtures and can be used for many other purposes besides lubrication, that is, the control of wear and friction. Modern technology places severe and varied demands on lubricants, so the selection and formulation of appropriate mixtures of hydrocarbons for the purposes of lubrication are skilled and complex processes. Most natural oils contain substances which can hinder their lubrication properties, but they also contain compounds essential to the lubrication process. Lubricants made from natural or mineral oils are partly refined and partly impure. The balance between impurity and purity is critical to the oxidation stability of the oil and it varies depending on the application of the lubricant. Chemicals which are deliberately added to an oil in order to improve its properties are called additives. Additives can radically change the properties of a lubricant and are essential to its overall performance. They also dictate specific characteristics of the lubricant such as corrosion tendency, foaming, clotting, oxidation, wear, friction and other properties.

There are two fundamental aspects of lubricant performance: achieving the required level of friction and wear rates, and maintaining these standards despite continuous degradation of the lubricant. Chemical reaction of the lubricant with atmospheric oxygen and water is inevitable since the lubricant is essentially a hydrocarbon. Additives present in the oil also deteriorate during operation since they react with the metallic parts of the machinery and with the environment. The degradation of the lubricant is inevitable and must be postponed until the required lifetime is achieved. In fact, a large part of lubricant technology is devoted to the preservation of lubricating oils when in use.

A typical lubricating oil is composed of 95% base stock and 5% additives. Base stock is the term used to describe plain mineral oil. The physical properties of an oil depend on its base stock. In most cases it is chemically inert. There are three sources of base stock: biological, mineral and synthetic. The oils manufactured from these sources exhibit different properties and they are suitable for different applications. For example:

· biological oils are suitable in applications where the risk of contamination must be reduced to a minimum, for example, in the food or pharmaceutical industry. They are usually applied to lubricate kilns, bakery ovens, etc. There can be two sources of this type of oil: vegetable and animal. Examples of vegetable oils are castor, palm and rape-seed oils while the examples of animal oils are sperm, fish and wool oils from sheep (lanolin).

· mineral oils are the most commonly used lubricants throughout industry. They are petroleum based and are used in applications where temperature requirements are moderate. Typical applications of mineral oils are to gears, bearings, engines, turbines, etc.

· synthetic oils are artificially developed substitutes for mineral oils. They are specifically developed to provide lubricants with superior properties to mineral oils. For example, temperature resistant synthetic oils are used in high performance machinery operating at high temperatures. Synthetic oils for very low temperature applications are also available.

Greases are not fundamentally different from oils. They consist of mineral or synthetic oil, but the oil is trapped in minute pockets formed by soap fibres which constitute the internal structure of the grease. Hence a grease is classified as 'mineral' or 'synthetic' according to the base stock used in its production. Greases have been developed especially to provide semi-permanent lubrication since the oil trapped in the fibrous structure is unable to flow away from the contacting surfaces. For this reason greases are widely used despite certain limitations in performance.

In this chapter the basic composition of mineral oils, synthetic oils and greases, such as their base stocks and additives, are described. Their characteristics, properties and typical applications are also outlined.

3.2 MINERAL OILS

Mineral oils are the most commonly used lubricants. They are manufactured from crude oil which is mined in various parts of the world. There are certain advantages and disadvantages of applying mineral oil to lubricate specific machinery, and these must be carefully considered when selecting a lubricant and designing a lubrication system. The cost of mineral oils is low and even with the rapid development of synthetic oils, solid lubricants and wear resistant polymers, their continued use in many industries seems certain.

Sources of Mineral Oils

The commonly accepted hypothesis about the origins of mineral oils is the fossil fuel theory. The theory states that the mineral oils are the result of decomposition of animal and plant matter in salt water [1]. According to the theory the remains of dead plants and animals were collected in sedimentary basins, especially in places where the rivers dump silt into the sea. Over time they were buried and compressed. Under these conditions the organic matter transformed into tar-like molecules called kerogen. As the temperature and pressure increased the kerogen gradually transformed into the complex hydrocarbon molecules which are the basic constituents of crude oil. When the temperature and pressure became sufficiently high methane was produced from the kerogen or crude oil and hence natural gas is often found together with crude oil. About 60% of the known world oil resources are in the Middle East, concentrated in 25 giant fields. It seems, according to conventional theory, that the Persian Gulf was a vast sink for plant and animal life for millions of years. Over the years, plants and animals deposited there were covered by impermeable layers which formed a sort of rock cap. In order for such a system to remain intact it must be left undisturbed, i.e., free of earthquakes, faultings, etc., for millions of years, and this creates some serious doubts in the validity of the fossil fuel hypothesis as the only source of mineral oils. To begin with it is quite difficult to believe that, in ancient times, most of the plant and animal life on Earth was concentrated in the Persian Gulf region. It is very unlikely that the Persian Gulf was free from earthquakes since it is known that most of the Middle East oil deposits lie along continental plate boundaries where the African, Eurasian and Arabian plates are pushing and pulling each other, and the probability for earthquakes occurring in this region is quite high in comparison to other regions. Interestingly, most of the rich oil deposits have been found along the most seismically active regions such as from Papua New Guinea through Indonesia and Burma to China. Despite these facts this theory is still widely accepted, perhaps because we do not have a valid, experimentally confirmed replacement.

There is another hypothesis about the origin of mineral oils suggested by Gold [2]. It has been known for some time that many hydrocarbons are present in meteorites and that these hydrocarbons cannot possibly originate from any plant or animal life. The hydrocarbons are also quite common on the other planets of the solar system. For example, Jupiter, Saturn, Uranus and Neptune have atmospheres rich in some forms of hydrocarbons. Even Titan, one of Saturn's moons, has large quantities of methane and ethane in its atmosphere. The new hypothesis suggested that, although some oil and gas may originate from biological sources, hydrocarbons on Earth originated from non-biological sources in the same way as on most of the other planets [2]. If the material from which the Earth was formed resembled some of the meteorites, then the Earth would release hydrocarbons when heated. The hydrocarbons would then accumulate under layers of rock and would generate very high pressures. This would lead to the migration of hydrocarbons through cracks and fissures in the Earth's crust. Although at high temperatures oil molecules break down to their most stable form, methane, at the very high pressures which occur several thousand metres below the surface of the Earth, some of the oil molecules would survive. The surviving hydrocarbons would migrate upwards along faults, deep rifts, continental plate boundaries and other fissures in the Earth's crust. Although the pressure would decrease, these places are also cooler, so the probability that the oil molecules would survive is very much higher. Some of the hydrocarbon molecules would dissolve, some of them would create or enrich coal deposits while some of them would be trapped under rock caps and create reservoirs. Some of the oil would be trapped about 3,000 [m] below the surface, and much of this has already been found. There seems to be a very strong correlation between fault and rift zones and the known reservoirs of oil and gas. There is also a strong pattern of trace elements occurring in the oil. For example, along the west coast of South America the oil is rich in vanadium, oils from the Persian Gulf, the Ural Mountains, and parts of West Africa have a constant ratio of nickel to vanadium. This seems to indicate that the origin of these oil deposits is deep within the Earth. According to the new hypothesis huge reservoirs of gas and oil are still waiting to be discovered. They are buried several thousand metres below the Earth's surface and an efficient deep drilling technology will be required to exploit them. This hypothesis, however plausible, has not yet been proved, but if true then there are many major oil and gas reservoirs yet to be discovered.

Manufacture of Mineral Oils

Crude oil exhibits a complex structure which is separated into a number of fractions by a distillation process which is called fractional distillation. The process of fractional distillation involves heating the crude oil to turn it into a vapour which is then passed through a tall vertical column (fractional tower) containing a number of trays at various levels. The vapour passes through the column and at each successive tray the temperature gradually drops. The fraction whose boiling point corresponds to the temperature at a particular tray will condense. In this manner the most volatile compounds will condense at the highest trays in the column while those with the highest boiling points condense at the lower trays. The condensed fractions are then tapped. There are certain temperature limits to which crude oil can be pre-heated. If the temperature is too high then some of the crude may decompose into coke and tarry matter. This problem is overcome by employing another distillation tower which operates at a lower pressure. By lowering the pressure the heavy fractions of crude can be vaporized at much lower temperatures. Thus in the manufacture of mineral oils and petroleum fuels distillation takes place at atmospheric pressure and also at significantly reduced pressures. At atmospheric pressures the following fractions of crude oil distillate are obtained in ascending order of boiling point: gas, gasoline, kerosene, naphtha, diesel oil, lubricating oil and residue. The unvaporized fraction will sediment at the bottom of the column as a residue. This unvaporized residue from the 'atmospheric column' is then placed in the 'vacuum column' and heated. At the lower distillation temperatures which result from using low pressures, the risk of decomposition is eliminated. The vapour condenses on subsequent trays and the distillation products are extracted by vacuum pumps. The following fractions of the remaining residue are obtained by this method in ascending order of boiling point: gas oil, lubricant fraction and short residue. The schematic diagram of a crude oil distillation process is shown in Figure 3.1.

Not all crude oils have to be treated in two stages. Depending on the origin, some of the crude oils are light enough to be heated to a temperature sufficient for their complete distillation at atmospheric pressure.

After the distillation, the lubricating oil fractions of the distillate are then subjected to several stages of refining and various treatments which result in a large variety of medical, cosmetic, industrial and automotive oils and lubricants. The refining process involves further distillation of impure lubricating oils and mixing with organic solvents for preferential leaching of impurities. The purpose of refining is to remove high molecular weight waxes, aromatic hydrocarbons and compounds containing sulphur and nitrogen. The waxes cause the oil to solidify or become near solid at inconveniently high temperatures, the aromatic compounds accentuate the decrease in viscosity of the oil with temperature and the sulphur or nitrogen compounds can cause corrosion of wearing surfaces, resulting in accelerated wear. They may also contribute to some other problems such as corrosion of seals. Filtration of the oil through absorbent clays and hydrogenation of the oil in the presence of a catalyst are applied at the later stages of refining. The lubricant may also be mixed with concentrated sulphuric acid as this is a very effective way of removing complex organic compounds as esters of sulphuric acid. This treatment, however, causes a severe waste disposal problem. For this reason, the sulphuric acid treatment is used only for special high purity oils, such as pharmaceutical oils. Klamann [3] and Dorinson and Ludema [4] give detailed discussions on lubricant oil refining processes which vary with the source of crude oil. As already underlined, the salient feature of refining is that crude oil is a variable and extremely complex mixture of hydrocarbons and that refining imposes only an approximate control on the final product. The objective of the process is not to produce a pure compound, but a product with specific characteristics which are desirable for a particular application.

It is possible to over-refine a lubricating oil, which does not happen very often in practice. In fact most lubricating oils have trace compounds deliberately left in. Many trace aromatic compounds are anti-oxidants, hence an over-refined oil is prone to rapid oxidation. Trace compounds, however, are usually a source of sludge and deposits on contacting surfaces so that a balance or optimization of refining is necessary [3]. In practice the crude oil and refining process are selected to give the desired type of lubricating oil.

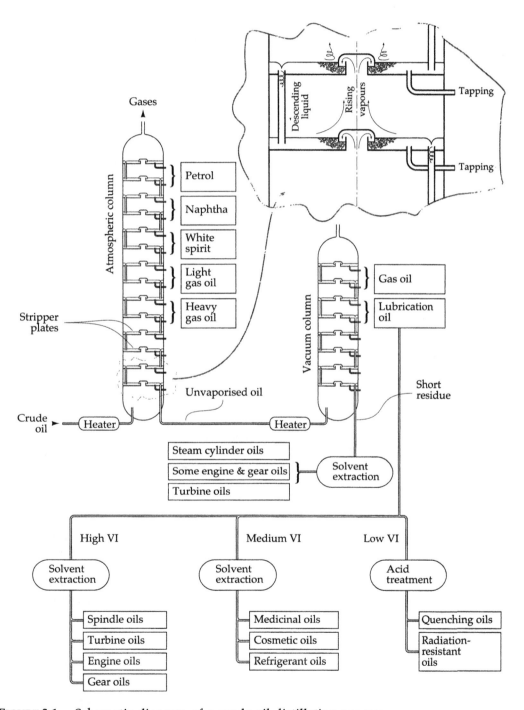

FIGURE 3.1   Schematic diagram of a crude oil distillation process.

Types of Mineral Oils

The structure of mineral oils is very complex. For example, a detailed analysis of crude oil revealed 125 different compounds of which only 45 have been analysed in detail [5]. An interesting consequence of this is that since it is not possible to give a precise analysis of mineral oil, wear and friction studies of lubricated contacts are being conducted in the presence of pure organic fluids of known composition such as hexadecane. The results obtained can then be compared between various research groups. The major part of mineral oils consists of hydrocarbons with approximately 30 carbon atoms in each molecule. The structure of each molecule is composed of several aliphatic (straight) chains and cyclic carbon chains bonded together. Almost any composition of cyclic and aliphatic chains may occur and a large number of the possible forms of the complex molecule are present in any single oil sample. The mineral oils are also impure. The impure nature of mineral oils results in a range of useful and harmful properties [5], e.g., trace compounds provide anti-oxidants and boundary lubrication properties but they also cause deposits which can impede lubrication. There are also many other compounds present in mineral oils such as waxes which are virtually useless and can easily be oxidized to form harmful organic acids. Special additives are needed to neutralize these waxes and related compounds.

Therefore, mineral oils differ from each other depending on the source of crude oil and refining process. The fundamental differences between mineral oils are based on:

- chemical forms,
- sulphur content,
- viscosity.

· Chemical Forms

There are three basic chemical forms of mineral oil:

- paraffinic,
- naphthenic,
- aromatic.

They originate from crudes from different sources and correspond to an exact chemical type. As shown in Figure 3.2 paraffinic implies straight chain hydrocarbons, naphthenic means cyclic carbon molecules with no unsaturated bonds and aromatic oils contain benzene type compounds. Oils are distinguished based on the relative proportions of paraffinic, naphthenic and aromatic components present.

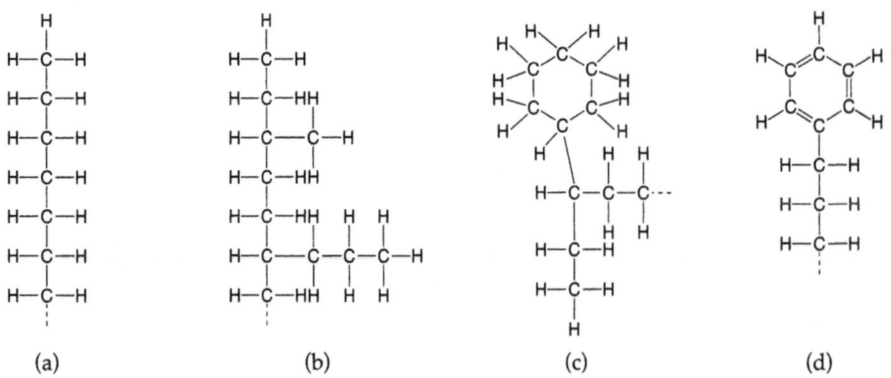

FIGURE 3.2   Types of mineral oils: a) straight paraffin, b) branched paraffin, c) naphthene, and d) aromatic.

The aromatic oil is present only as a minor component of naphthenic or paraffinic oils. The subtlety of the lubricant engineering definition of these terms is that the lubricant is named depending on which chemical type makes up its major proportion. For example, a paraffinic oil means that the majority of the hydrogen and carbon atoms are present as paraffinic chains. These paraffinic chains are then linked by carbon atoms bonded in a cyclic manner to form a more complex molecule. A naphthenic oil has much smaller paraffinic chains in each hydrocarbon molecule and most carbon is incorporated in cyclic molecules. There is also a limited quantity (about 20%) of simple paraffins (alkanes) present in the oil. The presence of one type or the other of these molecules determines some of the physical properties of the lubricants, i.e., pour point, viscosity index, pressure-viscosity characteristics, etc. For example, there are significant differences in viscosity-temperature characteristics and viscosity-pressure characteristics between paraffinic and naphthenic oils and care must be taken in distinguishing between them. Paraffinic oils are also generally more expensive since they require a few more stages of refining than naphthenic oils.

· *Sulphur Content*

Sulphur content in mineral oils varies, depending on the source of the crude oil and the refining process. Small amounts of sulphur in the oil are desirable to give good lubrication and oxidation properties. It has been demonstrated, for example, that between 0.1% and 1% of natural sulphur content ensures reduced wear [60]. On the other hand, too much sulphur is detrimental to the performance of the machinery, e.g., it may accelerate the corrosion of seals. Excess sulphur can be removed from oil by refining, but this can be expensive. The sulphur content varies with the source of crude oil and the range of concentration lies between 0% and 8%. For example, sulphur content of Pennsylvanian oil is <0.25%, Venezuelan ~2%, Middle East ~1%, Mexican 5%, etc.

· *Viscosity*

Mineral oils can also be classified by viscosity, which depends on the degree of refining. For commonly used mineral oils, viscosity varies from about 5 [cS] to 700 [cS]. For example, the viscosity of a typical spindle oil is about 20 [cS], engine oil between 30 and 300 [cS], bright stock about 600 [cS], etc.

3.3   SYNTHETIC OILS

Synthetic lubricants were originally developed early this century by countries lacking a reliable supply of mineral oil. These lubricants were expensive and initially did not gain general acceptance. The use of synthetic oils increased gradually, especially in more specialized applications for which mineral oils were inadequate. Despite many positive features such as availability and relatively low cost, mineral oils also have several serious defects, such as oxidation and viscosity loss at high temperatures, combustion or explosion in the presence of strong oxidizing agents and solidification at low temperatures. These effects are prohibitive in some specialized applications such as gas turbine engines where a high temperature lubricant is required, but occasionally very low temperatures must be sustained. In other applications such as vacuum pumps and jet engines, low vapour pressure lubricant is needed; in food processing and the pharmaceutical industry low toxicity lubricant is required, etc. In recent years the strongest demand has been for high performance lubricants, especially for applications in the aviation industry with high performance gas turbine engines. This led to the development of synthetic lubricants that can withstand high temperatures without decomposing and at the same time will provide a reduced fire hazard. The recent trend towards high operating temperatures of machinery has created a second and probably more durable period of interest in these lubricants.

Synthetic lubricants can generally be divided into two groups:

· fluids intended to provide superior lubrication at ambient or elevated temperatures, and

· lubricants for extremes of temperature or chemical attack.

There is also a clear distinction between exotic lubricants with high performance but high cost and more economical moderate performance lubricants. For example, the price of a halogen based synthetic lubricant reached $450/kg in 1987 which is close to the price of silver.

There are three basic types of synthetic lubricant currently in use:

· synthetic hydrocarbon lubricants,

· silicon analogues of hydrocarbons, and

· organohalogens.

All of the hundred or more specific types of synthetic lubricant available on the market conform to one of these broad categories. Phosphates, as in polyphenyl phosphate, deviate from the pattern as they are generally associated with simple hydrocarbons.

These three groups of synthetic lubricants have distinct characteristics which sustain the usefulness of this form of classification. These are:

· synthetic hydrocarbons which provide a lubricant that is similar in price to mineral oil but has superior performance,

· silicon analogues or silicones which are resistant to extremes of temperature and vacuum but do not provide good adsorption or extreme pressure lubrication (sometimes known as 'boundary characteristics') and are expensive,

· organohalogens which can offer effective lubrication by adsorption and extreme pressure lubrication mechanisms and resist extremes of temperature or chemical attack, but are also expensive.

Manufacturing of Synthetic Oils

In most cases synthetic hydrocarbon lubricants are produced from low molecular weight hydrocarbons which are derived from the 'cracking' of petroleum [1]. The process of cracking is performed in order to reduce the range of molecules present in the oil. Through the application of high pressures and catalysts large complex molecules present in the oil are decomposed to more simple, smaller and more uniform molecules. The low molecular weight hydrocarbons are then polymerized under carefully controlled conditions to produce fluids with the required low volatility and high viscosity. The polymerization is carefully limited otherwise a solid polymer results and, in strict technical terms, an oligomer as opposed to a polymer is produced. A prime example of this method of lubricant synthesis is the production of a polyolefin synthetic lubricant oil from olefins (alkenes).

Halogenated lubricants are also manufactured on a large scale; these are appropriate for low temperatures or where there is an extreme fire risk. These lubricants are made from ethylene and halogen compounds in a process of simultaneous halogenation and polymerization within a solvent [1]. Not all synthetic lubricants are produced by polymerization, some monomers, e.g., dibasic acid esters, are also useful for many applications.

Organohalogens and silicones are produced using catalysts. Organohalogens are manufactured by reacting hydrocarbon gas, i.e., methane and hydrogen chloride, under pressure and temperatures of about 250°C or more in the presence of a catalyst such as alumina gel or zinc chloride. During the process low molecular weight organohalogens (i.e., methyl-chloride) are formed which can later be polymerized, resulting in high molecular weight organohalogens. Silicones, on the other hand, are produced from methyl chloride (CH_3Cl) which is reacted with silicon in the presence of copper catalysts at 380°C to form dimethyl-silicon-chloride (($2CH_3)_2SiCl_2$). Secondary treatment with hydrochloric acid causes the removal of the chloride radicals to form a silicone. After neutralizing and dewatering the original stock the polymerization of silicones is then induced by alkali, resulting in the finished product. Chemical structures of the most common synthetic lubricants are shown in Table 3.1.

TABLE 3.1   Typical chemical structures of the most common synthetic lubricants.

HYDROCARBON SYNTHETIC LUBRICANTS	Polyalphaolefins e.g.	$(-CH_2-CH_2-CH_2-CH_2-)_n \cdots CH_2-CH_2-CH=CH_2$
	ESTERS E.G.	
	· Diesters e.g.	$C_8H_{17}-O-CO-C_8H_{16}-CO-O-C_8H_{17}$
	· Phosphate esters e.g.	$(CH_3-C_6H_4-O)_3P=O$
	· Silicate esters e.g.	$Si(O-C_8H_{17})_4$
	· Polyglycol esters e.g.	$CH_2-(-CH_2-O-CH_2)_n-CH_2$ with OH on each end
	· Fluoro esters e.g.	$F(CF_2)_4CH_2OOC(CF_2)_4F$
	· Fatty acid esters e.g.	$C_{13}H_{27}-OC(=O)-C_{18}H_{37}$
	· Neopentyl polyol esters e.g.	$CH_3-CH_2-C(CH_2-OOC-C_8H_{17})(CH_2-OOC-C_8H_{17})-OOC-C_8H_{17}$
	Cycloaliphatic e.g.	two cyclohexane rings, $CH_3-C(CH_3)-C(CH_3)-CH_3$
	Polyglycols e.g.	$OH-CH_3-CH_3-O-CH_3-CH_3-O \cdots CH_3-CH_3-OH$
SILICON ANALOGUES OF HYROCARBONS	Silicones e.g.	$CH_3-Si(CH_3)(CH_3)-[O-Si(CH_3)(CH_3)]_n-O-Si(CH_3)(CH_3)-CH_3$
	Silahydrocarbons e.g.	$(C_{12}H_{25})Si(C_6H_{13})_3$
ORGANOHALOGENS	Perfluoropolyethers e.g.	$CF_3-CF_2-O-CF_2-CF_3$
	Chlorofluorocarbons e.g.	$[-C(Cl)(F)-C(F)(F)-]_n$
	Chlorotrifluoroethylenes e.g.	$Cl-[-C(F)(F)-C(Cl)(F)-]_n-Cl$
	Perfluoropolyalkylethers e.g.	$F-[-C(F)(CF_3)-C(Cl)(F)-]_n-O-C(F)(F)-CF_3$

Hydrocarbon Synthetic Lubricants

There is an almost infinite variety of hydrocarbons that could be utilized as lubricants. The economics of production, however, severely restricts their range. The oils presently advocated as the optimum synthetic lubricants by various oil refiners are not necessarily ideal as lubricants, but they are relatively cheap to produce and therefore are economic for large volume applications such as engine oils. Engine oils constitute almost half the entire lubricating oil usage and there is a large profit to be made from a synthetic oil which costs only a little more than mineral oil but can improve engine performance, durability and prolong draining periods. Synthetic oils that can be classified as synthetic hydrocarbons are polyalphaolefins, esters, cyclo-aliphatics and polyglycols. Of course, the list is incomplete and future advances in refining and synthesis may extend it.

The oxidation stability of a synthetic hydrocarbon depends on the structure of the hydrocarbon chain. The bond energy of the **C-C** linkage (360 [MJ/kgmole]) is the fundamental limitation and higher oxidation stability can be achieved by applying various oxidation inhibitors. Oxidation stability can also be improved by replacing weakly bonded structures with branched hydrocarbons. The hydrocarbons can be optimized for their viscosity-temperature characteristics, low temperature performance and volatility.

· Polyalphaolefins

Polyalphaolefins are among the most promising general purpose synthetic lubricants. Olefins or alkenes are unsaturated hydrocarbons with the general formula $(-CH_2-)_n$. They consist of a straight carbon chain with an unsaturated carbon at one end of the chain. A typical example is polybutene. The presence of unsaturated carbons allows polymerization or oligomerization to form a lubricating oil. The preferred alkene is decene which produces an oil with a low minimum operating temperature (pour point). Higher molecular weight compounds such as dodecenes have a higher viscosity index but also a higher pour point. The viscosity of polydecenes can be varied from 0.3 [mPas] to 100 [mPas]. Their viscosity index is about 130 and pour point about -30°C [6]. Polydecenes are highly resistant to oxidation, have a low volatility due to the lack of small molecular weight substances and are not toxic or corrosive. These properties ensure the use of polydecenes as a general purpose synthetic lubricant.

· Polyphenyl Ethers

Polyphenyl ethers exhibit better boundary characteristics than silicone oils. They have very high oxidation and thermal stability, but are limited by poor viscosity-temperature characteristics. Thermal stability of these compounds is about 430°C and oxidation stability is also quite high at about 290°C. They are used as lubricants in aircraft hydraulic pumps.

· Esters

A very important group of synthetic hydrocarbons are the esters. They are produced by reacting alcohol with organic or inorganic acids. For applications such as lubrication, inorganic acids are widely used in their production. The linkages of esters are much more stable than those of typical hydrocarbons with their **C-C** bonds. The ester linkages have a much higher bond energy, thus they are more resistant to heat. Esters usually have good oxidation stability and excellent viscosity-temperature and volatility characteristics.

Dibasic Acid Esters (Diesters) have similar lubrication qualities to polydecenes, i.e., a high viscosity index and oxidation resistance. Dibasic acid esters can operate at higher temperatures than polydecenes and are used for applications where tolerance to heat is essential. Originally these oils were used in aircraft engines, but they have been gradually replaced by polyol esters. Polyol esters have an even higher operating temperature limit. Maximum operating temperatures for dibasic acid esters are around 200°C and for polyol esters close to 250°C.

<u>Phosphate Esters</u> have better thermal stability than diesters but they also have a high surface tension. They have excellent fire-resistant properties and are commonly used as hydraulic fluids for steam and gas turbines. A phosphate ester, tricresylphosphate (TCP), has good anti-wear properties and has been widely used as an anti-wear additive in many mineral and synthetic oils. Phosphate esters may also cause corrosive wear. Chlorine forms phosphorous oxychloride, which is used to manufacture phosphate esters, and entrained water may react with the residual chlorine to form corrosive agents [58]. They may cause only a small amount of corrosive wear but this is sufficient to disrupt the delicate hydraulic control systems. The cost of phosphate esters is so high that it prohibits their use as simple lubricants.

<u>Silicate Esters</u> have high thermal stability, low viscosity and relatively low volatility, but they have low resistance to the adverse effects of water. They are used as low temperature ordinance lubricants.

<u>Polyglycol Esters</u> have fair lubricating properties and are commonly used as hydraulic fluids.

<u>Fluoro Esters</u> have good oxidation stability characteristics, low flash and fire points and poor viscosity-temperature characteristics. They are used both as lubricants and as hydraulic fluids.

<u>Fatty Acid Esters</u> have moderately low volatility, low oxidation resistance and low thermal resistance. On the other hand, they have good boundary properties with metals and metal oxides. Since they cannot form large molecules, they are not particularly popular as lubricants in industrial applications but are commonly used as lubricants in most magnetic tapes and floppy disks.

<u>Neopentyl Polyol Esters</u> have volatility, oxidation stability and thermal stability superior to fatty acid esters. They are used as lubricants in gas turbine engines and as hydraulic fluids in supersonic aircraft.

· *Cycloaliphatics*

Cycloaliphatics are specialized oils specifically designed for the traction drives used in the machine tool, textile and computer hardware industries. In principle, traction drives allow continuously variable speed transmission without the need for gears at fixed speed ratios. Cyclic hydrocarbon molecules exhibit high pressure-viscosity coefficients which raise the limiting traction force in the elastohydrodynamic contact. The maximum traction power that can be transmitted across the contact determines the size of the unit. Therefore research efforts are concentrated on developing traction fluids which will allow transmission of higher forces and permit smaller traction drives for a given transmitted power [6].

· *Polyglycols*

Polyglycols were originally used as brake fluids but have now assumed importance as lubricants. The term 'polyglycol' is an abbreviation of the full chemical name 'polyalkylene glycol'. Certain types of polyglycols have a viscosity index greater than 200 and pour points less than -50°C. The pressure-viscosity coefficients of polyglycols are relatively low and the oxidation stability is inferior to other synthetic oils. Water soluble polyglycols tend to adsorb water and are mostly used as brake fluids. Polyglycols have distinct advantages as lubricants for systems operating at high temperatures such as furnace conveyor belts, where the polyglycol burns without leaving a carbonaceous deposit. Since the unburned polyglycol does not stain, it is also used as a lubricant in the textile industry.

Silicon Analogues of Hydrocarbons

Silicon analogues of hydrocarbons constitute a completely different branch of synthetic lubricants. They have been found to offer a significantly extended liquid temperature range and improved chemical stability. Two basic classes of compounds have attracted practical interest: silicones and silahydrocarbons. The silicones contain oxygen as well as silicon which distinguishes them from the silahydrocarbons which contain only silicon, carbon and hydrogen. These oils have similar but not identical qualities to lubricants or synthetic oils.

· Silicones

The most commonly used silicones are dimethyl, methyl phenyl and polymethyl silicones. Most of the silicones are chemically inert. They have excellent thermal and oxidation stability, good viscosity-temperature characteristics, low volatility, toxicity and surface tension. Their operating temperature range is between -50°C and 370°C, and the viscosity index of some lubricants is nearly 300, i.e., viscosity remains nearly constant. The fluids are available in a very wide range of viscosities from 0.1 [mPas] to 1 [Pas] at 25°C. Pressure-viscosity coefficients of silicones are also higher than those of mineral oils. Because of their chemical inertness they have poor boundary characteristics, especially with steel [7,8], but on the other hand they are effective as hydrodynamic lubricants. Their load capacity is quite low under thin film conditions. Four-ball tester seizure loads are about 0.1 of that of a mineral oil containing additives. Low solubility of most additives prevents any significant improvement in load capacity. When the methyl groups of dimethyl silicone fluid are replaced with hydrocarbon groups containing substituent fluorine atoms, the result is a lubricant with very much improved lubricating properties. Silicones are used in grease formulations for various space applications. Silicone oils can also be blended with high temperature thickeners to form heat-resistant greases. For example, lithium soaps are effective up to 200°C, carbon black and other solid lubricants can raise the operating temperature up to the decomposition temperature of the silicone, i.e., 370°C. The production cost of silicone lubricants is much higher than most of the other synthetics and this is reflected in the price. Silicones are usually employed in extreme operating temperatures where other lubricants fail to operate. They are widely used in military equipment.

· Silahydrocarbons

Although silahydrocarbons resemble silicones, as a silicon is substituted with hydrocarbon in these compounds, they are in fact different. They are synthesized from organometallics and silicontetrachloride and possess good oxidation and thermal stability as well as low volatility. They are, however, most resistant to thermal degradation in the absence of air or oxygen and have been specifically developed for aerospace hydraulic systems [61]. The operating range of temperatures for these lubricants is approximately from -40°C to +350°C [61]. They are also used as high temperature lubricants.

Organohalogens

Organohalogens and their related compounds, the halogenated hydrocarbons, are well established as lubricants which are stable against oxidation. Before these liquids were developed, devices such as air and oxygen compressors had to rely on pure sulphuric acid for lubrication. Sulphuric acid is an effective lubricant of steel but the practical difficulties of preventing mixing of the acid with moisture and contamination of the compressed gas are severe. Sulphuric acid is also very corrosive. Fluorine and chlorine, but not bromine, are used to develop compounds with desirable properties. Their oxidation and thermal stability are very good. They are used for applications over a wide temperature range and in various hostile environments, i.e., in a vacuum, under strong oxidation conditions, etc. The cost of these lubricants, however, is very high. In this group of synthetic lubricants the most commonly used are perfluoropolyethers, chlorofluorocarbons, chlorotrifluoroethylenes and perfluoropolyalkylethers.

· *Perfluoropolyethers*

These are among the most promising lubricants for high temperature applications. Perfluoropolyethers (PFPE) have very high oxidation stability (about 320°C) and thermal stability (about 370°C), low surface tension and are chemically inert. They are used in the formulation of greases for high temperature and high vacuum applications. Perfluoropolyethers are also used as hydraulic fluids, gas turbine oils and lubricants for computer hard disks [9]. In a computer hard disk lubricant is needed to control friction during starting and stopping of the disk where the hydrodynamic air film prevents wear of the diskette and head. The critical features of the system are an extremely light contact load, measured in the range of milliNewtons and a minute quantity of lubricant that binds tightly to the diskette to resist being thrown off by centripetal forces. The light load means that stiction, caused by surface tension forces, is of concern. This is different from typical mechanical contacts where frictional seizure is the prime limitation. Surface tension of the lubricant is therefore a critical lubricant's property in computer head-disk applications.

The small volume of lubricant compared to the wetted area on the disk means that the lubricant is extremely sensitive to chemical degradation whether it is purely oxidative or catalysed by the diskette surface. Usually perfluoropolyethers are used for this type of application but they are vulnerable to degradation catalysed by the worn diskette surface [65]. Recently phosphazenes have been considered for better control of stiction and reduced chemical degradation. It has been found that some phosphazenes can be blended with perfluoropolyethers as a form of additive to enhance the performance of the lubricant [66]. There are, however, significant problems of phase separation which means that the additives do not always stay perfectly mixed when in service. Also corrosive effects induced by PFPE lubricants may cause problems in future designs of computer hard disk drives.

PFPE lubricants such as Fomblin Z-DOL show a significant corrosive activity, especially when contaminated by water. Significant galvanic corrosion of iron, cobalt and nickel, but not of aluminium, chromium and titanium, was observed [68]. The lack of intact oxide films on metals, such as iron, cobalt and nickel, was attributed as the cause for this effect.

· *Chlorofluorocarbons*

In chlorofluorocarbon molecules, the hydrogen present in hydrocarbon compounds is replaced completely or in part by chlorine or fluorine. Chlorofluorocarbons are chemically inert and possess excellent oxidation and thermal stability. On the other hand, they have poor viscosity-temperature characteristics, high volatility and a high pour point. Although they are good lubricants, their applications are limited due to the very high production costs involved.

· *Chlorotrifluoroethylenes*

They are non-toxic and have good oxidation and thermal stability. They are available in a wide viscosity range from 0.1 [mPas] to 1 [Pas] which can be obtained by varying the molecular weight or carbon chain length of the compound. The viscosity index, however, of these lubricants is low, for example, for high viscosity grades it is about 27 [10].

· *Perfluoropolyalkylethers*

Perfluoropolyalkylethers have good oxidation and thermal stability, high viscosity index and a wide operating temperature range. Values of viscosity index about 200 are easily reached and the minimum operating temperature is about -60°C. They provide thin-film lubrication in applications where oxygen is absent, e.g., in a vacuum. It has been found that these lubricants decompose under sliding contact to form iron fluoride films on the worn metal surface [11,12]. Moderate friction coefficients of 0.1 have been obtained at a vacuum of 10^{-6} [Pa] where unlubricated metals would usually seize because conventional oils are vapourized under these conditions [12].

Cyclophosphazenes

Cyclophosphazenes are a possible new class of high temperature synthetic lubricants. They exhibit a combination of good chemical and thermal stability together with a low vapour pressure (so that the oil does not boil away in service). They also form protective films on the worn ferrous surfaces by a tribochemical reaction with iron resulting in films of fluorides and organic compounds containing oxygen, nitrogen, fluorine and phosphorus [69].

Some of the main characteristics of the typical synthetic lubricants are summarized in Table 3.2 [9].

Finally it has to be mentioned that most of the literature available on synthetic lubricants relates either to the manufacture or physical properties of these lubricants. There is very little impartial data on the performance of these lubricants. Much of the literature available is sales literature from commercial organizations and the evaluation of synthetic lubricants can be influenced by the perceived benefits of synthetic lubricants. An example of this trend can be found by comparing the data on a specific lubricant provided by an oil company research group [13] and the data provided by an engineering company [14]. The former describes synthetic oils as the solution to many engine lubrication problems, e.g., lubricant oil durability and fuel economy. The latter, however, found no increase in the frictional failure load of gears. Thus it should be remembered that the synthetic oils are not the panacea for all lubrication problems. They will definitely solve many problems related to oil oxidation and viscosity loss but will not affect the limitations of boundary lubrication.

3.4 EMULSIONS AND AQUEOUS LUBRICANTS

Water is an attractive extender of lubricating oils; cheap, good heat transfer characteristics and non-flammability are all useful attributes. Water by itself is a very poor lubricant but when mixed with oils to form emulsions or when mixed with water-soluble hydrocarbons to produce an aqueous solution, some useful lubricants can be developed. These liquids are used as coolants in metalworking where the combination of the lubricity of oil, high conductivity and the latent heat of water provide the optimum fluid for this application. Mining machinery is also lubricated by water-based fluids to minimize the risk of fire from leakage of lubricants. It has been observed that during the lubrication process by emulsions, water is excluded from the loaded contacts and as a result the performance of an emulsion is close to that of a pure mineral oil [15]. The most severe limitation of these lubricants is the temperature range at which they can successfully be applied. They are limited to the temperature range of water, which lies between the melting point of ice and the boiling point of water. This excludes these lubricants from many applications, for example, engine oils.

Manufacturing of Emulsions

Emulsions are produced by mixing water and oil with an emulsifier. An example of this relatively simple process, which usually occurs inadvertently, is when water contaminates a lubricating oil sump (most lubricating oils contain natural emulsifiers). The mixing must be sufficiently intense to disperse one of the liquids as a series of small droplets within the other liquid. About 1 - 10% by weight of emulsifier is added to stabilize the dispersed droplets and stop their coagulation. A 'water in oil' emulsion, commonly abbreviated to 'W/O', is a suspension of water droplets in oil. The converse, oil in water, contains oil droplets dispersed in water and is usually referred to as an 'O/W' emulsion. 'W/O' and 'O/W' emulsions have different lubrication characteristics. The 'W/O' emulsions are used as fire resistant hydraulic fluids, while the 'O/W' emulsions are suitable as metalworking coolants. A novel type of

TABLE 3.2   Some of the main characteristics of the typical synthetic lubricants (adapted from [9]).

Property	Mineral oils	Diesters	Neopentyl polyol esters	Phosphate esters	Silicate esters	Disiloxanes	Polyphenyl ethers			Perfluoro-polyethers	
							Phenyl methyl	4P-3E	5P-4E	Fomblin YR	Fomblin Z-25
Thermal stability [°C]	135	210	230	240	250	230	280	430	430	370	370
Kinematic viscosity [cSt] at -20°C	170	193	16	85	115	200	850			8000	1000
0°C	75	75	16	38	47	100	250			2500	440
40°C	19	13	15	11	12	33	74	70	363	515	150
100°C	5.5	3.3	4.5	4	4	11	25	6.3	13.1	35	150
200°C		1.1			1.3	3.8	22	1.4	2.1		41
Specific gravity at 20°C	0.86	0.90	0.96	1.09	0.89	0.93	1.03	1.18		1.92	1.87
Thermal conductivity [W/mK]	0.134	0.153		0.127				0.144	0.155	0.095	
Specific heat at 38°C [J/kgK]	1670	1925		1757	185			1423	1799	1004	837
Flash point [°C]	105	230	250	180	185	185	200	260	290	none	none
Pour point [°C]	-57	-60	-62	-57	-65	-70	-70	-7	+4	-30	-67
Oxidative stability [°C]								240	290	320	320
Vapour pressure at 20°C [Pa]	1.3×10^{-3} – 13.3	1.3×10^{-3}	1.3×10^{-4}	1.3×10^{-4}	1.3×10^{-4}	1.3×10^{-4}		6.67×10^{-5}	1.3×10^{-5}	1.3×10^{-6}	4×10^{-9}
Effect on metals	non-corrosive when pure	slightly corrosive with non-ferrous metals	corrosive to some non-ferrous metals	enhance corrosion in the presence of water		non-corrosive	non-corrosive	non-corrosive	non-corrosive	non-corrosive	non-corrosive
Effect on plastics	slight	may act as plasticiser	acts as plasticiser	solvent		slight	slight	slight	satisfactory	some softening when hot	some softening when hot
Resistance to attack by water	excellent	good	good	fair	good	poor	poor	very good	very good	excellent	excellent
Suitable rubbers	nitrile	nitrile, silicone	silicone	butyl	silicone	viton, nitrile, fluoro-silicone	viton, nitrile, fluoro-silicone	neoprene, viton	none: for very high temperatures	none: for very high temperatures	silicone

water-based emulsion involves suspensions of nanoparticles in water. It was found that titanium dioxide nanoparticles, with diameters in the range of 20-40 [nm] and with an adsorbed coating of oleic acid (cis-9-octadecenoic acid), perform as an effective lubricating agent when combined with a dispersant and added to water in concentrations around 0.5% by weight [70].

Despite some limitations, water-based fluids constitute an important and specialized form of lubricant.

Characteristics

The apparent viscosity of emulsions declines with increasing shear stress, and their viscosity index is usually high. 'W/O' emulsions have a high viscosity, several times that of the base oil. They exhibit an interesting behaviour in concentrated contacts operating in the elastohydrodynamic lubrication regime (EHL). The size of an EHL contact is comparable to the droplet size, or the volume of fluid within the contact is similar to the average droplet volume. This suggests that the elastohydrodynamic films generated would be unstable or fluctuate when an emulsion is used. This, however, is not confirmed experimentally, and in fact it is known that a low stability emulsion gives the best lubrication. It has been suspected for a long time that the emulsion is temporarily degraded at the EHL contact and releases oil for lubrication. In the work by Sakurai and Yoshida [15] it was suggested that the typically oleophilic metal surfaces drew oil into the EHL contact but excluded water. Measurements showed that EHL film thickness does not vary with water concentration and maintains a value close to that of the constituent mineral oil. The pressure-viscosity coefficient of water is negligibly small [16,17] so that without forming an entrapment of oil around the EHL contact, elastohydrodynamic lubrication would not be possible.

Although it is generally accepted that the temperature limit of emulsions is dictated by the boiling point of water, i.e., around 100°C at ambient pressure, the emulsions can effectively function at much higher temperatures. When a water-based emulsion is in contact with a surface significantly hotter than the ambient boiling point of water then film boiling or Leidenfrost boiling occurs. Thus when an emulsion is placed on a hot surface, its oil component is released and transferred across the vapour film reaching the hot surface and providing lubrication. In general, the boiling point of most oils is higher than that of water. At greatly elevated temperatures, the vapour film may become too thick or continuous in nature and the lubricant transfer ceases. At the temperature where the vapour film becomes established, known as the critical temperature, which is often more than 200°C, the emulsion effectively fails as a lubricant [71].

Apart from a limited temperature range emulsions exhibit poor storage capability and they may not only be degraded by oil oxidation but also by bacterial contamination of water.

Applications

Emulsions and aqueous solutions are mostly used as cutting fluids in the metal working industry and as fire resistant lubricants in the mining industry. Aqueous solutions of polyglycols are often used as fire resistant hydraulic oils with the added advantage of low viscosity and low pour points, e.g., -40°C. As a lubricant, however, polyglycol solutions offer only mediocre performance. The pressure-viscosity coefficient of a polyglycol solution is only 0.45×10^{-8} [Pa^{-1}] compared to 2.04×10^{-8} [Pa^{-1}] for a mineral oil [18]. Even small quantities of water can significantly reduce the pressure-viscosity coefficient. Thus the primary applications of these fluids are as fire resistant lubricants because even if all the water were evaporated from the lubricant, the polyglycol would burn only with difficulty.

3.5   GREASES

Greases are not simply very viscous lubricating oils. They are in fact mixtures of lubricating oils and thickeners. The thickeners are dispersed in lubricating oils in order to produce a stable colloidal structure or gel. Thus, a grease consists of oil constrained by minute thickener fibres. Since the oil is constrained and unable to flow it provides semi-permanent lubrication. For this reason, greases are widely used, despite certain limitations in performance. The most widespread application of greases is as low-maintenance, semi-permanent lubricants in rolling contact bearings and some gears. The grease may be packed into a bearing or gear set and left for a period of several months or longer before being replaced. Inaccessible wearing contacts, such as are found on caterpillar track assemblies or in agricultural machinery, are conveniently lubricated by this means. Low maintenance items are also suitable candidates for grease lubrication. The lubricating performance of greases is inferior to mineral oils except at low sliding speeds where some greases may be superior. Greases have to meet the same requirements as lubricating oils but with one extra condition, the grease must remain as a semi-solid mass despite high service temperatures. If the grease liquefies and flows away from the contact then the likelihood of lubrication failure rapidly increases. Furthermore, grease is unable to remove heat by convection as oil does, so unlike oil, it is not effective as a cooling agent. It also cannot be used at speeds as high as oil because frictional drag would cause overheating. The lifetime of a grease in service is often determined by the eventual loss of the semi-solid consistency to become either a liquid or a hard deposit.

Manufacturing of Greases

Greases are manufactured by adding alkali and fatty acid to a quantity of oil. The mixture is then heated and soap is formed from the alkali and fatty acid. After the reaction, the water necessary for soap formation is removed and the soap crystallizes. The final stages of manufacture involve mechanical working of the grease to homogenize the composition and allow blending in of additives and the remaining oil. Careful control of process variables is necessary to produce a grease of the correct consistency [3]. Several cycles of mixing and 'maturing' are often needed to obtain the required grease properties. Most greases are made by a batch process in large pots or reactors, but continuous production is gaining acceptance.

Composition

Greases always contain three basic active ingredients: a base mineral or synthetic oil, additives and thickener. For thickeners, metal soaps and clays are used. In most cases the mineral oil plays the most important role in determining the grease performance, but in some instances the additives and the thickener can be critical. The type and amount of thickener (typically 5 - 20%) have a critical effect on grease properties. Very often additives which are similar to those in lubricating oils are used. Sometimes fillers, such as metal oxides, carbon black, molybdenum disulphide, polytetrafluoroethylene, etc., are also added.

· Base Oils

Mineral oils are most often used as the base stock in grease formulation. About 99% of greases are made with mineral oils. Naphthenic oils are the most popular despite their low viscosity index. They maintain the liquid phase at low temperatures and easily combine with soaps. Paraffinic oils are poorer solvents for many of the additives used in greases, and with some soaps they may generate a weaker gel structure. On the other hand, they are more stable than naphthenic oils, hence are less likely to react chemically during grease formulation.

Synthetic oils are used for greases which are expected to operate in extreme conditions. The most commonly used are synthetic esters, phosphate esters, silicones and fluorocarbons. Synthetic base greases are designed to be fire resistant and to operate in extremes of temperature, low and high. Their most common applications are in high performance aircraft, missiles and in space. They are quite expensive.

Vegetable oils are also used in greases intended for the food and pharmaceutical industries, but even in this application their use is quite limited.

The viscosity of the base oil used in making a grease is important since it has some influence on the consistency, but the grease consistency is more dependent on the amount and type of thickener used.

· *Thickener*

The characteristics of a grease depend on the type of thickener used. For example, if the thickener can withstand heat, the grease will also be suitable for high temperature applications, if the thickener is water resistant the grease will also be water resistant, etc. Hence the grease type is usually classified by the type of thickener used in its manufacture. As there are two fundamental types of thickener that can be used in greases, the commercial greases are divided into two primary classes: soap and non-soap based.

Soap type greases are the most commonly produced. According to the principles of chemistry, in order to obtain soap it is necessary to heat some fats or oils in the presence of an alkali, e.g., caustic soda (NaOH). Apart from sodium hydroxide (NaOH) other alkali can be used in the reaction, as for example, lithium, calcium, aluminium, barium, etc. Fats and oils can be animal or vegetable and are produced from cattle, fish, castor bean, coconut, cottonseed, etc. The reaction products are soap, glycerol and water. Soaps are very important in the production of greases. The most commonly used soap type greases are calcium, lithium, aluminium, sodium and others (mainly barium).

In non-soap type greases inorganic, organic and synthetic materials are used as thickeners. Inorganic thickeners are in the form of very fine powders which have enough porosity and surface area to absorb oil. The most commonly used are the silica and bentonite clays. The powders must be evenly dispersed in the grease so either high-shear mechanical mixing or some special dispersing additives are required during grease formulation. Because of their structure these types of greases have no melting point, so their maximum operating temperature depends on the oxidation stability of the base oil and its inhibitor treatment. When properly formulated these greases can successfully be applied in high temperature applications. They are usually considered as multipurpose greases and are widely applied in rolling contact bearings and in the automotive industry. Synthetic and organic thickeners such as amides, anilides, arylureas and dies are stable over a wide temperature range and they give superior performance to soap based grease at high temperatures. They are used for special applications, such as military and aerospace use.

The thickeners form a soft, fibrous matrix of interlocking particles. The interlocking structure forms tiny pockets of about 10^{-6} [m] in which the oil is trapped. A diagram of the fibrous structure of a soap based grease is shown in Figure 3.3.

· *Additives*

The additives used in grease formulations are similar to those used in lubricating oils. Some of them modify the soap, others improve the oil characteristics. The most common additives include anti-oxidants, rust and corrosion inhibitors, tackiness, anti-wear and extreme pressure (EP) additives.

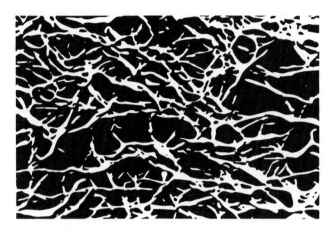

FIGURE 3.3   Diagram of the fibrous structure of a soap based grease (adapted from [4]).

Anti-oxidants must be selected to match the individual grease. Their primary function is to protect the grease during storage and extend the service life, especially in high temperature applications.

Rust and corrosion inhibitors are added to make the grease non-corrosive to bearings operating in machinery. The function of corrosion inhibitors is to protect the non-ferrous metals against corrosion whereas the function of rust inhibitors is to protect ferrous metals. Under wet or corrosive conditions the performance of most greases can be improved by a rust inhibitor. Most of the multipurpose greases contain these inhibitors.

Tackiness additives are sometimes added to impart a stringy texture and to increase the cohesion and adhesion of the grease to the surface. They are used, for example, in open gear lubricants.

Anti-wear and extreme pressure (EP) additives improve, in general, the load-carrying ability in most rolling contact bearings and gears. Extreme pressure additives react with the surface to form protective films which prevent metal to metal contact and the consequent scoring or welding of the surfaces. Although the EP additives are intended to improve the performance of a grease, in some cases the operating temperature is far too low for these additives to be useful. It has also been found that some thickening agents used in grease formulations inhibit the action of EP additives [19]. The additives most commonly used as anti-seize and anti-scuffing compounds are graphite and molybdenum disulphide.

Fillers

Fillers are sometimes used as fine solids in grease formulations to improve grease performance. Typical fillers are graphite, molybdenum disulphide, metal oxides and flakes, carbon black, talc and others. Graphite, for example, can minimize wear in sliding bearing surfaces, while molybdenum disulphide minimizes wear in gears. Zinc and magnesium oxide are used in the food processing industry since they neutralize acid. Metal flakes and powdered metals such as lead, zinc, tin and aluminium are used as anti-seize compounds in lubricants for pipe threads. Talc is used in die and drawing lubricants.

Lubrication Mechanism of Greases

Despite the practical importance of greases, there has been surprisingly little research into their lubrication mechanism. The question is, how do greases lubricate and what is the mechanism involved? The mechanism of oil lubrication might be hydrodynamic, elastohydrodynamic or boundary, depending on the operating conditions. The lubrication mechanism of greases, however, will be different since they have a different structure from oil. The structure of grease is gel-like or semi-solid. It is often assumed that grease acts as some sort of spongy reservoir for oil. It was thought for sometime that oil trapped between the soap fibres was slowly released into the interacting surfaces. The question of whether the grease bleeds oil in order to lubricate, or lubricates as one entity, is of critical importance to the understanding of the lubrication mechanism involved. Studies conducted disprove the oil bleeding model. Experiments were performed where different fluorescent colours were added to the soap thickener and to the oil of a grease. Mixing of the dyes was prevented by selecting a water-soluble dye for the thickener and an oil-soluble dye for the oil. Dispersal of the colours, red and blue, enabled observation of grease disintegration. Separation of the grease was not observed when it was used to lubricate a rolling bearing. After a few hours of operation, an equal amount of oil and thickener was found on the interacting surfaces [20]. It was therefore concluded that the bleeding of oil from the grease was not the principal mechanism of lubrication. It appears that the thickener, as well as the oil, takes part in the lubrication process and that grease as a whole is an effective lubricant.

In practice a large quantity of grease is applied to a system, despite the fact that only a very small amount of grease is needed for lubrication. The surplus of grease acts as a seal which prevents the lubricant from evaporating and from contamination, while also preventing the lubricant from migrating from the bearing. The surplus of lubricant also plays an important role as a reservoir from which grease feeds to the operating surfaces when needed [21]. It is thought that the following mechanism is acting: as the thickness of the lubricating film decreases there is an accompanying slight increase in generated frictional heat. As the temperature of grease in the vicinity of the contact increases, the grease expands and softens and more grease smears onto the interacting surfaces. This has been confirmed in an experiment where the oil and grease film thickness between gears has been measured. Contact voltage drop has been used in experiments to assess the operating film thickness [22]. It was found that when an oil was used as the lubricant, the contact resistance was relatively steady in comparison to the case when grease was used as the lubricant. This is shown in Figure 3.4 where the voltage drop for oil and grease is shown for two operating gears under load.

It is evident from Figure 3.4 that when grease is used as the lubricant, intermittent contact between gears occurs. The initial failure of the grease film causes the overall temperature to rise, eventually leading to softening and melting of the grease, resulting in the restoration of the lubricating film. Furthermore, when grease is used, the gear temperatures are usually higher despite lower loading (i.e., average contact load limit for oil is 2020 [kN/m] and for grease 1344 [kN/m]).

It was also found that the instability of a grease film increases the likelihood of gear failure by scuffing, and gear loading must be reduced by a factor of 0.7 compared to the equivalent load for a gear lubricated by a mineral oil [22].

Greases are commonly used in machinery operating under the elastohydrodynamic lubrication (EHL) regime, i.e., in rolling contact bearings and some gears. The question is, how does the grease behave in the EHL regime? Experiments revealed that the measured film thickness of grease under EHL conditions is greater initially than if the base oil contained in the grease were acting alone [23]. With continued running, however, the film thickness of the grease declines to about 0.6 of that of the base oil. The initial thick grease layer is rapidly removed by the rolling or sliding element and the lubrication is controlled by a thin viscous layer which is a mixture of oil and degraded thickener [67]. The decline in film thickness can only be explained in general terms of scarcity of grease in the contact. Grease is a semi-solid so that once expelled from the contact it probably returns only with difficulty. It has also been suggested that conveyance of oil by capillary action from the bulk grease to the wearing contact is possible [67]. However, there has been no detailed work conducted as yet to test this hypothesis.

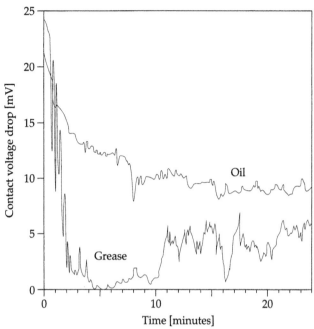

FIGURE 3.4   Fluctuations of oil film thickness between two gears one lubricated by oil and the other by grease (adapted from [22]).

The initial film thickness can be explained in terms of grease rheology [24]. The rheology of grease can be modelled by the Hershel-Bulkley equation:

$$\tau = \tau_p + (\eta_s \mathbf{du/dh})^n$$

where:

τ is the shear stress acting on the oil [Pa];

τ_p is the plastic flow stress [Pa];

η_s is the base oil dynamic viscosity [Pas];

du/dh   is the shear rate [s^{-1}];

n   is a constant.

The value of '**n**' is close to **1**. When '**n**' is exactly unity then the above equation reduces to the original Bingham equation which states that a fluid does not flow below a certain value of minimum shear stress, as shown in Figure 3.5. At high shear stresses, the fluid behaves as a Newtonian liquid. The Hershel-Bulkley equation usually gives good agreement with experiment. When used in the theoretical analysis of EHL and compared with experimental results, good agreement between theoretical and experimental data has been obtained. This is demonstrated in Figure 3.6, which shows the experimental EHL grease characteristics compared to the predicted theoretical values expressed as non-dimensional film thickness and speed [24].

It is generally assumed that, in the actual process of lubrication, the thickening agents are of secondary importance, but there is some evidence that thickeners have significant effects at low sliding speeds. There is surprisingly little published data on greases under these conditions. Some experimental work has been conducted to compare the effects of different thickeners on friction losses in journal bearings [25]. At high sliding speeds, all the lubricants tested provided very low friction, however, the minimum sliding speed to sustain low friction varied greatly between the lubricants. This is shown in Figure 3.7 where friction torque versus bearing speed is plotted.

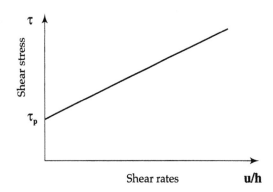

FIGURE 3.5   Bingham fluid.

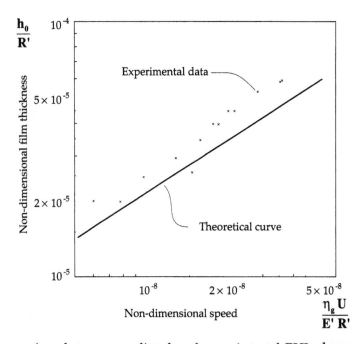

FIGURE 3.6   Comparison between predicted and experimental EHL characteristics of grease; h_0 is the minimum EHL film thickness [m], R' is the reduced radius of curvature [m], E' is the reduced Young's modulus [Pa], U is the surface velocity [m/s], η_g is the atmospheric grease viscosity [Pas] (adapted from [24]).

It can be seen that depending on the thickener, certain greases, in particular a lithium soap based grease, allow a very low friction level to persist even at very slow sliding speeds. On the other hand, the behaviour of some of the greases approximates that of the base oil, for example, aluminium and calcium soap based greases. Some more systematic research remains to be done in this area since the reported data is often contradictory. For example, it has also been found that calcium based grease shows a significant improvement in lubricating properties as compared to mineral oil [26].

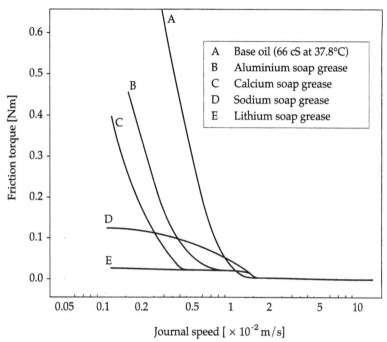

FIGURE 3.7   Low-speed journal bearing friction characteristics of various greases and a base oil [25].

Grease Characteristics

There are several performance characteristics of greases which are determined by well established procedures. The most commonly used in the characterization of greases are consistency, drop point, evaporation loss, oxidation stability, apparent viscosity, stability in storage and use, colour and odour.

· Consistency of Greases

Consistency or solidity is a measure of the hardness or shear strength of the grease. It is defined in terms of grease penetration depth by a standard cone under prescribed conditions of time and temperature (ASTM D-217, ASTM D-1403). A schematic diagram of a typical grease penetration apparatus is shown in Figure 3.8. The grease is placed in the cup and the surface is smoothed out to make it uniform and is maintained at a temperature of 25°C during the test. The cone tip is adjusted so it just touches the grease surface. The cone release mechanism is then activated and the cone is allowed to sink into the grease for 5 seconds. The indicator dial shows the penetration depth which is the measure of the consistency of the grease. The test is usually repeated at various temperatures and is used in conjunction with a standard grease-worker described in the next section. The consistency forms the basis for grease classification and its range is between **475** for a very soft grease and **85** for a very hard grease.

Although consistency is rather poorly defined it is a very important grease characteristic. The hardness of the grease must be sufficient so that it will remain as a solid lump adjacent to the sliding or rolling contact. This lump may be subjected to loads from centrifugal accelerations in rolling bearings and may also be subjected to frictional heat. However, if the grease is too hard 'channelling' may occur where the rolling or sliding element cuts a path through the grease and causes lubricant starvation. Excessively hard greases are also very difficult to pump and may cause blockage of the supply ducts to the bearings. Consistency of a grease also refers to the degree of aggregation of soap fibres. If the soap fibres are present as a tangled mass then the grease is said to be 'rough' and when the grease fibres have joined together to form larger fibres, the grease is said to be 'smooth'. Roughness or smoothness has a strong influence on the stable operation of rolling bearings [29]. If the grease is too smooth, then stable lumps of grease will never form in a rolling bearing during its operation. The grease will continue to slump and circulate in the bearing, and high operating temperatures and short grease life will result. The trade term for this problem is that the grease has failed to 'clear'. For some unknown reason a very rough grease will be expelled from the bearing and the bearing will rapidly wear out. A grease that is neither too rough nor too smooth usually gives the lowest operating temperatures and least wear.

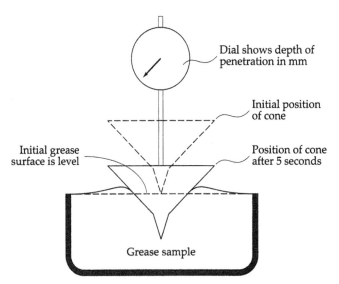

FIGURE 3.8   Schematic diagram of a typical penetration grease apparatus.

· *Mechanical Stability*

The consistency of a grease can change due to mechanical shearing. Even if at the beginning of the service grease possesses the optimum consistency for a particular application, mechanical working will damage the soap fibres and degrade the grease. Greases differ significantly in the level of damage they will incur due to mechanical working. For example, greases working in gear boxes and bearings or being pumped through pipes are subjected to shear. The changes in grease consistency depends on the stability of the grease structure. In some cases greases may become very soft, or even flow, but in most cases there is only slight softening or hardening of the grease. Consistency of the grease is often specified for worked and pre-worked conditions. The grease is worked in the test apparatus which consists of a container fitted with a perforated metal plate plunger which is actuated by a motor driven linkage. The schematic diagram of this apparatus is shown in Figure 3.9. There is a large clearance between the piston and the cylinder and the piston is perforated by a series of small holes. The piston is moved up and down and the grease is extruded through the holes and hence is subjected to shearing action. Usually the grease is worked through 60 double strokes of the piston and then the consistency is determined.

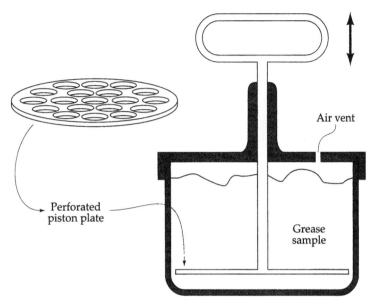

FIGURE 3.9   Schematic diagram of a grease-worker.

The consistency of greases made from several thickening agents has been measured after varying periods of mechanical working [30]. It was found that all greases were softened by mechanical working to some extent, but when calcium tallow soap was the thickening agent, little damage resulted. Lithium hydroxystearate and sodium tallow stearate suffered significant damage initially, but thereafter their consistency reached a stable value. Lithium stearate and aluminium stearate, however, showed a continuous progression in damage.

It was also found that if the grease in a rolling bearing fails to clear then the continued mechanical working of the grease makes the situation even worse. The high operating speeds of rolling bearings accelerate the mechanical degradation of grease and it is advisable to operate the bearing at slightly less than the maximum rated speed. A design level of 75% of maximum rated speed has been suggested [31].

· *Drop Point*

The drop point is the temperature at which a grease shows a change from a semi-solid to a liquid state under the prescribed conditions. The drop point is the maximum useful operating temperature of the grease. It can be determined in an apparatus in which the sample of grease is heated until a drop of liquid is formed and detaches from the grease (ASTM D-566, ASTM D-2265). The schematic diagram of a drop point test apparatus is shown in Figure 3.10. Although frequently quoted, drop point has only limited significance as a grease performance characteristic. Many other factors such as speed, load, evaporation losses, etc. determine the useful operating temperature range of the grease. Drop point is commonly used as a quality control parameter in grease manufacturing.

· *Oxidation Stability*

The oxidation stability of a grease (ASTM D-942) is the ability of the lubricant to resist oxidation. It is also used to evaluate grease stability during its storage. The base oil in grease will oxidize in the same way as a lubricating oil of a similar type. The thickener will also oxidize but is usually less prone to oxidation than the base oil. Oxidation stability of greases is measured in a test apparatus in which five grease dishes (4 grams each) are placed in an atmosphere of oxygen at a pressure of 758 [kPa]. The test is conducted at a temperature of 99°C and the pressure drop is monitored. The pressure drop indicates how much oxygen is being used to oxidize the grease. The schematic diagram of the grease oxidation stability apparatus is shown in Figure 3.11.

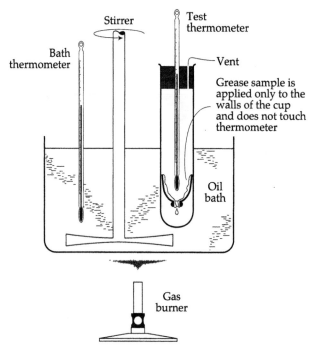

FIGURE 3.10   Schematic diagram of a drop point test apparatus.

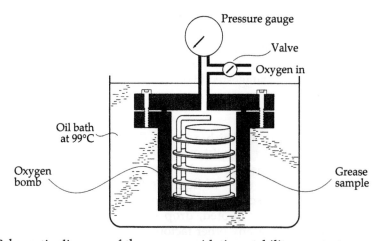

FIGURE 3.11   Schematic diagram of the grease oxidation stability apparatus.

Oxidized grease usually darkens and acidic products accumulate in the same manner as in a lubricating oil. Acidic compounds can cause softening of the grease, oil bleeding, and leakage resulting in secondary effects such as carbonization and hardening. In general the effects of oxidation in greases are more harmful than in oils.

· *Thermal Stability*

Greases cannot be heated above a certain temperature without starting to decompose. The temperature-life limits for typical greases are shown in Figure 3.12 [27]. The temperature limits for greases are determined by a number of grease characteristics such as oxidation stability, drop point and stiffening at low temperature.

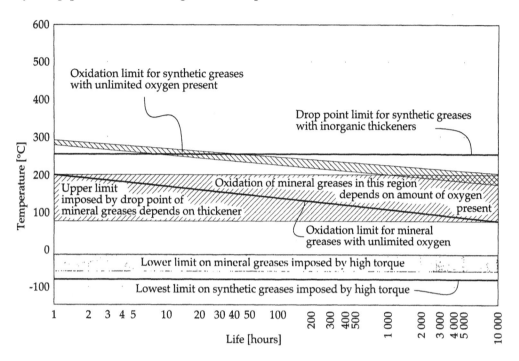

FIGURE 3.12   Temperature-life limits for typical greases [27].

· *Evaporation Loss*

As in oils, weight losses in greases due to evaporation can be quite significant. Volatile compounds and products of thermal degradation contribute to the losses, resulting in thickening of the lubricant, higher shear resistance and higher temperatures. The testing method involves placing the test sample in a heating bath and passing evaporating air over the sample's surface for 22 hours at temperatures ranging between 99°C and 150°C (ASTM D-972, ASTM D-2595). The percentage weight loss is then determined.

· *Grease Viscosity Characteristics*

Greases exhibit a number of similar characteristics to lubricating oils, e.g., they shear thin with increased shear rates, the apparent viscosity of a grease changes with the duration of shearing and grease consistency changes with temperature.

Apparent viscosity of a grease is the dynamic viscosity measured at the desired temperature and shear rate (ASTM D-1092, ASTM D-3232). Measurements are usually made in the temperature range between -53°C and 150°C in specially designed pressure viscometers. Apparent viscosity, defined as the ratio of shear stress to shear rate, is useful in predicting the grease performance at a specific temperature. It helps to predict the leakage, flow rate, and pressure drop in the system, the performance at low temperature and the pumpability. The apparent viscosity depends on the type of oil and the amount of thickener used in the grease formulation.

Shear thinning of greases is associated with the changes in the apparent viscosity of grease with increased shear rates. When shearing begins the grease's apparent viscosity is high but with increased rates of shearing it may drop to that of its base oil. An example of this non-Newtonian, pseudoplastic behaviour in calcium soap based greases is shown in Figure 3.13.

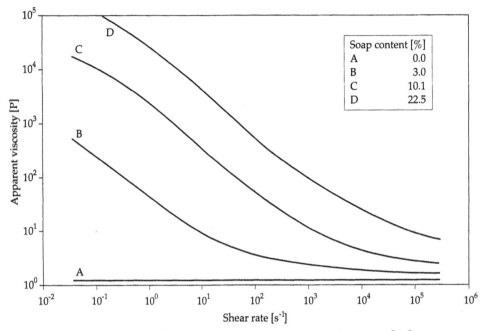

FIGURE 3.13 Non-Newtonian behaviour of calcium soap based greases [64].

Shear duration thinning of greases is associated with the changes which occur in the apparent viscosity of grease with the duration of shearing. As with oils, the greases which soften with duration of shearing and stiffen when shearing stops are called thixotropic. Depending on the type of grease a permanent softening or reverse effect of hardening can occur. In some applications this effect can be beneficial, in others it is detrimental. For example, the permanent softening of a small quantity of grease in rolling contact bearings will result in good lubrication, low friction and low contact temperatures. On the other hand, the softening of the main bulk of grease will result in its continuous circulation and high operating temperatures. Thixotropic greases are particularly useful where there is a leakage problem, for example, in a gear box. The grease in contact with the gears will be soft because of shearing, but outside the contact it will be stiffer and will not leak.

Grease consistency temperature relationship describes the changes in the grease consistency with temperature. As has already been mentioned in a previous chapter the viscosity of oil is very sensitive to temperature changes. Relatively small temperature variations may result in significant changes in viscosity. There are only relatively small changes in grease consistency with temperature until it reaches its drop point. At this temperature the grease structure breaks down and the grease becomes liquid. The variation in grease consistency, expressed in terms of penetration depth, with temperature for a sodium soap grease is shown in Figure 3.14 [21].

The structure of some non-soap greases will remain stable until the temperature rises to a point where either the base oil or the thickener decomposes. It has also been found that if a grease is heated above the drop point and then cooled it does not regain its grease-like consistency and its performance is unsatisfactory [21].

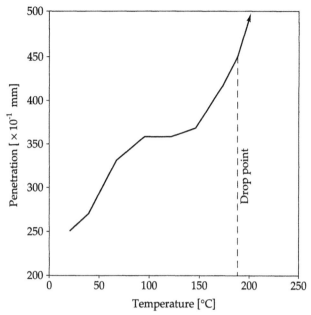

FIGURE 3.14 Variation in grease consistency, expressed in terms of penetration, with temperature for a sodium soap grease [21].

Classification of Greases

The most widely known classification of greases is related to their consistency and was established by the National Lubricating Grease Institute (NLGI). It classifies the greases into nine grades, according to their penetration depth, from the softest to the hardest [28], as shown in Table 3.3.

TABLE 3.3   NLGI grease classification [28].

NLGI grade	Worked (60 strokes) penetration range [$\times 10^{-1}$ mm] at 25°C		
000	445	-	475
00	400	-	430
0	355	-	385
1	310	-	340
2	265	-	295
3	220	-	250
4	175	-	205
5	130	-	160
6	85	-	115

Depending on the application a specific grease grade is selected. For example, soft greases, No. 000, 00, 0 and 1, are used in applications where low viscous friction is required, e.g., enclosed gears which are slow, small and have a tendency to leak oil. In open gears grease must effectively be retained on the gear surface and tacky or adhesive additives such as bitumen are used in its formulation to improve adhesion. Greases No. 0, 1 or 2 are used depending on the operating conditions such as speed, load and size of the gear. In rolling contact bearings greases No. 1, 2, 3 and 4 are usually used. The most commonly applied is No. 2. Harder greases are used in large bearings and in applications where there are problems associated with sealing and vibrations. They are also used for higher speed applications. In plain, slowly moving bearings (1 - 2 [m/s]) greases No. 1 and 2 are used. In general practice the most commonly used grease is Multipurpose Grease which is a grease No. 2 according to the NLGI classification, with aluminium or lithium soap thickeners.

The selection of a grease for a specific application mainly depends on the temperature at which the grease is expected to operate. For low temperature applications the important factor is the low-temperature limit of a specific grease, which is determined by the viscosity or pour point of the base oil. Examples of low temperature limits for selected greases are shown in Table 3.4 [21].

TABLE 3.4   Low temperature limits for selected greases [21].

Base oil	Thickener	Minimum temperature [°C]
Mineral oil	Calcium soap	-20
Mineral oil	Sodium soap	0
Mineral oil	Lithium soap	-40
Mineral oil	Bentonite clay	-30
Di-ester	Lithium soap	-75
Di-ester	Bentonite clay	-55
Silicone	Lithium soap	-55
Silicone	Dye	-75
Silicone	Silica	-50

The maximum operating temperature for a grease is limited by the drop point and the oxidation and thermal stability of the base oil and the other grease components. Typical properties together with the drop point values for selected greases are listed in Table 3.5 [63].

TABLE 3.5   Typical properties of selected greases [63].

Thickener	Drop point [°C]	Mecha-nical stability	Anti-wear	Water resista-nce	Thermal stability	Life	Anti-fretting	Churn-ing noise	Average relative cost
Sodium soap	185	medium	medium	fair-medium	fair	medium	fair	very quiet	1
Li/Ca mixed soap	185	good	medium	good-excellent	medium	medium-good	fair-medium	very quiet	1.4
Lithium complex	250	good-excellent	medium	good	good	good	poor	noisy	1.8
Calcium complex	240	fair-medium	good-excellent	medium	medium	medium	medium	noisy	1.5
Aluminium complex	250	good	poor	fair-medium	medium	medium	poor	noisy	1.6
Clay	>300	medium-good	poor-medium	good	medium-good	medium	poor-medium	noisy	1.5
Soap/clay mixed base	>300	good-excellent	medium	good-excellent	good	good	fair-medium	fair	1.9
Polyurea (di-urea)	270	excellent	excellent	good-excellent	excellent	excellent	medium	fair	2.5
Polyurea (tetra-urea)	260	fair-medium	excellent	good-medium	good	excellent	good	quiet	2

It is interesting to note that at temperatures above the drop point a grease may still provide effective lubrication but it will no longer be a grease since it will have changed its phase and become a liquid.

Environmental factors must also be considered in grease selection. Industries such as mining, pharmaceuticals, food processing, textiles, aero-space and others operate in specific environments where different types of greases are required. In some applications, due to their semi-solid nature, greases are essential. For example, in dirty environments such as mining, greases are ideal since they reduce the risk of fire and have good sealing properties. In the pharmaceutical and food industry they are widely applied because they seal against dirt and prevent leakages which might otherwise contaminate the product. The type of thickener and base oil that can be used in grease formulation is restricted and controlled in these industries, so that any accidental contamination of the product will not pose a health risk.

In aerospace applications, greases are expected to operate in extreme conditions. For example, aviation greases are expected to operate at the temperatures encountered by some of the high altitude military aircraft which range from -75°C to +200°C. Synthetic lubricants are used in these applications. In space, greases must have exceptionally low volatility to withstand high vacuum. Evaporation losses in space are controlled by specially designed seal systems.

Grease Compatibility

Two lubricating oils, provided that they are of the same type (i.e., mineral, silicone, silane, diester, etc.), should not present any problems with compatibility when mixed. The general rule, however, is that two greases should not be mixed, even if they are formulated from the same base oil and thickener, as this may lead to complete failure of the system [21]. The particular risk is that an oil added may dissolve or soften the thickener.

Degradation of Greases

Even though grease is prone to a greater number of degradation modes than oil, it is required to spend a greater period of time as a functioning lubricant. Grease remains packed within the rolling bearing, gear, etc., whereas oil is circulated from a sump. Grease failure often does not occur immediately but small changes in operating conditions, particularly temperature, may cause problems associated with grease degradation.

The modes of grease degradation are base oil oxidation, separation of oil from the thickening agent and breakdown of the thickening agent. Base oil oxidation proceeds in a similar manner to that already discussed for plain mineral oils. Separation of the oil and thickening agent, or 'bleeding', and breakdown of the thickening agent are peculiar to grease. Even in storage, where oil can be stored in a sealed container almost indefinitely, greases may separate, soften or harden or even become rancid as in the case of some soap thickened greases [21]. The composition and physical form of the soap control the likelihood of 'bleeding' or 'loss of consistency'. Loss of consistency means either that the grease has become too soft or too hard for the intended application or that the rheological and tribological characteristics have deteriorated.

The soap may be present in the oil as a tangled mass of fibres or as discrete crystals. It is only these fibres or crystals that prevent either the oil separating from the grease or the grease degenerating to a simple liquid. If a grease liquefies, this is called 'slumping' and is a major cause of grease failure. As mentioned earlier, the soap fibres are vulnerable to temperature and excessive mechanical working. Elevated temperature attacks the grease in two ways:

·   the base oil loses viscosity and therefore separates from the grease more readily,

·   the soap fibres melt, in some cases even at quite low temperatures.

If the soap fibres melt (or soften when there is no clear melting point), the grease disintegrates. Rolling bearings and gears can reach temperatures well in excess of 100°C during operation and special soaps, as opposed to the traditional calcium stearate, have been developed to meet these demands. An example is lithium hydroxy-stearate which does not soften up to 190°C, and other greases capable of withstanding even higher temperatures are also manufactured. The lifetime of any grease declines with temperature. For example, at 40°C the lifetime of a lithium hydroxy-stearate grease is approximately 20,000 hours, whereas at 140°C its lifetime is only 500 hours. Grease failure in these circumstances is caused by hardening of the grease and formation of deposits on bearing surfaces.

Most greases are reasonably resistant to damage by water despite their soap content. Whilst lithium and aluminium based greases are scarcely affected by water, sodium based greases are quite vulnerable to it. Calcium based greases, on the other hand, exhibit intermediate levels of water resistance.

3.6 LUBRICANT ADDITIVES

Lubricant additives are chemicals, nearly always organic or organometallic, that are added to oils in quantities of a few weight percent to improve the lubricating capacity and durability of the oil. This practice gained general acceptance in the 1940's and has since developed to provide an enormous range of additives. Specific purposes of lubricant additives are:

· improving the wear and friction characteristics by provision for adsorption and extreme pressure (EP) lubrication,

· improving the oxidation resistance,

· control of corrosion,

· control of contamination by reaction products, wear particles and other debris,

· reducing excessive decrease of lubricant viscosity at high temperatures,

· enhancing lubricant characteristics by reducing the pour point and inhibiting the generation of foam.

Carefully chosen additives are extremely effective in improving the performance of an oil. Perhaps for this reason, most additive suppliers maintain secrecy over the details of their products. One result of this secrecy is that the supplier and the user of the lubricant may only know that a particular oil contains a 'package' of additives and this can often impede analysis of lubricant failures. Another result is that large companies very often use many different brands of lubricants which are effectively the same or have similar properties and composition. This is quite costly to a company as a variety of lubricants must be stored and replaced from time to time. The secrecy surrounding additives also means that their formulation is partly an art rather than a purely scientific or technical process. The most common package of additives used in oil formulations contains anti-wear and EP lubrication additives, oxidation inhibitors, corrosion inhibitors, detergents, dispersants, viscosity improvers, pour point depressants and foam inhibitors. Sometimes other additives like dyes and odour improvers are also added to the oils.

Wear and Friction Improvers

Additives which improve wear and friction properties are probably the most important of all the additives used in oil formulations. Strictly speaking these chemicals are adsorption and extreme pressure (EP) additives and they control the lubricating performance of the oil. Performance enhancing properties of these additives are very important since, if oil lacks lubricating ability, excessive wear and friction will begin as soon as the oil is introduced into the machine. These additives can be divided into the following groups:

· adsorption or boundary additives,

· anti-wear additives,

· extreme pressure additives.

· *Adsorption or Boundary Additives*

The adsorption or boundary additives control the adsorption type of lubrication, and are also known in the literature as 'Friction Modifiers' [32] since they are often used to prevent slip-stick phenomena. The additives in current use are mostly the fatty acids and the esters and amines of the same fatty acids. They usually have a polar group (-OH) at one end of the molecule and react with the contacting surfaces through the mechanism of adsorption. The surface films generated by this mechanism are effective only at relatively low temperatures and loads. The molecules are attached to the surface by the polar group to form a carpet of molecules, as shown in Figure 3.15, which reduces friction and wear.

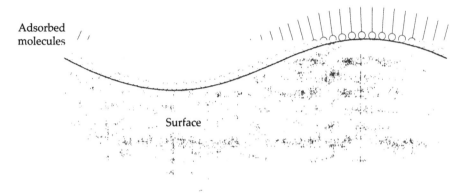

FIGURE 3.15 Adsorption lubrication mechanism by boundary additives.

The important characteristic of these additives is an unbranched chain of carbon atoms with sufficient length to ensure a stable and durable film. Specialized additives which combine adsorption or boundary properties with some other function such as corrosion protection are also in use [32]. Such additives are rarely described in detail in open literature, although the most frequently used are sulphurized fatty acid derivatives, phosphonic acids or N-acylated sarcosines [3]. Stearic acid derivatives such as methyl and ethyl stearates are also used. Adsorption or boundary additives are very sensitive to the effects of temperature. They lose their effectiveness at temperatures between 80°C and 150°C depending on the type of additive used. With increased temperature there is sufficient energy input to the surface for the additive to desorb. The critical temperature at which the additive is rendered ineffective can be manipulated by changing the additive's concentration, i.e., a higher concentration results in a higher critical temperature, but the cost is also increased.

· *Anti-Wear Additives*

In order to protect contacting surfaces at higher temperatures above the range of effectiveness of adsorption or boundary agents, anti-wear additives were designed and manufactured. There are several different types of anti-wear additives that are currently used in oil formulations. For example, in engine oils the most commonly used anti-wear additive is zinc dialkyldithiophosphate (ZnDDP), in gas turbine oils tricresylphosphate or other phosphate esters are used. Phosphorous additives are used where anti-wear protection at relatively low loads is required. These additives react with the surfaces through the mechanism of chemisorption, and the protective surface layer produced is much more durable than that generated by adsorption or boundary agents.

Common examples of these additives are zinc dialkyldithiophosphate, tricresylphosphate, dilaurylphosphate, diethylphosphate, dibutylphosphate, tributylphosphate and triparacresylphosphate. These additives are used in concentrations of 1% to 3% by weight.

<u>Zinc dialkyldithiophosphate (ZnDDP)</u> is a very important additive commonly used in engine oil formulations. It was originally developed as an anti-oxidant and detergent, but it was found later that this compound also acted as an anti-wear and mild extreme pressure additive. The term 'anti-wear' usually refers to wear reduction at moderate loads and temperatures whereas the term extreme pressure (EP) is reserved for high loads and temperatures. Although some authors recognize this additive as a mild EP additive, it is generally classified in the literature as an anti-wear additive. The chemical structure of ZnDDP is shown in Figure 3.16.

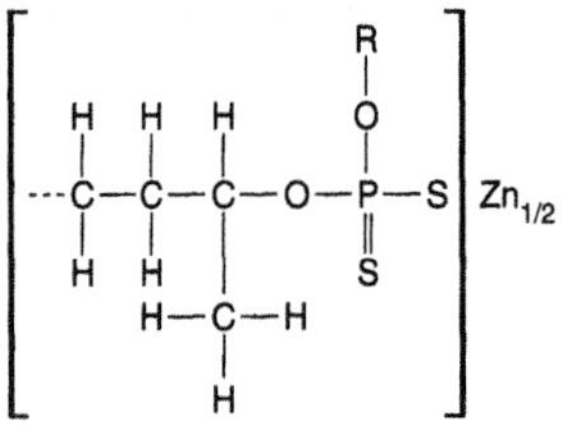

FIGURE 3.16 Chemical structure of zinc dialkyldithiophosphate.

By altering the side groups a series of related compounds can be obtained, an example of which is zinc diphenyldithiophosphate. These new compounds, however, are not as effective as ZnDDP in reducing wear and friction. The presence of zinc in ZnDDP plays an important role. The substitution of almost any other metal for zinc results in increased wear. For example, it was found that wear rates increased with various metals in the following order: cadmium, zinc, nickel, iron, silver, lead, tin, antimony and bismuth [56]. Cadmium gives the lowest wear rates but is far too toxic for practical applications. Interestingly, no definite explanation for the role of metals in the lubrication process by ZnDDP has yet been offered.

It is known that ZnDDP is a major source of sulphur and phosphorus presence in engine oils. Due to environmental pollution problems caused by sulphur and phosphorus, replacements for ZnDDP are being sought. Organic borates (exact composition appears to be still a commercial secret) appear to be promising substitutes for ZnDDP since they display good anti-oxidant and wear protection characteristics. However, organic borates cannot be used in combination with ZnDDP in the engine oil, since the calcium sulphonate component of the organic borate is antagonistic to ZnDDP [72].

Like many other lubricant additives, ZnDDP is usually not available in pure form and contains many impurities which affect lubrication performance to varying degrees [33]. The surface protective films which are formed as the result of action of ZnDDP act as the lubricant, reducing wear and friction between the two interacting surfaces. The lubrication mechanism of ZnDDP is quite complex as the additive has three interacting active elements, i.e., zinc, phosphorus and sulphur. Water and oxygen are also active elements, and their presence adds to the complexity of the lubrication mechanism. All of these elements and compounds might be involved in surface film formation, and our current understanding is that films of slightly different compositions can form depending on the operating conditions, contacting surfaces and the purity of ZnDDP. In one study the films formed were of the order of 10 [nm] and consisted of a matrix of zinc polyphosphate with inclusions of iron oxide and iron sulphide [34]. Varying contact stress levels affect the composition of films formed. An adsorbed film of ZnDDP was found in the unloaded parts of a wearing contact while much thinner film of zinc phosphates was found in loaded contact areas. Sliding contact or high temperature leads to a loss of the organic (alkane) component of the ZnDDP together with most of the sulphur leaving only a phosphate film [76]. It was also shown that when a purified ZnDDP was used at mild mechanical stresses at high temperature, or high stresses at ambient temperature of 25°C, these films consisted mainly of simple zinc/iron phosphates [76]. The presence of zinc polyphosphate tends to increase with increasing temperature.

It has also been suggested that the films might be formed by spontaneous decomposition of the additive on the worn surface since only a small amount of iron is found in the film [35]. Even the effective film thickness under operating conditions is a matter of controversy. In a different experiment the contact resistance measured between sliding surfaces lubricated by ZnDDP was found to be higher than expected. It indicated that a thicker surface film of perhaps 100 [nm] thickness was in place, which is much greater than when lubricated by surfactants which are boundary agents [33].

Care should be taken with the application of ZnDDP. This additive is most suitable for moderate loads and was initially applied to the valve train of an internal combustion engine, giving significant reduction in wear and friction [36]. For high load applications, ZnDDP may actually increase wear beyond that of a base oil [34]. It is also found that temperature can amplify these effects. This is demonstrated in Figure 3.17 [34] where the wear rates decreased with temperature at low loads for ZnDDP containing oils but the converse was true at high loads.

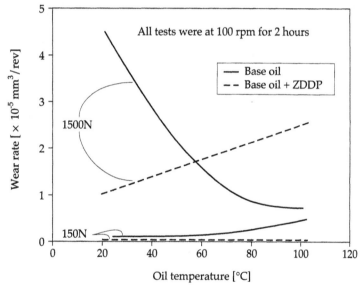

FIGURE 3.17 Influence of load and temperature on the effectiveness of ZnDDP on wear rates (adapted from [34]).

ZnDDP is a prime example of the empirical nature of much of the science of lubricant additive development. The problem of valve train wear and oil degradation in internal combustion engines was solved by applying ZnDDP many years ago. Scientific understanding and interpretation of the process have only recently become available.

Tricresylphosphate (TCP) has been used as an anti-wear additive for more than 50 years. Like ZnDDP, it functions by chemisorption to the operating surfaces, which is explained in detail in the chapter on 'Boundary and Extreme Pressure Lubrication'. It is very effective in reducing wear and friction at temperatures up to about 200°C. Beyond this temperature there is sufficient energy input to the surface for the chemisorbed films to desorb and it is believed that the compound will then form less effective, much weaker, thick phosphate films with limited load capacity [62].

Other anti-wear additives such as dilauryl phosphate, dibutylphosphate, diethylphosphate, tributylphosphate and triparacresylphosphate are also being used in lubricant formulation.

They function in the same manner as ZnDDP or TCP by producing chemisorbed surface films. Some of these additives, e.g., diethylphosphate, can even behave as a moderate EP additive, and these are discussed in the next section.

· *Extreme Pressure Additives*

These compounds are designed to react with metal surfaces under extreme conditions of load and velocity, i.e., slowly moving, heavily loaded gears. Under these conditions operating temperatures are high and the metal surfaces are hot. EP additives contain usually at least one aggressive non-metal such as sulphur, antimony, iodine or chlorine. They react with exposed metallic surfaces creating protective, low shear strength surface films, which reduce friction and wear. The reaction with the metallic surfaces is a form of mild corrosion, thus the additive concentration is critical. If the concentration of EP additive is too high then excessive corrosion may occur. If the concentration of EP additive is too low then the surfaces may not be fully protected and failure could result. EP additives, if they contain sulphur or phosphorus, may suppress oil oxidation but decomposition of these additives may occur at even moderate temperatures. Extended oil life at high temperatures is therefore not usually obtained by the addition of EP additives. Extreme pressure additives are not generally toxic but some early types were even poisonous, e.g., lead naphthenates.

There are several different types of extreme pressure additives currently added to oils. The most commonly used are dibenzyldisulphide, phosphosulphurized isobutene, trichlorocetane and chlorinated paraffin, sulphurchlorinated sperm oil, sulphurized derivatives of fatty acids and sulphurized sperm oil, cetyl chloride, mercaptobenzothiazole, chlorinated wax, lead naphthenates, chlorinated paraffinic oils and molybdenum disulphide. There are also other types of EP additives, e.g., tin based organochlorides, but these are not very popular because of toxicity and stability problems.

Dibenzyldisulphide is a mild EP additive which has sulphur positioned in a chain between two organic radicals as shown in Figure 3.18.

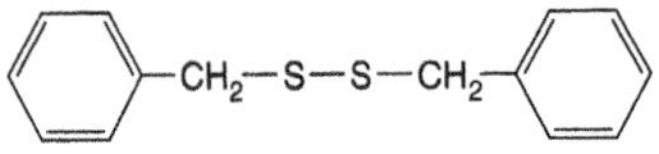

FIGURE 3.18   Structure of dibenzyldisulphide.

Examples of this type of additive are butylphenol disulphide and diphenyl disulphide. The specific type of hydrocarbon radical, e.g., diphenyl, provides a useful control of additive reactivity to minimize corrosion.

Trichlorocetane and chlorinated paraffin are powerful EP additives but they are also very corrosive, particularly when contaminated with water. They are applied in extreme situations of severe lubrication problems, e.g., screw cutting.

Paraffinic mineral oils and waxes can be chlorinated to produce EP additives. They are not very popular since the mineral oils are quite variable in their composition and usually a poorly characterized additive results from this procedure. Such additives may have very serious undesirable side effects, e.g., toxicity and corrosiveness.

Sulphurchlorinated sperm oil is an effective EP additive, but is becoming obsolete because of the increasing rarity of harvested sperm whale oil. It is still, however, used in heavy duty truck axles.

Sulphurized derivatives of fatty acids and sulphurized sperm oil provide a combination of extreme pressure and adsorption lubrication [38]. Sulphurization of fatty acids and sperm oil (which is a fatty material) produces a complex range of products so that names of individual products are not usually quoted. An early example is sulphurized lead naphthenate which has been used as an additive in hypoid gears. Although, in general, EP additives are not toxic, this particular additive is poisonous and largely for this reason it is gradually becoming obsolete. These additives can still be found in gear oils and cutting fluids for metalworking operations.

Molybdenum disulphide provides lubrication at high contact stresses. It functions by depositing a solid lubricant layer on the contacting surfaces. It is non-corrosive but is very sensitive to water contamination as water causes the additive to decompose.

Nanoparticle Additives

A new promising area in the lubricant additive developments involves nanoparticles. They are showing encouraging results as additives for conventional oils as well as for emulsions. Nickel oxythiomolybdate ($NiMoO_2S_2$) particles with an average diameter of 13 [nm], blended with a synthetic oil, pentaerythritoltetraester (PETE), displayed good anti-seizure characteristics and effective lubrication properties at temperatures beyond 300°C [73]. Lanthanum fluoride nanoparticles of average diameter of 6 [nm] blended with chemically pure paraffin oil were found to provide good anti-wear activity and load-carrying capacity when compared with ZnDDP during a four-ball wear test [74]. The addition of copper nanoparticles, with an average diameter around 80 [nm], to SAE30 motor oil provided a useful improvement in its lubricating characteristics [75]. Further promising developments in this area are expected.

Anti-Oxidants

· Oil Oxidation

Mineral oils inevitably oxidize during service and this causes significant increases in friction and wear which affects the performance of the machinery. The main effect of oxidation is a gradual rise in the viscosity and acidity of an oil. This effect is demonstrated in Figure 3.19 which shows the variation of viscosity and acidity of a mineral oil as a function of oxidation time [39].

It can be seen from Figure 3.19 that as the oxidation proceeds beyond 120 hours of operating time, there is a rapid rise in viscosity, increasing about 8 fold by 150 hours. A similar trend can also be observed with the acidity of the oil, expressed as Total Acid Number. A highly oxidized oil needs to be replaced since it causes power losses due to increased viscous drag and difficulties in pumping through the lubricant feed lines. It should be mentioned, however, that oxidation is not the only cause of viscosity increase in lubricating oils. Another cause is diesel soot. Elevated oil acidity can cause concentrated corrosion of certain machinery components such as seals and bearings. For example, lead, copper and cadmium are used in the bearing alloys of an internal combustion engine and they are particularly prone to corrosion. It is clear from Figure 3.19 that beyond a Total Acid Number of about 3 an oil needs to be replaced. The period of time prior to any drastic change in the lubricating properties of an oil represents its working life and is also referred to as the induction period.

The general mechanism of oil oxidation is believed to be a free-radical chain reaction. The identification of the precise mechanism, however, is hindered by the complexity and variability of mineral oil composition [3,40]. It is thought that a possible reaction mechanism of the initial stages of the oxidation process is as follows [3]:

$$\mathbf{R\text{-}H} \to \mathbf{R^\circ} + \mathbf{H^\circ} \qquad\qquad \text{(radical formation)}$$

$$\mathbf{R^\circ} + \mathbf{O_2} \to \mathbf{R\text{-}O\text{-}O^\circ} \qquad\qquad \text{(peroxide formation)}$$

$$\mathbf{R\text{-}O\text{-}O^\circ} + \mathbf{R\text{-}H} \to \mathbf{R\text{-}O\text{-}O\text{-}H} + \mathbf{R^\circ} \qquad\qquad \text{(propagation)}$$

where:

R-H	is a hydrocarbon;
R	is a radical;
R°	is a free radical;
H°	is a hydrogen ion;
R-O-O°	is a peroxide radical;
R-O-O-H	is an organic acid.

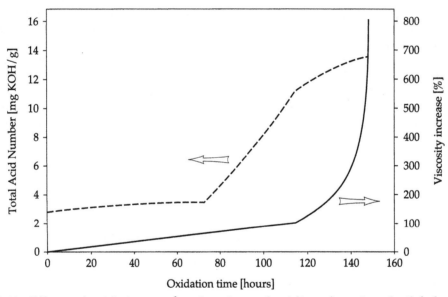

FIGURE 3.19 Effects of oxidation on the viscosity and acidity of a mineral oil (adapted from [39]).

The initial reaction is the formation of radicals which is essentially a thermally activated dissociation of the hydrocarbon molecules. In the second reaction peroxides are formed and this represents the first stage of oxidation, i.e., a direct reaction between hydrocarbon and dissolved oxygen. The propagation process is shown in the third reaction which illustrates how the oxidized hydrocarbons can exert a catalytic effect and hence greatly accelerate the oxidation process. The peroxide radical **R-O-O°** is very reactive and controls the reaction rate. By reacting with the hydrogen ion it produces a carboxyl acid **R-O-O-H**, and this provides the basis for accelerated oxidation [39]. Therefore each consequent oxidation stage leads to more products which are themselves capable of oxidation until eventually a large portion of the oil is oxidizing at any one time, as opposed to a trace quantity, and oil oxidation may proceed very rapidly beyond a critical point. The oxidation process is therefore self-accelerating. The end result is that the original hydrocarbons are converted to a series of carboxyl acids, ketones and alcohols which can then form higher molecular weight components by condensation. The high molecular weight components form sludge and deposits, which block the oil pathways.

The oxidation rates can be affected by temperature, metals in contact with oil, the amount of water and oxygen in the oil and the presence of ionizing radiation. The temperature especially has a profound effect on oxidation rates, which can be as much as tripled by a temperature rise of 10°C.

The presence of metallic wear debris in the oil can accelerate oxidation. This aspect of oil degradation has been known for a long time but the actual mechanism involved is still not fully understood. It has been shown that the oxidation of a thin layer of oil in a metal test-cup is influenced by the cup material [41]. Iron consistently accelerates the oxidation of oil, whereas when there are high concentrations of dissolved copper in the oil, oxidation is limited. It is found that copper shows an inhibiting effect when present in oil at concentrations as high as 2000 [ppm], but at the concentrations typical for used oils (i.e., ~ 100 [ppm]) there is no inhibiting effect and oxidation is accelerated [42]. Acceleration of oil oxidation by iron also depends on the form of contact between the iron and the oil. There are in fact several forms of contact possible between a metal and an oil. There can be contact between the oil and the adjacent metal surfaces, or there can be contact between suspended wear debris and dissolved metal. Each of these contacts affects oil oxidation to a varying degree. Some have already been investigated and others have not. For example, iron dissolved in oil exerts a far stronger pro-wear effect than particulate iron [43]. On the other hand, the relationship between suspended particulate iron and dissolved iron has not yet been investigated. Lead is also found to be easily dissolved in oil, but unlike copper or iron, it does not lead to the formation of high molecular weight insoluble products which cause sludging and formation of deposits on the interacting surfaces.

Oxidation rates depend on the quantity of oxygen dissolved in the oil. As the amount of oxygen in the oil can be increased by mixing, churning, etc., the lubrication systems should be designed in such a way that the oil is disturbed as little as possible.

Other factors can also contribute to the oxidation of oils or produce similar effects. In internal combustion engines blow-by gases and nitrous oxide can hasten the oxidation process [39] while nuclear radiation can cause a large increase in oil viscosity by inducing cross-links between hydrocarbon molecules [44]. Radiation also produces free radicals, thus increasing the rate of oxidation.

The problem of deposits formation on surfaces lubricated by mineral oils was recognized quite early, during the development of the internal combustion engine. Recent research in this area indicates that the probability of diesel engine scuffing is related to the critical temperature of lacquer formation [45]. It has been found that lacquer or deposit formation between piston rings and their adjacent grooves prevents their free movement. Scuffing then results because of the high contact forces between the ring and the cylinder wall. It is believed that the combined processes of oil oxidation and evaporation contribute to the formation of deposits. The rate of deposit formation depends on several factors, for example, rough surfaces accelerate deposit formation, the deposit forming tendency of mineral oils increases with the average molecular weight of the oil (high molecular weights giving the most deposits) and rate of deposition conforms to the Arrhenius law [46].

Oil oxidation can also affect the wear of mechanical components to varying degrees [47,57]. These effects are, however, poorly researched and understood. They are also unpredictable. It has been found, for example, that wear rates of various moving parts can be accelerated in a non-uniform manner by oxidation. The damage can then be concentrated on one component critical to the operation of the machinery.

Finally a limited amount of oxidation is not entirely detrimental to the oil, since initial oxidation products can provide thin-film lubrication [33]. This fact was recognized even before additives were introduced. The lubricating oils were deliberately oxidized before their application. An example of these effects is shown in Figure 3.20. It can be seen that mildly oxidized hexadecane gives better friction characteristics than pure hexadecane, especially above 90°C [33].

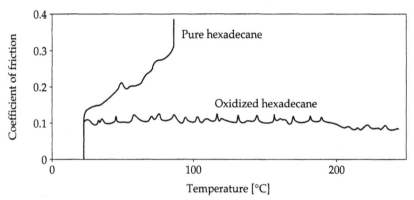

FIGURE 3.20   Effect of oxidation on the friction characteristics of hexadecane [33].

In general, mineral oils are vulnerable to chemical degradation in service and wear processes accentuate the severity of conditions. This limitation in fact provided one of the principal reasons for developing synthetic mineral oils. The move to more durable lubricants is a continuation of a long-term trend which began when mineral oils displaced lubricants such as castor and olive oil in the early 20th century because of their longer service life and availability.

· *Oxidation Inhibitors*

Most lubricating oils in present use contain anti-oxidant additives to delay the onset of severe oxidation of the oil. These are either natural anti-oxidants or artificially introduced additives that are able to suppress oxidation, and any differences in the oxidation resistance of oils largely depend on the presence of these inhibitors. Natural sulphur or nitrogen containing compounds which are present in mineral oils act as oxidation inhibitors by scavenging the radicals produced by the oxidation process. Sulphur based EP and anti-wear additives are also quite effective as anti-oxidants. Aside from their trace compounds, there are differences in oxidation resistance between particular oil types. For example, paraffinic oils usually have greater oxidation stability than naphthenic oils [3].

Widely used anti-oxidant additives are zinc dialkyldithiophosphate, metal deactivators, simple hydrocarbons such as phenol derivatives, amines and organic phosphates. Sulphur and phosphorus in elemental form or incorporated into organic compounds are also effective as anti-oxidants and anti-wear additives. They are sometimes added to oils (a very old practice) but are likely to cause corrosion problems or may precipitate and lose effectiveness as an additive. Anti-oxidants are usually added to the oil in very small quantities at a concentration of approximately 1% by weight.

Anti-oxidants can be classified into three basic categories:

· metal deactivators,

· radical inhibitors (or propagation inhibitors),

· peroxide decomposers.

Metal deactivators inhibit the acceleration of oil oxidation by entraining a metal such as iron and copper. These metals are the most commonly used materials for engineering machinery.

Metal deactivators function by chelation of the metal ions. Major sources of metal deactivators are derivatives of salicylic acid but they can also be derived from lecithin, phosphoric, acetic, citric and gluconic acids [3]. A specific example of this type of anti-oxidant is ethylenediaminetetraacetic acid. These compounds are added to the oil in very limited quantities of about 5 - 30 [mg/kg].

Radical inhibitors (or propagation inhibitors) function by neutralizing the peroxy radicals, and the anti-oxidation mechanism involved is shown below:

$$\mathbf{A\text{-}H + R\text{-}O\text{-}O^\circ \rightarrow A^\circ + R\text{-}O\text{-}O\text{-}H}$$ (production of hydroperoxides)

$$\mathbf{A^\circ + R\text{-}O\text{-}O^\circ \rightarrow A\text{-}O\text{-}O\text{-}R}$$ (termination of oxidation)

$$\mathbf{A^\circ + A^\circ \rightarrow 2A}$$ (deactivation of additives)

where:

R	is a radical;
R-O-O°	is a peroxide radical;
R-O-O-H	is a hydroperoxide;
A-H	is an additive;
A°	is an activated additive.

The first reaction shows the production of hydroperoxides which are generated by the reaction of the additives with peroxide radicals. The radical inhibitor usually consists of a hydrocarbon with a polarized or weakly bonded hydrogen atom. Unfortunately during this reaction organic acids are produced. In the next step the termination of oxidation takes place. The peroxy radicals are completely neutralized and form a relatively inert product. The third reaction illustrates the deactivation of the additive. The activated additives tend towards mutual neutralization, resulting in greater usage of the additives.

Examples of these additives are diarylamines, dihydroquinolenes and hindered phenols. These additives are also known in the literature as simple hydrocarbons. They are characterized by low volatility, which is an important feature since they can then only be used in very small quantities of 0.5 - 1% by weight, and furthermore their lifetime is very long.

Peroxide decomposers function by neutralizing the hydroperoxides which would otherwise accelerate the process of oxidation. An example of this additive is zinc dialkyldithiophosphate (ZnDDP) which functions by decomposing the peroxide radicals (R-O-O°) and hydroperoxides (R-O-O-H) formed during oil oxidation. This action prevents the acceleration of the oxidation process. ZnDDP is most frequently added to engine oils in small amounts of about ~1 - 2% by weight.

The decomposition of hydroperoxide causes gradual degradation and depletion of ZnDDP in the oil, reducing its efficiency as an additive. In one experiment the concentration of peroxides and ZnDDP in oxidizing hexadecane was measured by using the radio-isotope techniques [59]. It was found that as soon as a small quantity of ZnDDP was added to the lubricant it was converted to its immediate decomposition product di-isobutyldithiophosphoryl disulphide (DS), which, after an induction period, suppressed the further formation of peroxides. This is demonstrated in Figure 3.21 which shows the changes, with respect to time, in the concentration levels of hexadecane oxidation products (HOP), hydroperoxides (ROOH), ZnDDP and its immediate decomposition product such as di-isobutyldithiophosphoryl disulphide (DS).

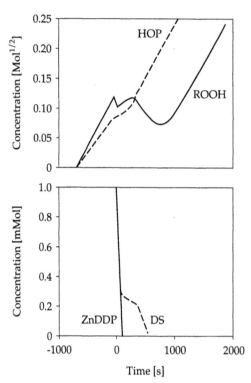

FIGURE 3.21   Inhibition of oil oxidation and simultaneous decomposition of anti-oxidant as measured by radio-isotopes (adapted from [59]).

The combination of organo-metallic and simple hydrocarbon (ash-less) anti-oxidants is often more effective than these two types used separately. Interestingly, ZnDDP does not even act as a true anti-oxidant, it simply decomposes the hydroperoxides which are the precursors of polymeric oxidation products and which accelerate wear by what appears to be a corrosive wear mechanism [59]. Instead of being prevented, the oxidation is converted to a milder, more acceptable form with a much reduced probability of rapidly accelerating oxidation by multiple reaction paths. ZnDDP is also found to delay both the induction time and the rate of deposit formation caused by oil oxidation [46]. This is different from ash-less anti-oxidants which prolong the induction time only slightly. The term 'anti-oxidants' thus refers to a diverse group of compounds with quite considerable differences in their functional characteristics.

Most known anti-oxidants will eventually decompose by oxidation during their working cycle. So in fact permanent protection of oil against oxidation at high temperatures is virtually impossible. Once the anti-oxidant is exhausted, the oil, which in most practical applications contains a high level of dissolved metal from wear, will rapidly oxidize. There are obviously considerable variations in oil oxidation rates, so that in some cases the period between depletion of anti-oxidant and breakdown of the oil will be quite long, whereas with some other oils there might be only a very short warning period before the lubricant fails entirely. Once the lubricant fails, the risk of machine failure is high.

Finally it has to be emphasized that the type of anti-oxidant used will depend on the application. For example, in systems where EP or anti-wear activity is essential and the oil life time is relatively short then ZnDDP or sulphur and phosphorus based EP additives give the best results. On the other hand, in applications which require a long lifetime from the oil at high temperatures, amine based inhibitors will be most suitable.

Corrosion Control Additives

In this category two groups of additives are distinguished in the literature: corrosion inhibitors and rust inhibitors. Corrosion inhibitors are used for non-ferrous metals (i.e., copper, aluminium, tin, cadmium, etc.) and are designed to protect their surfaces against any corrosive agents present in the oil. Rust inhibitors are needed for ferrous metals and their task is to protect ferrous surfaces against corrosion.

<u>Corrosion inhibitors</u> are used to protect the non-ferrous surfaces of bearings, seals, etc. against corrosive attack by various additives, especially those containing reactive elements such as sulphur, phosphorus, iodine, chlorine and oxidation products. Some of the oxidation products are very acidic and must be neutralized before they cause any damage to the operating parts of the machinery. The combination of corrosive additives, oxidation products, high temperature and very often water can make the corrosion attacks on non-ferrous metallic parts, which are used in almost every machine, very severe. The commonly used additives to control the corrosion of non-ferrous metals are benzotriazole, substituted azoles, zinc diethyldithiophosphate, zinc diethyldithiocarbamate and trialkyl phosphites. These act by forming protective films on the metallic surfaces.

<u>Rust inhibitors</u> are used to protect the ferrous components against corrosion. The main factors which contribute to accelerated corrosion attack of ferrous parts are oxygen dissolved in the oil and water. These can cause an electrolytic attack which may be even more accelerated with increased temperature. Rust inhibitors are usually long chain agents, which attach themselves to the surface, severely reducing the mobility of water, as shown in Figure 3.22. In some cases, two ends of the chain can be active, so the additive attaches itself to the surface with both ends, as demonstrated in Figure 3.22, and then less additive is needed to decrease the mobility of water. The commonly used additives which control the corrosion of ferrous metals are metal sulphonates (i.e., calcium, barium, etc.), amine succinates or other polar organic acids. The calcium and barium sulphonates are suitable for more severe corrosion conditions than the succinates and the other organic acids.

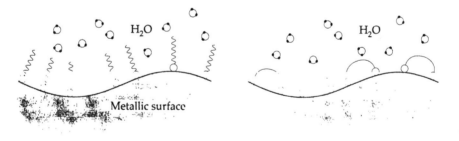

FIGURE 3.22 Operating mechanism of rust inhibitors.

Contamination Control Additives

With the introduction of the internal combustion engine a whole new class of additives has been developed. Engine oils are regularly exposed to fuel and combustion products which inevitably contribute to their contamination. Water also plays a major role since it accelerates the oxidation of the oil and may form an oil-water emulsion. When sulphur is present in the fuel, sulphurous or sulphuric acid is formed during combustion. If either of these compounds is dissolved by water then corrosion or corrosive wear of the engine will be accelerated. There can also be many other possible contaminants such as soot from inefficient fuel combustion, wear debris, unburned fuel, breakdown products of the base oil, corrosion products, dust from the atmosphere, organic debris from microbiological decomposition of the oil, etc. Without proper control of contamination, the oil will lose its lubricating capacity, become corrosive and will be unsuitable for service. Various additives have been developed to control the acidity of the products of sulphurous combustion of dirty fuel and to prevent agglomeration of soot from combustion and wear particles. The agglomeration of particles can be very destructive to engines since it blocks the oil supply pipelines or even the filters. Additives which prevent the development of all these detrimental effects are known in the literature as either 'detergents' or 'dispersants'. The latter term, however, is more accurate. The primary functions of these additives are:

· to neutralize any acids formed during the burning of fuel,

· to prevent lacquer and varnish formation on the operating parts of the engine,

· to prevent the flocculation or agglomeration of particles and carbon deposits which may choke the oil ways.

There are two types of dispersant: a mild dispersant and an over-based or alkaline dispersant.

Mild dispersants are often composed of simple hydrocarbons or ashless compounds (i.e., when the compound is burnt no oxides are left, since organic compounds burn to CO_2 and water). Mild dispersants are typically low molecular weight polymers of methacrylate esters, long chain alcohols, or polar vinyl compounds. The function of these additives is to disperse soot (carbon) and wear particles.

Over-based dispersants are calcium, barium or zinc salts of sulphonic, phenol or salicylic acids. Over-based means that an excess of alkali is used in the preparation of these additives. The additive is present in the mineral oil as a colloid. The alkaline prepared additive serves to neutralize any acid accumulated in the oil during service. Alkaline dispersants have a disadvantage in that they accelerate oil oxidation and therefore require the addition of an anti-oxidant to the oil. Commonly used dispersants are summarized in Table 3.6.

TABLE 3.6   Some commonly used dispersants.

Dispersant	Laquer and varnish prevention	Acid neutralisation	Coagulation prevention
Dispersants and over-based dispersants			
Calcium, barium or zinc salts of sulphonic, phenol or salicylic acid	Good, especially at high temperatures	Good	Fair
Carboxylic and salicylic type additives	Good	Fair	Poor (they can cause coagulation)
Mild dispersants (i.e., ashless compounds)			
Low weight polymers of methacrylate esters or long chain alcohols	Fair*	Weak	Good
Polar vinyl compounds	Fair*	Weak	Good
Amines (e.g., triethylene amine)	Fair*	Fair	Good
Saturated succinimide	Fair*	Weak	Good

* These additives are only effective at low temperatures

Dispersants usually work by stabilizing any colloid particles suspended in the oil as schematically shown in Figure 3.23.

There is a very quick and cheap way of checking whether the dispersants are still active in a particular oil. A drop of the oil is placed on a piece of blotting paper. It is preferable to use 'live oil' from the operating machinery. A 'greasy spot' will result. If the 'greasy spot' is evenly dispersed then the dispersants work correctly, but if there is a small black dot in the middle surrounded by the 'greasy spot' then the dispersants have been used up and the oil needs replacing.

Viscosity Improvers

These are additives which arrest the decline in oil viscosity with temperature and they are commonly known as viscosity index improvers. Viscosity improving additives are usually high molecular weight polymers which are dissolved in the oil and can change shape from spheroidal to linear as the temperature is increased. This effect is caused by a greater solubility of the polymer in the oil at higher temperatures and partly offsets the decline in base oil viscosity with temperature. The linear or uncoiled molecules cause a larger rise in viscosity in comparison to spheroidal or coiled molecules. Typical viscosity improvers are polymethacrylates in the molecular weight range between 10,000 and 100,000. It seems that linear polymer molecules with only a small number of side chains are the most effective. These additives are used in small concentrations of a few percent by weight in the base oil. They have been used for many years as an active ingredient of multigrade oils.

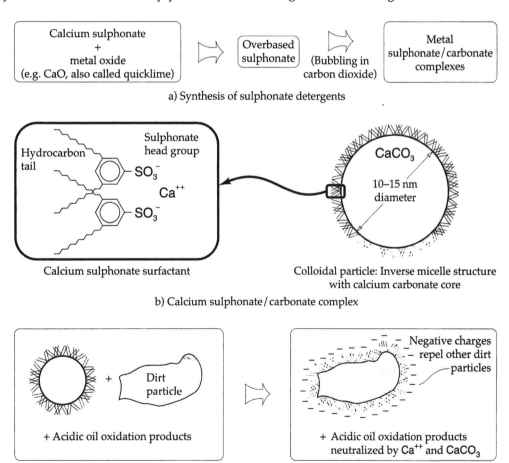

FIGURE 3.23 Synthesis and function of sulphonate detergents.

The main problem associated with these additives is that they are easily degraded by excessive shear rates and oxidation. Under high shear rates viscosity improvers can suffer permanent or temporary viscosity loss. Temporary viscosity loss results from the alignment of the polymer molecules under high shear rates and is reversible. On the other hand, permanent viscosity loss involves the breakdown of large polymer molecules under high shear rates and is irreversible. Viscosity improvers can usually raise the viscosity index of an oil from 110 to 150, but only at moderate shear rates and for a limited period of time. Oil oxidation can also contribute to the degradation or breakdown of the polymer molecules. Pressure-viscosity coefficients are not significantly affected by polymer viscosity improvers, although some minor effects have been reported in the literature. In general, they are relatively inert and do not interfere with the other additives, in particular ZnDDP [50]. Some of them may, however, affect the wear rates. It was found, for example, that polymethacrylate viscosity improvers with a molecular weight over 100,000 increased wear rates in comparison to plain mineral oil [54]. The acceleration in wear became even worse with increased contact load. To explain this, it has been suggested that the larger molecules of a viscosity improver can impede oil flow in the EHL contacts. Cam lobes of engines lubricated by polymer thickened oils were found to wear more rapidly than when plain oils were used [55]. At low contact loads, however, it was found that viscosity improvers actually reduced wear rates [54]. Factors which control the transition from the pro-wear to the anti-wear effect are, as yet, unknown and need to be investigated.

Pour Point Depressants

Pour point depressants are basically the same compounds as viscosity improvers. They prevent the generation of wax crystals at low temperatures by dislocating the wax structure. They are essential for low temperature operation in applications where the base stock is paraffinic. In cases when it is known that the oil operating temperature will never fall below 0°C they can be completely omitted from oil formulation.

Foam Inhibitors

The main task of foam inhibitors is to destabilize foam generated during the operation of the machinery. Usually long chain silicone polymers are used in very small quantities of about 0.05% to 0.5% by weight. The amount of additive used is quite critical, i.e., excessive amount of this additive is less effective.

Interference Between Additives

Interference between additives is a serious problem and the subject of continuing investigation. As mentioned earlier some of the anti-wear and EP additives can react with other additives and lose their effectiveness. For example, some of the fatty acids such as oleic acid or detergents and rust inhibitors can significantly suppress the lubricating action of ZnDDP. One of the problems which has hardly been investigated is the problem of chemical reaction between the lubricant and the fuel, or some other chemical. In some cases, zinc in ZnDDP can be replaced by some other element. A relatively harmless example is the substitution of zinc by lead in engine oils when leaded petrol is used. In this case, ZnDDP is converted to PbDDP which is a less efficient additive and causes increased wear of the engine [37]. The presence of zinc in ZnDDP can be used as a condition monitoring index, indicating when the level of ZnDDP has been depleted. When zinc is replaced by lead, the additive will still be present in the oil, but as a different compound, PbDDP, which is less efficient. Another example is the effect of ammonia on lubricants. In chemical plants, compression of ammonia is commonplace. In cases where an oil containing an acidic succinate corrosion inhibitor is contaminated by ammonia a serious sludging may occur which can lead to extensive plant damage [53]. Where there are several additives present in the oil along with an active contaminant, the number of possible interactions becomes very large and it is difficult to isolate the cause of any lubricant breakdown and consequent damage to the machinery.

Although dispersants are essential additives in oils, they unfortunately have a number of negative side effects which can be classified as 'additive interference'. It is well known that the dispersants accelerate the oxidation of oils and that an anti-oxidant must be included when these additives are used. Both the mild and over-based dispersants also have a strong effect on lubrication, in particular boundary and EP lubrication. ZnDDP and sulphur based additives may even be prevented from proper functioning because of dispersant interference. For example, the effect of dispersants on the coefficient of friction at various temperatures is shown in Figure 3.24. A hexadecane with a typical sulphur based additive, dibenzyl disulphide (DBDS), was initially used as the lubricating fluid and then common dispersants such as calcium sulphonate and n-octadecylamine (amine) were added [48].

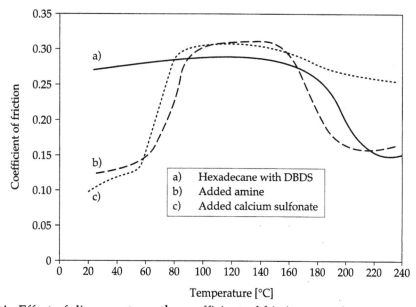

FIGURE 3.24   Effect of dispersants on the coefficient of friction at various temperatures [48].

It can be seen from Figure 3.24 that DBDS forms a lubricating film which results in a reduction of the coefficient of friction at about 160°C. Both dispersants reduce the coefficient of friction up to 80°C, but calcium sulphonate prevents the effective lubrication by DBDS beyond 160°C. The action of DBDS is thus inhibited. It was found that the increase in friction was paralleled by an ability of calcium sulphonate to prevent the formation of sulphide films on a steel substrate by DBDS [48].

A similar effect takes place between ZnDDP and dispersants. In this instance, however, the amine additives are the hindrance [49,50,51,52]. It is suggested that the amine forms a complex with ZnDDP preventing it from decomposing. Decomposition of ZnDDP is essential for the formation of lubricating films. In contrast, calcium sulphonate exerts only a very weak inhibition of the ZnDDP lubricant functions. It was found that almost all additives interfere with ZnDDP to some extent [50].

3.7 SUMMARY

The fundamental makeup of mineral and synthetic oils, emulsions and greases has been discussed in this chapter. A large variety of lubricating fluids are employed to lubricate machinery and new lubricants are continually being introduced to the market. The principles of lubricant selection, however, do not change and can be summarized as resistance to oxidation, wear and corrosion, maintenance of viscosity at high temperatures and provision for thin-film lubrication. Relatively few lubricants satisfy all of these criteria. It is thus a common practice to blend additives with a fluid or semi-solid lubricant to improve its properties. It must be emphasized, however, that additives are not the panacea for all lubrication problems. They can bring as many problems as they solve, for example, incompatibility with the base lubricant and with other additives has caused costly industrial failures.

Although specialized synthetic lubricants have been successfully replacing mineral oil in various applications for many years, general purpose synthetic lubricants have only recently been introduced on a large scale. They are generally more expensive but have better oxidation and thermal resistance than mineral oils. Performance data on synthetic lubricants is sparse, but will hopefully improve in the future. The present generation of synthetic lubricants is still imperfect and more development, particularly in the area of oxidation resistance, can be expected. There are also great advances in the production of a new generation of general purpose greases that can operate in more extreme environments, e.g., wide temperature range, vacuum, etc.

REVISION QUESTIONS

3.1   How can changes in refining techniques of lubricating oils affect the lubricating properties of the oil?

3.2   What is the 'base stock'?

3.3   Which of the three main chemical types of mineral oil - paraffinic, napthenic and aromatic - is preferred as a lubricant? State some reasons why.

3.4   For oils not used to lubricate food and drug processing machinery, what quality largely determines whether an oil is 'good' or 'bad'?

3.5   What is one of the main purposes of developing synthetic oils?

3.6   What problem occurs when oxygen compressors are lubricated by mineral oil?

3.7   Name a historical substitute for mineral oil in oxygen compressors.

3.8   Are pure, additive-free oils (base stocks) widely used in industry?

3.9   Why do engine oils have basic alkaline compounds added to them?

3.10   When should halogenated oils be used?

3.11   What percentage mass of a grease is soap?

3.12   Name a simple and effective rheological model of grease.

3.13   Under what conditions does the flow property of the grease revert to that of its constitutive mineral oil?

3.14   The soap in grease is present as a tangle of tiny fibres. When these fibres are too coarse what happens and why does it cause failure of the grease?

3.15   What was the purpose of developing a grease substitute for pantograph lubrication?

REFERENCES

1   G.P.K. Linkhammer and C.E. Lambert, Preservation of Organic Matter During Salinity Excursions, *Nature*, Vol. 339, 1989, pp. 271-274.

2   T. Gold, Terrestrial Sources of Carbon and Earthquake Outgassing, *Journal of Petroleum Geology*, Vol. 1, 1979, pp. 3-19.

3   D. Klamann, Lubricants and Related Products, Verlag Chemie, Weinheim, 1984, pp. 51-83.

4   A. Dorinson and K.C. Ludema, Mechanics and Chemistry in Lubrication, Elsevier, Amsterdam, 1985, pp. 472-500.

5   Y. Kimura and H. Okabe, An Introduction to Tribology, Youkandou Press, Tokyo 1982, pp. 69-82 (in Japanese).

6   A. Jackson, Synthetic Versus Mineral Fluids in Lubrication, Proc. Int. Tribology Conference, Melbourne, The Institution of Engineers, Australia, National Conference Publication No. 87/18, December, 1987, pp. 428-437.

7   D. Tabor and W.O. Winer, Silicone Fluids: Their Action as Boundary Lubricants, *ASLE Transactions*, Vol. 8, 1965, pp. 69-77.

8   G.V. Vinogradov, N.S. Nametkin and M.I. Nossov, Antiwear and Antifriction Properties of Polyorgano-Siloxanes and Their Mixtures with Hydrocarbons, *Wear*, Vol. 8, 1963, pp. 93-111.

9   B. Bhushan, Tribology and Mechanics of Magnetic Storage Devices, Springer-Verlag, 1990.

10   T.I. Fowle, Lubricants for Fluid Film and Hertzian Contact Conditions, *Proc. Inst. Mech. Engrs.*, Vol. 182, Pt. 3A, 1967-68, pp. 508-584.

11   D.J. Carre, Perfluoropolyalkylether Oil Degradation: Inference of FeF_3 Formation on Steel Surfaces Under Boundary Conditions, *ASLE Transactions*, Vol. 29, 1986, pp. 121-125.

12   S. Mori and W. Morales, Tribological Reactions of Perfluoroalkyl Polyether Oils with Stainless Steel Under Ultra-High Vacuum Conditions at Room Temperature, *Wear*, Vol. 132, 1989, pp. 111-122.

13   J.R. Lohuis and A.J. Harlow, Synthetic Lubricants for Passenger Car Diesel Engines, *Society of Automotive Engineers, Technical Paper*, 1985, No. 850564.

14   S. Ohkawa, Gear Load-Carrying Capacity of Synthetic Engine Oil, Proc. JSLE Int. Trib. Conf., 8-10 July 1985, Tokyo, Japan, pp. 497-502.

15   T. Sakurai and K. Yoshida, Tribological Behaviour of Dispersed Phase Systems, Proc. Int. Tribology Conference, Melbourne, The Institution of Engineers, Australia, National Conference Publication No. 87/18, December, 1987, pp. 110-115.

16   K.E. Bett and J.B. Cappi, Effect of Pressure on the Viscosity of Water, *Nature*, Vol. 207, 1965, pp. 620-621.

17   J. Wonham, Effect of Pressure on the Viscosity of Water, *Nature*, Vol. 215, 1967, pp. 1053-1054.

18   G. Dalmaz and M. Godet, Film Thickness and Effective Viscosity of Some Fire Resistant Fluids in Sliding Point Contacts, *Transactions ASME, Journal of Lubrication Technology*, Vol. 100, 1978, pp. 304-308.

19   H.B. Silver and I.R. Stanley, Effect of the Thickener on the Efficiency of Load Carrying Additives in Greases, *Tribology International*, Vol. 7, 1974, pp. 113-118.

20   R. O'Halloran, J.J. Kolfenbach and H.L. Leland, Grease Flow in Shielded Bearings, *Lubrication Engineering*, Vol. 14, 1958, pp. 104-107 and 117.

21   A.R. Lansdown, Lubrication, A Practical Guide to Lubricant Selection, Pergamon Press, 1982.

22   K. Fukunaga, Allowable Surface Durability in Grease Lubricated Gears, *Tribology Transactions*, Vol. 31, 1988, pp. 454-460.

23   S.Y. Poon, An Experimental Study of Grease in Elastohydrodynamic Lubrication, *Transactions ASME, Journal of Lubrication Technology*, Vol. 94, 1972, pp. 27-34.

24   W. Jonkisz and H. Krzeminski-Freda, The Properties of Elastohydrodynamic Grease Films, *Wear*, Vol. 77, 1982, pp. 277-285.

25   A.C. Horth, L.W. Sproule and W.C. Pattenden, Friction Reduction With Greases, *NLGI Spokesman*, Vol. 32, 1968, pp. 155-161.

26   D. Godfrey, Friction of Greases and Grease Components During Boundary Lubrication, *ASLE Transactions*, Vol. 7, 1964, pp. 24-31.

27   M.H. Jones and D. Scott, Industrial Tribology, The Practical Aspects of Friction, Lubrication and Wear, Elsevier, 1983.

28   SAE Standard, Automotive Lubricating Greases, SAE J310, August 1987.

29   A. Cameron, Principles of Lubrication, (J.F. Hutton, Lubricating Greases), Longmans, London 1966, pp. 521-541.

30   H.A. Woods and H.M. Trowbridge, Shell Roll Test for Evaluating Mechanical Stability, *NLGI Spokesman*, Vol. 19, 1955, pp. 26-27 and 30-31.

31   N.A. Scarlett, Use of Grease in Rolling Bearings, *Proc. Inst. Mech. Engrs.*, Vol. 182, Pt. 3A, 1967-1968, pp. 585-593.

32   A.G. Papay, Oil-Soluble Friction Reducers, Theory and Application, *Lubrication Engineering*, Vol. 39, 1983, pp. 419-426.

33 P. Cann, H.A. Spikes and A. Cameron, Thick Film Formation by Zinc Dialkyldithiophosphates, *ASLE Transactions*, Vol. 26, 1983, pp. 48-52.

34 S. Jahanmir, Wear Reduction and Surface Layer Formation by a ZDDP Additive, *Transactions ASME, Journal of Tribology*, Vol. 109, 1987, pp. 577-586.

35 H. Uetz, A. Khosrawi and J. Fohl, Mechanism of Reaction Layer Formation in Boundary Lubrication, *Wear*, Vol. 100, 1984, pp. 301-313.

36 F. Rounds, Contribution of Phosphorus to the Antiwear Performance of Zinc Dialkyldithiophosphates, *ASLE Transactions*, Vol. 28, 1985, pp. 475-485.

37 M. Kawamura, K. Fujita and K. Ninomiya, The Lubricating Properties of Used Engine Oil, *Wear*, Vol. 77, 1982, pp. 195-202.

38 A. Dorinson, Influence of Chemical Structures in Sulfurized Fats on Anti-Wear Behaviour, *ASLE Transactions*, Vol. 14, 1971, pp. 124-134.

39 S.M. Hsu, C.S. Ku and P.T. Pei, Oxidative Degradation Mechanisms of Lubricants, Aspects of Lubricant Oxidation, Editors, W.H. Stadtmiller and A.N. Smith, ASTM Tech. Publ. 916, ASTM Committee D-2 Symposium on Petroleum Products and Lubricants, Miami, Florida, 5 December 1983, pp. 27-48.

40 H.H. Zuidema, The Performance of Lubricating Oils, Reinhold Publ., New York, 1952, pp. 26-28.

41 D.B. Clark, E.E. Klaus and S.M. Hsu, The Role of Iron and Copper in the Oxidation Degradation of Lubricating Oils, *Lubrication Engineering*, Vol. 41, 1985, pp. 280-287.

42 E.E. Klaus, D.I. Ugwuzor, S.K. Naidu and J.L. Duda, Lubricant-Metal Interaction Under Conditions Simulating Automotive Bearing Lubrication, Proc. JSLE Int. Tribology Conf., 8-10 July, Japan, Elsevier, 1985, pp. 859-864.

43 T. Colclough, Role of Additives and Transition Metals in Lubricating Oil Oxidation, *Ind. Eng. Chem. Res.*, Vol. 26, 1987, pp. 1888-1895.

44 V.W. David and R. Irving, Effects of Nuclear Radiation on Hydrocarbon Oils, Greases and Some Synthetic Fluids, Proc. Conf. on Lubrication and Wear, Inst. Mech. Engrs., 1957, pp. 543-552.

45 S. Ohkawa, K. Seto, T. Nakashima and K. Takase, "Hot Tube Test" - Analysis of Lubricant Effect on Diesel Engine Scuffing, *Society of Automotive Engineers, Technical Paper*, 1984, No. 840262.

46 S. Tseregounis, J.A. Spearot and D.J. Kite, Formation of Deposits from Thin Films of Mineral Oil Base Stocks on Cast Iron, *Ind. Eng. Chem. Res.*, Vol. 26, 1987, pp. 886-894.

47 P.A. Willermet, S.K. Kandah and R.K. Jensen, The Influence of Auto-Oxidation on Wear Asymmetry With N-hexadecane, *ASLE Transactions*, Vol. 28, 1985, pp. 511-519.

48 H.A. Spikes and A. Cameron, Additive Interference in Dibenzyl Disulphide Extreme Pressure Lubrication, *ASLE Transactions*, Vol. 17, 1974, pp. 283-289.

49 F.G. Rounds, Some Effects of Amines on Zincdialkyldithiophosphate Antiwear Performance as Measured in 4-Ball Wear Tests, *ASLE Transactions*, Vol. 24, 1981, pp. 431-440.

50 F.G. Rounds, Additive Interactions and Their Effect on the Performance of a Zincdialkyldithiophosphate, *ASLE Transactions*, Vol. 21, 1978, pp. 91-101.

51 S. Shirahama, The Effects of Temperature and Additive Interaction on Valve Train Wear, Proc. JSLE. Int. Trib. Conf. 8-10 Tokyo, Japan, Elsevier, July 1985, pp. 331-336.

52 M.S. Hiomi, M. Tokashiki, H. Tomizawa, T. Nomura and T. Yamaji, Interaction Between Zincdialkyldithiophosphate and Amine, Proc. JSLE. Int. Trib. Conf., 8-10 Tokyo, Japan, Elsevier, July 1985, pp. 673-678.

53 D. Summers-Smith, The Unacceptable Face of Lubricating Additives, *Tribology International*, Vol. 11, 1978, pp. 318-320.

54 K. Yoshida, K. Hosonuma and T. Sakurai, Behaviour of Polymer-Thickened Oils in Lubricated Contacts, *Wear*, Vol. 98, 1984, pp. 63-78.

55 K. Yoshida, Effect of Sliding Speed and Temperature on Tribological Behaviour With Oils Containing a Polymer Additive or Soot, *Tribology Transactions*, Vol. 23, 1990, pp. 221-228.

56 K.G. Allum and E.S. Forbes, The Load Carrying Properties of Metal Dialkyldithiophosphates: The Effect of Chemical Structure, *Proc. Inst. Mech. Engrs.*, Vol. 183, Pt. 3P, 1968-1969, pp. 7-14.

57 J.J. Habeeb and W.H. Stover, The Role of Hydroperoxides in Engine Wear and the Effect of Zincdialkyldithiophosphates, *ASLE Transactions*, Vol. 30, 1987, pp. 419-426.

58 W.D. Phillips, The Electrochemical Erosion of Servo Valves by Phosphate Ester Fire-Resistant Hydraulic Fluids, *Lubrication Engineering*, Vol. 44, 1988, pp. 758-767.

59 M.D. Johnson, S. Korcek and M. Zinbo, Inhibition of Oxidation by ZDDP and Ashless Antioxidants in the Presence of Hydroperoxides at 160°C - Part II, *ASLE Transactions*, Vol. 29, 1986, pp. 136-140.

60 M.T. Benchaita, S. Gunsel and F.E. Lockwood, Wear Behaviour of Base Oil Fractions and Their Mixtures, *Tribology Transactions*, Vol. 33, 1990, pp. 371-383.

61 C.E. Snyder, L.J. Gschwender, C. Tamborski, G.J. Chen and D.R. Anderson, Synthesis and Characterization of Silahydrocarbons - A Class of Thermally Stable Wide-Liquid-Range Functional Fluid, *ASLE Transactions*, Vol. 25, 1982, pp. 299-308.

62 O.D. Faut and D.R. Wheeler, On the Mechanism of Lubrication by Tricresylphosphate - The Coefficient of Friction as a Function of Temperature for TCP on M-50 Steel, *ASLE Transactions*, Vol. 26, 1983, pp. 334-350.

63 R.H. Schade, Grease After Lithium, Proc. Int. Tribology Conference, Brisbane, The Institution of Engineers, Australia, National Conference Publication No. 90/14, December 1990, pp. 145-150.

64 M.H. Arveson, Flow of Petroleum Lubricating Greases, *Ind. Eng. Chem.*, Vol. 26, 1934, pp. 628-634.

65 Q. Zhao, H.J. Kang, L. Fu, F.E. Talke, D.J. Perettie and T.A. Morgan, Tribological Study of Phosphazene-Type Additives in Perfluoropolyether Lubricant for Hard Disk Applications, *Lubrication Engineering*, Vol. 55, 1999, pp. 16 - 21.

66 H. J. Kang, Q. Zhao, F.E. Talke, D.J. Perettie, B.M. Dekoven, T.A. Morgan, D.A. Fischer and S.M. Hsu, The Use of Cyclic Phosphazene Additives to Enhance the Performance of the Head Disk Interface, *Lubrication Engineering*, Vol. 55, 1999, pp. 22-27.

67 P.M.E. Cann, Thin-Film Grease Lubrication, *Proc. Inst. Mech. Eng., Part J, Journal of Engineering Tribology*, Vol. 213, 1999, pp. 405-416.

68 R.J. Greve, S.C. Langford and J.T. Dickinson, Oxidation and Reduction Reactions Responsible for Galvanic Corrosion of Ferrous and Reactive Metals in the Presence of a Perfluoropolyether Lubricant: Fomblin Z-DOL, *Wear*, Vol. 249, 2001, pp. 727-732.

69 W. Liu, Ch. Ye, Z. Zhang and L. Yu, Relationship Between Molecular Structures and Tribological Properties of Phosphazene Lubricants, *Wear*, Vol. 252, 2002, pp. 394-400.

70 Y. Gao, G. Chen, Y. Oli, Z. Zhang and Q. Xue, Study on Tribological Properties of Oleic Acid-Modified TiO_2 Nanoparticle in Water, *Wear*, Vol. 252, 2002, pp. 454-458.

71 K.R. Januszkiewicz, A.R. Riahi and S. Barakat, High Temperature Tribological Behaviour of Lubricating Emulsions, *Wear*, Vol. 256, 2004, pp. 1050-1061.

72 K. Varlot, M. Kasrai, G.M. Bancroft, E.S. Yamaguchi, P.R. Ryason and J. Igarashi, X-ray Adsorption Study of Antiwear Films Generated from ZDDP and Borate Micelles, *Wear*, Vol. 249, 2001, pp. 1029-1035.

73 P. Ye, X. Jiang, S. Li and S. Li, Preparation of $NiMoO_2S_2$ Nanoparticle and Investigation of Its Tribological Behavior as Additive in Lubricating Oils, *Wear*, Vol. 253, 2002, pp. 572-575.

74 J. Zhou, Z. Wu, Z. Zhang, W. Lu and H. Dang, Study on an Anti Wear and Extreme Pressure Additive of Surface Coated LaF_3 Nanoparticles in Liquid Paraffin, *Wear*, Vol. 249, 2001, pp. 333-337.

75 S. Tarasov, A. Kolubaev, S. Belyaev, M. Lerner and F. Tepper, Study of Friction Reduction by Nanocopper Additives to Motor Oil, *Wear*, Vol. 252, 2002, pp. 63-69.

76 A. Rossi, M. Eglin, F.M. Piras, K. Matsumoto and N.D. Spencer, Surface Analytical Studies of Surface-Additive Interactions, by Means of in Situ and Combinatorial Approaches, *Wear*, Vol. 256, 2004, pp. 578-584.

HYDRODYNAMIC LUBRICATION

4.1 INTRODUCTION

In the previous chapters the basic physical properties of lubricants, their composition and applications have been discussed. The fundamental question to be answered is: what causes a lubricant to lubricate? If some specific concepts of lubrication are formulated then the following questions become pertinent. What conditions must be fulfilled to fully separate two loaded surfaces in relative motion? How can we manipulate these conditions in order to minimize friction and wear? Is there only one or are there several mechanisms of lubrication? In what specific applications do they operate? What factors determine the classification of a load-carrying mechanism to a specific category? How do thrust and journal bearings operate? How can the design parameters for such bearings be estimated? How does cavitation or oil whirl affect bearing performance? Engineers are usually expected to know the answers to all of these questions.

In this chapter the basic principles of hydrodynamic lubrication will be discussed. The mechanisms of hydrodynamic film generation and the effects of operating variables such as velocity, temperature, load, design parameters, etc., on the performance of such films are outlined. This will be explained using bearings commonly found in many engineering applications as examples. Secondary effects in hydrodynamic lubrication such as viscous heating, compressible and non-Newtonian lubricants, bearing vibration and deformation will be described and their influence on bearing performance assessed.

4.2 REYNOLDS EQUATION

The serious appreciation of hydrodynamics in lubrication started at the end of the 19th century when Beauchamp Tower, an engineer, noticed that the oil in a journal bearing always leaked out of a hole located beneath the load. The leakage of oil was a nuisance so the hole was plugged first with a cork, which still allowed oil to ooze out, and then with a hard wooden bung. The hole was originally placed to allow oil to be supplied into the bearing to provide 'lubrication'. When the wooden bung was slowly forced out of the hole by the oil, Tower realized that the oil was pressurized by some as yet unknown mechanism. Tower then measured the oil pressure and found that it could separate the sliding surfaces by a hydraulic force [1]. At the time of Beauchamp Tower's discovery Osborne Reynolds and other theoreticians were working on a hydrodynamic theory of lubrication. By a most fortunate coincidence, Tower's detailed data was available to provide experimental confirmation of hydrodynamic lubrication almost at the exact time when Reynolds needed it. The result of this was a theory of hydrodynamic lubrication published in the Proceedings of the Royal Society by Reynolds in 1886 [2]. Reynolds provided the first analytical proof that a viscous liquid can physically separate two sliding surfaces by hydrodynamic pressure, resulting in low friction and theoretically zero wear.

At the beginning of the 20th century the theory of hydrodynamic lubrication was successfully applied to thrust bearings by Michell and Kingsbury and the pivoted pad bearing was developed as a result. The bearing was a major breakthrough in supporting the thrust of a ship propeller shaft and the load from a hydroelectric rotor. At the present level of technology, loads of several thousand tons are carried, at sliding speeds of 10 to 50 [m/s], in hydroelectric power stations. The operating surfaces of such bearings are fully separated by a lubricating film, so the friction coefficient is maintained at a very low level of about 0.005 and the failure of such bearings rarely occurs, usually only after faulty operation. Reynolds' theory explains the mechanism of lubrication through the generation of a viscous liquid film between the moving surfaces. The condition is that the surfaces must move, relatively to each other, with sufficient velocity to generate such a film. It was found by Reynolds and many later researchers that most of the lubricating effect of oil could be explained in terms of its relatively high viscosity. There are, however, some lubricating functions of an oil as opposed to other liquids which cannot be explained in terms of viscosity and these are described in more detail in Chapter 8 on 'Boundary and Extreme Pressure Lubrication'.

All hydrodynamic lubrication can be expressed mathematically in the form of an equation which was originally derived by Reynolds and is commonly known throughout the literature as the 'Reynolds equation'. There are several ways of deriving this equation. Since it is a simplification of the Navier-Stokes momentum and continuity equation it can be derived from this basis. It is, however, more often derived by considering the equilibrium of an element of liquid subjected to viscous shear and applying the continuity of flow principle.

There are two conditions for the occurrence of hydrodynamic lubrication:

- · two surfaces must move relatively to each other with sufficient velocity for a load-carrying lubricating film to be generated and,
- · surfaces must be inclined at some angle to each other, i.e., if the surfaces are parallel a pressure field will not form in the lubricating film to support the required load.

There are two exceptions to this last rule: hydrodynamic pressure can be generated between parallel stepped surfaces or the surfaces can move towards each other (these are special cases and are discussed later). The principle of hydrodynamic pressure generation between moving non-parallel surfaces is schematically illustrated in Figure 4.1.

It can be assumed that the bottom surface, sometimes called the 'runner', is covered with lubricant and moves with a certain velocity. The top surface is inclined at a certain angle to the bottom surface. As the bottom surface moves it drags the lubricant along it into the converging wedge. A pressure field is generated as otherwise there would be more lubricant entering the wedge than leaving it. Thus at the beginning of the wedge the increasing pressure restricts the entry flow and at the exit there is a decrease in pressure boosting the exit flow. The pressure gradient therefore causes the fluid velocity profile to bend inwards at the entrance to the wedge and bend outwards at the exit, as shown in Figure 4.1. The generated pressure separates the two surfaces and is also able to support a certain load. It is also possible for the wedge to be curved or wrapped around a shaft to form a journal bearing. If the wedge remains planar then a pad bearing is obtained. The entire process of hydrodynamic pressure generation can be described mathematically to enable accurate prediction of bearing characteristics.

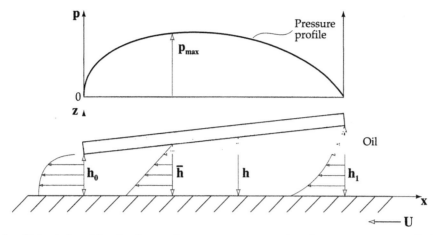

FIGURE 4.1   Principle of hydrodynamic pressure generation between non-parallel surfaces.

Simplifying Assumptions

In most engineering applications the controlling processes are too complicated to be easily described by exact mathematical equations. There are many interacting factors and variables in the real processes which make such a description extremely difficult, if not impossible. For example, with fluid mechanics in the early days of modelling, the terms of internal fluid friction were ignored. The mathematician John Newman observed sarcastically that these approximations have nothing to do with real fluids. It was like trying to study the flow of 'dry-water'. The situation dramatically changed with the introduction of computers so that mechanical systems could be studied in a more detailed fashion.

Similarly in hydrodynamics, several simplifying approximations have to be made before a mathematical description of the fundamental underlying mechanisms can be derived. All the simplifying assumptions necessary for the derivation of the Reynolds equation are summarized in Table 4.1 [3].

The Reynolds equation can now be conveniently derived by considering the equilibrium of an element (from which the expressions for fluid velocities can be obtained) and continuity of flow in a column.

Equilibrium of an Element

The equilibrium of an element of fluid is considered. This approach is frequently used in engineering to derive formulae in stress analysis, fluid mechanics, etc. Consider a small element of fluid from a hydrodynamic film shown in Figure 4.2. For simplicity, assume that the forces on the element are acting initially in the 'x' direction only.

Since the element is in equilibrium, forces acting to the left must balance the forces acting to the right, so

$$\mathbf{p}\,\mathbf{dy}\,\mathbf{dz} + (\tau_x + \frac{\partial \tau_x}{\partial \mathbf{z}}\mathbf{dz})\,\mathbf{dx}\,\mathbf{dy} = (\mathbf{p} + \frac{\partial \mathbf{p}}{\partial \mathbf{x}}\,\mathbf{dx})\,\mathbf{dy}\,\mathbf{dz} + \tau_x\,\mathbf{dx}\,\mathbf{dy}$$

(4.1)

which after simplifying gives:

$$\frac{\partial \tau_x}{\partial \mathbf{z}}\,\mathbf{dx}\,\mathbf{dy}\,\mathbf{dz} = \frac{\partial \mathbf{p}}{\partial \mathbf{x}}\,\mathbf{dx}\,\mathbf{dy}\,\mathbf{dz}$$

(4.2)

TABLE 4.1   Summary of simplifying assumptions in hydrodynamics.

	Assumption	Comments
1	Body forces are neglected	Always valid, since there are no extra outside fields of forces acting on the fluids with an exception of magnetohydrodynamic fluids and their applications.
2	Pressure is constant through the film	Always valid, since the thickness of hydrodynamic films is in the range of several micrometers. There might be some exceptions, however, with elastic films.
3	No slip at the boundaries	Always valid, since the velocity of the oil layer adjacent to the boundary is the same as that of the boundary.
4	Lubricant behaves as a Newtonian fluid	Usually valid with certain exceptions, e.g., polymeric oils.
5	Flow is laminar	Usually valid, except large bearings, e.g., turbines.
6	Fluid inertia is neglected	Valid for low bearing speeds or high loads. Inertia effects are included in more exact analyses.
7	Fluid density is constant	Usually valid for fluids when there is not much thermal expansion. Definitely not valid for gases.
8	Viscosity is constant throughout the generated fluid film	Crude assumption but necessary to simplify the calculations, although it is not true. Viscosity is not constant throughout the generated film.

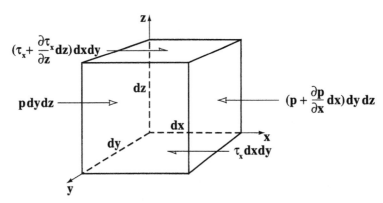

FIGURE 4.2   Equilibrium of an element of fluid from a hydrodynamic film; **p** is the pressure, τ_x is the shear stress acting in the 'x' direction.

Assuming that **dxdydz** $\neq$ **0** (i.e., non zero volume), both sides of equation (4.2) can be divided by this value and then the equilibrium condition for forces acting in the 'x' direction is obtained,

$$\frac{\partial \tau_x}{\partial z} = \frac{\partial p}{\partial x}$$

(4.3)

A similar exercise can be performed for the forces acting in the '**y**' (out of the page) direction, yielding the second equilibrium condition,

$$\frac{\partial \tau_y}{\partial z} = \frac{\partial p}{\partial y}$$

(4.4)

In the '**z**' direction since the pressure is constant through the film (Assumption 2) the pressure gradient is equal to zero:

$$\frac{\partial p}{\partial z} = 0$$

(4.5)

It should be noted that the shear stress in expression (4.3) is acting in the '**x**' direction while in expression (4.4) it is acting in the '**y**' direction, thus the values of the shear stress in these expressions are different.

Remembering the formula for dynamic viscosity discussed in Chapter 2, the shear stress 'τ' can be expressed in terms of dynamic viscosity and shear rates:

$$\tau_x = \eta \frac{u}{h} = \eta \frac{\partial u}{\partial z}$$

(4.6)

where:

   τ_x is the shear stress acting in the '**x**' direction [Pa].

Since '**u**' is the velocity along the '**x**' axis, the shear stress 'τ' is also acting along this direction. Along the '**y**' (out of the page) direction, however, the velocity is different and consequently the shear stress is different:

$$\tau_y = \eta \frac{v}{h} = \eta \frac{\partial v}{\partial z}$$

(4.7)

where:

   τ_y is the shear stress acting in the '**y**' direction [Pa];

   v   is the sliding velocity in the '**y**' direction [m/s].

Substituting (4.6) into (4.3) and (4.7) into (4.4), the equilibrium conditions for the forces acting in the '**x**' and '**y**' directions are obtained:

$$\frac{\partial p}{\partial x} = \frac{\partial}{\partial z}\left(\eta \frac{\partial u}{\partial z}\right)$$

(4.8)

$$\frac{\partial p}{\partial y} = \frac{\partial}{\partial z}\left(\eta \frac{\partial v}{\partial z}\right)$$

(4.9)

Equations 4.8 and 4.9 can now be integrated. Since the viscosity of the fluid is constant throughout the film (Assumption 8) and it is not a function of 'z' (i.e., $\eta \neq f(z)$), the process of integration is simple. For example, separating the variables in (4.8),

$$\frac{\partial p}{\partial x} \partial z = \partial \left(\eta \frac{\partial u}{\partial z} \right)$$

and integrating gives:

$$\frac{\partial p}{\partial x} z + C_1 = \eta \frac{\partial u}{\partial z}$$

Separating variables again,

$$\left(\frac{\partial p}{\partial x} z + C_1 \right) \partial z = \eta \partial u$$

and integrating again yields:

$$\boxed{\frac{\partial p}{\partial x} \frac{z^2}{2} + C_1 z + C_2 = \eta u}$$ (4.10)

Since there is no slip or velocity discontinuity between liquid and solid at the boundaries of the wedge (Assumption 3), the boundary conditions are:

$$u = U_2 \quad \text{at} \quad z = 0$$
$$u = U_1 \quad \text{at} \quad z = h$$

In the general case, there are two velocities corresponding to each of the surfaces 'U_1' and 'U_2'. By substituting these boundary conditions into (4.10) the constants 'C_1' and 'C_2' are calculated:

$$C_1 = (U_1 - U_2) \frac{\eta}{h} - \frac{\partial p}{\partial x} \frac{h}{2}$$
$$C_2 = \eta U_2$$

Substituting these into (4.10) yields:

$$\frac{\partial p}{\partial x} \frac{z^2}{2} + (U_1 - U_2) \frac{\eta z}{h} - \frac{\partial p}{\partial x} \frac{hz}{2} + \eta U_2 = \eta u$$

Dividing and simplifying gives the expression for velocity in the 'x' direction:

$$\boxed{u = \left(\frac{z^2 - zh}{2\eta} \right) \frac{\partial p}{\partial x} + (U_1 - U_2) \frac{z}{h} + U_2}$$ (4.11)

In a similar manner a formula for velocity in the 'y' direction is obtained.

$$v = \left(\frac{z^2 - zh}{2\eta}\right)\frac{\partial p}{\partial y} + (V_1 - V_2)\frac{z}{h} + V_2 \qquad (4.12)$$

The three separate terms in any of the velocity equations (4.11) and (4.12) represent the velocity profiles across the fluid film and they are schematically shown in Figure 4.3.

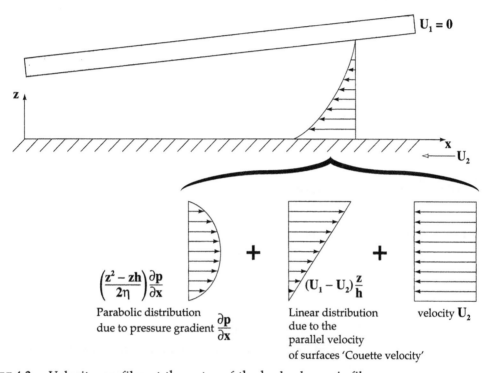

FIGURE 4.3   Velocity profiles at the entry of the hydrodynamic film.

Continuity of Flow in a Column

Consider a column of lubricant as shown in Figure 4.4. The lubricant flows into the column horizontally at rates of 'q_x' and 'q_y' and out of the column at rates of $(q_x + \frac{\partial q_x}{\partial x} dx)$ and $(q_y + \frac{\partial q_y}{\partial y} dy)$ per unit length and width, respectively. In the vertical direction the lubricant flows into the column at the rate of '$w_0 dx dy$' and out of the column at the rate of '$w_h dx dy$', where 'w_0' is the velocity at which the bottom of the column moves up and 'w_h' is the velocity at which the top of the column moves up.

The principle of continuity of flow requires that the influx of a liquid must equal its efflux from a control volume under steady conditions. If the density of the lubricant is constant (Assumption 7) then the following relation applies:

$$q_x dy + q_y dx + w_0 dx dy = \left(q_x + \frac{\partial q_x}{\partial x} dx\right)dy + \left(q_y + \frac{\partial q_y}{\partial y} dy\right)dx + w_h dx dy \qquad (4.13)$$

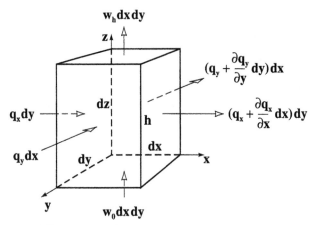

FIGURE 4.4   Continuity of flow in a column.

simplifying:

$$\frac{\partial q_x}{\partial x}dxdy + \frac{\partial q_y}{\partial y}dxdy + (w_h - w_0)dxdy = 0 \tag{4.14}$$

Since '$dxdy \neq 0$' equation (4.14) can be rewritten as:

$$\boxed{\frac{\partial q_x}{\partial x} + \frac{\partial q_y}{\partial y} + (w_h - w_0) = 0} \tag{4.15}$$

which is the equation of continuity of flow in a column.

Flow rates per unit length, 'q_x' and 'q_y', can be found from integrating the lubricant velocity profile over the film thickness, i.e.:

$$q_x = \int_0^h u\,dz \qquad \text{and} \tag{4.16}$$

$$q_y = \int_0^h v\,dz \tag{4.17}$$

substituting for 'u' from equation (4.11) yields:

$$q_x = \left| \left(\frac{z^3}{3} - \frac{z^2 h}{2}\right)\frac{\partial p}{2\eta \partial x} + (U_1 - U_2)\frac{z^2}{2h} + U_2 z \right|_0^h$$

which after simplifying gives the flow rate in the 'x' direction,

$$\boxed{q_x = -\frac{h^3}{12\eta}\frac{\partial p}{\partial x} + (U_1 + U_2)\frac{h}{2}} \tag{4.18}$$

Similarly the flow rate in the 'y' direction is found by substituting for 'v' from equation (4.12):

$$q_y = - \frac{h^3}{12\eta} \frac{\partial p}{\partial y} + (V_1 + V_2) \frac{h}{2}$$

(4.19)

Substituting now for flow rates into the continuity of flow equation (4.15):

$$\frac{\partial}{\partial x} \left[-\frac{h^3}{12\eta} \frac{\partial p}{\partial x} + (U_1 + U_2) \frac{h}{2} \right] + \frac{\partial}{\partial y} \left[-\frac{h^3}{12\eta} \frac{\partial p}{\partial y} + (V_1 + V_2) \frac{h}{2} \right] + (w_h - w_0) = 0$$

(4.20)

Defining $U = U_1 + U_2$ and $V = V_1 + V_2$ and assuming that there is no local variation in surface velocity in the 'x' and 'y' directions (i.e., $U \neq f(x)$ and $V \neq f(y)$) gives:

$$-\frac{\partial}{\partial x} \left(\frac{h^3}{12\eta} \frac{\partial p}{\partial x} \right) + \frac{U}{2} \frac{dh}{dx} - \frac{\partial}{\partial y} \left(\frac{h^3}{12\eta} \frac{\partial p}{\partial y} \right) + \frac{V}{2} \frac{dh}{dy} + (w_h - w_0) = 0$$

Further rearranging and simplifying yields the full Reynolds equation in three dimensions.

$$\frac{\partial}{\partial x} \left(\frac{h^3}{\eta} \frac{\partial p}{\partial x} \right) + \frac{\partial}{\partial y} \left(\frac{h^3}{\eta} \frac{\partial p}{\partial y} \right) = 6 \left(U \frac{dh}{dx} + V \frac{dh}{dy} \right) + 12(w_h - w_0)$$

(4.21)

Simplifications to the Reynolds Equation

It can be seen that the Reynolds equation in its full form is far too complex for practical engineering applications and some simplifications are required before it can conveniently be used. The following simplifications are commonly adopted in most studies:

· *Unidirectional Velocity Approximation*

It is always possible to choose axes in such a way that one of the velocities is equal to zero, i.e., $V = 0$. There are very few engineering systems, in which, for example, a journal bearing slides along a rotating shaft.

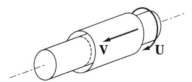

Assuming that $V = 0$ equation (4.21) can be rewritten in a more simplified form:

$$\frac{\partial}{\partial x} \left(\frac{h^3}{\eta} \frac{\partial p}{\partial x} \right) + \frac{\partial}{\partial y} \left(\frac{h^3}{\eta} \frac{\partial p}{\partial y} \right) = 6 U \frac{dh}{dx} + 12(w_h - w_0)$$

(4.22)

· *Steady Film Thickness Approximation*

It is also possible to assume that there is no vertical flow across the film, i.e., $w_h - w_0 = 0$. This assumption requires that the distance between the two surfaces remains constant during the operation. Some inaccuracy may result from this analytical simplification since most bearings usually vibrate and consequently the distance between the operating surfaces cyclically varies. Movement of surfaces normal to the sliding velocity is known as a squeeze film effect. Furthermore, in the case of porous bearings there is always some vertical flow of oil.

Assuming, however, that there is no vertical flow and $w_h - w_0 = 0$, equation (4.22) can be written in the form:

$$\frac{\partial}{\partial x}\left(\frac{h^3}{\eta}\frac{\partial p}{\partial x}\right) + \frac{\partial}{\partial y}\left(\frac{h^3}{\eta}\frac{\partial p}{\partial y}\right) = 6U\frac{dh}{dx}$$

(4.23)

· *Isoviscous Approximation*

For many practical engineering applications it is assumed that the lubricant viscosity is constant over the film, i.e., η = **constant**. This approach is known in the literature as the 'isoviscous' model where the thermal effects in hydrodynamic films are neglected. Thermal modification of lubricant viscosity does, however, occur in hydrodynamic films and must be considered in a more elaborate and accurate analysis which will be discussed later. Assuming that η = **constant** equation (4.23) can further be simplified:

$$\frac{\partial}{\partial x}\left(h^3\frac{\partial p}{\partial x}\right) + \frac{\partial}{\partial y}\left(h^3\frac{\partial p}{\partial y}\right) = 6U\eta\frac{dh}{dx}$$

(4.24)

This is in fact the most commonly quoted form of Reynolds equation throughout the literature.

· *Infinitely Long Bearing Approximation*

The simplified Reynolds equation (4.24) is two-dimensional and numerical methods are needed to obtain a solution. Thus, for a simple engineering analysis further simplifying assumptions are made.

It is assumed that the pressure gradient acting along the 'y' axis can be neglected, i.e., $\partial p/\partial y = 0$ and $h \neq f(y)$. It is therefore necessary to specify that the bearing is infinitely long in the 'y' direction. This approximation is known in the literature as the '**infinitely long bearing**' or simply '**long bearing approximation**', and is schematically illustrated in Figure 4.5. It can be said that the pressure gradient acting along the 'y' axis is negligibly small compared to the pressure gradient acting along the 'x' axis. This assumption reduces the Reynolds equation to a one-dimensional form which is very convenient for quick engineering analysis.

Since $\partial p/\partial y = 0$, the second term of the Reynolds equation (4.24) is also zero and equation (4.24) simplifies to:

$$\frac{\partial}{\partial x}\left(h^3\frac{\partial p}{\partial x}\right) = 6U\eta\frac{dh}{dx}$$

(4.25)

which can easily be integrated, i.e.:

$$h^3\frac{dp}{dx} = 6U\eta h + C$$

(4.26)

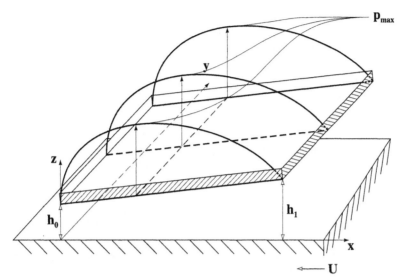

FIGURE 4.5   Pressure distribution in the long bearing approximation.

Now a boundary condition is needed to solve this equation and it is assumed that at some point along the film, pressure is at a maximum. At this point the pressure gradient is zero, i.e., **dp/dx = 0** and the corresponding film thickness is denoted as '$\bar{h}$'.

Thus the boundary condition is:

$$\frac{dp}{dx} = 0 \quad \text{at} \quad h = \bar{h}$$

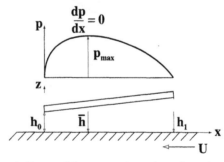

Substituting to (4.26) gives:

$$C = -6U\eta\bar{h}$$

and the final form of the one-dimensional Reynolds equation for the '**long bearing approximation**' is:

$$\boxed{\frac{dp}{dx} = 6U\eta\,\frac{h - \bar{h}}{h^3}} \qquad (4.27)$$

which is particularly useful in the analysis of linear pad bearings. Note that the velocity '**U**' in the convention assumed is negative, as shown in Figure 4.1.

· *Narrow Bearing Approximation*

Finally it is assumed that the pressure gradient acting along the '**x**' axis is very much smaller than along the '**y**' axis, i.e., $\partial p/\partial x \ll \partial p/\partial y$ as shown in Figure 4.6. This is known in the literature as a '**narrow bearing approximation**' or '**Ocvirk's approximation**' [3]. Actually this particular approach was introduced for the first time by an Australian, A.G.M. Michell, in 1905. It was applied to the approximate analysis of load capacity in a journal bearing [10]. A similar method was also presented by Cardullo [62]. Michell observed that the flow in a bearing of finite length was influenced more by pressure gradients perpendicular to the sides of the bearing than pressure gradients parallel to the direction of sliding. A formula for the hydrodynamic pressure field was derived based on the assumption that $\partial p/\partial x \ll \partial p/\partial y$. This work was severely criticized by other workers for neglecting the effect of pressure variation in the 'x' direction when equating for flow in the axial or 'y' direction and the work was ignored for about 25 years as a result of this initial unenthusiastic reception. Ocvirk and Dubois later developed the idea extensively in a series of excellent papers and Michell's approximation has since gained general acceptance.

The utility of this approximation became apparent as journal bearings with progressively shorter axial lengths were introduced into internal combustion engines. Advances in bearing materials allowed the reduction in bearing and engine size, and furthermore the reduction in bearing dimensions contributed to an increase in engine ratings. The axial length of the bearing eventually shrank to about half the diameter of the journal and during the 1950's this caused a reconsideration of the relative importance of the various terms in the Reynolds equation. During this period, Ocvirk realized the validity of considering the pressure gradient in the circumferential direction to be negligible compared to the pressure gradient in the axial direction. This approach later became known as the Ocvirk or narrow journal bearing approximation. An infinitely narrow bearing is schematically illustrated in Figure 4.6. The bearing resembles a well deformed narrow pad. Also the film geometry is similar to that of an 'unwrapped' film from a journal bearing, which will be discussed later.

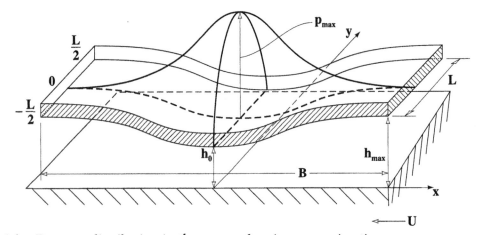

FIGURE 4.6   Pressure distribution in the narrow bearing approximation.

In this approximation since $L \ll B$ and $\partial p/\partial x \ll \partial p/\partial y$, the first term of the Reynolds equation (4.24) may be neglected and the equation becomes:

$$\frac{\partial}{\partial y}\left(h^3 \frac{\partial p}{\partial y}\right) = 6 U \eta \frac{dh}{dx} \qquad (4.28)$$

Also, since $h \neq f(y)$ then (4.28) can be further simplified,

$$\frac{d^2 p}{dy^2} = \frac{6 U \eta}{h^3} \frac{dh}{dx} \qquad (4.29)$$

Integrating once,

$$\frac{dp}{dy} = \frac{6 U \eta}{h^3} \frac{dh}{dx} y + C_1 \qquad (4.30)$$

and again gives:

$$p = \frac{6U\eta}{h^3} \frac{dh}{dx} \frac{y^2}{2} + C_1 y + C_2 \tag{4.31}$$

From Figure 4.6 the boundary conditions are:

$p = 0$ at $y = \pm L/2$ i.e., at the edges of the bearing and

$\dfrac{dp}{dy} = 0$ at $y = 0$ i.e., the pressure gradient is always zero along the central plane

of the bearing.

Substituting these into (4.30) and (4.31) gives the constants 'C_1' and 'C_2':

$$C_1 = 0$$

$$C_2 = \frac{-3U\eta}{h^3} \frac{dh}{dx} \frac{L^2}{4} \tag{4.32}$$

and the pressure distribution in a narrow bearing approximation is expressed by the formula:

$$\boxed{p = \frac{3U\eta}{h^3} \frac{dh}{dx} \left(y^2 - \frac{L^2}{4} \right)} \tag{4.33}$$

The infinitely long bearing approximation is acceptable when **L/B > 3** while the narrow bearing approximation can be used when **L/B < 1/3**. For the intermediate ratios of **1/3 < L/B < 3**, computed solutions for finite bearings are applied.

Bearing Parameters Predicted from Reynolds Equation

From the Reynolds equation most of the critical bearing design parameters such as pressure distribution, load capacity, friction force, coefficient of friction and oil flow are obtained by simple integration.

· Pressure Distribution

By integrating the Reynolds equation over a specific film shape described by some function **h = f(x,y)** the pressure distribution in the hydrodynamic lubricating film is found in terms of bearing geometry, lubricant viscosity and speed.

· Load Capacity

When the pressure distribution is integrated over the bearing area the corresponding load capacity of the lubricating film is found. If the load is varied then the film geometry will change to re-equilibrate the load and pressure field. The load that the bearing will support at a particular film geometry is:

$$W = \int_0^L \int_0^B p\, dx\, dy \tag{4.34}$$

The obtained load formula is expressed in terms of bearing geometry, lubricant viscosity and speed, hence the bearing operating and design parameters can be optimized to give the best performance.

· *Friction Force*

Assuming that the friction force results only from shearing of the fluid and integrating the shear stress 'τ' over the whole bearing area yields the total friction force operating across the hydrodynamic film, i.e.:

$$\mathbf{F} = \int_0^L \int_0^B \tau \, \mathbf{dx} \, \mathbf{dy}$$ (4.35)

The shear stress 'τ' is expressed in terms of dynamic viscosity and shear rates:

$$\tau = \eta \frac{\mathbf{du}}{\mathbf{dz}}$$

where **du/dz** is obtained by differentiating the velocity equation (4.11).

After substituting for 'τ' and integrating, the formula, expressed in terms of bearing geometry, lubricant viscosity and speed for friction force per unit length, is obtained. The derivation details of this formula are described later in this chapter.

$$\frac{\mathbf{F}}{\mathbf{L}} = \pm \int_0^B \frac{\mathbf{h}}{\mathbf{2}} \frac{\mathbf{dp}}{\mathbf{dx}} \, \mathbf{dx} - \int_0^B \frac{\mathbf{U}\eta}{\mathbf{h}} \, \mathbf{dx}$$

'+' and '−' refer to the upper and lower surface, respectively.

The '±' sign before the first term may cause some confusion as it appears that the friction force acting on the upper and the lower surface is different which apparently conflicts with the law of equal action and reaction forces. The balance of forces becomes much clearer when a closer inspection is made of the force distribution on the upper surface as schematically illustrated in Figure 4.7.

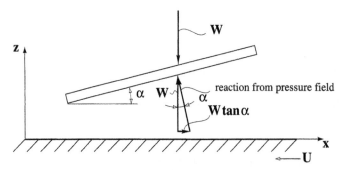

FIGURE 4.7   Load components acting on a hydrodynamic bearing.

It can be seen from Figure 4.7 that the reaction force from the pressure field acts in the direction normal to the inclined surface while a load is applied vertically. Since the load is at an angle to the normal there is a resulting component '**Wtanα**' acting in the opposite direction from the velocity. This is in fact the exact amount by which the frictional force acting on the upper surface is smaller than the force acting on the lower surface.

· *Coefficient of Friction*

The coefficient of friction is calculated from the load and friction forces:

$$\mu = \frac{\mathbf{F}}{\mathbf{W}} = \frac{\displaystyle\int_0^L \int_0^B \tau \,\mathbf{dx\,dy}}{\displaystyle\int_0^L \int_0^B \mathbf{p\,dx\,dy}} \tag{4.36}$$

Bearing parameters can then be optimized to give, for example, a minimum value of the coefficient of friction. This means, in approximate terms, minimizing the size of the bearing by allowing the highest possible hydrodynamic pressure. This is discussed in greater detail later as bearing optimization involves many other factors.

· *Lubricant Flow*

By integrating the flow expressions 'q_x' and 'q_y' (4.18) and (4.19) over the edges of the bearing, lubricant leakage out of the sides and ends of the bearing is found.

$$\mathbf{Q_x} = \int_0^L \mathbf{q_x\,dy}$$

$$\mathbf{Q_y} = \int_0^B \mathbf{q_y\,dx} \tag{4.37}$$

Lubricant flow is extremely important to the operation of a bearing since enough oil must be supplied to the hydrodynamic contact to prevent starvation and consequent failure. The flow formulae are expressed in terms of bearing parameters and the lubricant flow can also be optimized.

Summary

In summary it can be stated that the same basic analytical method is applied to the analysis of all hydrodynamic bearings regardless of their geometry.

Initially the bearing geometry, $\mathbf{h} = \mathbf{f(x,y)}$, must be defined and then substituted into the Reynolds equation. The Reynolds equation is then integrated to find the pressure distribution, load capacity, friction force and oil flow. The virtue of hydrodynamic analysis is that it is concise, simple, and the same procedure applies to all kinds of bearing geometries, i.e., linear bearings, step bearings, journal bearings, etc. The solution to the Reynolds equation becomes more complicated if other effects, such as heating, locally varying viscosity, elastic deformation, cavitation, etc., are introduced to the analysis. The basic method of analysis, however, remains unchanged. It is always necessary to start with a definition of the bearing geometry and to perform the integration procedure, taking into account extra terms and equations describing the additional effects that we wish to consider.

In the next section, some typical bearing geometries are considered and analysed. The one-dimensional Reynolds equation is used to study the linear pad and journal bearings since it provides qualitative indications of the effect of varying the controlling parameters such as load and speed. This approach is very useful as a method of rapid estimation and is widely applied in engineering analysis. For more exact treatments, the two-dimensional (2-D) Reynolds equation has to be employed. The solution of the 2-D Reynolds equation requires the application of numerical methods, and this will be discussed in the next chapter devoted to 'Computational Hydrodynamics'.

4.3 PAD BEARINGS

Pad bearings, which consist of a pad sliding over a smooth surface, are widely used in machinery to sustain thrust loads from shafts, e.g., from the propeller shaft in a ship. An example of this application is shown in Figure 4.8. The simple film geometry of these bearings, as compared to journal bearings, renders them a suitable example for introducing hydrodynamic bearing analysis.

Infinite Linear Pad Bearing

The infinite linear pad bearing, as already mentioned, is a pad bearing of infinite length normal to the direction of sliding. This particular bearing geometry is the easiest to analyse. It has been described in many books on lubrication theories [e.g., 3,4]. The basic procedures involved in the analysis are summarized in this section.

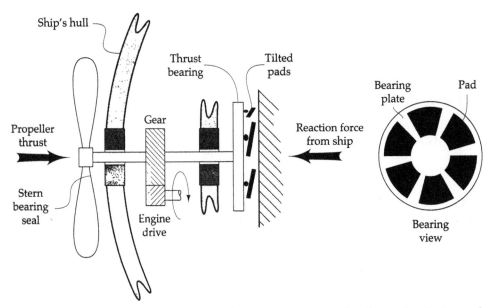

FIGURE 4.8   Example of a pad bearing application to sustain the thrust loads from the ship propeller shaft.

Consider an infinitely long linear wedge with **L/B > 3** as shown in Figure 4.9, where '**L**' and '**B**' are the pad dimensions normal to and parallel to the sliding direction, i.e., pad length and width, respectively. Assume that the bottom surface is moving in the direction shown, dragging the lubricant into the wedge which results in pressurization of the lubricant within the wedge. The inlet and the outlet conditions of the wedge are controlled by the maximum and minimum film thicknesses, 'h_1' and 'h_0', respectively.

· *Bearing Geometry*

As a first step in any bearing analysis the bearing geometry, i.e., **h = f(x)**, must be defined. The film thickness '**h**' in Figure 4.9 is expressed as a function:

$$h = h_0 + x \tan \alpha = h_0 + x \, \frac{h_1 - h_0}{B}$$

or simply:

$$h = h_0 \left(1 + \frac{h_1 - h_0}{h_0} \frac{x}{B} \right)$$

The term $(h_1 - h_0)/h_0$ is often known in the literature as the convergence ratio '**K**' [3,4]. The film geometry can then be expressed as:

$$h = h_0 \left(1 + \frac{Kx}{B} \right) \tag{4.38}$$

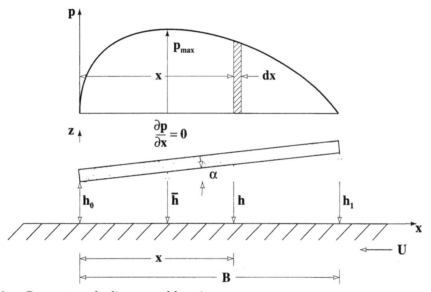

FIGURE 4.9　Geometry of a linear pad bearing.

· *Pressure Distribution*

As mentioned already, the pressure distribution can be calculated by integrating the Reynolds equation over the specific film geometry. Since the pressure gradient in the '**x**' direction is dominant, the one-dimensional Reynolds equation for the long bearing approximation (4.27) can be used for the analysis of this bearing.

$$\frac{dp}{dx} = 6U\eta \frac{h - \bar{h}}{h^3} \tag{4.27}$$

There are two variables '**x**' and '**h**' and the equation can be integrated with respect to '**x**' or '**h**'. Since it does not really matter with respect to which variable the integration is performed we choose '**h**'. Firstly one variable is replaced by the other. This can be achieved by differentiating (4.38) which gives '**dx**' in terms of '**dh**':

$$dx = \frac{B}{Kh_0} dh \tag{4.39}$$

Substituting into (4.27) yields:

$$\frac{dp}{\dfrac{B}{Kh_0}\,dh} = 6U\eta\,\frac{h-\bar{h}}{h^3}$$

and after simplifying and separating variables:

$$\frac{Kh_0}{6U\eta B}\,dp = \frac{h-\bar{h}}{h^3}\,dh \tag{4.40}$$

which is the differential formula for pressure distribution in this bearing. Equation (4.40) can be integrated to give:

$$\frac{Kh_0}{6U\eta B}\,p = -\frac{1}{h} + \frac{\bar{h}}{2h^2} + C \tag{4.41}$$

The boundary conditions, taken from the bearing's inlet and outlet, are (Figure 4.9):

$$\begin{aligned} p &= 0 \quad \text{at} \quad h = h_0 \\ p &= 0 \quad \text{at} \quad h = h_1 \end{aligned} \tag{4.42}$$

Substituting into (4.41) the constants '$\bar{h}$' and 'C' are:

$$\bar{h} = \frac{2h_0 h_1}{h_1 + h_0}$$

$$C = \frac{1}{h_1 + h_0} \tag{4.43}$$

The maximum film thickness 'h_1', can also be expressed in terms of the convergence ratio 'K':

$$K = \frac{h_1 - h_0}{h_0}$$

Thus:

$$h_1 = h_0(K+1) \tag{4.44}$$

Substituting into (4.43) the constants '$\bar{h}$' and 'C' in terms of 'K' are:

$$\bar{h} = 2h_0\frac{(K+1)}{(K+2)}$$

$$C = \frac{1}{h_0(K+2)} \tag{4.45}$$

Substituting into (4.41) gives:

$$\frac{Kh_0}{6U\eta B}\,p = -\frac{1}{h} + \frac{h_0}{h^2}\frac{(K+1)}{(K+2)} + \frac{1}{h_0(K+2)}$$

or:

$$p = \frac{6U\eta B}{Kh_0}\left(-\frac{1}{h} + \frac{h_0}{h^2}\frac{(K+1)}{(K+2)} + \frac{1}{h_0(K+2)}\right) \tag{4.46}$$

Note that the velocity 'U', in the convention assumed, is negative, as shown in Figure 4.1.

It is useful to find the pressure distribution in the bearing expressed in terms of bearing geometry and operating parameters such as the velocity 'U' and lubricant viscosity 'η'. A convenient method of finding the controlling influence of these parameters is to introduce non-dimensional parameters. In bearing analysis non-dimensional parameters such as pressure and load are used. Equation (4.46) can be expressed in terms of a non-dimensional pressure, i.e.:

$$p^* = \frac{h_0}{K}\left(-\frac{1}{h} + \frac{h_0}{h^2}\frac{(K+1)}{(K+2)} + \frac{1}{h_0(K+2)}\right) \tag{4.47}$$

where the non-dimensional pressure 'p*' is:

$$p^* = \frac{h_0^2}{6U\eta B}p \tag{4.48}$$

It is clear that hydrodynamic pressure is proportional to sliding speed 'U' and bearing width 'B' for a given value of dimensionless pressure and proportional to the reciprocal of film thickness squared. If a quick estimate of hydrodynamic pressure is required to check, for example, whether the pad material will suffer plastic deformation, a representative value of dimensionless pressure can be multiplied by the selected values of sliding speed, viscosity and bearing dimensions to yield the necessary information.

· *Load Capacity*

The total load that a bearing will support at a specific film geometry is obtained by integrating the pressure distribution over the specific bearing area.

$$W = \int_0^L\int_0^B p\,dx\,dy$$

This can be re-written in terms of load per unit length:

$$\frac{W}{L} = \int_0^B p\,dx$$

and substituting for 'p', equation (4.46), yields:

$$\frac{W}{L} = \frac{6U\eta B}{Kh_0}\int_0^B\left(-\frac{1}{h} + \frac{h_0}{h^2}\frac{(K+1)}{(K+2)} + \frac{1}{h_0(K+2)}\right)dx \tag{4.49}$$

Again there are two variables in (4.49), 'x' and 'h', and one has to be replaced by the other before the integration can be performed. Substituting (4.39) for 'dx',

$$\frac{W}{L} = \frac{6U\eta B}{Kh_0}\frac{B}{Kh_0}\int_{h_0}^{h_1}\left(-\frac{1}{h} + \frac{h_0}{h^2}\frac{(K+1)}{(K+2)} + \frac{1}{h_0(K+2)}\right)dh$$

and integrating yields:

$$\frac{W}{L} = \frac{6U\eta B}{Kh_0}\frac{B}{Kh_0}\left|\left(-\ln h - \frac{h_0}{h}\frac{(K+1)}{(K+2)} + \frac{h}{h_0(K+2)}\right)\right|_{h_0}^{h_0(K+1)}$$

$$\boxed{\frac{W}{L} = \frac{6U\eta B^2}{K^2 h_0^2}\left(-\ln(K+1) + \frac{2K}{K+2}\right)} \tag{4.50}$$

Equation (4.50) is the total load per unit length the bearing will support expressed in terms of the bearing's geometrical and operating parameters. In terms of the non-dimensional load 'W*' equation (4.50) can be expressed as:

$$\boxed{W^* = \frac{1}{K^2}\left(-\ln(K+1) + \frac{2K}{K+2}\right)} \tag{4.51}$$

where:

$$\boxed{W^* = \frac{h_0^2}{6U\eta B^2 L}W} \tag{4.52}$$

Bearing geometry can now be optimized to give maximum load capacity. By differentiating (4.51) and equating to zero an optimum value for 'K' is obtained which is:

K= 1.2

for maximum load capacity for the bearing geometry analysed. The inlet 'h_1' and the outlet 'h_0' film thickness can then be adjusted to give the maximum load capacity. From (4.38) it can be seen that the maximum load capacity occurs at a ratio of inlet and outlet film thicknesses of:

$$\frac{h_1}{h_0} = 2.2$$

· *Friction Force*

The friction force generated in the bearing due to the shearing of the lubricant is obtained by integrating the shear stress 'τ' over the bearing area (eq. 4.35):

$$F = \int_0^L\int_0^B \tau\, dx\, dy$$

The friction force per unit length is:

$$\frac{F}{L} = \int_0^B \tau dx$$

As already mentioned, shear stress is defined in terms of dynamic viscosity and shear rate:

$$\tau = \eta \frac{du}{dz}$$

where **du/dz** is obtained by differentiating the velocity equation (4.11).

In the bearing considered, the bottom surface is moving while the top surface remains stationary, i.e.:

$$U_1 = 0 \text{ and } U_2 = U$$

thus the velocity equation (4.11) is:

$$u = \left(\frac{z^2 - zh}{2\eta}\right)\frac{\partial p}{\partial x} - U\frac{z}{h} + U$$

Differentiating gives the shear rate:

$$\frac{du}{dz} = \left(2z - h\right)\frac{1}{2\eta}\frac{dp}{dx} - \frac{U}{h} \tag{4.53}$$

and substituting yields the friction force per unit length:

$$\frac{F}{L} = \int_0^B \left[\left(z - \frac{h}{2}\right)\frac{dp}{dx} - \frac{U\eta}{h}\right] dx \tag{4.54}$$

The friction force on the lower moving surface, as explained already, is greater than on the upper stationary surface. At the moving surface $z = 0$ (as shown in Figure 4.9), hence the acting frictional force per unit length is:

$$\frac{F}{L} = \int_0^B \left(-\frac{h}{2}\frac{dp}{dx} - \frac{U\eta}{h}\right) dx$$

or:

$$\frac{F}{L} = -\int_0^B \frac{h}{2}\frac{dp}{dx} dx - \int_0^B \frac{U\eta}{h} dx \tag{4.55}$$

The first part of the above equation must be integrated by parts. According to the theorems of integration, the general mathematical formula to integrate by parts is:

$$\int a\, db = ab - \int b\, da$$

So if,

$$a = \frac{h}{2} \quad \text{and} \quad db = \frac{dp}{dx} dx$$

then:

$$da = \frac{1}{2}dh \quad and \quad b = \int \frac{dp}{dx}dx = p$$

substituting:

$$-\int_0^B \frac{h}{2}\frac{dp}{dx}dx = -\left(\left|\frac{h}{2}p\right|_0^B - \int_0^B \frac{1}{2}pdh \right)$$

Since $p = 0$ at $x = 0$ and at $x = B$ (Figure 4.9) the term $\left|\frac{h}{2}p\right|_0^B$ also equals zero. In the remaining term variables are replaced before integration and substituting for 'dh' (eq. 4.39) gives:

$$-\int_0^B \frac{h}{2}\frac{dp}{dx}dx = 0 + \int_0^B \frac{1}{2}p\frac{Kh_0}{B}dx = \frac{Kh_0}{2B}\int_0^B pdx$$

Thus the first term of equation (4.55) is:

$$-\int_0^B \frac{h}{2}\frac{dp}{dx}dx = \frac{Kh_0}{2B}\frac{W}{L} \tag{4.56}$$

Integrating the second term of equation (4.55):

$$\int_0^B \frac{U\eta}{h}dx = \int_0^B \frac{U\eta}{h_0\left(1 + \frac{Kx}{B}\right)}dx = \frac{U\eta}{h_0}\int_0^B \frac{dx}{\left(1 + \frac{Kx}{B}\right)}$$

hence:

$$\int_0^B \frac{U\eta}{h}dx = \frac{U\eta B}{h_0 K}\ln(1 + K) \tag{4.57}$$

Substituting (4.56) and (4.57) into (4.55) the expression for friction force per unit length for a linear pad bearing is obtained:

$$\frac{F}{L} = \frac{Kh_0}{2B}\frac{W}{L} - \frac{U\eta B}{h_0 K}\ln(1 + K) \tag{4.58}$$

Note that calculating the friction force for the upper surface, i.e., for $z = h$, and subtracting from equation (4.58) yields $Kh_0W/BL = W\tan\alpha/L$. Substituting for 'W' (eq. 4.50),

$$W = \frac{6U\eta B^2 L}{K^2 h_0{}^2}\left(-\ln(K + 1) + \frac{2K}{K + 2}\right)$$

and simplifying:

$$\boxed{\frac{F}{L} = \frac{U\eta B}{h_0}\left(\frac{6}{K + 2} - \frac{4\ln(K + 1)}{K}\right)} \tag{4.59}$$

In a similar manner to load and pressure, frictional force is expressed in terms of the bearing's geometrical and operating parameters. In terms of the non-dimensional friction force 'F*' equation (4.59) is given by:

$$F^* = \frac{6}{K+2} - \frac{4\ln(K+1)}{K}$$

(4.60)

where:

$$F^* = \frac{h_0}{U\eta BL} F$$

(4.61)

The bearing geometry can now be optimized to give a minimum friction force, but it is more useful to optimize the bearing to find the minimum coefficient of friction since this provides the most efficient bearing geometry for any imposed load.

· *Coefficient of Friction*

By definition the coefficient of friction is expressed as a ratio of the friction and normal forces acting on the surface:

$$\mu = \frac{F}{W} = \frac{F/L}{W/L}$$

(4.62)

substituting for F/L and W/L and simplifying:

$$\mu = \frac{Kh_0}{B}\left[\frac{3K - 2(K+2)\ln(K+1)}{6K - 3(K+2)\ln(K+1)}\right]$$

(4.63)

As was performed with load and friction, 'μ' can also be expressed in a so-called non-dimensional or normalized form. In precise terms 'μ' is already non-dimensional but the purpose here is to find a general parameter which is independent of basic bearing characteristics such as load and size. Therefore 'μ' is defined entirely in terms of other non-dimensional parameters:

$$\mu^* = K\left[\frac{3K - 2(K+2)\ln(K+1)}{6K - 3(K+2)\ln(K+1)}\right]$$

(4.64)

where:

$$\mu^* = \frac{B}{h_0}\mu$$

(4.65)

The optimum bearing geometry which gives a minimum value of coefficient of friction can now be calculated. Differentiating (4.64) with respect to 'K' and equating to zero gives:

K = 1.55

which is the optimum convergence ratio for a minimum coefficient of friction.

As stated previously, the maximum load capacity occurs at **K = 1.2** but the minimum coefficient of friction is obtained when **K = 1.55**. In bearing design there must consequently be a compromise and '**K**' is chosen between these two values, i.e., **1.2 < K < 1.55** to give the optimum performance. This is evident when plotting 'μ^*' and '$6W^*$' (known as the load coefficient) against '**K**' as shown in Figure 4.10.

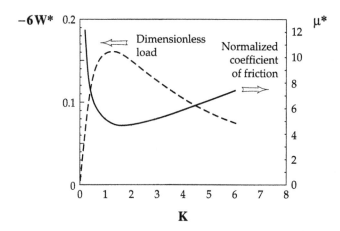

FIGURE 4.10   Variation of load capacity and coefficient of friction with a convergence ratio in a linear pad bearing.

It is quite easy to see what coefficient of friction can be anticipated in a linear pad bearing. For example, for a **0.1** [m] bearing width, a film thickness of **0.1** [mm] is typical. The minimum value of 'μ^*' is approximately **5** and the ratio B/h_0 is **1000** in this case, therefore the real value of the coefficient of friction is μ = **0.005** which is an extremely small value. Hydrodynamic lubrication is one of the most efficient means known of reducing friction and the associated power loss.

·   *Lubricant Flow Rate*

Lubricant flow rate is an important design parameter since enough lubricant must be supplied to the bearing to fully separate the surfaces by a hydrodynamic film. If an excess of lubricant is supplied, however, then secondary frictional losses such as churning of the lubricant become significant. This effect can ever overweigh the direct bearing frictional power loss. Precise calculation of lubricant flow is necessary to prevent overheating of the bearing from either lack of lubricant or excessive churning.

Since the bearing is infinitely long it can be assumed that there is no side leakage (in the '**y**' direction), i.e.:

$$q_y = 0$$

Hence the lubricant flow in the bearing is obtained by integrating the flow per unit length 'q_x' over the length of the bearing:

$$Q_x = \int_0^L q_x dy \qquad\qquad (4.66)$$

substituting for 'q_x' (eq. 4.18):

$$Q_x = \int_0^L \left(-\frac{h^3}{12\eta}\frac{\partial p}{\partial x} + \frac{Uh}{2} \right) dy \qquad (4.67)$$

The boundary conditions shown in Figure 4.9 are:

$$\frac{dp}{dx} = 0 \quad \text{at} \quad h = \bar{h} \quad \text{(point of maximum pressure)} \qquad (4.68)$$

substituting into (4.67) the flow is:

$$Q_x = \int_0^L \frac{U\bar{h}}{2}\, dy \qquad (4.69)$$

substituting for '$\bar{h}$' (eq. 4.45):

$$Q_x = \int_0^L \frac{U}{2} 2h_0\left(\frac{K+1}{K+2}\right) dy$$

and simplifying yields the lubricant flow per unit length:

$$\boxed{\frac{Q_x}{L} = Uh_0\left(\frac{K+1}{K+2}\right)} \qquad (4.70)$$

Lubricant flow is therefore determined by sliding speed and film geometry but not by viscosity or length in the direction of sliding. In real bearings, however, 'K' and 'h_0' are usually indirectly affected by oil viscosity and the length in the direction of sliding. For example, for a typical high speed pad bearing $U = 10$ [m/s], $h_0 = 0.1$ [mm] and 'K' is approximately 1.5. This gives a lubricant flow of **0.0007** [m²/s] (flow per unit length) or **0.7** [litres/sm]. If the bearing length 'L' is 0.2 [m] then **0.14** [litres/s] of lubricant is required to maintain lubrication.

Infinite Rayleigh Step Bearing

In 1918 Lord Rayleigh discovered a method of introducing a fixed variation in the lubricant film thickness without the use of tilting [5]. His new design moved away from the well established trend that lubricant film thickness variation can only be produced by tilting the pad. Rayleigh introduced a film geometry where a step divided the film into two levels of film thickness. The geometry of the Rayleigh step bearing is shown in Figure 4.11. This film geometry was advocated as simpler to manufacture than arrangements which allowed very small controlled angles of tilt.

The inlet and the outlet conditions are controlled by the maximum and minimum film thicknesses 'h_1', and 'h_0', respectively.

In this bearing there are two surfaces parallel to the bottom surface which divide the lubricant film into two zones as shown in Figure 4.11. The pressure gradients generated in each of the zones are constant, i.e., **dp/dx = constant**. This condition shortens the analysis considerably since the pressure gradients can be written directly from Figure 4.11.

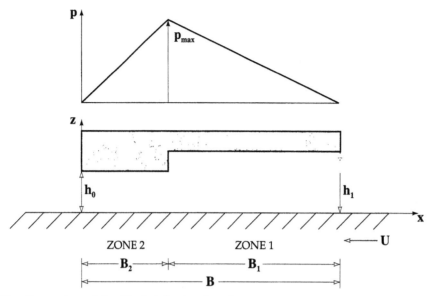

FIGURE 4.11 Geometry of the Rayleigh step bearing.

ZONE 1 $\left(\dfrac{dp}{dx}\right)_1 = -\dfrac{p_{max}}{B_1}$ (4.71)

ZONE 2 $\left(\dfrac{dp}{dx}\right)_2 = \dfrac{p_{max}}{B_2}$ (4.72)

Note that physically, for the configuration shown in Figure 4.11, $\left(\dfrac{dp}{dx}\right)_1$ is positive while $\left(\dfrac{dp}{dx}\right)_2$ is negative. In the entry zone the oil flow per unit length (eq. 4.18) into the bearing is (note that the velocity 'U' is negative):

$$q_1 = -\frac{h_1^3}{12\eta}\left(\frac{dp}{dx}\right)_1 - \frac{Uh_1}{2}$$

substituting for pressure gradient (eq. 4.71) gives flow into the bearing:

$$q_1 = \frac{h_1^3}{12\eta}\frac{p_{max}}{B_1} - \frac{Uh_1}{2}$$ (4.73)

On the other hand, in the exit zone the lubricant flow per unit length is:

$$q_2 = -\frac{h_0^3}{12\eta}\left(\frac{dp}{dx}\right)_2 - \frac{Uh_0}{2}$$

substituting for pressure gradient (eq. 4.72):

$$q_2 = -\frac{h_0^3}{12\eta}\frac{p_{max}}{B_2} - \frac{Uh_0}{2}$$ (4.74)

For continuity of flow:

$$q_1 = q_2$$

Thus:

$$\frac{h_1^3}{12\eta}\frac{p_{max}}{B_1} - \frac{Uh_1}{2} = -\frac{h_0^3}{12\eta}\frac{p_{max}}{B_2} - \frac{Uh_0}{2}$$

Simplifying and rearranging gives:

$$p_{max} = \frac{6U\eta(h_1 - h_0)}{\left(\dfrac{h_1^3}{B_1} + \dfrac{h_0^3}{B_2}\right)} \qquad (4.75)$$

Load capacity per unit length is simply the area under the pressure triangle:

$$\frac{W}{L} = \frac{1}{2}p_{max}B \qquad (4.76)$$

Frequently in the literature 'p_{max}' is quoted as 'p_s' for the step. Both ratios h_1/h_0 and B_1/B_2 can vary and it was found that these bearings give the maximum load capacity when the following ratios are selected [4]:

$$\frac{h_1}{h_0} = 1.87 \quad \text{and} \quad \frac{B_1}{B_2} = 2.588$$

An interesting story is that Lord Rayleigh first tried his bearing using two pennies (19th century coinage ~25 [mm] in diameter) [3]. Grooves were cut with a file and on one penny, the recessed areas were produced by etching with nitric acid. The other contact surface was left flat. Prepared in such a manner, the bearing worked!

The principal advantage of these bearings is that they give higher load capacity than linear pads. At their optimum configuration the load coefficient is **6W* = 0.206** as opposed to **0.1602** for infinite linear pad bearings while the coefficient of friction is almost the same. Despite the distinct advantages of other bearing types the Rayleigh step profile is still used in thrust and pad bearings. The principal reason for this practice is the ease of manufacture of the Rayleigh step as compared to the pivoted Michell pad in particular. Whereas the Michell pad requires an elaborate system of pivots, the Rayleigh step can be made by applying relatively simple machining techniques or even by covering one-half of a plane surface by protective tape and then exposing the whole surface to sandblasting or chemical etching. When the protective covering is removed, a completed Rayleigh step bearing is obtained.

The disadvantage of the Rayleigh bearing, however, is that as the step wears out then the hydrodynamic pressure falls and the bearing ceases to function as required. For bearings of finite length, the lubricant leaks more easily to the sides of the bearing than for a linear sloping pad which results in a lower load capacity. In other words at, e.g., **L/B = 1**, the Rayleigh pad has a lower value of '**W***' than the linear sloping pad, despite the fact that the opposite is the case at **L/B » 1**. To obtain higher efficiency from a Rayleigh bearing it is necessary to introduce side lands on the edges of the bearing [6]. An example of this modification is shown in Figure 4.12.

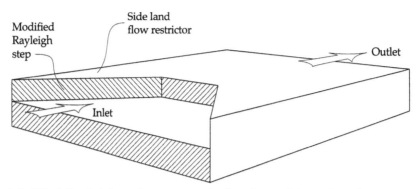

FIGURE 4.12   Modified Rayleigh pad geometry for bearings of finite length.

Other Wedge Geometries of Infinite Pad Bearings

Many different wedge shapes have been analysed and tried. Some of these designs were successful and applied in practice but most of them were destined to remain undisturbed on the shelf. The geometries of wedges most commonly applied in practice are briefly described below.

· *Tapered Land Wedge*

An example of the tapered land wedge is shown in Figure 4.13. At the end of the bearing a flat, called a '**land**', is machined. This is a very practical design since it accommodates the wear that would occur on a completely tapered wedge when the bearing decelerates to stop or accelerates from rest.

The film geometry is similar to both a linear and a Rayleigh pad bearing. Thus the bearing must be treated in two sections: the section with a taper first and then the parallel section, this is analogous to the Rayleigh step. The film geometry in the tapered section (for $x > B_2$) is described by:

$$h = h_0 \left[1 + \frac{K(x - B_2)}{B_1} \right]$$

The load capacity is strongly dependent on the amount of taper [3] and it was found that the optimum bearing configuration for maximum load capacity is achieved for the ratios $B_1/B = 0.8$ (i.e., $B_1/B_2 = 4$) and $h_1/h_0 = 2.25$ [4]. In this bearing, the combination of two geometries of linear and Rayleigh pad bearings results in the load coefficient of $6W^* = 0.192$ which is slightly lower than that for the step bearing and higher than that for the linear pad bearing.

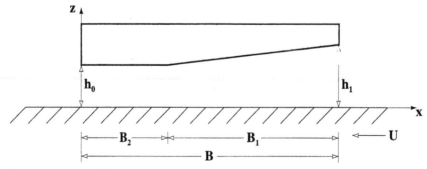

FIGURE 4.13   Geometry of the tapered land bearing.

· *Parabolic Wedge*

This particular film geometry is employed in the piston rings of combustion engines. The circumference of a piston ring is usually very much greater than either of its dimensions in the direction of travel so that the infinitely long bearing approximation is very appropriate. An example of the parabolic wedge is shown in Figure 4.14 while the pressure profile of an unbounded parabolic bearing is shown in Figure 4.15.

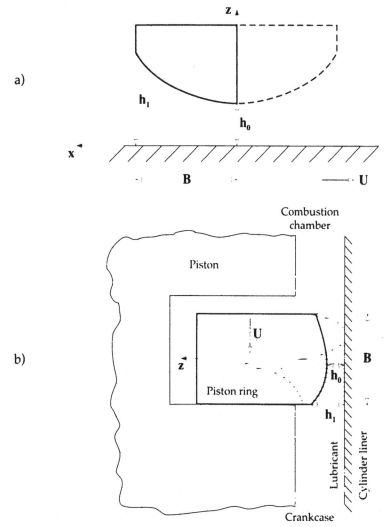

FIGURE 4.14   Geometry of the parabolic wedge bearing (a) and an example of its application in piston ring (b) [65].

The film geometry is described by the equation:

$$h = h_0 + (1 + x/B_c)^n(h_1 - h_0)$$

where:

  n   is a constant and equals **2** for a simple parabolic profile;

  B$_c$ is a characteristic width which is usually, but not always, equal to the bearing width '**B**' [m];

x is the distance along the 'x' axis starting from the minimum film thickness [m].

In this bearing there is no specific inlet and instead, beyond a certain distance, film thickness is so large that it becomes irrelevant to the pressure field close to the minimum film thickness. More information about the parabolic pressure profile can be found elsewhere [55].

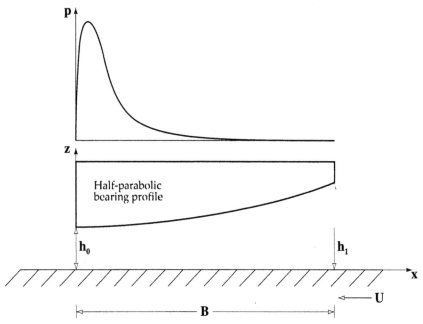

FIGURE 4.15   Pressure profile in a parabolic wedge bearing.

The parabolic profile has the advantage that it tends to be self-perpetuating under wear since the piston ring tends to rock inside its groove during reciprocating movement and causes preferential wear of the edges of the ring. If the wear is well advanced, or the edges of the piston ring have been deliberately rounded, then the starting point of hydrodynamic pressure generation cannot be precisely determined. Under these conditions, the model of parabolic film profile is very appropriate.

·   *Parallel Surface Bearings*

The low friction obtained during operation of parallel surface bearings appears to contradict the Reynolds equation since no wedge or step has been included in the bearing. It was found by Beauchamp Tower in 1891 that a bearing could be made of two flat parallel surfaces with one of them having four radial grooves cut in it [16]. The surfaces were parallel with no apparent wedge, so theoretically they should not support any load under sliding without severe wear and friction. Low friction and negligible wear was, however, obtained. Research conducted later showed that the thermal distortions of the bearing surfaces were sufficiently large to form a lubricating wedge [3,4]. These types of bearings are still used to support small intermittent loads. Thermal distortion of the bearing surface is the result of a temperature gradient between the relatively hot sliding surfaces and the cooler outer surfaces of the bearing. Since most bearing materials have considerable thermal expansion, curvature or 'crowning' of the bearing surfaces result. The distorted profile enables hydrodynamic pressure generation to occur. The principle is illustrated in Figure 4.16.

It is also possible for a parallel surface bearing to deform to produce a hydrodynamic wedge without any thermal deformation. A nominally parallel surface bearing consisting of a cantilever supported rigidly at one end is an effective bearing [17]. A maximum dimensionless load capacity '6W*' of **0.16** can be obtained by this means.

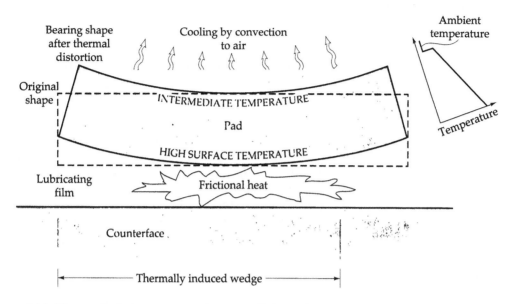

FIGURE 4.16 Thermal deformation of a parallel surface bearing to allow hydrodynamic lubrication.

· *Spiral Groove Bearing*

When the step is curved around a circular boundary, a very useful form of bearing results which is known as the spiral groove bearing and is shown in Figure 4.17.

As illustrated in Figure 4.17, two forms of the spiral groove bearing exist; in one the centre of the spiral has the lower film thickness and is known as the '**closed form**', in the other the centre of the spiral has the larger film thickness and is known as the '**open form**'.

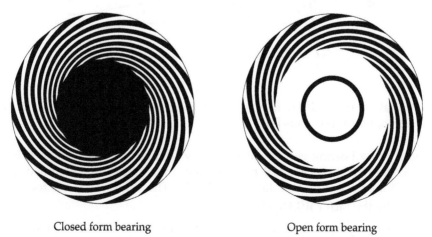

Closed form bearing   Open form bearing

FIGURE 4.17 Film geometry of spiral groove bearings (dark areas have lower film thickness).

The spiral groove bearing is often constructed in pairs of opposing spirals to allow reverse rotation with positive pressure generation and load capacity. The theory of spiral groove thrust bearings is discussed in detail elsewhere [12].

Spiral groove bearings are effective and relatively cheap alternatives for use as thrust bearings. When made from silicon carbide ceramic these bearings were found to work reliably in the presence of abrasive slurry which also acted as a lubricant [13].

Finite Pad Bearings

As indicated earlier the long bearing approximation provides adequate estimates of load capacity and friction for the ratios of **L/B > 3**. The bearings with a ratio **1/3 < L/B < 3** are called finite bearings. For these bearings all the important parameters such as pressure, load capacity, friction force and lubricant flow are usually computed by numerical methods. In certain limited cases, however, it is possible to derive analytical expressions of load capacity, friction force, etc. for finite bearings. A great deal of intellectual effort was expended on this task before computers were available. The disadvantage of the analytical approach is that it is impossible at present to incorporate additional factors such as lubricant heating. In the next chapter, a widely used numerical method, called the Finite Difference method, is introduced and its applications to bearing analysis are illustrated by examples. At this stage, however, it is helpful to consider the application of data generated by numerical bearing analysis.

In the literature computed data are presented in the forms of graphs or data sheets. An example is shown in Figure 4.18 [4] where the load coefficient '**6W***' is plotted against the convergence ratio '**K**' for various **L/B** ratios for rectangular linear pads.

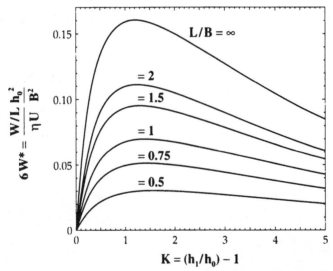

FIGURE 4.18 Variation of load capacity with convergence ratio for various **L/B** ratios for rectangular linear pads [4].

The load capacity of a bearing is then calculated by finding the appropriate **L/B** value or by interpolation where necessary. For **L/B** ratios greater than **2**, it can be assumed that values of '**6W***' for **L/B = 3** are very close to values for **L/B = ∞**. The value of load is found from '**6W***' by multiplying it by the factor $B^2LU\eta / h_0^2$.

Pads are usually employed in thrust bearings. They can also be found in pivoted pad journal bearings which are often used in machine tool applications. In thrust bearings the pads are usually not square since this would be impractical. The collar is circular and the pads are part of a circle. They are called sector-shape pads and were analysed by Pinkus [3,7] in 1958. The analysis is much more complex than that of rectangular pads. In practical engineering cases, however, it is usually sufficiently accurate to assume that the pad is of rectangular shape, since the error of this approximation is less than 10%. More information on these and more specialized pads can be found in the literature [e.g., 7-9].

It can be seen from Figure 4.18 that for the lower **L/B** ratios the load capacity of the bearing is less sensitive to changes in the h_1/h_0 ratios, i.e., the bearing is more stable. The continuous changes of h_1/h_0 ratios with load pose the greatest problem in this type of bearing. The way in which this problem has been overcome will be discussed in the next section.

EXAMPLE

Calculate the maximum load capacity for a square pad of B = 0.1 [m] side dimension. Assume a sliding speed of U = 10 [m/s], lubricant viscosity η = 0.05 [Pas] and minimum film thickness $h_0 = 10^{-4}$ [m].

From Figure 4.18;   **6W* = 0.07** hence $\mathbf{W = 0.07 \dfrac{0.1^3 \times 10 \times 0.05}{\left(10^{-4}\right)^2} = 3.5}$ [kN]

Pivoted Pad Bearing

The pivoted pad bearing allows the angle of tilt to vary with load as this has been found to improve the load capacity of the bearing. The problem of a limited load capacity caused by a fixed level of tilt can be illustrated by considering changes in load capacity when the load and 'W*' are increased progressively from zero. Until the pad bearing reaches its optimum h_1/h_0 or 'K' ratios, the load capacity balances the applied load. If the load is increased further then 'K' will increase as the tilt ($h_1 - h_0$) becomes greater than h_0. Non-dimensional load 'W*' then starts to decline since 'K' shifts towards the right as can be seen from Figure 4.10. The real load, however, as opposed to 'W*', may still rise in theory because of the strong multiplying effect of 'h_0^{-2}' but in practice there are other factors such as distortion of the pad and heating of the lubricant caused by the more intensive shearing at thin film thicknesses which render the decline in 'W*' with 'K' much more severe than shown in Figure 4.10.

The changes in 'W*' with 'K' posed a very serious engineering problem since it was not possible to run the bearing at the optimal h_1/h_0 ratio. The problem was eventually solved by an Australian engineer, A.G.M. Michell. During his work as consulting engineer he saw the limitations of conventional thrust bearings in which metal-to-metal contact frequently occurred. The existing designs of thrust bearings were also complicated and large in size.

For example, a thrust bearing for a single propeller ship could have ten or more collar bearings. There were obvious difficulties in maintaining the close tolerances to ensure a uniform contact pressure on the collars. Consequently the bearings were oversized, noisy, inefficient and not particularly reliable.

Michell came up with an ingenious solution which was a major breakthrough in lubrication science viz the pivoted pad bearing [10]. His design required only one thrust collar on a ship propeller shaft instead of the ten previously required. This simplification enabled a considerable reduction in noise and fuel consumption and less space was needed for the bearing. Michell patented his bearing in 1905 and a working model was installed in pumps at Cohuna on the Murray River in 1907 [11].

In 1910, Kingsbury independently patented a similar bearing in the United States. The only difference was that the pads of his bearing were pivoted centrally whereas in the Michell bearing they were offset as shown schematically in Figure 4.19. Also Kingsbury's approach was empirical, lacking the rigour and elegance of Michell's mathematical analysis [11].

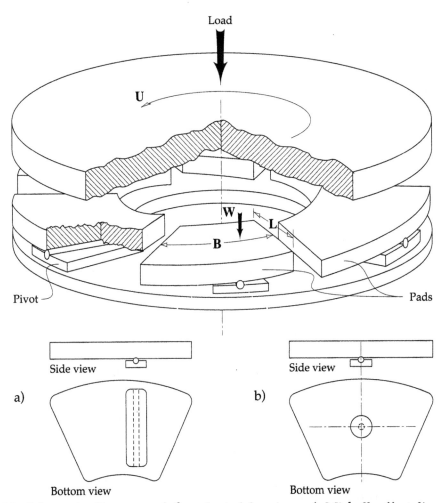

FIGURE 4.19 Schematic diagrams of the pivoted bearings a) Michell offset line pivot, b) Kingsbury button point pivot (adapted from [67]).

Kingsbury's design, however, made allowance for misalignment between mobile and stationary sides of the bearing by the use of a point pivot rather than a linear pivot.

The greatest advantage of pivoted pad bearings over fixed pad bearings is that the ratio h_1/h_0 always remains the same, whatever the load. These bearings self-adjust their film thickness geometry with load to give optimum performance. The pivot should be placed in the centre of pressure, otherwise the bearing becomes unstable. The centre of pressure and hence the pivot position can easily be found by taking moments about the trailing edge of the bearing as shown in Figure 4.20.

The moment of force about the bearing outlet is:

$$WX = \int_0^L \int_0^B px\,dx\,dy$$

or per unit length:

$$\frac{WX}{L} = \int_0^B px\,dx \qquad (4.77)$$

substituting for 'p' (eq. 4.46), 'x' (from eq. 4.38) and 'dx' (eq. 4.39) gives:

$$\frac{WX}{L} = \frac{6U\eta B}{Kh_0}\frac{B^2}{K^2h_0^2}\int_{h_0}^{h_1}\left(-\frac{1}{h} + \frac{h_0}{h^2}\frac{(K+1)}{(K+2)} + \frac{1}{h_0(K+2)}\right)(h-h_0)dh \tag{4.78}$$

After substituting for **W/L** (eq. 4.50) into (4.78) and manipulating the equation, the position of pivot from the trailing edge of the bearing is found.

$$\frac{X}{B} = 1 - \frac{2(3+K)(1+K)\ln(1+K) - K(6+5K)}{2K[(2+K)\ln(1+K) - 2K]} \tag{4.79}$$

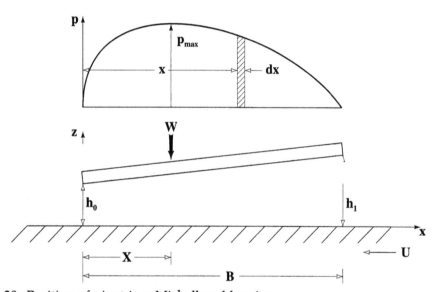

FIGURE 4.20   Position of pivot in a Michell pad bearing.

It can be seen that the position of the pivot depends only on the convergence ratio 'K' or the ratio of film thicknesses. Thus the film thickness ratio 'K' remains constant once the position of the pivot is decided. Some calculated values of the pivot position for various convergence ratios are shown in Table 4.2. It can be seen from Table 4.2 that for an infinite linear pad bearing at its optimum h_1/h_0 ratio of **2.2** the pivot position is **0.422B** from the trailing edge.

Inlet Boundary Conditions in Pad Bearing Analysis

Finally, the question of pad boundary conditions has to be discussed in more detail. It is usually assumed for the purpose of analysis that the inlet pressure to a pad bearing is either zero or identical to atmospheric pressure. This assumption ignores the possibility of an inertial pressure rise at the bearing inlet. Such an inertial pressure rise is schematically shown in Figure 4.21 and is the result of deceleration or acceleration of the lubricant when it flows over or around the bearing pad.

This pressure rise is often referred to as the 'Ram Effect' and can be significant at high sliding speeds. The pressure rise has been measured experimentally [14] and modelled by computational methods [15]. The pressure rise can be estimated from the following equation:

TABLE 4.2   Pivot position for various 'K' ratios.

$\dfrac{X}{B}$	K	$\dfrac{h_1}{h_0}$
0	0	1
0.482	0.2	1.2
0.466	0.4	1.4
0.453	0.6	1.6
0.442	0.8	1.8
0.431	1.0	2.0
0.422	1.2	2.2
0.414	1.4	2.4
0.406	1.6	2.6
0.399	1.8	2.8
0.393	2.0	3.0
0.387	2.2	3.2
0.381	2.4	3.4
0.376	2.6	3.6
0.371	2.8	3.8
0.366	3.0	4.0

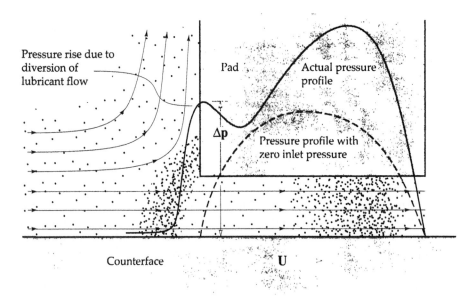

FIGURE 4.21   Stagnation pressure rise at the inlet to a pad bearing.

$$\frac{\Delta p}{\rho U^2} = \frac{A}{Re} + B$$

where:

Δp is the pressure rise [Pa];

ρ is the lubricant density [kg/m³];

U   is the sliding speed [m/s];

Re   is the Reynolds number defined as **Re = h / ρUh₁**;

A, B   are constants.

A graph of the dimensionless pressure rise versus the reciprocal of the Reynolds number is shown in Figure 4.22 for two bearing geometries.

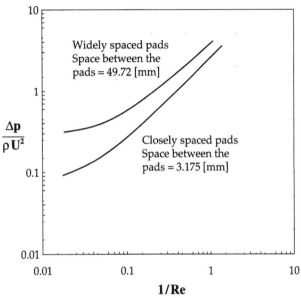

FIGURE 4.22   Dimensionless bearing inlet pressure rise versus the reciprocal of the Reynolds number [15].

It can be seen that for a bearing with a standard geometry, e.g., a planar wall normal to the direction of sliding and widely spaced pads, the pressure increase is high, particularly at low Reynolds' numbers. On the other hand, for the bearing with a 45° chamfer at the inlet and a smaller distance between the pads this effect is reduced. The difference in pressure is the result of the varying mode of lubricant flow at the bearing inlet. This effect is a prime example of the important role the bearing inlet conditions play in controlling the bearing characteristics. Examples of measured constant values of '**A**' and '**B**' can be found elsewhere [15].

The effect of a positive inlet pressure is to raise the load capacity of both fixed inclination pads and pivoted pads [63]. In some cases, the increase in load capacity is predicted to be quite large, e.g., a doubling of load capacity compared to that calculated with the simple assumption of zero inlet pressure. Unless suction is deliberately applied, the inlet pressure is always positive and, in effect, provides a margin of safety for the design of pad bearings based on classical hydrodynamic theory.

4.4   CONVERGING-DIVERGING WEDGES

In the previous sections an infinite linear pad bearing was considered where the geometry was merely a converging wedge. In many engineering applications, however, such as in journal bearings, the situation can be a little more complicated; the wedge can initially converge and then, after reaching a minimum, diverge. The question is: what will then happen? How would this affect the generated pressure and consequently the load capacity? The overall approach to the analysis in such cases is unaffected. The first step is to define the film geometry and then the pressure distribution, load capacity and other important parameters can be obtained by integrating the Reynolds equation. This problem can be illustrated by initially considering the secant wedge. It has been suggested that this wedge is the most useful for teaching purposes because of its simplicity [4]. It also gives a clear introduction to the more complicated geometry of journal bearings.

Consider a converging-diverging wedge described by a secant function such as that illustrated in Figure 4.23.

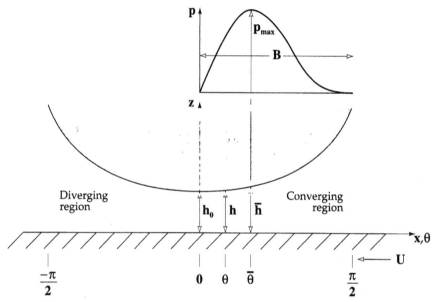

FIGURE 4.23 Geometry of a secant wedge. Note the negative velocity 'U'.

Bearing Geometry

The wedge profile is described by the function:

$$h = h_0 \sec \theta \tag{4.80}$$

where:

$$\theta = \frac{x}{B} \frac{\pi}{2} \tag{4.81}$$

and the minimum film thickness 'h_0' occurs when $\theta = 0$.

Pressure Distribution

The Reynolds equation for the long bearing approximation (4.27) can be used since the pressure gradient acting along the 'x' axis is dominant:

$$\frac{dp}{dx} = 6 U \eta \frac{h - \bar{h}}{h^3}$$

substituting for '**h**' and '$\overline{\mathbf{h}}$' which is defined as:

$$\overline{\mathbf{h}} = \mathbf{h_0} \sec \overline{\theta}$$

gives:

$$\frac{\mathbf{dp}}{\mathbf{dx}} = 6\mathbf{U}\eta \, \frac{\mathbf{h_0}(\sec\theta - \sec\overline{\theta})}{\mathbf{h_0}^3 \sec^3\theta}$$

In a similar manner to the previously analysed linear bearing, '$\overline{\theta}$' gives the position at which the maximum pressure occurs at the corresponding film thickness '$\overline{\mathbf{h}}$'. Differentiating (4.81):

$$\mathbf{dx} = \frac{2\mathbf{B}}{\pi} \, \mathbf{d\theta} \tag{4.82}$$

and substituting, gives

$$\mathbf{dp} = \frac{6\mathbf{U}\eta}{\mathbf{h_0}^2} \frac{2\mathbf{B}}{\pi} \left(\frac{1}{\sec^2\theta} - \frac{\sec\overline{\theta}}{\sec^3\theta} \right) \mathbf{d\theta}$$

which can be integrated. Remembering the standard integrals:

$$\int \frac{1}{\sec^2\theta} \, \mathbf{d\theta} = \int \cos^2\theta \, \mathbf{d\theta} = \frac{1}{2}(\theta + \sin\theta\cos\theta) \quad \text{and}$$

$$\int \frac{1}{\sec^3\theta} \, \mathbf{d\theta} = \int \cos^3\theta \, \mathbf{d\theta} = \frac{1}{3}(\sin\theta\cos^2\theta + 2\sin\theta)$$

the pressure distribution in the secant wedge is given by:

$$\mathbf{p} = \frac{6\mathbf{U}\eta}{\mathbf{h_0}^2} \frac{2\mathbf{B}}{\pi} \left[\frac{1}{2}(\theta + \sin\theta\cos\theta) - \frac{\sec\overline{\theta}}{3}(\sin\theta\cos^2\theta + 2\sin\theta) + \mathbf{C} \right] \tag{4.83}$$

which can then be expressed in terms of a non-dimensional pressure:

$$\boxed{\mathbf{p^*} = \frac{2}{\pi} \left[\frac{1}{2}(\theta + \sin\theta\cos\theta) - \frac{\sec\overline{\theta}}{3}(\sin\theta\cos^2\theta + 2\sin\theta) + \mathbf{C} \right]} \tag{4.84}$$

where:

$$\boxed{\mathbf{p^*} = \frac{\mathbf{h_0}^2}{6\mathbf{U}\eta\mathbf{B}} \mathbf{p}} \tag{4.85}$$

There are two constants '$\overline{\theta}$' and '**C**' and some boundary conditions are needed in order to determine them. Throughout the literature three sets of boundary conditions known as the Full-Sommerfeld, Half-Sommerfeld and Reynolds are widely quoted, and these are described below.

· *Full-Sommerfeld Boundary Condition*

The Full-Sommerfeld condition [18] is perhaps the most obvious and simplest of the boundary conditions. It assumes that the pressure is equal to zero at the edges of the wedge, i.e.:

$$p = 0 \quad \text{at} \quad \theta = \pm \frac{\pi}{2}$$

substituting into (4.83) the constants '**C**' and '**sec$\bar{\theta}$**' can be determined:

$$C = 0$$

$$\sec \bar{\theta} = \frac{3}{8}\, \pi = 1.1781 \tag{4.86}$$

It can be seen from (4.86) that the pressure reaches its maximum at $\bar{\theta} = 31.92°$ and substituting this into (4.84) yields:

$$p^* = \frac{2}{\pi} \left[\frac{1}{2}\, (\theta + \sin\theta\cos\theta) - \frac{\pi}{8}\, (\sin\theta\cos^2\theta + 2\sin\theta) \right] \tag{4.87}$$

The pressure distribution for the Full-Sommerfeld condition is shown in Figure 4.24.

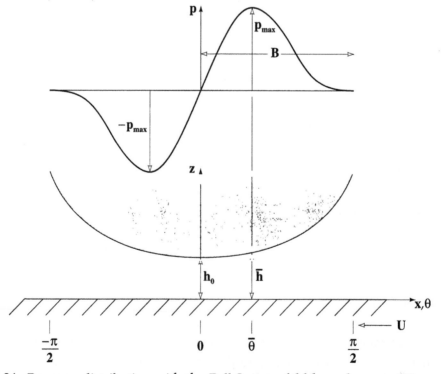

FIGURE 4.24 Pressure distribution with the Full-Sommerfeld boundary condition.

It can be seen from Figure 4.24 that the Full-Sommerfeld condition is unlikely to apply to real fluids. There is a large negative pressure in the diverging region which is the mirror image of the pressure distribution in the converging region. As discussed later, large negative pressures are physically unrealistic. Furthermore, because of these opposing negative and positive pressures the predicted load capacity is zero. On the other hand, it has been shown that the hydrodynamic lubrication film is very efficient under such geometries and is capable of supporting a load. Hence some other boundary condition should apply.

· *Half-Sommerfeld   Boundary   Condition*

To allow for the reality that large negative pressures do not exist in a diverging region because of the limitations of lubricants, a simple model has been adopted which states that the predicted negative pressures are in fact equal to zero, i.e.:

$$p = 0 \quad \text{for} \quad x \le 0$$

and the Half-Sommerfeld boundary condition is:

$$p = 0 \quad \text{at} \quad \theta = \frac{\pi}{2}$$

$$p = 0 \quad \text{at} \quad \theta = 0 \tag{4.88}$$

$$p = 0 \quad \text{at} \quad -\frac{\pi}{2} < \theta < 0$$

Substituting this condition gives a pressure distribution similar to the Full-Sommerfeld with the only difference being that in the diverging region the pressure remains constant and equal to zero as shown in Figure 4.25.

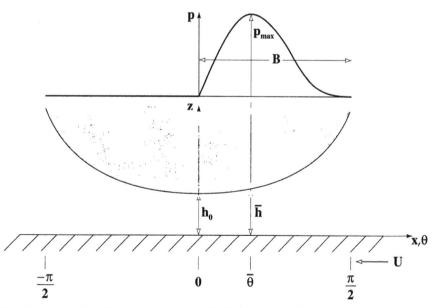

FIGURE 4.25   Pressure distribution with the Half-Sommerfeld boundary condition.

From the engineering viewpoint the Half-Sommerfeld boundary condition is very simple and easy to apply. Its physical basis, however, is erroneous since discontinuity of flow at the boundary between the zero and the non-zero pressure regions is implied. The discrepancy between the flow of lubricant out of the non-zero pressure region and into the zero pressure region is analysed below.

Flow rate per unit length into the wedge is (eq. 4.18):

$$q_x = - \frac{h^3}{12\eta} \frac{dp}{dx} + U\frac{h}{2}$$

since:

$$\frac{dp}{dx} = 0 \quad at \quad \bar{h}$$

then the flow rate into the bearing is given by:

$$q_{x_{in}} = \frac{U\bar{h}}{2} = \frac{Uh_0 \sec\bar{\theta}}{2}$$

substituting for 'sec$\bar{\theta}$' (eq. 4.86) yields:

$$\boxed{q_{x_{in}} = 1.1781 \frac{Uh_0}{2}} \tag{4.89}$$

Since in the diverging region pressure is continuously equal to zero then:

$$\frac{dp}{dx} = 0 \quad for \ all \quad x \le 0$$

This means that:

$$\frac{dp}{dx} = 0 \quad at \quad h_0$$

and the flow out of the bearing in the diverging region is:

$$\boxed{q_{x_{out}} = \frac{Uh_0}{2}} \tag{4.90}$$

For continuity of flow:

$$q_{x_{in}} = q_{x_{out}}$$

but this is clearly not the case, since:

$$1.1781 \frac{Uh_0}{2} \ne \frac{Uh_0}{2}$$

Despite the lack of flow continuity, the Half-Sommerfeld boundary condition is used in some engineering calculations, as the errors introduced are small.

Summarizing, both conditions analysed so far are not physically realistic, since one leads to predictions of large negative pressures and the other to a discontinuity of flow. A more exact solution is required before accurate estimations of load capacity, bearing friction and lubricant flow rate can be obtained.

· *Reynolds Boundary Condition*

The solution to the problem was suggested by Reynolds who simply stated that there are no negative pressures and that at the boundary between zero and non-zero pressure the following condition should apply:

$$p = \frac{dp}{dx} = 0$$

Reynolds assumed that in the diverging region, at film thickness '$\overline{h}$' (located at '$-\overline{\theta}$') the pressure gradient **dp/dx** and the pressure are equal to zero. In this region the lubricating film starts to divide into streamers of lubricant and air spaces and $\overline{h} = h_{cav}$ (where the subscript '**cav**' stands for cavitation). As the film thickness continues to increase, the proportion of space occupied by lubricant streamers is correspondingly reduced. The balance between streamer volume and volume of air space is determined by the condition that the lubricant flow remains constant within the zero pressure region. The Reynolds boundary condition for an infinitely long bearing is shown in Figure 4.26 which is a plan view of the hydrodynamic pressure field.

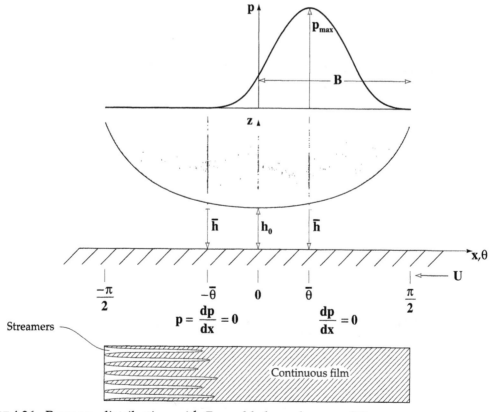

FIGURE 4.26   Pressure distribution with Reynolds boundary condition.

It can be seen from Figure 4.26 that the Reynolds boundary condition assumes:

$$p = 0 \quad \text{at} \quad \theta = \frac{\pi}{2}$$

$$p = 0 \quad \text{at} \quad \theta = -\bar{\theta} \quad \text{when} \quad \frac{dp}{dx} = 0 \tag{4.91}$$

Substituting the above boundary condition into equation (4.83) yields the constants 'C' and 'sec$\bar{\theta}$':

$$C = -0.03685$$

$$\sec \bar{\theta} = 1.1228 \tag{4.92}$$

Substituting into (4.84) gives the non-dimensional pressure for the Reynolds boundary condition:

$$p^* = \frac{2}{\pi} \left[\frac{1}{2} (\theta + \sin\theta \cos\theta) - 0.3743 (\sin\theta \cos^2\theta + 2\sin\theta) - 0.03685 \right] \tag{4.93}$$

It can also be found from (4.92) that the maximum pressure is reached at $\bar{\theta} = 27.05°$ and is zero at $\bar{\theta} = -27.05°$ (Figure 4.26).

Load Capacity

As already mentioned the total load that the bearing will support is obtained by integrating over the pressure field and the load per unit length is:

$$\frac{W}{L} = \int_0^B p \, dx$$

Substituting for 'dx' (4.82):

$$\frac{W}{L} = \frac{2B}{\pi} \int_{\theta_1}^{\theta_2} p \, d\theta \tag{4.94}$$

where 'θ_1' and 'θ_2' mark the extent of the pressure field, which depends on the boundary condition applied. Substituting for 'p' (4.83):

$$\frac{W}{L} = \frac{6U\eta}{h_0^2} \frac{2B}{\pi} \frac{2B}{\pi} \int_{\theta_1}^{\theta_2} \left[\frac{1}{2} (\theta + \sin\theta \cos\theta) - \frac{\sec\bar{\theta}}{3} (\sin\theta \cos^2\theta + 2\sin\theta) + C \right] d\theta$$

Integrating yields:

$$\frac{W}{L} = \frac{6U\eta B^2}{h_0^2} \frac{4}{\pi^2} \left| \frac{\theta^2}{4} + \frac{\sin^2\theta}{4} + \frac{\sec\bar{\theta}}{3} \left(\frac{\cos^3\theta}{3} + 2\cos\theta \right) + C\theta \right|_{\theta_1}^{\theta_2} \tag{4.95}$$

which is the total load per unit length that the bearing will support. In terms of non-dimensional load (4.95) is expressed as:

$$W^* = \frac{4}{\pi^2} \left| \frac{\theta^2}{4} + \frac{\sin^2\theta}{4} + \frac{\sec\bar{\theta}}{3}\left(\frac{\cos^3\theta}{3} + 2\cos\theta\right) + C\theta \right|_{\theta_1}^{\theta_2}$$

(4.96)

where:

$$W^* = \frac{h_0^2}{6U\eta B^2 L} W$$

Substituting for the appropriate values of 'θ' defining the extent of the pressure field and constants 'C' and '$\sec\bar{\theta}$', the non-dimensional load can easily be determined for all the conditions analysed. The results obtained are summarized in Table 4.3.

TABLE 4.3   Extent of pressure fields and non-dimensional loads for Full-Sommerfeld, Half-Sommerfeld and Reynolds boundary conditions.

	The extent of the pressure field		Constants		W*
	θ_1	θ_2	$\sec\bar{\theta}$	C	
Full - Sommerfeld	−90°	90°	1.1781	0	0
Half - Sommerfeld	0°	90°	1.1781	0	0.020041
Reynolds	−27.05°	90°	1.1228	−0.03685	0.028438

It can be seen from Table 4.3 that with the Full-Sommerfeld boundary condition there is no resultant load. The reason for this becomes clear after examining Figure 4.24. The positive pressure field is counterbalanced by a negative pressure field, leaving only a turning moment. It can also be noticed that the pressure field and therefore the load found for the Reynolds boundary condition are higher than for the Full and Half-Sommerfeld boundary conditions. This is shown in Figure 4.27 [4] where the pressure distributions, for all three conditions, are superimposed.

The Reynolds boundary condition gives the most accurate results of all the boundary conditions presented in this section, but other more specialised conditions have been suggested and will be discussed in Chapter 5 on 'Computational Hydrodynamics'. Although it is relatively accurate, the Reynolds boundary condition is still an approximation to the transition from full fluid flow to cavitated flow. It is also relatively difficult to apply in analytical solutions of realistic film geometries. The mathematics can become quite involved especially when locating the position of the boundary. Thus, as mentioned already, the Half-Sommerfeld boundary condition can be employed since it gives acceptable estimates of load capacity. The Reynolds boundary condition is, however, easily incorporated into numerical solutions of hydrodynamic pressure fields as shown in the next chapter on 'Computational Hydrodynamics' and for this reason is the most widely used boundary condition.

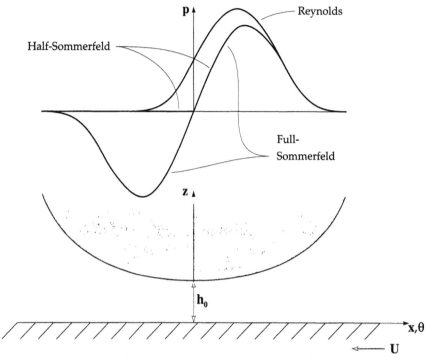

FIGURE 4.27 Pressure fields for Full-Sommerfeld, Half-Sommerfeld and Reynolds boundary conditions [4].

4.5 JOURNAL BEARINGS

Journal bearings are very common engineering components and are used in almost all types of machinery. Combustion engines and turbines virtually depend on journal bearings to obtain high efficiency and reliability. A journal bearing consists of a shaft rotating within a stationary bush. The hydrodynamic film which supports the load is generated between the moving surfaces of the shaft and the bush.

There are two basic aspects of journal bearing analysis. The first refers to the basic analysis of journal bearing load capacity, friction and lubricant flow rate as a function of load, speed and any other controlling parameters. The second aspect of journal bearings relates to practical or operational problems, such as methods of lubricant supply, bearing designs to suppress vibration and cavitation or to allow for misalignment, and frictional heating of the lubricant. These aspects are introduced in this chapter and continued in the next chapter.

Evaluation of the Main Parameters

The same method of analysis applies to journal bearings as to the previously described examples of linear and converging-diverging wedges. The film geometry is defined and then the Reynolds equation is applied to find the pressure field and load capacity. The geometry of the journal bearing is shown in Figure 4.28.

· Bearing Geometry

In the same manner as previous cases, bearing geometry is defined in the first step of the analysis. Consider the triangle $O_s O_B A$ from Figure 4.28 which is shown in detail in Figure 4.29.

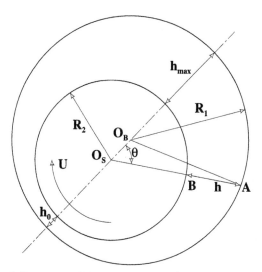

FIGURE 4.28   Geometry of the journal bearing; R_1 is the radius of the bush, R_2 is the radius of the shaft, O_B is the centre of the bush, O_S is the centre of the shaft.

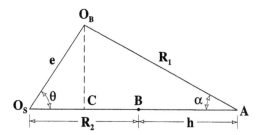

FIGURE 4.29   Details of geometry for the evaluation of film shape in journal bearings; **e** is the eccentricity (i.e., distance $O_S O_B$ between the axial centres of shaft and bush during the bearing's operation) [m], **h** is the film thickness [m].

It should be noted that the angle 'α' is very small. From inspection of the triangle $O_S O_B A$ it can be written:

$$\mathbf{O_s A = O_s C + CA = O_s B + BA} \quad \text{or}$$

$$\mathbf{O_s A = e\cos\theta + R_1\cos\alpha = R_2 + h}$$

thus:

$$\mathbf{h = e\cos\theta + R_1\cos\alpha - R_2} \tag{4.97}$$

applying the sine rule gives:

$$\frac{\mathbf{e}}{\mathbf{\sin\alpha}} = \frac{\mathbf{R_1}}{\mathbf{\sin\theta}} \quad \text{and}$$

$$\mathbf{\sin\alpha = \frac{e}{R_1}\sin\theta}$$

Remembering that:

$$\sin^2\alpha + \cos^2\alpha = 1$$

and substituting for 'sinα' yields:

$$\cos\alpha = \sqrt{1 - \sin^2\alpha} = \sqrt{1 - \left(\frac{e}{R_1}\right)^2 \sin^2\theta}$$

Since $e/R_1 \ll 1$ then:

$$\cos\alpha \approx 1 \tag{4.98}$$

Substituting into (4.97) yields:

$$h = e\cos\theta + R_1 - R_2 = e\cos\theta + c$$

where:

    c   is the clearance, i.e., the difference between the radii of bush and shaft $(R_1 - R_2)$ [m].

or:

$$\boxed{h = c(1 + \varepsilon\cos\theta)} \tag{4.99}$$

where:

    ε   is the eccentricity ratio, i.e., the ratio of eccentricity to clearance (e/c).

Equation (4.99) gives a description of the film shape in journal bearings to within 0.1% accuracy [3].

· *Pressure Distribution*

In most journal bearings where the axial length is less than the shaft diameter, the pressure gradient along the 'y' axis is much larger than the pressure gradient along the 'x' axis (circumferential direction). The narrow bearing approximation can therefore be used. This approximation gives accurate results for $L/D < 1/3$. The one-dimensional Reynolds equation for the narrow bearing approximation is given by (4.33).

$$p = \frac{3U\eta}{h^3}\frac{dh}{dx}\left(y^2 - \frac{L^2}{4}\right)$$

where 'L' is the length of the bearing along the 'y' axis. Substituting 'x' for angular displacement times radius gives:

$$x = R\theta$$

differentiating:

$$dx = R d\theta$$

and substituting gives:

$$p = \frac{3U\eta}{h^3 R}\frac{dh}{d\theta}\left(y^2 - \frac{L^2}{4}\right)$$ (4.100)

Differentiating (4.99):

$$dh = -c\varepsilon\sin\theta\,d\theta$$

rearranging:

$$\frac{dh}{d\theta} = -c\varepsilon\sin\theta$$ (4.101)

and substituting for '**h**' and '**dh/dθ**' to (4.100) yields the pressure distribution in a narrow journal bearing:

$$p = \frac{3U\eta\varepsilon\sin\theta}{Rc^2(1+\varepsilon\cos\theta)^3}\left(\frac{L^2}{4} - y^2\right)$$ (4.102)

It can be seen that this equation complies implicitly with the Full-Sommerfeld and Half-Sommerfeld boundary conditions since:

$$p = 0 \quad \text{at} \quad \theta = 0, \pi \text{ and } 2\pi$$

· *Load Capacity*

The total load that the bearing will support is found by integrating the pressure around the bearing. In the early literature, the Half-Sommerfeld condition was used for load calculations, i.e., the negative pressures in one half of the bearing were discounted. Load is usually calculated from two components: one acting along the line of shaft and bush centres and a second component perpendicular to the first. This method allows calculation of the angle between the line of centres and the load line. As will be shown later, the shaft does not deflect co-directionally with the load but instead always moves at an angle to the load line. The angle is known as the '**attitude angle**' and results in the position of minimum film thickness lying some distance from where the load line intersects the shaft and bush. The load components and pressure field of a journal bearing are shown in Figure 4.30.

To analyse and derive expressions for the load components '**W$_1$**' and '**W$_2$**' consider a small element of area **Rdθdy** where the '**y**' axis is normal to the plane of the diagram in Figure 4.30 and hence is invisible. The axis is shown in an 'unwrapped film' in Figure 4.31. Hydrodynamic pressure fields of journal bearings are often shown in 'unwrapped form' where the film is shown as a plane surface in plan view. This is equivalent to taking a sheet of paper and rolling it to form a cylinder. In cylindrical form, the true shape of the hydrodynamic film is represented when unwrapped or flat. In this way the form most convenient for film visualization is obtained.

The increment of force exerted by the hydrodynamic pressure on the element of area is **pRdθdy** and this force is resolved into two components:

· **pRcosθdθdy**   acting along the line of shaft and bush centres and

· **pRsinθdθdy**   acting in the direction normal to the line of centres.

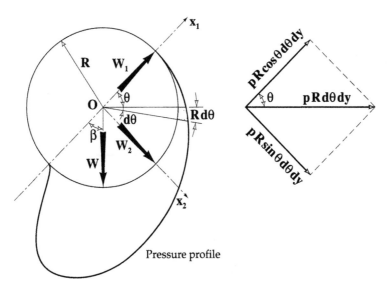

FIGURE 4.30 Load components and pressure field acting in a journal bearing.

Thus the load component acting along the line of centres is expressed by:

$$W_1 = \int_0^\pi \int_{-\frac{L}{2}}^{\frac{L}{2}} pR\cos\theta\, d\theta\, dy \tag{4.103}$$

similarly the component acting in the direction normal to the line of centres is:

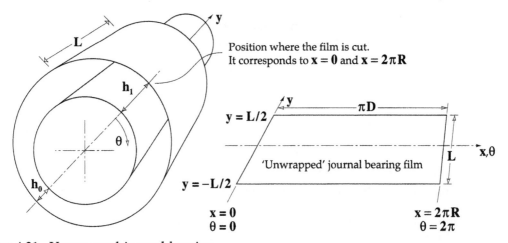

FIGURE 4.31 Unwrapped journal bearing.

$$W_2 = \int_0^\pi \int_{-\frac{L}{2}}^{\frac{L}{2}} pR\sin\theta\, d\theta\, dy \tag{4.104}$$

Substituting for 'p' (4.102) and separating variables gives:

$$W_1 = \int_0^\pi \int_{-\frac{L}{2}}^{\frac{L}{2}} \frac{3U\eta\varepsilon R\sin\theta\cos\theta}{Rc^2(1+\varepsilon\cos\theta)^3}\left(\frac{L^2}{4}-y^2\right)d\theta\,dy = \frac{3U\eta\varepsilon}{c^2}\int_0^\pi \int_{-\frac{L}{2}}^{\frac{L}{2}} \frac{\sin\theta\cos\theta}{(1+\varepsilon\cos\theta)^3}\,d\theta\left(\frac{L^2}{4}-y^2\right)dy$$

$$W_2 = \int_0^\pi \int_{-\frac{L}{2}}^{\frac{L}{2}} \frac{3U\eta\varepsilon R\sin^2\theta}{Rc^2(1+\varepsilon\cos\theta)^3}\left(\frac{L^2}{4}-y^2\right)d\theta\,dy = \frac{3U\eta\varepsilon}{c^2}\int_0^\pi \int_{-\frac{L}{2}}^{\frac{L}{2}} \frac{\sin^2\theta}{(1+\varepsilon\cos\theta)^3}\,d\theta\left(\frac{L^2}{4}-y^2\right)dy$$

The individual integrals can be evaluated separately from each other and they are:

$$\int_0^\pi \frac{\sin\theta\cos\theta}{(1+\varepsilon\cos\theta)^3}\,d\theta = -\frac{2\varepsilon}{(1-\varepsilon^2)^2}$$

$$\int_0^\pi \frac{\sin^2\theta}{(1+\varepsilon\cos\theta)^3}\,d\theta = \frac{\pi}{2(1-\varepsilon^2)^{3/2}}$$

$$\int_{-\frac{L}{2}}^{\frac{L}{2}}\left(\frac{L^2}{4}-y^2\right)dy = \frac{L^3}{6}$$

Substituting yields:

$$W_1 = -\frac{U\eta L^3\varepsilon^2}{c^2(1-\varepsilon^2)^2} \tag{4.105}$$

$$W_2 = \frac{U\eta\varepsilon\pi L^3}{4c^2(1-\varepsilon^2)^{3/2}} \tag{4.106}$$

The total load that the bearing will support is the resultant of the components 'W_1' and 'W_2':

$$W = \sqrt{W_1^{\,2} + W_2^{\,2}} \tag{4.107}$$

Substituting for 'W_1' and 'W_2' gives the expression for the total load that the bearing will support:

$$\boxed{W = \frac{U\eta\varepsilon L^3}{c^2(1-\varepsilon^2)^2}\frac{\pi}{4}\sqrt{\left(\frac{16}{\pi^2}-1\right)\varepsilon^2+1}} \tag{4.108}$$

It can be seen that in a similar fashion to the other bearings analysed, the total load is expressed in terms of the geometrical and operating parameters of the bearing. Equation (4.108) can be rewritten in the form:

$$\frac{Wc^2}{LU\eta R^2}\frac{4R^2}{L^2} = \frac{\pi\varepsilon}{(1-\varepsilon^2)^2}(0.621\varepsilon^2+1)^{0.5} \tag{4.109}$$

Introducing a variable 'Δ':

$$\Delta = \frac{W}{LU\eta}\left(\frac{c}{R}\right)^2$$

(4.110)

which is also known as the 'Sommerfeld Number' or 'Duty Parameter', equation (4.109) becomes:

$$\Delta\left(\frac{D}{L}\right)^2 = \frac{\pi\varepsilon}{(1-\varepsilon^2)^2}(0.621\varepsilon^2+1)^{0.5}$$

(4.111)

where:

D = 2R   is the shaft diameter [m].

The Sommerfeld Number is a very important parameter in bearing design since it expresses the bearing load characteristic as a function of eccentricity ratio. Computed values of Sommerfeld number 'Δ' versus eccentricity ratio 'ε' are shown in Figure 4.32 [3]. The curves were computed using the Reynolds boundary condition which is the more accurate. Data for long journal bearings which cannot be calculated from the above equations are also included. The data are also based on a bearing geometry where 180° of bearing sector on the unloaded side of the bearing has been removed. Removal of the bearing shell at positions where hydrodynamic pressure is negligible is a convenient means of reducing friction and the bearings are known as partial arc bearings. The effect on load capacity is negligible except at extremely small eccentricity ratios. An engineer can find from Figure 4.32 a value of Sommerfeld number for a specific eccentricity and **L/D** ratio and then the bearing and operating parameters can be selected to give an optimum performance. It is usually assumed that the optimum value of eccentricity ratio is close to:

$$\varepsilon_{optimum} = 0.7$$

Higher values of eccentricity ratio are prone to shaft misalignment difficulties; lower values may cause shaft vibration and are associated with higher friction and lubricant temperature.

If the surface speed of the shaft is replaced by the angular velocity of the shaft then the left hand side of the graph shown in Figure 4.32 can be used. When the shaft angular velocity is expressed in revolutions per second [rps] then the modified Sommerfeld parameter becomes **S** = πΔ. Since:

$$U = 2\pi RN$$

substituting into equation (4.110) gives:

$$\Delta = \frac{W}{L\eta 2\pi RN}\left(\frac{c}{R}\right)^2$$

Introducing 'P' [4]:

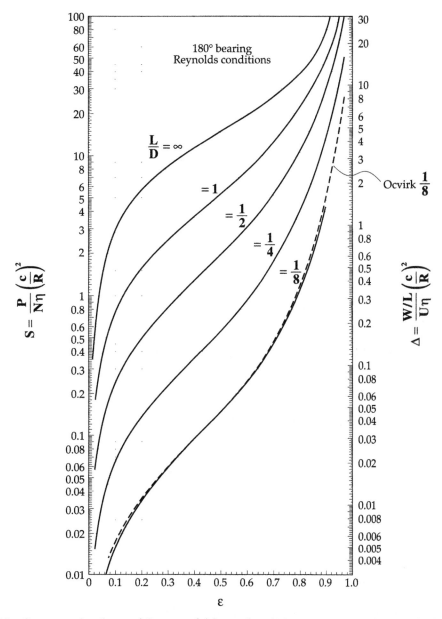

FIGURE 4.32   Computed values of Sommerfeld number 'Δ' versus eccentricity ratio 'ε' [3].

$$P = \frac{W}{2LR}$$

$$\Delta = \frac{P}{N\eta\pi}\left(\frac{c}{R}\right)^2$$

Thus:

$$\boxed{S = \Delta\pi = \frac{P}{N\eta}\left(\frac{c}{R}\right)^2}$$

(4.112)

It can also be seen from Figure 4.30 that the attitude angle 'β' between the load line and the line of centres can be determined directly from the load components 'W₁' and 'W₂' from the following relation:

$$\tan\beta = -\frac{W_2}{W_1}$$

Substituting for 'W₁' and 'W₂' yields:

$$\boxed{\tan\beta = \frac{\pi}{4}\frac{(1-\varepsilon^2)^{1/2}}{\varepsilon}}$$

(4.113)

· *Friction Force*

The friction force can be calculated by integrating the shear stress 'τ' over the bearing area:

$$F = \int_0^L\int_0^B \tau\, dx\, dy = \int_0^L\int_0^B \eta\frac{du}{dz}\, dx\, dy$$

In journal bearings, the bottom surface is stationary whereas the top surface, the shaft, is moving, i.e.:

$$U_1 = U \quad \text{and} \quad U_2 = 0$$

which is the opposite case from linear pad bearings. Thus the velocity equation (4.11) becomes:

$$u = \left(\frac{z^2 - zh}{2\eta}\right)\frac{\partial p}{\partial x} + U\frac{z}{h}$$

Differentiating with respect to 'z' gives the shear rate:

$$\frac{du}{dz} = \left(2z - h\right)\frac{1}{2\eta}\frac{dp}{dx} + \frac{U}{h}$$

After substituting, the expression for friction force is obtained:

$$F = \int_0^L\int_0^B \left[\left(z - \frac{h}{2}\right)\frac{dp}{dx} + \frac{U\eta}{h}\right] dx\, dy$$

(4.114)

In the narrow bearing approximation it is assumed that $\partial p/\partial x \approx 0$ since $\partial p/\partial x \ll \partial p/\partial y$ and (4.114) becomes:

$$F = \int_0^L\int_0^B \frac{U\eta}{h}\, dx\, dy$$

(4.115)

and the friction force on the moving surface, i.e., the shaft, is given by:

$$F = \int_0^B \frac{U\eta L}{h} dx \tag{4.116}$$

Substituting for '**h**' from (4.99) and '**dx = Rdθ**' gives:

$$F = \int_0^\pi \frac{U\eta LR}{c(1 + \varepsilon\cos\theta)} d\theta = \frac{U\eta LR}{c} \int_0^\pi \frac{d\theta}{(1 + \varepsilon\cos\theta)}$$

and integrating yields:

$$\boxed{F = \frac{2\pi\eta ULR}{c} \frac{1}{(1 - \varepsilon^2)^{0.5}}} \tag{4.117}$$

which is the friction in journal bearings at the surface of the shaft for the Half-Sommerfeld condition.

It can be seen from equation (4.117) that when:

· the shaft and bush are concentric then:

$$e = 0 \quad \text{and} \quad \varepsilon = 0$$

and the value of the second term of equation (4.117) becomes unity. The equation now reduces to the first term only. This is known as '**Petroff friction**' since it was first published by Petroff in 1883 [3].

· the shaft and bush are touching then:

$$e = c \quad \text{and} \quad \varepsilon = 1$$

which causes infinite friction according to the model of hydrodynamic lubrication. In practice the friction may not reach infinitely high values if the shaft and bush touch but the friction will be much higher than that typical of hydrodynamic lubrication. It is also true that as the eccentricity ratio approaches unity, the friction coefficient rises. The second term of (4.117) is known as the '**Petroff multiplier**'. Figure 4.33 shows the relationship between the calculated Petroff multiplier and the eccentricity ratio for infinitely long 360° journal bearings [8]. The calculated values are higher than those predicted from $(1 - \varepsilon^2)^{-0.5}$ since the effects of pressure on the shear stress of the lubricant are not included in equation (4.117). The effect of cavitation, i.e., the zero pressure region, does have a significant effect on friction and this together with pressure effects are discussed in the next chapter on 'Computational Hydrodynamics'.

· *Coefficient of Friction*

The coefficient of friction of a bearing is calculated once the load and friction forces are known:

$$\mu = \frac{F}{W}$$

As can be seen from equation (4.108) or from Figure 4.32 the load capacity rises sharply with an increase in eccentricity ratio. Friction force is relatively unaffected by changes in

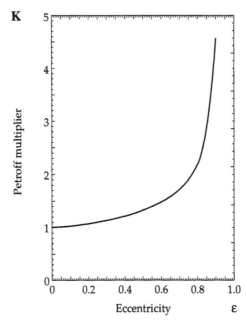

FIGURE 4.33   Relationship between Petroff multiplier and eccentricity ratio for infinitely long 360° bearings [8].

eccentricity ratio until an eccentricity ratio of about **0.8** is reached. Although the operation of bearings at the highest possible levels of Sommerfeld number and eccentricity ratio will allow minimum bearing dimensions and oil consumption, the optimum value of the eccentricity ratio, as already mentioned, is approximately $\varepsilon = 0.7$. Interestingly the optimal ratio of maximum to minimum film thickness for journal bearings is much higher than for pad bearings as is shown below:

at $\theta = 0$ where film thickness is a maximum, $h_1 = c (1 + \varepsilon)$ and

at $\theta = \pi$ where film thickness is a minimum, $h_0 = c (1 - \varepsilon)$

so that the optimal inlet/outlet film thickness ratio for journal bearings is $\dfrac{h_1}{h_0} = \dfrac{1 + \varepsilon}{1 - \varepsilon} = \dfrac{1 + 0.7}{1 - 0.7} = 5.67$. This ratio is higher than for linear pad bearings for which it is equal to **2.2**. There is a noticeable discrepancy in optimum ratios of maximum to minimum film thickness but strictly speaking these two ratios are not comparable. In the case of linear pad bearings classical theory predicts a maximum load capacity while for journal bearings there is no maximum theoretical capacity, instead a limit is imposed by theoretical considerations. When cavitation effects are ignored, the friction coefficient for a bearing with the Half-Sommerfeld condition is:

$$\mu = \frac{8Rc(1 - \varepsilon^2)^{1.5}}{L^2\varepsilon(0.621\varepsilon^2 + 1)^{0.5}} \tag{4.118}$$

It needs to be mentioned that in real bearings contamination of the lubricating oil by particles of, for example, sand or other minerals, may alter the friction coefficient from values calculated using the above formula. It has been shown that particles smaller than the minimum film thickness can affect the friction coefficient by interlocking between particles trapped on opposing surfaces [68]. In dry bearings a large wear particle can also sporadically be present. Such a particle can become trapped between the shaft and journal and the usual consequence of this is rapidly rising friction or seizure of the bearing. A remedy for such problems is to provide recesses for the escape of wear debris by machining grooves on the shaft or bush surface [69].

· *Lubricant Flow Rate*

For narrow bearings, the flow equation (4.18) is simplified since $\partial p/\partial x \approx 0$ and is expressed in the form:

$$q_x = \frac{Uh}{2} \tag{4.119}$$

and the lubricant flow in the bearing is:

$$Q_x = \int_0^L q_x dy = \int_0^L \frac{Uh}{2} \, dy = \frac{UhL}{2}$$

Substituting for '**h**' from (4.99), gives the flow in the bearing:

$$\boxed{Q_x = \frac{UL}{2} c(1 + \varepsilon\cos\theta)} \tag{4.120}$$

In order to prevent the depletion of lubricant inside the bearing, the lubricant lost due to side leakage must be compensated for. The rate of lubricant supply can be calculated by applying the boundary inlet-outlet conditions to equation (4.120). From a diagram of the unwrapped journal bearing film shown in Figure 4.34 it can be seen that the oil flows into the bearing at $\theta = 0$ and $h = h_1$ and out of the bearing at $\theta = \pi$ and $h = h_0$.

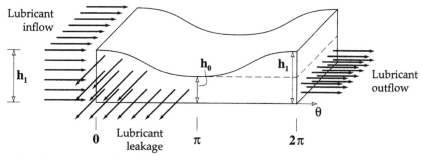

FIGURE 4.34 Unwrapped oil film in a journal bearing.

Substituting the above boundary conditions into (4.120) it is found that the lubricant flow rate into the bearing is:

$$Q_1 = \frac{UL}{2} c(1 + \varepsilon)$$

and the lubricant flow rate out of the bearing is:

$$Q_0 = \frac{UL}{2} c(1 - \varepsilon)$$

The rate at which lubricant is lost due to side leakage is:

$$Q = Q_1 - Q_0$$

and thus:

$$Q = UcL\varepsilon \tag{4.121}$$

Lubricant must be supplied at this rate to the bearing for sustained operation. If this requirement is not met, '**lubricant starvation**' will occur.

For long bearings and eccentricity ratios approaching unity, the effect of hydrodynamic pressure gradients becomes significant and equation (4.121) loses accuracy. Lubricant flow rates for some finite bearings as a function of eccentricity ratio are shown in Figures 4.35 and 4.36 [8]. The data are computed using the Reynolds boundary condition, values for a 360° arc or complete journal bearing are shown in Figure 4.35 and similar data for a 180° arc or partial journal bearing are shown in Figure 4.36.

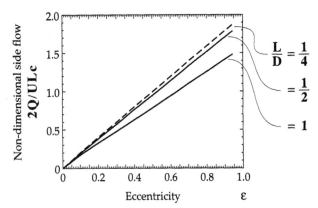

FIGURE 4.35 Lubricant leakage rate versus eccentricity ratio for some finite 360° bearings [8].

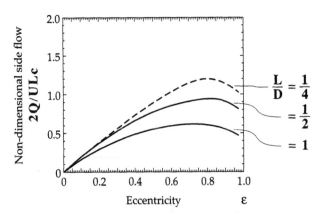

FIGURE 4.36 Lubricant leakage rate versus eccentricity ratio for some finite 180° bearings [8].

Practical and Operational Aspects of Journal Bearings

Journal bearings are commonly incorporated as integral parts of various machinery with a wide range of design requirements. Thus there are some problems associated with practical implementation and operation of journal bearings. For example, in many practical applications the lubricant is fed under pressure into the bearing or there are some critical resonant shaft speeds to be avoided. The shaft is usually misaligned and there are almost always some effects of cavitation for liquid lubricants. Elastic deformation of the bearing will certainly occur but this is usually less significant than for pad bearings. All of these issues will affect the performance of a bearing to some extent and allowance should be made during the design and operation of the bearing. Some of these problems will be addressed in this section and some will be discussed later in the next chapter on 'Computational Hydrodynamics'.

· Lubricant Supply

In almost all bearings, a hole and groove are cut into the bush at a position remote from the point directly beneath the load. Lubricant is then supplied through the hole to be distributed over a large fraction of the bearing length by the groove. Ideally, the groove should be the same length as the bearing but this would cause all the lubricant to leak from the sides of the groove. As a compromise the groove length is usually about half the length of the bearing. Unless the groove and oil hole are deliberately positioned beneath the load there is little effect of groove geometry on load capacity. Circumferential grooves in the middle of the bearing are useful for applications where the load changes direction but have the effect of converting a bearing into two narrow bearings. These grooves are mostly used in crankcase bearings where the load rotates. Typical groove shapes are shown in Figure 4.37. The edges of grooves are usually recessed to prevent debris accumulating.

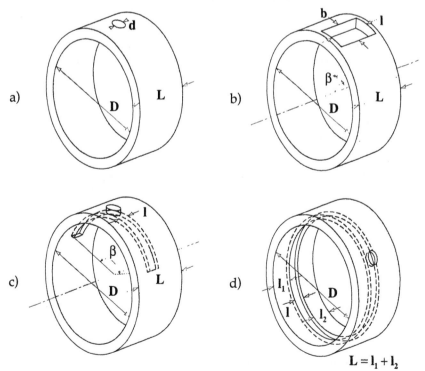

FIGURE 4.37 Typical lubricant supply grooves in journal bearings: a) single hole, b) short angle groove, c) large angle groove and d) circumferential groove (adapted from [19]).

The idealized lubricant supply conditions assumed previously for load capacity analysis do not cause significant error except for certain cases such as the circumferential groove. The calculation of lubricant flow from grooves requires computation for accurate values and is described in the next chapter. Only a simple method of estimating lubricant flow is described in this section. With careful design, grooves and lubricant holes can be more than just a means of lubricant supply but can also be used to manipulate friction levels and bearing stability.

Lubricant can be supplied to the bearing either pressurized or unpressurized. The advantage of unpressurized lubricant supply is that it is simpler, and for many small bearings a can of lubricant positioned above the bearing and connected by a tube is sufficient for several hours operation. The bearing draws in lubricant efficiently and there is no absolute necessity for pressurized supply. Pressurization of lubricant supply does, however, provide certain advantages which are:

· high pressure lubricant can be supplied close to the load line to suppress lubricant heating and viscosity loss. This practice is known as '**cold jacking**',

· for large bearings, pressurized lubricant supply close to the load line prevents shaft to bush contact during starting and stopping. This is a form of hydrostatic lubrication,

· lubricant pressurization can be used to modify vibrational stability of a bearing,

· cavitation can be suppressed if the lubricant is supplied to a cavitated region by a suitably located groove. Alternatively the groove can be enlarged so that almost all of the cavitated region is covered, which prevents cavitation within it.

For design purposes it is necessary to calculate the flow of lubricant through the groove. It is undesirable to try to force the bearing to function on less than the lubricant flow dictated by hydrodynamic lubrication since the bearing can exert a strong suction effect on the lubricant in such circumstances. When the bearing is rotating, the movement of the shaft entrains any available fluid into the clearance space. It is not possible for the bearing to rotate at any significant speed without some flow through the groove or supply hole. If lubricant flow is restricted then suction may cause the lubricant to cavitate in the supply line which causes pockets of air to pass down the supply line and into the bearing or the groove may become partially cavitated. When the latter occurs there is no guarantee that the lubricant flow from the groove will remain stable, and instead lubricant may be released in pulses. In either case, the hydrodynamic lubrication would suffer periodic failure with severe damage to the bearing.

There are two components of total flow 'Q' from a groove or supply hole into a bearing: the net Couette flow 'Q_c' due to the difference in film thickness between the upstream and downstream side of the groove/hole and the imposed flow 'Q_p' from the externally pressurized lubricant, i.e.:

$$Q = Q_c + Q_p$$

An expression for the net Couette flow is:

$$Q_c = 0.5Ul(h_d - h_u) \qquad (4.122)$$

where:

Q_c is the net Couette flow [m³/s];

U is the sliding velocity [m/s];

l is the axial width of the groove/hole [m];

$\mathbf{h_d}$ is the film thickness on the downstream side of the groove/hole [m], as shown in Figure 4.38;

$\mathbf{h_u}$ is either the film thickness on the upstream side of the groove or the film thickness at the position of cavitation if the bearing is cavitated [m], as shown in Figure 4.38.

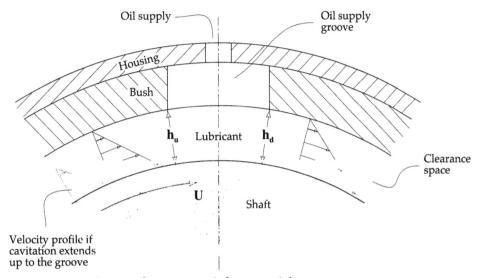

FIGURE 4.38   Couette flow at the entry and the exit of the groove.

Note that '$\mathbf{h_d}$' depends on the position at which the groove is located and can be calculated from the bearing geometry. On the other hand, when cavitation occurs a generous estimate for '$\mathbf{h_u}$' is the minimum film thickness, i.e., $\mathbf{h_u = h_0 = c(1 - \varepsilon)}$. The net Couette flow is the minimum flow of lubricant that should pass through the groove/hole even if the lubricant supply is not pressurized. If this flow is not maintained then the problems of suction and intermittent supply described above will occur.

However, even the net Couette flow may not be sufficient to prevent starvation of lubricant particularly if the groove/hole is small compared to the bearing length. For small grooves/holes and for circumferential grooves, pressurization of lubricant is necessary for correct functioning of the bearing. In fact the Couette flow in bearings with circumferential grooves is equal to zero, i.e., $\mathbf{Q_c = 0}$. The pressurized flow of lubricant from a groove has been summarized in a series of formulae [19]. These formulae supersede earlier estimates of pressurized flow [3] which contain certain inaccuracies. Formulae for pressurized flow from a single circular oil hole, rectangular feed groove (small angular extent), rectangular feed groove (large angular extent) and a circumferential groove are summarized in Table 4.4 [19]. Coefficients '$\mathbf{f_1}$' and '$\mathbf{f_2}$' required or the calculations of lubricant flow from a rectangular groove of large angular extent are determined from the chart shown in Figure 4.39.

The grooves are centred on the load line but positioned at 180° to the point where the load vector intersects the shaft and bush. The transition between 'large angular extent' and 'small angular extent' depends on the **L/D** ratio; e.g., for **L/D = 1, 180°** is the transition point whereas for **L/D ≤ 0.5** the limit is at **270°**. For angular extents greater than **90°** it is recommended, however, that both calculation methods be applied to check accuracy.

It should be noted that the pressurized flow of large angular extent bearings is significantly influenced by eccentricity so that it is necessary to calculate the value of this parameter first.

TABLE 4.4   Formulae for the calculation of lubricant flow through typical grooves (adapted from [19]).

Type of oil feed	Pressurized oil flow
Single circular hole $(d_h < L/2)$	$Q_p = 0.675 \dfrac{p_s h_g^3}{\eta} \left(\dfrac{d_h}{L} + 0.4 \right)^{1.75}$
Single rectangular groove with small angular extent $(\beta < 5°)$	$Q_p = \dfrac{p_s}{\eta} \left[\dfrac{h_g^3}{3} \left(\dfrac{1.25 - 0.25\,(l/L)}{(L/l - 1)^{0.333}} \right) + \dfrac{h_g^3}{3} \left(\dfrac{b/L}{1 - l/L} \right) \right]$
Single rectangular groove with large angular extent $(5° < \beta < 180°)$	$Q_p = \dfrac{c^3 p_s}{\eta} \left[\left(\dfrac{1.25 - 0.25\,(l/L)}{6\,(L/l - 1)^{0.333}}\, f_1 \right) + \left(\dfrac{D/L}{6\,(1 - l/L)}\, f_2 \right) \right]$
Circumferential groove $(360°)$	$Q_p = \dfrac{\pi D c^3 p_s}{3\eta} \dfrac{(1 + 1.5\varepsilon^2)}{(L - l)}$

where:

Q_p is the pressurized lubricant flow from the hole or groove [m³/s];

p_s is the oil supply pressure [Pa];

η is the dynamic viscosity of the lubricant [Pas];

h_g is the film thickness at the position of the groove [m];

c   is the radial clearance [m];

d_h is the diameter of the hole [m];

L   is the axial length of the bearing [m] (In the case of bearings with a circumferential groove it is the sum of two land lengths as shown in Figure 4.37. Note that in this case the bearing is split into two bearings.)

l   is the axial length of the groove [m];

b   is the width of the groove in the sliding direction [m];

D   is the diameter of the bush [m];

ε is the eccentricity ratio;

f_1, f_2 are the coefficients determined from Figure 4.39.

For small grooves/holes, the lubricant supply pressure may be determined from the amount of pressurized flow required to compensate for the difference between Couette flow and the lubricant consumption of full hydrodynamic lubrication. At very low eccentricities some excess flow may be required to induce replenishment of lubricant since the hydrodynamic lubricant flow rate declines to zero with decreasing eccentricity. If this precaution is not applied, progressive overheating of the lubricant and loss of viscosity may result particularly as low eccentricity is characteristic of high bearing speed, e.g., 10,000 [rpm] [20].

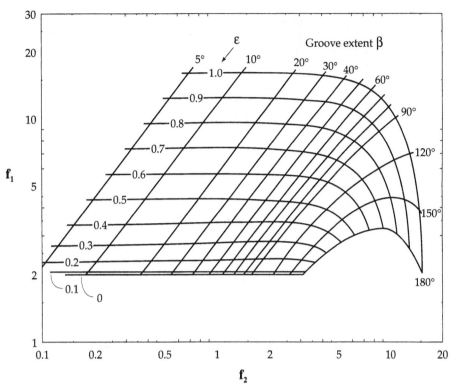

FIGURE 4.39   Parameters for calculation of pressurized oil flow from grooves (adapted from [19]).

· *Cavitation*

As discussed already, large negative pressures in the hydrodynamic film are predicted when surfaces move apart or mutually sliding surfaces move in a divergent direction. For gases, a negative pressure does not exist and for most liquids a phenomenon known as cavitation occurs when the pressure falls below atmospheric pressure. The reason for this is that most liquids contain dissolved air and minute dirt particles. When the pressure becomes sub-atmospheric, bubbles of previously dissolved air nucleate on pits, cracks and other surface irregularities on the sliding surfaces and also on dirt particles. It has been shown that very clean fluids containing a minimum of dissolved gas can support negative pressures but this has limited relevance to lubricants which are usually rich in wear particles and are regularly aerated by churning. If there is a significant drop in pressure, the operating temperature can be sufficient for the lubricant to evaporate. The lubricant vapour accumulates in the bubbles and their sudden collapse is the cause of most cavitation damage. The critical difference between 'gaseous cavitation', i.e., cavitation involving bubbles of dissolved air, and 'vaporous cavitation' is that with the latter, sudden bubble collapse is possible. When a bubble collapses against a solid surface very high stresses, reaching 0.5 [GPa] in some cases, are generated and this will usually cause wear. Wear caused by vaporous cavitation progressively damages the bearing until it ceases to function effectively. The risk of vaporous cavitation occurring increases with elevation of bearing speeds and loads [21]. Cavitation in bearings is also referred to as 'film rupture' but this term is old fashion and is usually avoided.

Cavitation occurs in liquid lubricated journal bearings, in elastohydrodynamics and in applications other than bearings such as propeller blades. In journal bearings, cavitation causes a series of 'streamers' to form in the film space. The lubricant feed pressure has some ability to reduce the cavitation in the area adjacent to the groove [22], as shown in Figure 4.40.

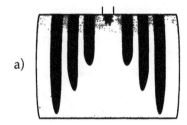
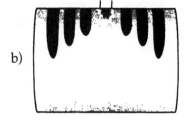

FIGURE 4.40 Cavitation in a journal bearing: a) oil fed under low pressure and b) oil fed under high pressure (adapted from [22]).

Large lubricant supply grooves can be used to suppress negative hydrodynamic film pressures and so prevent cavitation. This practice is similar to using partial arc bearings and has the disadvantage of raising the lubricant flow rate and the precise location of the cavitation front varies with eccentricity. This means that cavitation might only be prevented for a restricted range of loads and speeds. In practice it is very difficult to avoid cavitation completely with the conventional journal bearing.

· *Journal Bearings With Movable Pads*

Multi-lobe bearings consist of a series of Michell pads arranged around a shaft as a substitute for a journal bearing. Figure 4.41 shows a schematic illustration of multi-lobe bearings incorporating pivoted pads and self-aligning pads.

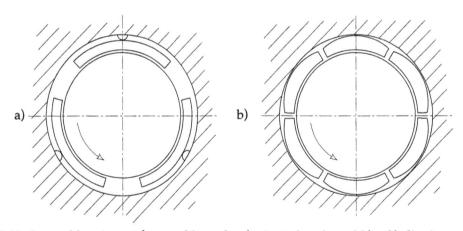

FIGURE 4.41 Journal bearing with movable pads: a) pivoted pads and b) self-aligning pads.

The number of pads can be varied from two to almost any number, but in practice, two, three or four pads are usually chosen for pivoted pad designs [23]. The pads can also be fitted with curved backs to form self-aligning pads which eliminates the need for pivots. The rolling pads are simpler to manufacture than pivoted pads and do not suffer from wear of the pivots. The reduction in the number of parts allows a larger number of pads to be used with the self-aligning pad design and bearings with up to six pads have been manufactured [24].

The adoption of pads ensures that all hydrodynamic pressure generation occurs between surfaces that are converging in the direction of sliding motion. This practice ensures the prevention of cavitation and associated problems. There is a further advantage discussed in more detail later and this is a greater vibrational stability. The method of analysis of this bearing type is described elsewhere [23,24] and is not fundamentally different from the treatment of Michell pads already presented.

· *Journal Bearings Incorporating a Rayleigh Step*

The Rayleigh step is used to advantage in journal bearings as well as in pad bearings. As with the spiral groove thrust bearing, a series of Rayleigh steps are used to form a 'grooved bearing'. A bearing design incorporating helical grooves terminating against a flat surface was introduced by Whipple [3,25]. This design is known as the 'viscosity plate'. An alternative design where two series of helical grooves of opposing helix face each other is also used in practical applications and is known as the 'herring bone' bearing. The herring bone and viscosity plate bearings are illustrated in Figure 4.42. The analysis of these bearings, also known as 'spiral groove' bearings, is described in detail elsewhere [12].

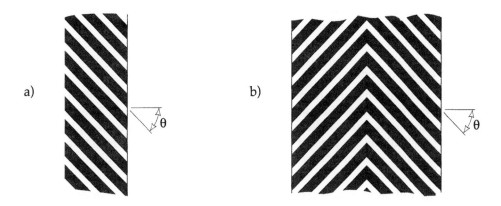

FIGURE 4.42 Examples of grooved bearings: a) viscosity plate bearing and b) herring bone bearing (adapted from [4]).

This type of bearing is suitable for use as a gas-lubricated journal bearing operating at high speed. The grooves can be formed by the sand blasting method which avoids complicated machining of the helical grooves. A 9 [mm] journal diameter bearing was tested to 350,000 [rpm] [26]. The bearing functioned satisfactorily provided that the expansion of the shaft by centrifugal stress and thermal expansion was closely controlled. In the design of these bearings the accurate assessment of the deformation of the bearing is critical and unless it is precisely calculated, by e.g. the finite element method, it is possible for bearing clearances during operation to become so small that contact between the shaft and the bush may occur.

· *Oil Whirl or Lubricant Caused Vibration*

Oil whirl is the colloquial term describing hydrodynamically induced vibration of a journal bearing. This can cause serious problems in the operation of journal bearings and must be considered during the design process. Oil whirl is characterized by severe vibration of the shaft which occurs at a specific speed. There is also another form of bearing vibration known as 'shaft whip' which is caused by the combined action of shaft flexibility and bearing vibration characteristics. Although it may appear unlikely that a liquid such as oil would cause vibration, according to the hydrodynamic theory discussed previously, a change in load on the bearing is always accompanied by a finite displacement. This constitutes a form of mechanical stiffness or spring constant and when combined with the mass of the shaft, vibration is the natural result. A rotating shaft nearly always provides sufficient exciting force due to small imbalance forces. For engineering analysis it is essential to know the critical speed at which oil whirl occurs and avoid it during operation. It has been found that severe whirl occurs when the shaft speed is approximately twice the bearing critical frequency. The question is, what is this critical frequency and how can it be estimated? The answer to this question and most bearing vibration problems is found by numerical analysis.

A complete analysis of bearing vibration is very complex as non-linear stiffness and damping coefficients are involved. Two types of analysis are currently employed. The first provides a means of determining whether unstable vibration will occur and is based on linearized stiffness and damping coefficients. These coefficients are accurate for small stable vibrations and a critical shaft speed is found by this method. A full discussion of the linearized method is given in the chapter on 'Computational Hydrodynamics' as computation of the stiffness and damping coefficients is required. The second method provides an exact analysis of bearing motion under specific levels of load, speed and vibrating mass. Exact non-linear coefficients of stiffness and damping are computed and applied to an equation of motion for the shaft to find the shaft acceleration. A notional small exciting displacement is applied to the shaft and the subsequent motion of the shaft is then traced by a Runge-Kutta or similar stepwise progression technique using the acceleration as original data [3]. A hammer blow on the shaft or bearing is a close physical equivalent of the initial displacement. The motion of the shaft centre is known as the shaft trajectory or orbit. Figure 4.43 shows an example of a computed shaft centre trajectory.

The data are in non-dimensional form so that the maximum range of shaft movement is equal to **1** which corresponds to the radial clearance in real dimensions. The circle defines the limit of possible shaft movement without contacting the bush. It can be seen from Figure 4.43 that when stable oscillations are present the shaft centre rapidly converges to a fixed position, whereas when unstable oscillations occur the shaft centre remains mobile for an indefinite period.

The purpose of the full analysis of shaft motion is to check whether the shaft merely wanders around the bush centre without approaching the bush too closely. If the vibration is unstable then a very large spiral trajectory results. This in practice leads to bearing failure because the very small clearances between shaft and bush at the extremes of vibration amplitude cannot be maintained and would lead to shaft/bush contact. In many cases, however, it is found that contact between shaft and bush does not occur despite indications of unstable vibration from the linearized method. The reason for this is the large change in stiffness and damping coefficients as the shaft moves from the equilibrium load position.

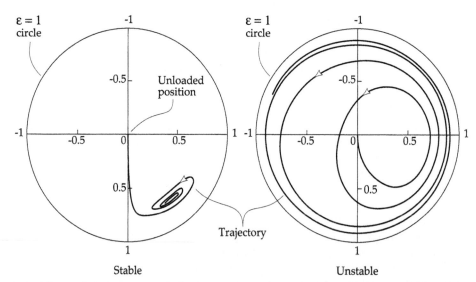

FIGURE 4.43 Example of computed shaft trajectories in journal bearings; stable condition, i.e., declining spiral trajectory, and unstable condition, i.e., self-propagating spiral trajectory (adapted from [51]).

Vibrational data are often collated into a stability diagram which shows the transition between stable and unstable vibration as a function of eccentricity ratio and the load parameter which is defined as:

$$P = \frac{2F}{Mc\omega^2}$$

(4.123)

where:

P	is the stability parameter;
F	is the static load on the bearing [N];
M	is the vibrating mass [kg];
c	is the radial clearance [m];
ω	is the angular velocity of the bearing [rad/s].

The vibrating mass is the mass of the shaft and connected rotating mass, e.g., a turbine rotor. The factor of two in the definition of '**P**' arises from the need for two bearings to support one vibrating mass.

A stability diagram is illustrated schematically in Figure 4.44 as a graph of the transitional value of '**P**' separating stability from instability as a function of eccentricity.

Transition values of '**P**' are also included for various sizes of grooves where size is defined by the subtended angle of the groove. The groove geometry consists of two grooves positioned at 90° to the load line. It can be seen that for large eccentricities, i.e., ε > 0.8, the bearing is stable at all levels of load and exciting mass. For all other values of eccentricity, unstable vibration is likely to occur when **P < 0.2**. Despite many studies of bearing geometry to optimize vibration stability this value does not appear to decline much below **0.2** for bearings with monolithic bushes and may be used as an estimate of stability. Multipad journal bearings have much better resistance to vibration because of the intrinsic stability of the Michell pad [23,24].

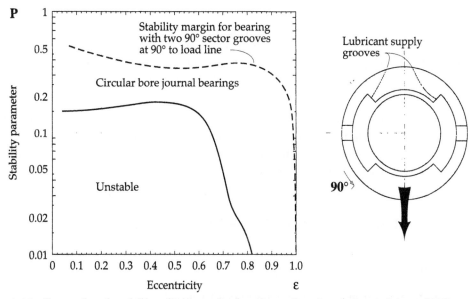

FIGURE 4.44 Example of stability diagram for bearing vibration (adapted from [27]).

Factors such as grooves, misalignment and elastic deformation have a strong (usually negative) influence on vibrational stability and are the subject of continuing study [26,27]. Large angular extent grooves, e.g., 90° extent, are particularly deleterious to stability. An accepted solution of bearing vibration problems is to apply specially designed bearings with an anti-whirl configuration. The basic principle in these designs is to destroy the symmetry of a plain journal bearing which encourages vibration. Although many anti-whirl configurations have already been patented, no solution has yet been found that completely eliminates oil whirl. A recently developed solution is to apply multi-lobed bearings. Some of the typical anti-whirl geometries of plain journal bearings are shown in Figure 4.45.

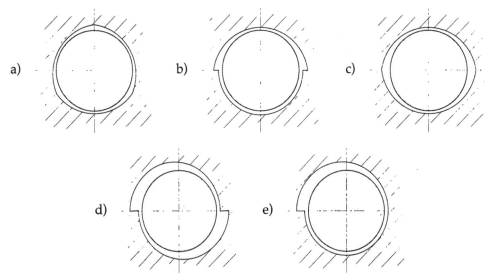

FIGURE 4.45 Typical anti-whirl bearing geometries: a) three-lobed, b) half-lemon, c) lemon, d) displaced and e) spiral (adapted from [4]).

· *Rotating Load*

In the analysis presented so far, only steady loads, acting in a fixed direction, have been considered. There are, however, many practical engineering applications where the load rotates around the bearing. A prime example of this can be found in the internal combustion engine where the load vector rotates in tandem with the working cycle. The issue is, what effect will this have on bearing performance?

Consider that the load rotates around the bearing with some angular velocity 'ω_L' and the shaft rotates with an angular velocity 'ω_S'. To visualize the effect of the load vector movement, it is helpful to consider velocities relative to the load vector, i.e., add '$-\omega_L$' to the shaft and bush velocities as shown in Figure 4.46.

The effective surface velocity 'U' can be determined by inspecting Figure 4.46, i.e.:

$$\mathbf{U} = \mathbf{U_1} + \mathbf{U_2} = \mathbf{R}(\omega_S - \omega_L + (-\omega_L)) = \mathbf{R}(\omega_S - 2\omega_L)$$

where:

$\mathbf{R}$ is the radius of the shaft [m];

ω_L is the angular velocity of the load vector [rad/s];

ω_S is the angular velocity of the shaft [rad/s].

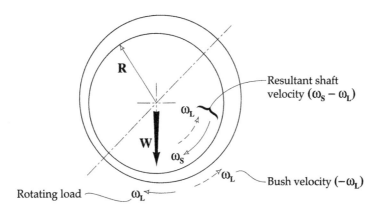

FIGURE 4.46   Angular velocities in a bearing with a rotating load (adapted from [4]).

It is evident from the above relationship that when the surface velocity 'U' is equal to zero then:

$$\omega_L = 0.5\omega_S$$

This relationship gives the condition which should be avoided when operating bearings with a rotating load. If the angular velocity of the rotating load is half the angular velocity of the rotating shaft then the total surface velocity is zero. When this occurs, wedge-type hydrodynamic lubrication ceases and only squeeze film hydrodynamic lubrication is viable. Squeeze film lubrication offers only temporary protection so that only short periods of load vector rotating at half the shaft speed can be tolerated. Failure to observe this rule may cause bearing seizure.

The load capacity of a journal bearing subjected to a rotating load is conveniently summarized as a plot of the ratio of rotating and static load capacities versus ratio of load and shaft angular velocities. A simplified version of the graph originally derived by Burwell [29] is shown in Figure 4.47.

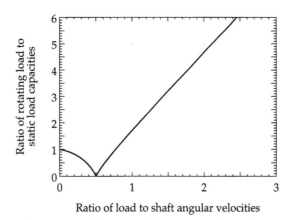

FIGURE 4.47   Relative load capacity of a journal bearing subjected to rotating loads [29].

It can be seen that at low angular velocities of the load, rotation has a detrimental effect on load capacity. There is zero load capacity when the load angular velocity is half the shaft angular velocity. This characteristic of load capacity corresponds to the model of rotating load described above. Load capacity rapidly recovers when half shaft-speed is exceeded so that at an angular velocity ratio of '**1**', the rotating load capacity is greater than the static load capacity. The angular velocity ratio of '**1**' corresponds to forces produced by shaft imbalance so it can be concluded that imbalance forces are relatively unlikely to cause bearing failure.

· *Tilted Shafts*

In practical applications, shafts are not usually aligned parallel to the bearing axis. Even if the shaft is accurately aligned during assembly, the load on the shaft causes bending and tilting of the shaft in a bearing. The critical minimum film thickness will occur at the edge of the bearing, as shown in Figure 4.48.

The critical film thickness for tilted shafts will in general be considerably less than for parallel shafts. The basic parameter to describe the tilt of the shaft is the tilt ratio which is defined as:

$$t = \frac{m}{c}$$

where:

t	is the tilt ratio or non-dimensional tilt;
m	is the distance between the axes of the tilted and non-tilted shaft measured at the edges of the bearing [m];
c	is the radial clearance [m].

To calculate the minimum film thickness, the loss in film thickness due to misalignment is added to the eccentricity. Assuming that minimum film thickness occurs along the load line:

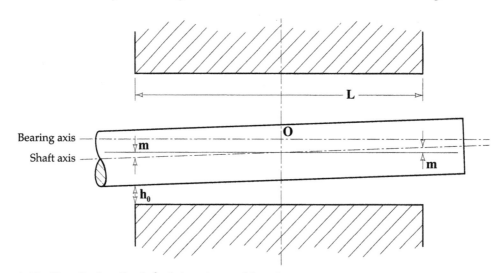

FIGURE 4.48   Detail of a tilted shaft in a journal bearing.

$$h_0 = c(1 - \varepsilon\cos\beta) - m$$

where:

β   is the attitude angle.

In most cases of heavily loaded shafts, the attitude angle is small and its cosine can be approximated by unity.

To calculate the effect of misalignment on bearing geometry, the Reynolds equation is applied to the journal bearing with a film geometry modified by misalignment. The main effect of shaft tilting is to shift the point of support (centre of hydrodynamic pressure) towards the minimum film thickness, which increases the maximum hydrodynamic pressure and affects the stability threshold of bearing vibration [30,31]. Values of maximum hydrodynamic pressure and stability threshold can be calculated for specified amounts of misalignment by applying the computer programs described in the next chapter on 'Computational Hydrodynamics'.

· *Partial Bearings*

In real bearings, it can be advantageous for the bush not to encircle the shaft completely. If the load is acting in an approximately constant direction, then only part of a bearing arc is often employed. The most common bearings of this type are 180° arc bearings, although narrower arcs are also in use. The main advantage of partial bearings is that they have a lower viscous drag and hence lower frictional power losses. Cavitation is also suppressed. Partial arc bearings can be analysed by the same Reynolds equation and film geometry as full journal bearings, the only difference lying in the entry and exit boundary conditions. In the full 360° bearing the entry condition is:

$$\mathbf{h}_1 = \mathbf{c}(1 + \varepsilon) \quad \text{at} \quad \theta = 0$$

whereas in the partial bearing:

$$\mathbf{h}_1{}' = \mathbf{c}(1 + \varepsilon\cos\theta) \quad \text{at} \quad \theta = \theta_1$$

as shown in Figure 4.49.

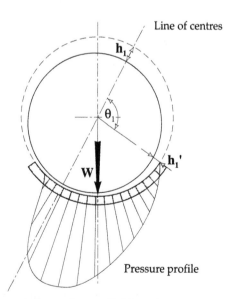

FIGURE 4.49 Schematic representation of a partial bearing.

The practical analysis of such bearings is discussed in the next chapter. Some results for the numerical solutions of various arcs are shown elsewhere [3,32]. The effect of arc on load capacity is very small unless eccentricities as low as **0.3** are considered and very narrow arcs such as 90° are chosen. In these circumstances, load capacity can be less than half that of the equivalent 360° arc bearing.

· *Elastic Deformation of the Bearing*

The interacting surfaces of the bearing and the shaft will deform elastically under load. It is very difficult to prevent elastic deformation and the hydrodynamic pressure field is inevitably affected by the imposed changes in film geometry. The first recorded example of the modification of hydrodynamic pressure by elastic deformation was provided (unknowingly) by Beauchamp Tower [1] with his pressure profile measured from an actual bearing. Reynolds cited Tower's experimental data as evidence in support of a model of hydrodynamic lubrication between perfectly rigid surfaces. Almost a century later, however, it was found that Tower's pressure profile corresponded to that expected from a deformed bearing [33]. The effect of deformation was to bend the bearing shell, resulting in a relatively flat pressure profile which declined sharply at the edges of the bearing. The pressure profile and film geometry are illustrated schematically in Figure 4.50.

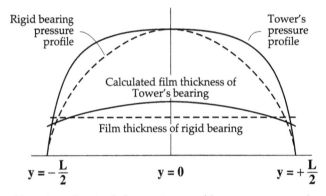

FIGURE 4.50   Effect of bearing elastic deformation on film geometry and pressure profile.

Distortion of the film geometry by elastic deformation becomes more significant with increasing size of bearings. Elastic deformation of the surfaces affects the lubricant film geometry which, in turn, influences all the other bearing parameters such as pressure distribution, load capacity, friction losses and lubricant flow rate. The effect of elastic deformation on the hydrodynamic pressure field is to reduce the peak pressure and generate a more widely distributed pressure profile. Elastic deformation can also improve the vibrational stability of a bearing [34] so that there is no particular need to minimize deformation during the design of a bearing. To calculate load capacity and the other parameters for a deformable bearing requires computation since simultaneous solution of the Reynolds equation and elastic deformation equations is required. A simple example of such an analysis (for a Michell pad) is provided in the next chapter on 'Computational Hydrodynamics'.

· *Infinitely Long Approximation in Journal Bearings*

In the analysis presented so far, it has been assumed that a bearing is '**narrow**' or $\partial p/\partial y \gg \partial p/\partial x$. It is possible to assume the contrary and analyse an '**infinitely long bearing**' where $\partial p/\partial y \ll \partial p/\partial x$. The application of the infinite length or 'long approximation' to the analysis of journal bearings requires more complicated mathematics than the narrow approximation. The values of load capacity provided by this analysis are only applicable to bearings with **L/D > 3**. For any bearings narrower than this, unrealistically high predictions for the load capacity of the bearing are obtained. The '**infinitely long approximation**' is therefore of limited practical value since bearings as long as **L/D > 3** are prone to misalignment. For interested readers, the analysis of an infinitely long journal bearing is given elsewhere [3,4].

4.6 THERMAL EFFECTS IN BEARINGS

It has been assumed so far that the lubricant viscosity remains constant throughout the hydrodynamic film. This is a crude approximation which allowed the derivation and algebraic solution of the Reynolds equation. In practice, the bearing temperature is raised by frictional heat and the lubricant viscosity varies accordingly. As illustrated in Chapter 2, a temperature rise as small as 25°C can cause the lubricant viscosity to collapse to 20% of its original value. The direct effects of heat in terms of lubricant hydrodynamic pressure, load capacity, friction and power losses can readily be imagined. More pernicious still are the indirect effects of thermal distortion on the bearing geometry which can distort a film profile from the intended optimum to something far less satisfactory. Most bearing materials also have a maximum temperature limit for safe operation. This maximum temperature must be allowed for in design calculations. When all these factors are taken into consideration, it becomes clear that thermal effects play a major role in bearing operation and cannot be ignored.

In general there are two approaches to the problem:

·　isoviscous method with 'effective viscosity',

·　rigorous analysis with a locally varying viscosity in the lubricant film.

As is usually the case, one method (the former) is relatively simple but inaccurate while the other is more accurate but complicated to apply. In fact, the analysis with locally varying viscosity has only recently become available, while the 'effective viscosity' methods have persisted for decades.

Before introducing the analysis of thermally modified hydrodynamics and thermal effects in bearings, the fundamental heat transfer mechanisms are discussed.

Heat Transfer Mechanisms in Bearings

Heat in bearings is generated by viscous shearing in the lubricant and is released from the bearings by either conduction from the lubricant to the surrounding structure or convection. These two mechanisms may act simultaneously or one mechanism may be dominant. To demonstrate the mechanism of heat transfer, consider the simplest possible film geometry, i.e., two parallel surfaces, as shown in Figure 4.51.

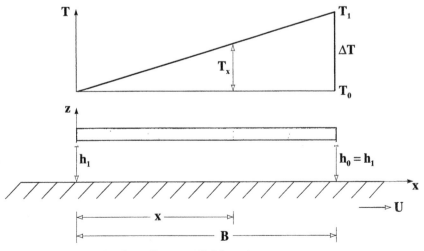

FIGURE 4.51　Temperature rise in a flat parallel bearing.

It is assumed that the temperature rises linearly across the film from zero to 'ΔT' at the exit, so at any point 'x', the surface temperature 'T_x' is:

$$T_x = \frac{\Delta T}{B} \, x$$

The temperature gradient across the film is assumed to be linear (which is not always so in real bearings) and is:

$$\frac{T_x}{h} = \frac{\Delta T}{B} \frac{x}{h}$$

· *Conduction*

According to the principles of heat transfer and thermodynamics, the conduction of heat is calculated from the integration of temperature gradient over the specific bearing geometry, i.e.:

$$H_{cond} = \int_0^B K \frac{\Delta T}{B} \frac{x}{h} \, dx$$

where:

H_{cond}	is the conducted heat per unit length [W/m];
K	is the thermal conductivity of the oil [W/mK];
ΔT	is the temperature rise [K];
B	is the width of the bearing [m];
h	is the hydrodynamic film thickness [m].

The ratio of film thickness to bearing dimensions in almost all bearings is such that conduction in the plane of the lubricant film is of negligible significance.

Since the surfaces are parallel, $h \neq f(x)$ and integrating gives:

$$\boxed{H_{cond} = \Delta T \, \frac{KB}{2h}} \tag{4.124}$$

· *Convection*

The heat removed by the lubricant flow can be calculated from the continuity condition:

$$H_{conv} = \text{mass flow} \times \text{specific heat} \times \text{average temperature rise}$$

Since the surfaces are parallel the pressure gradient $\partial p/\partial x = 0$ and the flow rate along the 'x' axis is (eq. 4.18):

$$q_x = \frac{Uh}{2}$$

Multiplying this term by the lubricant density 'ρ' gives the mass flow per unit length. The average temperature rise of the lubricant is '$\Delta T/2$' since it is assumed that the temperature increases linearly from entry to exit of the bearing. The convected heat is calculated from:

$$H_{conv} = \frac{Uh\rho}{2}\,\sigma\,\frac{\Delta T}{2}$$ (4.125)

where:

H_{conv} is the convected heat per unit length [W/m];

σ is the specific heat of the lubricant [J/kgK];

ρ is the density of the lubricant [kg/m³];

U is the surface velocity [m/s].

· *Conducted/Convected Heat Ratio*

From the above equations the ratio of conducted to convected heat can be calculated to determine which of the two mechanisms of heat removal from the bearing is the more significant. Combining (4.124) and (4.125) the ratio is given by:

$$\frac{H_{cond}}{H_{conv}} = \left(\frac{K}{\rho\sigma}\right)\frac{2B}{Uh^2}$$ (4.126)

where:

$K/\rho\sigma = \chi$ is the thermal diffusivity of the fluid [m²/s].

Typical values of density, specific heat, thermal conductivity and thermal diffusivity are shown in Table 2.7, Chapter 2.

EXAMPLE

Find the ratio of conducted to convected heat in a journal bearing of diameter D = 0.1 [m] and length L = 0.157 [m] which operates at 3000 [rpm]. The hydrodynamic film thickness is h = 0.0001 [m]. The bearing is lubricated by mineral oil of thermal diffusivity $\chi = 0.084 \times 10^{-6}$ [m²/s]. Since D = 0.1 [m], B = πD/2 = 0.157 [m] thus

$$\frac{H_{cond}}{H_{conv}} = 0.084 \times 10^{-6}\,\frac{2 \times 0.157}{3000\,(\pi \times 0.1/60)\,0.0001^2} = 0.168$$

It is clear from the above example that for journal bearings the amount of convected heat is quite significant. When the film thickness of lubricant is much smaller or a highly conductive fluid such as mercury is used as a lubricant, then the ratio of conducted to convected heat will be different. Hydrodynamic lubrication at much thinner film thicknesses is discussed in the chapter on 'Elastohydrodynamic Lubrication'. The significance of the ratio of conducted to convected heat for hydrodynamic bearings is that convection must be included in the equations of heat transfer in a hydrodynamic film. This condition renders the numerical analysis of the heat transfer equations much more complicated than would otherwise be the case, as is demonstrated in the next chapter.

Another ramification of the above result is that conductive heat transfer is still significant although it is often the smaller component of overall heat transfer. Most hydrodynamic bearings operate under a condition between adiabatic and isothermal heat transfer. Adiabatic heat transfer can be modelled by a perfectly insulating shaft and bush. In this case, all the heat is transferred by the lubricant as convection. Isothermal heat transfer represents a bearing made of perfectly conductive material which maximizes heat transfer by conduction in the lubricant. The adiabatic model gives the lowest load capacity since the highest possible lubricant temperatures are predicted. The isothermal model conversely predicts the maximum attainable load capacity. Combined solution of the two models provides valid upper and lower limits of load capacity. If a more accurate estimate of load capacity is required then it is necessary to estimate heat transfer coefficients to the surrounding bearing structure. This is a very complex task and is still under investigation [35]. This topic is discussed further in the chapter on 'Computational Hydrodynamics'.

As is probably very clear here, exact analysis of thermal effects in bearings is a demanding task and most designers of bearings have used the 'effective viscosity' methods even though they are at least partly based on supposition.

Isoviscous Thermal Analysis of Bearings

A simple method of estimating the loss in load capacity due to frictional heat dissipation is to adopt an isoviscous model. It is assumed that the lubricant viscosity is lowered by frictional heating to a uniform value over the whole film. Viscosity may vary with time during operation of the bearing but its value remains uniform throughout the lubricating film. An 'effective temperature' is introduced with a corresponding 'effective viscosity' which is used to calculate load capacity. Two methods are available to find the effective temperature and viscosity, an 'iterative method' which requires computation and a 'constant flow method' which can be executed on a pocket calculator. These are discussed below.

· Iterative Method

The iterative method is effective and accurate in finding the value of effective viscosity. The standard procedure is conducted in the following stages:

· An effective bearing temperature is initially assumed for the purposes of iteration. The assumed value must lie between the inlet temperature 'T_{inlet}' and the maximum temperature 'T_{max}' of the bearing material, i.e.:

$$T_{inlet} < T_{eff,s1} < T_{max}$$

where:

$T_{eff,s1}$ is the effective temperature at the start 's' of iteration, first cycle [°C].

The maximum temperature is usually set by the manufacturer and for most bearing materials 'T_{max}' is about 120 [°C]. For computing purposes, the initial value for '$T_{eff,s1}$' is usually assumed as equal to the inlet temperature.

· the corresponding effective viscosity '$\eta_{eff,s1}$' is found from '$T_{eff,s1}$' using the ASTM viscosity chart or applying the appropriate viscosity-temperature law, e.g., Vogel equation.

· for a given film geometry and effective viscosity, bearing parameters such as friction force 'F' and lubricant flow rate 'Q' can now be calculated.

· the values of 'F' and 'Q' are used to calculate the new effective temperature. This will be different from the previous value unless they happen to coincide. The effective temperature is calculated from [4]:

$$T_{eff} = T_{inlet} + k\Delta T \qquad (4.127)$$

where:

k   is an empirical constant with a value of **0.8** giving good agreement between theory and experiment [4];

ΔT is the frictionally induced temperature rise dependent on '**F**' and '**Q**' [°C].

The frictionally induced temperature rise is found from the following argument. The heat generated in the bearing is:

$$\mathbf{H = FU}$$

At equilibrium, the heat generated by friction balances the heat removed by convection assuming an adiabatic bearing, thus:

$$FU = Q\rho\sigma\frac{\Delta T}{2}$$

Rearranging this gives:

$$\Delta T = \frac{2FU}{Q\rho\sigma}$$

Substituting into (4.127) yields:

$$T_{eff} = T_{inlet} + \frac{1.6FU}{Q\rho\sigma} \qquad (4.128)$$

A new effective temperature is then calculated from (4.128).

·   for the new effective temperature '$T_{eff,n1}$' ('**n**' denotes new) a corresponding effective viscosity is then found from, for example, the ASTM chart or Vogel equation.

·   if the difference between the new effective viscosity and the former effective viscosity is less than a prescribed limit, then the iteration is terminated. If the difference in viscosities is still too large the new viscosity value '$\eta_{eff,s2}$' is assumed and the procedure is continued until the required convergence is achieved. A relaxation factor is usually incorporated at this stage (see program listed in Appendix) to prevent unstable iteration.

The iteration procedure is summarized in a flowchart shown in Figure 4.52, and a computer program '**SIMPLE**' written in Matlab to perform this analysis for narrow journal bearings is listed in the Appendix.

· *Constant Flow Method*

The constant flow method is simpler than the formal iterative method and does not require a computer. In journal bearings, lubricant flow remains approximately constant between eccentricity ratios **0.6 < ε < 0.95** as can be seen in Figure 4.36. It was found experimentally that in this range the lubricant flow can be approximated by the formula [3,4]:

$$Q = 0.3\left(2 - \frac{L}{D}\right)UcL$$

The friction is approximated by the 'Petroff friction' (eq. 4.117):

$$F = \frac{2\pi\eta ULR}{c}$$

Substituting into (4.128) yields:

$$T_{eff} = T_{inlet} + 10.67\frac{\pi\eta UR}{c^2\rho\sigma(2 - L/D)} \tag{4.129}$$

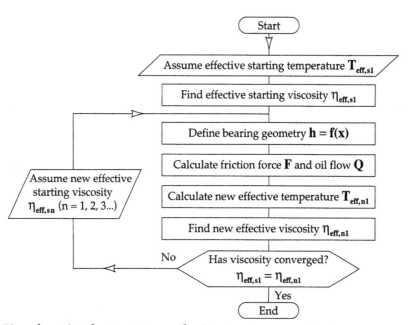

FIGURE 4.52 Flowchart for the iterative method in isoviscous analysis.

The solution can now be obtained by applying only one equation which is easily programmed into a pocket calculator. As with the iterative method it is necessary to perform some iteration since an initial value of effective temperature '$T_{eff,s1}$' has to be assumed. The corresponding viscosity '$\eta_{eff,s1}$' is found from the ASTM chart or Vogel equation as described previously. It is important to choose a sensible value of guessed initial effective temperature, i.e., between the inlet temperature and bearing material limit. This value is then substituted into (4.129) which in turn produces a new value of the effective temperature '$T_{eff,n1}$'. A new corresponding viscosity '$\eta_{eff,n1}$' is then found and (4.129) re-applied. This procedure is repeated until satisfactory convergence is obtained.

Non-Isoviscous Thermal Analysis of Bearings With Locally Varying Viscosity

The assumption that lubricant viscosity is uniform across the film is in fact erroneous and inevitably causes inaccuracy. This error can be quite large especially when the lubricant viscosity has been severely reduced by frictional heating. Lubricant viscosity varies with position in the film, both in the plane of sliding and normal to the direction of sliding. An example of computed film temperatures inside a journal bearing operating at an eccentricity ratio of **0.8** and **L/D = 1** is shown in Figure 4.53. The figure shows a temperature field through a radial section parallel to the load line and an axial section through the midplane of the bearing. The temperature distribution was calculated for a bearing speed of 2,000 [rpm], lubricant inlet temperature of 33°C and ambient temperature of 23°C [36].

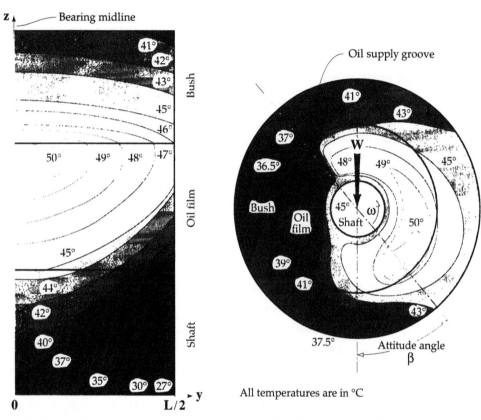

FIGURE 4.53 Example of computed temperature distribution in a hydrodynamic bearing (adapted from [36]).

It can be seen from Figure 4.53 that the hottest part of the lubricant film is at the centre of the bearing close to the minimum film thickness. The temperature at this location is 50°C which is 27°C higher than ambient temperature and causes a large viscosity loss in most lubricating oils. On the other hand, the shaft tends to a uniform temperature with angular position because it is rotating at a high speed. Bearing temperatures, even at the coolest point, are higher than the lubricant inlet temperature. Frictional heat accumulates in the bearing and there is a large temperature rise from the initial level to ensure sufficient dissipation of heat by convection or conduction through the external bearing structure. The uniformity of shaft temperature ensures a temperature difference across the lubricant film since the temperature of the bush varies with angular position. The temperatures of both shaft and bush vary in the axial direction. The temperature characteristic of a journal bearing is clearly non-uniform and all 'effective viscosity' methods are, at best, approximations to a complex problem.

The solution to the problem of calculating the pressure field and load capacity with variable viscosity involves the simultaneous solution of a variable viscosity form of the Reynolds equation and a heat transfer equation for the lubricant film. This is clearly beyond the scope of analytical solution and numerical methods are almost exclusively employed. The solution method which is generally referred to as 'thermohydrodynamics' is discussed in the next chapter. Thermal effects render invalid many of the predictions of classical 'isoviscous hydrodynamics', e.g., that the load capacity is proportional to surface speed, and constitute the prime reason why the viscosity index of an oil is such an important property for the maintenance of viscosity at high localized operating temperatures, as was discussed in Chapter 2.

Multiple Regression in Bearing Analysis

The multiple regression method is very useful in finding the correlation between variables, and also in expressing one variable, selected as a dependent variable, in terms of all the other variables which are independent variables. Any variable present in a particular process can be selected as a dependent variable and expressed in terms of the other variables. Some form of approximating function is assumed, and polynomials or exponential functions are the most frequently used. Correlation coefficients between variables and coefficients of approximation are usually computed. The technique is used for analysing experimental data and has also been applied to bearing analysis [39]. Theoretical data from several hundred bearings were analysed by multiple regression resulting in a set of equations which directly provide information about the design and performance parameters of journal bearings. The equations are given in an exponential form, i.e.:

$$z = Cv_1^a v_2^b v_3^c \dots v_n^m$$

where:

z — is the dependent variable;

C — is the calculated regression constant;

$v_1, v_2, \dots, v_n$ — are the independent variables, n =1,2,3,...;

$a, b, \dots, m$ — are the calculated exponents.

The dependent and independent variables together with calculated constants and exponents are shown in Table 4.5 [39].

For example, if the load capacity is required to be calculated for a specific journal bearing, then the following equation from Table 4.5, row 1, can be used:

$$W = 2.7861 \times 10^1 \, v_{37.8°C}^{-1.1} \, v_{93.3°C}^{2.46} L^{2.515} D^{0.563} N^{0.528} c^{-1.09} T_S^{-0.383} \left(\frac{\varepsilon}{1-\varepsilon^2}\right)^{1.385}$$

or row 7:

$$W = 1.7575 \times 10^4 \, v_{37.8°C}^{-0.793} \, v_{93.3°C}^{2.033} L^{2.596} D^{1.042} N^{0.884} c^{-1.51} T_{mean}^{-2.02} \left(\frac{\varepsilon}{1-\varepsilon^2}\right)^{1.108}$$

One of the equations is expressed in terms of lubricant supply temperature and the other in terms of mean lubricant temperature. Both, however, should give a similar result. The results predicted by these equations are acceptable and can be used in quick engineering analysis.

TABLE 4.5 Multiple regression relationships between journal bearing design and performance parameters [39]. (Note that the temperature is in degrees Fahrenheit [°F], i.e., $T°F = 1.8 \times T°C + 32$).

N°	Dependent parameter	Curve constant	$\upsilon_{37.8°C}$ [cS]	$\upsilon_{93.3°C}$ [cS]	L [m]	D [m]	N [rps]	c [m]	T_s [°F]	T_{mean} [°F]	$1+\ln W^*$	W [N]	$\frac{\varepsilon}{1-\varepsilon^2}$	$1-\varepsilon$
1	**W** [N]	2.7861×10^1	-1.100	2.460	2.512	0.563	0.528	-1.090	-0.383	-	-	-	1.385	-
2	**H** [W]	3.9307×10^3	-0.706	1.577	0.477	2.240	1.287	0.249	-0.204	-	1.324	-	-	-
3	ε	1.2666×10^{-2}	0.536	1.120	-1.050	-0.578	-0.217	0.476	0.214	-	-	0.422	-	-
4	$\mathbf{Q_s}$ [kg/s]	1.4791×10^3	0.524	-1.070	0.212	1.381	0.821	1.457	0.276	-	1.699	-	-	-
5	$\mathbf{T_{max}}$ [°F]	3.8608×10^0	0.137	-0.063	0.024	0.387	0.272	-0.311	0.081	-	-0.011	-	-	-
6	$\mathbf{T_{max}\!-\!T_s}$ [°F]	2.0528×10^{-2}	-0.783	1.730	-0.367	0.881	0.496	-0.690	-0.348	-	-	0.162	-	-
7	**W** [N]	1.7575×10^4	-0.793	2.033	2.596	1.042	0.884	-1.510	-	-2.020	-	-	1.108	-
8	**H** [W]	2.7915×10^3	-0.579	1.530	0.873	2.500	1.642	-0.225	-	-1.470	0.659	-	-	-
9	ε	1.0516×10^2	0.399	-1.040	1.372	-0.539	-0.458	0.765	-	0.962	-	0.535	-	-
10	$\mathbf{T_{max}}$ [°F]	2.1918×10^{-1}	0.097	-0.055	-0.064	-0.314	0.170	-0.176	-	0.426	0.194	-	-	-
11	$\mathbf{T_{max}}$ [°F]	1.7580×10^{-1}	0.005	0.134	0.010	-0.302	0.190	-0.244	-	0.396	-	-	-	-0.142

Bearing Inlet Temperature and Thermal Interaction between Pads of a Michell Bearing

Frictional heat not only affects the load capacity of a bearing by directly influencing the viscosity but also leads to thermal interaction between adjacent bearings. For example, Michell pads are used in a combination of several pads distributed around a circle to form a larger thrust bearing. According to isoviscous theory a minimum of space should be left between the pads to maximize load bearing area. However, experimental measurements have revealed an optimum pad coverage fraction for maximum load capacity [37]. A certain amount of space is required between the pads for the hot lubricant discharged from one pad to be replaced by cool lubricant before entrainment in the following pad. In practice, the replacement of lubricant is never perfect and a phenomenon known as 'hot oil carry over' is almost inevitable. This phenomenon is illustrated schematically in Figure 4.54.

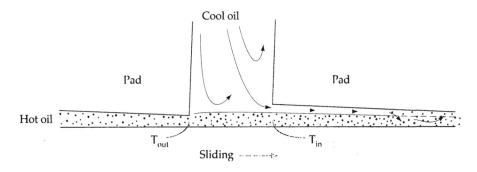

FIGURE 4.54 'Hot oil carry over' in a multiple pad bearing.

It was found from boundary layer theory that the lubricant inlet temperature can be calculated from [38]:

$$\mathbf{T}_{inlet} = \mathbf{T}_s(1 - m) / (1 - 0.5m) + 0.5\mathbf{T}_{outlet}m/(1 - 0.5m) \qquad (4.130)$$

where:

$\mathbf{T}_{inlet}$ is the lubricant inlet temperature [°C];

$\mathbf{T}_s$ is the lubricant supply temperature [°C];

$\mathbf{T}_{outlet}$ is the lubricant outlet temperature [°C];

$\mathbf{m}$ is the hot oil carry over coefficient.

The hot oil carry over coefficient is a function of sliding speed and space between adjacent pads. For a small gap width of 5 [mm] between pads, 'm' has a value of **0.8** at 20 [m/s] and **0.7** at 40 [m/s]. For a large gap width of 50 [mm] between pads, 'm' has a value of **0.55** at 20 [m/s] and **0.5** at 40 [m/s]. The minimum value of hot oil carry over coefficient occurs at approximately 40 [m/s] with a sharp rise in 'm' beyond this speed [38].

The effect of 'hot oil carry over' in a multipad thrust bearing was found to be sufficiently strong to ensure that individual pad temperatures were reduced when the number of pads was reduced from 8 to 3 [37]. This reduction in temperature occurred despite the greater concentration of frictional power dissipation per pad at constant load and speed. Even when the ratio of combined pad area to area swept by the pads was lowered to less than 35% the bearing still functioned efficiently. It is therefore unnecessary to fit a large number of closely spaced pads in a high-speed thrust bearing because this merely allows hot lubricant to recirculate almost indefinitely. The number of pads can be reduced for the same load capacity, thus achieving considerable economies in the manufacture of the bearing. This is illustrated in Figure 4.55.

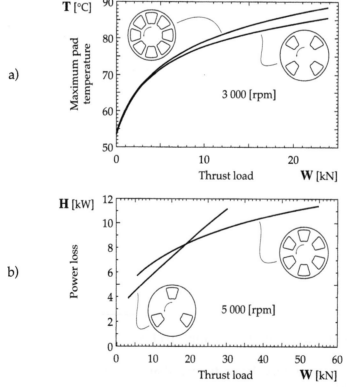

FIGURE 4.55   Effect of pad number on the performance of a thrust bearing: a) bearing temperature and b) power loss (adapted from [37] and [64]).

It can be seen that bearing temperature is reduced especially at high loads when four pads are used instead of eight. The reduction in bearing temperature coincides with improved load capacity. Removal of pads can raise load capacity under certain conditions, as shown in Figure 4.55, and this is the result of a diminished loss of lubricant operating viscosity.

4.7   LIMITS OF HYDRODYNAMIC LUBRICATION

As has been implied throughout this chapter, hydrodynamic lubrication is only effective when an appreciable sliding velocity exists. A sliding velocity of 1 [m/s] is typical of many bearings. As the sliding velocity is reduced the film thickness also declines to maintain the pressure field. This process is very effective as pressure magnitudes are proportional to the square of the reciprocal of film thickness. Eventually though the film thickness will have diminished to such a level that the small high points or asperities on each surface will come into contact. Contact between asperities causes wear and elevated friction. This condition, where the hydrodynamic film still supports most of the load but cannot prevent some contact between the opposing surfaces, is known as 'partial hydrodynamic lubrication'. When the sliding speed is reduced still further the hydrodynamic lubrication fails completely and solid contact occurs. A lubricant may still, however, influence the coefficient of friction and wear rate to some degree, as is discussed in subsequent chapters. Original research into the limits of hydrodynamic lubrication was performed early in the 20th century by Stribeck [40] and Gumbel [41]. The limits of hydrodynamic lubrication are summarized in a graph shown in Figure 4.56.

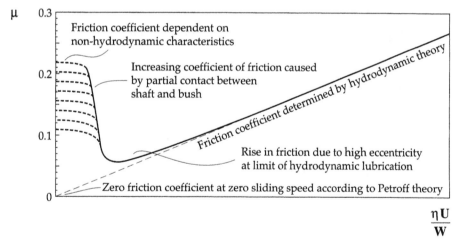

FIGURE 4.56 Schematic diagram of changes in friction coefficient at the limits of hydrodynamic lubrication.

When the friction measurements from a journal bearing were plotted on a graph against a controlling parameter defined as '$\eta U/W$' it was found that, for all but very small sliding speeds, friction 'μ' was proportional to the above parameter which is known as the 'Stribeck number'. When a critically low value of this parameter was reached, the friction rose from values of about 0.01 to much higher levels of 0.1 or more. The rapid change in the coefficient of friction represents the termination of hydrodynamic lubrication. Later work revealed that hydrodynamic lubrication persists until the largest asperities are separated by only a few nanometres of fluid. It was found that a minimum film thickness of more than twice the combined roughness of the opposing surfaces ensures full hydrodynamic lubrication of perfectly flat surfaces [3]. With the level of surface roughness attainable today on machined surfaces, a minimum film thickness of a few micrometres could thus be acceptable. In fact for small bearings, e.g., a journal bearing of 80 [mm] diameter, it is possible to use twice the combined roughness as a minimum limit for film thickness. On the other hand, for large hydrodynamic bearings, larger clearances are usually selected because of the great difficulty in ensuring that such a small minimum film thickness is maintained over the entire bearing surface. Even if the bearing surfaces are machined accurately, elastic or thermal deflection would almost certainly cause contact between the bearing surfaces. If contact between sliding surfaces occurs then, particularly at high speeds of, e.g., 10 [m/s], the dramatic increase in frictional power dissipation can cause overheating of the lubricant and possibly seizure of the bearing. Most hydrodynamic bearings, particularly the larger bearings, are designed to operate at film thicknesses well above the estimated transition point between fully hydrodynamic lubrication and wearing contact because:

· the transition loads and speeds are difficult to specify accurately,

· bearing failure is almost inevitable if hydrodynamic lubrication is allowed to fail even momentarily.

4.8 HYDRODYNAMIC LUBRICATION WITH NON-NEWTONIAN FLUIDS

Most fluids in use as lubricants either have a rheology that cannot be described as Newtonian or are modified by additives to cause deviations from Newtonian behaviour. All fluids have a non-zero density and therefore the hydrodynamic equations should ideally include the effect of inertia and at high bearing speeds, turbulent flow can also occur. Some of these flow effects are deleterious to bearing performance but others can be beneficial. The Reynolds equation presented so far does not include the characteristics of complex fluids and the refinement of hydrodynamic lubrication theory to model complex fluids is a subject of many current research projects. Some of the problems associated with complex fluids in hydrodynamics are outlined in this section.

Turbulence and Hydrodynamic Lubrication

In most bearings operating at moderate speeds, laminar flow of lubricant prevails but at high speed, the lubricant flow becomes turbulent and this affects the load capacity and particularly the friction coefficient of the bearing. Turbulent flow in a hydrodynamic bearing can be modelled by introducing 'turbulence coefficients' into the Reynolds equation. An example of a modified Reynolds equation used in the analysis of Michell pad bearings is in the form [24]:

$$\frac{\partial}{\partial x}\left(\frac{G_x h^3}{\eta}\frac{\partial p}{\partial x}\right) + \frac{R^2}{L^2}\frac{\partial}{\partial y}\left(\frac{G_y h^3}{\eta}\frac{\partial p}{\partial y}\right) = \frac{1}{2}\frac{\partial h}{\partial x} \qquad (4.131)$$

where:

G_x and G_y are the turbulence coefficients defined as:

$$G_x = \frac{1}{12(1 + 0.00116\,Re_h^{0.916})}$$

$$G_y = \frac{1}{12(1 + 0.00120\,Re_h^{0.854})}$$

Re_h is the local Reynolds number defined as:

$$Re_h = hU\rho / \eta$$

R   is the radius of a bearing [m];

h   is the film thickness [m];

U   is the surface velocity [m/s];

L   is the length of the bearing [m];

ρ is the oil density [kg/m^3];

η is the oil viscosity [Pas].

The solution to this equation is usually obtained by computation since the 'turbulence coefficients' are functions of film thickness. At the high speeds where turbulence occurs, heating effects are quite significant and must be incorporated in the solution.

It was found that the onset of turbulence in the bearing results in a slight increase in load capacity and marginal effect on stiffness coefficients (i.e., the variation in hydrodynamic load with film thickness change), which can slightly alter the limiting speed before bearing vibration occurs (i.e., vibration stability threshold). The main effect of turbulence is a large increase in bearing friction. The friction coefficient in a turbulent bearing is about 80% larger than if laminar flow prevails [24,42]. To suppress turbulent flow additives in the form of oil soluble macromolecules, e.g., polymethylmethacrylate, are used, but they are eventually degraded by prolonged shearing [42].

Hydrodynamic Lubrication With Non-Newtonian Lubricants

Common examples of non-Newtonian lubricants are the multigrade oils with polymer viscosity index improvers described in Chapter 3. There are two forms of deviation from ideal Newtonian rheology which have been studied in terms of the effect on hydrodynamic lubrication. The most common features of lubricating oil rheology are 'shear thinning' and 'viscoelasticity'.

When incorporating non-Newtonian lubricant behaviour into hydrodynamics it is necessary to derive a controlling equation from a basic rheological formula relating shear stress to shear rate. The introduction of special coefficients into Reynolds equation to allow for this effect, in a similar manner to that conducted when considering turbulent flow, is inaccurate as an analytical method [43]. This is because of the variation of the effective viscosity with shear rate which depends on the position in the film. A detailed description of the analysis and examples of the shear-thinning effect on standard lubricating oils is provided elsewhere [43]. A shear-thinning non-Newtonian oil provides only a reduced load capacity as compared to a Newtonian oil with the same apparent viscosity at near zero shear rate. An example of the effect of shear thinning on bearing load capacity is shown in Figure 4.57.

It can be seen from Figure 4.57 that the effect of shear thinning is most pronounced at high eccentricity ratios where shear rates are greatest. For example, the load capacity of the Newtonian lubricant at an eccentricity ratio of 0.6 is only equalled by the shear-thinning lubricant at an eccentricity ratio of about 0.7 and there is an even greater difference at higher eccentricities. This means that a Newtonian lubricant may provide a sharper increase in load capacity with eccentricity ratio than non-Newtonian lubricants. At extremes of eccentricity, i.e., high loads, the reduced load capacity compared to an equivalent Newtonian fluid may be critical to bearing survival.

In contrast, viscoelasticity, which is the combination of elasticity and viscosity in a fluid, generally has a positive effect on load capacity. It has often been reported that the addition of viscoelastic additives to lubricating oils raises the load capacity of journal bearings and reduces any wear that might occur. A review of the work published on the effects of viscoelastic lubricants in hydrodynamics is summarized elsewhere [44]. The strongest effect is in the load capacity improvement at high eccentricity ratios, which tends to prevent bearing failure under extremes of load [45]. Viscoelastic lubricants can also raise the threshold of vibration instability in hydrodynamic bearings with stiff shafts [44]. The substitution of a viscoelastic lubricant may be a simple solution to bearing vibration.

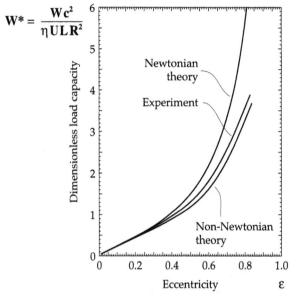

$$W^* = \frac{Wc^2}{\eta ULR^2}$$

FIGURE 4.57   Comparison of load capacity versus eccentricity ratio for a shear-thinning lubricating oil and an equivalent Newtonian lubricating oil [43].

Inertia Effects in Hydrodynamics

Fluid inertia, like turbulence, becomes significant at high bearing speeds and alters many bearing performance characteristics. It is found that although cavitation is suppressed by fluid inertia [46-50] in journal bearings, the effect on load capacity is quite small for high eccentricity ratios, $\varepsilon > 0.55$ [50]. Inertia effects become significant when the reduced Reynolds number is greater than unity, i.e., $Re^* > 1$. The reduced Reynolds number is defined as:

$$Re^* = (\rho U h_0/\eta)(h_0/R) = \frac{\rho U h_0^2}{\eta R} \qquad (4.132)$$

where:

h_0 is the minimum film thickness [m];

R is the radius of the bearing [m].

For the analysis of bearings with significant inertia effects a modified form of the Reynolds equation is used. The one-dimensional Reynolds equation, which includes an inertia term, is in the form:

$$\frac{\partial}{\partial x}\left(h^3 \frac{\partial p}{\partial x}\right) = \eta U\left[6\frac{dh}{dx} + h_0 Re^* \frac{dH}{dx}\right] \qquad (4.133)$$

where:

H is a parameter allowing for inertia effects [49].

'**H(x)**' is defined entirely in terms of higher powers of '**h**', '**dh/dx**', '$\partial p/\partial x$' and '$\partial^2 p/\partial x^2$' obtained from the standard Reynolds equation which ignores inertia effects. The full form of the expression for '**H(x)**' can be found elsewhere [49].

The effect of inertia in high speed turbulent journal bearings is to raise the vibration stability threshold until it is almost identical to that predicted by laminar theory [51].

The influence of inertia is fundamental to the operation of 'squeeze-film dampers'. A squeeze film damper is a journal bearing which is fitted around a rolling contact bearing. These systems are used in aircraft engines where rolling bearings prevent bearing failure in the event of interruption to the lubricant supply. The journal bearing or 'squeeze film damper' is used to provide damping of the shaft vibrations which are completely unaffected (not suppressed) by the presence of the rolling bearings. A schematic diagram of a squeeze film damper is shown in Figure 4.58.

At the high speeds which are typical of the operation of squeeze dampers, the journal bearing creates higher friction than the rolling bearing so that it does not usually rotate. There are always, however, imbalance forces and these form a rotating load vector with a high angular velocity. The rolling bearing housing is therefore prevented from making contact with the external structure and tends to stay positioned concentrically with the outer bearing. The stability of this bearing is analysed by applying the Reynolds equation to a journal bearing operating at a low eccentricity ratio. It is found that unless fluid inertia effects are included, gross inaccuracies result in calculated values of vibration stability [52,53]. If the limit of journal bearing stability against vibration is exceeded, the squeeze damper will vibrate, negating its intended purpose.

Compressible Fluids

In the analysis conducted so far it has been assumed that the fluid density is constant. The assumption is nearly always valid with liquid lubricants since although liquids expand thermally during bearing operation, causing a reduction in density, this is a relatively small effect and can be neglected. The situation is entirely different for bearings lubricated by gases, i.e., gas bearings. The density of gases can vary significantly with pressure and temperature.

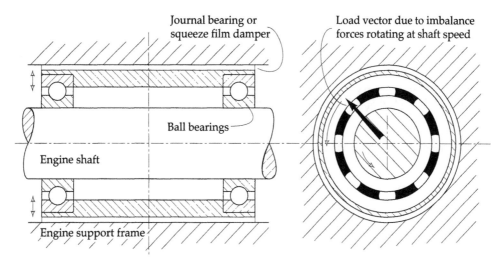

FIGURE 4.58 Schematic diagram of the squeeze film damper.

A modified Reynolds equation is used for the analysis of bearings lubricated by fluids with variable density. As already mentioned, the term **12(w_h - w_0)** in the Reynolds equation (4.22) describes the vertical flow across the film, or in other words it describes the rate of change in the film thickness, i.e.:

$$12(w_h - w_0) = 12\frac{dh}{dt}$$

Thus the Reynolds equation is in the form:

$$\frac{\partial}{\partial x}\left(\frac{h^3}{\eta}\frac{\partial p}{\partial x}\right) + \frac{\partial}{\partial y}\left(\frac{h^3}{\eta}\frac{\partial p}{\partial y}\right) = 6U\eta\frac{\partial h}{\partial x} + 12\eta\frac{\partial h}{\partial t} \qquad (4.134)$$

The continuity of mass flow with variable density requires that:

$$\frac{\partial}{\partial x}\left(\rho u\right) + \frac{\partial}{\partial y}\left(\rho v\right) + \frac{\partial}{\partial z}\left(\rho w\right) + \frac{\partial\rho}{\partial t} = 0$$

Thus the Reynolds equation (4.134), incorporating variable density but still assuming constant viscosity, can be written in the form:

$$\frac{\partial}{\partial x}\left(h^3\rho\frac{\partial p}{\partial x}\right) + \frac{\partial}{\partial y}\left(h^3\rho\frac{\partial p}{\partial y}\right) = 6U\eta\frac{\partial(h\rho)}{\partial x} + 12\eta\frac{\partial(h\rho)}{\partial t} \qquad (4.135)$$

According to the gas laws:

$$pV^\gamma \quad \text{or} \quad \frac{p}{\rho^\gamma} = \text{constant}$$

For isothermal conditions $\gamma = 1$ and thus the pressure 'p' is proportional to density 'ρ':

$$p \; \alpha \; \rho$$

This relation enables the replacement of density with pressure and equation (4.135) becomes:

$$\frac{\partial}{\partial x}\left(h^3 p\frac{\partial p}{\partial x}\right) + \frac{\partial}{\partial y}\left(h^3 p\frac{\partial p}{\partial y}\right) = 6U\eta\frac{\partial(hp)}{\partial x} + 12\eta\frac{\partial(hp)}{\partial t} \qquad (4.136)$$

Introducing the non-dimensional parameters [2]:

$$\bar{p} = \frac{p}{p_a} \qquad \Rightarrow \qquad p = p_a\bar{p}$$

$$h^* = \frac{h}{c} \qquad \Rightarrow \qquad h = ch^*$$

$$x^* = \frac{x}{B} \qquad \Rightarrow \qquad x = Bx^*$$

$$y^* = \frac{y}{L} \qquad \Rightarrow \qquad y = Ly^*$$

where:

p_a is the external ambient pressure [Pa];

c   is the radial clearance [m];

B   is the width of the bearing [m];

L   is the length of the bearing [m];

and substituting into equation (4.136) yields:

$$\frac{\partial}{\partial(Bx*)}\left[(ch*)^3(p_a\bar{p})\frac{\partial(p_a\bar{p})}{\partial(Bx*)}\right] + \frac{\partial}{\partial(Ly*)}\left[(ch*)^3(p_a\bar{p})\frac{\partial(p_a\bar{p})}{\partial(Ly*)}\right] = 6U\eta\frac{\partial(ch*p_a\bar{p})}{\partial(Bx*)} + 12\eta\frac{\partial(ch*p_a\bar{p})}{\partial t}$$

or:

$$\frac{c^3p_a{}^2}{B^2}\left[\frac{\partial}{\partial x*}\left(h*^3\bar{p}\frac{\partial\bar{p}}{\partial x*}\right) + \frac{B^2}{L^2}\frac{\partial}{\partial y*}\left(h*^3\bar{p}\frac{\partial\bar{p}}{\partial y*}\right)\right] = \frac{6\eta Ucp_a}{B}\left[\frac{\partial(h*\bar{p})}{\partial x*} + \frac{2B}{U}\frac{\partial(h*\bar{p})}{\partial t}\right]$$

which simplifies to:

$$\frac{\partial}{\partial x*}\left(h*^3\bar{p}\frac{\partial\bar{p}}{\partial x*}\right) + \frac{B^2}{L^2}\frac{\partial}{\partial y*}\left(h*^3\bar{p}\frac{\partial\bar{p}}{\partial y*}\right) = \frac{6\eta UB}{c^2p_a}\left[\frac{\partial(h*\bar{p})}{\partial x*} + \frac{2B}{U}\frac{\partial(h*\bar{p})}{\partial t}\right] \qquad (4.137)$$

The non-dimensional parameter:

$$\frac{6\eta UB}{c^2p_a} = \lambda \qquad\qquad\qquad (4.138)$$

is known in the literature as the Harrison number in memory of the person who first derived these equations [54]. The Harrison number is also called the compressibility number [3]. It is apparent that the Harrison number is a function of bearing design and operating parameters as well as ambient pressure.

The modified Reynolds equation (4.137) can now be used for analysis of bearings lubricated with compressible fluids. The procedure is the same as outlined in the previous sections, i.e., define film geometry and boundary conditions and then solve the Reynolds equation. For example, assuming an infinitely long bearing, i.e., $B/L \Rightarrow 0$, running under steady conditions where $\partial h/\partial t = 0$, equation (4.137) becomes:

$$\frac{\partial}{\partial x*}\left(h*^3\bar{p}\frac{\partial\bar{p}}{\partial x*}\right) = \lambda\frac{\partial(h*\bar{p})}{\partial x*}$$

which can now be integrated over a specified bearing geometry. More detailed analysis of such bearings is discussed in [55].

Compressible Hydrodynamic Lubrication in Gas Bearings

Compressible hydrodynamic lubrication is typically found in gas bearings. The lubrication of these bearings is characterized by extremely small film thicknesses which can restrict the free movement of individual gas molecules. The mean free path of molecules in air at atmospheric pressure is 0.064 [μm] and this is reduced with higher operating pressures.

When the film thickness is comparable to this value, then the gas begins to be subjected to rarefaction and loses its continuity as a film. It may be asked, do similar processes happen with liquids at very thin film thicknesses? The lubrication mechanism for liquids, in these cases, is fundamentally different from classical hydrodynamics, introduced so far, and is discussed in subsequent chapters. Returning to gases, a parameter known as the Knudsen number is used to measure the degree of gas rarefaction at thin film thicknesses. The Knudsen number is the ratio of the mean free path of gas molecules to the available range of movement, i.e.:

$$K_n = \frac{\gamma}{h} \tag{4.139}$$

where:

K_n is the Knudsen number;

γ is the mean free path of a molecule [m], typically $\gamma = 0.064$ [μm] for air at standard temperature and pressure;

h is the film thickness [m].

When the Knudsen number is in the range between $0.01 < K_n < 0.1$, which is typical of many small bearings and corresponds to a film thickness in a range between 0.64 [μm] $< h < 6.4$ [μm] in atmospheric air, then the 'slip flow regime' applies. In this regime a thin layer of gas close to the solid surface loses its fluid characteristic and is capable of 'slipping' against the surface. In terms of the controlling equations, this means that the effective viscosity of the gas declines with its proximity to the surface. The formation of a rarefied gas layer close to the sliding surface and the occurrence of 'slip' is illustrated in Figure 4.59.

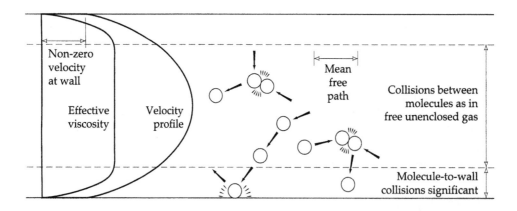

FIGURE 4.59   Rarefaction and surface-proximal viscosity loss in gas films at high Knudsen numbers.

It is possible to model this phenomenon with another modified form of the Reynolds equation. This equation is too complex and specialized to describe here, and more information is available elsewhere [56,57]. The consequence of rarefaction effects is that gases (unlike liquids) provide a declining load capacity as film thickness is reduced and that a compressible gas bearing is far more likely to fail under excess load than a liquid lubricated bearing. A graph of the relative load capacity of an inclined pad bearing against Knudsen number is shown in Figure 4.60. The relative load capacity 'W_n' is defined as the calculated load capacity for a given Knudsen number divided by the load capacity at zero Knudsen number. The initial load capacities are calculated from a non-dimensionalized Reynolds equation, which removes the effect of variation in bearing clearance on load capacity. A reduction in bearing clearance is in fact the only means available to increase the Knudsen number, and as clearance becomes smaller, load capacity increases. The data presented in Figure 4.60 clearly show that the increase in load capacity is diminished by rarefaction effects.

The major application of thin gas film bearings lies in data recording heads for computers. Air is used as the lubricant for the recording heads which are designed to be separated from the magnetic recording disc by a hydrodynamic film. The need for high recording densities in magnetic discs necessitates the smallest possible air film thickness between the head and the disc. A typical film thickness is around 1 [μm] which is in the range of Knudsen number where load capacity sharply decreases with declining film thickness. The decline in load capacity means that the bearing can have a significant negative stiffness (i.e., anomalously low stiffness) and is prone to 'head crash', which is when the recording head contacts and wears the disc.

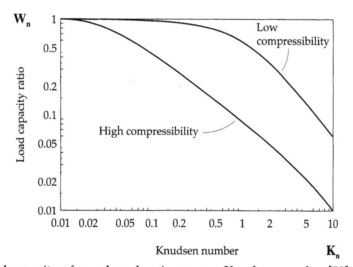

FIGURE 4.60   Load capacity of a pad gas bearing versus Knudsen number [58].

4.9   REYNOLDS EQUATION FOR SQUEEZE FILMS

In the treatment of hydrodynamics presented so far, little mention has been made of load capacity under a time dependent film thickness, e.g., cases where $\partial h/\partial t \neq 0$. Squeeze film is a term denoting a hydrodynamic film that sustains a negative $\partial h/\partial t$, i.e., when the opposing surfaces are being squeezed together. An extremely useful characteristic of squeeze films is that they provide increased load capacity (although temporary) when a bearing is suddenly subjected to an abnormally high load. This feature is essential to the reliability of crankcase bearings which must withstand transient combustion forces. A further aspect of squeeze films is that the squeeze film force is always opposite in direction to the motion of either bearing surface. This is a form of damping and, as will be shown in the next chapter on 'Computational Hydrodynamics', squeeze film forces contribute to the vibrational stability of a bearing. To analyse squeeze film forces, the term $\partial h/\partial t$ is kept in the Reynolds equation and is given precedence over the film geometry term $\partial h/\partial x$. The Reynolds equation with the squeeze term is in the form (eq. 4.134):

$$\frac{\partial}{\partial x}\left(\frac{h^3}{\eta}\frac{\partial p}{\partial x}\right) + \frac{\partial}{\partial y}\left(\frac{h^3}{\eta}\frac{\partial p}{\partial y}\right) = 6U\frac{\partial h}{\partial x} + 12\frac{\partial h}{\partial t}$$

Assuming an isoviscous lubricant and zero entraining velocity this equation becomes:

$$\frac{\partial}{\partial x}\left(h^3\frac{\partial p}{\partial x}\right) + \frac{\partial}{\partial y}\left(h^3\frac{\partial p}{\partial y}\right) = 12\eta\frac{\partial h}{\partial t} \qquad (4.140)$$

This equation defines the hydrodynamic pressure field when the wedge effect is absent, e.g., when the load vector rotates as mentioned previously. It can be integrated in terms of a specified bearing geometry to provide load capacity, maximum pressure or any other required bearing characteristic in terms of $\partial h/\partial t$. The 'squeeze time' which means the time required for film thickness to decline to some critical minimum value can also be determined by integrating $\partial h/\partial t$ with respect to time.

To illustrate the principles involved, an example consisting of two long parallel plates squeezed together as shown in Figure 4.61 is considered.

Pressure Distribution

For two parallel and infinitely long plates:

$$h \neq f(x) \quad \text{and} \quad \frac{\partial p}{\partial y} = 0$$

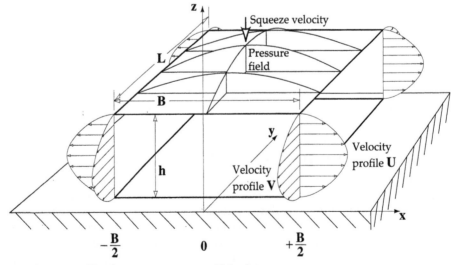

FIGURE 4.61 Squeeze film between two parallel plates.

and equation (4.140) reduces to:

$$\frac{d^2 p}{dx^2} = \frac{12\eta}{h^3}\frac{dh}{dt}$$

Integrating once:

$$\frac{dp}{dx} = \frac{12\eta}{h^3} \frac{dh}{dt} x + C_1 \tag{4.141}$$

and again:

$$p = \frac{6\eta}{h^3} \frac{dh}{dt} x^2 + C_1 x + C_2 \tag{4.142}$$

Boundary conditions are, from Figure 4.61:

$$p = 0 \quad \text{at} \quad x = \pm\frac{B}{2} \quad \text{at the edges of the plates}$$

and the pressure gradient:

$$\frac{dp}{dx} = 0 \quad \text{at} \quad x = 0$$

Substituting into (4.141) and (4.142) yields:

$$C_1 = 0$$

$$C_2 = -\frac{6\eta B^2}{4h^3} \frac{dh}{dt}$$

Substituting for 'C_1' and 'C_2' into (4.142) gives the pressure distribution as a function of **dh/dt**:

$$\boxed{p = \frac{6\eta}{h^3} \frac{dh}{dt} \left(x^2 - \frac{B^2}{4} \right)} \tag{4.143}$$

where the term **dh/dt** is negative for a positive squeeze direction as shown in Figure 4.61.

Load Capacity

The load that the plates can support, or more exactly the force separating the plates, can be obtained by integrating the pressure distribution over the bearing area:

$$W = \int_0^{1.} \int_{-B/2}^{B/2} p \, dx \, dy$$

The load per unit length is:

$$\frac{W}{L} = \int_{-B/2}^{B/2} p \, dx$$

substituting for '**p**' from (4.143):

$$\frac{W}{L} = \int_{-B/2}^{B/2} \frac{6\eta}{h^3} \frac{dh}{dt} \left(x^2 - \frac{B^2}{4}\right) dx$$

and integrating yields:

$$\boxed{\frac{W}{L} = -\frac{\eta B^3}{h^3} \frac{dh}{dt}}$$ (4.144)

For finite plates, equation (4.144) is modified by a computed coefficient 'β' shown in Table 4.6 for different **B/L** ratios [4]. Hence the equation of load capacity for finite plates as a function of **dh/dt** becomes:

$$\boxed{\frac{W}{L} = -\beta \frac{\eta B^3}{h^3} \frac{dh}{dt}}$$ (4.145)

TABLE 4.6   Computed values of 'β' coefficient for load capacity of finite parallel plates [4].

B/L	1	5/6	2/3	1/2	2/5	1/3	1/4	1/5	1/10	0
β	0.421	0.498	0.580	0.633	0.748	0.790	0.845	0.874	0.937	1

Squeeze Time

The time necessary for the film thickness between parallel plates to change between specified limits, the squeeze time, is found by rearranging (4.144):

$$dt = -\frac{\eta B^3}{h^3} \frac{L}{W} dh$$

Noting that for many problems, the load does not vary with time, i.e., **W ≠ f(t)**, and integrating accordingly yields:

$$\Delta t = -\frac{\eta B^3 L}{W} \int_{h_1}^{h_2} \frac{dh}{h^3}$$

and finally:

$$\boxed{\Delta t = \frac{\eta B^3 L}{2W} \left(\frac{1}{h_2^2} - \frac{1}{h_1^2}\right)}$$ (4.146)

Similarly for finite parallel plates, the squeeze time is given by:

$$\boxed{\Delta t = \beta \frac{\eta B^3 L}{2W} \left(\frac{1}{h_2^2} - \frac{1}{h_1^2}\right)}$$ (4.147)

and for flat circular plates:

$$\Delta t = \frac{3\pi\eta R^4}{4W}\left(\frac{1}{h_2^2} - \frac{1}{h_1^2}\right)$$

(4.148)

where:

    Δt is the time required for the film thickness to decline from 'h_1' to 'h_2' [s];

    R is the radius of a circular plate [m];

    h_1 is the initial film thickness [m];

    h_2 is the final film thickness [m];

    W is the load [N].

In systems where squeeze film lubrication persists, the minimum film thickness is usually determined by the combined surface roughnesses of the opposing faces and, in the case of large bearings, by elastic deformation.

Cavitation and Squeeze Films

In a system where positive and negative squeeze occurs in the presence of a liquid lubricant, cavitation is almost inevitable. Cavitation affects squeeze film forces by the formation of compressible bubbles in an otherwise incompressible lubricant. Bubbles can also persist or grow in size by coalescence until the squeeze characteristics of the system are fundamentally changed. The persistence of bubbles even under temporarily positive lubricant pressure is due to the much slower rate of bubble dissolution as compared to the rate of bubble formation. The process of bubble accretion under oscillating squeeze is illustrated in Figure 4.62.

The overall effect is to significantly reduce the load capacity from the values calculated by hydrodynamic theory assuming no cavitation effects [59,60]. A model of oscillating squeeze films with cavitation and bubble formation is described elsewhere [59]. The analysis of bubble formation and its effect on load capacity is beyond the scope of the material discussed in this chapter. A graphical comparison between dimensionless load capacity of an oscillating squeeze film with and without cavitation is shown in Figure 4.63. The graph shows the load capacity of a squeeze bearing, i.e., stationary shaft and bush with rotating load, in which the lubricant is subjected to cyclic cavitation as illustrated in Figure 4.62.

The effect of cavitation on load capacity can be divided into three regimes based on eccentricity ratio. The eccentricity ratio in this instance denotes the amplitude of shaft movement at any given position as the rotating load ensures that the shaft is in continuous movement. This mode of bearing operation is different from the more usual quasi-static position of the shaft centre under uni-directional load. It can be seen from Figure 4.63 that at low eccentricities, cavitation does not exert any significant effect but at eccentricities ranging from low to medium values bubble formation is significant and load capacity drops accordingly. At very high eccentricities, the bubbles become highly compressed which significantly raises load capacity.

Microscopic Squeeze Film Effects between Rough Sliding Surfaces

In the Reynolds equation presented so far, it has been assumed for the purposes of analysis that the bearing surfaces are perfectly smooth. In most practical applications, this assumption is very accurate despite the roughness present on any machined surface. When the film thickness is reduced to a level where the depth of an asperity or the measured amplitude of roughness is comparable to the film thickness, then transient squeeze velocities may occur when large asperities of opposing surfaces pass over each other. This concept of transient squeeze between rough surfaces in close contact is illustrated in Figure 4.64.

The effect of transient squeeze on load capacity is still controversial and it is not quite clear whether load capacity is reduced or increased [57]. Both forms of squeeze, positive and negative, occur during asperity approach and separation so the net effect on load capacity is small except for gases. In gases, the relationship between compression and pressure ensures that the pressure increase due to positive squeeze is larger than the pressure decrease due to expansion during negative squeeze. There is, as a result, a net load carrying effect for gases [3].

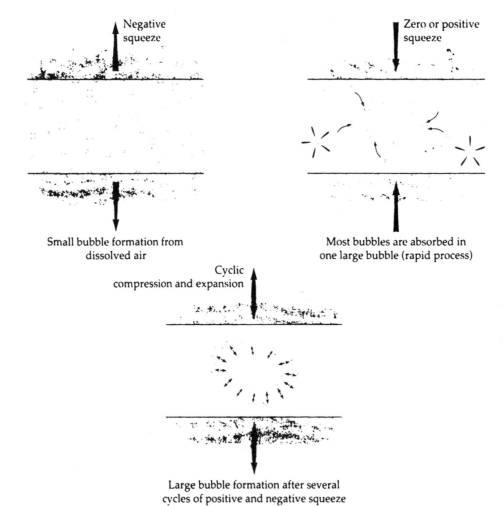

FIGURE 4.62 Mechanism of bubble accretion under oscillating squeeze.

4.10 POROUS BEARINGS

Porous bearings are intended to provide hydrodynamic lubrication without the need for an external lubricant supply. These bearings consist of a cylindrical bush made of compacted metal powder which has been impregnated with a lubricant. A typical lubricant used is mineral oil and the metal powder is usually bronze. The operating principle is that oil is drawn out of the unloaded parts of the bearing by hydrodynamic suction and then returned

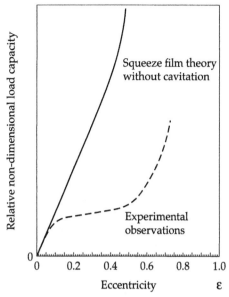

FIGURE 4.63   Effect on load capacity of cavitation in an oscillating squeeze film [66].

to the bearing in the loaded area where hydrodynamic pressure is positive. Very little, however, is known about the mechanism of oil recirculation around the bush during operation. The hydrodynamic lubrication of porous bearings can be solved by incorporating a leakage term 'K_p' into the right-hand side of the standard Reynolds equation:

$$\frac{\partial}{\partial x}\left(h^3\frac{\partial p}{\partial x}\right) + \frac{\partial}{\partial y}\left(h^3\frac{\partial p}{\partial y}\right) = \eta\left[6U\frac{\partial h}{\partial x} + K_p\left(\frac{\partial p}{\partial z}\right)_0\right]$$

where:

K_p is a constant which depends on the permeability of the porous bearing;

$(\partial p/\partial z)_0$ is the pressure gradient in the porous material immediately beneath the bearing surface.

The full solution of this equation for narrow bearings is provided elsewhere [61].

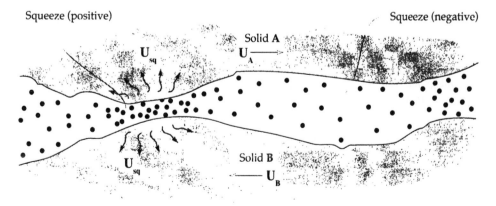

FIGURE 4.64   Transient squeeze between rough surfaces in close contact.

A basic disadvantage of porous bearings as predicted by hydrodynamic theory is that the load at $\varepsilon = 1$ is finite which is in contrast to infinite load for a plain journal bearing. The finite load capacity ensures that it is relatively easy to cause seizure in a porous bearing by overload compared to a plain bearing. As a result, porous bearings have suffered declining interest in recent years.

4.11   SUMMARY

The theory of hydrodynamic lubrication has been presented to demonstrate how a basic property of all liquids, such as viscosity, can be used to produce cheap, reliable bearings that operate with low friction and wear. Like many important scientific principles, chance observation played an important role in the recognition of hydrodynamic action as a basic mechanism of bearing lubrication. The complete separation of sliding surfaces by a liquid film under full hydrodynamic lubrication can allow bearings to operate indefinitely without any wear. Any liquid or gas can be used for this form of lubrication, provided that no chemical attack of the bearing occurs. The disadvantage of hydrodynamic lubrication is that a non-zero sliding or 'squeeze' velocity is required before load capacity is obtained. Some damage to bearings during starting or stopping is inevitable because of this condition. There is also the risk of a large rise in friction and possible bearing seizure if the limits of hydrodynamic lubrication are exceeded by excessive load or insufficient speed. A further problem is that vibration induced by hydrodynamic instability may occur during operation at high speeds and this should always be carefully controlled. Despite these deficiencies, hydrodynamic lubrication is the preferred form of lubrication in most bearing systems.

REVISION QUESTIONS

4.1   Briefly describe the principle behind hydrodynamic lubrication.

4.2   Why is hydrodynamic lubrication so vital to machine design?

4.3   What basic principle of fluid mechanics dictates that a pressure must be generated when a liquid is drawn by a sliding surface into a tapered space?

4.4   What is the principal mechanism of lubrication between an oil lubricated roll and the rolled steel sheet?

4.5   Suggest a practical reason why the Michell pad is superior to rolling bearings for support of rotor load in a hydroelectric power station.

4.6   Why is the Michell pad supported on a pivot?

4.7   What level of friction coefficient is attained by a pad bearing?

4.8   What is the average oil film thickness in a pad bearing?

4.9   Why do elastic and thermal distortions of the pivoted pads have such an adverse effect on load carrying capacity?

4.10   What is the disadvantage of the offset pivot compared to centrally pivoted pads?

4.11   The centrally pivoted pad is suitable for load carrying in both directions of rotation, while the off-centre pad is unsuitable. The centrally pivoted pad is, however, inferior to the off-centre pad in one important aspect, what is this?

4.12   Give a definition of 'eccentricity ratio'.

4.13   Of all the parameters controlling hydrodynamic lubrication, which parameter is most easily optimised and therefore left as the design variable?

4.14   Why is oil so useful as a hydrodynamic lubricant?

4.15   Is hydrodynamic lubrication confined to oils?

4.16   How much of the frictional heat is absorbed by the oil?

4.17   Name the phenomenon that causes difficulty in cooling the lubricating oil at high operating speeds of a pad bearing.

4.18   Given that hydrodynamic lubrication provides low friction and virtually indefinite bearing life, why is it limited to specialized applications?

4.19   Churning of oil by pads causes the oil to overheat and also to become bubbly. Why does this bubbliness reduce the load carrying capacity of the pads?

4.20   How are friction and wear at 'startup' eliminated?

4.21   An infinitely long linear wedge bearing of length $L = 100$ [mm] and width $B = 20$ [mm] is operating under an inlet film thickness of $h_0 = 20$ [μm]. Assuming constant velocity $U = 10$ [m/s] and constant oil viscosity $\eta = 3.5 \times 10^{-3}$ [Pas]:

•   Find the maximum load the bearing can support,

•   Find the non-dimensional load,

•   Calculate the outlet film thickness. (Ans. $W = 560$ [N], $W^* = 0.02676$, $h_1 = 44 \times 10^{-6}$ [m])

4.22   For the infinitely long Rayleigh step bearing shown in Figure 4.11, find the relationship between the total load supported W, minimum film thickness h_0 and step height Δh. Assume that $B_1 = B_2 = B/2$. (Ans.
$$\frac{W}{L} = \frac{3UB^2\eta\Delta h}{2\left[(h_0 + \Delta h)^3 + h_0^3\right]})$$

4.23   For the infinitely long Rayleigh step bearing shown in Figure 4.11, calculate the load and friction per unit length and coefficient of friction. Assume that $B_1 = 0.6B$, $B_2 = 0.4B$, $B = 0.2$ [m], $h_0 = 15$ [μm], height of the step is $\Delta h = 5$ [μm], lubricant viscosity is $\eta = 5 \times 10^{-3}$ [Pas] and surface velocity is $U = 10$ [m/s]. (Ans. $W/L = 1.378$ [MN/m], $F/L = 719.89$ [N/m], $\mu = 0.001$)

4.24   A single square linear wedge thrust pad is required to support a load of 0.4 [MN] and give minimum film thickness of not less than 50×10^{-6} [m] at a temperature of 50.5°C. The lubricant has a viscosity of 19.5 [cS] at 100°C, a viscosity index of 95 and density (independent of temperature) of 870 [kg/m^3]. The sliding velocity is 50 [m/s]. Assuming an isoviscous lubricating film and assuming an optimal convergence ratio:

•   calculate the minimum pad dimensions needed to fulfil the above requirements, and

•   estimate the friction force and coefficient of friction generated at the moving surface. (Ans. L & B = 133 [mm], F = 1.605 [kN], $\mu = 0.004$)

4.25   A journal bearing of 80 [mm] diameter and 20 [mm] length is operating at a speed of 3500 [rpm] and load of 1 [kN]. The oil effective viscosity is 9.8 [cS], oil density is $\rho = 900$ [kg/m^3] and the bearing operating temperature is 80°C. Assume clearance/radius ratio $c/R = 0.0015$.

•   estimate the amount of side leakage,

•   estimate the friction force and the coefficient of friction,

•   calculate the bearing maximum pressure for these conditions. (Ans. Q = 1.3×10^{-5} [m^3/s], F = 16.67 [N], $\mu = 0.017$, $p_{max} = 3$ [MPa])

4.26   A journal bearing of 0.06 [m] diameter, 0.015 [m] length and 10 [μm] clearance operating at 1000 [rpm] supports a load of 10 [kN]. Assuming isothermal conditions and optimal eccentricity ratio of $\varepsilon = 0.7$, find the oil viscosity needed to lubricate this bearing. (Ans. $\eta = 0.043$ [Pas])

4.27   For the bearing in problem 4.26 calculate oil flow and friction. (Ans. F = 533.3 [N])

4.28   An engine bearing of 80 [mm] diameter, 10 [mm] length and clearance to radius ratio of 10^{-3}, lubricated by the oil with viscosity $\eta = 0.01$ [Pas], is operating under a load of 1.5 [kN] at a speed of 5000 [rpm]. Select the filter pore size to limit wear on this bearing. (Ans. 50 [μm])

4.29   A 150 [mm] diameter shaft carrying a load of 10 [kN] and rotating at 1000 [rpm] is supported by a 40 [mm] long journal bearing. The bearing is lubricated with **SAE 20W/50** of density $\rho = 900$ [kg/m^3] and specific heat $\sigma = 1650$ [J/kgK]. Assume that the bearing clearance is $c = 100$ [μm] and the inlet oil temperature is $T_{in} = 50$°C. Using the iteration method calculate the operating temperature of the bearing. Discuss the results obtained. (Ans. $T_{eff} = 77$°C)

4.30   For the bearing specified in problem 4.29, calculate the operating temperature using a constant flow method. Compare the results obtained by the constant flow and iteration methods (problem 4.29). (Ans. $T_{eff} = 76$°C)

4.31   Two journal bearings of 0.08 [m] diameter and 0.04 [m] length are used to support a rotor weighing 20 [kN] at a speed of 6000 [rpm]. The viscosity of the oil supplied to the bearing is $\upsilon_s = 22$ [cS]. Within the bearing the oil viscosity changes according to the equation:

$$\eta = \eta_s e^{-\gamma \Delta T_{eff}}$$

where:

ΔT_{eff} is the effective temperature rise and 'γ is **0.035** per K. The specific heat of the oil is **1900** [J/kgK].

Assume that:

*   the shaft runs concentric in the bearing,

*   the average temperature rise of the oil escaping the bearing is $(T_{max} - T_{inlet})/2$,

*   the effective viscosity is at $T_{eff} = T_{inlet} + 0.8(T_{max} - T_{inlet})$,

*   all heat generated is absorbed by the oil flow,

*   oil flow through the bearing is $UcL/2$,

Find the algebraic expression for the temperature rise $\Delta T = T_{max} - T_{inlet}$ and find the maximum allowable radial clearance so that ΔT does not exceeds **60K**. (Ans. $\Delta T = \dfrac{4\pi \upsilon UD}{c^2 \sigma}$, $c = 3.015 \times 10^{-5}$ [m])

REFERENCES

1   Beauchamp Tower, First Report on Friction Experiments, *Proc. Inst. Mech. Engrs.*, Nov. 1883, pp. 632-659.

2   O. Reynolds, On the Theory of Lubrication and its Application to Mr Beauchamp Tower's Experiments Including an Experimental Determination of the Viscosity of Olive Oil, *Phil. Trans., Roy. Soc. London*, Vol. 177 (i), 1886, pp. 157-234.

3   A. Cameron, Principles of Lubrication, Longmans Green and Co. Ltd., London, 1966.

4   A. Cameron, Basic Lubrication Theory, Ellis Horwood Ltd., London, 1981.

5   Lord Rayleigh, Notes on the Theory of Lubrication, *Phil. Mag.*, 1918, No. 35, pp. 1-12.

6   R.C.R. Johnston and C.F. Kettleborough, An Experimental Investigation into Stepped Thrust Bearings, *Proc. Inst. Mech. Engrs.*, Vol. 170, 1956, pp. 511-520.

7   O. Pinkus, Solution of the Tapered-Land Sector Thrust Bearing, *Transactions ASME*, Vol. 80, 1958, pp. 1510-1516.

8   O. Pinkus and B. Sternlicht, Theory of Hydrodynamic Lubrication, McGraw-Hill, New York, 1961.

9   A.A. Raimondi, The Influence of Longitudinal and Transverse Profile on the Load Capacity of Pivoted Pad Bearings, *ASLE Transactions*, Vol. 3, 1960, pp. 265-276.

10 A.G.M. Michell, The Lubrication of Plane Surfaces, *Zeitschrift fur Mathematik und Physik*, Vol. 52, 1905, pp. 123-137.

11 A.W. Roberts, Bulk Materials Handling - A Key Discipline of Engineering, *Inst. of Engineers Aust., Mech. Eng. Trans.*, Vol. ME14, 1989, pp. 84-96.

12 E.A. Muijdermann, Spiral Groove Bearings, Springer-Verlag, Berlin, 1966.

13 Y. Kimura, N. Osada and K. Sasaki, Characteristics of Ceramic Spiral Groove Thrust Bearings under Lubricating Condition with Low Viscosity Liquid (Part 1), *JSLE Transactions*, Vol. 33, 1988, pp. 790-797.

14 W. Lewicki, Theory of Hydrodynamic Lubrication in Parallel Sliding, *Engineer*, Vol. 200, 1955, pp. 939-941.

15 D.D. Heckelman and C.M. McC. Ettles, Viscous and Inertial Pressure Effects at the Inlet to a Bearing Film, *STLE Transactions*, Vol. 31, 1988, pp. 1-5.

16 Beauchamp Tower, Reports of the Research Committee on Friction, No. 4, Experiments on Pivot Friction, *Proc. Inst. Mech. Engrs.*, 1891, pp. 111-140.

17 A.H. Bennett and C.M. McC. Ettles, A Self Acting Parallel Surface Thrust Bearing, *Proc. Inst. Mech. Engrs.*, London, Vol. 182, Pt. 3N, 1968, pp. 141-146.

18 A. Sommerfeld, Zur hydrodynamischen Theorie der Schmiermittelreibung, *Zeitschrift der Mathematik und Physik*, Vol. 40, 1904, pp. 97-155.

19 F.A. Martin, Feed-Pressure Flow in Plain Journal Bearings, *ASLE Transactions*, Vol. 26, 1983, pp. 381-392.

20 D.F. Wilcock, Influence of Feed Groove Pressure and Related Lubricant Flow on Journal Bearing Performance, *Tribology Transactions*, Vol. 31, 1988, pp. 397-403.

21 R.W. Wilson, Cavitation Damage in Plain Bearings, Proc. 1st Leeds-Lyon Symp. on Tribology, Cavitation and Related Phenomena in Lubrication, editors, D. Dowson, C.M. Taylor and M. Godet, Sept. 1974, Inst. Mech. Engrs. Publ., London, 1975, pp. 177-184.

22 J.A. Cole and C.J. Hughes, Oil Flow and Film Extent in Complete Journal Bearings, *Proc. Inst. Mech. Engrs.*, Vol. 170, 1956, pp. 499-510.

23 R.D. Flack and R.F. Lanes, Effects of Three-Lobe Bearing Geometries on Rigid-Rotor Stability, *ASLE Transactions*, Vol. 25, 1982, pp. 221-228.

24 M. Mikami, M. Kumagai, S. Uno and H. Hashimoto, Static and Dynamic Characteristics of Rolling-Pad Journal Bearings in Super-Laminar Flow Regime, *Transactions ASME, Journal of Tribology*, Vol. 110, 1988, pp. 73-79.

25 R.T.P. Whipple, Theory of the Spiral Grooved Thrust Plate with Liquid or Gas Lubricant, Report T/R 622, Atomic Energy Establishment, 1951.

26 A.K. Molyneaux and M. Leonhard, The Use of Spiral Groove Gas Bearings in a 350,000 rpm Cryogenic Expander, *Tribology Transactions*, Vol. 32, 1989, pp. 197-204.

27 M. Akkok and C.M.McC. Ettles, The Effect of Load and Feed Pressure on Whirl in a Grooved Journal Bearing, *ASLE Transactions*, Vol. 22, 1979, pp. 175-184.

28 M. Akkok and C.M.McC. Ettles, The Effect of Grooving and Bore Shape on the Stability of Journal Bearings, *ASLE Transactions*, Vol. 23, 1980, pp. 431-441.

29 J.T. Burwell, The Calculated Performance of Dynamically Loaded Sleeve Bearings, *Journal Appl. Mech.*, Vol. 69, 1947, pp. 231-245.

30 A.J. Smalley and H. McCallion, The Effect of Journal Misalignment on the Performance of a Journal Bearing under Steady Running Conditions, *Proc. Inst. Mech. Engrs.*, Vol. 181, Pt. 3B, 1966-1967, pp. 45-54.

31 J.R. Stokley and R.R. Donaldson, Misalignment Effects in 180° Partial Journal Bearings, *ASLE Transactions*, Vol. 12, 1969, pp. 216-226.

32 T. Someya (editor), Journal Bearing Data-book, Springer Verlag, Berlin, 1989.

33 C.M.McC. Ettles, M. Akkok and A. Cameron, Inverse Hydrodynamic Methods Applied to Mr. Beauchamp Tower's experiments of 1885, *Transactions ASME, Journal of Lubrication Technology*, Vol. 102, 1980, pp. 172-181.

34 S.C. Jain, R. Sinhasan and S.C. Pilli, A Study on the Dynamic Response of Compliant Shell Journal Bearings, *Tribology Transactions*, Vol. 32, 1989, pp. 297-304.

35 C.M.McC Ettles, Transient Thermoelastic Effects in Fluid Film Bearings, *Wear*, Vol. 79, 1989, pp. 53-71.

36 R. Boncompain and J. Frene, Thermohydrodynamic Analysis of a Finite Journal Bearing's Static and Dynamic Characteristics, Proc. 6th Leeds-Lyon Symp. on Tribology, Thermal Effects in Tribology, Sept. 1979, editors, D. Dowson, C.M. Taylor, M. Godet and D. Berthe, Inst. Mech. Engrs. Publ., London, 1980, pp. 33-41.

37 P.B. Neal, Some Factors Influencing the Operating Temperature of Pad Thrust Bearings, Proc. 6th Leeds-Lyon Symp. on Tribology, Thermal Effects in Tribology, Sept. 1979, editors, D. Dowson, C.M. Taylor, M. Godet and D. Berthe, Inst. Mech. Engrs. Publ., London, 1980, pp. 137-142.

38 C.M.McC. Ettles and S. Advani, The Control of Thermal and Elastic Effects in Thrust Bearings, Proc. 6th Leeds-Lyon Symp. on Tribology, Thermal Effects in Tribology, Sept. 1979, editors, D. Dowson, C.M. Taylor, M. Godet and D. Berthe, Inst. Mech. Engrs. Publ., London, 1980, pp. 105-116.

39 C.M.McC. Ettles, *Private Communication*, University of London, 1978.

40 R. Stribeck, Die Wesentlich Eigenschaften der Gleit-und-Rollen-lager, *Zeitschrift des Vereines deutscher Ingenieure*, Vol. 46, 1902, No. 38, pp. 1341-1348, pp. 1432-1438, also No. 39, pp. 1463-1470.

41 L. Gumbel, Das Problem der Lagerreibung, *Monatsblatter, Berliner Bezirks Verein Deutscher Ingenieure (V.D.I)*, No. 5, 1914, pp. 97-104 and 109-120.

42 H. Fukayama, M. Tanaka and Y. Hori, Friction Reduction in Turbulent Journal Bearings by Highpolymers, *Transactions ASME, Journal of Lubrication Technology*, Vol. 102, 1980, pp. 439-444.

43 B.A. Gecim, Non-Newtonian Effects of Multigrade Oils on Journal Bearing Performance, *Tribology Transactions*, Vol. 33, 1990, pp. 384-394.

44 A. Mukherjee and A.M. Rao Dasary, Experimental Study of Rotor Bearing Systems Influenced by Dilute Viscoelastic Lubricants, *Tribology International*, Vol. 21, 1988, pp. 109-115.

45 A. Harnoy, An Analysis of Stress Relaxation in Elastico-Viscous Fluid Lubrication of Journal Bearings, *Transactions ASME, Journal of Lubrication Technology*, Vol. 100, 1978, pp. 287-295.

46   V.N. Constantinescu, On the Influence of Inertia Forces in Turbulent and Laminar Self-Acting Film, *Transactions ASME, Journal of Lubrication Technology*, Vol. 92, 1970, pp. 473-481.

47   B.E. Launder and M. Leschziner, Flow in Finite-Width Thrust Bearings Including Thrust Effects 1 - Laminar Flow, *Transactions ASME, Journal of Lubrication Technology*, Vol. 100, 1978, pp. 330-338.

48   J.A. Tichy and S.H. Chen, Plane Slider Bearing Load Due to Fluid Inertia - Experiment and Theory, *Transactions ASME, Journal of Lubrication Technology*, Vol. 107, 1985, pp. 32-38.

49   H.I. You and S.S. Lu, Inertia Effect in Hydrodynamic Lubrication With Film Rupture, *Transactions ASME, Journal of Tribology*, Vol. 109, 1987, pp. 87-90.

50   H.I. You and S.S. Lu, The Effect of Fluid Inertia on the Operating Characteristics of a Journal Bearing, *Transactions ASME, Journal of Tribology*, Vol. 110, 1988, pp. 499-502.

51   H. Hashimoto, S. Wada and M. Sumitomo, The Effects of Fluid Inertia Forces on the Dynamic Behaviour of Short Journal Bearings in Superlaminar Flow Regime, *Transactions ASME, Journal of Tribology*, Vol. 110, 1988, pp. 539-547.

52   J.A. Tichy, Measurements of Squeeze-Film Bearing Forces and Pressures, Including the Effect of Fluid Inertia, *ASLE Transactions*, Vol. 28, 1985, pp. 520-526.

53   L.A. San Andres and J.M. Vance, Effect of Fluid Inertia on Squeeze-Film Damper Forces for Small-Amplitude Circular-Centered Motions, *ASLE Transactions*, Vol. 30, 1987, pp. 63-68.

54   W.J. Harrison, The Hydrodynamical Theory of Lubrication with Special Reference to Air as a Lubricant, *Trans. Cambridge Phil. Soc.*, Vol. 22, 1913, pp. 39-54.

55   W.A. Gross, Fluid Film Lubrication, John Wiley, New York, 1980.

56   A. Burgdorfer, The Influence of the Molecular Mean Free Path on the Performance of Hydrodynamic Gas Lubricated Bearings, *Transactions ASME, Journal of Basic Engineering*, Vol. 81, 1959, pp. 94-100.

57   B. Bhushan and K. Tonder, Roughness-Induced Shear- and Squeeze-Film Effects in Magnetic Recording, Parts 1 and 2, *Transactions ASME, Journal of Tribology*, Vol. 111, 1989, Part 1, pp. 220-227, Part 2, pp. 228-237.

58   Y. Mitsuya, Stokes Roughness Effects on Hydrodynamic Lubrication, Part 2 - Effects under Slip Flow Boundary Conditions, *Transactions ASME, Journal of Tribology*, Vol. 108, 1986, pp. 159-166.

59   S. Haber and I. Etsion, Analysis of an Oscillatory Oil Squeeze Film Containing a Central Gas Bubble, *ASLE Transactions*, Vol. 28, 1985, pp. 253-260.

60   D.W. Parkins and W.T. Stanley, Characteristics of an Oil Squeeze Film, *Transactions ASME, Journal of Lubrication Technology*, Vol. 104, 1982, pp. 497-503.

61   V.T. Morgan and A. Cameron, Mechanism of Lubrication in Porous Metal Bearings, Proceedings Conf. on Lubrication and Wear, Inst. Mech. Engrs., London, 1957, pp. 151-157.

62   F.E. Cardullo, Some Practical Deductions from the Theory of the Lubrication of Cylindrical Bearings, *Transactions ASME*, Vol. 52, 1930, pp. 143-153.

63   Cz. M. Rodkiewicz, K.W. Kim and J.S. Kennedy, On the Significance of the Inlet Pressure Build-up in the Design of Tilting-Pad Bearings, *Transactions ASME, Journal of Tribology*, Vol. 112, 1990, pp. 17-22.

64   J.A. Cole, Experimental Investigation of Power Loss in High Speed Plain Thrust Bearings, Proceedings Conf. on Lubrication and Wear, Inst. Mech. Engrs., London, 1957, pp. 158-163.

65   D. Dowson, Non-Steady State Effects in EHL, New Directions in Lubrication, Materials, Wear and Surface Interactions, Tribology in the 80's, edited by W.R. Loomis, Noyes Publications, Park Ridge, New Jersey, USA, 1985.

66   J.A. Cole and C.J. Hughes, Visual Study of Film Extent in Dynamically Loaded Complete Journal Bearings, Lubrication and Wear Conf., *Proc. Inst. Mech. Engrs.*, 1957, pp. 147-149.

67   E.R. Booser, CRC Handbook of Lubrication, Volume II, CRC Press, Boca Raton, Florida, 1984.

68   M. Tomimoto, Experimental Verification of a Particle Induced Friction Model in Journal Bearings, *Wear*, Vol. 254, 2003, pp. 749-762.

69   M. Mosleh, N. Saka and N.P. Suh, A Mechanism of High Friction in Dry Sliding Bearings, *Wear*, Vol. 252, 2002, pp. 1-8.

5

COMPUTATIONAL

HYDRODYNAMICS

5.1 INTRODUCTION

The differential equations which arose from the theories of Reynolds and later workers rapidly exceeded the capacity of analytical solution. For many years some heroic attempts were made to solve these equations using specialized and obscure mathematical functions but this process was tedious and the range of solutions was limited. A gap or discrepancy always existed between what was required in the engineering solutions to hydrodynamic problems and the solutions available. Before numerical methods were developed, analogue methods, such as electrically conductive paper, were experimented with as a means of determining hydrodynamic pressure fields. These methods became largely obsolete with the advancement of numerical methods to solve differential equations. This change radically affected the general understanding and approach to hydrodynamic lubrication and other subjects, e.g., heat transfer. It is now possible to incorporate in the numerical analysis of the bearing common features such as heat transfer from a bearing to its housing. The application of traditional, analytical methods would require to assume that the bearing is either isothermal or adiabatic. Numerical solutions to hydrodynamic lubrication problems can now satisfy most engineering requirements for prediction of bearing characteristics and improvements in the quality of prediction continue to be found. In engineering practice problems such as what is the maximum size of the groove to reduce friction before lubricant leakage becomes excessive or how does bending of the pad affect the load capacity of a bearing, need to be solved.

In this chapter the application of numerical analysis to problems encountered in hydrodynamic lubrication is described. A popular numerical technique, the '**finite difference method**', is introduced and its application to the analysis of hydrodynamic lubrication is demonstrated. The steps necessary to obtain solutions for different bearing geometries and operating conditions are discussed. Based on the example of the finite journal bearing it is shown how fundamental characteristics of the bearing, e.g., the rigidity of the bearing, the intensity of frictional heat dissipation and its lubrication regime, control its load capacity.

5.2 NON-DIMENSIONALIZATION OF THE REYNOLDS EQUATION

Non-dimensionalization is the substitution of all real variables in an equation, e.g., pressure, film thickness, etc., by dimensionless fractions of two or more real parameters. This process extends the generality of a numerical solution. A basic disadvantage of a numerical solution is that data are only provided for specific values of controlling variables, e.g., one value of friction force for a particular combination of sliding speed, lubricant viscosity, film thickness and bearing dimensions. Analytical expressions, on the other hand, are not limited to any specific values and are suited for providing data for general use, for example, they can be incorporated in an optimization process to determine the optimum lubricant viscosity. A computer program would have to be executed for literally thousands of cases to provide a comprehensive coverage of all the controlling parameters. The benefit of non-dimensionalization is that the number of controlling parameters is reduced and a relatively limited data set provides the required information on any bearing.

The Reynolds equation (4.24) is expressed in terms of film thickness '**h**', pressure '**p**', entraining velocity '**U**' and dynamic viscosity '**η**'. Non-dimensional forms of the equation's variables are:

$$\mathbf{h^*} = \frac{\mathbf{h}}{\mathbf{c}}$$

$$\mathbf{x^*} = \frac{\mathbf{x}}{\mathbf{R}}$$

$$\mathbf{y^*} = \frac{\mathbf{y}}{\mathbf{L}} \qquad\qquad (5.1)$$

$$\mathbf{p^*} = \frac{\mathbf{p c^2}}{\mathbf{6 U \eta R}}$$

where:

h	is the hydrodynamic film thickness [m];
c	is the bearing radial clearance [m];
R	is the bearing radius [m];
L	is the bearing axial length [m];
p	is the pressure [Pa];
U	is the bearing entraining velocity [m/s], i.e., $\mathbf{U} = (\mathbf{U_1} + \mathbf{U_2})/2$;
η	is the dynamic viscosity of the bearing [Pas];
x, y	are hydrodynamic film co-ordinates [m].

The Reynolds equation in its non-dimensional form is:

$$\boxed{\frac{\partial}{\partial \mathbf{x^*}}\left(\mathbf{h^{*3}}\frac{\partial \mathbf{p^*}}{\partial \mathbf{x^*}}\right) + \left(\frac{\mathbf{R}}{\mathbf{L}}\right)^2 \frac{\partial}{\partial \mathbf{y^*}}\left(\mathbf{h^{*3}}\frac{\partial \mathbf{p^*}}{\partial \mathbf{y^*}}\right) = \frac{\partial \mathbf{h^*}}{\partial \mathbf{x^*}}} \qquad (5.2)$$

All terms in equation (5.2) are non-dimensional apart from '**R**' and '**L**' which are only present as a non-dimensional ratio.

Although any other scheme of non-dimensionalization can be used, this particular scheme is the most popular and convenient. For planar pads, '**R**' is substituted by the pad width '**B**' in the direction of sliding.

5.3 THE VOGELPOHL PARAMETER

The Vogelpohl parameter was developed to improve the accuracy of numerical solutions of the Reynolds equation and was introduced by Vogelpohl [1] in the 1930's. The Vogelpohl parameter '**M_v**' is defined as follows:

$$M_v = p^* h^{*1.5}$$

(5.3)

Substitution into the non-dimensional form of Reynolds equation (5.2) yields the 'Vogelpohl equation':

$$\frac{\partial^2 M_v}{\partial x^{*2}} + \left(\frac{R}{L}\right)^2 \frac{\partial^2 M_v}{\partial y^{*2}} = FM_v + G$$

(5.4)

where parameters 'F' and 'G' for journal bearings are as follows:

$$F = \frac{0.75\left[\left(\frac{\partial h^*}{\partial x^*}\right)^2 + \left(\frac{R}{L}\right)^2 \left(\frac{\partial h^*}{\partial y^*}\right)^2\right]}{h^{*2}} + \frac{1.5\left[\frac{\partial^2 h^*}{\partial x^{*2}} + \left(\frac{R}{L}\right)^2 \frac{\partial^2 h^*}{\partial y^{*2}}\right]}{h^*}$$

(5.5)

$$G = \frac{\left(\frac{\partial h^*}{\partial x^*}\right)}{h^{*1.5}}$$

(5.6)

The Vogelpohl parameter facilitates computing by simplifying the differential operators of the Reynolds equation, and furthermore it does not show high values of higher derivatives in the final solution, i.e., $d^n M_v/dx^{*n}$ where $n > 2$, unlike the dimensionless pressure 'p^*'. This is because where there is a sharp increase in 'p^*' close to the minimum of hydrodynamic film thickness 'h^*', 'M_v' remains at moderate values. Large values of higher derivatives cause significant truncation error in numerical analysis. The characteristics of 'M_v' and 'p^*' for a journal bearing at an eccentricity of 0.95 are shown in Figure 5.1.

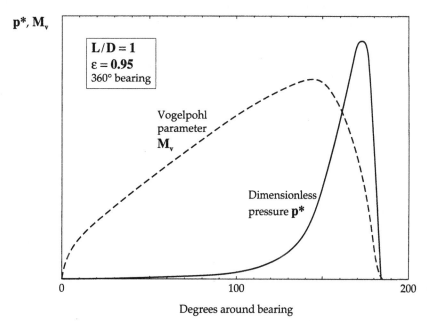

FIGURE 5.1   Variation of dimensionless pressure and the Vogelpohl parameter along the centre plane of a journal bearing [4].

It can be seen from Figure 5.1 that the introduction of the Vogelpohl parameter does not complicate the boundary conditions in the Reynolds equation, since wherever **p* = 0**, also **M_v = 0** (zero values of '**h**', i.e., solid to solid contact, are not included in the analysis). As discussed later in this chapter, wherever cavitation occurs, the gradient of '**M_v**' adjacent and normal to the cavitation front is zero like that of '**p***'.

Numerical solutions of the Reynolds equation are obtained in terms of '**M_v**' and values of '**p***' found from the definition $M_v/h^{*1.5} = p^*$.

5.4 FINITE DIFFERENCE EQUIVALENT OF THE REYNOLDS EQUATION

Journal and pad bearing problems are usually solved by a 'finite difference' method, although a 'finite element' method has also been employed [2]. The finite difference method is based on approximating a differential quantity by the difference between function values at two or more adjacent nodes. For example, the finite difference approximation to $\partial M_v/\partial x^*$ is given by:

$$\left(\frac{\partial M_v}{\partial x^*}\right)_i \approx \frac{M_{v,i+1} - M_{v,i-1}}{2\,\delta x^*} \tag{5.7}$$

where the subscripts **i-1** and **i+1** denote positions immediately behind and in front of the central position '**i**' and '**δx***' is the step length between nodes. A similar expression results for the second differential $\partial^2 M_v/\partial x^{*2}$. This expression can be found according to the principle illustrated in Figure 5.2.

The second differential $\partial^2 M_v/\partial x^{*2}$ is found by subtracting the expression for $\partial M_v/\partial x^*$ at the **i-0.5** nodal position from the **i+0.5** nodal position and dividing by δx*, i.e.:

$$\left(\frac{\partial^2 M_v}{\partial x^{*2}}\right)_i \approx \frac{\left(\frac{\partial M_v}{\partial x^*}\right)_{i+0.5} - \left(\frac{\partial M_v}{\partial x^*}\right)_{i-0.5}}{\delta x^*} \tag{5.8}$$

where:

$$\left(\frac{\partial M_v}{\partial x^*}\right)_{i+0.5} \approx \frac{M_{v,i+1} - M_{v,i}}{\delta x^*}$$

$$\left(\frac{\partial M_v}{\partial x^*}\right)_{i-0.5} \approx \frac{M_{v,i} - M_{v,i-1}}{\delta x^*}$$

substituting into (5.8) yields:

$$\left(\frac{\partial^2 M_v}{\partial x^{*2}}\right)_i \approx \frac{M_{v,i+1} + M_{v,i-1} - 2M_{v,i}}{(\delta x^*)^2}$$

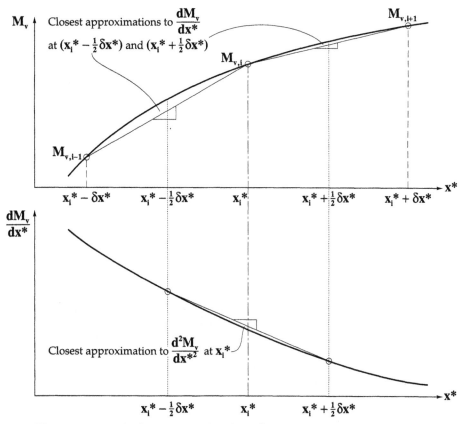

FIGURE 5.2   Illustration of the principle for the derivation of the finite difference approximation of the second derivative of a function.

The finite difference equivalent of $(\partial^2 M_v/\partial x^{*2} + \partial^2 M_v/\partial y^{*2})$ is found by considering the nodal variation of 'M_v' in two axes, i.e., the 'x' and 'y' axes. A second nodal position variable is introduced along the 'y' axis, the 'j' parameter. The expressions for $\partial M_v/\partial y^*$ and $\partial^2 M_v/\partial y^{*2}$ are exactly the same as the expressions for the 'x' axis but with 'i' substituted by 'j'. The coefficients of 'M_v' at the 'i'-th node and adjacent nodes required by the Reynolds equation which form a 'finite difference operator' are usually conveniently illustrated as a 'computing molecule' as shown in Figure 5.3.

The finite difference operator is convenient for computation and does not create any difficulties with boundary conditions. When the finite difference operator is located at the boundary of a solution domain, special arrangements may be required with imaginary nodes outside of the boundary. The solution domain is the range over which a solution is applicable, i.e., the dimensions of a bearing. There are more complex finite difference operators available based on longer strings of nodes but these are difficult to apply because of the requirement for nodes outside of the solution domain and are rarely used despite their greater accuracy. The terms 'F' and 'G' can be included with the finite difference operator to form a complete equivalent of the Reynolds equation. The equation can then be rearranged to provide an expression for '$M_{v,i,j}$' i.e.:

$$M_{v,i,j} = \frac{C_1\left(M_{v,i+1,j} + M_{v,i-1,j}\right) + \left(\dfrac{R}{L}\right)^2 C_2\left(M_{v,i,j+1} + M_{v,i,j-1}\right) - G_{i,j}}{2C_1 + 2C_2 + F_{i,j}} \tag{5.9}$$

where:

$$C_1 = \frac{1}{\delta x^{*2}}$$

$$C_2 = \frac{1}{\delta y^{*2}}$$

This expression forms the basis of the finite difference method for the solution of the Reynolds equation. Its solution gives the required nodal values of 'M_v'.

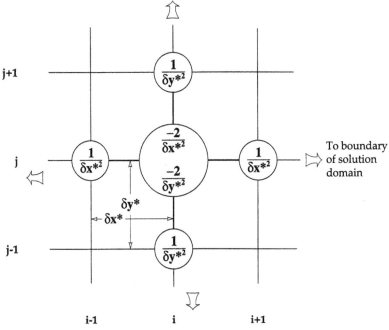

FIGURE 5.3   Finite difference operator and nodal scheme for numerical analysis of the Reynolds equation.

Definition of Solution Domain and Boundary Conditions

After establishing the controlling equation, the next step in numerical analysis is to define the boundary conditions and range of values to be computed. For the journal or pad bearing, the boundary conditions require that 'p^*' or 'M_v' is zero at the edges of the bearing and also that cavitation can occur to prevent negative pressures occurring within the bearing. The range of 'x^*' is between 0 - 2π (360° angle) for a complete bearing or some smaller angle for a partial arc bearing. The range of 'y^*' is from -0.5 to +0.5 if the mid-line of the bearing is selected as a datum. A domain of the journal bearing where symmetry can be exploited to cover either half of the bearing area, i.e., from $y^* = 0$ to $y^* = 0.5$, or the whole bearing area is shown in Figure 5.4. Nodes on the edges of the bearing remain at a pre-determined zero value while all other nodes require solution by the finite difference method. When symmetry is exploited to solve for only a half domain, it should be noted that nodes on the mid-line of the bearing are also variable and the finite difference operator requires an extra column of nodes outside the solution domain, as zero values along the edge of the solution domain cannot be assumed. This extra column is generated by adopting node values from the column one step from the mid-line on the opposite side. In analytical terms this is achieved by setting:

$$\mathbf{M}_{v,i,jnode+1} = \mathbf{M}_{v,i,jnode-1} \tag{5.10}$$

where '**jnode**' is the number of nodes in the '**j**' or '**y***' direction.

A split domain reduces the number of nodes but when analyzing a non-symmetric or misaligned bearing then a domain covering the complete bearing area is necessary. The domain in this case is the complete bearing with limits of '**y***' from **-0.5** to **+0.5** and the mid-line boundary condition vanishes.

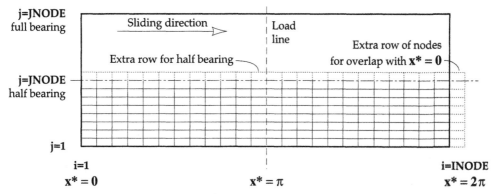

FIGURE 5.4   Nodal pressure or Vogelpohl parameter domains for finite difference analysis of hydrodynamic bearings.

Calculation of Pressure Field

It is possible to apply the direct solution method to calculate the pressure field but this is quite complex. In practice the pressure field in a bearing is calculated by the iteration procedure, which is discussed in the next section.

Calculation of Dimensionless Friction Force and Friction Coefficient

Once the pressure field has been found, it is possible to calculate the friction force and friction coefficient from the film thickness and pressure gradient data. As discussed already in Chapter 4, the frictional force operating across the hydrodynamic film is calculated by integrating the shear stress 'τ' over the bearing area, i.e.:

$$\mathbf{F} = \int_0^L \int_0^{2\pi R} \tau \, \mathbf{dx} \, \mathbf{dy} \tag{5.11}$$

where the shear stress 'τ' is given by:

$$\tau = \frac{\eta U}{h} + \frac{h}{2} \frac{dp}{dx} \tag{5.12}$$

where:

τ is the shear stress [Pa];

η is the dynamic viscosity of the lubricant [Pas];

U is the entraining velocity [m/s];

h   is the hydrodynamic film thickness [m];

p   is the hydrodynamic pressure [Pa];

F   is the friction force [N];

x   is the distance in the direction of sliding [m];

y   is the distance normal to the direction of sliding [m].

In a manner similar to the computation of pressure, the equation for friction force can be expressed in terms of non-dimensional quantities. From (5.1) $h = h^*c$, $x = x^*R$ and $p = p^*(6U\eta R)/c^2$ and substituting into (5.12) yields:

$$\tau = \frac{\eta U}{c}\frac{1}{h^*} + \frac{ch^*}{2}\frac{6U\eta R}{c^2}\frac{1}{R}\frac{dp^*}{dx^*} = \left(\frac{U\eta}{c}\right)\left(\frac{1}{h^*} + 3h^*\frac{dp^*}{dx^*}\right) \tag{5.13}$$

Substituting for 'x' and 'y' from (5.1) the '$\tau dxdy$' in terms of non-dimensional quantities is:

$$\tau\, dx\, dy = \tau\, dx^*\, dy^* RL \tag{5.14}$$

Substituting (5.14) and (5.13) into (5.11) results in an expression for frictional force in terms of non-dimensional quantities:

$$\begin{aligned} F = \int_0^L\int_0^{2\pi R} \tau\, dx\, dy &= RL\int_0^1\int_0^{2\pi} \tau\, dx^*\, dy^* \\ &= \frac{RL\eta U}{c}\int_0^1\int_0^{2\pi}\left(\frac{1}{h^*} + 3h^*\frac{dp^*}{dx^*}\right)dx^*\, dy^* \end{aligned} \tag{5.15}$$

It can be seen from equation (5.15) that the non-dimensional shear stress 'τ^*' is expressed as:

$$\tau^* = \frac{1}{h^*} + 3h^*\frac{dp^*}{dx^*} \tag{5.16}$$

and equation (5.15) can be re-written in the following form:

$$F = \frac{RL\eta U}{c}\int_0^1\int_0^{2\pi} \tau^*\, dx^*\, dy^* = F^*\left(\frac{RL\eta U}{c}\right) \tag{5.17}$$

The coefficient of friction is calculated by dividing the friction force by the load. A similar quantity is also found when the dimensionless friction is divided by the dimensionless load. Thus in journal bearings the friction coefficient is the ratio of circumferential friction force divided by the load:

$$\mu = \frac{F}{W} = \frac{\displaystyle\int_0^L\int_0^{2\pi R} \tau\, dx\, dy}{\displaystyle\int_0^L\int_0^{2\pi R} p\, dx\, dy} \tag{5.18}$$

where:

μ is the coefficient of friction;

W   is the bearing load [N].

Load on a journal bearing is often expressed as:

$$W = \int_0^L \int_0^{2\pi R} -\cos(x^*)\, p\, dx\, dy \tag{5.19}$$

where the term **-cos(x*)** arises from the fact that load supporting pressure is located close to **x* = π** or **cos(x*) = -1**. Any pressure close to **x* = 0** merely imposes an extra load on the bearing since it acts in the direction of the load. The negative sign refers to the fact that the load vector does not coincide with the position of maximum film thickness.

Expressing equation (5.19) in terms of non-dimensional quantities yields:

$$W = \left(\frac{6U\eta R}{c^2}\right) RL \int_0^1 \int_0^{2\pi} -\cos(x^*)\, p^*\, dx^*\, dy^* = W^* \left(\frac{6R^2 LU\eta}{c^2}\right) \tag{5.20}$$

Substituting (5.17) and (5.20) into (5.18) gives the expression for coefficient of friction:

$$\mu = \frac{F}{W} = \frac{\left(\dfrac{RL\eta U}{c}\right)}{\left(\dfrac{6R^2 LU\eta}{c^2}\right)} \frac{F^*}{W^*} = \left(\frac{c}{6R}\right)\left(\frac{F^*}{W^*}\right) \tag{5.21}$$

hence:

$$\frac{F^*}{W^*} = \left(\frac{6R}{c}\right)\mu \tag{5.22}$$

The presence of cavitation in the bearing adds some complication to the calculation of the coefficient of friction. Within the cavitated region, the proportion of clearance space between shaft and bush that is filled by lubricant is:

$$\frac{h_{cav}^*}{h^*}$$

where:

h_{cav}^* is the dimensionless film thickness at the cavitation front;

h^* is the dimensionless film thickness at a specified position downstream of the cavitation front.

The average or 'effective' coefficient of friction is proportional to the lubricant filled fraction of the clearance space and within the cavitated region '**p***' and '**dp*/dx***' are equal to zero. The symbol for 'effective' dimensionless shear stress is 'τ_e^*'. Assuming a simple proportionality between fluid filled volume and total shear force, an average value of 'τ_e^*' that allows for zero shear stress between streamers of lubricant is given by:

$$\tau_e^* = \frac{h_{cav}^*}{h^{*2}}$$ (5.23)

This value of dimensionless shear stress is included in the integral for dimensionless friction force (eq. 5.17) with no further modification.

Values of h^*, $\partial h^*/\partial x^*$, $\partial h^*/\partial y^*$ and $\partial^2 h^*/\partial x^{*2}$ are also required in computation and the expressions for these are:

$$h^* = y^* t \cos(x^*) + \varepsilon \cos(x^* - \beta) + 1$$ (5.24)

where:

  ε is the eccentricity ratio;

  t is the misalignment factor;

  β is the attitude angle.

Note that the variation in 'h^*' due to misalignment is dependent on 'x^*' whereas the variation in 'h^*' due to eccentricity is also controlled by the attitude angle.

The derivatives of 'h^*' are found by direct differentiation of (5.24), i.e.:

$$\frac{dh^*}{dx^*} = -y^* t \sin(x^*) - \varepsilon \sin(x^* - \beta)$$ (5.25)

$$\frac{dh^*}{dy^*} = t \cos(x^*)$$ (5.26)

$$\frac{d^2h^*}{dx^{*2}} = -y^* t \cos(x^*) - \varepsilon \cos(x^* - \beta)$$ (5.27)

Numerical Solution Technique for Vogelpohl Equation

The nodal values of 'M_v' are conveniently arranged in a matrix with 'i' and 'j' as the column and row ordinates. The coefficients in equation (5.9) can also be organized into a 'sparse' matrix with all coefficients lying close to the main diagonal. It is therefore possible to solve equation (5.9) by matrix inversion but this requires elaborate computation. Programming is greatly simplified when iterative solution methods are applied. The Gauss-Seidel iterative method is used in this chapter. All node values are assigned an initial zero value and the finite difference equation (5.9) is repeatedly applied until convergence is obtained.

5.5 NUMERICAL ANALYSIS OF HYDRODYNAMIC LUBRICATION IN IDEALIZED JOURNAL AND PARTIAL ARC BEARINGS

A numerical solution to the Reynolds equation for the full and partial arc journal bearings is necessary for the calculation of pressure distribution, load capacity, lubricant flow rate and friction coefficient when the bearings are neither 'infinitely long' nor 'infinitely narrow'. This condition is valid for bearings with **L/D** ratio in the range **1/3 < L/D < 3**, where '**L**' is the bearing length and '**D**' is the bearing diameter. Equation (5.9) is solved numerically in order to find the dimensionless pressure field corresponding to equation (5.2) and the other important bearing parameters. An example of the flowchart of the computer program 'PARTIAL' for the analysis of a partial arc or full 360°, isothermal, rigid and non-vibrating journal bearing is shown in Figure 5.5 while the full listing of the program with description is provided in the Appendix. The program provides a solution for aligned and misaligned journal bearings. Misalignment has a pronounced effect on bearing characteristics but cannot be modelled by either infinitely long or narrow bearing theories. Numerical methods help to overcome this problem.

Example of Data from Numerical Analysis, the Effect of Shaft Misalignment

The computed solution to the classical Reynolds equation as applied to journal bearings has been comprehensively used to obtain basic information for bearing design. An example of this data was shown in Chapter 4 for the 360° journal bearing (Figure 4.32). Tables of data for load and attitude angle as a function of eccentricity, **L/D** ratio and partial arc angle can be found elsewhere [e.g., 4,5,6]. A computer program 'PARTIAL' for the analysis of a partial arc bearing is listed in the Appendix. The program calculates the dimensionless load, attitude angle, Petroff multiplier and dimensionless friction coefficient for a specified angle of partial arc bearing, **L/D**, eccentricity and misalignment ratios. The solution is based on an isoviscous model of hydrodynamic lubrication with no elastic deflection of the bearing.

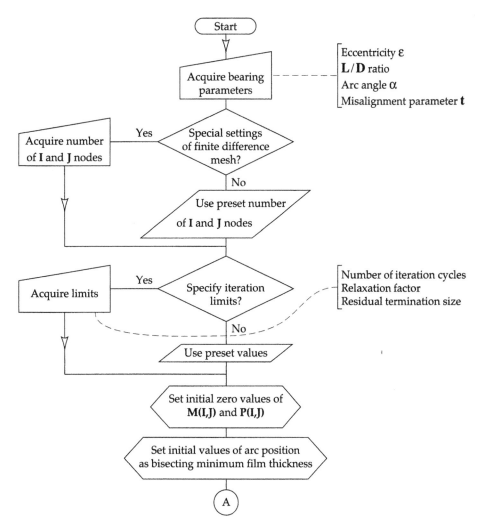

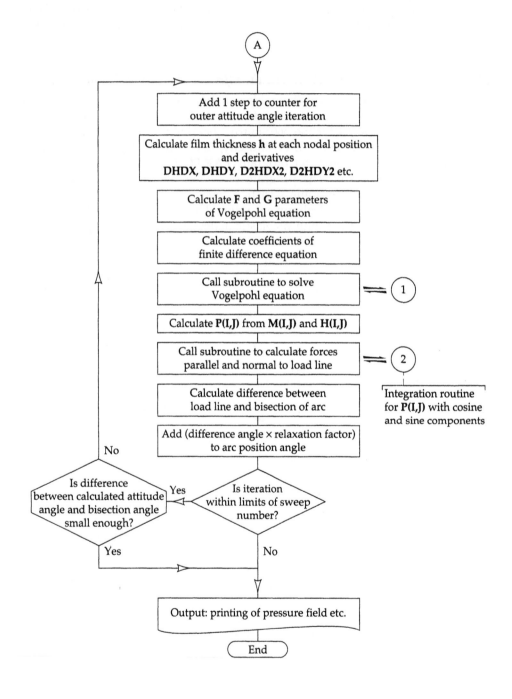

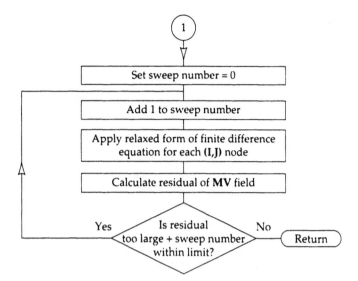

FIGURE 5.5 Flowchart of the program for idealized, isothermal rigid and non-vibrating journal bearing.

The effect of misalignment is very important in bearing design since it has a pronounced influence on maximum hydrodynamic pressure. In Figure 5.6 this effect is illustrated for a **120°** arc bearing, **L/D = 1** and an eccentricity ratio of **0.7**.

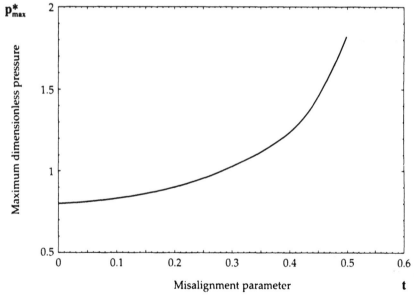

FIGURE 5.6 Effect of misalignment on maximum hydrodynamic pressure in a partial arc bearing.

It can be seen that the maximum pressure is more than doubled as misalignment increases from 0 to **0.5**. The limiting value of misalignment before contact occurs between the shaft and the bush is **0.6** for a value of eccentricity ratio of **0.7**. This is based on equation (5.24) which implies that the sum of eccentricity and half the misalignment must be less than **1** for no solid contact.

This effect can also be demonstrated easily by comparing the pressure fields for perfectly aligned and misaligned bearings. The computed pressure field for a perfectly aligned **120°** partial arc bearing, **L/D = 1** at an eccentricity ratio of **0.7** is shown in Figure 5.7 and the computed pressure field for the same bearing with a misalignment parameter **t = 0.5** is shown in Figure 5.8. The data are obtained by executing program '**PARTIAL**' and are arranged in a perspective view to show the entire profile rather than just a section through '**x***' or '**y***'. All pressures are presented as percentages of the peak pressure.

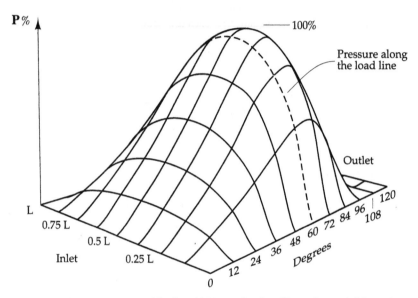

FIGURE 5.7   Computed pressure profile for 120° perfectly aligned partial bearing.

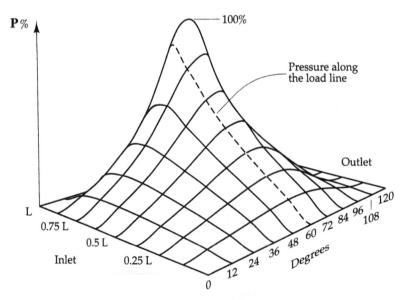

FIGURE 5.8   Computed pressure field for 120° misaligned partial bearing.

It can be seen from Figures 5.7 and 5.8 that the misaligned bearing shows a skewed pressure field with a relatively high pressure close to the minimum film thickness, whereas the perfectly aligned bearing gives the more uniform distribution of pressure. The maximum pressure for the perfectly aligned bearing is about 40% of the maximum pressure under extreme misalignment for the conditions considered.

The computed data also show that load capacity is hardly affected by misalignment apart from a rise in load capacity immediately prior to shaft contact. The accepted view that misalignment has no significant effect on load capacity [4] is confirmed but the aforementioned data show that other critical parameters are controlled by misalignment. For instance, certain soft bearing materials, e.g., babbits, may suffer damage from excessive hydrodynamic pressure.

The data generated by the program 'PARTIAL' are in dimensionless form but for practical applications the dimensioned quantities are usually required. The following example shows how these dimensioned values can be found.

EXAMPLE

For a 120° partial arc bearing with $L/D = 1$, eccentricity ratio $\varepsilon = 0.7$ and misalignment parameter $t = 0.4$ find the load capacity, the maximum pressure and minimum film thickness. Assume that the bearing dimensions are $R = 0.1$ [m], $L = 0.2$ [m], $c = 0.0002$ [m] and that the bearing entraining velocity is $U = 10$ [m/s] and the dynamic viscosity of the lubricant is $\eta = 0.05$ [Pas].

· *Solution*

Executing the program 'PARTIAL' with a mesh size of **11** columns in the 'x^*' direction and **9** rows in the 'y^*' direction yields: dimensionless load $W^* = 0.6029$ and maximum dimensionless pressure $p^*_{max} = 1.2684$.

Since the load 'W' and pressure 'p' expressed in non-dimensional terms are:

$$W = W^* \left(\frac{6R^2 LU\eta}{c^2} \right)$$

$$p = p^* \left(\frac{6RU\eta}{c^2} \right)$$

substituting the bearing data yields:

$$W = 0.6029 \left(\frac{6 \times 0.1^2 \times 0.2 \times 10 \times 0.05}{0.0002^2} \right) = 90.435 \text{ [kN]}$$

$$p_{max} = 1.2684 \left(\frac{6 \times 0.1 \times 10 \times 0.05}{0.0002^2} \right) = 9.513 \text{ [MPa]}$$

The difference in lubricant film thickness or shaft clearance along the load line from one side of the bearing to the other (i.e., the distance between the axes of the tilted and non-tilted shaft measured at the edges of the bearing) is:

$$m = tc$$

Substituting gives:

$$m = 0.4 \times 0.0002 = 8 \times 10^{-5} \text{ [m]}$$

The minimum film thickness has two components: one due to eccentricity (as described in Chapter 4, i.e., $h_{min,ecc} = c(1 - \varepsilon)$) and one due to misalignment and can be estimated from the formula:

$$h_{min} = h_{min,ecc} - 0.5tc = c(1 - \varepsilon) - 0.5tc$$

Misalignment is calculated from the centre of the bearing, hence the term '$0.5tc$'. Strictly speaking, an exact calculation should allow for the angle between the film thickness variation due to misalignment and that due to eccentricity, i.e., the attitude angle. The approximate method underestimates the minimum clearance and therefore provides a small margin of error biased towards reliable bearing operation. Substituting for 'c', 'ε' and 't' yields:

$$h_{min} = 0.0002 \times (1 - 0.7) - 0.5 \times 0.4 \times 0.0002 = 2 \times 10^{-5} \; [m]$$

It should also be mentioned that the small value of clearances in the bearing illustrates the limitations of increasing the potential load capacity of a bearing by reducing the nominal clearance. Reduction in clearance also increases the maximum film pressures to a level where most bearing materials fail. For example, if the clearance is reduced by a factor of **10**, then the peak pressure is increased by a factor of **100** to **951.3** [MPa].

The likelihood of misalignment in real bearings is one of the major factors preventing the measurement of shaft load from a limited number of film pressure measurements. Unless the misalignment is accurately measured, no assumptions can be made about the pressure profile. This condition impedes what would be a very convenient means of load monitoring on large journal bearings.

The effect of misalignment on friction coefficient can also be tested. The effect is relatively mild and is limited to about 10% decline in friction coefficient as misalignment reaches its maximum value before shaft and bush contact.

5.6 NUMERICAL ANALYSIS OF HYDRODYNAMIC LUBRICATION IN A REAL BEARING

The numerical analysis of the ideal bearing described in the previous section was compiled many years ago. Present research is mostly directed to modelling effects such as heating and elastic distortion in bearings. Heating and elastic distortion diminish the load capacity of a bearing compared to the predictions of the classical Reynolds theory. The prevailing trends towards higher speeds and loads have heightened the need for accurate predictions of load capacity. Realistic modelling of the method of lubricant supply, i.e., the grooves and feed-holes, requires detailed computation of the cavitation and reformation fronts. It also cannot always be assumed that the lubricant is supplied from a groove equal in length to the bearing and located directly above the load vector. Calculation of vibrational instability, which is known as 'oil whirl' or 'oil whip', also depends on numerical solutions. In the following sections the computation methods allowing for these effects will be discussed.

5.6.1 THERMOHYDRODYNAMIC LUBRICATION

There are many examples where liquids can be heated by intense viscous shearing, e.g., a hydrodynamic brake or viscous damper. This process occurs in a high speed bearing and the loss of viscosity due to heating causes a significant loss of bearing load capacity. An example of this effect is illustrated in Figure 5.9 where the pressure field in a pad bearing predicted by both the isoviscous (Reynolds) model and the isothermal model is shown.

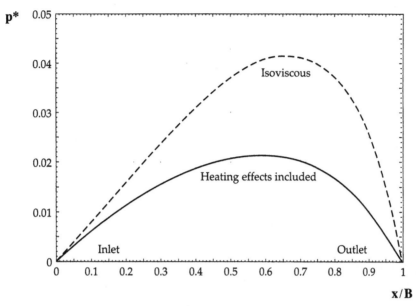

FIGURE 5.9   Pressure field in a pad bearing predicted by the isoviscous and isothermal models of hydrodynamic lubrication [8].

It can be seen from Figure 5.9 that the isoviscous pressure field at any point is almost twice that of the isothermal pressure field. The ratio between corresponding load capacities, in this case, is also approximately **2 : 1**.

The analysis of a hydrodynamic lubrication problem with allowance for viscous heating is commonly known as thermohydrodynamics. Highly precise, more realistic solutions of thermohydrodynamic problems include effects such as heat transfer to the bearing housing and forced cooling of a bearing. The rise in temperature may also cause thermally induced distortion of the film geometry [9], further complicating the analysis. In this section a simplified example of thermohydrodynamic analysis for a rigid pad bearing which is either perfectly adiabatic or isothermal is presented. Thermohydrodynamic analysis of a journal bearing involves the same principles as for pad bearings but with the complication of cavitation and reformation boundaries. This problem is, however, beyond the scope of this book.

In thermohydrodynamic analysis, real variables are used instead of non-dimensional variables. Non-dimensionalization of thermohydrodynamics has not yet gained widespread usage because of the complexity and the lack of general agreement on one particular scheme of non-dimensionalization.

Governing Equations and Boundary Conditions in Thermohydrodynamic Lubrication

In the analysis of thermohydrodynamic lubrication the equation allowing for heating effects, the '**energy equation**', is used. The energy equation is a simplified form of the heat transfer equation commonly used in fluid mechanics.

Terms present in the heat transfer equation can be eliminated by an order of magnitude analysis to yield only the convection terms and the conduction normal to the plane of the film. The characteristics of pressure and shear stress equilibrium in the oil film are unchanged except that the viscosity is variable in all three dimensions, i.e., viscosity also varies through the film thickness. Earlier solutions by, for example, Vohr [10] assumed that viscosity was constant through the oil film thickness but this has since been found to be inaccurate. The standard form of the Reynolds equation is therefore not adequate for thermohydrodynamic analysis.

· *Governing Equations in Thermohydrodynamic Lubrication for a One-Dimensional Bearing*

The energy equation for a one-dimensional bearing is given by [8]:

$$\frac{\partial}{\partial x}\left(\rho u c_p T\right) + \frac{\partial}{\partial z}\left(\rho w c_p T\right) - K\frac{\partial^2 T}{\partial z^2} = \eta\left(\frac{\partial u}{\partial z}\right)^2 \qquad (5.28)$$

where:

u	is the velocity of the lubricant in the direction of sliding [m/s];
ρ	is the density of the lubricant [kg/m³];
c_p	is the specific heat of the lubricant [J/kgK] (For the purposes of computation, notation 'c_p' is used in this chapter for specific heat instead of 'σ');
K	is the thermal conductivity of the lubricant [W/mK];
T	is the operating temperature of the lubricant [K];
w	is the lubricant velocity normal to the horizontal plane of the oil film (i.e., in the '**z**' direction) [m/s];
x	is the co-ordinate in the direction of sliding [m];
z	is the co-ordinate normal to the horizontal plane of the lubricant film [m].

Changes in lubricant viscosity with temperature can be calculated from, for example, the Vogel equation (Table 2.1):

$$\eta = a e^{b/(T-c)} \qquad (5.29)$$

where:

η	is the dynamic viscosity of the lubricant at the required temperature '**T**' [Pas];
a, b, c	are constants.

Changes in lubricant density with temperature due to thermal expansion can be evaluated from, for example, the exponential law:

$$\rho = \rho_0 e^{\zeta(T-T_0)} \qquad (5.30)$$

where:

ρ	is the density of the lubricant at the required temperature '**T**' [kg/m³];
ρ_0	is the density of the lubricant at the reference temperature '**T₀**' [kg/m³];

ζ is an exponent of density-temperature dependence (typically $\zeta = 0.001$) [K^{-1}].

For oils, thermal expansion is not significant and is usually neglected.

As discussed already in Chapter 4, for a one-dimensional bearing, the equilibrium condition for forces acting in the 'x' direction is given by:

$$\boxed{\frac{\partial p}{\partial x} - \frac{\partial}{\partial z}\left(\eta\frac{\partial u}{\partial z}\right) = 0}$$

(5.31)

Since pressure is constant through the hydrodynamic film thickness, the pressure gradient in the 'z' direction is equal to zero:

$$\boxed{\frac{\partial p}{\partial z} = 0}$$

(5.32)

and the continuity of flow requires that:

$$\boxed{\frac{\partial u}{\partial x} + \frac{\partial w}{\partial z} = 0}$$

(5.33)

Integrating equation (5.31) with respect to 'z' yields the expression for 'u' at a particular value of 'x' and 'z' in the oil film. This is the one-dimensional form of the Navier-Stokes equation, i.e.:

$$u_z = \frac{dp}{dx}\int_0^z \frac{z}{\eta}\,dz + A\int_0^z \frac{dz}{\eta} + U$$

(5.34)

where 'U' is the bearing velocity [m/s] and 'A' is given by:

$$A = \frac{\left(-\dfrac{dp}{dx}\displaystyle\int_0^h \frac{z}{\eta}\,dz\right) - U}{\displaystyle\int_0^h \frac{dz}{\eta}}$$

(5.35)

In the bearing considered the bottom surface is moving whereas the top surface is stationary, i.e.:

$u = U$ at $z = 0$ and

$u = 0$ at $z = h$

The solution for 'u' is found from the constant value of lubricant flow along the one-dimensional bearing. Since there is no side leakage in one-dimensional bearings the continuity of flow in the 'x' direction is maintained. In algebraic terms this means that the integral of lubricant velocity across the film thickness is constant, i.e.:

$$\frac{\partial}{\partial x}\left(\int_0^h u\,dz\right) = 0 \tag{5.36}$$

The velocity 'u' can be expressed in terms of parameters 'M' and 'N' which are composed of the integrals of viscosity 'η' [8], i.e.:

$$u = M\frac{dp}{dx} + NU \tag{5.37}$$

where the terms 'M' and 'N' are given by:

$$M = \int_0^z \frac{z}{\eta}\,dz - \frac{\int_0^h \frac{z}{\eta}\,dz \int_0^z \frac{dz}{\eta}}{\int_0^h \frac{dz}{\eta}} \tag{5.38}$$

$$N = 1 - \frac{\int_0^z \frac{dz}{\eta}}{\int_0^h \frac{dz}{\eta}} \tag{5.39}$$

Equations 5.34 and 5.35 can be rearranged to form terms that are products of 'dp/dx' or 'U', i.e.:

$$\int_0^h u\,dz = \frac{dp}{dx}\int_0^h M\,dz + U\int_0^h N\,dz \tag{5.40}$$

Substituting (5.40) into (5.36) gives the continuity of flow condition as an explicit equation for dp/dx, i.e.:

$$\boxed{\frac{d^2p}{dx^2}\int_0^h M\,dz + \frac{dp}{dx}\left(\frac{d}{dx}\int_0^h M\,dz\right) + U\frac{d}{dx}\int_0^h N\,dz = 0} \tag{5.41}$$

If the values of 'M' and 'N' are calculated from a known or assumed viscosity field, then a one-dimensional elliptic equation for pressure is formed which can be solved iteratively.

The continuity equation (5.33) is more conveniently solved by differentiating with respect to 'z' [8] which gives:

$$\frac{\partial^2 w}{\partial z^2} = -\frac{\partial}{\partial z}\left(\frac{\partial u}{\partial x}\right) \tag{5.42}$$

This form of the continuity equation facilitates solution by iteration of the equivalent finite difference equation (assuming that 'u' is already known).

To summarize the application of equations (5.37), (5.40) and (5.42), a known or assumed viscosity field is used to solve for pressure based on equation (5.41). The lubricant velocity in the direction of sliding is found from equation (5.37) and the lubricant velocity normal to the sliding direction from equation (5.42). Finally, based on this information a fluid temperature field is found.

· *Thermohydrodynamic Equations for the Finite Pad Bearing*

For a bearing of finite length a very similar analysis to the two-dimensional case can be applied [10]. The extra dimension, the 'y' axis, does not introduce any qualitative differences to the analysis. The energy equation in its three-dimensional form is:

$$\frac{\partial}{\partial x}\left(\rho u c_p T\right) + \frac{\partial}{\partial y}\left(\rho v c_p T\right) + \frac{\partial}{\partial z}\left(\rho w c_p T\right) - K\frac{\partial^2 T}{\partial z^2} = \eta\left[\left(\frac{\partial u}{\partial z}\right)^2 + \left(\frac{\partial v}{\partial z}\right)^2\right] \tag{5.43}$$

where:

    v   is the lubricant velocity in the 'y' direction (i.e., the direction normal to sliding) [m/s].

The equation for pressure is derived from the same principles as described in Chapter 4 for the infinitely long bearing. Equations (5.31) and (5.32) are applied and, in addition, an equation for pressure equilibrium along the 'y' axis is introduced, i.e.:

$$\frac{\partial p}{\partial y} = \frac{\partial}{\partial z}\left(\eta\frac{\partial v}{\partial z}\right) \tag{5.44}$$

with the boundary conditions:

    v = 0   at   **z = h**   and   **z = 0**

This leads to an expression for 'v' as a function of **dp/dy**, based on reasoning similar to that used in deriving equations (5.37), (5.38) and (5.39), i.e.:

$$v = \frac{dp}{dy}\left[\int_0^z \frac{z}{\eta}\,dz - \frac{\left(\int_0^z \frac{dz}{\eta}\right)\left(\int_0^h \frac{z}{\eta}\,dz\right)}{\int_0^h \frac{dz}{\eta}}\right] \tag{5.45}$$

Substituting for 'M' equation (5.45) simplifies to:

$$v = M\frac{dp}{dy} \tag{5.46}$$

An expression for 'p' in terms of derivatives with respect to 'x' and 'y' axes can now be derived. This expression is comparable to the isothermal Reynolds equation. The continuity of flow condition is given by:

$$\frac{\partial}{\partial x}\int_0^h u\,\partial z + \frac{\partial}{\partial y}\int_0^h v\,\partial z = 0 \tag{5.47}$$

Substituting (5.40) and (5.46) into (5.47) yields:

$$\frac{\partial}{\partial x}\left(\frac{\partial p}{\partial x}\int_0^h M\partial z + U\int_0^h N\partial z\right) + \frac{\partial}{\partial y}\left(\frac{\partial p}{\partial y}\int_0^h M\partial z\right) = 0 \tag{5.48}$$

This equation can be re-arranged to give:

$$\left(\frac{\partial^2 p}{\partial x^2} + \frac{\partial^2 p}{\partial y^2}\right)\int_0^h M\partial z + \left(\frac{\partial p}{\partial x}\frac{\partial}{\partial x} + \frac{\partial p}{\partial y}\frac{\partial}{\partial y}\right)\int_0^h M\partial z + U\frac{\partial}{\partial x}\int_0^h N\partial z = 0 \tag{5.49}$$

The equation in this form can also be solved by Gaussian iteration of equivalent finite difference equations.

Once the pressure field has been found, it is then possible to find values of 'w' from the three-dimensional continuity equation:

$$\frac{\partial w}{\partial z} + \frac{\partial v}{\partial y} + \frac{\partial u}{\partial x} = 0 \tag{5.50}$$

According to the method proposed by Ettles [9] this can be differentiated with respect to 'z' giving:

$$\frac{\partial^2 w}{\partial z^2} + \frac{\partial}{\partial z}\left(\frac{\partial v}{\partial y} + \frac{\partial u}{\partial x}\right) = 0 \tag{5.51}$$

When values of 'u' and 'v' are specified, this equation reduces to a one-dimensional equation for 'w' which can be solved iteratively.

· *Boundary Conditions*

The boundary conditions for the thermohydrodynamic bearing can cause as much difficulty as the controlling equations. Although the ideal bearings, in broad terms, can be classified as either isothermal or adiabatic, real bearings are neither perfectly isothermal nor perfectly adiabatic. The boundary conditions for adiabatic and isothermal bearings are illustrated in Figure 5.10.

For an isothermal bearing, all bearing surfaces remain at a fixed temperature, whereas for an adiabatic bearing, the temperature gradient of the lubricant becomes zero close to the static surface. The lack of temperature gradient close to the pad surface is caused by the absence of heat conduction from the lubricant to the pad. The adiabatic boundary condition can result in much higher temperatures than for the isothermal bearing. Fixed surface temperatures are more convenient for computation since this enables the boundary temperature nodes of the solution domain to be assigned a pre-determined value. For either isothermal or adiabatic bearings, the temperature of the rotating surface remains constant for most practical levels of sliding velocity. At the bearing outlet, the temperature gradient in the sliding direction, i.e., $\partial T/\partial x$ declines to zero as the lubricant leaves the control volume with an unchanged temperature.

The bearing inlet has a variable boundary condition which depends on the direction of lubricant flow at the bearing inlet. If there is no 'backflow', i.e., reverse flow of lubricant, at the inlet as illustrated in Figure 5.10, then the lubricant maintains a pre-determined inlet temperature. Although the phenomenon of 'hot oil carry over' described in Chapter 4 may also take place, it is not included in the model because of the analytical complexity it entails. If backflow occurs then the concept of an inlet temperature becomes invalid and it is necessary to assume that continuity of temperature gradient is maintained across the inlet boundary. In terms of computation, if backflow occurs then it is necessary to iterate at the inlet boundary and create an additional array of temperature nodes with values based on linear extrapolation from inside the bearing. Boundary conditions for the three-dimensional bearing, i.e., finite bearing, are similar to the two-dimensional bearing (i.e., width and film thickness) with the addition that at the sides of the bearing $\partial T/\partial y = 0$. Since the lubricant always flows outwards from the sides, provision for backflow at the sides is not required.

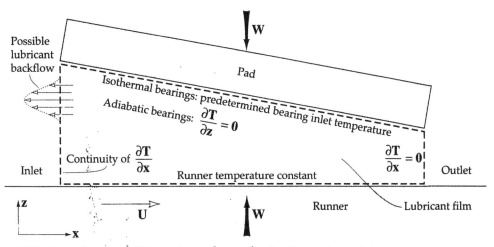

FIGURE 5.10 Boundary conditions for a thermohydrodynamic pad.

Finite Difference Equations for Thermohydrodynamic Lubrication

The complexity of the governing equations of thermohydrodynamic lubrication precludes a direct substitution of finite difference equivalents as was performed in the isoviscous case. The particular problem is that the energy equation contains convection terms which enforce a distinction between 'down-stream' and 'up-stream' in terms of lubricant flow. A finite difference method must therefore include a way of enforcing directionality of lubricant flow and this has to be specifically written into the computing procedure. The provision of a finite difference mesh that allows for varying film thickness also necessitates modification of the finite difference equations.

In order to solve this problem, a 'control volume' method has been introduced and adopted by several workers [e.g., 7,8]. A control volume mesh of the oil film for static and moving surfaces of the bearing was devised. An example of a control volume mesh composed of trapezia with the dimensions normal to the plane of the lubricant film varying in proportion to the lubricant film thickness is illustrated in Figure 5.11.

The grid scheme shown in Figure 5.11 is two dimensional. Other researchers [e.g., 10] have developed three-dimensional schemes which are more accurate but far more demanding of computing time. The edges or faces of each control volume (which is a quadrilateral) are given a code for use in the finite difference equations. Many researchers use points of the compass, i.e., '**North**', '**South**', '**East**' and '**West**', to indicate the four directions of relevance. These directions are denoted by the subscripts 'N', 'S', 'E' and 'W'. 'North' and 'South' directions are also referred to as '**Top**' and '**Bottom**' and are denoted by the subscripts 'T' and 'B', respectively. The centre of the control volume is commonly referred to as a '**Pole**' and is denoted by a subscript '**p**'. An example of the control volume mesh with this coding is shown in Figure 5.12. The dimensions of the mesh are individually labelled to allow the asymmetry of the control volume to be incorporated in the controlling equations.

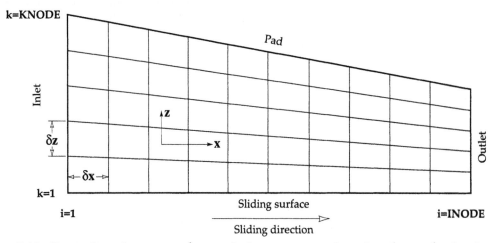

FIGURE 5.11 Control volume mesh used for computation in thermohydrodynamic lubrication.

The control volume equations can be found in [8]:

$$(a_p + E)T_p = a_E T_E + a_W T_W + a_N T_N + a_S T_S + b + E T_{p,old} \tag{5.52}$$

where:

$$a_E = -\frac{\Delta Z_E (\rho u)_e c_p}{2} \tag{5.53}$$

$$a_W = \frac{\Delta Z_W (\rho u)_w c_p}{2} \tag{5.54}$$

$$a_N = \frac{K \Delta X}{\Delta Z_p} - \frac{\Delta X (\rho w)_n c_p}{2} \tag{5.55}$$

$$a_S = \frac{K \Delta X}{\Delta Z_p} + \frac{\Delta X (\rho w)_s c_p}{2} \tag{5.56}$$

$$a_p = a_E + a_W + a_N + a_S - S_p \Delta X \Delta Z_p \tag{5.57}$$

$$b = S_c \Delta X \Delta Z_p \tag{5.58}$$

$T_{p,old}$ is the temperature value from the preceding iteration cycle;
K is the thermal conductivity of the lubricant [W/mK];
S_c is the temperature independent component of viscous heating [W/m³];
S_p is the temperature dependent component of viscous heating [W/m³K].

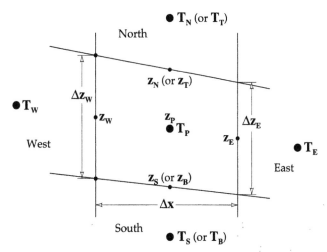

FIGURE 5.12 Coded mesh of the control volume used in thermohydrodynamics [8].

The parameter 'E' is used to enforce flow directionality or convection into the controlling finite difference equation (5.52). There are two values of 'E' and the selection is based on the difference between lubricant flow 'in' and 'out' of the control volume.

$$E = E_1 \text{ if } E_1 > 0 \quad \text{and} \quad E = 0 \text{ if } E_1 \leq 0 \tag{5.59}$$

$$E_1 = |a_E| + |a_W| + |a_N| + |a_S| - |a_P| \tag{5.60}$$

The inequality in equation (5.59) allows for reversal of flow which otherwise causes numerical instability.

The subscripts for all the mass flow terms, e.g., $(\rho w)_n$, are lowercase, denoting that an average between velocities at the central and peripheral node is taken; for example: $w_n = 0.5(w_N + w_P)$.

'S_p' and 'S_c' are terms representing viscous heating where allowance is made for the strong influence of temperature on viscosity. These two quantities are derived from the basic expression for viscous heating which is caused by the shearing of the lubricant:

$$S = \eta \left(\frac{\partial u}{\partial z} \right)^2 \tag{5.61}$$

where:

S   is the intensity of viscous heating $[W/m^3]$;

The controlling equation for 'S_p' and 'S_c' is based on the assumption of a linear dependence of the heat source term on temperature:

$$S = S_c + S_p T_p \tag{5.62}$$

The precise forms of 'S_p' and 'S_c' are given by:

$$S_p = \frac{S}{\eta_p} \frac{d\eta}{dT_{p,old}} \tag{5.63}$$

$$S_c = S\left(1 - \frac{T_p}{\eta_p}\frac{d\eta}{dT_{p,old}}\right) \tag{5.64}$$

where all terms are as calculated from the previous sweep of iteration for temperature and are referred to as 'old' values.

The exponential viscosity law can be written as (Table 2.1):

$$\eta_p = \eta_0 e^{-\gamma T_p} \tag{5.65}$$

where:

η_p is the predicted dynamic viscosity of the lubricant [Pas];

η_0 is the dynamic viscosity of the lubricant at some reference temperature 'T_0' [Pas];

γ is an exponent of viscosity-temperature dependence (typically $\gamma = 0.05$) [K^{-1}].

or rearranged as:

$$\frac{d\eta_p}{dT_p} = -\gamma\eta_p \tag{5.66}$$

Substituting (5.66) into (5.63) and (5.64) yields:

$$\boxed{S_p = -\gamma S} \tag{5.67}$$

$$\boxed{S_c = S\left(1 - \frac{T_p}{\eta_p}\frac{d\eta}{dT_{p,old}}\right) = S\left(1 + \gamma T_p\right)} \tag{5.68}$$

Treatment of Boundary Conditions in Thermohydrodynamic Lubrication

As mentioned already, the boundary conditions necessary when viscous heating is modelled are considerably more complicated than in the isoviscous case.

The finite difference equations presented are arranged so that they allow solution by methods appropriate to elliptic differential equations. This means that if iteration is used, the direction or order in which nodes are iterated does not affect the solution. If the equations were of a parabolic type then it would be necessary to iterate in the down-stream direction, i.e., apply a marching procedure but when reverse flow occurs this method generally fails. A marching procedure is a process of establishing nodal values in a specified sequence.

The temperature boundary conditions at the interfaces of the hydrodynamic film vary according to the heat transfer mode of the bearing. It is thus necessary to modify the finite difference mesh to provide a means of solving the thermohydrodynamic equations for the specified boundary conditions. An example of the modified mesh is shown in Figure 5.13. The mesh can be applied to solve both isothermal and adiabatic cases.

If the isothermal bearing is studied, then the boundary conditions at the sliding surfaces simplify to a fixed temperature for the boundary nodes. Iteration is then confined to interior nodes without any need for extra arrays of imaginary nodes apart from at the outlet and inlet.

On the other hand, if an adiabatic bearing is to be analyzed, then the boundary condition at the pad surface changes to an unknown pad temperature but with a zero temperature gradient normal to the plane of the lubricant film. In this case it is necessary to invoke an imaginary array of temperature nodes above the pad surface with values of temperature maintained equal to the adjacent pad surface temperature. Iteration then includes temperature nodes at the interface between the pad and the hydrodynamic film. Even for an adiabatic pad, the temperature on the pad at the bearing inlet which involves just one node remains the same as the lubricant inlet temperature.

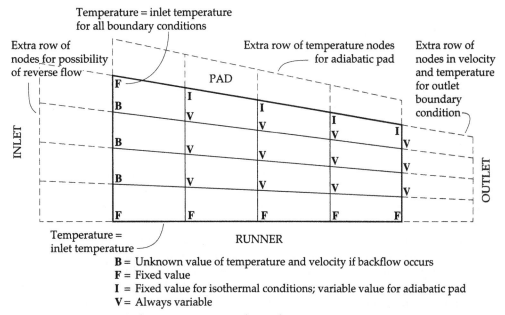

B = Unknown value of temperature and velocity if backflow occurs
F = Fixed value
I = Fixed value for isothermal conditions; variable value for adiabatic pad
V = Always variable

FIGURE 5.13 Example of the modified mesh with accessories for boundary conditions in thermohydrodynamic lubrication.

Provided that 'reverse flow' does not occur, the temperatures at the bearing inlet are equal to the lubricant supply temperature. Temperatures at the outlet are unknown and can be calculated by applying the boundary condition $\partial T/\partial x = 0$. The condition $\partial T/\partial x = 0$ relates to the very slow change in temperature by cooling once the lubricant leaves the bearing exit. There will only be a negligible variation in 'T' with respect to 'x' compared to changes within the bearing where heat generation occurs. This condition can be accommodated by supplying an extra column of nodes with temperatures equal to the adjacent node's outlet temperature. The iteration procedure then includes the extra nodes at the bearing outlet. Wherever reverse flow at the inlet of the oil film occurs, temperatures are iterated on the boundary with the assumption that $\partial u/\partial x$ and $\partial T/\partial x$ remain constant across the boundary. This condition is met by another column of nodes up-stream of the bearing inlet which maintain the values of temperature and velocity calculated by linear extrapolation from node temperatures inside the bearing.

Computer Program for the Analysis of an Infinitely Long Pad Bearing in the Case of Thermohydrodynamic Lubrication

A computer program '**THERMAL**' for the analysis of both isothermal and adiabatic infinitely long pad bearings is listed and described in the Appendix. For bearings that are neither isothermal nor adiabatic, an estimation of the effects of bearing heat transfer can be deduced from a comparison of data from the adiabatic and isothermal conditions, which represent lower and upper limits of load capacity, respectively. The program is based on a two-level iteration in temperature and pressure and its flowchart is shown in Figure 5.14.

An initial constant temperature field equal to the oil inlet temperature is assumed and a pressure solution is calculated from the resulting viscosity field. A new temperature field is then derived from the viscous shearing terms created by the pressure field. This new temperature field is then used to produce a second viscosity field which completes the first cycle of iteration. This iteration cycle is repeated until adequate convergence in the pressure field between successive iterations of temperature is reached. On completion of the iteration, pressure is integrated with respect to distance to obtain film force per unit length and the data is then printed to complete the program.

Example of the Analysis of an Infinitely Long Pad Bearing in the Case of Thermohydrodynamic Lubrication

The computer program 'THERMAL' described in the previous section provides a means of calculating the reduction in load capacity of a bearing due to heating effects. To demonstrate this effect the load capacity of a typical industrial bearing operating under conditions similar to those studied by Ettles [9] was analyzed by this program. The bearing parameters were chosen as typical of an industrial bearing.

The selected values of controlling parameters were as follows: bearing width (i.e., length in the direction of sliding) **0.1** [m], maximum film thickness **10^{-4}** [m], minimum film thickness **5×10^{-5}** [m], lubricant viscosity temperature coefficient **0.05**, lubricant specific heat **2000** [J/kgK], lubricant density **900** [kg/m^3] and lubricant thermal conductivity **0.15** [W/mK]. Two values of viscosity were considered, **0.05** [Pas] and **0.5** [Pas] at the bearing inlet temperature of **50°C**. The performance of the bearing was studied over a range of sliding speeds from **1** to **100** [m/s] for the lower viscosity and **0.3** to **20** [m/s] for the higher viscosity. Sliding speed values used for computation were **0.3, 1, 3, 10, 20, 30** and **100** [m/s]. For higher speeds the computing time required to obtain convergence was unfortunately far too long for practical use.

The calculated temperature distributions within an isothermal and adiabatic bearing are illustrated in Figure 5.15. The temperature fields were obtained for a lubricant inlet viscosity of **0.5** [Pas] and bearing sliding speed of **10** [m/s]. It can be seen from Figure 5.15 that the maximum temperature occurs at the outlet of the bearing.

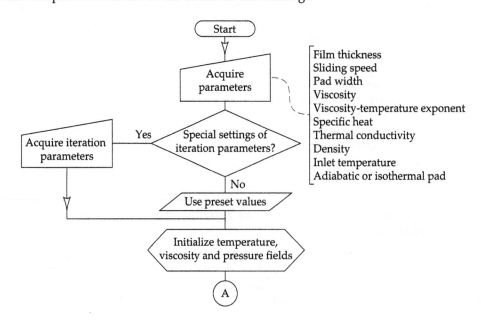

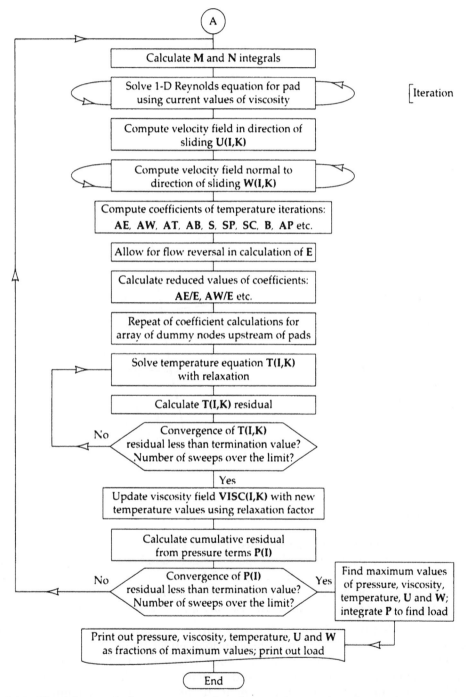

FIGURE 5.14 Flowchart of the program for the analysis of a thermohydrodynamic pad bearing.

A strong effect of pad heat transfer on the temperatures inside the lubricant film is clear. The maximum temperature in the isothermal bearing is **71**°C, compared to **116**°C for the adiabatic bearing. The location of the maximum temperature is also different for these bearings. For the isothermal bearing the peak occurs close to the middle of the bearing. A small decline in temperature beyond this maximum is due to improved thermal conduction with reduced film thickness. The location of the peak temperature in the adiabatic bearing is at the downstream end of the pad at the interface with the lubricant. The lubricant is progressively heated to higher temperatures as it passes down the bearing and the pad surface becomes very hot as it is remote from any source of cooling.

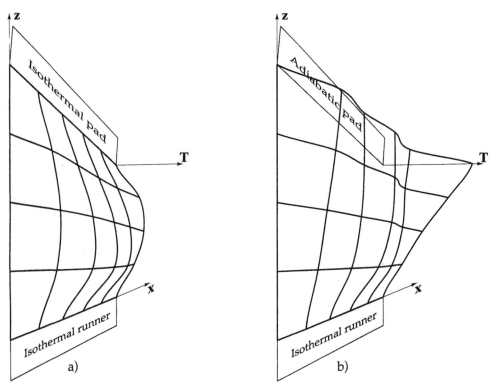

FIGURE 5.15 Computed temperature field in isothermal and adiabatic pad bearing at high sliding speed.

The dependence between bearing load (defined as load per unit length divided by the product of sliding speed and viscosity) and sliding speed for both adiabatic and isothermal bearing at two different viscosity levels is shown in Figure 5.16. Defining the bearing load as load per unit length divided by the product of sliding speed and viscosity allowed for comparison of the heating effects on lubricants of different viscosity and at various sliding speeds.

At low sliding speeds close to **1** [m/s], the load parameter converges to a common value of about **640,000** [dimensionless]. This indicates that load under these conditions is proportional to the product of sliding speed and viscosity, which agrees well with the isoviscous theory of hydrodynamic lubrication. As the sliding speed is increased, the load parameter declines and heating effects are gradually becoming evident. It can be seen from Figure 5.16 that the threshold sliding speed at which decreases in load capacity from the isoviscous level become significant is lowered by the higher lubricant viscosity. At high sliding speeds, the rise in lubricant viscosity may not provide as large an increase in load capacity as might be expected. It can also be seen that an isothermal bearing has a higher load capacity than an adiabatic bearing. Improvements in cooling of a real bearing can therefore bring an improvement in load capacity.

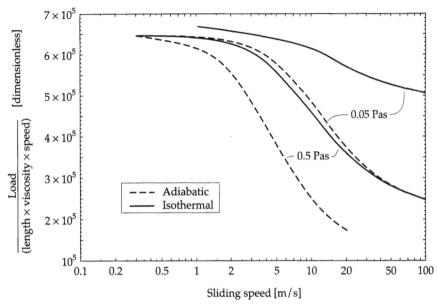

FIGURE 5.16   Computed effect of lubricant heating on relative load capacity of a pad bearing.

5.6.2   ELASTIC DEFORMATIONS IN A PAD BEARING

Almost all plain bearings operate with very small clearances and a requirement of nearly flat sliding surface. All bearings are made of material with a finite elastic modulus; if they deform or bend there may be a significant deviation from the optimum surface geometry considerably affecting the bearing performance. Pad bearings are particularly vulnerable to this phenomenon which is known in the literature as '**crowning**'. In a Michell bearing, the pad bends about the pivot point to form a curved or crowned shape which has a much lower load capacity than a rigid pad. This effect can become extreme at small film thicknesses, where even very limited deflections due to bending may severely distort the film geometry.

To illustrate this problem a one-dimensional pad has been selected as an example since the relevant elastic deflections can be found from simple bending theory. The two-dimensional case would require the analysis of deflections in a plate which is far more complex [9]. Elastic deflections combined with thermohydrodynamic effects have also been analysed and a strong interaction between these effects has been found [9,11].

An example of a computer program '**DEFLECTION**' for analysis of an elastically deforming one-dimensional pivoted Michell pad bearing is listed and described in the Appendix. Thermal effects, although significant, have been omitted in the program because of limitations of computing speed. The controlling equations of the bearing are the isoviscous Reynolds equation and the elastic deformation equation:

$$\boxed{\frac{d^2z'}{dx^2} = \frac{M'}{EI}}$$

(5.69)

where:

z' is the deflection of the pad in the 'z' direction [m];

    M'   is the local bending moment [Nm];

    E   is the elastic modulus of the pad material [Pa];

    I   is the second moment of area of the pad [m^4].

The pad is modelled as an infinitely long plate of uniform thickness so that 'I' (in terms of second moment of area per unit length) is a constant. The bearing load is assumed to be supported at the pivot. The pivot is located at the calculated centroid of hydrodynamic pressure. A two level iteration procedure is used in this analysis. The isoviscous hydrodynamic pressure field is first computed by iteration and then the bending moments are found and the resulting pad deflection calculated. The hydrodynamic pressure field is then re-iterated and a new series of pad deflections is found. The process is repeated until the pad deflections converge to sufficient accuracy.

Hydrodynamic pressure is found from a finite difference equivalent of the one-dimensional isoviscous Reynolds equation. The one-dimensional isoviscous Reynolds equation (4.25) can be written as:

$$\frac{d}{dx}\left(U_0 h\right) - \frac{1}{6\eta}\frac{d}{dx}\left(h^3\frac{dp}{dx}\right) = 0 \tag{5.70}$$

or as:

$$6\eta\, U_0\frac{dh}{dx} - h^3\frac{d^2p}{dx^2} - 3h^2\frac{dh}{dx}\frac{dp}{dx} = 0 \tag{5.71}$$

The finite difference equivalent of this equation rearranged to give an expression for the nodal pressure value is:

$$P_i = \frac{0.5(P_{i+1} + P_{i-1}) + 0.75(P_{i+1} - P_{i-1})\delta x\left(\frac{dh}{dx}\right)_i}{h_i} - \frac{3\eta\, U_0\delta x^2\left(\frac{dh}{dx}\right)_i}{h_i^3} \tag{5.72}$$

The finite difference equation (5.72) forms the basis of the iteration for pressure. Since cavitation due to extreme elastic deflection is also possible, even in a pad bearing, whenever this occurs the negative pressures are set to zero.

The bending deflection equation is applied with the following boundary conditions:

    ·   the bending moment 'M'' and shear force normal to the pad 'S' are equal to zero at both ends of the pad bearing,

    ·   the pad is balanced at the pivot point and there are no other forms of support to the pad,

    ·   pad deflection and deflection slope **dz'/dx** are zero at the pivot.

With these conditions, for $x < x_c$ where 'x_c' is the position of the pressure centroid, the expressions for 'S' and 'M'' are:

$$S = \int_0^x p\,dx \tag{5.73}$$

$$M' = \int_0^x S\,dx \tag{5.74}$$

For $x > x_c$ the expressions for 'S' and 'M''' can be written as:

$$S = \int_0^x p \, dx - \int_0^B p \, dx \qquad (5.75)$$

$$M' = \int_0^x S \, dx - (x - x_c) \int_0^B p \, dx \qquad (5.76)$$

where '**B**' is the width of the pad [m].

The deflection of the pad for all 'x' is found by integrating of (5.69) twice with respect to 'z' and is given by:

$$z' = \int_0^x \left(\int_0^x M' \, dx \right) dx + C_1 x + C_2 \qquad (5.77)$$

The constants 'C_1' and 'C_2' are:

$$C_1 = - \int_0^{x_c} M' \, dx \qquad (5.78)$$

$$C_2 = x_c \int_0^{x_c} M' \, dx - \int_0^{x_c} \left(\int_0^x M' \, dx \right) dx \qquad (5.79)$$

Computer Program for the Analysis of an Elastically Deforming One-Dimensional Pivoted Michell Pad Bearing

The flowchart of the computer program for the analysis of an elastically deforming one-dimensional pivoted Michell pad bearing is shown in Figure 5.17. A two level iteration in pressure and elastic deflection is conducted in order to determine the hydrodynamic pressure of a deformable pad bearing.

Effect of Elastic Deformation of the Pad on Load Capacity and Film Thickness

The computer program 'DEFLECTION' described earlier can provide useful information for the mechanical design of hydrodynamic bearings. For instance, the effect of pad thickness and elastic modulus of pad material on the load capacity can be assessed with the aid of this program. The effect of pad thickness on load capacity is demonstrated as an example of possible applications of this program.

It is of practical importance to know how thick the bearing pad should be to provide sufficient rigidity for a particular size of bearing and nominal film thickness. A reduction in the hydrodynamic film thickness can increase load capacity but at the same time it also increases bearing sensitivity to elastic distortion. Optimization of bearing characteristics is therefore essential to the design process. The computed load capacity of a bearing of 1 [m] pad width, lubricated by a lubricant of 1 [Pas] viscosity versus pad thickness is shown in Figure 5.18. The Young's modulus of the pad's material is 207 [GPa]. The hydrodynamic film thickness is 2 [mm] at the inlet and 1 [mm] at the outlet of the pad.

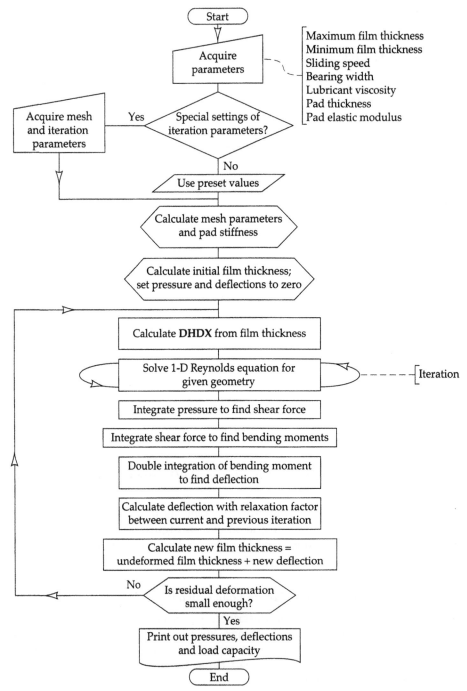

FIGURE 5.17 Flowchart of program to compute load capacity of an elastically deforming one-dimensional pivoted pad bearing.

It can be seen from Figure 5.18 that as the pad thickness is reduced from **200** [mm] to **30** [mm] the load capacity declines by **70%**. When the thickness of the pad is **200** [mm], then the load capacity is identical to that of the rigid pad. At the pad thickness of **100** [mm], load capacity is only reduced by about **10%** as compared to a rigid pad. It can thus be concluded that **100** [mm] is close to the optimum pad thickness for this particular bearing. The relationships between the film thickness and pressure for pads of **100** [mm] and **30** [mm] thickness are shown in Figures 5.19 and 5.20, respectively.

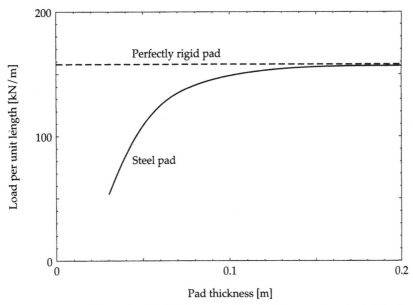

FIGURE 5.18 Computed effect of pad thickness on the load capacity of a Michell pad bearing.

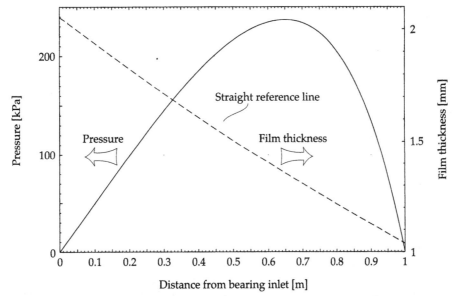

FIGURE 5.19 Effect of elastic deflection on the hydrodynamic film thickness and pressure profile for a **100** [mm] thick pad.

It can be seen from Figure 5.19 that with a pad thickness of **100** [mm], elastic deflection is small and the pressure field is essentially the same as for a rigid bearing. When the pad thickness is reduced to **30** [mm], however, the geometry of the bearing is distorted from a tapered wedge to a converging-diverging film profile. For this pad thickness, the divergence beyond the minimum film thickness is sufficiently small so that cavitation does not occur.

The pressure profile does, however, shift forward along with the pivot point as shown in Figure 5.20. With further reduction in either the pad thickness or the film thickness, cavitation occurs and this causes a severe reduction in load capacity. Cavitation causes the effective load bearing area to shrink and the load capacity declines even if specific hydrodynamic pressures remain high.

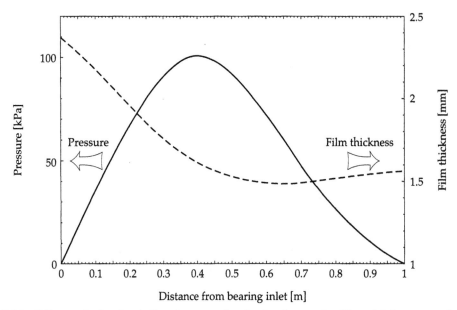

FIGURE 5.20   Effect of elastic deflection on the hydrodynamic film thickness and pressure profile for a **30** [mm] thick pad.

From the example presented it is clear that the performance of pad bearings depends on the use of high modulus materials and thick bearing sections. Low modulus materials such as polymers, although attractive as bearing materials, would require a metal backing for all bearings with the exception of very small pad sizes.

5.6.3   CAVITATION AND FILM REFORMATION IN GROOVED JOURNAL BEARINGS

Cavitation occurs in liquid lubricated journal bearings to suppress any negative pressures that would otherwise occur. In the numerical analysis of complete journal bearings, as opposed to partial arc bearings, cavitation and reformation must be included in the numerical model. For partial arc bearings, it can be assumed that the inlet side of the bearing is fully flooded and cavitation is usually limited to a small area down-stream of the load vector. Full 360° journal bearings are lubricated through the lubricant supply holes or grooves, and the cavitation and reformation fronts that form around the grooves or holes control the load capacity of the bearing. For the standard configuration of two grooves positioned perpendicular to the load line, a cavitation front forms down-stream of each groove and a reformation front is located up-stream of each groove. This is illustrated in Figure 5.21 which shows the cavitation and reformation fronts on an 'unwrapped' lubricant film.

A method of predicting the location of the cavitation and reformation fronts is required for numerical analysis of the grooved bearing. The cavitation front can be determined by applying the Reynolds condition that all negative pressures generated during computations are set to zero. It is found that if this rule is applied then not only are all the negative pressures removed, but also the gradient of pressure normal to the cavitation front is zero as predicted by Reynolds.

The reformation front creates considerably more difficulty in numerical modelling. The boundary condition is based on mass conservation between the lubricant flow from the cavitated and fully reformed oil film. One of the first analyses of reformation fronts was performed by Elrod [12] based on a model of cavitation and film reformation which had been developed by Jakobsson and Floberg [13] and Olsson [14]. The Jakobsson-Floberg-Olsson model provides boundary conditions which satisfy the continuity condition for any geometry of reformation fronts, and is applied in the computer program 'GROOVE' for the analysis of a 360° journal bearing. This particular model is the most appropriate amongst available models for the solution of heavily loaded bearings. At light bearing loads, however, other models may be more suitable. The program 'GROOVE' is listed in the Appendix.

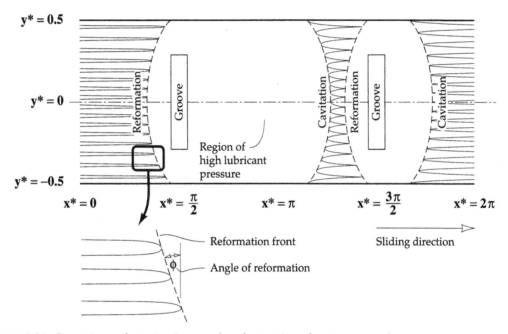

FIGURE 5.21 Location of cavitation and reformation fronts around grooves in a centrally loaded 360° journal bearing.

At the reformation front the following equation of mass conservation applies.

$$h_c^* = h_r^* - h_r^{*3}\left[\frac{\partial p^*}{\partial x^*} + \left(\frac{R}{L}\right)^2 \tan\phi \frac{\partial p^*}{\partial y^*}\right]$$

(5.80)

where:

h_c^* is the film thickness at the cavitation front directly up-stream (i.e., with the same 'x*' position);

h_r^* is the film thickness at the reformation front;

ϕ is the angle of the adjacent section of the reformation front.

The angle 'ϕ' of the reformation front is shown in Figure 5.21. When the reformation front is aligned to be parallel to 'y*' then $\phi = 0$ and when the front is parallel to 'x*' then $\phi = \pi/2$. In descriptive terms, equation (5.80) states that flow from a cavitation front which is constant and equal to the Couette flow at this front should be equal to the increased Couette flow at the reformation front minus the backwards pressure flow from the reformed oil film. This condition imposes a significant positive pressure gradient at the reformation front because of the step change from cavitated to full flow. This principle of film reformation is illustrated in Figure 5.22.

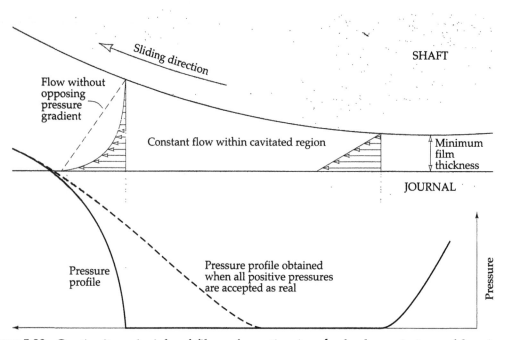

FIGURE 5.22 Continuity principle of film reformation in a hydrodynamic journal bearing.

The reformation condition may be applied to every positive pressure generated down-stream of the cavitation front as an inequality based on finite difference approximations to the pressure gradients. Positive pressures which do not satisfy this inequality are set to zero until a front is established. A similar method was developed by Dowson et al. [15]. If the reformation condition is not applied in the analysis then the extent of the non-cavitated lubricant film will be overestimated, causing an imbalance between lubricant flow from the grooves and lubricant flow out of the bearing. This creates a risk of underestimating the lubricant consumption. The effect of lubricant starvation on the extent of the load-bearing film would also be underestimated. Re-arranging equation (5.80) to isolate the pressure terms gives:

$$\frac{h_r^* - h_c^*}{h_r^{*3}} = \frac{\partial p^*}{\partial x^*} + \left(\frac{R}{L}\right)^2 \tan\phi \frac{\partial p^*}{\partial y^*} \qquad (5.81)$$

According to the finite difference method the terms $\partial p^*/\partial x^*$ and $\partial p^*/\partial y^*$ can be approximated by:

$$\frac{\partial p^*}{\partial x^*} \approx \frac{P_{i+1,j}^* - P_{i,j}^*}{\delta x^*} \qquad (5.82)$$

$$\frac{\partial p^*}{\partial y^*} \approx \frac{P^*_{i,j+1} - P^*_{i,j}}{\delta y^*} \tag{5.83}$$

These terms, when substituted into the modified form of the reformation condition (5.80), give an expression for the pressure gradient at the reformation front in terms of three nodal pressures. In the computer program '**GROOVE**' the modified reformation condition with finite difference equivalents of pressure gradients is applied as an inequality, i.e.:

$$\frac{h^*_r - h^*_c}{h^{*3}_r} \leq \frac{P^*_{i+1,j} - P^*_{i,j}}{\delta x^*} + \frac{(P^*_{i,j+1} - P^*_{i,j})\left(\frac{R}{L}\right)^2 \tan\phi}{\delta y^*} \tag{5.84}$$

Most of the difficulty in applying this condition arises because of the two following problems:

· devising criteria to discriminate between positive pressures which should be subjected to this condition from the pressures which are remote from the reformation front,

· deducing a value of '$\tan\phi$' from a front of unknown geometry.

Methods adopted in the program '**GROOVE**' will be described in greater detail in the next section.

The pressure field between the cavitation and the reformation fronts is found by the same Vogelpohl equations in finite difference form which were applied earlier in the computer program '**PARTIAL**'. Grooves are incorporated into the bearing by assigning a groove pressure to nodes within the grooves and excluding these from iteration. The overlap at $x^* = 2\pi$ and $x^* = 0$ is allowed for by creating an extra row of nodes at $x^* = 2\pi + \delta x^*$, where 'δx^*' is the mesh spacing in the 'x^*' direction. This approach enables iteration up to $x^* = 2\pi$ with values of 'M_v' at $x^* = 0$ being set equal to the values at $x^* = 2\pi$ at the end of each iteration sweep.

The flow from the grooves is computed as well as the flow from the sides of the bearing. This provides a check on the accuracy of calculations since the total groove flow should equal the side flow. Perfect equality between these flows is unlikely because of truncation errors in the finite difference scheme, thus the discrepancy between the flows provides an indication of the precision of the computed results. The flow terms are given by the following expressions:

$$\boxed{Q^*_{side} = -\left(\frac{R}{L}\right)^2 \int_0^{2\pi} h^{*3}\left(\frac{\partial p^*}{\partial y^*}\right)dx^*} \tag{5.85}$$

$$\boxed{Q^*_{axial} = \int_0^1 \left(h^* - h^{*3}\frac{\partial p^*}{\partial x^*}\right)dy^*} \tag{5.86}$$

where:

Q^*_{side} is the lubricant flow normal to the direction of sliding [non-dimensional];

Q^*_{axial} is the lubricant flow parallel to the direction of sliding [non-dimensional];

Computer Program for the Analysis of Grooved 360° Journal Bearings

The example of a computer program 'GROOVE' for the analysis of a grooved 360° journal bearing is listed and described in the Appendix. The program incorporates the solution procedure for the Vogelpohl parameter, used already in the program for partial bearing analysis, as well as other procedures to define groove geometry and to apply the film reformation condition. A flowchart of the program is shown in Figure 5.23.

An additional procedure for determining whether a positive nodal pressure belongs to a true pressure field or should be checked by the reformation condition is used and the detailed flowchart of the procedure is shown in Figure 5.24.

Example of the Analysis of a Grooved 360° Journal Bearing

The computer program 'GROOVE' described in the previous section provides a means of calculating the lubricant flow from a bearing and testing the effect of groove geometry and lubricant supply pressure on load capacity. Rapid estimations of load capacity can be found from an analysis of an equivalent partial arc bearing, i.e., a partial arc that fits in the space between the grooves. A large number of nodes is required to accurately estimate the pressure gradients around the grooves which results in longer computing time.

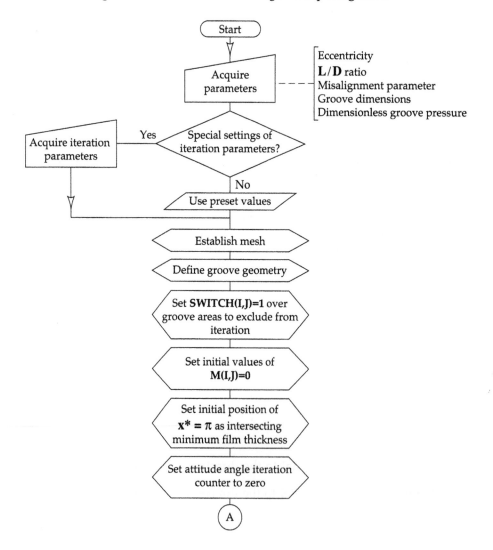

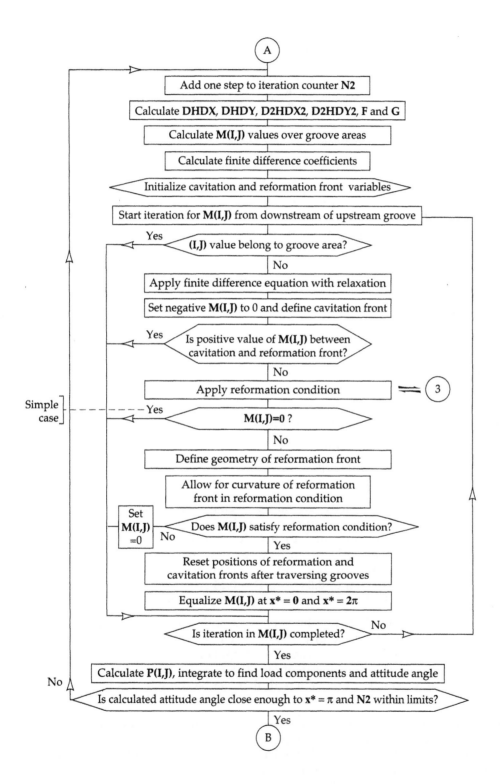

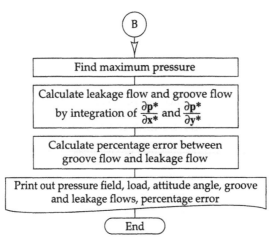

FIGURE 5.23 Flowchart of a computer program for the analysis of grooved 360° journal bearings.

Real lubricant flow is found from the dimensionless quantities by applying the following relation:

$$Q = \frac{Q^*LUc}{2} \tag{5.87}$$

where:

Q   is the lubricant flow [m³/s];

Q*   is the non-dimensional lubricant flow;

L   is the length of the bearing [m];

U   is the entraining velocity [m/s];

c   is the radial clearance [m].

The plots of non-dimensional load capacity and oil flow versus relative length of the grooves for **L/D = 0.25** and **L/D = 1** are shown in Figures 5.25 and 5.26, respectively. The eccentricity ratio is assumed to be constant and equal to **0.8**. It is also assumed that there is no misalignment and the subtended angle of each groove is **36°**. The mesh assumed for computation has **21** nodes in the '**x***' direction and **11** nodes in the '**y***' direction. The pre-set dimensionless groove pressure is equal to **0.05** for **L/D = 0.25** and to **0.2** for **L/D = 1**, giving a ratio of groove pressure to peak hydrodynamic pressure of approximately **0.1** in both cases.

The prime feature of the computed data is the very weak influence of groove geometry on load capacity. This means that unless the groove is so small as to impose extreme lubricant starvation on the bearing, hydrodynamic lubrication remains effective. The limitations imposed by the node density prevent investigation of the minimum groove size to cause lubricant starvation since the smallest groove length that can be computed is given by three step lengths in the '**J**' direction.

For the wider bearing with **L/D = 1** side flow remains almost constant until a relative groove length (i.e., ratio of groove length to axial length of the bearing) of **0.5** is reached. It is therefore possible to fit wide grooves with relative length equal to **0.5** into **L/D = 1** bearings to reduce the Petroff friction force without incurring an increased lubricant pumping power loss. Petroff friction is negligible inside the groove because of the large clearance within the groove. With the narrower bearing of **L/D = 0.25**, side flow rises sharply with groove length

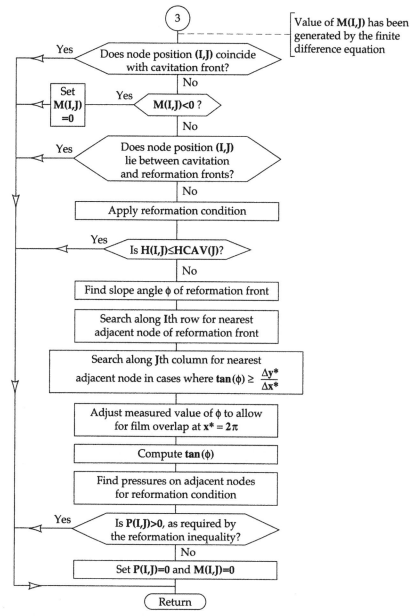

FIGURE 5.24 Detailed flowchart of procedure to apply reformation condition.

which implies that the groove length should be made as small as possible. The data obtained also reveals that the distribution of flow between the up-stream and down-stream grooves is not equal. At the eccentricity ratio of **0.8** selected for this example, the up-stream groove consumed more lubricant with a flow rate of typically **80%** of total side flow. Although at lower eccentricities the difference in groove flows becomes smaller, in almost all cases the up-stream groove has the larger flow. The calculation of lubricant flow around the boundary of a groove requires a very high density of nodes. When the pressure gradient at the up-stream edge of a groove is very high, a pressure drop from full groove pressure to zero within one mesh length may be less than the pressure gradient that would be formed in a real bearing. In such cases, the truncation error of flow calculation becomes significant. The side flow is not subject to such problems and accurate values are computed in most cases.

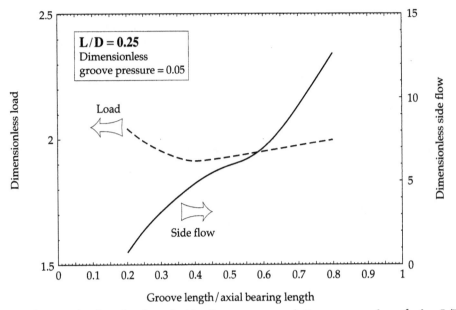

FIGURE 5.25 Dimensionless load and side flow versus relative groove length for **L/D = 0.25**, p^*_{groove} = **0.05** and groove subtended angle **36°**.

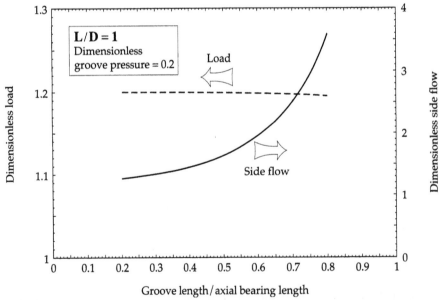

FIGURE 5.26 Dimensionless load and side flow versus relative groove length for **L/D = 1**, p^*_{groove} = **0.2** and groove subtended angle **36°**.

Lubricant flow in bearings can be quite large. It can be seen from Figure 5.25 that for a **36°** subtended groove angle and **0.6** relative groove length, the dimensionless flow is about **6.8**.

Assuming that the bearing entraining velocity is **U = 10** [m/s], bearing length **L = 0.2** [m] and the radial clearance of the bearing is **c = 0.0004** [m], then from equation (5.87) the value of flow '**Q**' is:

$$Q = 0.5 \times 6.8 \times 0.2 \times 10 \times 0.0004 = 2.72 \times 10^{-3} \text{ [m}^3\text{/s]} = 2.72 \text{ [litres/s]}$$

It is evident that in some cases hydrodynamic bearings can require large flow rates of lubricant, and accurate estimates of side and groove flow are essential information in bearing design.

The shape of the cavitation and reformation fronts can also provide information on the adequacy of lubricant supply to the bearing. An example of the effect of groove on pressure distribution in a bearing of **L/D = 1**, eccentricity ratio of **0.7**, is shown in Figures 5.27 and 5.28. The relative groove length is equal to **0.2** and groove subtended angle is **72°**. The perfectly aligned case is shown in Figure 5.27 whereas the case with an extreme misalignment of **0.5** is shown in Figure 5.28.

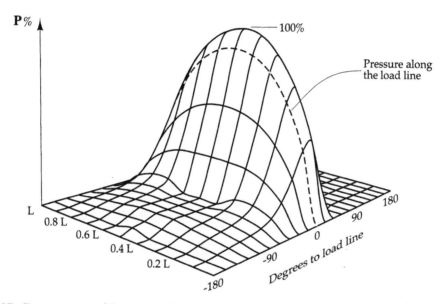

FIGURE 5.27 Pressure profile of grooved perfectly aligned journal bearing of **L/D = 1** and eccentricity ratio of **0.7** (not to scale).

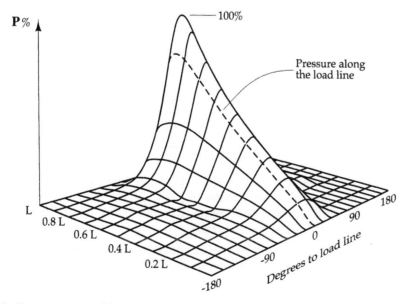

FIGURE 5.28 Pressure profile of grooved misaligned journal bearing of **L/D = 1** and eccentricity ratio of **0.7** (not to scale).

Misalignment has surprisingly little effect on the location of the reformation front or even the cavitation front compared to its effect on the pressure peak. This feature is probably due to the large film thickness in the cavitated regions of the hydrodynamic film which ensures a small relative change in film thickness with misalignment. Lubricant flow rates are also relatively unaffected by misalignment which renders unlikely the possibility of lubricant starvation with increasing shaft misalignment. The high values of maximum pressure, however, are undesirable.

5.6.4 VIBRATIONAL STABILITY IN JOURNAL BEARINGS

As discussed already in Chapter 4, hydrodynamic bearings are prone to a vibrational instability known as 'oil whirl'. Vibration characteristics of a hydrodynamic film can be modelled by a series of stiffness and damping coefficients. These coefficients can be computed from the solutions of the Reynolds equation. Vibration analysis of hydrodynamic bearings can be directed to the computation of the shaft trajectory in a vibrating bearing. This approach, however, involves a rigorous analysis of bearing instability and is a specialized task requiring extensive computing. A much simpler mode of analysis for practical engineering applications is discussed in this section. In this approach the limiting shaft speed at the onset of vibration is calculated using the Routh-Hurwitz criterion of stability. The criterion provides a conservative estimate of the shaft speed at which some level of sustained vibration occurs. It has often been found that at moderate shaft speeds, shaft vibration may occur but it is limited to a finite and safe amplitude. On the other hand, at higher speeds, there is no limit to the amplitude of vibration and the shaft will oscillate in ever wider trajectories until it touches the bush which inevitably results in destruction of the bearing.

In order to analyze shaft trajectories, the non-linear variation in stiffness and damping coefficients with shaft position must be included in the analysis. The advantage of the Routh-Hurwitz method is that only infinitesimal amplitudes of vibration are considered which allow the use of linearized stiffness and damping coefficients. The linearized Routh-Hurwitz analysis of bearing vibration and the computation method are described in the following sections.

Determination of Stiffness and Damping Coefficients

Stiffness and damping coefficients are obtained by including in the Reynolds equation the effect of small displacements and squeeze velocities. Stiffness and damping coefficients are calculated from the change in pressure integral by dividing the changes by the displacement and squeeze velocity, respectively. Magnitudes of displacements and squeeze velocities are held at small values in order to minimize inaccuracy due to non-linear variation of film forces. A cartesian coordinate system aligned with the direction of bearing load, shown in Figure 5.29, is established and values of stiffness and damping coefficients normal and co-directional with the load line are then computed.

Four stiffness coefficients relating to the range of possible bearing movements, 'K_{xx}', 'K_{yy}', 'K_{xy}' and 'K_{yx}', and four damping coefficients, 'C_{xx}', 'C_{yy}', 'C_{xy}' and 'C_{yx}', are required for vibration analysis. To find these coefficients the effect of small displacements on hydrodynamic pressure integral must be analyzed.

Shaft displacements are modelled in the Reynolds equation in terms of their effect on **dh/dx**. It is convenient to use non-dimensional forms of shaft displacement in terms of the radial bearing clearance, i.e.:

$$\boxed{\frac{\Delta x}{c} = \Delta x^*}$$

(5.88)

where:

Δx is the displacement of the shaft centre in the 'x' direction [m];

c   is the radial clearance of the bearing [m];

Δx^* is the non-dimensional displacement.

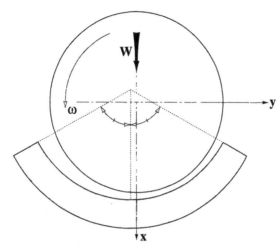

FIGURE 5.29 Journal bearing coordinate configuration for vibration analysis.

A similar relationship applies to 'Δy', the displacement in the 'y' direction. The equation for **dh*/dx*** is given in the following form according to basic geometrical principles:

$$\boxed{\frac{\partial h^*}{\partial x^*} = \left(\frac{\partial h^*}{\partial x^*}\right)_{static} + \frac{\partial}{\partial x^*}\left[\Delta x^* \cos(x^*) + \Delta y^* \sin(x^*)\right]}$$

(5.89)

where:

x^* refers to the film ordinate around the bearing;

$\left(\dfrac{\partial h^*}{\partial x^*}\right)_{static}$ is the variation in film thickness for the static case.

The modified forms of '**h***' and $\partial^2 h^*/\partial x^{*2}$ which are required for the Vogelpohl equation follow the scheme already described and are given by:

$$h^* = h^*_{static} + \Delta x^* \cos(x^*) + \Delta y^* \sin(x^*)$$

(5.90)

$$\frac{\partial^2 h^*}{\partial x^{*2}} = \left(\frac{\partial^2 h^*}{\partial x^{*2}}\right)_{static} + \frac{\partial^2}{\partial x^{*2}}\left[\Delta x^* \cos(x^*) + \Delta y^* \sin(x^*)\right]$$

(5.91)

The Vogelpohl equation (5.4) is then solved in terms of the modified forms of '**h***' and its derivatives, i.e., $\partial h^*/\partial x^*$, $\partial h^*/\partial y^*$, etc. Non-dimensional stiffness coefficients are defined as :

$$K^* = \frac{Kc}{W}$$

(5.92)

where:

    K^* is the non-dimensional stiffness;

    K is the real stiffness (note, in this section 'K' denotes the stiffness) [N/m];

    c is the radial clearance of the bearing [m];

    W is the bearing load [N].

This form of non-dimensionalization can be shown to be equivalent to:

$$K^* = \frac{\Delta W^*}{\Delta x^* W^*_{static}}$$

(5.93)

Since 'δx^*' is very small then:

$$W^* \approx W^*_{static}$$

In other words, non-dimensional stiffness coefficients are equal to the change in non-dimensional load divided by the product of non-dimensional displacement and static non-dimensional load. The change in load 'ΔW^*' is calculated from the total load found by integration of the hydrodynamic pressure field with the displacement parameters included, and the static load, i.e.:

$$\Delta W^* = W^* - W^*_{static}$$

(5.94)

In exact terms, only the change in film force along the 'x' or 'y' axis is calculated, not the change in the total load. For example, 'K^*_{xx}' stiffness is calculated according to the following equation, i.e.:

$$K^*_{xx} = \frac{\Delta W^*_x}{\Delta x^* W^*}$$

(5.95)

where 'ΔW^*_x' is the load change in the 'x' direction (i.e., first index denotes the axis along which the deflection occurs, while the second index denotes the axis of the force).

Similarly, stiffness 'K^*_{yx}' is given by:

$$K^*_{yx} = \frac{\Delta W^*_y}{\Delta x^* W^*}$$

(5.96)

where 'ΔW^*_x' is the load change in the 'y' direction. A similar convention applies for stiffnesses 'K^*_{yy}' and 'K^*_{xy}'.

Damping coefficients are found by adding appropriate squeeze terms to the Reynolds equation. A non-dimensional squeeze term is defined as:

$$w^* = \frac{w}{c\omega}$$

(5.97)

where:

w is the squeeze velocity [m/s];

c is the radial clearance of the bearing [m];

ω is the angular velocity of the shaft [rad/s].

and the non-dimensional form of the Reynolds equation with squeeze terms is given by:

$$\frac{\partial}{\partial x^*}\left(h^{*3}\frac{\partial p^*}{\partial x^*}\right) + \left(\frac{R}{L}\right)^2\frac{\partial}{\partial y^*}\left(h^{*3}\frac{\partial p^*}{\partial y^*}\right) = \frac{\partial h^*}{\partial x^*} + 2w^*$$

(5.98)

The squeeze velocity is not constant around the hydrodynamic film but varies in a sinusoidal manner similar to the displacements. An expression for the dimensionless squeeze velocity at any position on the hydrodynamic film in terms of squeeze velocities along the 'x' and 'y' axes is given by:

$$w^* = w_x^*\cos(x^*) + w_y^*\sin(x^*)$$

(5.99)

The squeeze term 'w*' can be included in the parameter 'G' of the Vogelpohl equation, i.e.:

$$\frac{\partial^2 M_v}{\partial x^{*2}} + \left(\frac{R}{L}\right)^2\frac{\partial^2 M_v}{\partial y^{*2}} = FM_v + G$$

$$= \frac{FM_v + \dfrac{\partial h^*}{\partial x^*} + 2w^*}{h^{*1.5}}$$

(5.100)

Damping coefficients are computed in a similar manner to the stiffness coefficients, i.e., an arbitrary infinitesimal squeeze velocity is applied to cause a change in the pressure integral. The non-dimensional damping coefficient is defined in a similar manner to the non-dimensional stiffness coefficient, i.e.:

$$C^* = C\left(\frac{c\omega}{W}\right)$$

(5.101)

where:

C^* is the non-dimensional damping coefficient;

C is the real damping coefficient [Ns/m].

Expressing (5.101) in terms of non-dimensional quantities gives the non-dimensional damping coefficient, i.e.:

$$C^* = \frac{\Delta W^*}{w^* W^*}$$

(5.102)

and a specific damping coefficient, e.g., 'C_{xx}^*', is calculated according to:

$$C_{xx}^* = \frac{\Delta W_x^*}{w_x^* W^*}$$

(5.103)

After determining all the necessary values of stiffness and damping coefficients the vibrational stability of a bearing can be evaluated. There are various theories of bearing vibrational analysis, and the obtained stiffness and damping coefficients can be used in any of these methods. A very useful theory for vibrational analysis of a journal bearing was developed by Hori [7]. In this theory a simple disc of a mass 'm' mounted centrally on a shaft supported by two journal bearings is considered. The disc tends to vibrate in the 'x' and 'y' directions which are both normal to the shaft axis. The configuration is shown in Figure 5.30.

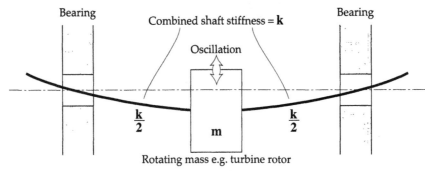

FIGURE 5.30 Hori's model for journal bearing vibration analysis.

There are two sources of disc deflection in this model: the shaft can bend and the two bearings are of finite stiffness which allows translation of the shaft. This system was analyzed by Newton's second law of motion to provide a series of equations relating the acceleration of the rotor in either the 'x' or the 'y' direction to the mass of the disc, shaft and bearing stiffnesses, and bearing damping coefficients. The description of this analysis can be found elsewhere [7]. The equations of motion of the disc can be solved to produce shaft trajectory but this is not often required since the most important information resulting from the analysis is the limiting shaft speed at the onset of bearing vibration. The limiting shaft speed is derived from the Routh-Hurwitz criterion which provides the following expression for the '**threshold speed of self-excited vibration**' or the '**critical frequency**' as it is often called:

$$\omega_c^{*2} = \frac{A_1 A_3 A_5^2}{(A_1^2 + A_2 A_5^2 - A_1 A_4 A_5)(A_5 + \gamma A_1)}$$

(5.104)

where:

   A_1, A_2, A_5 are the dimensionless stiffness and damping products;

   ω_c^* is the dimensionless bearing critical frequency.

The bearing critical frequency is also given by:

$$\omega_c^* = \frac{\omega_c}{(g/c)^{0.5}}$$

(5.105)

where:

ω_c is the angular speed of the shaft [rad/s];

g   is the acceleration due to gravity [m/s^2];

c   is the radial clearance of the bearing [m].

and the 'γ' parameter is expressed by:

$$\gamma = \frac{W}{kc} \tag{5.106}$$

where:

W   is the weight on the shaft [N];

k   is the stiffness of the shaft [N/m].

Since the 'γ' parameter is independent of bearing geometry it must be specified before commencing computing of a solution to equation (5.104).

The '**A**' terms relate to stiffness and damping coefficients in the following manner [7]:

$$A_1 = K_{xx}^* C_{yy}^* - K_{xy}^* C_{yx}^* - K_{yx}^* C_{xy}^* + K_{yy}^* C_{xx}^* \tag{5.107}$$

$$A_2 = K_{xx}^* K_{yy}^* - K_{xy}^* K_{yx}^* \tag{5.108}$$

$$A_3 = C_{xx}^* C_{yy}^* - C_{xy}^* C_{yx}^* \tag{5.109}$$

$$A_4 = K_{xx}^* + K_{yy}^* \tag{5.110}$$

$$A_5 = C_{xx}^* + C_{yy}^* \tag{5.111}$$

The analysis is completed with the calculation of the non-dimensional critical frequency 'ω_c^*'.

Computer Program for the Analysis of Vibrational Stability in a Partial Arc Journal Bearing

An example of a computer program '**STABILITY**' for analysis of vibrational stability in a partial arc journal bearing is listed and described in the Appendix and its flowchart is shown in Figure 5.31. The program computes the limits of bearing vibrational stability.

The Vogelpohl equation is solved by the same method described for the program '**PARTIAL**'. Although the program '**STABILITY**' specifically refers to partial arc bearings, a similar program could be developed for grooved bearings since the principles applied are the same.

Example of the Analysis of Vibrational Stability in a Partial Arc Journal Bearing

Comprehensive tables of a perfectly aligned bearing can be found elsewhere [7]. Of considerable practical interest, however, is the effect of shaft misalignment on bearing critical frequency. The computed results of the effect of shaft misalignment on critical frequency of a **120°** partial arc bearing, **L/D = 1**, eccentricity ratio **0.7** and dimensionless exciter mass **0.1** are shown in Figure 5.32. A mesh density of **11** rows in both '**x***' and '**y***' directions was applied in computation.

It can be seen from Figure 5.32 that there is a decline in critical frequency with increasing misalignment. However, at extreme values of misalignment the critical frequency rises as a result of the sharp increase in the principal stiffness coefficient 'K_{xx}^*'.

In practical bearing systems where misalignment is inevitable, operating the bearing at speeds very close to the critical speed as predicted from the perfectly aligned condition is not recommended. For example, if the value of radial clearance is **0.0002** [m] and **g = 9.81** [m/s²], then the conversion factor from non-dimensional to real frequency according to equation (5.105) is equal to:

$$(g/c)^{0.5} = (9.81/0.0002)^{0.5} = 221.5 \ [Hz]$$

The calculated difference between the minimum dimensionless critical speed for the bearing with a misalignment parameter of **t = 0.2** and a perfectly aligned bearing is:

$$\omega_{c,misaligned}^* - \omega_{c,aligned}^* = 2.2647 - 1.8591 = 0.4056$$

which makes the difference in the angular speed of the shaft about:

$$221.5 \times 0.4056 = 89.6 \ [Hz] \ \text{or} \ [rad/s]$$

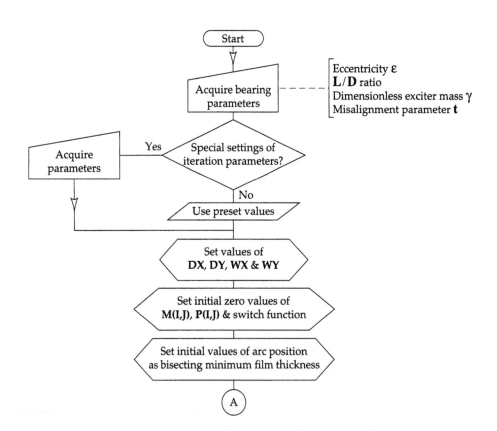

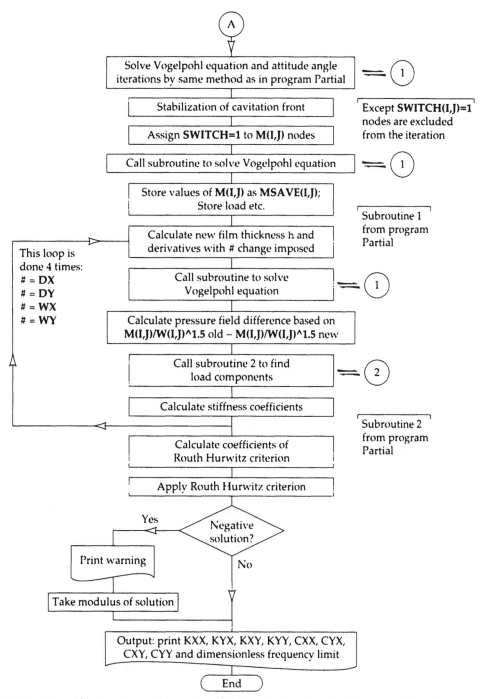

FIGURE 5.31 Flowchart of program for the analysis of vibration stability in partial arc bearings.

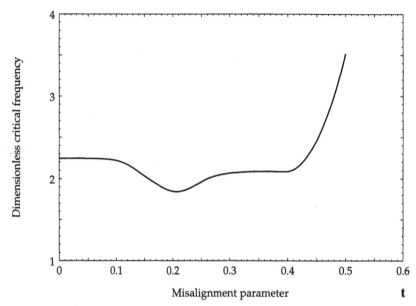

FIGURE 5.32 Effect of shaft misalignment on bearing dimensionless critical frequency.

5.7 SUMMARY

Numerical analysis has allowed models of hydrodynamic lubrication to include closer approximations to the characteristics of real bearings than the original idealized analytical solutions. Some adaptations of numerical models presented in current research literature have been introduced in this chapter to illustrate the potential of computational methods. The strong influence of secondary effects such as lubricant heating and bearing deformation on load capacity is shown, together with possible methods of controlling the negative effects these have on bearing performance. The scope of numerical analysis is continually being extended. With increases in the speed of computing it may become possible to perform the simultaneous analysis of several different effects on bearing performance, e.g., the combined effect of heating, deformation and misalignment. The finite difference method used in numerical analysis is versatile and simple to apply, but is also relatively inaccurate. Newer methods of devising numerical equivalents of differential equations are being increasingly adopted. However, the fundamental principles of numerical analysis outlined in this chapter remain unaltered.

REVISION QUESTIONS

5.1   What are the advantages and disadvantages of computational hydrodynamics as compared to classical hydrodynamics, list and discuss.

5.2   Apply the thermo-hydrodynamic program to oils of increasing thermal conductivity under, for example, constant sliding speed, and discuss the results.

5.3   What is the difference between the isothermal and adiabatic bearing models used in the program 'THERMAL'?

5.4   Do the isothermal or adiabatic models used in program 'THERMAL' give an accurate representation of thermal conditions operating in a bearing?

5.5   Apply the elastic pad program to various common engineering materials of varying elastic modulus. Discuss the data obtained.

5.6   Apply external pressure to the bearing for the grooved cavitation program, as in a bearing for sub-sea operation, and observe the effect on cavitation. Note: this question would require re-writing some lines of the program where the negative pressures are set to zero, so it is only suitable for more advanced students.

5.7   Investigate how the reduced groove size and reduced groove pressure affect the load capacity. Is there a critical groove size before the load capacity is significantly affected?

5.8   Can we ignore viscosity variation across the film (the 'z' axis) in a thermo-hydrodynamic bearing?

5.9   Polymers have good sliding characteristics against metals. Could a polymer be used to make an effective tilted pad bearing?

5.10   Why would one need to calculate the flow rate in a grooved bearing?

5.11   Why would one need to study the effect of groove size on bearing performance?

5.12   How does misalignment alter the static and dynamic characteristics of a journal bearing?

5.13   Suggest some future developments in computational hydrodynamics, especially in regard to the increased speed of computing.

5.14   Suggest a 'stretched goal' or long-term objective for computational hydrodynamics. Hint: What is a basic assumption about the lubricant film for all the programs?

REFERENCES

1   G. Vogelpohl, Beitraege zur Kenntnis der Gleitlagerreibung (Contributions to Study of Journal Bearing Friction), Ver. Deutsch. Ing., Forschungsheft, Vol. 386, 1937, pp. 1-28.

2   M.M. Reddi and T.Y. Chu, Finite Element Solution of the Steady-State Incompressible Lubrication Problem, *Transactions ASME, Journal of Lubrication Technology*, Vol. 92, 1970, pp. 495-503.

3   J.F. Booker and K.K. Huebner, Application of Finite Element Methods to Lubrication, an Engineering Approach, *Transactions ASME, Journal of Lubrication Technology*, Vol. 94, 1972, pp. 313-323.

4   A. Cameron, Principles of Lubrication, Chapter by M.R. Osborne on Computation of Reynolds' Equation, Longmans, London, 1966, pp. 426-439.

5   A.A. Raimondi and J. Boyd, A Solution for the Finite Journal Bearing and its Application to Analysis and Design, *ASLE Transactions*, Vol. 1, 1958, pp. 159-209.

6   A. Cameron, Principles of Lubrication, Longmans, London, 1966, pp. 305-340.

7   T. Someya (editor), Journal-Bearing Data-Book, Springer Verlag, Berlin, Heidelberg, 1989.

8   A.J. Colynuck and J.B. Medley, Comparison of Two Finite Difference Methods for the Numerical Analysis of Thermohydrodynamic Lubrication, *Tribology Transactions*, Vol. 32, 1989, pp. 346-356.

9   C.M.Mc. Ettles, Transient Thermoelastic Effects in Fluid Film Bearings, *Wear*, Vol. 79, 1982, pp. 53-71.

10   J.H. Vohr, Prediction of the Operating Temperature of Thrust Bearings, *Transactions ASME, Journal of Lubrication Technology*, Vol. 103, 1981, pp. 97-106.

11   S.M. Rohde and K.P. Oh, A Thermoelastohydrodynamic Analysis of a Finite Slider Bearing, *Transactions ASME, Journal of Lubrication Technology*, Vol. 97, 1975, pp. 450-460.

12   H.G. Elrod, A Cavitation Algorithm, *Transactions ASME, Journal of Lubrication Technology*, Vol. 103, 1981, pp. 350-354.

13   B. Jakobsson and L. Floberg, The Finite Journal Bearing Considering Vaporization, Chalmers Tekniska Hoegskolas Madlinar, Vol. 190, 1957, pp. 1-116.

14   K.O. Olsson, Cavitation in Dynamically Loaded Journal Bearings, Chalmers University of Technology, 1965, Goteborg.

15   D. Dowson, A.A.S. Miranda and C.M. Taylor, Implementation of an Algorithm Enabling the Determination of Film Rupture and Reformation Boundaries in a Liquid Film Bearing, Proc. 10th Leeds-Lyon Symp. on Numerical and Experimental Methods in Tribology, Sept. 1983, editors: D. Dowson, C.M. Taylor, M. Godet and D. Berthe, Butterworths, 1984, pp. 60-70.

HYDROSTATIC LUBRICATION

6.1 INTRODUCTION

In hydrostatic lubrication the bearing surfaces are fully separated by a lubricating film of liquid or gas forced between the surfaces by an external pressure. The pressure is generated by an external pump instead of by viscous drag as is the case with hydrodynamic lubrication. As long as a continuous supply of pressurized lubricant is maintained, a complete film is present even at zero sliding speed. Hydrostatic films usually have a considerable thickness reaching 100 [μm] and therefore prevent contact between the asperities of even the roughest surfaces. This ensures a complete absence of sticking friction. Furthermore, the friction generated by viscous shear of the lubricant decreases to zero at zero sliding speed. Hydrostatic bearings can support very large masses and allow them to be moved from their stationary positions with the use of minimal force. These extraordinary features of zero static friction and high load capacity were applied, for example, in the 5.08 [m] diameter Mount Palomar telescope and in many radar installations. With other types of bearing, starting friction is inevitable and can cause distortion and damage to large structures. This problem is critical to the design of large telescopes which rely on extreme accuracy of telescope positioning.

Hydrostatic bearings have a wide range of characteristics and need to be carefully controlled for optimum effect. The following questions summarize the potential problems that an engineer or tribologist might confront. If it is possible to generate films similar to hydrodynamic films, how can these films be controlled and produced when needed? What are the practical applications of this type of lubrication? What are the critical design parameters of hydrostatic bearings? What is the bearing stiffness and how can it be controlled? The engineer or tribologist should know how to find the answer to all these questions.

According to available records, the first hydrostatic bearing was invented in 1851 by Girard [1,2] who employed a bearing fed by high pressure water for a system of railway propulsion. Since then there have been a number of patents and publications dealing with different design aspects and incorporating various features. Some of these designs introduced genuine improvements but the majority merely introduced complexity rather than simplicity and are destined to be forgotten.

As well as the true hydrostatic bearing, hybrid bearings have also been developed. These are hydrodynamic bearings assisted by an externally pressurized lubricant supply.

In this chapter, the mechanism of film generation in hydrostatic bearings together with methods of calculating basic bearing operational and design parameters are discussed. Commonly used methods of controlling the bearing stiffness are also outlined.

6.2   HYDROSTATIC BEARING ANALYSIS

The analysis of hydrostatic bearings is much simpler than the analysis of hydrodynamic bearings. It is greatly simplified by the condition that the surfaces of these bearings are parallel.

Flat Circular Hydrostatic Pad Bearings

Consider, as an example, a flat circular hydrostatic pad bearing with a central recess as shown in Figure 6.1 [2].

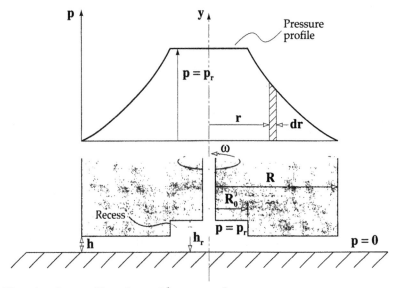

FIGURE 6.1   Flat circular pad bearing with a central recess.

· *Pressure distribution*

The pressure distribution can be calculated by considering the lubricant flow in a bearing. For a bearing supplied with lubricant under pressure, the flow rate given by equation (4.18) becomes:

$$q_x = -\frac{h^3}{12\eta}\frac{\partial p}{\partial x}$$

Since the bearing is circular, the flow through the elemental ring at radius 'r' is:

$$Q = -\frac{h^3}{12\eta}\frac{\partial p}{\partial r}\,2\pi r \tag{6.1}$$

rearranging and integrating yields (surfaces are parallel, i.e., $h \neq f(r)$):

$$p = -\frac{6\eta Q}{\pi h^3}\int\frac{dr}{r} = -\frac{6\eta Q}{\pi h^3}\ln r + C \tag{6.2}$$

Boundary conditions from Figure 6.1 are:

$$\mathbf{p = 0} \quad \text{at} \quad \mathbf{r = R}$$

Substituting into equation (6.2) yields the constant '**C**':

$$C = \frac{6\eta Q}{\pi h^3} \ln R \tag{6.3}$$

Hence the pressure distribution for this type of bearing in terms of lubricant flow, bearing geometry and lubricant viscosity is given by:

$$\boxed{p = \frac{6\eta Q}{\pi h^3} \ln\left(\frac{R}{r}\right)} \tag{6.4}$$

· *Lubricant Flow*

By rearranging equation (6.4), the lubricant flow, i.e., the minimum amount of lubricant required from the pump to maintain film thickness '**h**' in a bearing, is obtained:

$$Q = \frac{\pi h^3 p}{6\eta} \frac{1}{\ln(R/r)}$$

Since at $\mathbf{r = R_0}$, $\mathbf{p = p_r}$ then:

$$\boxed{Q = \frac{\pi h^3 p_r}{6\eta} \frac{1}{\ln(R/R_0)}} \tag{6.5}$$

where:

$\mathbf{p_r}$	is the recess pressure [Pa];
$\mathbf{h}$	is the lubricant film thickness [m];
η	is the lubricant dynamic viscosity [Pas];
$\mathbf{R}$	is the outer radius of the bearing [m];
$\mathbf{R_0}$	is the radius of the recess [m];
$\mathbf{Q}$	is the lubricant flow [m³/s].

It can be seen that by merely substituting for flow (eq. 6.5), the pressure distribution (eq. 6.4) is expressed only in terms of the recess pressure and bearing geometry, i.e.:

$$\boxed{p = p_r \frac{\ln(R/r)}{\ln(R/R_0)}} \tag{6.6}$$

· *Load Capacity*

The total load supported by the bearing can be obtained by integrating the pressure distribution over the specific bearing area:

$$W = \int_A p\, dA = \iint_A p\, dx\, dy$$

It can be seen from the pressure distribution shown in Figure 6.1 that for the bearing considered, the expression for total load is composed of two terms: one related to the recess area and the other to the bearing load area. The general integral for load shown above can therefore be conveniently divided into two integrals:

$$W = \int_0^{2\pi}\int_0^{R_0} p_r r\, d\theta\, dr + \int_0^{2\pi}\int_{R_0}^{R} p r\, d\theta\, dr$$

Since the recess pressure is constant the expression is reduced to:

$$W = p_r \pi R_0^2 + 2\pi \int_{R_0}^{R} p r\, dr$$

Substituting for pressure equation (6.6),

$$W = p_r \pi R_0^2 + 2\pi p_r \frac{1}{\ln(R/R_0)} \int_{R_0}^{R} r \ln\!\left(\frac{R}{r}\right) dr$$

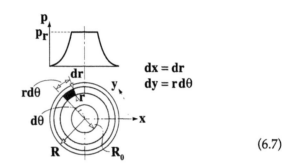

(6.7)

Integrating by parts and substituting yields:

$$W = p_r \pi R_0^2 + 2\pi p_r \frac{1}{\ln(R/R_0)} \left|\frac{r^2}{2}\ln\!\left(\frac{R}{r}\right) + \frac{r^2}{4}\right|_{R_0}^{R}$$

After simplifying, the expression for the total load that the bearing can support is:

$$\boxed{W = \frac{p_r \pi}{2}\left[\frac{R^2 - R_0^2}{\ln(R/R_0)}\right]}$$

(6.8)

where:

W is the bearing load capacity [N].

· *Friction Torque*

The frictional resistance of a rotating hydrostatic circular pad bearing consists only of friction torque which is usually very small and can be calculated from:

$$T = \int_A r\, dF$$

(6.9)

In a similar manner to the expression for load, the expression for total torque has two components: one related to the recess area and the other to the bearing load area:

$$T = T_r + T_l$$

where:

T   is the total friction torque [Nm];

T_r is the friction torque acting on the recess area [Nm];

T_l is the friction torque acting on the bearing load area [Nm].

The bearing surfaces are parallel which ensures a uniform film thickness, **h = constant**, hence the velocity equation (4.11) becomes:

$$u = U\frac{z}{h}$$

It should be noted that in this bearing the upper surface is moving while the bottom surface remains stationary. This is the opposite situation from the hydrodynamic pad bearing discussed previously. Thus the following conditions for velocity apply:

$$U_1 = U \quad \text{and} \quad U_2 = 0$$

Differentiation of the velocity equation gives the shear rate:

$$\frac{du}{dz} = \frac{U}{h}$$

Thus the shear stress is:

$$\tau = \eta\frac{du}{dz} = \eta\frac{U}{h}$$

The friction force in its differential form 'dF', also has two components: one referring to the recess area and the other to the bearing load (land) area, i.e.:

$$dF = \eta\frac{U}{h_r}r\,d\theta\,dr + \eta\frac{U}{h}r\,d\theta\,dr \qquad\qquad (6.10)$$

where:

h   is the hydrostatic film thickness in the bearing load area [m];

h_r is the hydrostatic film thickness in the recess area [m].

Substituting (6.10) into (6.9) gives:

$$T = \int_0^{2\pi}\int_0^{R_0} \eta\frac{U}{h_r}r^2\,d\theta\,dr + \int_0^{2\pi}\int_{R_0}^{R} \eta\frac{U}{h}r^2\,d\theta\,dr$$

Expressing 'U' in terms of revolutions per second:

$$U = 2\pi r n$$

where:

n   is the speed of the bearing [rev/s].

and substituting gives:

$$T = \int_0^{2\pi} \int_0^{R_0} \eta \frac{2\pi n}{h_r} r^3 d\theta dr + \int_0^{2\pi} \int_{R_0}^{R} \eta \frac{2\pi n}{h} r^3 d\theta dr$$

Assuming constant viscosity and velocity and integrating yields:

$$T = \frac{\pi^2 \eta n}{h_r} R_0^4 + \frac{\pi^2 \eta n}{h} (R^4 - R_0^4) \qquad (6.11)$$

where:

T   is the torque needed to rotate the bearing [Nm].

Since in practical applications the recess depth 'h_r' is at least 16 to 20 times the bearing film thickness 'h' [2], the first term of equation (6.11), which is related to the recess area, is very small and may be neglected.

· *Friction Power Loss*

The friction power loss which is transmitted through the operating surfaces can be calculated from:

$$H_f = T\omega = 2T\pi n$$

where:

ω   is the bearing angular velocity, $\omega = 2\pi n$, [rad/s];

H_f is the friction power loss in the bearing [W].

Substituting for torque (eq. 6.11) yields:

$$H_f = 2\pi^3 \eta n^2 \left[\frac{R_0^4}{h_r} + \frac{(R^4 - R_0^4)}{h} \right] \qquad (6.12)$$

It has been suggested that in practice this equation gives slightly underestimated results due to the neglect of the recess effects such as flow recirculation [2,7]. To allow for these effects it has been proposed to increase the recess component in equation (6.12) by a factor of four [2,7]. Very often, however, in practical applications an allowance for these effects is made by treating the whole bearing area as the bearing load area [2] and hence equation (6.12) becomes:

$$H_f = \frac{2\pi^3 \eta n^2 R^4}{h}$$

Non-Flat Circular Hydrostatic Pad Bearings

Flat hydrostatic bearings are only suitable for supporting a load normal to the plane of contact. In some mechanical systems it is, however, very convenient to support oblique loads while allowing rotation and non-flat circular pad bearings are suitable for this purpose. Examples of non-flat bearings used in mechanical equipment are bearings based on a conical or hemi-spherical shape. The typical geometries of these bearings are shown in Figure 6.2.

Non-flat circular pad bearings can be analysed in the same manner as already discussed for flat circular pads. For example, the geometry of the conical bearing is shown in Figure 6.3 and the following analysis is applicable.

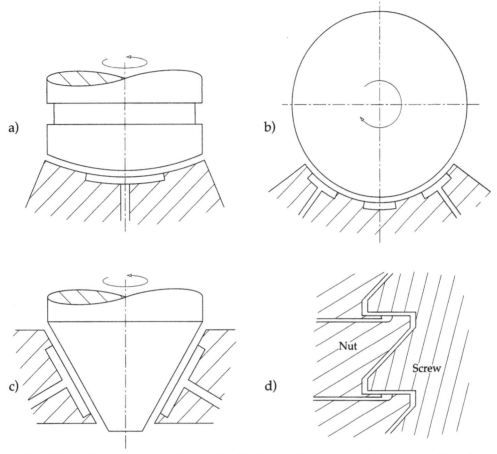

FIGURE 6.2   Typical geometries of non-flat hydrostatic bearings: a) footstep, b) spherical, c) conical and d) hydrostatic screw thread.

Over the flat part of the bearing surface, hydrostatic pressure is equal to the supply pressure while a nearly linear decrease in pressure prevails in the conical bearing region. The pressure profile is then very similar to the flat pad bearing pressure profile already discussed. In the conical bearing, pressure does show a fully asymptotic profile outside the constant pressure region because the bearing radius 'r' increases at a slower rate with respect to distance travelled by the escaping fluid. The calculation procedure of bearing operating parameters such as pressure, load, flow and friction force is the same as for flat circular pad bearings. An allowance has to be made, however, for the bearing geometry as shown in the following sections.

· *Pressure Distribution*

In this bearing, the lubricant flow through an elemental ring at radius 'r' is given by the equation:

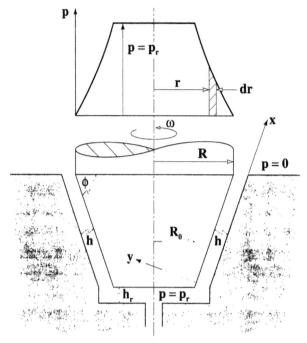

FIGURE 6.3   Film geometry of conical hydrostatic bearing.

$$Q = -\frac{h^3}{12\eta}\frac{dp}{dx}2\pi r$$

where:

$$dx = \frac{dr}{\cos\phi} \quad \text{and} \quad dy = rd\theta$$

substituting and integrating gives:

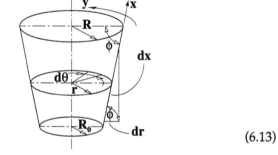

$$p = \frac{6\eta Q}{\pi h^3 \cos\phi}\ln\left(\frac{R}{r}\right)$$

(6.13)

· *Lubricant Flow*

Rearranging (6.13) and substituting for boundary conditions, i.e., $r = R_0$ at $p = p_r$ gives the lubricant flow:

$$Q = \frac{\pi h^3 p_r \cos\phi}{6\eta}\frac{1}{\ln(R/R_0)}$$

(6.14)

As was described for flat circular pad bearings, substitution of the flow expression into the pressure equation (6.13) gives the pressure distribution in terms of recess pressure and bearing geometry:

$$p = p_r\frac{\ln(R/r)}{\ln(R/R_0)}$$

· *Load Capacity*

The total load capacity that the bearing can support is found by integration of pressure over the bearing area:

$$W = \int_0^{2\pi}\int_0^{R_0} p_r r d\theta \frac{dr}{\cos\phi} + \int_0^{2\pi}\int_{R_0}^{R} p r d\theta \frac{dr}{\cos\phi}$$

Substituting for 'p' and integrating give the bearing load capacity '**W**':

$$W = \frac{p_r \pi}{2\cos\phi}\left[\frac{R^2 - R_0^2}{\ln(R/R_0)}\right] \tag{6.15}$$

· *Friction Torque*

The friction torque is calculated from equation (6.9), according to the same principles as outlined for flat circular pad bearings. The friction force in its differential form '**dF**' also has recess and bearing load components, i.e.:

$$dF = \eta \frac{U}{h_r} r d\theta dr + \eta \frac{U}{h} r d\theta \frac{dr}{\cos\phi}$$

where:

    h   is the hydrostatic film thickness in the bearing load area [m];

    h$_r$ is the hydrostatic film thickness in the recess area [m].

Expressing 'U' in terms of revolutions per second 'n' and substituting into (6.9) yield:

$$T = \int_0^{2\pi}\int_0^{R_0} \eta \frac{2\pi n}{h_r} r^3 d\theta dr + \int_0^{2\pi}\int_{R_0}^{R} \eta \frac{2\pi n}{h}\frac{r^3}{\cos\phi} d\theta dr$$

Assuming no change in viscosity 'η' and rotational velocity '**n**' and integrating give:

$$T = \frac{\pi^2 \eta n}{h_r} R_0^4 + \frac{\pi^2 \eta n}{h}\frac{(R^4 - R_0^4)}{\cos\phi} \tag{6.16}$$

where:

    T   is the torque needed to rotate the bearing [Nm].

Similarly as with flat circular pad bearings, the recess component in these bearings is usually very small and may be omitted for elementary estimates of friction torque.

· *Friction Power Loss*

The friction power loss which is transmitted through the operating surfaces for a conical hydrostatic bearing is given by:

$$H_f = 2\pi^3 \eta n^2\left[\frac{R_0^4}{h_r} + \frac{(R^4 - R_0^4)}{h\cos\phi}\right] \tag{6.17}$$

As discussed for flat circular pad bearings, the allowance for recess effects in practical applications can be made by treating the recess area as a bearing load area and equation (6.17) becomes:

$$H_f = \frac{2\pi^3 \eta\, n^2 R^4}{h\cos\phi}$$

The procedure described earlier can be applied in a similar manner to any other bearing geometry, e.g., square pads, rectangular pads, etc. More information about these bearings can be found elsewhere [2].

6.3 GENERALIZED APPROACH TO HYDROSTATIC BEARING ANALYSIS

In hydrostatic bearing analysis, load and flow are frequently expressed for simplicity in terms of non-dimensional parameters. Examples of this treatment are given below for flat circular pads and flat square pads.

Flat Circular Pad Bearings

The load capacity and the lubricant flow rate are expressed in terms of non-dimensional load and flow multiplied by a non-dimensional scale factor.

$$\boxed{W = A\,p_r \overline{A} = \pi R^2 p_r \overline{A}}\tag{6.18}$$

$$\boxed{Q = \frac{p_r h^3}{\eta}\,\overline{B}}\tag{6.19}$$

where:

A	is the total pad area [m²];
A and **B**	are non-dimensional load and flow coefficients defined as:

$$\overline{A} = \frac{1}{2R^2}\left[\frac{R^2 - R_0^2}{\ln(R/R_0)}\right]$$

$$\overline{B} = \frac{\pi}{6}\,\frac{1}{\ln(R/R_0)}$$

Non-dimensional coefficients $\overline{A}$ and $\overline{B}$ are usually plotted against bearing dimensions as shown in Figure 6.4 [2]. The bearing analysis and design are now greatly simplified since for any given R/R_0 ratio, the non-dimensional coefficients $\overline{A}$ and $\overline{B}$ can easily be found and load capacity and lubricant flow calculated. It can be noted that another parameter **H** is included and its significance is explained in the next section.

Flat Square Pad Bearings

The load capacity and lubricant flow rate in flat square pad bearings can be expressed in a similar manner to the circular pad bearings, i.e., in terms of non-dimensional coefficients:

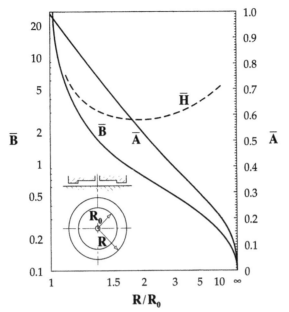

FIGURE 6.4   Design coefficients for flat circular pad bearings [2].

$$W = A\,p_r\overline{A} = B^2 p_r\overline{A}$$

$$Q = \frac{p_r h^3}{\eta}\,\overline{B}$$

where:

A	is the total pad area;
B	is the width of the bearing;
$\overline{A}$ and $\overline{B}$	are non-dimensional load and flow coefficients.

The coefficients $\overline{A}$ and $\overline{B}$ are plotted against bearing dimensions as shown in Figure 6.5 [2]. It can be seen that the ratio of recess dimension to pad dimension **C/B** significantly affects load capacity and lubricant flow rate.

6.4   OPTIMIZATION OF HYDROSTATIC BEARING DESIGN

The parameters of a hydrostatic bearing, such as bearing area, recess area, lubricant flow rate, etc., can be varied to achieve maximum stiffness, maximum load capacity for a given oil flow or minimum pumping power. Since the bearing is almost entirely under external control, it is possible to regulate the characteristics of these bearings to a far greater extent than, for example, those of hydrodynamic bearings.

Minimization of Power

As can be seen from Figure 6.1, if the recess in the bearing is made almost as large as the bearing diameter, then supply pressure is maintained over virtually the entire area of the bearing. This would ensure a higher load capacity than with a smaller recess but with the disadvantage of requiring a very high rate of lubricant supply pumping power.

The total power required is the sum of friction power and the pumping power, i.e.:

$$H_t = H_f + H_p$$

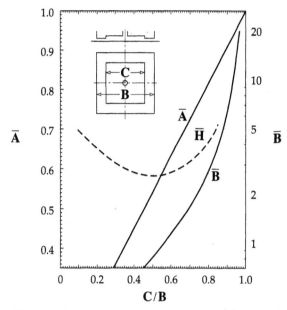

FIGURE 6.5   Design coefficients for square pad bearings; **C** is the width of the recess [2].

Pumping power 'H_p' is defined as the product of the lubricant flow '**Q**' and the recess pressure 'p_r', i.e.:

$$H_p = Qp_r$$

The total power describes the rate at which the friction and pumping energies are converted into heat in the bearing. The heat dissipation rate is the product of mass flow rate, specific heat and temperature, thus:

$$H_t = Q\rho\sigma\Delta T$$

where:

Q   is the lubricant flow [m³/s];

ρ is the density of the lubricant [kg/m³];

σ is the specific heat of the lubricant [J/kgK];

ΔT is the temperature rise [°C].

For most applications, however, the sliding speed is low so that the total power required 'H_t' and therefore the temperature rise 'ΔT' are small. Possible hydrodynamic effects that may occur at low operating speeds are also negligible. At high speeds, however, hydrodynamic effects become significant and the whole bearing design must be reconsidered.

Ratio of friction power to pumping power 'ζ' (i.e., $\zeta = H_f/H_p$) can be used as a measure of the proportion of the hydrodynamic effects to the hydrostatic effects. If the bearing is not rotating then $H_f = 0$ and $\zeta = 0$, and a purely hydrostatic load support mechanism is operating. As 'ζ' increases the hydrodynamic effects become more significant. The bearings operating with $\zeta \geq$ **1** are considered as 'high speed bearings' and with $\zeta \ll$ **1** as 'low speed bearings' [2]. When $\zeta \geq$ **3** then the hydrostatic and hydrodynamic effects on load are of the same order [2]. The 'ζ' factor is used as a design parameter in bearing optimization for minimum power and maximum load.

· *Low Speed Recessed Bearings*

In low speed bearings $H_f \approx 0$, thus $H_t \approx H_p$ and $\zeta = 0$. The bearing geometry can be optimized so that at a given total bearing area, film thickness, viscosity and applied load the pumping power is a minimum. This is achieved by calculating pumping power to load ratio, i.e.:

$$\frac{H_t}{W} = \frac{H_p}{W} = \frac{Qp_r}{W}$$ (6.20)

Substituting for load (6.18) and flow (6.19), the ratio of pumping power to load expressed in terms of non-dimensional coefficients becomes:

$$\frac{H_p}{W} = \frac{p_r\left(\dfrac{p_r h^3}{\eta}\overline{B}\right)W}{(\pi R^2 p_r \overline{A})^2} = \frac{Wh^3}{\pi^2 R^4 \eta}\frac{\overline{B}}{\overline{A}^2}$$

Thus:

$$\boxed{\frac{H_p}{W} = \frac{Wh^3}{\pi^2 R^4 \eta}\overline{H}}$$ (6.21)

The values of the '$\overline{H}$' coefficient for flat circular and square pad bearings can be found from Figures 6.4 and 6.5, respectively [2].

· *High Speed Recessed Bearings*

The effects of viscous shear due to the relative motion between the surfaces may become significant if the bearing is forced to operate at high speeds. Since the ratio of friction power to pumping power is expressed as:

$$\zeta = H_f/H_p$$

The total power in terms of 'ζ' and 'H_p' is:

$$H_t = (1 + \zeta)H_p$$

Substituting into (eq. 6.20) gives the power to load ratio for high speed bearings:

$$\frac{H_t}{W} = \frac{(1 + \zeta)H_p}{W} = \frac{(1 + \zeta)Qp_r}{W}$$

rearranging yields:

$$\boxed{\frac{(1 + \zeta)H_p}{W} = \frac{(1 + \zeta)Wh^3}{\pi^2 R^4 \eta}\overline{H}}$$ (6.22)

For any value of 'ζ' it is still necessary to find a minimum value of the '$\overline{H}$' parameter to optimize the bearing geometry for maximum load and minimum power.

Other parameters such as viscosity and bearing clearance can also be optimized and the details are given elsewhere [2]. For example, the lubricant viscosity can be optimized by calculating power losses and load capacity for a range of viscosities while maintaining all the other parameters at required design levels. The optimum clearance is obtained when the power ratio $\zeta = 3$ [2]. It has also been shown that the bearing gives the optimum performance when the power ratio 'ζ' is between $1 \leq \zeta \leq 3$ [2]. The effect of optimization is relatively small since the difference between most and least effective procedures is only about 15% of total bearing power consumption [2,8].

More information about hydrostatic bearing design methods can be found elsewhere [2,4,9].

Control of Lubricant Film Thickness and Bearing Stiffness

It can be seen from the derived equations that film thickness has no direct effect on hydrostatic pressure field (i.e., equation 6.4), which is in direct contrast to hydrodynamic lubrication. According to the theory presented so far, film thickness is free to vary and the bearing has no stiffness in the direction of load. In real hydrostatic bearings, there is a distinct stiffness which can reach very high values. Stiffness in fact depends on the pressure-flow characteristic of the lubricant supply as is shown below.

The load and flow equations can be rearranged so that load is expressed in terms of flow. For example, by rearranging equation (6.8), 'p_r' is obtained in terms of load:

$$p_r = \frac{2W}{\pi} \frac{\ln(R/R_0)}{R^2 - R_0^2}$$

Substituting this equation into the flow equation (6.5) and simplifying give the load in terms of flow, i.e.:

$$W = 3\eta (R^2 - R_0^2) \frac{Q}{h^3} \tag{6.23}$$

When the above expression is examined, it is clear that the changes in load are related to changes in lubricant flow, film thickness, bearing geometry and lubricant viscosity. Assuming that the bearing geometry and lubricant viscosity remain constant, load still depends on changes in flow and film thickness or vice versa. It can be seen from equation (6.23) that if there is a sudden increase in load the lubricating film may collapse, and if there is sudden decrease in load the lubricating film will increase and the bearing may become unstable. By controlling the flow rate, it is possible to prevent both film collapse after a precipitate increase in load and bearing instability caused by a sudden reduction in load. In practical applications, the lubricant flow into the bearing is controlled by the following methods:

· constant delivery pump,
· capillary,
· orifice,
· pressure sensors.

The parameter of bearing stiffness is introduced to describe variations in film thickness with imposed load. If the load is increased by an amount 'ΔW' then the film thickness is reduced by a corresponding amount '$-\Delta h$' (negative for squeeze, i.e., to allow for direction of load). The stiffness is defined as the limit of the ratio:

$$\lambda = -\lim_{\Delta W \Rightarrow 0} \frac{\Delta W}{\Delta h} = -\frac{dW}{dh}$$

Thus the stiffness is a measure of the force necessary to produce changes in film thickness. Each of the lubricant flow control methods employed in a bearing will give a different stiffness, i.e., a different response to the load changes as shown below.

· *Stiffness With Constant Flow Method*

Lubricant is delivered at a constant rate of '**Q**' to the bearing when a constant delivery pump or constant flow valve is used. The constant flow valve has a variable orifice controlled by the flow itself. A pump is rarely used to supply just a single bearing since this arrangement is impractical and expensive.

The lubricant flow required from the pump to maintain the film thickness '**h**' is given by equation (6.19):

$$Q = \frac{p_r h^3}{\eta} \overline{B}$$

Rearranging this equation gives the recess pressure in terms of lubricant flow rate:

$$p_r = \frac{Q\eta}{h^3\overline{B}}$$

Substituting this into equation (6.18) defines the load in terms of lubricant flow rate:

$$W = \frac{AQ\eta\overline{A}}{h^3\overline{B}}$$

Differentiation of this expression with respect to film thickness enables the stiffness for constant flow to be obtained:

$$\lambda = -\frac{dW}{dh} = \frac{3AQ\eta\overline{A}}{h^4\overline{B}}$$

hence:

$$\boxed{\lambda = \frac{3W}{h}}$$

(6.24)

It can be seen from the above analysis that the bearing stiffness controlled by a constant flow method depends on bearing temperature, i.e., the viscosity term present in the stiffness equation. The prime benefit of this mode of bearing control is that a constant lubricant flow rate is maintained which is extremely important in large multiple bearing systems. The disadvantage is that the stiffness cannot be varied to suit design requirements. High bearing stiffness is essential to the accuracy of precision machine tools. For this reason, more elaborate methods of lubricant supply have been applied to hydrostatic bearings.

· *Stiffness With Capillary Restrictors*

A capillary or narrow diameter tube can be introduced between the main lubricant supply line and the bearing recess as shown in Figure 6.6 as a means of controlling lubricant flow. The capillary diameter and length are selected so that the required bearing stiffness is obtained. The theory of capillary controlled stiffness is outlined below.

The flow rate through a straight capillary is given by the Hagen-Poiseuille equation. This equation applies to the tubes with high 'l/d' ratios, i.e., **l/d > 100** and a Reynolds number less that 2000, i.e., **Re < 2000**:

$$Q = \frac{p_s - p_r}{k_c \eta}$$

where:

      Q is the flow rate through the capillary [m³/s];

      p_s is the lubricant supply pressure [Pa];

      p_r is the bearing recess pressure [Pa];

      η is the lubricant dynamic viscosity [Pas];

      k_c is the capillary constant.

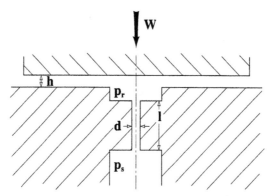

FIGURE 6.6   Flat circular pad bearing with capillary controlled flow.

The capillary constant 'k_c' can be calculated from:

$$k_c = \frac{128 \times l}{\pi d^4}$$

where:

      l is the capillary length [m];

      d is the capillary diameter [m].

Equating the flow through the capillary to the total lubricant flow (eq. 6.19) gives:

$$\frac{p_s - p_r}{k_c \eta} = \frac{p_r h^3}{\eta} \overline{B}$$

Rearranging:

$$p_r = \frac{p_s}{1 + k_c h^3 \overline{B}}$$

Substituting into the load expression (eq. 6.18):

$$W = \frac{A\,p_s\overline{A}}{1 + k_c h^3 \overline{B}}$$

Differentiating with respect to film thickness gives the stiffness with a capillary restrictor [2]:

$$\lambda = \frac{3W}{h}\,\frac{1}{1 + 1/(k_c h^3 \overline{\overline{B}})} \tag{6.25}$$

A comparison of equations (6.24) and (6.25) reveals that a capillary restrictor gives a lower stiffness than a constant flow device. It is also possible with this method of controlling the lubricant flow to vary stiffness to suit specific design requirements as opposed to the fixed values of stiffness provided by the constant flow device. It can also be noted that the lubricant viscosity is eliminated when equating the flow through the capillary to the total lubricant flow. Hence the bearing stiffness in this case is independent of temperature since the effect of viscosity on flow through the bearing is balanced by the viscosity effect on flow through the capillary.

· *Stiffness With an Orifice*

In some applications an orifice is fitted to the lubricant supply line as shown in Figure 6.7. Although this device is very simple it controls lubricant flow quite effectively and allows some freedom in the selection of load and stiffness characteristics.

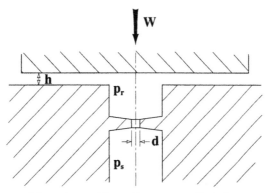

FIGURE 6.7   Flat circular pad bearing with orifice controlled flow.

The flow rate through an orifice of diameter '**d**' is:

$$Q = \frac{\pi d^2}{2}\left(\frac{p_s - p_r}{2\rho}\right)^{1/2} C_d$$

where:

Q	is the flow rate through the orifice [m³/s];
d	is the diameter of the orifice [m];
p_s	is the lubricant supply pressure [Pa];
p_r	is the bearing recess pressure [Pa];
ρ	is the lubricant density [kg/m³];
C_d	is the discharge coefficient.

Equating flow through the orifice to the total lubricant flow through the bearing (6.19):

$$\frac{\pi d^2}{2}\left(\frac{p_s - p_r}{2\rho}\right)^{1/2} C_d = \frac{p_r h^3}{\eta}\,\overline{B}$$

Rearranging:

$$p_r = \frac{\pi d^2 \eta}{2h^3 \overline{\overline{B}}}\left(\frac{p_s - p_r}{2\rho}\right)^{1/2} C_d$$

Substituting into the expression for load (6.18) gives:

$$W = \frac{A \pi d^2 \eta \overline{A}}{2h^3 \overline{B}}\left(\frac{p_s - p_r}{2\rho}\right)^{1/2} C_d$$

The bearing stiffness with an orifice is found by differentiating with respect to film thickness 'h':

$$\lambda = -\frac{dW}{dh} = \frac{3A\pi d^2 \eta \overline{A}}{2h^4 \overline{\overline{B}}}\left(\frac{p_s - p_r}{2\rho}\right)^{1/2} C_d$$

Hence:

$$\boxed{\lambda = \frac{3W}{h}}$$

(6.26)

which is the same expression for stiffness as found in the case of the constant flow method. Although the orifice is easy to apply it has one major deficiency. The flow characteristics of an orifice can easily be altered by erosive wear from contaminant particles in the lubricant. The stiffness and flow characteristics will gradually vary with time unless a very clean lubricant is used. The bearing stiffness with this method of lubricant flow control may be affected by the operating temperature. Similar effects will also be observed when the constant flow method is applied. The orifices are in general more compact than capillaries and give fractionally greater stiffnesses [2]. Although the traditionally quoted values of discharge coefficient are between **0.6** and **0.65**, detailed testing has demonstrated that there could be some significant variations from this value [10]. For example, in cases when either the flow rate of the lubricant is low or the orifice diameter and lubricant viscosity are large then the value of 'C_d' is less than **0.6** and may even be as low as **0.4** [10].

· *Stiffness With Pressure Sensors*

In more sophisticated systems, pressure sensors can be used to provide feedback of recess pressure for control purposes. By applying these devices a hydrostatic bearing of even infinite stiffness can be designed and built. Most pressure sensors are based on a diaphragm which deflects under increased hydrostatic pressure to allow a greater flow of lubricant. The most common type of diaphragm valve is the Mohsin [5] valve shown in Figure 6.8.

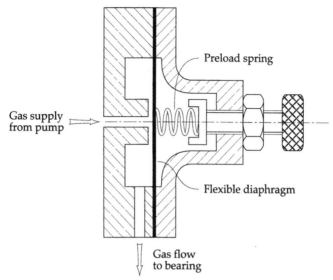

Gas supply
from pump

Preload spring

Flexible diaphragm

Gas flow
to bearing

FIGURE 6.8   Diaphragm pressure sensor valve developed by Mohsin [5].

There are two basic limitations to diaphragm bearings: it is necessary to adjust the spring force on the diaphragm to match the bearing stiffness and high stiffness is only maintained up to a certain load where bearing deflection rises sharply with any further increase in load. More detailed information about pressure sensing valves can be found elsewhere [2].

6.5   AEROSTATIC BEARINGS

In some applications a gas is used to lubricate the bearings and these are known as aerostatic bearings. The mechanism of film generation is the same as in liquid bearings. Gas lubricated bearings offer some advantages such as:

· gas viscosity increases with temperature, thus reducing heating effects during over-load or abnormal operating conditions,

· some gases are chemically stable over a wider temperature range than hydrocarbon lubricants,

· a non-combustible gas eliminates the fire hazard associated with hydrocarbons,

· if air is selected as the hydrostatic lubricant, then it is not necessary to purchase or recycle the lubricant,

· gases can offer greater cleanliness and non-toxicity than fluid lubricants.

The viscosities of some gases used in aerostatic bearings are shown in Table 6.1.

TABLE 6.1   Viscosities of various gases at 20°C and 0.1 [MPa] pressure.

Gas	Viscosity [Pas]
Hydrogen	8.80×10^{-6}
Helium	1.96×10^{-5}
Nitrogen	1.76×10^{-5}
Oxygen	2.03×10^{-5}
Carbon Dioxide	1.47×10^{-5}
Air	1.82×10^{-5}

Aerostatic bearings are used in the precision machine tool industry, metrology, computer peripheral devices and in dental drills, where the air used in the bearing also drives the drill. They are particularly useful for high speed applications and where precision is required since very thin film thicknesses are possible. The main disadvantage, however, is that the load capacity of gas lubricated bearings or aerostatic bearings is much lower than the load capacity of fluid lubricated bearings of the same size. The most commonly used gas is air, but other gases such as carbon dioxide and helium have been used in specialized systems, e.g., nuclear technology.

The analysis of aerostatic bearings is very similar to liquid hydrostatic bearings. The main difference, however, is that the gas compressibility is now a distinctive feature and has to be incorporated into the analysis. Since the pressures generated in these bearings are much lower than in liquid lubricated bearings, ambient pressure cannot be neglected and is also included in the analysis. For example, assuming isothermal behaviour of the gas, the main performance parameters of the aerostatic flat circular pad bearings can be calculated from the following formulae [6].

Pressure Distribution

The pressure distribution is found from a simplified form of the compressible Reynolds equation for radial coordinates with angular symmetry and negligible sliding velocity. The derivation of the pressure distribution can be found elsewhere [5] and the final form of the equation for pressure at any given radial position between the recess and bearing edge is:

$$p = p_s \left[1 - \frac{\ln(r/R_0)}{\ln(R/R_0)} \right] \left[1 - \left(\frac{p_a}{p_r} \right)^2 \right]^{0.5}$$

(6.27)

where:

p_s is the supply pressure [Pa];

p_r is the recess pressure [Pa];

p_a is the ambient pressure [Pa];

R is the outer radius of the bearing [m];

R_0 is the radius of the recess [m];

r is the radius from the centre of the circular thrust bearing [m].

Despite the apparent complexity of the above expression, the pressure distribution approximates to a linear decline of pressure with distance from the recess.

Gas Flow

The gas flow in flat circular aerostatic bearings is given by the expression:

$$Q = \frac{\pi h^3}{12 \eta \ln(R/R_0)} \frac{p_r^2 - p_a^2}{p_r}$$

(6.28)

where:

h is the film thickness [m];

η is the gas dynamic viscosity [Pas];

Q   is the gas flow (i.e., amount of gas required to maintain the film thickness 'h' in a bearing) [m³/s].

Load Capacity

The expression for load capacity is found by integrating the pressure distribution over the bearing area. The resulting expression is very complex and is available elsewhere [5]. To estimate the load capacity it is useful to introduce a proportionality constant between the load capacity and nominal load based on recess pressure and bearing area, i.e.:

$$K^* = \frac{W}{p_r \pi R^2}$$

(6.29)

where:

W   is the bearing load capacity [N];

p_r is the recess pressure [Pa];

R   is the outer radius of the bearing [m].

Values of 'K*' as a function of the ratio of bearing radius to recess radius are shown in Figure 6.9.

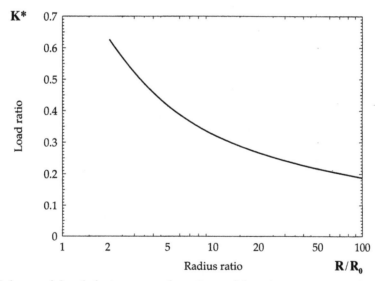

FIGURE 6.9   Values of load factor as a function of bearing geometry for a circular pad aerostatic thrust bearing.

Friction Torque

The friction torque or friction coefficient of a gas bearing is not usually considered because its value is extremely small unless the bearing is operating at a very high sliding speed. If it is necessary to estimate a friction torque, a Petroff approximation can be used. In other words, the shear rate should be multiplied by the gas viscosity and bearing area. The viscosity of most gases is not greatly affected by temperature so that the ambient temperature value of viscosity can safely be used even if the gas leaves the bearing at a high temperature. At high sliding speeds, turbulent flow of gas in the bearing may occur and this should be checked by computing the local Reynolds number based on film thickness. The friction torque for flat circular pad bearings can be calculated from equation (6.11) provided that laminar flow prevails.

EXAMPLE

Calculate the load capacity of a 0.1 [m] radius flat circular aerostatic bearing with a radius ratio of 2 and a recess pressure of 1 [MPa]. From Figure 6.9, K* = 0.63. The load capacity is then:

$$\mathbf{W} = \pi \times 10^6 \times 0.1^2 \times 1 \times 0.63 = 19\,792 \text{ [N]}.$$

Power Loss

The power loss for most slow sliding speed gas bearings is determined by the pumping power required since the friction power loss is extremely small. As mentioned already the pumping power is the product of gas flow rate and the total pressurization required (including losses in capillaries and supply lines). If the friction power loss is significant then this should be added to the pumping power loss to give the total power loss. The friction power loss for flat circular pad bearings can be calculated from equation (6.12).

6.6   HYBRID BEARINGS

Hybrid bearings function by the combined action of hydrostatic and hydrodynamic lubrication. A bearing technology resembling a hybrid bearing was described in Chapter 4 where a pressurized lubricant supply is used to prevent metallic contact during starting or stopping of a Michell pad bearing. The principle of augmenting hydrodynamic lubrication with a hydrostatic effect or vice versa has been developed further than this limited application. The distinction between a true 'hybrid bearing' and the pressurized lubricant supply to a journal bearing is that in the latter case, the purpose of the extra supply is to supply cool lubricant into the hottest part of the bearing.

An important difference between hybrid and hydrostatic bearings is the absence of recesses in the hybrid bearings. Recesses cause reduced hydrodynamic pressures in the loaded parts of the bearing which are where hydrostatic gas or liquid outlets are usually positioned. As discussed previously, lubricant supply outlets are usually located remote from the loaded part of the bearing for efficient hydrodynamic lubrication. An example of a hybrid journal bearing is shown in Figure 6.10.

Where several lubricant outlets are used it is important to avoid interconnection of the supply lines. If the supply lines are connected then recirculating flow of lubricant will occur which reduces the hydrodynamic pressure.

The basic parameters of these bearings such as pumping power and size are designed as for a hydrostatic bearing and any hydrodynamic effect which improves bearing performance is regarded as a bonus [2].

6.7   STABILITY OF HYDROSTATIC AND AEROSTATIC BEARINGS

Hydrostatic bearings are subject to vibrational instability particularly under variable loads or where a gas is used as the lubricant. The mechanism causing vibrational instability is the same as discussed already in hydrodynamic lubrication, i.e., a resonance dependent on the stiffness and damping coefficients of the load-carrying film and the coupled mass. 'Oil whirl' can also occur in hydrostatic and hybrid journal bearings at high speeds in a similar manner to hydrodynamic bearings. Most high-speed externally pressurized bearings are aerostatic and the analysis of the corresponding vibrational stability is highly specialized. For the practical prevention of vibrational instability it is important to minimize the depth of the recess in a hydrostatic bearing since this increases the storage capacity of energy in the bearing, particularly so if gas is used. The depth of the recess should only be a small multiple of the design film thickness, e.g., only 10 to 20 times larger but not 100 times larger. A high bearing stiffness can raise the critical vibration speed in these bearings. The bearing stiffness may be increased by supplying a gas under higher pressure. The method of gas or liquid supply is also important, as restrictors and capillary compensation are associated with vibration problems. More information on bearing stability can be found in Rowe [2] and Mohsin [5].

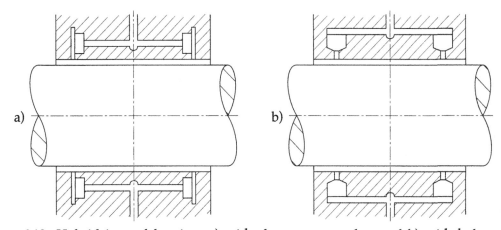

FIGURE 6.10   Hybrid journal bearing: a) with slots as gas outlets and b) with holes as gas outlets.

6.8   SUMMARY

Hydrostatic lubrication provides complete separation of sliding surfaces to ensure zero or negligible wear and very low friction. Hydrostatic lubrication is based on the same physical principles as hydrodynamic lubrication but has certain fundamental differences. There is no friction force at infinitesimal sliding speeds unlike hydrodynamic lubrication which is a uniquely useful characteristic in the design and operation of precision control systems. The disadvantage of hydrostatic lubrication is a complete reliance on an external pressurized supply of lubricant which means that the pump must be reliable and the supply lines free of dirt that might block the flow of lubricant. Hydrostatic lubrication with a gas, which is known as aerostatic lubrication, can provide very low friction even at extremely high sliding speeds because of the low viscosity of gases. Quasi-ideal characteristics of zero wear and friction are obtained with hydrostatic or aerostatic lubrication at low to medium contact stresses but a more complicated technology, e.g., the application of an external high pressure pump, is required in comparison to other forms of lubrication. Bearing stiffness in these bearings can also be manipulated more easily than with other types of bearings to suit specific design requirements.

REVISION QUESTIONS

6.1   What is the main difference between hydrodynamic and hydrostatic lubrication?

6.2   Are hydrostatic bearings effective at high sliding speeds?

6.3   How would you expect contaminated lubricant to affect a hydrostatic bearing?

6.4   Are hydrostatic bearings resistant to high temperatures?

6.5 Can hybrid hydrostatic and hydrodynamic bearings be developed and used to advantage?

6.6 Give examples of operating conditions under which the application of hydrostatic bearings would be necessary or highly desirable.

6.7 List advantages and disadvantages of gas-lubricated bearings over oil-lubricated bearings.

6.8 A hydrostatic circular pad bearing operating under a minimum oil film thickness of **50** [μm] supports a vertical load of **50** [kN] at a shaft speed of **2000** [rpm]. The lubricant's viscosity at the operating temperature is **8.5** [cP], the recess oil pressure is **2** [MPa] and the external pressure outside the bearing is $\mathbf{p_0}$ **= 0**. Assuming that $\mathbf{D/D_0 = 3}$, calculate:

- the bearing dimensions,

- the rate of oil flow through the bearing,

- the power losses '**H**' due to viscous friction in the bearing and

- the coefficient of friction (assume that the friction force is acting at the mean radius of the bearing area). (Ans. $D_0 = 0.095$ [m], $D = 0.282$ [m], $Q = 1.402 \times 10^{-5}$ [m^3/s], $H = 3.69$ [kW], $\mu = 0.004$)

6.9 For the bearing specified in problem 6.8 calculate the capillary dimensions necessary to maintain the minimum film thickness. The pump supply pressure is **3** [MPa]. (Assume **l/d = 100**, where '**l**' is the length while '**d**' is the diameter of the capillary). (Ans. d = 0.8 [mm], l = 0.079 [m])

6.10 A pump capable of maintaining **20** [MPa] of pressure is supplying oil of viscosity **80** [cP] to a circular hydrostatic thrust bearing of an outside diameter of **0.1** [m] at a rate of **0.4x10^{-5}** [m^3/s], generating a steady film thickness of **30** [μm]. The bearing supports a weight of **60** [kN] and rotates at **1000** [rpm]. Calculate a suitable recess radius and pumping power. (Ans. $R_0 = 0.0285$ [m], H = 50.89 [W])

6.11 For the bearing specified in problem 6.10 calculate the percentage in load increase that would cause a **20%** decrease in film thickness. (Ans. 95%)

6.12 A circular hydrostatic bearing with an outside diameter of **150** [mm], recess diameter of **60** [mm] and rotating at **500** [rpm] supports a load of **50** [kN]. The bearing is lubricated by a mineral oil of density $\rho =$ **900** [kg/m^3] delivered to the bearing by a constant flow rate delivery pump operating at **50×10^{-9}** [m^{3}s^{-1}]. The operating temperature of the bearing is **30°C**.

- Calculate the recess pressure and

- Select an SAE oil such that the film thickness will never be less than **10** [μm]. (Ans. $p_r = 6.173$ [MPa], SAE20)

6.13 A circular pad hydrostatic bearing has an outside radius of **0.1** [m] and circular recess radius of **0.02** [m]. The viscosity of the lubricant used is **30** [cP]. The oil is supplied to the bearing by a pump at a constant flow rate of **20** [l/min]. Calculate the load capacity for a uniform film thickness of **100** [μm], and the percentage increase in load capacity if the film thickness is reduced by **10%**.

For the same bearing assume now that the oil is supplied by a capillary restrictor, such that the film thickness is still **100** [μm] and that the pressure drops across the capillary from **60** [MPa] (pump supply pressure '$\mathbf{p_s}$') to **30** [MPa] (recess pressure '$\mathbf{p_r}$'). Calculate the load carried by the film again and the percentage increase in load if the film thickness is reduced by **10%**. (Ans. W = 288 [kN] and 395 [kN], 37.17%, W = 281 [kN], 15.66%)

REFERENCES

1 L.D. Girard, Hydraulique Appliqué, Nouveau Système de Locomotion Sur les Chemins de Fer, Paris, Bachelier, 1852.

2 W. B. Rowe, Hydrostatic and Hybrid Bearing Design, Butterworths, 1983.

3 A. Cameron, Basic Lubrication Theory, Ellis Horwood Ltd, 1981.

4 J.P. O'Donoghue and W.B. Rowe, Hydrostatic Bearing Design, *Tribology*, Vol. 2, 1969, pp. 25-71.

5 M.E. Mohsin, The Use of Controlled Restrictors for Compensating Hydrostatic Bearings, Advances in Machine Tool Design and Research, Pergamon Press, Oxford, 1962.

6 W.A. Gross, Fluid Film Lubrication, John Wiley and Sons, 1980.

7 J.N. Shinkle and K.G. Hornung, Frictional Characteristics of Liquid Hydrostatic Journal Bearings, *Transactions ASME, Journal of Basic Engineering*, Vol. 87, 1965, pp. 163-169.

8   H. Opitz, Pressure Pad Bearings, Conf. Lubrication and Wear, Fundamentals and Application to Design, London, 1967, *Proc. Inst. Mech. Engrs.*, Vol. 182, Pt. 3A, 1967-1968, pp. 100-115.

9   R. Bassani and B. Piccigallo, Hydrostatic Lubrication: Theory and Practice, Elsevier, Amsterdam, 1992.

10   J.K. Scharrer and R.I. Hibs Jr., Flow Coefficients for the Orifice of a Hydrostatic Bearing, *Tribology Transactions*, Vol. 33, 1990, pp. 543-550.

E L A S T O H Y D R O D Y N A M I C

L U B R I C A T I O N

7.1 INTRODUCTION

Elastohydrodynamic lubrication can be defined as a form of hydrodynamic lubrication where the elastic deformation of the contacting bodies and the changes of viscosity with pressure play fundamental roles. The influence of elasticity is not limited to second-order changes in load capacity or friction as described for pivoted pad and journal bearings. Instead, the deformation of the bodies has to be included in the basic model of elastohydrodynamic lubrication. The same refers to the changes in viscosity due to pressure.

The existence of elastohydrodynamic lubrication was suspected long before it could be proved or described using specific scientific concepts. The lubrication mechanisms in conformal contacts such as those encountered in hydrodynamic and hydrostatic bearings were well described and defined and the reasons for their effectiveness were well understood. However, the mechanism of lubrication operating in highly loaded non-conformal contacts, such as those which are found in gears, rolling contact bearings, cams and tappets, although effective, was poorly understood. The wear rates of these devices were very low which implied the existence of films sufficiently thick to separate the opposing surfaces, yet this conclusion was in direct contradiction to the calculated values of hydrodynamic film thicknesses. The predicted values of film thickness were so low that it was inconceivable for the contacting surfaces to be separated by a viscous liquid film. In fact, the calculated film thicknesses suggested that the surfaces were lubricated by films only one molecule in thickness. In experiments specifically designed to permit only lubrication by monomolecular films, much higher wear rates and friction coefficients were obtained. This apparent contradiction between the empirical observation of effective lubrication and the limits of known lubrication mechanisms could not be explained for a considerable period of time. The entire problem acquired an aura of mystery and many elaborate experiments and theories were developed as a result. From the viewpoint of an engineer, the answers to the questions of what controls the lubrication mechanism and how it can be optimized are very important, since heavily loaded point contacts are often found, and provision for effective lubrication of these contacts is critical.

In the 1940's a substantial amount of work was devoted to resolving elastohydrodynamics and the first realistic model which provided an albeit approximate solution for elastohydrodynamic film thickness was proposed by Ertel and Grubin. The work was published by Grubin in 1949 [1]. It was found that the combination of three effects:

hydrodynamics, elastic deformation of the metal surfaces and the increase in the viscosity of oil under extreme pressures are instrumental to this mechanism. This lubrication regime is referred to in the literature as elastohydrodynamic lubrication which is commonly abbreviated as EHL or EHD. At this stage, it should be realized that elastohydrodynamic lubrication is effectively limited to oils as opposed to other viscous liquids because of the pressure-viscosity dependence. It was shown theoretically that under conditions of intense contact stress a lubricating oil film can be formed. The lubricated contacts in which these three effects take place are said to be operating elastohydrodynamically, which effectively means that the contacting surfaces deform elastically under the hydrodynamic pressure generated in the layer of lubricating film. The lubricating films are very thin, in the range of 0.1 to 1 [μm], but manage to separate the interacting surfaces, resulting in a significant reduction of wear and friction. Although this regime generally operates between non-conforming surfaces, it can also occur under certain circumstances in the contacts classified as conformal such as highly loaded journal and pad bearings which have a significant component of contact and bending deformation. However, enormous loads are required for this to occur and very few journal or pad bearings operate under these conditions. Significant progress has been made towards a complete understanding of the mechanism of elastohydrodynamic lubrication. The pioneering work in this field was conducted by Martin (1916) [2], followed by Grubin (1949) [1] and was continued by Dowson and Higginson [3], Crook [4], Cameron and Gohar [5] and others.

In this chapter the fundamental mechanisms of film generation in elastohydrodynamic contacts, together with the methods for calculating the minimum film thickness in rolling bearings and gears, will be outlined. Some particular characteristics of elastohydrodynamic contacts such as traction and flash temperature will be discussed, along with the methods of their evaluation.

7.2   CONTACT STRESSES

From elementary mechanics it is known that two contacting surfaces under load will deform. The deformation may be either plastic or elastic depending on the magnitude of the applied load and the material's hardness. In many engineering applications, for example, rolling contact bearings, gears, cams, seals, etc., the contacting surfaces are non-conformal hence the resulting contact areas are very small and the resulting pressures are very high. From the viewpoint of machine design it is essential to know the values of stresses acting in such contacts. These stresses can be determined from the analytical formulae, based on the theory of elasticity, developed by Hertz in 1881 [6-8]. Hertz developed these formulae during his Christmas vacation in 1880 when he was 23 years old [7].

Simplifying Assumptions to Hertz's Theory

Hertz's model of contact stress is based on the following simplifying assumptions [6]:

- ·   the materials in contact are homogeneous and the yield stress is not exceeded,
- ·   contact stress is caused by the load which is normal to the contact tangent plane which effectively means that there are no tangential forces acting between the solids,
- ·   the contact area is very small compared with the dimensions of the contacting solids,
- ·   the contacting solids are at rest and in equilibrium,
- ·   the effect of surface roughness is negligible.

Subsequent refinements of Hertz's model by later workers have removed most of these assumptions, and Hertz's theory forms the basis of the model of elastohydrodynamic lubrication.

Stress Status in Static Contact

Consider two bodies in contact under a static load and with no movement relative to each other. Since there is no movement between the bodies, shearing does not occur at the interface and therefore the shear stress acting is equal to zero. According to the principles of solid mechanics, the planes on which the shear stress is zero are called the principal planes. Thus the interface between two bodies in a static contact is a principal plane on which a principal stress 'σ_1' is the only stress acting. It is also known from solid mechanics that the maximum shear stress occurs at 45° to the principal plane, as shown in Figure 7.1.

If the contact load is sufficiently high then the maximum shear stress will exceed the yield stress of the material, i.e., $\tau_{max} > k$, and plastic deformation takes place. Material will then deform (slip) along the line of action of maximum shear stress. The maximum shear stress 'τ_{max}', also referred to in the literature as '$\tau_{45°}$' since it acts on planes inclined at 45° to the interface (in static contacts), is given by:

$$\tau_{max} = \tau_{45°} = \pm k = \pm \left(\frac{\sigma_1 - \sigma_3}{2} \right) \tag{7.1}$$

The stresses 'σ_x', 'σ_z' and 'τ' vary with depth below the interface. An example of the stress field beneath the surface of two parallel cylinders in static contact is shown in Figure 7.2 [9].

For a circular contact, the subsurface stress field is very similar [9]. It can be seen from Figure 7.2 that the maximum shear stress '$\tau_{45°}$' occurs at some depth below the surface. This depth depends on load and is therefore related to the contact area. In a circular contact, for example, the maximum shear stress occurs at approximately **0.6a**, where '**a**' is the radius of the contact area.

Stress Status in Lubricated Rolling and Sliding Contacts

In a lubricated rolling contact, the contact stresses are affected by the lubricating film separating the opposing surfaces and the level of rolling and sliding. The hydrodynamic film generated under these conditions and the relative movement of the surfaces cause significant changes to the original static stress distribution and will be discussed later.

Rolling, in general, results in an increase in contact area and a subsequent modification of the Hertzian stress field in both dry and lubricated conditions. The most critical influence on subsurface stress fields, however, is exerted by sliding. To illustrate the effect of sliding on the stress distribution, consider two bodies in contact with some sliding occurring between them. Frictional forces are the inevitable result of sliding and cause a shear stress to act along the interface, as shown in Figure 7.3.

The frictional stress acting at the interface is balanced by rotating the planes of principal stresses through an angle 'ϕ' from their original positions when frictional forces are absent. The magnitude of the angle 'ϕ' depends on the frictional stress μq acting at the interface according to the relation:

$\phi = 1/2\cos^{-1}(\mu q/k)$

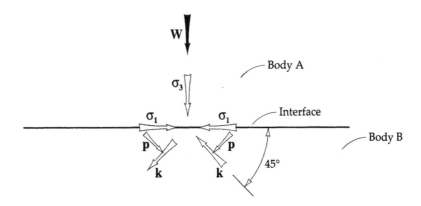

FIGURE 7.1   Stress status in a static contact; σ_1, σ_3 are the principal stresses, **p** is the hydrostatic pressure and **k** is the shear yield stress of the material.

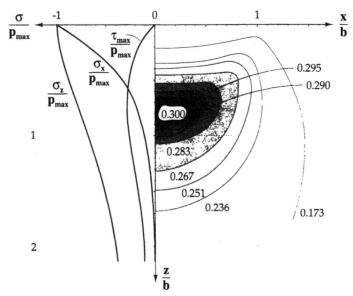

FIGURE 7.2   Subsurface stress field for two cylinders in static contact; p_{max} is the maximum contact pressure and **b** is the half width of the contact rectangle [9].

The variation with depth below the interface of the principal shear stress 'τ_{max}' for a cylinder and the plane on which it slides is shown in Figure 7.4. The contours show the principal shear stress due to the combined normal pressure and tangential stress for a coefficient of friction $\mu = 0.2$ [9]. It can clearly be seen that as friction force increases, the maximum shear stress moves towards the interface. Thus there is a gradual increase in shear stress acting at the interface as the friction force increases. This phenomenon is very important in crack formation and the subsequent surface failure and will be discussed later.

7.3   CONTACT BETWEEN TWO ELASTIC SPHERICAL OR SPHEROIDAL BODIES

Elastic bodies in contact deform and the contact geometry, load and material properties determine the contact area and stresses. The contact geometry depends on whether the contact occurs between surfaces which are both convex or a combination of flat, convex and concave. In this section the methods of defining the contact geometry and calculating the contact stresses and deformations are described.

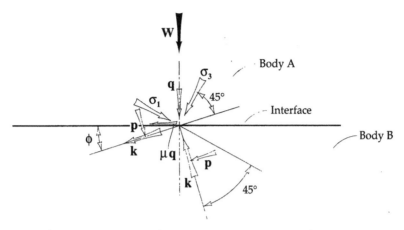

FIGURE 7.3   Stresses in a contact with sliding; σ_1 and σ_3 are the principal stresses, p is the hydrostatic pressure, k is the shear yield stress of the material, μ is the coefficient of friction, q is the stress normal to the interface or compressive stress due to load and ϕ is the angle by which the planes of principal stress are rotated from the corresponding zero friction positions to balance the frictional stress.

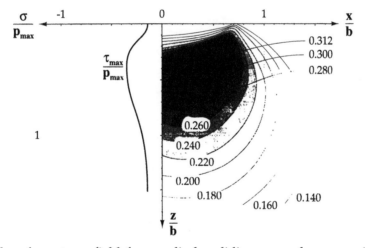

FIGURE 7.4   Subsurface stress field for a cylinder sliding on a plane; p_{max} is the maximum contact pressure and b is the half width of the contact rectangle [9].

Geometry of Contacting Elastic Bodies

The shape of the contact area depends on the shape (curvature) of the contacting bodies. For example, point contacts occur between two balls, line contacts occur between two parallel cylinders and elliptical contacts, which are most frequently found in many practical engineering applications, occur when two cylinders are crossed, or a moving ball is in contact with the inner ring of a bearing, or two gear teeth are in contact. The curvature of the bodies can be convex, flat or concave. It is defined by convention that **convex surfaces** possess a **'positive curvature'** and **concave surfaces** have a **'negative curvature'** [7]. The following general rule can be applied to distinguish between these surfaces: **if the centre of curvature**

lies within the solid then the curvature is positive, if it lies outside the solid then the curvature is negative. This distinction is critical in defining the parameter characterizing the contact geometry which is known as the reduced radius of curvature.

· *Two Elastic Bodies with Convex Surfaces in Contact*

The configuration of two elastic bodies with convex surfaces in contact was originally considered by Hertz in 1881 and is shown in Figure 7.5.

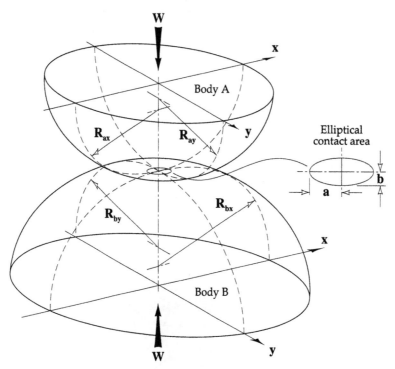

FIGURE 7.5   Geometry of two elastic bodies with convex surfaces in contact.

The reduced radius of curvature for this case is defined as:

$$\frac{1}{R'} = \frac{1}{R_x} + \frac{1}{R_y} = \frac{1}{R_{ax}} + \frac{1}{R_{bx}} + \frac{1}{R_{ay}} + \frac{1}{R_{by}}$$

(7.2)

where:

$$\frac{1}{R_x} = \frac{1}{R_{ax}} + \frac{1}{R_{bx}}$$

$$\frac{1}{R_y} = \frac{1}{R_{ay}} + \frac{1}{R_{by}}$$

R_x is the reduced radius of curvature in the 'x' direction [m];

R_y is the reduced radius of curvature in the 'y' direction [m];

R_{ax} is the radius of curvature of body 'A' in the 'x' direction [m];

R_{ay} is the radius of curvature of body 'A' in the 'y' direction [m];

R_{bx} is the radius of curvature of body 'B' in the 'x' direction [m];

R_{by} is the radius of curvature of body 'B' in the 'y' direction [m].

Convention

The method of arranging the 'x' and 'y' coordinates plays an important role in the calculation of contact parameters. It is important to locate the 'x' and 'y' coordinates so that the following condition is fulfilled:

$$\boxed{\frac{1}{R_x} \geq \frac{1}{R_y}} \tag{7.3}$$

When this rule is applied, the coordinate 'x' determines the direction of the semiminor axis of the contact and the coordinate 'y' determines the direction of the semimajor axis [7]. If $1/R_x = 1/R_y$ then there is a circular contact and when $1/R_x < 1/R_y$ then it is necessary to transpose the directions of the coordinates, i.e., 'x' becomes 'y' and vice versa and 'R_x' becomes 'R_y'.

· *Two Elastic Bodies with One Convex and One Flat Surface in Contact*

The geometry of a contact between a flat surface and a convex surface is shown in Figure 7.6. The reduced radius of curvature for this case is defined according to (7.2) as:

$$\frac{1}{R'} = \frac{1}{R_x} + \frac{1}{R_y} = \frac{1}{R_{ax}} + \frac{1}{R_{bx}} + \frac{1}{R_{ay}} + \frac{1}{R_{by}}$$

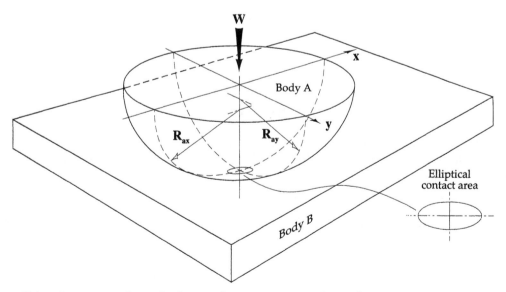

FIGURE 7.6   Geometry of two bodies with one convex and one flat surface in contact.

Since one of the contacting surfaces is a plane then it has infinite radii of curvature, i.e.:

$$R_{bx} = R_{by} = \infty$$

and the reduced radius of curvature becomes:

$$\frac{1}{R'} = \frac{1}{R_x} + \frac{1}{R_y} = \frac{1}{R_{ax}} + \frac{1}{R_{ay}}$$

(7.4)

· *Two Elastic Bodies with One Convex and One Concave Surface in Contact*

The contact geometry between a convex and a concave surface is shown in Figure 7.7.

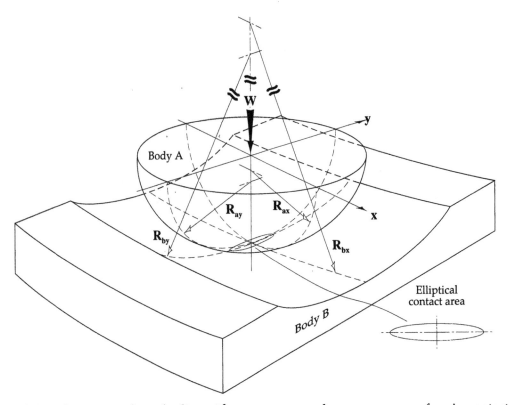

FIGURE 7.7   Geometry of two bodies with one convex and one concave surface in contact.

Body 'B' has a concave surface and according to the convention its curvature is negative, i.e., 'R_{bx}' and 'R_{by}' are negative. The reduced radius of curvature for this contact according to equation (7.2) becomes:

$$\frac{1}{R'} = \frac{1}{R_x} + \frac{1}{R_y} = \frac{1}{R_{ax}} - \frac{1}{R_{bx}} + \frac{1}{R_{ay}} - \frac{1}{R_{by}}$$

(7.5)

or

$$\frac{1}{R'} = \left(\frac{1}{R_{ax}} + \frac{1}{R_{ay}}\right) - \left(\frac{1}{R_{bx}} + \frac{1}{R_{by}}\right)$$

where:

$$\frac{1}{R_x} = \frac{1}{R_{ax}} - \frac{1}{R_{bx}}$$

$$\frac{1}{R_y} = \frac{1}{R_{ay}} - \frac{1}{R_{by}}$$

Contact Area, Pressure, Maximum Deflection and Position of the Maximum Shear Stress

The evaluation of contact parameters is essential in many practical engineering applications. The most frequently used contact parameters are:

· the contact area dimensions,

· the maximum contact pressure, often called the Hertzian stress,

· the maximum deflection at the centre of the contacting surfaces,

· the position of the maximum shear stress under the surface.

The contact area depends on the geometry of the contacting bodies, load and material properties. In most cases, the contact area is enveloped by an ellipse such as in the case of two cylinders crossed at an angle ≠ 90°. A circular contact area is found between two balls in contact or when two cylinders are crossed at 90°. The contact area between two parallel cylinders is enclosed by a narrow rectangle. Contact pressures and deflections also depend on the geometry of the contacting bodies.

Since the formulae for contact parameters were originally developed by Hertz, terms such as Hertzian contacts or Hertzian stresses can frequently be found in the literature. The solution suggested by Hertz requires the determination of the ellipticity parameter together with complete elliptic integrals. This is usually accomplished by numerical procedures or charts. Numerical iterative procedures were suggested by Hamrock and Anderson in 1973 [10,11]. The simplified equations for the determination of these parameters were developed later in 1977 by Brewe and Hamrock [12,11]. The formulae for the calculations of contact parameters are summarized in this section. More specialized and detailed information about the above parameters can be found elsewhere [e.g., 7,9,11,13].

· *Contact between Two Spheres*

The contact area between two spheres is enveloped by a circle. The formulae for the main contact parameters of two spheres in contact, shown in Figure 7.8, are summarized in Table 7.1.

TABLE 7.1   Formulae for contact parameters between two spheres.

Contact area dimensions	Maximum contact pressure	Average contact pressure	Maximum deflection	Maximum shear stress
$a = \left(\dfrac{3WR'}{E'}\right)^{1/3}$ circle a	$p_{max} = \dfrac{3W}{2\pi a^2}$ Hemispherical pressure distribution	$p_{average} = \dfrac{W}{\pi a^2}$	$\delta = 1.0397\left(\dfrac{W^2}{E'^2 R'}\right)^{1/3}$	$\tau_{max} = \dfrac{1}{3} p_{max}$ at a depth of $z = 0.638a$

where:

a	is the radius of the contact area [m];
W	is the normal load [N];
p	is the contact pressure (Hertzian stress) [Pa];
δ	is the total deflection at the centre of the contact (i.e., $\delta = \delta_A + \delta_B$; where '$\delta_A$' and '$\delta_B$' are the maximum deflections of body 'A' and 'B', respectively) [m];
τ	is the shear stress [Pa];
z	is the depth under the surface where the maximum shear stress acts [m];
E'	is the reduced Young's modulus [Pa];
R'	is the reduced radius of curvature [m].

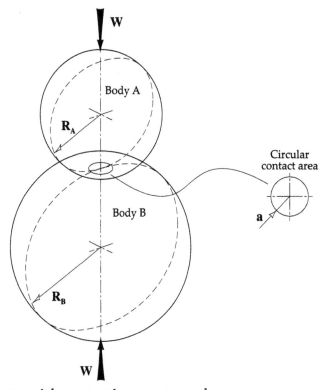

FIGURE 7.8   Geometry of the contact between two spheres.

The reduced Young's modulus is defined as:

$$\frac{1}{E'} = \frac{1}{2}\left[\frac{1-\upsilon_A^{2}}{E_A} + \frac{1-\upsilon_B^{2}}{E_B}\right]$$
(7.6)

where:

υ_A and υ_B	are the Poisson's ratios of the contacting bodies 'A' and 'B', respectively;
E_A and E_B	are the Young's moduli of the contacting bodies 'A' and 'B', respectively.

For example, reduced Young's modulus for contact between steel spheres of $\upsilon_{steel} = 0.3$ and $E_{steel} = 2.1 \times 10^{11}$ [Pa] is: **E' = 2.308 × 10¹¹ [Pa].**

It can be noted that for the spheres:

$$\mathbf{R_{ax} = R_{ay} = R_A} \quad \text{and} \quad \mathbf{R_{bx} = R_{by} = R_B}$$

where:

$\mathbf{R_A}$ and $\mathbf{R_B}$ are the radii of the spheres 'A' and 'B', respectively.

Substituting into equation (7.2) gives:

$$\boxed{\frac{1}{R'} = \frac{1}{R_x} + \frac{1}{R_y} = \frac{1}{R_A} + \frac{1}{R_B} + \frac{1}{R_A} + \frac{1}{R_B} = 2\left(\frac{1}{R_A} + \frac{1}{R_B}\right)} \tag{7.7}$$

where:

$$\frac{1}{R_x} = \frac{1}{R_y} = \frac{1}{R_A} + \frac{1}{R_B}$$

It should be noted that in some publications the reduced radius for the contact between two spheres is defined as:

$$\frac{1}{R'} = \frac{1}{R_A} + \frac{1}{R_B}$$

and consequently the formulae for the contact area dimension 'a' and the maximum deflection 'δ' are presented in the slightly altered form:

$$\mathbf{a} = \left(\frac{3WR'}{2E'}\right)^{1/3}, \quad \delta = 1.31\left(\frac{W^2}{E'^2R'}\right)^{1/3}$$

EXAMPLE

Find the contact parameters for two steel balls. The normal force is W = 5 [N], the radii of the balls are $R_A = 10 \times 10^{-3}$ [m] and $R_B = 15 \times 10^{-3}$ [m]. The Young's modulus for both balls is $E = 2.1 \times 10^{11}$ [Pa] and the Poisson's ratio of steel is $\upsilon = 0.3$.

· *Reduced Radius of Curvature*

Since $R_{ax} = R_{ay} = R_A = 10 \times 10^{-3}$ [m] and $R_{bx} = R_{by} = R_B = 15 \times 10^{-3}$ [m] the reduced radii of curvature in the 'x' and 'y' directions are:

$$\frac{1}{R_x} = \frac{1}{R_{ax}} + \frac{1}{R_{bx}} = \frac{1}{10 \times 10^{-3}} + \frac{1}{15 \times 10^{-3}} = 166.67 \qquad \boxed{\Rightarrow R_x = 6 \times 10^{-3}[m]}$$

$$\frac{1}{R_y} = \frac{1}{R_{ay}} + \frac{1}{R_{by}} = \frac{1}{10 \times 10^{-3}} + \frac{1}{15 \times 10^{-3}} = 166.67 \qquad \boxed{\Rightarrow R_y = 6 \times 10^{-3}[m]}$$

Note that $1/R_x = 1/R_y$, i.e., condition (7.3) is satisfied (circular contact), and the reduced radius of curvature is:

$$\frac{1}{R'} = \frac{1}{R_x} + \frac{1}{R_y} = 166.67 + 166.67 = 333.34 \qquad \boxed{\Rightarrow R' = 3 \times 10^{-3}\,[\text{m}]}$$

· *Reduced Young's Modulus*

$$\frac{1}{E'} = \frac{1}{2}\left[\frac{1-\upsilon_A^2}{E_A} + \frac{1-\upsilon_B^2}{E_B}\right] = \frac{1}{2}\left[\frac{1-0.3^2}{2.1\times10^{11}} + \frac{1-0.3^2}{2.1\times10^{11}}\right] \qquad \boxed{\Rightarrow E' = 2.308 \times 10^{11}\,[\text{Pa}]}$$

· *Contact Area Dimensions*

$$a = \left(\frac{3WR'}{E'}\right)^{1/3} = \left(\frac{3\times5\times(3\times10^{-3})}{2.308\times10^{11}}\right)^{1/3} \qquad \boxed{= 5.799 \times 10^{-5}\,[\text{m}]}$$

· *Maximum and Average Contact Pressures*

$$p_{max} = \frac{3W}{2\pi a^2} = \frac{3\times5}{2\pi(5.799\times10^{-5})^2} \qquad \boxed{= 709.9\,[\text{MPa}]}$$

$$p_{average} = \frac{W}{\pi a^2} = \frac{5}{\pi(5.799\times10^{-5})^2} \qquad \boxed{= 473.3\,[\text{MPa}]}$$

· *Maximum Deflection*

$$\delta = 1.0397\left(\frac{W^2}{E'^2 R'}\right)^{1/3} = 1.0397\left(\frac{5^2}{(2.308\times10^{11})^2 3\times10^{-3}}\right)^{1/3} \qquad \boxed{= 5.6 \times 10^{-7}\,[\text{m}]}$$

· *Maximum Shear Stress*

$$\tau_{max} = \frac{1}{3}\,p_{max} = \frac{1}{3}\,709.9 \qquad \boxed{= 236.6\,[\text{MPa}]}$$

· *Depth at which Maximum Shear Stress Occurs*

$$z = 0.638a = 0.638 \times (5.799\times10^{-5}) \qquad \boxed{= 3.7 \times 10^{-5}\,[\text{m}]}$$

· *Contact between a Sphere and a Plane Surface*

The contact area between a sphere and a plane surface, as shown in Figure 7.9, is also circular. The contact parameters for this configuration can be calculated according to the formulae summarized in Table 7.1.

The radii of curvature of a plane surface are infinite and symmetry of the sphere applies so that $R_{bx} = R_{by} = \infty$ and $R_{ax} = R_{ay} = R_A$. The reduced radius of curvature according to (7.2) is therefore given by:

$$\boxed{\frac{1}{R'} = \frac{1}{R_x} + \frac{1}{R_y} = \frac{1}{R_A} + \frac{1}{\infty} + \frac{1}{R_A} + \frac{1}{\infty} = \frac{2}{R_A}} \qquad (7.8)$$

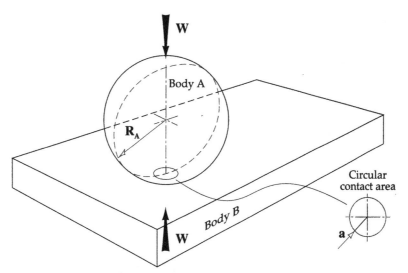

FIGURE 7.9   Contact between a sphere and a flat surface.

where:

$$R_x = R_y = R_A$$

EXAMPLE

Find the contact parameters for a steel ball on a flat steel plate. The normal force is $W = 5$ [N], the radius of the ball is $R_A = 10 \times 10^{-3}$ [m], the Young's modulus for ball and plate is $E = 2.1 \times 10^{11}$ [Pa] and the Poisson's ratio is $\upsilon = 0.3$.

· *Reduced Radius of Curvature*

Since the radii of the ball and the plate are $R_{ax} = R_{ay} = 10 \times 10^{-3}$ [m] and $R_{bx} = R_{by} = \infty$ [m], respectively, the reduced radii of curvature in '**x**' and '**y**' directions are:

$$\frac{1}{R_x} = \frac{1}{R_{ax}} + \frac{1}{R_{bx}} = \frac{1}{10 \times 10^{-3}} + \frac{1}{\infty} = 100 \qquad \boxed{\Rightarrow R_x = 0.01 \, [\text{m}]}$$

$$\frac{1}{R_y} = \frac{1}{R_{ay}} + \frac{1}{R_{by}} = \frac{1}{10 \times 10^{-3}} + \frac{1}{\infty} = 100 \qquad \boxed{\Rightarrow R_y = 0.01 \, [\text{m}]}$$

Condition (7.3), i.e., $1/R_x = 1/R_y$ is satisfied (circular contact), and the reduced radius of curvature is:

$$\frac{1}{R'} = \frac{1}{R_x} + \frac{1}{R_y} = 100 + 100 = 200 \qquad \boxed{\Rightarrow R' = 5 \times 10^{-3} \, [\text{m}]}$$

· *Reduced Young's Modulus*

$$\boxed{E' = 2.308 \times 10^{11} \, [\text{Pa}]}$$

· *Contact Area Dimensions*

$$a = \left(\frac{3WR'}{E'}\right)^{1/3} = \left(\frac{3 \times 5 \times (5 \times 10^{-3})}{2.308 \times 10^{11}}\right)^{1/3} \qquad \boxed{= 6.88 \times 10^{-5}\,[\text{m}]}$$

· *Maximum and Average Contact Pressures*

$$p_{max} = \frac{3W}{2\pi a^2} = \frac{3 \times 5}{2\pi(6.88 \times 10^{-5})^2} \qquad \boxed{= 504.4\,[\text{MPa}]}$$

$$p_{average} = \frac{W}{\pi a^2} = \frac{5}{\pi(6.88 \times 10^{-5})^2} \qquad \boxed{= 336.2\,[\text{MPa}]}$$

· *Maximum Deflection*

$$\delta = 1.0397\left(\frac{W^2}{E'^2 R'}\right)^{1/3} = 1.0397\left(\frac{5^2}{(2.308 \times 10^{11})^2 5 \times 10^{-3}}\right)^{1/3} \quad \boxed{= 4.7 \times 10^{-7}\,[\text{m}]}$$

· *Maximum Shear Stress*

$$\tau_{max} = \frac{1}{3}p_{max} = \frac{1}{3}504.4 \qquad \boxed{= 168.1\,[\text{MPa}]}$$

· *Depth at which Maximum Shear Stress Occurs*

$$z = 0.638a = 0.638 \times (6.88 \times 10^{-5}) \qquad \boxed{= 4.4 \times 10^{-5}\,[\text{m}]}$$

· *Contact between Two Parallel Cylinders*

The contact area between two parallel cylinders is circumscribed by a narrow rectangle. The geometry of parallel cylinders in contact is shown in Figure 7.10 and the formulae for the main contact parameters are summarized in Table 7.2.

TABLE 7.2   Formulae for contact parameters between two parallel cylinders.

Contact area dimensions	Maximum contact pressure	Average contact pressure	Maximum deflection	Maximum shear stress
$b = \left(\frac{4WR'}{\pi l E'}\right)^{1/2}$ rectangle $2b$ $\mid\leftarrow 2l \rightarrow\mid$	$p_{max} = \frac{W}{\pi b l}$ Elliptical pressure distribution	$p_{average} = \frac{W}{4bl}$	$\delta = 0.319\left(\frac{W}{E'l}\right)$ $\times\left[\frac{2}{3} + \ln\left(\frac{4R_A R_B}{b^2}\right)\right]$	τ_{max} $= 0.304\,p_{max}$ at a depth of $z = 0.786b$

where:

b   is the half width of the contact rectangle [m];

l   is the half length of the contact rectangle [m];

R'   is the reduced radius of curvature for the two parallel cylinders in contact [m]. For the cylinders: $R_{ax} = R_A$, $R_{ay} = \infty$, $R_{bx} = R_B$, $R_{by} = \infty$ where 'R_A' and 'R_B' are the radii of the cylinders '**A**' and '**B**', respectively.

Substituting into equation (7.2) yields:

$$\frac{1}{R'} = \frac{1}{R_x} + \frac{1}{R_y} = \frac{1}{R_A} + \frac{1}{R_B} + \frac{1}{\infty} + \frac{1}{\infty} = \frac{1}{R_A} + \frac{1}{R_B}$$

(7.9)

where:

$$\frac{1}{R_x} = \frac{1}{R_A} + \frac{1}{R_B} \quad \text{and} \quad \frac{1}{R_y} = 0$$

The rest of the parameters are as defined for Table 7.1.

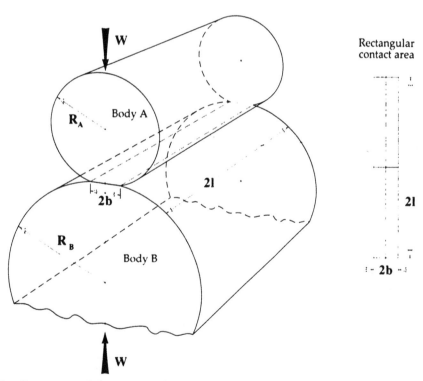

FIGURE 7.10   Geometry of the contact between two parallel cylinders.

EXAMPLE

Find the contact parameters for two parallel steel rollers. The normal force is W = 5 [N], radii of the rollers are $R_A = 10 \times 10^{-3}$ [m] and $R_B = 15 \times 10^{-3}$ [m], Young's modulus for both rollers is $E = 2.1 \times 10^{11}$ [Pa] and the Poisson's ratio is $\upsilon = 0.3$. The length of both rollers is $2l = 10 \times 10^{-3}$ [m].

· *Reduced Radius of Curvature*

Since the radii of the cylinders are $R_{ax} = R_A = 10 \times 10^{-3}$ [m], $R_{ay} = \infty$ and $R_{bx} = R_B = 15 \times 10^{-3}$ [m], $R_{by} = \infty$, respectively, the reduced radii of curvature in the '**x**' and '**y**' directions are:

$$\frac{1}{R_x} = \frac{1}{R_{ax}} + \frac{1}{R_{bx}} = \frac{1}{10 \times 10^{-3}} + \frac{1}{15 \times 10^{-3}} = 166.67 \qquad \boxed{\Rightarrow R_x = 6 \times 10^{-3}\,[\text{m}]}$$

$$\frac{1}{R_y} = \frac{1}{R_{ay}} + \frac{1}{R_{by}} = \frac{1}{\infty} + \frac{1}{\infty} = 0 \qquad \boxed{\Rightarrow R_y = \infty\,[\text{m}]}$$

Since $1/R_x > 1/R_y$ condition (7.3) is satisfied and the reduced radius of curvature is:

$$\frac{1}{R'} = \frac{1}{R_x} = 166.67 \qquad \boxed{\Rightarrow R' = 6 \times 10^{-3}\,[\text{m}]}$$

· *Reduced Young's Modulus*

$$\boxed{E' = 2.308 \times 10^{11}\,[\text{Pa}]}$$

· *Contact Area Dimensions*

$$b = \left(\frac{4WR'}{\pi l E'}\right)^{1/2} = \left(\frac{4 \times 5 \times (6 \times 10^{-3})}{\pi \times (5 \times 10^{-3}) \times (2.308 \times 10^{11})}\right)^{1/2} \qquad \boxed{= 5.75 \times 10^{-6}\,[\text{m}]}$$

· *Maximum and Average Contact Pressures*

$$p_{max} = \frac{W}{\pi b l} = \frac{5}{\pi \times (5.75 \times 10^{-6}) \times (5 \times 10^{-3})} \qquad \boxed{= 55.4\,[\text{MPa}]}$$

$$p_{average} = \frac{W}{4 b l} = \frac{5}{4 \times (5.75 \times 10^{-6}) \times (5 \times 10^{-3})} \qquad \boxed{= 43.5\,[\text{MPa}]}$$

· *Maximum Deflection*

$$\delta = 0.319\left[\frac{W}{E'l}\right]\left[\frac{2}{3} + \ln\left(\frac{4 R_A R_B}{b^2}\right)\right]$$

$$= 0.319\left[\frac{5}{(2.308 \times 10^{11}) \times (5 \times 10^{-3})}\right]\left[\frac{2}{3} + \ln\left(\frac{4 \times (10 \times 10^{-3}) \times (15 \times 10^{-3})}{(5.75 \times 10^{-6})^2}\right)\right]$$

$$\boxed{= 2.40 \times 10^{-8}\,[\text{m}]}$$

· *Maximum Shear Stress*

$$\tau_{max} = 0.304\,p_{max} = 0.304 \times 55.4 \qquad \boxed{= 16.8\,[\text{MPa}]}$$

· *Depth at which Maximum Shear Stress Occurs*

$$z = 0.786\,b = 0.786 \times (5.75 \times 10^{-6}) \qquad \boxed{= 4.5 \times 10^{-6}\,[\text{m}]}$$

· *Contact between Two Crossed Cylinders with Equal Diameters*

The contact area between two cylinders with equal diameters crossed at 90° is bounded by a circle. This configuration is frequently used in wear experiments since the contact parameters can easily be determined. The contacting cylinders are shown in Figure 7.11 and the contact parameters can be calculated according to the formulae summarized in Table 7.1.

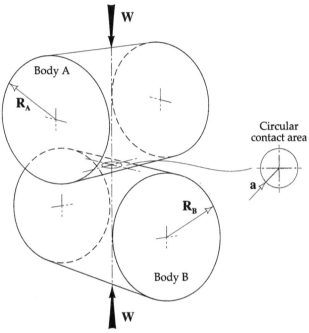

FIGURE 7.11 Geometry of the contact between two cylinders of equal diameters with axes perpendicular.

Since $R_A = R_B$ then in this configuration $R_{ax} = \infty$, $R_{ay} = R_A$, $R_{bx} = R_B$ and $R_{by} = \infty$. The reduced radius according to (7.2) is given by:

$$\frac{1}{R'} = \frac{1}{R_x} + \frac{1}{R_y} = \frac{1}{\infty} + \frac{1}{R_B} + \frac{1}{R_A} + \frac{1}{\infty} = \frac{2}{R_A}$$ (7.10)

which is the same as for a sphere on a plane surface.

If the cylinders are crossed at an angle other than 0° or 90°, i.e., their axes are neither parallel nor perpendicular, then the contact area is enclosed by an ellipse. Examples of the analysis of such cylindrical contacts can be found in the specialized literature [14]. The formulae for evaluation of parameters of elliptical contacts are described next.

EXAMPLE

Find the contact parameters for two steel wires of the same diameter crossed at 90°. This configuration is often used in fretting wear studies. The normal force is W = 5 [N], radii of the wires are $R_A = R_B = 1.5 \times 10^{-3}$ [m], the Young's modulus for both wires is E = 2.1 × 10^{11} [Pa] and the Poisson's ratio is $\upsilon = 0.3$.

· *Reduced Radius of Curvature*

Since the radii of the wires are $R_{ax} = \infty$, $R_{ay} = R_A = 1.5 \times 10^{-3}$ [m], and $R_{bx} = R_B = 1.5 \times 10^{-3}$ [m], $R_{by} = \infty$, respectively, the reduced radii of curvature in the 'x' and 'y' directions are:

$$\frac{1}{R_x} = \frac{1}{R_{ax}} + \frac{1}{R_{bx}} = \frac{1}{\infty} + \frac{1}{1.5 \times 10^{-3}} = 666.67 \qquad \boxed{\Rightarrow R_x = 0.0015 \,[m]}$$

$$\frac{1}{R_y} = \frac{1}{R_{ay}} + \frac{1}{R_{by}} = \frac{1}{1.5 \times 10^{-3}} + \frac{1}{\infty} = 666.67 \qquad \boxed{\Rightarrow R_y = 0.0015 \,[m]}$$

Since $1/R_x = 1/R_y$ condition (7.3) is satisfied and the reduced radius of curvature is:

$$\frac{1}{R'} = \frac{1}{R_x} + \frac{1}{R_y} = 666.67 + 666.67 = 1333.34 \qquad \boxed{\Rightarrow R' = 7.5 \times 10^{-4} \,[m]}$$

· *Reduced Young's Modulus*

$$\boxed{E' = 2.308 \times 10^{11} \,[Pa]}$$

· *Contact Area Dimensions*

$$a = \left(\frac{3WR'}{E'}\right)^{1/3} = \left(\frac{3 \times 5 \times (7.5 \times 10^{-4})}{2.308 \times 10^{11}}\right)^{1/3} \qquad \boxed{= 3.65 \times 10^{-5} \,[m]}$$

· *Maximum and Average Contact Pressures*

$$p_{max} = \frac{3W}{2\pi a^2} = \frac{3 \times 5}{2\pi(3.65 \times 10^{-5})^2} \qquad \boxed{= 1791.9 \,[MPa]}$$

$$p_{average} = \frac{W}{\pi a^2} = \frac{5}{\pi(3.65 \times 10^{-5})^2} \qquad \boxed{= 1194.6 \,[MPa]}$$

· *Maximum Deflection*

$$\delta = 1.0397\left(\frac{W^2}{E'^2 R'}\right)^{1/3}$$

$$= 1.0397\left(\frac{5^2}{(2.308 \times 10^{11})^2 \times (7.5 \times 10^{-4})}\right)^{1/3} \qquad \boxed{= 8.9 \times 10^{-7} \,[m]}$$

· *Maximum Shear Stress*

$$\tau_{max} = \frac{1}{3}\,p_{max} = \frac{1}{3}\,1791.9 \qquad \boxed{= 597.3 \,[MPa]}$$

· *Depth at which Maximum Shear Stress Occurs*

$$z = 0.638\,a = 0.638 \times (3.65 \times 10^{-5}) \qquad \boxed{= 2.3 \times 10^{-5} \,[m]}$$

· *Elliptical Contact between Two Elastic Bodies, General Case*

Elliptical contacts are found between solid bodies which have different principal relative radii of curvature in orthogonal planes. Examples of this are encountered in spherical bearings and gears. The contact area is described by an ellipse. An illustration of this form of contact is shown in Figure 7.5 and the formulae for the main contact parameters are summarized in Table 7.3.

TABLE 7.3   Formulae for contact parameters between two elastic bodies; elliptical contacts, general case.

Contact area dimensions	Maximum contact pressure	Average contact pressure	Maximum deflection	Maximum shear stress
$a = k_1 \left(\dfrac{3WR'}{E'} \right)^{1/3}$ $b = k_2 \left(\dfrac{3WR'}{E'} \right)^{1/3}$ ellipse $\quad$ b $\overset{\downarrow}{\underset{\uparrow}{\ominus}}$ $\rightarrowtail$ a $\leftarrowtail$	$p_{max} = \dfrac{3W}{2\pi ab}$ Elliptical pressure distribution	$p_{average} = \dfrac{W}{\pi ab}$	$\delta = 0.52 k_3 \left(\dfrac{W^2}{E'^2 R'} \right)^{1/3}$	$\tau_{max} = k_4 p_{max}$ $\approx 0.3 p_{max}$ at a depth of $z = k_5 b$

where:

a	is the semimajor axis of the contact ellipse [m];
b	is the semiminor axis of the contact ellipse [m];
R'	is the reduced radius of curvature [m];
k_1, k_2, k_3, k_4, k_5	are the contact coefficients.

The other parameters are as defined previously. Contact coefficients can be found from the charts shown in Figures 7.12 and 7.13 [13]. In Figure 7.12 the coefficients 'k_1', 'k_2' and 'k_3' are plotted against the 'k_0' coefficient which is defined as:

$$k_0 = \frac{\left[\left(\dfrac{1}{R_{ax}} - \dfrac{1}{R_{ay}} \right)^2 + \left(\dfrac{1}{R_{bx}} - \dfrac{1}{R_{by}} \right)^2 + 2 \left(\dfrac{1}{R_{ax}} - \dfrac{1}{R_{ay}} \right) \left(\dfrac{1}{R_{bx}} - \dfrac{1}{R_{by}} \right) \cos 2\phi \right]^{1/2}}{\left(\dfrac{1}{R_{ax}} + \dfrac{1}{R_{ay}} + \dfrac{1}{R_{bx}} + \dfrac{1}{R_{by}} \right)}$$

where:

φ   is the angle between the plane containing the minimum principal radius of curvature of body '**A**' and the plane containing the minimum principal radius of curvature of body '**B**'. For example, for a wheel on a rail contact φ = 90° while for parallel cylinders in contact φ = 0°.

The remaining contact coefficients 'k_4' and 'k_5' are plotted against the k_2/k_1 ratio as shown in Figure 7.13.

A very useful development in the evaluation of contact parameters is due to Hamrock and Dowson [7]. The method of linear regression by the least squares method has been applied to derive simplified expressions for the elliptic integrals required for the stress and deflection calculations in Hertzian contacts. The derived formulae apply to any contact and eliminate the need to use numerical methods or charts such as those shown in Figures 7.12 and 7.13. The formulae are summarized in Table 7.4. Although they are only approximations, the differences between the calculated values and the exact predictions from the Hertzian analysis are very small. This can easily be demonstrated by applying these formulae to the previously considered examples, with the exception of the two parallel cylinders. In this case the contact is described by an elongated rectangle and these formulae cannot be used. In general, these equations can be used in most of the practical engineering applications.

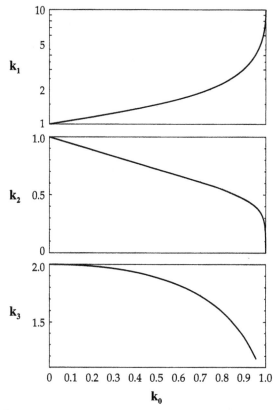

FIGURE 7.12   Chart for the determination of the contact coefficients 'k_1', 'k_2' and 'k_3' [13].

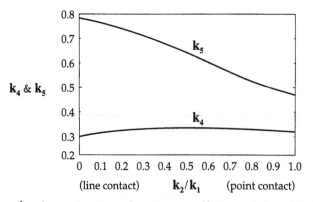

FIGURE 7.13   Chart for the determination of contact coefficients 'k_4' and 'k_5' [13].

TABLE 7.4   Approximate formulae for contact parameters between two elastic bodies [7].

Contact area dimensions	Maximum contact pressure	Maximum deflection	Simplified elliptical integrals
$a = \left(\dfrac{6\bar{k}^2\bar{\varepsilon}WR'}{\pi E'}\right)^{1/3}$ $b = \left(\dfrac{6\bar{\varepsilon}WR'}{\pi \bar{k}E'}\right)^{1/3}$ ellipse	$p_{max} = \dfrac{3W}{2\pi ab}$ Average contact pressure $p_{average} = \dfrac{W}{\pi ab}$	$\delta = \bar{\xi}\left[\left(\dfrac{4.5}{\bar{\varepsilon}R'}\right)\left(\dfrac{W}{\pi\bar{k}E'}\right)^2\right]^{1/3}$	$\bar{\varepsilon} = 1.0003 + \dfrac{0.5968\,R_x}{R_y}$ $\bar{\xi} = 1.5277 + 0.6023\ln\left(\dfrac{R_y}{R_x}\right)$ **Ellipticity parameter** $\bar{k} = 1.0339\left(\dfrac{R_y}{R_x}\right)^{0.636}$

where:

$\bar{\varepsilon}$ and $\bar{\xi}$ are the simplified elliptic integrals;

$\bar{k}$ is the simplified ellipticity parameter. The exact value of the ellipticity parameter is defined as the ratio of the semiaxis of the contact ellipse in the transverse direction to the semiaxis in the direction of motion, i.e., $k = a/b$. The differences between the ellipticity parameter '$\bar{k}$' calculated from the approximate formula, Table 7.4, and the ellipticity parameter calculated from the exact formula, $k = a/b$, are very small [7].

The other parameters are as defined already.

EXAMPLE

Find the contact parameters for a steel ball in contact with a groove on the inside of a steel ring (as shown in Figure 7.7). The normal force is W = 50 [N], radius of the ball is $R_{ax} = R_{ay} = R_A = 15 \times 10^{-3}$ [m], the radius of the groove is $R_{bx} = 30 \times 10^{-3}$ [m] and the radius of the ring is $R_{by} = 60 \times 10^{-3}$ [m]. The Young's modulus for both ball and ring is E = 2.1 × 10^{11} [Pa] and the Poisson's ratio is $\upsilon = 0.3$.

·   *Reduced Radius of Curvature*

Since the radii of the ball and the grooved ring are $R_{ax} = 15 \times 10^{-3}$ [m], $R_{ay} = 15 \times 10^{-3}$ [m] and $R_{bx} = -30 \times 10^{-3}$ [m] (concave surface), $R_{by} = -60 \times 10^{-3}$ [m] (concave surface), respectively, the reduced radii of curvature in the 'x' and 'y' directions are:

$$\frac{1}{R_x} = \frac{1}{R_{ax}} + \frac{1}{R_{bx}} = \frac{1}{15 \times 10^{-3}} + \frac{1}{-30 \times 10^{-3}} = 33.33 \qquad \boxed{\Rightarrow R_x = 0.03\,[m]}$$

$$\frac{1}{R_y} = \frac{1}{R_{ay}} + \frac{1}{R_{by}} = \frac{1}{15 \times 10^{-3}} + \frac{1}{-60 \times 10^{-3}} = 50.0 \qquad \boxed{\Rightarrow R_y = 0.02\,[m]}$$

Since $1/R_x < 1/R_y$ condition (7.3) is not satisfied. According to the convention it is necessary to transpose the directions of the coordinates, so 'R_x' and 'R_y' become:

$$R_x = 0.02\ [m] \quad \text{and} \quad R_y = 0.03\ [m]$$

and the reduced radius of curvature is:

$$\frac{1}{R'} = \frac{1}{R_x} + \frac{1}{R_y} = 50.0 + 33.33 = 83.33 \qquad \boxed{\Rightarrow R' = 0.012\,[m]}$$

· *Reduced Young's Modulus*

$$\boxed{E' = 2.308 \times 10^{11}\,[Pa]}$$

· *Contact Coefficients*

The angle between the plane containing the minimum principal radius of curvature of the ball and the plane containing the minimum principal radius of the ring is:

$$\phi = 0°$$

The contact coefficients are:

$$k_0 = \frac{\left[\left(\dfrac{1}{R_{ax}} - \dfrac{1}{R_{ay}}\right)^2 + \left(\dfrac{1}{R_{bx}} - \dfrac{1}{R_{by}}\right)^2 + 2\left(\dfrac{1}{R_{ax}} - \dfrac{1}{R_{ay}}\right)\left(\dfrac{1}{R_{bx}} - \dfrac{1}{R_{by}}\right)\cos 2\phi\right]^{1/2}}{\left(\dfrac{1}{R_{ax}} + \dfrac{1}{R_{ay}} + \dfrac{1}{R_{bx}} + \dfrac{1}{R_{by}}\right)}$$

$$= \frac{\left[\left(\dfrac{1}{15 \times 10^{-3}} - \dfrac{1}{15 \times 10^{-3}}\right)^2 + \left(\dfrac{1}{-60 \times 10^{-3}} - \dfrac{1}{-30 \times 10^{-3}}\right)^2 \right. }{\left(\dfrac{1}{15 \times 10^{-3}} + \dfrac{1}{15 \times 10^{-3}} + \dfrac{1}{-60 \times 10^{-3}} + \dfrac{1}{-30 \times 10^{-3}}\right)}$$

$$\left. + 2\left(\dfrac{1}{15 \times 10^{-3}} - \dfrac{1}{15 \times 10^{-3}}\right)\left(\dfrac{1}{-60 \times 10^{-3}} - \dfrac{1}{-30 \times 10^{-3}}\right)\cos 0°\right]^{1/2}$$

$$= \frac{16.67}{83.33} \qquad \boxed{= 0.2}$$

From Figure 7.12, for $k_0 = 0.2$:

$$k_1 = 1.17, \quad k_2 = 0.88 \quad \text{and} \quad k_3 = 1.98$$

and from Figure 7.13 where $k_2/k_1 = 0.88/1.17 = 0.75$, the other constants have the following values:

$$k_4 = 0.33 \quad \text{and} \quad k_5 = 0.54$$

· *Contact Area Dimensions*

$$a = k_1\left(\frac{3WR'}{E'}\right)^{1/3} = 1.17\left(\frac{3 \times 50 \times 0.012}{2.308 \times 10^{11}}\right)^{1/3} \qquad \boxed{= 2.32 \times 10^{-4}\,[m]}$$

$$b = k_2\left(\frac{3WR'}{E'}\right)^{1/3} = 0.88\left(\frac{3 \times 50 \times 0.012}{2.308 \times 10^{11}}\right)^{1/3} \qquad \boxed{= 1.75 \times 10^{-4}\,[m]}$$

· *Maximum and Average Contact Pressures*

$$p_{max} = \frac{3W}{2\pi ab} = \frac{3 \times 50}{2\pi (2.32 \times 10^{-4}) \times (1.75 \times 10^{-4})} \qquad \boxed{= 588.0 \,[\text{MPa}]}$$

$$p_{average} = \frac{W}{\pi ab} = \frac{50}{\pi (2.32 \times 10^{-4}) \times (1.75 \times 10^{-4})} \qquad \boxed{= 392.0 \,[\text{MPa}]}$$

· *Maximum Deflection*

$$\delta = 0.52 k_3 \left(\frac{W^2}{E'^2 R'}\right)^{1/3} = 0.52 \times 1.98 \left(\frac{50^2}{(2.308 \times 10^{11})^2 0.012}\right)^{1/3} \qquad \boxed{= 1.6 \times 10^{-6} \,[\text{m}]}$$

· *Maximum Shear Stress*

$$\tau_{max} = k_4 p_{max} = 0.33 \times 588.0 \qquad \boxed{= 194.0 \,[\text{MPa}]}$$

· *Depth at which Maximum Shear Stress Occurs*

$$z = k_5 b = 0.54 \times (1.75 \times 10^{-4}) \qquad \boxed{= 9.5 \times 10^{-5} \,[\text{m}]}$$

It can easily be found that the Hamrock-Dowson approximate formulae (Table 7.4) give very similar results, e.g.:

· *Ellipticity Parameter*

$$\bar{k} = 1.0339 \left(\frac{R_y}{R_x}\right)^{0.636} = 1.0339 \left(\frac{0.03}{0.02}\right)^{0.636} \qquad \boxed{= 1.3380}$$

· *Simplified Elliptical Integrals*

$$\bar{\varepsilon} = 1.0003 + \frac{0.5968 R_x}{R_y} = 1.0003 + \frac{0.5968 \times 0.02}{0.03} \qquad \boxed{= 1.3982}$$

$$\bar{\xi} = 1.5277 + 0.6023 \ln\left(\frac{R_y}{R_x}\right) = 1.5277 + 0.6023 \ln\left(\frac{0.03}{0.02}\right) \qquad \boxed{= 1.7719}$$

· *Contact Area Dimensions*

$$a = \left(\frac{6\bar{k}^2 \bar{\varepsilon} W R'}{\pi E'}\right)^{1/3} = \left(\frac{6 \times 1.3380^2 \times 1.3982 \times 50 \times 0.012}{\pi (2.308 \times 10^{11})}\right)^{1/3} \qquad \boxed{= 2.32 \times 10^{-4} \,[\text{m}]}$$

$$b = \left(\frac{6\bar{\varepsilon} W R'}{\pi \bar{k} E'}\right)^{1/3} = \left(\frac{6 \times 1.3982 \times 50 \times 0.012}{\pi \times 1.3380 \times (2.308 \times 10^{11})}\right)^{1/3} \qquad \boxed{= 1.73 \times 10^{-4} \,[\text{m}]}$$

· *Maximum and Average Contact Pressures*

$$p_{max} = \frac{3W}{2\pi ab} = \frac{3 \times 50}{2\pi(2.32 \times 10^{-4}) \times (1.73 \times 10^{-4})} \qquad \boxed{= 594.8\,[\text{MPa}]}$$

$$p_{average} = \frac{W}{\pi ab} = \frac{50}{\pi(2.32 \times 10^{-4}) \times (1.73 \times 10^{-4})} \qquad \boxed{= 396.5\,[\text{MPa}]}$$

Maximum Deflection

$$\delta = \bar{\xi}\left[\left(\frac{4.5}{\bar{\varepsilon}R'}\right)\left(\frac{W}{\pi\bar{k}E'}\right)^2\right]^{1/3}$$

$$= 1.7719\left[\left(\frac{4.5}{1.3982 \times 0.012}\right)\left(\frac{50}{\pi 1.3380 \times (2.308 \times 10^{11})}\right)^2\right]^{1/3} \quad \boxed{= 1.6 \times 10^{-6}\,[\text{m}]}$$

When comparing the results obtained by the Hertz theory and the Hamrock-Dowson approximation it is apparent that the differences between the results obtained by both methods are very small. Errors due to the approximation on reading values of contact coefficients from Figures 7.12 and 7.13 may contribute significantly to the difference.

The benefits of applying the Hamrock-Dowson formulae to the evaluation of contact parameters are demonstrated by the simplification of the calculations without any compromise in accuracy. Hence the Hamrock-Dowson formulae can be used with confidence in most practical engineering applications.

Total Deflection

In some practical engineering applications, such as rolling bearings, the rolling element is squeezed between the inner and the outer ring and the total deflection is the sum of the deflections between the element and both rings, i.e.:

$$\delta_T = \delta_o + \delta_i \tag{7.11}$$

where:

δ_T is the total combined deflection between the rolling element and the inner and outer rings [m];

δ_o is the deflection between the rolling element and the outer ring [m];

δ_i is the deflection between the rolling element and the inner ring [m].

According to the formula from Table 7.4, the maximum deflections for the inner and outer conjunctions can be written as:

$$\delta_i = \bar{\xi}_i\left[\left(\frac{4.5}{\bar{\varepsilon}_i R_i'}\right)\left(\frac{W}{\pi\bar{k}_i E'}\right)^2\right]^{1/3}$$

$$\tag{7.12}$$

$$\delta_o = \bar{\xi}_o\left[\left(\frac{4.5}{\bar{\varepsilon}_o R_o'}\right)\left(\frac{W}{\pi\bar{k}_o E'}\right)^2\right]^{1/3}$$

where 'i' and 'o' are the indices referring to the inner and outer conjunction, respectively. Note that each of these conjunctions has a different contact geometry resulting in a different reduced radius 'R''', ellipticity parameter 'k' and simplified integrals 'ξ' and 'ϵ'.

Introducing coefficients which are a function of the contact geometry and material properties, i.e.:

$$\overline{K}_i = \pi \overline{k}_i E' \left(\frac{\overline{\epsilon}_i R_i'}{4.5 \overline{\xi}_i^3} \right)^{1/2}$$

$$\overline{K}_o = \pi \overline{k}_o E' \left(\frac{\overline{\epsilon}_o R_o'}{4.5 \overline{\xi}_o^3} \right)^{1/2}$$

(7.13)

The deflections can be written as:

$$\delta_i = \left(\frac{W}{\overline{K}_i} \right)^{2/3}$$

$$\delta_o = \left(\frac{W}{\overline{K}_o} \right)^{2/3}$$

and

$$\delta_T = \left(\frac{W}{\overline{K}_T} \right)^{2/3}$$

Substituting into equation (7.11) yields:

$$\left(\frac{W}{\overline{K}_T} \right)^{2/3} = \left(\frac{W}{\overline{K}_i} \right)^{2/3} + \left(\frac{W}{\overline{K}_o} \right)^{2/3}$$

(7.14)

By rearranging the above expression the coefficient 'K_T' for the total combined deflection, in terms of the 'K_i' and 'K_o' coefficients, can be obtained [7], i.e.:

$$\frac{1}{\overline{K}_T} = \left[\left(\frac{1}{\overline{K}_i} \right)^{2/3} + \left(\frac{1}{\overline{K}_o} \right)^{2/3} \right]^{3/2}$$

(7.15)

It should be realized that the deflections and furthermore the pressures resulting from different loads cannot be superimposed. This is because Hertzian deflections are not linear functions of load.

7.4   ELASTOHYDRODYNAMIC LUBRICATING FILMS

The term elastohydrodynamic lubricating film refers to the lubricating oil which separates the opposing surfaces of a concentrated contact. The properties of this minute amount of oil, typically 1 [μm] thick and 400 [μm] across for a point contact, and which is subjected to extremes of pressure and shear, determine the efficiency of the lubrication mechanism under rolling contact.

Effects Contributing to the Generation of Elastohydrodynamic Films

The three following effects play a major role in the formation of lubrication films in elastohydrodynamic lubrication:

· the hydrodynamic film formation,

· the modification of the film geometry by elastic deformation,

· the transformation of the lubricant's viscosity and rheology under pressure.

All three effects act simultaneously and cause the generation of elastohydrodynamic films.

· *Hydrodynamic Film Formation*

The geometry of interacting surfaces in Hertzian contacts contains converging and diverging wedges so that some form of hydrodynamic lubrication occurs. The basic principles of hydrodynamic lubrication outlined in Chapter 4 apply, but with some major differences. Unlike classical hydrodynamics, both contact geometry and lubricant viscosity are a function of hydrodynamic pressure. It is therefore impossible to specify precisely a film geometry and viscosity before proceeding to solve the Reynolds equation. Early attempts by Martin [2] were made, for example, to estimate the film thickness in elastohydrodynamic contacts using a pre-determined film geometry, and erroneously thin film thicknesses were predicted.

· *Modification of Film Geometry by Elastic Deformation*

For all materials whatever their modulus of elasticity, the surfaces in a Hertzian contact deform elastically. The principal effect of elastic deformation on the lubricant film profile is to interpose a central region of quasi-parallel surfaces between inlet and outlet wedges. This geometric effect is shown in Figure 7.14 where two bodies, i.e., a flat surface and a ball, in elastic contact are illustrated. The contact is shown in one plane and the contact radii are '∞' and '**R**' for the flat surface and ball, respectively.

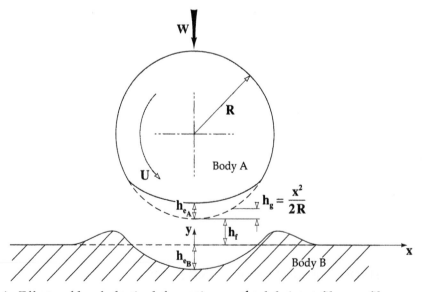

FIGURE 7.14   Effects of local elastic deformation on the lubricant film profile.

The film profile in the '**x**' direction is given by [15]:

$$h = h_f + h_e + h_g$$

where:

h_f is constant [m];

h_e is the combined elastic deformation of the solids [m], i.e., $h_e = h_{e_A} + h_{e_B}$;

h_g is the separation due to the geometry of the undeformed solids [m], i.e., for the
        ball on a flat plate shown in Figure 7.14 $h_g = x^2/2R$;

R is the radius of the ball [m].

· *Transformation of Lubricant Viscosity and Rheology under Pressure*

The non-conformal geometry of the contacting surfaces causes an intense concentration of load over a very small area for almost all Hertzian contacts of practical use. When a liquid separates the two surfaces, extreme pressures many times higher than those encountered in hydrodynamic lubrication are inevitable. Lubricant pressures from 1 to 4 [GPa] are found in typical machine elements such as gears. As previously discussed in Chapter 2, the viscosity of oil and many other lubricants increases dramatically with pressure. This phenomenon is known as piezoviscosity. The viscosity-pressure relationship is usually described by a mathematically convenient but approximate equation known as the Barus law:

$$\eta_p = \eta_0 e^{\alpha p}$$

where:

η_p is the lubricant viscosity at pressure 'p' and temperature 'θ' [Pas];

η_0 is the viscosity at atmospheric pressure and temperature 'θ' [Pas];

α is the pressure-viscosity coefficient [m²/N].

As an example of the radical effect of pressure on viscosity, it has been reported that at contact pressures of about 1 [GPa], the viscosity of mineral oil may increase by a factor of 1 million (10^6) from its original value at atmospheric pressure [15].

With sufficiently hard surfaces in contact, the lubricant pressure may rise to even higher levels and the question of whether there is a limit to the enhancement of viscosity becomes pertinent. The answer is that indeed there are constraints where the lubricant loses its liquid character and becomes semi-solid. This aspect of elastohydrodynamic lubrication is the focus of present research and is discussed later in this chapter. For now, however, it is assumed that the Barus law is exactly applicable.

Approximate Solution of Reynolds Equation with Simultaneous Elastic Deformation and Viscosity Rise

An approximate solution for elastohydrodynamic film thickness as a function of load, rolling speed and other controlling variables was put forward by Grubin and was later superseded by more exact equations. Grubin's expression for film thickness is, however, relatively accurate and the same basic principles that were originally established have been applied in later work. For these reasons, Grubin's equation is derived in this section to illustrate the principles of how the elastohydrodynamic film thickness is determined.

The derivation of the film thickness equation for elastohydrodynamic contacts begins with the one-dimensional form of the Reynolds equation without squeeze effects (i.e., 4.27):

$$\frac{dp}{dx} = 6U\eta\left(\frac{h - \bar{h}}{h^3}\right)$$

where the symbols follow the conventions established in Chapter 4 and are:

p is the hydrodynamic pressure [Pa];

U is the surface velocity [m/s];

η is the lubricant viscosity [Pas];

h is the film thickness [m];

$\bar{h}$ is the film thickness where the pressure gradient is zero [m];

x is the distance in direction of rolling [m].

Substituting into the Reynolds equation the expression for viscosity according to the Barus law yields:

$$\frac{dp}{dx} = 6U\eta_0 e^{\alpha p}\left(\frac{h - \bar{h}}{h^3}\right) \qquad (7.16)$$

To solve this equation, Grubin introduced an artificial variable, known as the 'reduced pressure', defined as:

$$q = \frac{1}{\alpha}\left(1 - e^{-\alpha p}\right) \qquad (7.17)$$

Differentiating gives:

$$\frac{dq}{dx} = e^{-\alpha p}\frac{dp}{dx}$$

When this term is substituted into the Reynolds equation (7.16), a separation of pressure and film thickness is achieved:

$$\frac{dq}{dx} = 6U\eta_0\left(\frac{h - \bar{h}}{h^3}\right) \qquad (7.18)$$

Two independent controlling variables, i.e., 'x' and 'h', however, still remain and replacement of either of these variables by the other (since $x = f(h)$) is required for the solution. The argument used to achieve this reduction in unknown variables is perhaps the most original and innovative part of Grubin's analysis.

Grubin observed that at the inlet of the EHL contact, the contact pressure rises very sharply as predicted by Hertzian contact theory. If a hydrodynamic film is established, then the hydrodynamic pressure should also rise sharply at the inlet. This sharp rise in pressure can be approximated as a step jump to some value in pressure comparable to the peak Hertzian contact pressure. If this pressure is assumed to be large enough then the term $e^{-\alpha p} \ll 1$ and it can be seen from equation (7.17) that $q \approx 1/\alpha$. Grubin reasoned that since the stresses and the deformations in the EHL contacts were substantially identical to Hertzian, the opposing surfaces must almost be parallel and thus the film thickness is approximately uniform within the contact. Inside the contact therefore, the film thickness $h = $ **constant** so that $h = \bar{h}$. Since '$\bar{h}$' occurs where 'p_{max}' takes place, Grubin deduced that there must be a sharp increase in pressure in the inlet zone to the contact as shown in Figure 7.15. It therefore follows that according to this model $q \approx 1/\alpha = $ **constant**, $dq/dx = 0$ and $h = \bar{h}$ within the contact.

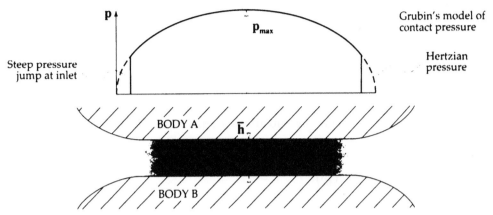

FIGURE 7.15 Grubin's approximation to film thickness within an EHL contact.

A formal expression for '**q**' is found by integrating (7.18);

$$q = 6U\eta_0 \int_{h_\infty}^{h_1} \left(\frac{h - \bar{h}}{h^3}\right) dx \qquad (7.19)$$

where:

h_1 is the inlet film thickness to the EHL contact [m];

h_∞ is the film thickness at a distance 'infinitely' far from the contact [m].

Since **q** ≈ **1/α** the above equation (7.19) can be written in the form:

$$q = \frac{1}{\alpha} = 6U\eta_0 \int_{h_\infty}^{h_1} \left(\frac{h - \bar{h}}{h^3}\right) dx \qquad (7.20)$$

After replacing one variable with another (i.e., expressing '**x**' in terms of '**h**'), this integral is solved numerically by assuming that the values of film thickness '**h**' are equal to the distance separating the contacting dry bodies plus the film thickness within the EHL contact. The constant of integration is zero for the selected limits of this integral since at any position remote from the contact, **p = 0** and therefore **q = 0**. The following approximation was calculated numerically for the integral as applied to a line contact:

$$\int_{h_\infty}^{h_1} \left(\frac{h - \bar{h}}{h^3}\right) dx = 0.131 \left(\frac{W}{LE'R'}\right)^{-0.625} \left(\frac{b}{R'^2}\right) \left(\frac{\bar{h}}{R'}\right)^{1.375} \qquad (7.21)$$

where:

R' is the reduced radius of curvature [m];

E' is the reduced Young's modulus [Pa];

L is the full length of the EHL contact, i.e., $L = 2l$, [m];

b is the half width of the EHL contact [m];

h is the film thickness where the pressure gradient is zero, i.e., Grubin's EHL film thickness as shown in Figure 7.15 [m];

W   is the contact load [N].

Rearranging (7.20) gives:

$$\int_{h_\infty}^{h_1}\left(\frac{h-\bar{h}}{h^3}\right)dx = \frac{1}{6U\eta_0\alpha} \tag{7.22}$$

The integral term is then eliminated by substituting equation (7.22) into equation (7.21), i.e.:

$$1.275\frac{R'^2}{bU\eta_0\alpha}\left(\frac{W}{LE'R'}\right)^{0.625} = \left(\frac{\bar{h}}{R'}\right)^{-1.375} \tag{7.23}$$

Expressing equation (7.23) as a unit power of $\bar{h}/R'$ yields:

$$\left(\frac{\bar{h}}{R'}\right) = 1.193\left(\frac{R'^2}{bU\eta_0\alpha}\right)^{-0.7273}\left(\frac{W}{LE'R'}\right)^{-0.4545} \tag{7.24}$$

Substituting for contact width 'b' the Hertzian contact formula (Table 7.2) yields a more convenient expression for routine film thickness calculation. The expression for 'b' (Table 7.2) is:

$$b = \left(\frac{4WR'}{\pi lE'}\right)^{1/2} = \left(\frac{8WR'}{\pi LE'}\right)^{1/2}$$

Substituting into (7.24) gives Grubin's expression for film thickness in the elastohydrodynamic linear contact, i.e.:

$$\left(\frac{\bar{h}}{R'}\right) = 1.657\left(\frac{U\eta_0\alpha}{R'}\right)^{0.7273}\left(\frac{W}{LE'R'}\right)^{-0.0909} \tag{7.25}$$

It can be seen that all the variables are combined in dimensionless groups making the interpretation of the irrational exponents easier.

Grubin was able to demonstrate with the above expression that oil films with sufficient thickness to separate typical engineering surfaces existed in concentrated line contacts. The values of film thickness provided by this approximate formula are surprisingly accurate. The relative effects of load, rolling velocity and pressure-viscosity dependence are shown in terms of indices that correspond closely to more exact analyses. The comparatively weak effect of load should be noted which explains the high load capacity of elastohydrodynamic films. More advanced solutions of the elastohydrodynamic film thickness equation involve the two-dimensional Reynolds equations and more sophisticated inlet conditions. Grubin also assumed that the contact was 'fully flooded', i.e., the rolling elements moved in a bath of oil. More exact work has allowed for the effect of oil shortage in the contact and thermal effects at high speeds. The exact analysis of elastohydrodynamic lubrication involves a simultaneous iterative numerical solution of the equations describing hydrodynamic film formation, elastic deformation and piezoviscosity in a lubricated Hertzian contact. These are the same fundamental equations described earlier but they are solved directly without any analytical simplifications. The numerical procedures and mathematics involved are described in detail elsewhere [7,11].

Pressure Distribution in Elastohydrodynamic Films

In a static contact, the pressure distribution is hemispherical or ellipsoidal in profile according to classical Hertzian theory. The pressure field will change, however, when the surfaces start moving relative to each other in the presence of a piezoviscous lubricant such as oil. Relative motion between the two surfaces causes a hydrodynamic lubricating film to be generated which modifies the pressure distribution to a certain extent. The greatest changes to the pressure profile occur at the entry and exit regions of the contact. The combined effect of rolling and a lubricating film results in a slightly enlarged contact area. Consequently at the entry region, the hydrodynamic pressure is lower than the value for a dry Hertzian contact. This has been demonstrated in numerous experiments. The opposing surfaces within the contact are almost parallel and planar and film thickness is often described in this region by the central film thickness 'h_c'. The lubricant experiences a precipitous rise in viscosity as it enters the contact followed by an equally sharp decline to ambient viscosity levels at the exit of the contact. To maintain continuity of flow and compensate for the loss of lubricant viscosity at the contact exit, a constriction is formed close to the exit. The minimum film thickness 'h_0' is found at the constriction as shown in Figure 7.16. The minimum film thickness is an important parameter since it controls the likelihood of asperity interaction between the two surfaces. Viscosity declines even more sharply at the exit than at the entry to the contact. A large pressure peak is generated next to the constriction on the upstream side, and downstream the pressure rapidly declines to less than dry Hertzian values. The peak pressure is usually larger than the maximum Hertzian contact pressure and diminishes as the severity of lubricant starvation increases and dry conditions are approached [7]. The size and the steepness of the pressure peak depend strongly on the lubricant's pressure-viscosity characteristics.

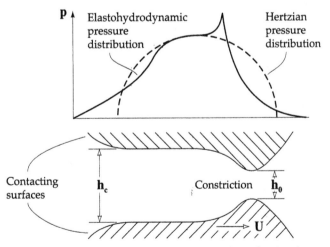

FIGURE 7.16 Hydrodynamic pressure distribution in an elastohydrodynamic contact; h_c is the central film thickness and h_0 is the minimum film thickness.

The end constriction to the EHL film is even more distinctive for a 'point' contact, e.g., two steel balls in contact. In this case the contact is circular and the end constriction has to be curved in order to fit into the contact boundary. This effect is known as the 'horseshoe' constriction and is shown later in Figure 7.22 which illustrates a plan view of the EHL film (as opposed to the side view shown in Figure 7.16). The minimum film thickness in a point contact is found at both ends of the 'horseshoe' and at these locations the film thickness is only about **60%** of its central value.

Elastohydrodynamic Film Thickness Formulae

The exact analysis of elastohydrodynamic lubrication by Hamrock and Dowson [7,16] provided the most important information about EHL. The results of this analysis are the formulae for the calculation of the minimum film thickness in elastohydrodynamic contacts. The formulae derived by Hamrock and Dowson apply to any contact, such as point, linear or elliptical, and are now routinely used in EHL film thickness calculations. They can be used with confidence for many material combinations including steel on steel even up to maximum pressures of 3-4 [GPa] [11]. The numerically derived formulae for the central and minimum film thicknesses, as shown in Figure 7.16, are in the following form [7]:

$$\frac{h_c}{R'} = 2.69 \left(\frac{U \eta_0}{E'R'} \right)^{0.67} \left(\alpha E' \right)^{0.53} \left(\frac{W}{E'R'^2} \right)^{-0.067} \left(1 - 0.61 e^{-0.73k} \right) \tag{7.26}$$

$$\frac{h_0}{R'} = 3.63 \left(\frac{U \eta_0}{E'R'} \right)^{0.68} \left(\alpha E' \right)^{0.49} \left(\frac{W}{E'R'^2} \right)^{-0.073} \left(1 - e^{-0.68k} \right) \tag{7.27}$$

where:

h_c is the central film thickness [m];

h_0 is the minimum film thickness [m];

U is the entraining surface velocity [m/s], i.e., $U = (U_A + U_B)/2$, where the subscripts 'A' and 'B' refer to the velocities of bodies 'A' and 'B', respectively;

η_0 is the viscosity at atmospheric pressure of the lubricant [Pas];

E' is the reduced Young's modulus (7.6) [Pa];

R' is the reduced radius of curvature [m];

α is the pressure-viscosity coefficient [m^2/N];

W is the contact load [N];

k is the ellipticity parameter defined as: **k = a/b**, where 'a' is the semiaxis of the contact ellipse in the transverse direction [m] and 'b' is the semiaxis in the direction of motion [m].

As mentioned already, the approximate value of the ellipticity parameter can be calculated with sufficient accuracy from:

$$\bar{k} = 1.0339 \left(\frac{R_y}{R_x} \right)^{0.636}$$

where:

R_x and R_y are the reduced radii of curvature in the 'x' and 'y' directions, respectively.

It can be seen that for line contacts $k = \infty$ and for point contact **k = 1**. It has been shown that the above EHL film thickness equations are applicable for 'k' values between **0.1** and ∞ [17].

The non-dimensional groups in equations (7.26) and (7.27) are frequently referred to in the literature as:

- the non-dimensional film parameter

$$\mathbf{H} = \frac{\mathbf{h}}{\mathbf{R'}}$$

- the non-dimensional speed parameter

$$\mathbf{U} = \left(\frac{\mathbf{U}\eta_0}{\mathbf{E'R'}}\right)$$

- the non-dimensional materials parameter

$$\mathbf{G} = (\alpha\mathbf{E'})$$

- the non-dimensional load parameter

$$\mathbf{W} = \left(\frac{\mathbf{W}}{\mathbf{E'R'^2}}\right)$$

- the non-dimensional ellipticity parameter

$$\mathbf{k} = \frac{\mathbf{a}}{\mathbf{b}}$$

Effects of the Non-Dimensional Parameters on EHL Contact Pressures and Film Profiles

The changes in the non-dimensional parameters have varying effects on the EHL film thicknesses and pressures. To demonstrate these effects, Hamrock and Dowson allowed one specific parameter to vary while holding all the other parameters constant [7].

· Effect of the Speed Parameter

As would be expected from the need for relative movement to generate a hydrodynamic pressure field, the speed parameter has a strong effect on EHL. The influence of the speed parameter 'U' on the pressure and film thickness profiles is shown in Figure 7.17. The pressure and film profiles are calculated for $\mathbf{k} = 6$, $\mathbf{W} = 7.371 \times 10^{-7}$ and $\mathbf{G} = 4.522 \times 10^3$ [7].

It can be seen that in the inlet region there is a gradual increase in pressure with speed and a corresponding decline in pressure in the outlet region of the Hertzian contact area. The effect of elevated speed is to radically distort the pressure profile from the Hertzian form to the profile of a sharply pointed peak. This change in pressure profile increases the maximum contact pressure for a given load which may cause damage to the underlying material. When the speed parameter is reduced, the pressure profile reverts to the Hertzian form, but with a pressure peak at the exit constriction. The effect of the speed parameter on the film thickness profile is to (a) increase film thickness, (b) reduce the proportion of contact area where the two surfaces are virtually parallel and (c) increase the proportion of contact area covered by the exit constriction. The first effect, i.e., increase in the film thickness, is the most significant, while the importance of the other effects is unclear. It is evident that the film thickness varies considerably with speed, which illustrates the dominant effect of the non-dimensional speed parameter on the minimum film thickness in elastohydrodynamic contacts.

These findings have been confirmed experimentally by many researchers. The experiments usually demonstrated a remarkable agreement with theory. The pressure distribution, position of the pressure peak and film profile could be accurately and effectively predicted at a particular velocity and load. There was, however, some discrepancy concerning the height of the pressure peak since the measured peak was very much smaller than that predicted by theory. This was eventually rectified by introducing the lubricant compressibility into the calculations which resulted in a reduction in the pressure spike [18].

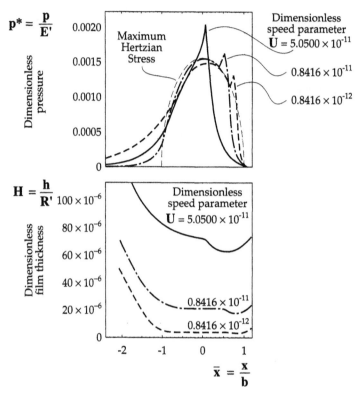

FIGURE 7.17 Effects of speed parameter 'U' on the pressure and film thickness in an EHL contact; **b** is the semiaxis of the contact ellipse in the direction of motion [7].

· *Effect of the Materials Parameter*

In general terms, the type of materials used will determine the regime of hydrodynamic lubrication, whether it is true EHL or some other variant. For example, substituting rubber for steel reduces the contact stress sufficiently to preclude the pressure dependent viscosity rise found in EHL. It is, however, difficult to show the effect of small variations of the materials parameter on EHL since the dimensioned parameters defining the materials parameter, such as the reduced Young's modulus, are also included in the non-dimensional load and speed parameters. The minimum film thickness as a function of the material properties and these other parameters can be written as [7]:

$$H_{min} \; \alpha \; G^{0.45}$$

· *Effect of the Load Parameter*

Load also has a strong effect on film thickness in general and more importantly on the minimum film thickness at the exit constriction. Figure 7.18 shows the effect of varying load parameter on hydrodynamic pressure and film thickness for constant values of ellipticity, speed parameter and materials parameter: $k = 6$, $U = 1.683 \times 10^{-12}$, $G = 4.522 \times 10^{3}$ [7].

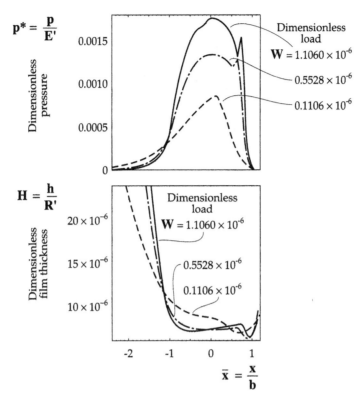

$$p^* = \frac{p}{E'}$$

$$H = \frac{h}{R'}$$

$$\bar{x} = \frac{x}{b}$$

FIGURE 7.18   Effects of load parameter on pressure and film thickness in EHL contacts; **b** is as defined previously [7].

It can be seen that as the load is increased, hydrodynamic pressure becomes almost completely confined inside the nominal Hertzian contact area. This effect is so strong that with an increase in load, pressure outside the contact area, i.e., at the inlet, actually declines. The increase in load also causes an increase in film thickness between the inlet and exit constriction which is a re-entrant profile. This feature is attributed to lubricant compressibility [7].

It is evident that the central film thickness declines with load till a certain level where film thickness becomes virtually independent of load. This is a very useful feature of EHL but it should also be noted that the minimum film thickness at the constriction does not decline significantly with increased load.

· *Effect of the Ellipticity Parameter*

Ellipticity has a strong effect on the hydrodynamic pressure profile and film thickness. Figure 7.19 shows pressure and film thickness profiles for 'k' ranging from **1.25** to **6** for the following values of the non-dimensional controlling parameters: $U = 1.683 \times 10^{-12}$, $W = 1.106 \times 10^{-7}$ and $G = 4.522 \times 10^3$ [7]. The profile is shown for a section codirectional with the rolling velocity.

The pressure 'spike' is predicted for $k = 1.25$ and 2.5 but not for $k = 6$. The film thickness appears to increase in proportion to '**k**' and this trend is due to the relative widening of the contact which enhances the generation of hydrodynamic pressure for a given film thickness by preventing side leakage of lubricant. The re-entrant form of the film profile when $k = 1.25$ is attributed to lubricant compressibility. When the compressibility is considered, the local film thickness is reduced by an amount corresponding to the change in fluid volume with pressure.

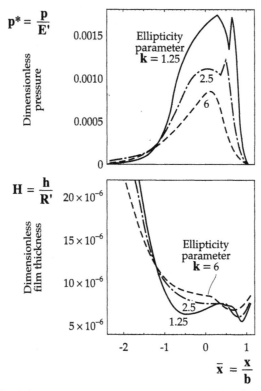

FIGURE 7.19 Effect of ellipticity parameter on pressure and film thickness in an EHL contact; **b** is as defined previously [7].

Lubrication Regimes in EHL - Film Thickness Formulae

Although the EHL film thickness equations (7.26) and (7.27) apply to most of the elastohydrodynamic contacts, there may be some practical engineering applications where more precise formulae can be used. For example, in heavily loaded contacts where the elastic deformations and changes in viscosity with pressure are significant, equations (7.26) and (7.27) give accurate film thickness predictions. However, there are other engineering applications, such as very lightly loaded rolling bearings where the elastic and viscosity effects are small, yet the contacts are classified as elastohydrodynamic. The magnitude of elastic deformation and changes in lubricant viscosity depend mostly on the applied load and the Young's modulus of the material. Depending on the values of load and material properties, the changes in film geometry and lubricant viscosity can be either more or less pronounced. In general, four well defined lubrication regimes are distinguished in full-film elastohydrodynamics [7]. Each of these regimes is characterized by the operating conditions and the properties of the material. Accurate equations for minimum film thickness have been developed for each of these regimes which are:

· isoviscous-rigid body (comparable to classical hydrodynamics),

· piezoviscous-rigid body,

· isoviscous-elastic body,

· piezoviscous-elastic body (as discussed in this chapter).

In engineering calculations it is important to first assess which EHL regime applies to the contact or mechanical component under study and then apply the appropriate equation to determine the minimum film thickness.

It has also been observed in the literature that the set of non-dimensional parameters used in the film thickness formulae (7.26 and 7.27), i.e., '**H**', '**U**', '**G**', '**W**' and '**k**', can be reduced by one parameter without any loss of generality. New non-dimensional parameters expressed in terms of those already defined have been suggested [7]:

· the non-dimensional film parameter (new) $\hat{H} = H\left(\dfrac{W}{U}\right)^2$

· the non-dimensional viscosity parameter (new) $G_V = \dfrac{GW^3}{U^2}$

· the non-dimensional elasticity parameter (new) $G_E = \dfrac{W^{8\,3}}{U^2}$

· the non-dimensional ellipticity parameter (unchanged) $k = \dfrac{a}{b}$

The utility of this simplification of controlling parameters is that it enables identification of the operating parameters and also facilitates the construction of a chart defining the regimes of elastohydrodynamic lubrication.

The film thickness formulae for the four regimes of EHL mentioned above are presented below starting with the simplest case of isoviscous-rigid body.

· *Isoviscous-Rigid*

In the isoviscous-rigid regime, elastic deformations are small and can be neglected. The maximum film pressure is too low to significantly increase the lubricant viscosity. This regime is typically found in very lightly loaded rolling bearings.

The non-dimensional minimum and central film thickness can be calculated from the formula [7]:

$$\hat{H}_{min} = \hat{H}_c = 128\,\alpha_a\,\lambda_b{}^2\left[0.131\tan^{-1}\left(\frac{\alpha_a}{2}\right) + 1.683\right]^2 \qquad (7.28)$$

where:

$\hat{H}_{min}$ is the non-dimensional minimum film thickness;

$\hat{H}_c$ is the non-dimensional central film thickness;

α_a and λ_b are coefficients which can be calculated from:

$$\alpha_a = \frac{R_B}{R_A} \approx 0.955k$$

$$\lambda_b = \left(1 + \frac{0.698}{k}\right)^1$$

k   is the ellipticity parameter as previously defined.

It can be seen that the minimum film thickness is only a function of the geometry of the contact.

· *Piezoviscous-Rigid*

In the piezoviscous-rigid regime, the elastic deformations are very small and can be neglected but the film pressures are sufficiently high to significantly increase the lubricant viscosity inside the contact. This regime is typically found in moderately loaded cylindrical tapered rollers and some piston rings and cylinder liners. The non-dimensional minimum and central film thickness for this regime can be calculated from the equation [7]:

$$\hat{H}_{min} = \hat{H}_c = 1.66 G_V^{2/3}(1 - e^{-0.68k})$$

(7.29)

· *Isoviscous-Elastic*

In the isoviscous-elastic regime of EHL, the elastic deformations of contacting surfaces make a considerable contribution to the thickness of the generated film. The film pressures are either too low to raise the lubricant viscosity or else the lubricant viscosity is relatively insensitive to pressure. A prime example of such a lubricant is pure water (but not necessarily an aqueous solution of another substance). This regime is typically found between contacting solids with low Young's moduli, e.g., human joints, seals, tyres, etc.

The non-dimensional minimum and central film thickness can be calculated from the following equations [7]:

$$\hat{H}_{min} = 8.70 G_E^{0.67}(1 - 0.85 e^{-0.31k})$$

(7.30)

$$\hat{H}_c = 11.15 G_E^{0.67}(1 - 0.72 e^{-0.28k})$$

(7.31)

· *Piezoviscous-Elastic*

Under piezoviscous-elastic conditions, film thickness is controlled by the combined action of elastic deformation and viscosity elevation as discussed previously. This regime is a form of fully developed elastohydrodynamic lubrication typically encountered in rolling bearings, gears, cams and followers, etc.

The non-dimensional minimum and central film thicknesses can be calculated from the formulae [7]:

$$\hat{H}_{min} = 3.42 G_V^{0.49} G_E^{0.17}(1 - e^{-0.68k})$$

(7.32)

$$\hat{H}_c = 3.61 G_V^{0.53} G_E^{0.13}(1 - 0.61 e^{-0.73k})$$

(7.33)

Identification of the Lubrication Regime

As mentioned earlier, it is important to identify which lubrication regime a specific machine component is operating in before applying a film thickness equation. Hamrock and Dowson produced a map of lubrication regimes [7,19] to simplify this identification. An example of this map for a value of the ellipticity parameter, **k = 1**, is shown in Figure 7.20 [7].

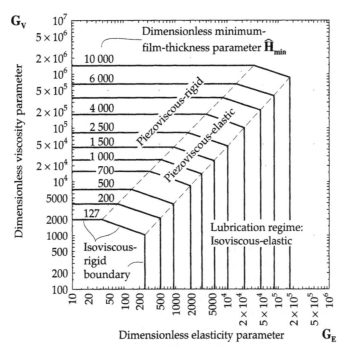

FIGURE 7.20   Map of lubrication regimes for an ellipticity parameter **k =1** [7].

Elastohydrodynamic Film Thickness Measurements

Various elastohydrodynamic film thickness measurement techniques have been developed over the years. These can be generally classified as electrical resistance, capacitance, X-ray, mechanical and optical interferometry methods.

The electrical resistance method involves measuring the electrical resistance of the lubricating film. The method is useful for the detection of lubricating films, but there are some problems associated with the assessment of the film thickness. The resistance is almost zero when metal-to-metal contact is established between the asperities of opposite surfaces and then increases in quite a complex manner with the thickness of lubricating film. The method is primarily used in detecting the breakdown of lubricating films in contact. Many difficulties arise in the evaluation of film thickness and the method is rather unreliable.

The electrical capacitance method involves the measurement of the electrical capacity of the lubricating film. The film thickness can be estimated to reasonable accuracy by this method. A major problem associated with this method is that the dielectric constant of the lubricating oil varies with temperature and pressure. The constant must be determined before the measurements of film thickness. The electrical capacitance method was pioneered by Crook [20] who measured the film thickness between steel rollers and refined later by Dyson et al. [21].

The X-ray method involves passing an X-ray beam through the lubricated contact between two surfaces. Since the lubricant scarcely absorbs the X-rays whereas the absorption by the metallic contacting bodies is very strong, the differences in film thickness can be detected. The technique was originally developed by Sibley et al. [22,23]. In the experiments conducted, the X-ray beam was shone along the tangent plane between two lubricated rolling discs and the film thickness was evaluated from the radiation intensity measurements of the emerging beam. The problems in applying this technique are principally associated with maintaining the parallelism of the beam to the common tangent of the contacting surfaces and with the calibration of film thickness [7].

The mechanical methods involve the measurements of differences in strain caused by elastohydrodynamic films. Strain gauges are used for measurements. The method was developed by Meyer and Wilson [24] to measure the EHL film thickness in a ball bearing. The main advantage of this method is that it can be used for EHL film thickness evaluation in real operating machinery. The other methods usually require simulation of the EHL contact in an experimental apparatus.

The optical interferometry method of elastohydrodynamic film thickness measurement was first pioneered by Kirk [25] and Cameron and Gohar [26]. In its original form, the method utilizes a steel ball which is driven in nominally pure rolling by a glass disc as shown in Figure 7.21. The disc is coated on one side with an approximately 10 [nm] thick semi-reflecting layer of chromium. When the disc is rotated in the presence of lubricant an elastohydrodynamic film is formed between the ball and the disc. White light is shone through the contact between the glass disc and the steel ball. The semi-reflecting chromium layer applied to the surface of the disc reflects off some of the light while some light passes through the lubricant and is reflected off the steel ball. The intensity of the two reflected beams is similar and they will either constructively or destructively interfere to produce an interference pattern, resulting in a graduation of colours depending on film thickness. Since the elastohydrodynamic film thickness is of the same order as the wavelength of visible light, it can be used to measure the generated elastohydrodynamic film thickness. The interference pattern is reflected back through the objective to the viewing port of the microscope. The corresponding optical film thickness is determined from the colours of the optical interference pattern and the real film thickness found after dividing the optical film thickness by the refractive index of the fluid.

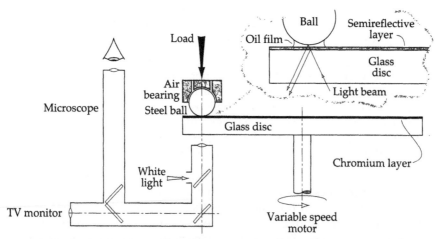

FIGURE 7.21 Schematic diagram of the apparatus for measurement of the EHL film thickness by the optical interferometry technique.

A schematic representation of the observed image for a point contact is shown in Figure 7.22. The 'horseshoe' shaped constriction found in the elastohydrodynamic film is clearly evident.

One of the great advantages of this method is that a 'contour map' of the contact can be obtained from the image [26].

The method of interferometric film measurement, however, has to be calibrated before its application, i.e., the film thickness corresponding to a particular fringe colour must be known. Calibration can be performed by using sodium monochromatic light to illuminate the contact. There are distinct advantages in using monochromatic light for calibration. As the film thickness increases, there are corresponding phase changes in the observed interference fringes. Since the sodium light used is monochromatic there is no graduation of colours and as the phase change occurs there is a corresponding change only from black to yellow and vice versa. When white light is used the phase change would be manifested by a colour change from yellow through red and blue to green, etc. (the colour change cycle will repeat). The phase changes found with monochromatic light can be related to the film thickness since the change in colour occurs at every $\lambda/4$ increase in the film thickness (where 'λ' is the wavelength of sodium light, i.e., $\lambda = 0.59$ [μm]). A calibration curve between film thickness and the corresponding phase change is obtained. The technique of obtaining this calibration curve is described elsewhere [27]. During the calibration process, the ball is placed on a stationary glass disc covered with oil. When stationary, the observed image is dark in the centre and the zero order fringe defines the Hertzian contact diameter. The speed of the disc is then slowly increased until a phase change occurs and all the dark fringes in the contact turn to yellow and then the yellow fringes turn dark, etc. The measurements are now conducted with white light since the change in colours (phase change) are already related to corresponding changes in the film thickness. In this manner a calibration curve, allocating a specific film thickness to a particular colour, is obtained.

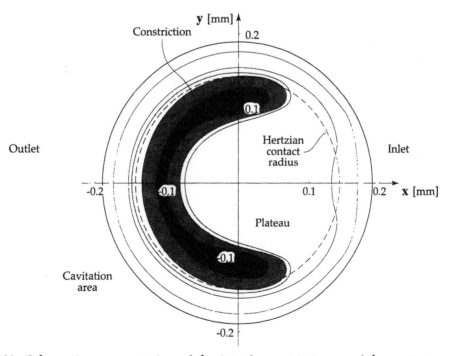

FIGURE 7.22 Schematic representation of the interferometric image of the contact area under EHL conditions [69].

It may be noticed that the optical interferometry method of elastohydrodynamic film thickness measurement also allows for the accurate measurements of the pressure-viscosity coefficient at high shear rates. Since the variables such as film thickness 'h_c', load 'W', velocity 'U', reduced radius of curvature 'R''', lubricant viscosity at atmospheric pressure 'η_0', reduced Young's modulus 'E''' and the ellipticity parameter 'k' are known then the pressure-viscosity coefficient 'α' can be calculated from equation (7.26). The major advantage of using this technique in the evaluation of pressure-viscosity coefficient is that the coefficient is being determined at the realistically high shear rates which operate in EHL contacts. High pressure viscometers which are usually used in determining the values of pressure-viscosity coefficient give its values at very low shear rates of about $10^2 - 10^3$ [s⁻¹] . Since the shear rates found in elastohydrodynamic contacts are very high, in the range of $10^6 - 10^7$ [s⁻¹], this method gives more realistic estimates of the pressure-viscosity coefficient.

The major limitation of the optical interferometry method is that one of the contacting bodies must be transparent. This restriction limits optical interferometry as essentially a laboratory technique.

To summarize, although it is possible to measure EHL film thickness accurately in a laboratory apparatus simulating real contacts, the measurement of EHL film thickness in practical engineering machinery is very difficult and accurate results are almost impossible to obtain.

7.5 MICRO-ELASTOHYDRODYNAMIC LUBRICATION AND MIXED OR PARTIAL EHL

In the evaluation of EHL film thickness it has been assumed that the contacting surfaces lubricated by elastohydrodynamic films are flat. In practice, however, the surfaces are never flat, they are rough, covered by features of various shapes, sizes and distribution. The question arises of how the surface roughness affects the mechanism of elastohydrodynamic film generation. For example, it has been reported that many engineering components operate successfully, without failure, with a calculated minimum film thickness of the same order as the surface roughness [28]. However, the question is: how are these surfaces lubricated?

If the surface asperities are of the same height as the elastohydrodynamic film thickness then one may wonder whether there is any separation at all between the surfaces by a lubricating film. For example, EHL film thickness is often found to be in the range of 0.2 - 0.4 [μm] which is similar to the surface roughness of ground surfaces.

Local film variation as a function of local surface roughness is perhaps best characterized by a parameter proposed by Tallian [29]. The ratio of the minimum film thickness to the composite surface roughness of two surfaces in contact is defined as:

$$\lambda = \frac{h_0}{(\sigma_A^2 + \sigma_B^2)^{0.5}} \tag{7.34}$$

where:

h_0 is the minimum film thickness [m];

σ_A is the RMS surface roughness of body 'A' [m];

σ_B is the RMS surface roughness of body 'B' [m];

λ is the parameter characterizing the ratio of the minimum film thickness to the composite surface roughness.

Measured values of 'λ' have been found to correlate closely with the limits of EHL and the onset of damage to the contacting surfaces. A common form of surface damage is surface fatigue where spalls or pits develop on the contacting surfaces and prevent smooth rolling or sliding. It is also possible for wear, i.e., surface material uniformly removed from the contacting surface, to occur when EHL is inadequate. The rapidity of pitting and spalling or simple wear is described in terms of a fatigue life which is the number of rolling/sliding contacts till pitting is sufficient to prevent smooth motion between the opposing surfaces. The relationship between 'λ' and fatigue life is shown in Figure 7.23.

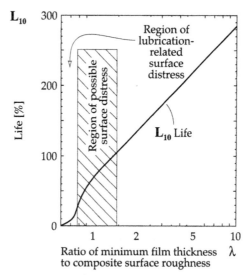

FIGURE 7.23   Effects of minimum film thickness and composite surface roughness on contact fatigue life [29].

It has been found that if 'λ' is less than **1**, surface smearing or deformation accompanied by wear can occur. When 'λ' is between **1** and **1.5** surface distress is possible. The term 'surface distress' means that surface glazing and spalling will occur. When the surface has been 'glazed', it is assumed that the original surface roughness has been suppressed by extreme plastic deformation of the asperities. For the values of 'λ' between **1.5** and **3** some glazing of the surface may occur; however, this glazing will not impair bearing operation or result in pitting. At values about **3** or greater minimal wear can be expected with no glazing. When 'λ' is greater than **4**, full separation of the surfaces by an EHL film can be expected.

As mentioned already it has been found that a great percentage of machine elements operate quite well even though λ ≈ **1**, i.e., in the region of 'possible surface distresses'. This would suggest that in order for lubrication to be effective, elastic deformation which flattens the asperities occurs and elastohydrodynamic lubrication is established between the asperities. This poorly understood process where asperities are somehow prevented from contacting each other is known as '**micro-elastohydrodynamic lubrication**' or '**micro-EHL**'.

Partial or Mixed EHL

In many instances of EHL, direct contact between the deformed asperities will still occur despite the presence of micro-EHL. If the lubricating film separating the surfaces is such that it allows some contact between the deformed asperities, then this type of lubrication is considered in the literature as '**mixed**' or '**partial lubrication**'. The contact load is shared between the contacting asperities and the film when mixed or partial lubrication prevails. The theory describing the mechanism of partial elastohydrodynamic lubrication was developed by Johnson, Greenwood and Poon [30]. It was found that during partial lubrication, the average surface separation between two rough surfaces is about the same as predicted for smooth surfaces. It has also been found that the average asperity pressure depends on the composite (RMS) surface roughness 'σ' and, since the mean separation of rough surfaces approximately equals the minimum film thickness 'h_0', the number of contacting asperities is also a function of 'λ', i.e., $λ = h_0/σ$ [11,30]. In the central part of the EHL film, the asperity pressure is nearly uniform. From the simple examination of the EHL film thickness equations (7.26) and (7.27) it is clear that the film thickness is almost independent of load. Thus the asperity pressure must also be load independent. With the increasing load, the contact area increases and consequently the number of asperity contacts increases [31]. The number of asperities deforming plastically depends on the plasticity index and the 'λ' parameter. According to the classical Greenwood-Williamson model the plasticity index is defined as [31]:

if $\left(\dfrac{E'}{H}\right) \times \left(\dfrac{σ^*}{r}\right)^{0.5} < 0.6$ elastic contact and

if $\left(\dfrac{E'}{H}\right) \times \left(\dfrac{σ^*}{r}\right)^{0.5} > 1$ plastic contact

where:

E'   is the composite Young's modulus [Pa]. Note that the composite Young's modulus differs from the reduced Young's modulus (eq. 7.6) by a factor of **2**, i.e.:

$$\frac{1}{E'} = \frac{1 - υ_A^2}{E_A} + \frac{1 - υ_B^2}{E_B} \tag{7.35}$$

H   is the hardness of the deforming surface [Pa];

σ*   is the standard deviation of the surface peak height distribution [m];

r   is the asperity radius, constant in this model [m].

In this theory, however, the effects of asperity interaction, which could affect the mechanism of lubrication by raising the entry temperature to the contact, were not considered [30]. More information on plasticity index can be found in Chapter 10.

The shape of the asperities, not just their size compared to the EHL film thickness, is believed to be important. In a model by Tallian and McCool [32] it was assumed that the shape of asperity peaks is prismatic with a rounded tip whereas Johnson et al. [30] assumed hemispherical shape of the peaks. It has been found that the 'sharp peaks', i.e., asperities with high slope or low radius sustained a higher proportion of the contact load than 'flat peaks', i.e., asperities with a low slope or large radius. An improved surface finish enables a diminished fraction of contact load supported by the asperities and the likelihood of a perfect elastohydrodynamic film is enhanced. When surfaces are polished to an extreme smoothness, however, a contrary trend to lowered load capacity is probable. It has often been observed in engineering practice that if the surface is too smooth, e.g., with a surface roughness of 0.001 [μm] R_a, then there is a risk of sudden seizure. In this instance it is commonly believed that small asperities play a useful role as a reservoir for the lubricant by entrapment between asperities. Under extremes of contact pressures the trapped lubricant can be expelled by asperity deformation to provide a final reserve of lubricating oil. The effect of surface roughness on partial EHL is illustrated in Figure 7.24.

Another characteristic feature of this lubrication regime is a progressive change in contact geometry and surface roughness because of wear occurring. The primary effect of lubrication is to alter the distribution of wear within the contact and create a wedge shaped film geometry [70]. It has been found that the plane of the wear scar on the ball slid against a steel disc is tilted relative to the plane of sliding. Under dry sliding conditions this 'tilt' of the wear scar does not occur, i.e., wear scar remains parallel to the worn surface. The 'tilt' is also not observed for very smooth surfaces. The amount of wear that produces this 'tilt' is limited to a depth which is approximately equal to the original surface roughness. At high levels of surface roughness, the EHL film sustains a reduction in minimum film thickness and 'tilt' formation is obscured or prevented by rapid wear of the contact [70].

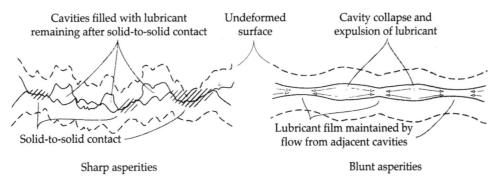

FIGURE 7.24   Effect of roughness and asperity shape on survival of EHL films.

Micro-Elastohydrodynamic Lubrication

Micro-EHL is a poorly understood lubrication concept that has only recently been invoked to explain the survival of heavily loaded concentrated contacts.

A lubrication mechanism acting when two surfaces have a relative separation of $\lambda \approx 1$ was proposed by Sayles et al. [33]. It was suggested that on the contacting surfaces there are features of the surface waviness which exhibit wavelengths of the same order but shorter than the contact width. On these features a much finer random surface texture of very much shorter wavelength is superimposed, as shown in Figure 7.25. These surface features may deform elastically under EHL pressures to conform with similar features on the other contacting surface. Surfaces with wavy features are found in most practical applications.

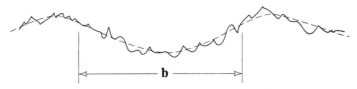

FIGURE 7.25   Surface texture of the contacting surfaces; **b** is the semiaxis of the contact ellipse in the direction of motion.

The size of an asperity or a protuberance from the surface has a strong influence on the load required for plastic deformation. In simple terms, when an asperity becomes smaller, its corresponding radius of curvature must also be reduced whatever the shape of the asperity. Applying the Hertzian theory of contact stresses, the load required to generate a constant stress in an asperity declines sharply with diminished radius of curvature. For a surface composed of asperities with a range of radii of curvature, a combined occurrence of limited elastic deformation and severe plastic deformation is possible. Returning to the concept of surface wavelength, an approximate proportionality between surface wavelength and radius of curvature can be assumed and this observation was developed further in the Sayles model as is described later.

The wavelengths smaller than the contact width constitute the surface features that can be deformed elastically, while the fine surface texture forms the smaller features which can be deformed plastically or even partially removed during metal-to-metal contact. It is thought that the elastic deformation of surface features forces them to conform to the opposing surface and allows full EHL or micro-EHL lubrication to prevail despite the low λ values, e.g., λ ≈ 1. Experiments conducted on an optical interferometry rig seem to confirm this theory [34]. Studies of contacts generated between a glass disc and a steel ball deliberately roughened by laser irradiation revealed that the elastic deformations taking place between the wave features, which were smaller then the width of the contact, seemed to play a major role in inhibiting the metal-to-metal contacts under EHL conditions. In such cases it is plausible that micro-EHL functions between these elastically deforming surface features. The directionality of roughness is also significant [35]. Micro-EHL is favoured by alignment of the grooves or 'lay' of a surface roughness normal to the direction of rolling or sliding since this arrangement creates a series of microscopic wedges. It was also found that lubricating oils containing additives in high concentrations, such as ZnDDP, tend to influence the measured EHL film thickness. ZnDDP, in particular, was found to increase film thickness at low rolling speeds where it is believed there is sufficient time between successive contacts for a protective film to form on the worn surface [71].

The possible mechanism of micro-EHL is shown in Figure 7.26 where the asperities are separated by transient squeeze films. The high viscosity of oil in an EHL contact would ensure that such squeeze forces are large enough to deform and flatten the asperities.

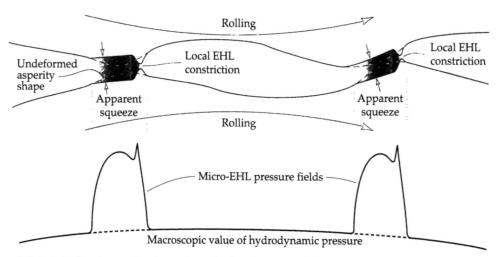

FIGURE 7.26 Mechanism of micro-elastohydrodynamic lubrication.

There are many different models of the micro-elastohydrodynamic lubrication regime [e.g., 36-40] in which the lubrication film between a single asperity and a smooth surface in rolling, sliding and even collision between the asperities is considered.

An example of such analysis is shown in the work of Houpert and Hamrock [41] where the problem of a single asperity (surface bump) of approximately half the Hertzian contact width, passing through a rolling-sliding line contact, was considered. The results of the numerical simulation are shown in Figure 7.27. It can be seen that under very high pressures, the shape of the bump and the pressure profile change. A large pressure spike is formed on the bump traversing the contact.

When the surface is covered with a series of small bumps and other imperfections there will be a number of corresponding pressure peaks superimposed on the smooth macroscopic pressure distribution as these surface features pass through the contact representing the micro-EHL pressure disturbances. The size of these pressure peaks depends on the asperity wavelength and height. Studies of the elastohydrodynamic lubrication of surfaces with such wavy features seem to indicate that such pressure ripples can indeed develop on a nominally smooth elastohydrodynamic pressure distribution [11,42]. If the local pressure variation is sufficiently large then elastic asperity deformation takes place and the micro-elastohydrodynamic lubricating films are generated, inhibiting contact between the asperities. The generated local pressures can significantly affect the stress distribution underneath the deforming asperities which can influence wear (i.e., contact fatigue). It was found that under practical loads the localized stress directly under the surface defect can often exceed the yield stress of the material [11,43].

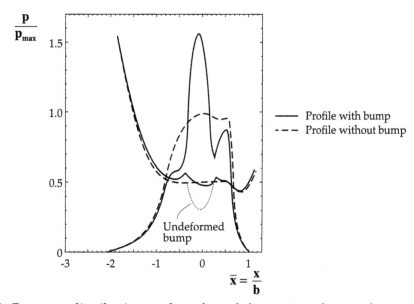

FIGURE 7.27   Pressure distribution and surface deformation of a single asperity passing through the EHL contact, where $W = 2.5 \times 10^{-5}$, $U = 1.3 \times 10^{-11}$, $G = 8000$, slide to roll ratio $U_2/U_1 - 1 = 10$, depth of the bump 1 [μm] and width of the bump 0.5 [μm] [41].

The phenomenon of micro-EHL is a very important research topic of many current and future studies. The development of an accurate model of micro-EHL is fundamental to tribology since it relates to the lubrication of real, rough surfaces.

7.6   SURFACE TEMPERATURE AT THE CONJUNCTION BETWEEN CONTACTING SOLIDS AND ITS EFFECT ON EHL

Surface temperature has a strong effect on EHL, as is the case with hydrodynamic lubrication. Elevated temperatures lower the lubricating oil viscosity and usually decrease the pressure-viscosity coefficient 'α'. A reduction in either of these parameters will reduce the EHL film thickness which may cause lubricant failure. Excessively high temperatures may also interfere with some auxiliary mechanisms of lubrication necessary for the stable functioning of partial EHL. Lubrication mechanisms auxiliary to partial EHL involve monomolecular films and are discussed in Chapter 8 on 'Boundary and Extreme Pressure Lubrication'. The maximum contact temperature is of particular engineering interest, especially in predicting problems associated with excessive surface temperatures which may lead to transitions in the lubrication mechanisms, changes in the wear rates through structural changes in the surface layers and the consequent failure of the machinery.

Calculation of Surface Conjunction Temperature

EHL is almost always found in concentrated contacts and in order to estimate the temperature rise during sliding contact, it is convenient to model the contact as a point or localized source of heat as a first approximation. In more detailed work, the variation of temperature within the contact is also considered, but this is essentially a refinement only. Since the intense release of frictional heat occurs over the small area of a concentrated contact, the resulting frictional temperatures within the contact are high, even when outside temperatures are close to ambient.

The temperature at the interface between contacting and mutually sliding solids is known as the '**surface conjunction temperature**'. It is possible to calculate this temperature by applying the laws of energy conservation and heat transfer. Most of the energy dissipated during the process of friction is converted into heat, resulting in a significant local surface temperature rise. For any specific part of the sliding surface, frictional temperature rises are of very short duration and the temperatures generated are called '**flash temperatures**'. From the engineering viewpoint it is important to know the expected values of these temperatures since they can severely affect not only EHL but also wear and dry friction through the formation of oxides, production of metallurgically transformed surface layers, alteration of local geometry caused by thermal expansion effects or even surface melting [44]. As well as the transient 'flash temperatures' there is also a steady state '**flash temperature rise**' at the sliding contact. When the contact is efficiently lubricated, the transient flash temperatures are relatively small and are superimposed on a large, steady-state temperature peak. In dry friction, or where lubrication failure is imminent, the transient flash temperatures may become larger than the steady-state component [45].

The flash temperature theory was originally formulated by Blok in 1937 [46] and developed further by Jaeger in 1944 [47] and Archard in 1958 [48]. The theory provides a set of formulae for the calculation of flash temperature for various velocity ranges and contact geometries.

According to Blok, Jaeger and Archard's theory, the flash temperature is the temperature rise above the temperature of the solids entering the contact which is called the '**bulk temperature**'. The maximum contact temperature has therefore two components: the bulk temperature of the contacting solids and the maximum flash temperature rise, i.e.:

$$T_c = T_b + T_{fmax} \qquad\qquad (7.36)$$

where:

T_c is the maximum surface contact temperature [°C];

T_b is the bulk temperature of the contacting solids before entering the contact [°C];

T_{fmax} is the maximum flash temperature [°C].

Evaluation of the flash temperature is basically a heat transfer problem where the frictional heat generated in the contact is modelled as a heat source moving over the surface [46,47]. The following simplifying assumptions are made for the analysis:

· thermal properties of the contacting bodies are independent of temperature,

· the single area of contact is regarded as a plane source of heat,

· frictional heat is uniformly generated at the area of the contact,

· all heat produced is conducted into the contacting solids,

· the coefficient of friction between the contacting solids is known and attains some steady value,

· a steady state condition (i.e., $\partial T / \partial t = 0$, the temperature is steady over time) is attained.

Some of these assumptions appear to be dubious. For example, the presence of the lubricant in the contact will affect the heat transfer characteristics. Although most of the heat produced will be conducted into the solids, a portion of it will be convected away by the lubricant, resulting in cooling of the surfaces. An accurate value of the coefficient of friction is very difficult, if not impossible, to obtain. The friction coefficient is dependent on the level of the heat generated as well as many other variables such as the nature of the contacting surfaces, the lubricant used and the lubrication mechanism acting. Even when an experimental measurement of the friction coefficient is available, in many cases the friction coefficient continually varies over a wide range. It is therefore necessary to calculate temperatures using a minimum and maximum value of friction coefficient. Temperatures at the beginning of sliding movement should also be considered since flash temperatures do not form instantaneously. Flash temperatures tend to stabilize within a very short sliding distance but the gradual accumulation of heat in the surrounding material and consequent slow rise in bulk temperature should not be overlooked.

Not withstanding these assumptions, the analysis gives temperature predictions which, although not very precise, are a good indication of the temperatures that might be expected between the operating surfaces.

As already mentioned, flash temperature calculations are based on the assumption that heat generated at the rate of:

$$q = Q/A$$

where:

Q   is the generated heat [W];

A   is the contact area [m²].

is conducted to the solids. The frictional heat generated is expressed in terms of the coefficient of friction, load and velocity, i.e.:

$$Q = \mu W |U_A - U_B|$$

where:

μ is the coefficient of friction;

W   is the normal load [N];

U_A is the surface velocity of the solid 'A' [m/s];

U_B is the surface velocity of the solid 'B' [m/s].

There is no single algebraic equation giving the flash temperature for the whole range of surface velocities. A non-dimensional measure of the speed at which the 'heat source' moves across the surface called the '**Peclet number**' has been introduced as a criterion allowing the differentiation between various speed regimes. The Peclet number is defined as [47]:

$$L = Ua/2\chi$$

where:

L   is the Peclet number;

U   is the velocity of a solid ('A' or 'B') [m/s];

a   is the contact dimension [m], (i.e., contact radius for circular contacts, half width of the contact square for square contacts and the half width of the rectangle for linear contacts);

χ is the thermal diffusivity [m²/s], i.e., $\chi = K/\rho\sigma$ where:

K   is the thermal conductivity [W/mK];

ρ is the density [kg/m³];

σ is the specific heat [J/kgK].

The Peclet number is an indicator of the heat penetration into the bulk of the contacting solid, i.e., it describes whether there is sufficient time for the surface temperature distribution of the contact to diffuse into the stationary solid. A higher Peclet number indicates a higher surface velocity for constant material characteristics.

Since all frictional heat is generated in the contact, the contact is modelled and treated as a heat source in the analysis. Flash temperature equations are derived, based on the assumption that the contact area moves with some velocity 'U' over the flat surface of a body 'B' as shown in Figure 7.28.

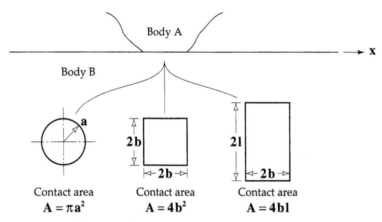

FIGURE 7.28   Geometry of the circular, square and linear contacts.

The heat transfer effects vary with the Peclet number as shown schematically in Figure 7.29. The following velocity ranges, defined by their Peclet number, are considered in flash temperature analysis:

     L < 0.1   one surface moves very slowly with respect to the other. There is enough time for the temperature distribution of the contact to be established in the stationary body. In this case, the situation closely approximates to steady state conduction [44],

     0.1 < L < 5   intermediate region. One surface moves faster with respect to the other and a slowly moving heat source model is assumed,

     L > 5   one surface moves fast with respect to the other and is modelled by a fast moving heat source. There is insufficient time for the temperature distribution of the contact to be established in the stationary body and the equations of linear heat diffusion normal to the surface apply [44]. The depth to which the heat penetrates into the stationary body is very small compared to the contact dimensions.

Flash temperature equations are given in terms of the heat supply over the contact area, the velocity and the thermal properties of the material. They are derived based on the assumption that the proportion of the total heat flowing into each contacting body is such that the average temperature over the contact area is the same for both bodies. The flash temperature equations were developed by Blok and Jaeger for linear and square contacts [46,47] and by Archard for circular contacts [48].

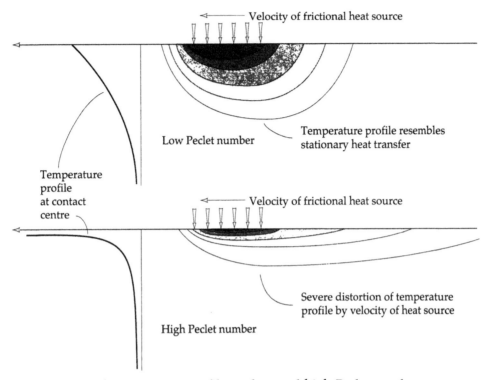

FIGURE 7.29   Frictional temperature profiles at low and high Peclet numbers.

· *Flash Temperature in Circular Contacts*

In developing the flash temperature formulae for a circular contact, it was assumed that the portion of the surface in contact is of height approximately equal to the contact radius 'a'. The temperature at the distance 'a' from the surface is considered as a bulk temperature 'T_b' of the body. This can be visualised as a cylinder of height equal to its radius with one end in contact and the other end maintained at the bulk temperature of the body. The geometry of the contact is shown in Figure 7.28.

Average and maximum flash temperature formulae for circular contacts and various velocity ranges are summarized in Table 7.5. The average flash temperature corresponds to the steady-state component of flash temperature, while the maximum value includes the transient component. The maximum flash temperature occurs when the maximum load is concentrated at the smallest possible area, i.e., when the load is carried by a plastically deformed contact [48].

· *Flash Temperature in Square Contacts*

Flash temperature equations for square contacts have been developed by Jaeger [47]. Although square contacts are rather artificial the formulae might be of use in some applications. The geometry of the contact is shown in Figure 7.28. The formulae for various velocity ranges are summarized in Table 7.6. Constants 'C_1' and 'C_2' required in flash temperature calculations for the intermediate velocity range are determined from the chart shown in Figure 7.30 [47].

TABLE 7.5   Average and maximum flash temperature formulae for circular contacts.

Peclet number	Average flash temperature T_{f_a}	Maximum flash temperature $T_{f_{max}}$
L < 0.1	$T_{f_a} = 0.5 NL = \dfrac{\pi}{4} \dfrac{qa}{K}$ or $T_{f_a} = 0.25 \dfrac{\mu W \lvert U_A - U_B \rvert}{Ka}$	$T_{f_{max}} = 0.25 N'L'$ or $T_{f_{max}} = 0.222 \dfrac{\mu U}{K} \left(p_y W \right)^{0.5}$
0.1 < L < 5	$T_{f_a} = 0.5\alpha NL = \alpha \dfrac{\pi}{4} \dfrac{qa}{K}$ or $T_{f_a} = 0.25\alpha \dfrac{\mu W \lvert U_A - U_B \rvert}{Ka}$ α ranges from **0.85** at **L = 0.1** to **0.35** at **L = 5**	$T_{f_{max}} = 0.25\beta N'L'$ or $T_{f_{max}} = 0.222\beta \dfrac{\mu U}{K} \left(p_y W \right)^{0.5}$ β ranges from **0.95** at **L = 0.1** to **0.50** at **L = 5**
L > 5	$T_{f_a} = 0.435 NL^{0.5} = \dfrac{\pi}{3.251} \dfrac{q}{K} \left(\dfrac{\chi a}{U} \right)^{0.5}$ or $T_{f_a} = 0.308 \dfrac{\mu W \lvert U_A - U_B \rvert}{Ka} \left(\dfrac{\chi}{Ua} \right)^{0.5}$	$T_{f_{max}} = 0.435 \gamma N'L'^{0.5}$ or $T_{f_{max}} = 0.726 \gamma \mu p_y \left(\dfrac{U}{K \rho \sigma} \sqrt{\dfrac{W}{p_y}} \right)^{0.5}$ or in general $T_{f_{max}} = 1.64 T_{f_a}$ γ ranges from **0.72** at **L = 5** to **0.92** at **L = 100**. For **L > 100**, $\gamma = 1$

where:

T_{fa}	is the average flash temperature [°C];
T_{fmax}	is the maximum flash temperature [°C];
μ	is the coefficient of friction;
W	is the normal load [N];
p_y	is the flow or yield stress of the material [Pa];
U_A, U_B	are the surface velocities of solid 'A' and solid 'B', respectively [m/s];
U	is the velocity of solid 'A' or 'B';
a	is the radius of the contact circle [m] (Figure 7.28);
χ	is the thermal diffusivity, $\chi = K/\rho\sigma$, [m²/s];
K	is the thermal conductivity [W/mK];
ρ	is the density [kg/m³];
σ	is the specific heat [J/kgK];
α, β, γ	are coefficients;
L	is the Peclet number; $L = Ua/2\chi = Ua\rho\sigma/2K$;

N   is the variable [°C], defined as:

$$N = \pi q / \rho \sigma U$$

where:

$$q = Q/\pi a^2 = \mu W |U_A - U_B| / \pi a^2$$

is the rate of heat supply per unit area (circular) [W/m²];

L'   is the variable defined as:

$$L' = \frac{U}{2\chi}\left(\frac{W}{\pi p_y}\right)^{0.5}$$

N'   is the variable [°C], defined as:

$$N' = \pi \mu p_y / \rho \sigma$$

TABLE 7.6   Average and maximum flash temperature formulae for square contacts.

Peclet number	Average flash temperature T_{f_a}	Maximum flash temperature $T_{f_{max}}$
L < 0.1	$T_{f_a} = 0.946\dfrac{qb}{K}$ or $T_{f_a} = 0.237\dfrac{\mu W |U_A - U_B|}{Kb}$	$T_{f_{max}} = 1.122\dfrac{qb}{K}$ or $T_{f_{max}} = 0.281\dfrac{\mu W |U_A - U_B|}{Kb}$
0.1 < L < 5	$T_{f_a} = C_1\dfrac{2}{\pi}\dfrac{\chi q}{KU}$ or $T_{f_a} = 0.159 C_1\dfrac{\mu W |U_A - U_B|}{Kb}\left(\dfrac{\chi}{Ub}\right)$ C_1 from Figure 7.30.	$T_{f_{max}} = C_2\dfrac{2}{\pi}\dfrac{\chi q}{KU}$ or $T_{f_{max}} = 0.159 C_2\dfrac{\mu W |U_A - U_B|}{Kb}\left(\dfrac{\chi}{Ub}\right)$ C_2 from Figure 7.30.
L > 5	$T_{f_a} = 1.064\dfrac{q}{K}\left(\dfrac{\chi b}{U}\right)^{0.5}$ or $T_{f_a} = 0.266\dfrac{\mu W |U_A - U_B|}{Kb}\left(\dfrac{\chi}{Ub}\right)^{0.5}$	$T_{f_{max}} = \dfrac{2q}{K}\left(\dfrac{2\chi b}{\pi U}\right)^{0.5}$ or $T_{f_{max}} = 0.399\dfrac{\mu W |U_A - U_B|}{Kb}\left(\dfrac{\chi}{Ub}\right)^{0.5}$

where:

b   is the half width of the contact square [m] (Figure 7.28);

L   is the Peclet number; $L = Ub/2\chi$;

q   is the rate of heat supply per unit area (square) [W/m²];

$$q = Q/4b^2 = \mu W |U_A - U_B| / 4b^2$$

The other variables are as already defined.

· *Flash Temperature in Line Contacts*

Flash temperature formulae for line contacts for various velocity ranges are summarized in Table 7.7 [47]. They are applicable in many practical cases such as gears, roller bearings, cutting tools, etc. The contact geometry is shown in Figure 7.28 and constants 'C_3' and 'C_4' required in flash temperature calculations for the intermediate velocity range are determined from the chart shown in Figure 7.30 [47].

TABLE 7.7   Average and maximum flash temperature formulae for line contacts.

Peclet number	Average flash temperature T_{f_a}	Maximum flash temperature $T_{f_{max}}$
$L < 0.1$	$T_{f_a} = \dfrac{4\chi q}{\pi KU}\left(-2.303L\log_{10}2L + 1.616L\right)$ or $T_{f_a} = 0.318\dfrac{\mu W\,\lvert U_A - U_B\rvert}{Kl}\left(\dfrac{\chi}{Ub}\right)$ $\times\left(-2.303L\log_{10}2L + 1.616L\right)$	$T_{f_{max}} = \dfrac{4\chi q}{\pi KU}\left(-2.303L\log_{10}L + 1.116L\right)$ or $T_{f_{max}} = 0.318\dfrac{\mu W\,\lvert U_A - U_B\rvert}{Kl}\left(\dfrac{\chi}{Ub}\right)$ $\times\left(-2.303L\log_{10}L + 1.116L\right)$
$0.1 < L < 5$	$T_{f_a} = C_3\dfrac{2}{\pi}\dfrac{\chi q}{KU}$ or $T_{f_a} = 0.159\,C_3\dfrac{\mu W\,\lvert U_A - U_B\rvert}{Kl}\left(\dfrac{\chi}{Ub}\right)$ C_3 from Figure 7.30.	$T_{f_{max}} = C_4\dfrac{2}{\pi}\dfrac{\chi q}{KU}$ or $T_{f_{max}} = 0.159\,C_4\dfrac{\mu W\,\lvert U_A - U_B\rvert}{Kl}\left(\dfrac{\chi}{Ub}\right)$ C_4 from Figure 7.30.
$L > 5$	$T_{f_a} = 1.064\dfrac{q}{K}\left(\dfrac{\chi b}{U}\right)^{0.5}$ or $T_{f_a} = 0.266\dfrac{\mu W\,\lvert U_A - U_B\rvert}{Kl}\left(\dfrac{\chi}{Ub}\right)^{0.5}$	$T_{f_{max}} = \dfrac{2q}{K}\left(\dfrac{2\chi b}{\pi U}\right)^{0.5}$ or $T_{f_{max}} = 0.399\dfrac{\mu W\,\lvert U_A - U_B\rvert}{Kl}\left(\dfrac{\chi}{Ub}\right)^{0.5}$

where:

b is the half width of the contact rectangle [m] (Figure 7.28);

l is the half length of the contact rectangle [m] (Figure 7.28);

L is the Peclet number; $L = Ub/2\chi$;

q is the rate of heat supply per unit area (rectangle) [W/m²];

$q = Q/4bl = \mu W\lvert U_A - U_B\rvert/4bl$

The other variables are as already defined.

It can be seen from Tables 7.5 and 7.6 that the average flash temperature equations for circular and square contacts are identical apart from a small difference in the proportionality constant. The shape of the contact, with the exception of elongated contacts, has a small effect on flash temperature and the average flash temperature formulae for square sources can be used for most irregular shapes of sources [47].

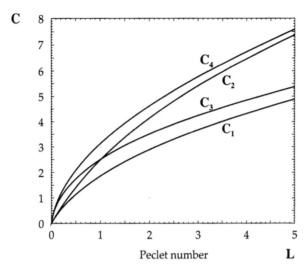

FIGURE 7.30 Diagram for evaluation of constants 'C' required in flash temperature calculations for the intermediate velocity range [47].

True Flash Temperature Rise

The heat generated in frictional contacts is divided between the contacting solids. The proportion of the total heat flowing to each body is determined on the basis that the average surface temperature is the same for both bodies [48]. A simple way of estimating the true temperature rise in the contact is to initially assume that all the heat generated is supplied to body 'A'. The appropriate flash temperature equation for a given speed and contact geometry conditions is then selected and a flash temperature 'T_{fA}' calculated. The next step in the calculation procedure is to assume that all the heat generated is transferred to body 'B'. The appropriate flash temperature equation for this second model is then selected and the corresponding flash temperature 'T_{fB}' calculated. It should be noted that for each body, the flash temperature equations adequate for their speed conditions must be selected. For example, both surfaces of meshing gears move with high velocity, thus equations for Peclet number **L > 5** apply to both bodies 'A' and 'B'. On the other hand, in some applications where only one of the surfaces moves fast and the other moves very slowly, the equations for **L > 5** and **L < 0.1** should be applied consecutively. In such cases the flash temperatures are calculated assuming that initially all frictional energy is conducted to the moving surface and then that all frictional energy is conducted to the stationary surface. The true flash temperature rise must be the same for both solids in contact and is calculated from:

$$1/T_f = 1/T_{fA} + 1/T_{fB} \tag{7.37}$$

For example, for two fast moving surfaces in line contact, the maximum temperature rise of the conjunction 'T_{fmaxc}' is given by the expression:

$$1/T_{fmaxc} = 1/T_{fmaxA} + 1/T_{fmaxB} \tag{7.38}$$

Substituting the expressions for maximum flash temperature from Table 7.7 for **L > 5**, the commonly used equations for maximum temperature rise in line contacts are obtained, i.e.:

$$T_{fmaxc} = \frac{1.11\mu W |U_A - U_B|}{(2l)(2b)^{0.5}[(K\rho\sigma U)_A^{0.5} + (K\rho\sigma U)_B^{0.5}]} \tag{7.39}$$

If the contacting solids are of the same material then their thermal constants are also the same and the above equation can be written as:

$$T_{f_{maxc}} = \frac{1.11\mu W \left|U_A^{0.5} - U_B^{0.5}\right|}{(2l)(2b)^{0.5}(K\rho\sigma)^{0.5}} \tag{7.40}$$

or in terms of the reduced radius of curvature 'R″' and Young's modulus 'E″' as:

$$T_{f_{maxc}} = \frac{0.62\mu \left|U_A^{0.5} - U_B^{0.5}\right|}{(K\rho\sigma)^{0.5}} \left(\frac{W}{2l}\right)^{0.75}\left(\frac{E'}{R'}\right)^{0.25} \tag{7.41}$$

where:

E' is the reduced Young's modulus [Pa]. For two solids of the same material:

$$E' = E_A/(1 - \upsilon_A{}^2) = E_B/(1 - \upsilon_B{}^2)$$

R' is the reduced radius of curvature of the undeformed surfaces [m].

or in terms of maximum contact pressure as:

$$T_{f_{maxc}} = \frac{2.45\mu p_{max}{}^{1.5}\left|U_A^{0.5} - U_B^{0.5}\right|}{(K\rho\sigma)^{0.5}} \left(\frac{R'}{E'}\right)^{0.5} \tag{7.42}$$

where:

p_{max} is the maximum contact pressure [Pa].

Although the procedure outlined does not always provide precise values of the temperature distribution over the entire interface between contacting solids, it greatly facilitates the physical interpretation of frictional temperatures. For example, consider a pin-on-disc machine with the pin and the disc manufactured from the same material. Since the pin is stationary, low speed conditions of heat transfer apply, whereas high speed conditions apply to the disc. The interfacial temperature of the disc will be very much lower than that of the pin since the disc is constantly presenting fresh cool material to the interface. Hence the temperature distribution at the interface will be mostly determined by the heat flow equations in the disc [44,49]. Another example of this effect which is more closely related to EHL is the difference in frictional temperatures between a large and a small gear wheel when meshed together.

The maximum flash temperature rise is located towards the trailing region of the contact and its location depends on the Peclet number as shown in Figure 7.31 [47].

The maximum flash temperature distribution for high speed conditions in circular contacts is shown in Figure 7.32 [44]. It can be seen that the maximum temperature is about $T_{fmax} = 1.64T_{fa}$ and occurs at the centre of the trailing edge of the contact.

It should be noted that the heat source considered in the analysis was treated as uniform, i.e., the frictional energy generated is uniformly distributed over the contact area. It has been found that for the non-uniform heat sources arising from the Hertzian pressure distribution, the value of 'q_{max}' is almost unaffected by the non-uniform distribution of 'q' [44,49]. However, it was found that the maximum temperature for a circular contact is increased by 16% compared to the uniform heat source, and its location is moved inward from the trailing edge [44].

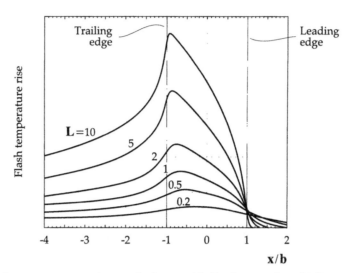

FIGURE 7.31   Flash temperature rise variations with Peclet number [47].

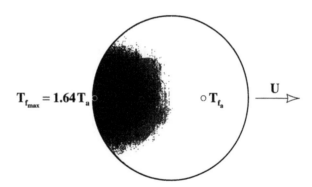

FIGURE 7.32   Flash temperature distribution in circular contacts for the high speed condition (adapted from [44]).

As already mentioned the calculated flash temperature rise must be added to the bulk temperature to obtain the temperature of the conjunction, i.e.:

$$\mathbf{T_c} = \mathbf{T_b} + \mathbf{T_{fmax}}$$

The above equation is valid if the bulk temperatures of the two solids entering the contact are the same. If the bulk temperatures of the contacting solids differ then a new adjusted value of bulk temperature must be calculated and substituted into the above equation. The adjusted value of bulk temperature is evaluated from the formula [51]:

$$\mathbf{T_{b_{new}}} = \frac{1}{2}\left(\mathbf{T_{b_A}} + \mathbf{T_{b_B}}\right) + \frac{1}{2}\left(\frac{\mathbf{n-1}}{\mathbf{n+1}}\right)\left(\mathbf{T_{b_A}} - \mathbf{T_{b_B}}\right) \tag{7.43}$$

where:

n   is a constant calculated from:

$$\left(\frac{\mathbf{U_A\rho_A\sigma_AK_A}}{\mathbf{U_B\rho_B\sigma_BK_B}}\right)^{0.5}$$

T_{bnew} is the new adjusted bulk temperature [°C];

T_{bA} is the bulk temperature of body A [°C];

T_{bB} is the bulk temperature of body B [°C].

The other variables are as already defined.

For $0.2 \leq n \leq 5$ the average bulk temperature can be calculated with sufficient accuracy from:

$$T_{bnew} = 0.5 \times (T_{bA} + T_{bB}) \qquad (7.44)$$

Flash temperature under the surface drops rapidly with depth. For example, for **L = 2** at a depth of **15** [μm] below the surface, the temperature drops to about **0.1** of its initial value at the rear end of the source. Thus any thermal stresses or thermally induced microstructure transformations caused by flash temperature fluctuations are very superficial [47].

EXAMPLE

Find the maximum flash temperature rise for two steel rollers of radii $R_A = 10 \times 10^{-3}$ [m] and $R_B = 15 \times 10^{-3}$ [m], working under a load of 5 [kN] and rotating at different speeds of $U_A = 2$ [m/s] and $U_B = 1$ [m/s]. Assume that the thermal and material properties of the rollers are the same and are: Young's modulus $E = 2.1 \times 10^{11}$ [Pa], Poisson's ratio $\upsilon = 0.3$, thermal conductivity K = 46.7 [W/mK], density $\rho = 7800$ [kg/m³] and specific heat $\sigma = 460$ [J/kgK]. The length of both rollers is 2l = 10×10^{-3} [m]. The bulk temperature of the rollers is $T_{bA} = T_{bB} = 80°C$ and the coefficient of friction is $\mu = 0.2$.

· *Reduced Radius of Curvature*

Since the dimensions of the rollers are the same as in the example already considered the reduced radius of curvature is:

$$\boxed{R' = 6 \times 10^{-3} \, [m]}$$

· *Reduced Young's Modulus*

$$\boxed{E' = 2.308 \times 10^{11} \, [Pa]}$$

· *Half Width of the Contact Rectangle (Table 7.2)*

$$b = \left(\frac{4WR'}{\pi l E'}\right)^{0.5} = \left(\frac{4 \times (5 \times 10^3) \times (6 \times 10^{-3})}{\pi \times (5 \times 10^{-3}) \times (2.308 \times 10^{11})}\right)^{0.5} \qquad \boxed{= 1.82 \times 10^{-4} \, [m]}$$

· *Thermal Diffusivity*

$$\chi = \frac{K}{\rho\sigma} = \frac{46.7}{7800 \times 460} \qquad \boxed{= 13.02 \times 10^{-6} \, [m^2/s]}$$

· *Peclet Number*

$$L_A = \frac{U_A b}{2\chi} = \frac{2 \times (1.82 \times 10^{-4})}{2 \times (13.02 \times 10^{-6})} \qquad \boxed{= 13.98}$$

$$L_B = \frac{U_B b}{2\chi} = \frac{1 \times (1.82 \times 10^{-4})}{2 \times (13.02 \times 10^{-6})} \qquad \boxed{= 6.99}$$

Note that L_A and $L_B > 5$.

· *True Maximum Flash Temperature of the Conjunction*

$$T_{f_{maxc}} = \frac{1.11\mu W |U_A^{0.5} - U_B^{0.5}|}{(2l)(2b)^{0.5}(K\rho\sigma)^{0.5}}$$

$$= \frac{1.11 \times 0.2 \times (5 \times 10^3) |\sqrt{2} - \sqrt{1}|}{(10 \times 10^{-3}) \times (2 \times 1.82 \times 10^{-4})^{0.5}(46.7 \times 7800 \times 460)^{0.5}} \boxed{= 186.2\ [°C]}$$

This value is added to the bulk temperature of the rollers in order to obtain the true maximum temperature of the conjunction, i.e.:

$$T_c = T_b + T_{fmaxc} = 80° + 186.2° \qquad \boxed{= 266.2\ [°C]}$$

The same result is obtained when applying the two other flash temperature formulae 7.41 and 7.42. For example, calculating flash temperature in terms of the maximum contact pressure equation (7.42) yields:

$$p_{max} = \frac{W}{\pi b l} = \frac{5 \times 10^3}{\pi (1.82 \times 10^{-4}) \times (5 \times 10^{-3})} \qquad \boxed{= 1748.96\ [MPa]}$$

hence:

$$T_{f_{maxc}} = \frac{2.45 \times 0.2 \times (1748.96 \times 10^6)^{1.5} |\sqrt{2} - \sqrt{1}|}{(46.7 \times 7800 \times 460)^{0.5}} \left(\frac{6 \times 10^{-3}}{2.308 \times 10^{11}}\right)^{0.5}$$

$$\boxed{= 184.91\ [°C]}$$

Frictional Temperature Rise of Lubricated Contacts

All surface lubricating films can affect the surface temperatures through changes in the coefficient of friction. The temperature of the substrate is usually not affected by these films as long as the presence of a film does not affect friction. However, when a lubricating film is relatively thick and the lubricant has sufficiently low thermal conductivity the conjunction temperature can be significantly altered by the presence of a lubricating film.

In elastohydrodynamic contacts the surfaces are separated by thin, low thermal conductivity films. The oil viscosity in these contacts varies with pressure from a low value at the entry side to a maximum at the centre to a low value at the exit. Consequently the force needed to shear the lubricating film will also vary along the contact. Thus in the middle of the lamellar elastohydrodynamic film heat is generated at a greater rate since the viscosity attains its highest value. In this region the rates of heat generation are proportional to viscosity [48] and are highest at the centre of the film. Variations in rates of heat generation will obviously affect the temperature distribution on the surfaces but due to a slow temperature response to these variations this effect will be small. The main effect, however, is in the increase of the maximum temperature in the oil film and this maximum temperature has the tendency to move towards the centre of the parallel film [48]. Experimental measurements of temperature within an EHL film by infra-red spectroscopy have confirmed these theoretical predictions of a temperature maximum at the centre of the EHL film [52,53]. Since the middle of the parallel elastohydrodynamic film is where heat is generated and dissipated at the greatest rate, then the heat distribution between two solids depends on their thermal properties and the EHL film thickness. In effect, two surfaces can have different temperatures as long as they are separated by a film. Their temperatures will be the same if the film disappears and the separation ceases to exist. The temperature profiles of a contact lubricated by an EHL film and a dry sliding contact are shown schematically in Figure 7.33.

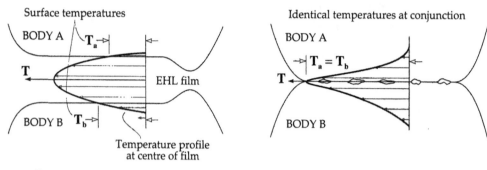

FIGURE 7.33 Temperature profiles in an EHL contact and in a dry contact.

The difference in temperature between the centre of the lubricant film and the surfaces can be as large as 60°C or even higher despite the extreme thinness of an EHL film which is usually only 0.1 to 1 [μm] thick. The high temperatures occurring in the EHL contacts can explain why the EHL film, once formed, can fail, resulting, for example, in scuffing in gears. The temperatures within the EHL conjunction may be high enough for the lubricant to decompose and cause lubricating film failure. It is possible to discriminate between the temperature of the lubricating oil and the surface temperatures of a contacting solid in an EHL contact by the difference in emissivity between the oil and the metal or sapphire surfaces. An example of the temperature fields of the contacting surface and the peak oil temperature [52] is shown in Figure 7.34. The measurements were taken at the maximum contact stress of 1.05 [GPa], sliding speed 1.4 [m/s] and a naphthenic mineral oil was used as the lubricant.

The most important difference between the temperature field of the surface of the contacting solid (a steel ball) and the lubricant oil temperature field is the much greater variability of the latter. The surface temperature varies at a near uniform gradient with position from a temperature maximum close to the exit restriction of the EHL contact. In contrast, the lubricant temperature field reveals a number of peaks or 'hot spots' distributed along the exit constriction of the EHL contact. The high lubricant temperature may be the result of intense viscous heat generation in the oil which has a much lower heat capacity/unit volume than steel. Under these experimental conditions the EHL film is on the point of sustaining thermally induced collapse or 'burn out' and this mode of failure may be the fundamental limit to all EHL films.

Even when the EHL film is not subjected to high levels of sliding and load, thermal effects can still be significant. At high rolling speeds, viscous shear heating at the inlet will cause the lubricant to enter the EHL contact with reduced viscosity and a corresponding lowered film thickness [54]. The relation between film thickness and rolling speed as predicted by the Dowson-Higginson formulae (7.26 and 7.27) is an overestimate of the film thickness at high rolling speeds. Errors in the Dowson-Higginson formulae become significant at film thicknesses higher than 1 [μm] and rolling speeds greater than 10 [m/s] [54]. However, a maximum or limiting film thickness was not observed even for the highest of rolling speeds.

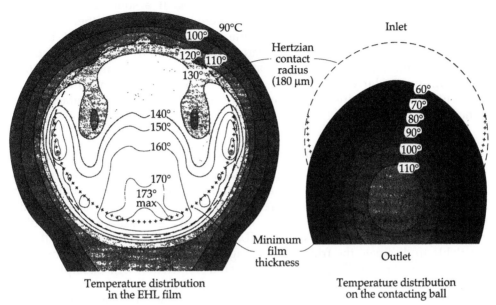

FIGURE 7.34 Experimental measurements of surface temperature and oil temperature in an EHL contact (adapted from [52]).

Mechanism of Heat Transfer within the EHL Film

Frictional heat is transmitted through lubricating oil film mainly by convection and conduction [55]. Other forms of heat transfer such as by bubble nucleation during boiling may occur in EHL contacts operating at a high sliding speed and load but these mechanisms of heat transfer have not yet been observed. As mentioned in Chapter 4, the balance between convection and conduction depends largely on the lubricating oil film thickness. An example of an approximate calculation of the relative importance of convection and conduction is shown below.

EXAMPLE

Find the ratio of convected to conducted heat in a rolling contact bearing operating with a surface velocity of U = 15.71 [m/s]. The contact width between the inner race and a roller is B = 0.0001 [m] and the film thickness is h = 0.5 [μm]. The bearing is lubricated by mineral oil of thermal diffusivity $\chi = 0.084 \times 10^{-6}$ [m²/s]. Substituting these values into equation (4.126) yields:

$$\frac{H_{cond}}{H_{conv}} = 0.084 \times 10^{-6} \frac{2 \times 0.0001}{15.71(0.5 \times 10^{-6})^2} = 4.278$$

This result implies that conduction is a far more significant mechanism than convection for thin EHL films as opposed to the much thicker lubricant films found in journal or pad bearings. Some reductions in EHL film temperatures may therefore be achieved by supplying lubricants with conductivities much greater than conventional oils.

Effect of Surface Films on Conjunction Temperatures

Surface films of solids such as metal oxides can also affect the surface temperature to a varying degree, depending on their thermal properties. If a solid layer is a good heat conductor then the surface temperature will be lowered, whereas if its thermal conductivity is low relative to the bulk material then the surface temperature will be increased. Significant modification of frictional temperature rises are, however, only found when the thickness of solid film material is much greater than molecular dimensions [47,48]. Films of solid material, particularly oxides, are also instrumental in controlling the friction coefficient, but this aspect of frictional temperatures in dry or lubricated contacts is discussed in later chapters.

Measurements of Surface Temperature in the EHL Contacts

The temperature in EHL contacts has recently been measured by infra-red spectroscopy. One of the contacting surfaces was made of material transparent to infra-red radiation and hard enough to sustain high Hertzian contact stresses. Sapphire or diamond has been used for this purpose [56,57] in an arrangement shown schematically in Figure 7.35. The high shear stresses and shear rates prevailing in an EHL contact and the relatively small volume of liquid available to dissipate the frictional heating ensure that high contact temperatures are reached even when very little power is dissipated by friction. An example of a surface temperature rise above bulk inlet temperature, measured by infra-red spectroscopy through the centre of a sphere-on-plane contact, is shown in Figure 7.36 [56]. The lubricant used in the experiments was a synthetic perfluoroether tested at sliding speeds ranging from 0.5 to 2 [m/s] and a maximum contact pressure of 1.05 [GPa] [56].

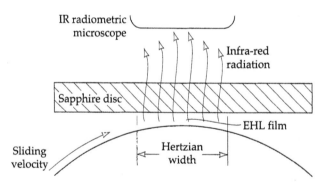

FIGURE 7.35 Schematic diagram of the apparatus for the determination of surface temperature profile by infra-red spectroscopy [56].

A temperature rise approaching 80°C at the centre of the contact was found even under the relatively mild conditions of the test. However, the exact mechanism by which contact temperatures can prevent effective EHL is still poorly understood [56-58].

A fundamental limitation of the infra-red spectroscopic measurement of surface temperature is the need for a special transparent window to function as one of the contacting surfaces. This requirement precludes common engineering materials such as steel from being used for both contacting surfaces. A conventional thermocouple embedded in the contacting material is unsuitable for the measurements because the temperature rise is confined to the surface. The only way that a thermocouple can be used with accuracy is if a lamellar thermocouple is attached to the surface. A lamellar thermocouple is made by depositing on the surface successive thin films approximately 0.1 [μm] thick of insulants and two metals. The specialized form of thermocouple required to measure flash temperatures is illustrated schematically in Figure 7.37.

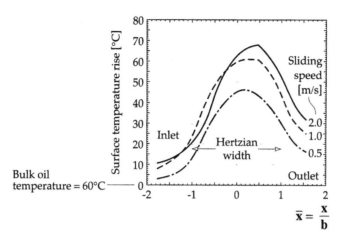

FIGURE 7.36 Surface temperature profiles within an EHL contact determined by infra-red spectroscopic measurements [56].

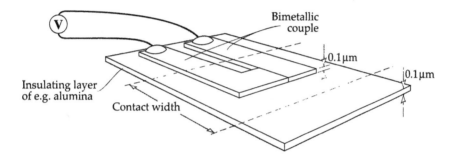

FIGURE 7.37 Lamellar thermocouple suitable for the measurement of flash temperature.

The lamellar thermocouple requires elaborate coating equipment for its manufacture and is not very durable against wear. The measurement of surface temperature can be a very difficult experimental task and for most studies it is more appropriate to estimate it by calculating the range of surface temperatures that may be found in the particular EHL contact.

7.7   TRACTION AND EHL

Traction is the application of frictional forces to allow the transmission of mechanical energy rather than its dissipation. The most common example of the distinction between traction and friction is the contact between a wheel and a road. When the wheel rolls without skidding, traction is obtained and the frictional forces available enable propulsion of a vehicle. When skidding occurs, the same frictional forces will now dissipate any mechanical energy applied to the wheel. Thus the difference between traction and friction is in the way that the mechanical energy is processed, e.g., in the case of traction this energy is transmitted between the contacting bodies (i.e., one body is driving another) whereas with friction it is dissipated. Traction can also be applied to lubricated contacts despite the relatively low coefficients of friction involved. EHL contacts can provide sufficiently high traction to be used as interfaces for variable speed transmissions. Unique features of variable speed transmissions such as infinitely variable output speed, almost a constant torque over the speed range and low noise make them particularly attractive for applications in computers, machine tools and the textile industry, or even in motor vehicles. A lubricated contact is selected to suppress wear which would otherwise shorten the lifetime of the transmission. The operating principles of these transmissions are shown schematically in Figure 7.38.

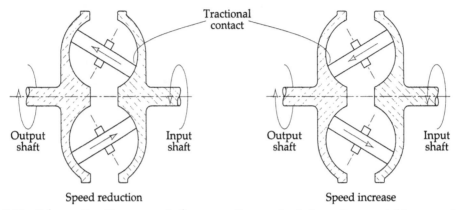

FIGURE 7.38 Schematic diagram of the operating principles of a variable speed toroidal transmission.

The level of traction in an EHL contact also accelerates wear of the corresponding rolling/sliding elements and traction should therefore be controlled wherever possible. This topic is discussed further in later chapters.

It is possible to analyze the traction force in an EHL contact and obtain a good agreement between theory and experiment although the models involved are fairly complex [55,59]. A basic simplification used in most analyses of traction is to assume a uniform film thickness inside the EHL contact and ignore the end constriction. This is a film geometry similar to Grubin's original model of EHL. When traction is applied, there is a small but non-zero sliding speed between the contacting surfaces. This non-zero sliding speed is inevitable since all the tractional force in an EHL contact is the result of viscous shear. The envisaged simplified film geometry and velocity profiles of the sheared lubricant are shown in Figure 7.39.

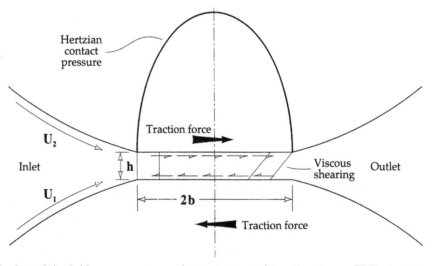

FIGURE 7.39 Simplified film geometry and generation of traction in an EHL contact.

From an elementary analysis of the relationship between shear rate and shear stress in a Newtonian fluid, discussed in Chapter 2, it can be seen that the traction force is a product of contact area, local viscosity and velocity difference between the surfaces divided by film thickness, i.e.:

$$\mathbf{F} = \eta \times \mathbf{A} \times \Delta \mathbf{U/h} \tag{7.45}$$

For the purposes of argument it is assumed either that the local viscosity remains constant in the EHL contact or that its average value can be found. In more refined analyses, however, the local variation of viscosity is included. Under conditions of constant load, geometry and lubricant characteristics, the contact area, local viscosity and film thickness remain almost invariant. A '**coefficient of traction**' is obtained by dividing the traction force by load, i.e.:

$$\mu_T = \mathbf{F/W} \tag{7.46}$$

where:

μ_T is the traction coefficient;

$\mathbf{F}$ is the traction force [N];

$\mathbf{W}$ is the contact load [N].

Substituting for traction force (7.45) yields:

$$\mu_T = \eta \mathbf{A} \Delta \mathbf{U} \,/\, \mathbf{hW} = \kappa \times \Delta \mathbf{U} \tag{7.47}$$

where:

η is the dynamic viscosity of the lubricant [Pas];

$\mathbf{A}$ is the contact area [m^2];

$\Delta \mathbf{U}$ is the surface velocity (i.e., velocity difference between the contacting surfaces) [m/s];

$\mathbf{h}$ is the film thickness [m];

κ is constant defined as: $\kappa = \eta \mathbf{A} \,/\, \mathbf{hW}$ [s/m].

The velocity difference is often normalized as a coefficient which is obtained by division with the larger velocity. This coefficient is known in the literature as the '**slide to roll ratio**' and is defined as:

$$\Delta \mathbf{U/U} = (\mathbf{U_A} - \mathbf{U_B})/\mathbf{U_A}$$

where:

$\Delta \mathbf{U}$ is the velocity difference [m];

$\mathbf{U_A}, \mathbf{U_B}$ are the surface velocities of body '$\mathbf{A}$' and '$\mathbf{B}$', respectively [m].

The relationship between the traction coefficient and the slide to roll ratio then is:

$$\mu_T = \kappa' \times \Delta \mathbf{U/U_A}$$

where:

κ' is the coefficient defined as: $\kappa' = \kappa \mathbf{U_A}$.

The velocity difference between the contacting surfaces is usually extremely small and for EHL traction systems a single 'velocity' value is often given in the literature instead of accurate values of '$\mathbf{U_A}$' and '$\mathbf{U_B}$'.

This extremely simple and approximate analysis predicts a straight proportionality between traction coefficient and slide to roll ratio and this is verified experimentally for low levels of slide to roll ratio. At higher slide to roll ratios this proportionality is lost and an entirely different pattern is observed. A schematic representation of the relationship between traction coefficient and slide to roll ratio is shown in Figure 7.40.

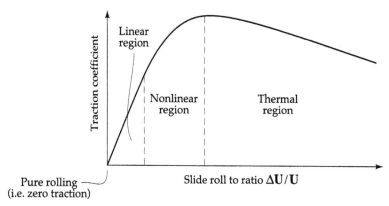

FIGURE 7.40 Typical traction curve (adapted from [67]).

The relationship between traction coefficient and slide to roll ratio is initially linear but later reaches a maximum value of traction coefficient beyond which there is a gradual decline in traction coefficient. In general, the peak traction coefficient occurs at about 0.1 or 10% of slide to roll ratio. For most lubricants, however, the traction peak is at about 1% of slide to roll ratio. Traction beyond slide to roll ratios of 10% is not usually considered for use in technology. The traction characteristic is usually divided into three regions: linear, nonlinear and thermal region. The extent of the linear region depends strongly on pressure [67]. In the nonlinear region, non-Newtonian lubricant rheology is the controlling factor while at high slide to roll ratios in the thermal region, viscous heating of the lubricant by intense shearing is the most significant influence. Most of the complexity in modelling traction originates from the inclusion of thermal and non-Newtonian effects.

A Simplified Analysis of Traction in the EHL Contact

The analysis of traction is very complicated. The general approach to traction problems illustrated in this section is based on a simplified example.

Consider two cylinders in contact such as are shown in Figure 7.10. The cylinders are rotating with the velocities 'U_A' and 'U_B' in the presence of a lubricant. As shown already in Chapter 4 the friction force generated between the cylinders can be calculated from equation (4.35):

$$\mathbf{F} = \int_0^L \int_{-b}^{b} \tau \, \mathbf{dx} \, \mathbf{dy}$$

where:

    b   is the half width of the contact rectangle [m];

    L   is the length of the contact rectangle [m].

In terms of frictional force per unit length, the double integral (4.35) is reduced to a single integral:

$$\frac{F}{L} = \pm \int_{-b}^{b} \frac{h}{2} \frac{dp}{dx} dx - \int_{-b}^{b} \frac{(U_A - U_B)\eta}{h} dx \qquad (7.48)$$

Two terms are present in the above equation. The first relates to the asymmetry of the pressure field and the second is a true traction term. The first term is present even when there is no sliding (i.e., $U_A = U_B$) and does not contribute to traction, it is in fact the rolling resistance of the EHL contact. The origin of rolling resistance in EHL is unclear, it may be the result of energy dissipation in compressing the lubricant or it may be a result of elastic hysteresis in the deformation of the contacting bodies (or differences in the velocity gradient in the centre and convergent inlet region of the EHL film). There are also other probable sources of rolling resistance. The second term in equation (7.48) relates directly to traction and is by far the larger term. It can be noticed that it only becomes relevant when sliding (i.e., $U_A \neq U_B$) in the contact takes place. Hence the traction force per unit length in sliding is given by:

$$\frac{F_{sl}}{L} = \int_{-b}^{b} \frac{(U_A - U_B)\eta}{h} dx \qquad (7.49)$$

where:

    F_{sl} is the traction force in sliding [N].

Assuming that in the EHL contact considered the surfaces are parallel (i.e., $h \neq f(x)$) and the sliding velocity is constant, equation (7.49) can be simplified to:

$$\frac{F_{sl}}{L} = \frac{U_A - U_B}{h} \int_{-b}^{b} \eta \, dx \qquad (7.50)$$

The viscosity of the lubricant, however, changes with the pressure. So in order to solve this equation two other equations, one describing the pressure distribution and the other describing the viscosity-pressure relationship, must be applied. The simplest solution is obtained by assuming the Barus pressure-viscosity dependence:

$$\eta = \eta_0 e^{\alpha p}$$

and the Hertzian equation for contact stress [55], i.e.:

$$p = p_{max} \left(1 - \frac{x^2}{b^2}\right)^{1/2}$$

Substituting these equations into (7.50) yields:

$$\frac{F_{sl}}{L} = \frac{U_A - U_B}{h} \int_{-b}^{b} \eta_0 \exp\left[\alpha p_{max}\left(1 - \frac{x^2}{b^2}\right)^{1/2}\right] dx \qquad (7.51)$$

Even in this very simplified approach the integral obtained for traction force is quite difficult to solve and apart from that it also suffers from some inaccuracy. In more accurate numerical work, the Roelands relationship between pressure and viscosity (described in Chapter 2) is often employed. Furthermore the analysis so far does not include non-Newtonian or thermal effects which further complicate the mathematics involved. For example, non-Newtonian behaviour of the lubricant has been modelled by applying an Eyring relation between shear stress and shear rate given by [59]:

$$\eta \frac{du}{dh} = \tau_0 \sinh\left(\frac{\tau}{\tau_0}\right) \qquad (7.52)$$

where:

  η is the reference, Newtonian, viscosity [Pas];

  du/dh   is the shear rate [s^{-1}];

  τ_0 is the reference shear stress acting on a lubricant (when Newtonian) [Pa];

  τ is the actual shear stress acting on a lubricant [Pa].

It can be seen from this equation that at low shear rates when $\tau \approx \tau_0$ the Newtonian law applies since equation (7.52) reduces to:

$$\tau = \eta \frac{du}{dh}$$

The increase in shear rate greatly complicates the determination of the traction force since 'τ' from a formula modelling the non-Newtonian behaviour of the lubricant (e.g., 7.52) is substituted into (4.35) and the outlined procedure for traction force determination repeated. In such cases more refined equations describing the pressure field, viscosity-pressure relationship and thermal effects are also incorporated.

Thermal effects are modelled by a modified form of the flash temperature theory, described in the previous section, where the heat is assumed to be generated through an EHL film of finite thickness and released to the contacting surfaces by conduction. In the simple theory of flash temperature, the frictional heat source was assumed to be planar without any thickness. A significant temperature variation through the film thickness is contained in the refined model of traction [59] and the analysis is extremely complex. A reasonable agreement between experiment and theory has, however, been found.

Non-Newtonian Lubricant Rheology and EHL

Even a simple mineral oil can reveal non-Newtonian characteristics at the extreme conditions of shear found in EHL. There are two aspects to the problem: EHL by non-Newtonian lubricants and the phenomenon of complex rheology under extreme conditions.

Non-Newtonian lubricants are, in most cases, mineral oils blended with 'viscosity index improvers' (VI improvers), described in Chapter 3. Under the conditions of intense shearing found in an EHL contact, molecular alignment and temporary viscosity loss occurs [60,61]. Molecular alignment means that the long linear molecules of VI improvers are forced to lie parallel to the rolling/sliding surfaces within the EHL contact. The effectiveness of VI improvers at postponing the decline in viscosity with temperature is greatly reduced by this process. Molecular alignment and viscosity loss was observed at the inlet to the EHL contact so that even before the VI enhanced lubricant had entered the EHL contact, the effect of the VI improver was already nullified [61]. As well as VI improvers, oils contain a wide range of other additives and these may affect lubricant rheology in EHL contacts by as yet unknown mechanisms. A study of a refined mineral oil blended with a range of additives such as dispersants, anti-wear additives and friction modifiers found that the EHL film thickness increased significantly for all additives tested [62]. The measured film thicknesses were in the range between 0.2 [μm] and 1 [μm] which is too large for monomolecular films to have any effect on modifying the lubricant rheology. A form of additive antagonism was observed when the film thickness increase obtained from an anti-wear additive was eliminated by the addition of detergent to the lubricating oil. It is possible that the combination of high levels of pressure, shear rate and the extreme thinness of an EHL film ensures that lubricant rheology is easily affected by small amounts of dissolved or colloidal material.

Even a plain mineral oil which shows Newtonian rheology when measured in a conventional viscometer reveals non-Newtonian characteristics within an EHL film. Two basic concepts that frequently appear in the literature are '**glass transition**' and '**limiting shear stress**'. The 'glass transition' is a term used to describe a change from the liquid state to an amorphous solid or glassy state under the conditions of extreme pressure and shear rate as found in EHL [63,64]. The effect of pressure is to raise the temperature where the reverse glass transition occurs, i.e., from glassy to liquid state. In most oils the transition temperature at atmospheric pressure is close to 0°C while at pressures of about 1 [GPa], the transition is closer to 100°C [63]. The lubricating oil therefore transforms to the glassy state even though the temperatures in the EHL contact can be relatively high. The most direct effect of the glass transition is on traction rather than film thickness [63]. The difference between a solid and a liquid is that the former has a finite shear stress while the latter can, in theory, sustain an infinite shear stress. In terms of traction, the glass transition in lubricating oils implies that the oil in an EHL contact will reach a limiting shear stress, which in turn limits the maximum traction coefficient. This aspect of lubricant rheology is modelled in the theory of traction by exponential functions to relate viscous stress and shear rate. Glass transition has relatively little influence on EHL film thickness unless it affects the EHL inlet conditions which virtually control the film thickness [63,64].

It is also possible that the lubricant slips at the interface between a liquid and a solid surface in apparent contradiction of some basic principles of fluid mechanics [65]. A discontinuity in the velocity profile was detected in a ball on plane contact during mixed sliding and rolling of the ball but not during pure rolling [65]. The concept of 'lubricant slip' under sliding is illustrated schematically in Figure 7.41. It is not clear, however, whether this phenomenon relates directly to the glass transition, but this is clear confirmation of lubricant failure by quasi-solid shear.

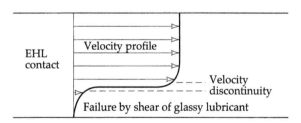

FIGURE 7.41 Lubricant slip or velocity discontinuity in an EHL contact.

Non-uniform shear by the lubricant or 'limiting shear stress' has been considered as the cause of failure of EHL films under sliding [66] but the effect of heating, which is inevitable under sliding conditions, was not included in these models and no definite conclusion was found. The limiting shear stress of an oil is believed to decline significantly with increased temperature [56] and this characteristic cannot be neglected in any model of EHL with sliding present.

EHL between Meshing Gear Wheels

From the viewpoint of practical engineering an important EHL contact takes place between the lubricated teeth of opposing gears. As is the case with rolling bearings, it is essential to maintain an adequate EHL film thickness to prevent wear and pitting of the gear teeth. The same fundamental equations for EHL film thickness described for a simple Hertzian contact also apply for gears. However, before applying the formulae for contact parameters and minimum film thickness it is necessary to define reduced radius of curvature, contact load and surface velocity for a specific gear. The contact geometry is illustrated in Figure 7.42.

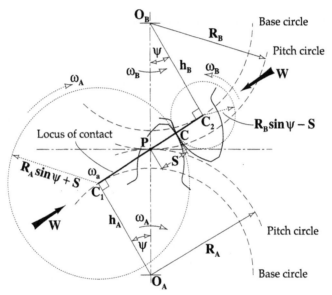

FIGURE 7.42   Contact geometry of meshing involute gear teeth.

The surface contact velocity is expressed as:

$$U = \frac{U_A + U_B}{2} = \frac{\omega_A R_A \sin\psi + \omega_B R_B \sin\psi}{2}$$

where:

R_A, R_B are the pitch circle radii of the driver and follower, respectively [m];

ψ is the pressure angle (acute angle between contact normal and the common tangent to the pitch circles);

ω_A, ω_B are the angular velocities of the driver and follower, respectively [rad/s].

Since:

$$\frac{R_A}{R_B} = \frac{\omega_B}{\omega_A}$$

Then the contact surface velocity is:

$$U = \omega_A R_A \sin\psi = \omega_B R_B \sin\psi \qquad (7.53)$$

Assuming that the total load is carried by one tooth only then, from Figure 7.42, the contact load in terms of the torque exerted is given by:

$$W = \frac{T_B}{h_B} = \frac{T_B}{R_B \cos\psi} \qquad (7.54)$$

where:

W is the total load on the tooth [N];

h_B is the distance from the centre of the follower to interception of the locus of the contact with its base circle, i.e., $h_B = R_B \cos\psi$ [m];

T_B is the torque exerted on the follower [Nm].

The torque exerted on the driver and the follower expressed in terms of the transmitted power is calculated from:

$$T_A = \frac{H}{\omega_A} = 9.55 \frac{H}{N_A}$$

$$T_B = \frac{H}{\omega_B} = 9.55 \frac{H}{N_B}$$

where:

N_A, N_B are the rotational speeds of the driver and follower, respectively [rps];

H is the transmitted power [kW].

Substituting into (7.54) yields the contact load. The minimum and central EHL film thicknesses can then be calculated from formulae (7.26) and (7.27).

The line from 'C_1' to 'C_2' (Figure 7.42) is the locus of the contact and it can be seen that the distance 'S' between the gear teeth contact and the pitch line is continuously changing with the contact position during the meshing cycle of the gears. It is thus possible to model any specific contact position on the tooth surface of an involute gear by two rotating circular discs of radii ($R_A \sin\Psi + S$) and ($R_B \sin\Psi - S$) as shown in Figure 7.42. This idea is applied in a testing apparatus generally known as a '**twin disc**' or '**two disc**' machine shown schematically in Figure 7.43. Since the gear tooth contact is closely simulated by the two rotating discs, these machines are widely used to model gear lubrication and wear and in selecting lubricants or materials for gears. It is much cheaper and more convenient experimentally to use metal discs instead of actual gears for friction and wear testing. The wear testing virtually ensures the destruction of the test specimens and it is far easier to inspect and analyse a worn disc surface than the recessed surface of a gear wheel.

It may also be apparent that the fixed dimensions of the discs only allow modelling of one particular position in the contact cycle. Of particular importance to friction and wear studies is the increasing amount of sliding as the contact between opposing gear teeth moves away from the line of shaft centres. The radii of curvature also vary with position of gear teeth so that the 'two-disc' test rig is not entirely satisfactory and another model gear apparatus such as the 'Ryder gear tester' may be necessary for some studies. A recently developed test-apparatus where two contacting discs are supplied with additional movement of their corresponding shafts allows a much closer, more realistic simulation of the entire gear tooth contact cycle [68].

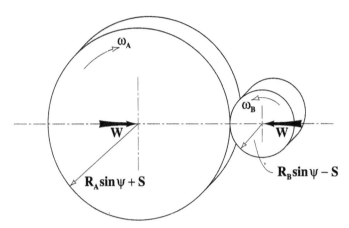

FIGURE 7.43 Schematic diagram of a 'two disc' machine used to simulate rolling/sliding contact in meshing gears, i.e.: for **S = 0** pure rolling and for **S ≠ 0** rolling/sliding in EHL contact; **S** is the distance between the pitch line and the gear teeth contact [m].

7.8 SUMMARY

A fundamental lubrication mechanism involved in highly loaded concentrated contacts was discussed in this chapter. The remarkable efficiency of elastohydrodynamic lubrication in preventing solid to solid contact even under extreme contact stresses prevents the rapid destruction of many basic mechanical components such as rolling bearings or gears. EHL is, however, mostly confined to mineral or synthetic oils since it is essential that the lubricant is piezo-viscous. The mechanism of EHL involves a rapid change in the lubricant from a nearly ideal liquid state outside of the contact to an extremely viscous or semi-solid state within the contact. This transformation allows the lubricant to be drawn into the contact by viscous drag while generating sufficient contact stress within the contact to separate the opposing surfaces. If a simple solid, i.e., a fine powder, is supplied instead, there is no viscous drag to entrain the powder and consequently only poor lubrication results. A non-piezo-viscous lubricant simply does not achieve the required high viscosity within the contact necessary for the formation of the lubricating film. The formulae for the calculation of the EHL film thickness are relatively simple and are based on load, velocity, dimensions and elastic modulus of the contacting materials. As well as providing lubrication of concentrated contacts, the EHL mechanism can be used to generate traction, i.e., where frictional forces enable power transmission. A unique combination of high tractive force with minimal wear, reduced noise levels, infinitely variable output speed and an almost constant torque over the speed range can be obtained by this means.

REVISION QUESTIONS

7.1   Under what level of pressure does elastohydrodynamic lubrication occur?

7.2   Can water sustain elastohydrodynamic lubrication?

7.3   Why is the interaction between a lubricant with pressure dependent viscosity and elastic deformation of a sphere or disc contacting a plane of such great practical significance?

7.4   For the dimensionless speed parameter $U = 10^{-11}$, dimensionless material parameter $G = 5000$ and dimensionless load parameter $W = 3 \times 10^{-5}$, calculate the h_0/R' ratio for a line contact. If $R' = 0.03$ [m], calculate h_0 using the Dowson-Hamrock equation. (Ans. 1.67×10^{-5} and 5×10^{-7} [m])

7.5   What would be the surface roughness required to sustain the EHL film thickness of Question 7.4? What kind of machining process would be needed to produce this surface finish? (Ans. $\sigma = 0.12$ [μm])

7.6   Can a mild steel or equivalent material be used in EHL contacts?

7.7   What is the lowest ratio of minimum film thickness to combined surface roughness 'λ' that ensures a long operating life for lubricated rolling bearings?

7.8   Why should the 'λ' value be of practical and economic significance?

7.9   Are changes in phase of the oil thought to occur in the EHL contact?

7.10   Name two causes of deviation from the ideal EHL film thickness formula which occur at high rolling and sliding speeds.

7.11   Name an important advantage of rolling bearings over plain bearings.

7.12   A steel ball starts rolling from rest down a steel plate inclined at **45°** to the horizontal, as shown below. The plate is covered by a thin oil film of viscosity η_0 = **50** [cP] and pressure viscosity coefficient α = **2.2×10⁻⁸** [m²/N]. The diameter of the ball is **d = 3×10⁻² [m]**, Young's modulus is **E = 2.1×10¹¹ [Pa]**, Poisson's ratio is υ = **0.3** and density is ρ = **7800** [kg/m³]. What is the minimum thickness of the elastohydrodynamic film after the ball has rolled **1** [cm] and **1** [m]? Assume no sliding. (Ans. h_0 = 0.059 & 0.281 [μm])

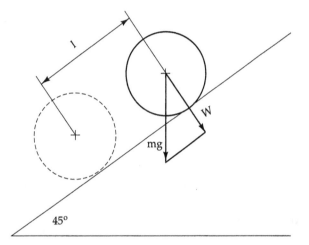

7.13   The highly loaded cam, shown below, has a minimum radius of **10** [mm], is **20** [mm] wide and is subjected to a load of **4** [kN]. The cam rotates against a convex stationary follower that has a radius of **40** [mm]. The sliding speed at the point of contact is approximately **0.1** [m/s]. The cam is lubricated with oil that has a dynamic viscosity of η_0 = **1.44** [Pas] at the operating temperature and a pressure-viscosity coefficient α = **2.8×10⁻⁸** [Pa⁻¹]. The material of both cam and follower has a modulus of elasticity **E = 2.07×10¹¹ [Pa]** and Poisson's ratio υ = **0.3**. Estimate the surface roughness of the cam and the follower necessary to prevent the occurrence of scuffing. (Ans. σ = 0.0768 [μm])

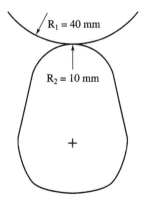

7.14   Assuming the coefficient of friction μ = **0.05** for problem 7.13, calculate the oil temperature increase after **1** [min] of operation. For simplicity, assume that all heat remains in the cam, follower and oil, that the coefficient of friction is independent of temperature, that the temperature of the cam, follower and oil is the same and that the total heat capacity of the assembly is **C = 200** [J/K]. Assume that the cam and follower are adiabatic. (Ans. $\Delta T = 6°C$)

REFERENCES

1   A.N. Grubin, Fundamentals of the Hydrodynamic Theory of Lubrication of Heavily Loaded Cylindrical Surfaces, in Investigation of the Contact Machine Components, Kh.F. Ketova, ed. Translation of Russian Book No. 30, Central Scientific Institute for Technology and Mechanical Engineering, Moscow, 1949.

2   H.M. Martin, Lubrication of Gear Teeth, *Engineering, London*, Vol. 102, 1916, pp. 119-121.

3   D. Dowson and G.R. Higginson, Elastohydrodynamic Lubrication, Pergamon Press, Oxford, 1977.

4   A.W. Crook, Elastohydrodynamic Lubrication of Rollers, *Nature*, Vol. 190, 1961, p. 1182.

5   A. Cameron and R. Gohar, Optical Measurement of Oil Film Thickness under Elasto-hydrodynamic Lubrication, *Nature*, Vol. 200, 1963, pp. 458-459.

6   H. Hertz, Uber die Beruhrung Fester Elasticher Korper, (On the Contact of Elastic Solids), *J. Reine und Angewandte Mathematik*, Vol. 92, 1881, pp. 156-171.

7   B.J. Hamrock and D. Dowson, Ball Bearing Lubrication, The Elastohydrodynamics of Elliptical Contacts, John Willey & Sons, 1981.

8   H. Hertz, Miscellaneous Papers by H. Hertz, Jones & Schott (eds), Macmillan, London, 1986.

9   K.L. Johnson, Contact Mechanics, Cambridge University Press, 1985.

10   B.J. Hamrock and W.J. Anderson, Analysis of an Arched Outer-Race Ball Bearing Considering Centrifugal Forces, *Transactions ASME, Journal of Lubrication Technology*, Vol. 95, 1973, pp. 265-276.

11   R. Gohar, Elastohydrodynamics, Ellis Horwood Limited, 1988.

12   D.E. Brewe and B.J. Hamrock, Simplified Solution for Elliptical Contact Deformation Between Two Elastic Solids, *Transactions ASME, Journal of Lubrication Technology*, Vol. 99, 1977, pp. 485-487.

13   F.T. Barwell, Bearing Systems, Principles and Practice, Oxford University Press, 1979.

14   Engineering Sciences Data Unit, Stress and Strength Sub-Series, Contact Stresses, Vol. 9, No. 78035, 1985.

15   H. Christensen, The Oil Film in a Closing Gap, *Proc. Roy. Soc., London*, Series A, Vol. 266, 1962, pp. 312-328.

16   B.J. Hamrock and D. Dowson, Isothermal Elastohydrodynamic Lubrication of Point Contacts, Part III - Fully Flooded Results, *Transactions ASME, Journal of Lubrication Technology*, Vol. 99, 1977, pp. 264-276.

17   K.A. Kaye and W.O. Winer, An Experimental Evaluation of the Hamrock and Dowson Minimum Film Thickness Equation for Fully Flooded EHD Point Contact, *Transactions ASME, Journal of Lubrication Technology*, Vol. 103, 1981, pp. 284-294.

18   D. Dowson and A.V. Whitaker, A Numerical Procedure for the Solution of the Elastohydrodynamic Problems of Rolling and Sliding Contacts Lubricated by Newtonian Fluid, *Proc. Inst. Mech. Engrs.*, London, Vol. 180, Pt. 3B, 1965-1966, pp. 57-71.

19   B.J. Hamrock and D. Dowson, Minimum Film Thickness in Elliptical Contacts for Different Regimes of Fluid-Film Lubrication, Proc. 5th Leeds-Lyon Symp. on Tribology, Elastohydrodynamics and Related Topics, editors: D. Dowson, C.M. Taylor, M. Godet and D. Berthe, Sept. 1978, Inst. Mech. Engrs. Publ., London, 1979, pp. 22-27.

20   A.W. Crook, The Lubrication of Rollers, Part I, *Phil. Trans. Roy. Soc., London*, Series A, Vol. 250, 1958, pp. 387-409.

21   A. Dyson, H. Naylor and A.R. Wilson, The Measurement of Oil Film Thickness in Elastohydrodynamic Contacts, *Proc. Inst. Mech. Engrs.*, Vol. 180, Pt. 3B, 1965, pp. 119-134.

22   L.B. Sibley, J.C. Bell, F.K. Orcutt and C.M. Allen, A Study of the Influence of Lubricant Properties on the Performance of Aircraft Gas Engine Rolling Contact Bearings, WADD Technical Report, 1960, pp. 60-189.

23   L.B. Sibley and A.E. Austin, An X-Ray Method for Measuring Thin Lubricant Films Between Rollers, *ISA Transactions*, Vol. 3, 1962, pp. 237-243.

24   D.R. Meyer and C.C. Wilson, Measurement of Elastohydrodynamic Oil Film Thickness and Wear in Ball Bearings by the Strain Gage Method, *Transactions ASME, Journal of Lubrication Technology*, Vol. 93, 1971, pp. 224-230.

25   A.T. Kirk, Hydrodynamic Lubrication of Perspex, *Nature*, Vol. 194, 1962, pp. 965-966.

26   A. Cameron and R. Gohar, Theoretical and Experimental Studies of the Oil Film in Lubricated Point Contacts, *Proc. Roy. Soc., London*, Series A, Vol. 291, 1966, pp. 520-536.

27   N. Thorp and R. Gohar, Oil Film Thickness and Shape for Ball Sliding in a Grooved Raceway, *Transactions ASME, Journal of Lubrication Technology*, Vol. 94, 1972, pp. 199-210.

28   D. Dowson, Recent Developments in Studies of Fluid Film Lubrication, Proc. Int. Tribology Conference, Melbourne, The Institution of Engineers, Australia, National Conference Publication No. 87/18, December, 1987, pp. 353-359.

29   T.E. Tallian, On Competing Failure Modes in Rolling Contact, *ASLE Transactions*, Vol. 10, 1967, pp. 418-439.

30   K.L. Johnson, J.A. Greenwood and S.Y. Poon, A Simple Theory of Asperity Contact in Elastohydrodynamic Lubrication, *Wear*, Vol. 19, 1972, pp. 91-108.

31   J.A. Greenwood and J.B.P. Williamson, Contact of Nominally Flat Surfaces, *Proc. Roy. Soc., London*, Series A, Vol. 295, 1966, pp. 300-319.

32   T.E. Tallian and J.I. McCool, An Engineering Model of Spalling Fatigue Failure in Rolling Contact, II. The Surface Model, *Wear*, Vol. 17, 1971, pp. 447-461.

33   R.S. Sayles, G.M.S. deSilva, J.A. Leather, J.C. Anderson and P.B. Macpherson, Elastic Conformity in Hertzian Contacts, *Tribology International*, Vol. 14, 1981, pp. 315-322.

34   G.M.S. De Silva, J.A. Leather and R.S. Sayles, The Influence of Surface Topography on Lubricant Film Thickness in EHD Point Contact, Proc. 12th Leeds-Lyon Symp. on Tribology, Mechanisms and Surface Distress: Global Studies of Mechanisms and Local Analyses of Surface Distress Phenomena, editors: D. Dowson, C.M. Taylor, M. Godet and D. Berthe, Sept. 1985, Inst. Mech. Engrs. Publ., London, 1986, pp. 258-272.

35   N. Patir and H.S. Cheng, Effect of Surface Roughness Orientation on the Central Film Thickness in EHD Contacts, Proc. 5th Leeds-Lyon Symp. on Tribology, Elastohydrodynamics and Related Topics, editors: D. Dowson, C.M. Taylor, M. Godet and D. Berthe, Sept. 1978, Inst. Mech. Engrs. Publ., London, 1979, pp. 15-21.

36   H.S. Cheng, On Aspects of Microelastohydrodynamic Lubrication, Proc. 4th Leeds-Lyon Symp. on Tribology, Surface Roughness Effects in Lubrication, editors: D. Dowson, C.M. Taylor, M. Godet and D. Berthe, Sept. 1977, Inst. Mech. Engrs. Publ., London, 1978, pp. 71-79.

37   X. Ai and L. Zheng, A General Model for Microelastohydrodynamic Lubrication and its Full Numerical Solution, *Transactions ASME, Journal of Tribology*, Vol. 111, 1989, pp. 569-576.

38   P. Goglia, T.F. Conry and C. Cusano, The Effects of Surface Irregularities on the Elastohydrodynamic Lubrication of Sliding Line Contacts, Parts 1 and 2, *Transactions ASME, Journal of Tribology*, Vol. 106, 1984, Part 1, pp. 104-112, Part 2, pp. 113-119.

39   C.C. Kweh, H.P. Evans and R.W. Snidle, Micro-Elastohydrodynamic Lubrication of an Elliptical Contact With Transverse and 3-D Sinusoidal Roughness, *Transactions ASME, Journal of Tribology*, Vol. 111, 1989, pp. 577-584.

40   L. Chang and M.N. Webster, A Study of Elastohydrodynamic Lubrication of Rough Surfaces, *Transactions ASME, Journal of Tribology*, Vol. 113, 1991, pp. 110-115.

41   L.G. Houpert and B.J. Hamrock, EHD Lubrication Calculation Used as a Tool to Study Scuffing, Proc. 12th Leeds-Lyon Symp. on Tribology, Mechanisms and Surface Distress: Global Studies of Mechanisms and Local Analyses of Surface Distress Phenomena, editors: D. Dowson, C.M. Taylor, M. Godet and D. Berthe, Sept. 1985, Inst. Mech. Engrs. Publ., London, 1986, pp. 146-155.

42   K.P. Baglin, EHD Pressure Rippling in Cylinders Finished With a Circumferential Lay, *Proc. Inst. Mech. Engrs*, Vol. 200, 1986, pp. 335-347.

43   B. Michau, D. Berthe and M. Godet, Influence of Pressure Modulation in Line Hertzian Contact on the Internal Stress Field, *Wear*, Vol. 28, 1974, pp. 187-195.

44   J.F. Archard and R.A. Rowntree, The Temperature of Rubbing Bodies, Part 2, The Distribution of Temperature, *Wear*, Vol. 128, 1988, pp. 1-17.

45   F.P. Bowden and D. Tabor, Friction and Lubricating Wear of Solids, Part 1, Oxford: Clarendon Press, 1964.

46   H. Blok, Theoretical Study of Temperature Rise at Surfaces of Actual Contact Under Oiliness Lubricating Conditions, General Discussion on Lubrication, *Inst. Mech. Engrs, London*, Vol. 2, 1937, pp. 222-235.

47   J.C. Jaeger, Moving Sources of Heat and the Temperature at Sliding Contacts, *Proc. Roy. Soc., N.S.W.*, Vol. 76, 1943, pp. 203-224.

48   J.F. Archard, The Temperature of Rubbing Surfaces, *Wear*, Vol. 2, 1958/59, pp. 438-455.

49   F.E. Kennedy, Thermal and Thermomechanical Effects in Dry Sliding, *Wear*, Vol. 100, 1984, pp. 453-476.

50 H. Blok, The Postulate About the Constancy of Scoring Temperature, Interdisciplinary Approach to Lubrication of Concentrated Contacts, P.M. Ku (ed.), Washington DC, Scientific and Technical Information Division, NASA, 1970, pp. 153-248.

51 T.A. Stolarski, Tribology in Machine Design, Heineman Newnes, 1990.

52 V.K. Ausherman, H.S. Nagaraj, D.M. Sanborn and W.O. Winer, Infrared Temperature Mapping in Elastohydrodynamic Lubrication, *Transactions ASME, Journal of Lubrication Technology*, Vol. 98, 1976, pp. 236-243.

53 V.W. King and J.L. Lauer, Temperature Gradients Through EHD Films and Molecular Alignment Evidenced by Infrared Spectroscopy, *Transactions ASME, Journal of Lubrication Technology*, Vol. 103, 1981, pp. 65-73.

54 A.R. Wilson, An Experimental Thermal Correction for Predicted Oil Film Thickness in Elastohydrodynamic Contacts, Proc. 6th Leeds-Lyon Symp. on Tribology, Thermal Effects in Tribology, Sept. 1979, editors: D. Dowson, C.M. Taylor, M. Godet and D. Berthe, Inst. Mech. Engrs. Publ., London, 1980, pp. 179-190.

55 J.L. Tevaarwerk, Traction Calculations Using the Shear Plane Hypothesis, Proc. 6th Leeds-Lyon Symp. on Tribology, Thermal Effects in Tribology, Sept. 1979, editors: D. Dowson, C.M. Taylor, M. Godet and D. Berthe, Inst. Mech. Engrs. Publ., London, 1980, pp. 201-215.

56 H.A. Spikes and P.M. Cann, The Influence of Sliding Speed and Lubricant Shear Stress on EHD Contact Temperatures, *Tribology Transactions*, Vol. 33, 1990, pp. 355-362.

57 W.O. Winer and E.H. Kool, Simultaneous Temperature Mapping and Traction Measurements in EHD Contacts, Proc. 6th Leeds-Lyon Symp. on Tribology, Thermal Effects in Tribology, Sept. 1979, editors: D. Dowson, C.M. Taylor, M. Godet and D. Berthe, Inst. Mech. Engrs. Publ., London, 1980, pp. 191-200.

58 T.A. Dow and W. Kannel, Evaluation of Rolling/Sliding EHD Temperatures, Proc. 6th Leeds-Lyon Symp. on Tribology, Thermal Effects in Tribology, Sept. 1979, editors: D. Dowson, C.M. Taylor, M. Godet and D. Berthe, Inst. Mech. Engrs. Publ., London, 1980, pp. 228-240.

59 K.L. Johnson and J.A. Greenwood, Thermal Analysis of an Eyring Fluid in Elastohydrodynamic Traction, *Wear*, Vol. 61, 1980, pp. 353-374.

60 J.L. Lauer and Y-J. Ahn, Lubricants and Lubricant Additives Under Shear Studied Under Operating Conditions by Optical and Infra Red Spectroscopic Methods, *Tribology Transactions*, Vol. 31, 1988, pp. 120-127.

61 P.M. Cann and H.A. Spikes, In Lubro Studies of Lubricants in EHD Contacts Using FITR Absorption Spectroscopy, *Tribology Transactions*, Vol. 34, 1991, pp. 248-256.

62 F.L. Snyder, J. L. Tevaarwerk and J. A. Schey, Effects of Oil Additives on Lubricant Film Thickness and Traction, SAE Tech. Paper No. 840263, 1984.

63 S. Bair and W.O. Winer, Some Observations in High Pressure Rheology of Lubricants, *Transactions ASME, Journal of Lubrication Technology*, Vol. 104, 1982, pp. 357-364.

64 M. Alsaad, S. Bair, D.M. Sanborn and W.O. Winer, Glass Transitions in Lubricants: Its Relation to Elastohydrodynamic Lubrication (EHD), *Transactions ASME, Journal of Lubrication Technology*, Vol. 100, 1978, pp. 404-417.

65 M. Kaneta, H. Nishikawa and K. Kameishi, Observation of Wall Slip in Elastohydrodynamic Lubrication, *Transactions ASME, Journal of Tribology*, Vol. 112, 1990, pp. 447-452.

66 K.L. Johnson and J.G. Higginson, A Non-Newtonian Effect of Sliding in Micro-EHL, *Wear*, Vol. 128, 1988, pp. 249-264.

67 K.L. Johnson and J.L. Tevaarwerk, Shear Behaviour of Elastohydrodynamic Oil Films, *Proc. Roy. Soc., London*, Series A, Vol. 356, 1977, pp. 215-236.

68 E. Van Damme, Surface Engineering, Gear Wear Simulations, Proc. International Tribology Conference, Melbourne, 1987, The Institution of Engineers, Australia, National Conference Publication No. 87/18, December, 1987, pp. 391-396.

69 C.A. Foord, W.C. Hammann and A. Cameron, Evaluation of Lubricants Using Optical Elastohydrodynamics, *ASLE Transactions*, Vol. 11, 1968, pp. 31-43.

70 P.L. Wong, P.Huang, W. Wang and Z. Zhang, Effect of Geometry Change of Rough Point Contact Due to Lubricated Sliding Wear on Lubrication, *Tribology Letters*, Vol. 5, 1998, pp. 265-274.

71 C. Bovington, Elastohydrodynamic Lubrication: a Lubricant Industry Perspective, *Proc. Inst. Mech. Eng., Part J, Journal of Engineering Tribology*, Vol. 213, 1999, pp. 417-426.

BOUNDARY AND EXTREME PRESSURE LUBRICATION

8.1 INTRODUCTION

In many practical applications there are cases where the operating conditions are such that neither hydrodynamic nor EHL lubrication is effective. The questions then are: how are the interacting machine components lubricated and what is the lubrication mechanism involved? The models of lubrication which are thought to operate under such conditions are discussed in this chapter. The traditional name for this type of lubrication is '**boundary lubrication**' or '**boundary and extreme-pressure lubrication**'. Neither of these terms describes accurately the processes at work since they were conceived long before any fundamental understanding of the mechanisms was available. Several specialized modes of lubrication such as adsorption, surface localized viscosity enhancement, amorphous layers and sacrificial films are commonly involved in this lubrication regime to ensure the smooth-functioning and reliability of machinery. The imprecise nature of present knowledge about these modes or mechanisms of lubrication contrasts with their practical importance. Many vital items of engineering equipment such as steel gears, piston-rings and metal-working tools depend on one or more of these lubrication modes to prevent severe wear or high coefficients of friction and seizure.

Boundary and EP lubrication is a complex phenomenon. The lubrication mechanisms involved can be classified in terms of relative load capacity and limiting frictional temperature as shown in Table 8.1, and they will be described in this chapter.

These lubrication mechanisms are usually controlled by additives present in the oil. Since the cost of a lubricant additive is usually negligible compared to the value of the mechanical equipment, the commercial benefits involved in this type of lubrication can be quite large.

In general, boundary and EP lubrication involves the formation of low friction, protective layers on the wearing surfaces. One exception is when the surface-localized viscosity enhancement takes place. The occurrence of surface-localized viscosity enhancement, however, is extremely limited as is explained in the next section.

The operating principle of the boundary lubrication regime can perhaps be best illustrated by considering the coefficient of friction. In simple terms the coefficient of friction 'μ' is defined as the ratio of frictional force '**F**' and the load applied normal to the surface '**W**', i.e.:

$$\mu = F/W \tag{8.1}$$

TABLE 8.1   Categories of boundary and EP lubrication.

Temperature	Load	Lubrication mechanisms
Low	Low	Viscosity enhancement close to contacting surface, not specific to lubricant.
	High	Friction minimization by coverage of contacting surfaces with adsorbed mono-molecular layers of surfactants.
High	Medium	Irreversible formation of soap layers and other viscous materials on worn surface by chemical reaction between lubricant additives and metal surface.
		Surface-localized viscosity enhancement specific to lubricant additive and basestock.
		Formation of amorphous layers of finely divided debris from reaction between additives and substrate metal surface.
	High	Reaction between lubricant additives and metal surface.
		Formation of sacrificial films of inorganic material on the worn surface preventing metallic contact and severe wear.

Since the contacting surfaces are covered by asperities, 'dry' contact is established between the individual asperities and the 'true' total contact area is the sum of the individual contact areas between the asperities. Assuming that the major component of the frictional force is due to adhesion between the asperities (other effects, e.g., ploughing, are negligible), then the expression for frictional force '**F**' can be written as:

$$\mathbf{F} = \mathbf{A}_t \tau$$

where:

$\mathbf{F}$ is the frictional force [N];

$\mathbf{A}_t$ is the true contact area [m^2];

τ is the effective shear stress of the material [Pa].

Applied load can be expressed in terms of contact area, i.e.:

$$\mathbf{W} = \mathbf{A}_t \mathbf{p}_y$$

where:

$\mathbf{p}_y$ is the plastic flow stress of the material (related to the indentation hardness) [Pa].

        $\mathbf{p}_y \approx 3\,\sigma_y$, where '$\sigma_y$' is the yield strength of the material [Pa].

Substituting for '**F**' and '**W**' to (8.1) yields:

$$\mu = \tau/\mathbf{p}_y \qquad\qquad\qquad (8.2)$$

This simple model explains the rationale behind boundary lubrication. It can be seen from equation (8.2) that in order to obtain a low coefficient of friction, material of low shear strength and high hardness is required. These requirements are clearly incompatible. However, if a low shear-strength layer can be formed on a hard substrate then low coefficients of friction can be achieved. Thus, in general terms, the fundamental principles behind boundary and EP lubrication involve the formation of low shear-strength lubricating layers on hard substrates. It is evident that, since with most materials the ratio of 'τ' and '$\mathbf{p}_y$' does not vary greatly, changing the material type has little effect on friction.

8.2   LOW TEMPERATURE - LOW LOAD LUBRICATION MECHANISMS

For a very large range of sliding speeds and loads, classical hydrodynamic lubrication prevails in a lubricated contact. As the sliding speed is reduced, hydrodynamic lubrication reaches its limit where the hydrodynamic film thickness declines until eventually the asperities of the opposing surfaces interact. This process was originally investigated by Stribeck and has already been discussed in Chapter 4.

At low speeds, under certain conditions, contact between opposing surfaces can be prevented by the mechanism involving surface-localized viscosity enhancement. In other words, a thin layer of liquid with an anomalously high viscosity can form on the contacting surfaces. Hydrodynamic lubrication or quasi-hydrodynamic lubrication then persists to prevent solid contact and severe wear. In such cases linear molecules of a hydrocarbon align themselves normally to the contacting surfaces to form a lubricating, protective layer as shown in Figure 8.1. Since the molecules are polar the opposite ends are attracted to form pairs of molecules which are subsequently incorporated into the viscous surface layer. At the interface with the metallic substrate the attractive force of the free end of the molecules to the substrate is sufficient to firmly bond the entire layer.

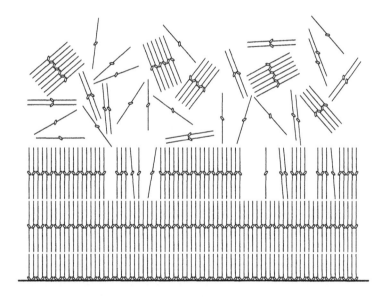

FIGURE 8.1   Low-temperature, low-load mechanism of lubrication [1].

It has been found that linear molecules are more effective than other hydrocarbons in preventing solid contact. The variation in film thickness between parallel discs as a function of the square root of squeeze time for paraffinic oil and cyclohexane is shown in Figure 8.2 [2]. According to the theory of hydrodynamic lubrication described in Chapter 4, there is a linear decline in film thickness with square root of squeeze time but as can be seen from Figure 8.2 this linearity is soon lost. The MS-20 oil contains paraffinic molecules which are approximately linear and this allows for the formation and persistence of a thicker film than for cyclohexane. Cyclohexane is a non-linear molecule which impedes the linear alignment of molecules and therefore the resulting film is less effective in preventing solid contact.

The effectiveness of this mechanism of lubrication is limited to low temperatures and low loads. The data shown in Figure 8.2 was obtained at contact pressures of 0.4 [MPa], and further work revealed that at contact pressures beyond 2 [MPa] the residual film thickness is very small [2]. Since in many contacts pressures in the range of 1 [GPa] are quite common, the disadvantages of this lubrication mechanism are obvious. The temperature also has a pronounced effect on these films. It was found that even relatively low operating temperatures of about 50°C can result in severe decline in the film thickness [2]. Since the practical applications in which this mode of boundary lubrication occurs are extremely limited, the topic does not incite much technological interest and has consequently been neglected by most researchers.

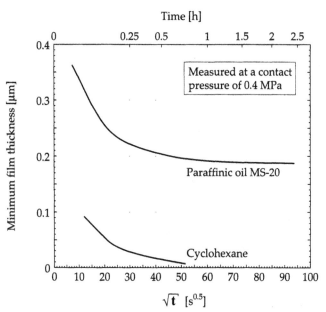

FIGURE 8.2   Detection of permanent films formation as evidence of a surface-proximal layer of aligned molecules [2].

8.3   LOW TEMPERATURE - HIGH LOAD LUBRICATION MECHANISMS

The lubrication mechanism acting at low temperature and high load is of considerable practical importance. It is generally known as 'adsorption lubrication'. This mechanism of lubrication is quite effective with contact pressures up to 1 [GPa] and relatively low surface temperatures between 100 and 150°C. Adsorption lubrication is different from hydrodynamic, EHL or even the viscous layer described in the previous section in that the opposing contact surfaces are not separated by a thick layer of fluid. A mono-molecular layer separates the contacting surfaces and this layer is so thin that the mechanics of asperity contact are identical to that of dry surfaces in contact. This mono-molecular layer is formed by adsorption of the lubricant or, more precisely, lubricant additives on the worn surface. The lubricating effect or friction reduction is caused by the formation of a low shear strength interface between the opposing surfaces.

It can be seen from equation (8.2) that the role of adsorption lubrication is to reduce the effective shear stress 'τ' at the interface without affecting the plastic flow stress 'p_y' of the substrate. This is achieved by the formation of an adsorbed film on the surface which introduces a plane of weakness parallel to the plane of sliding. This principle is illustrated in Figure 8.3 which shows a schematic comparison of contact between dry unlubricated solids and solids with a lubricant film on asperity peaks.

If the film is thin, then any structural weakness in the direction of the contact load will be compensated by the substrate. The shear stress anisotropy or low shear stress in the plane of sliding is obtained by inducing a discontinuity in intermolecular bonding between opposite sides of the sliding interface. At all locations other than the interface, bonding between atoms even of different substances, e.g., film material and substrate, is relatively strong. The characteristics of adsorbed layers, in particular of polar organic substances, allow this system to form on metallic surfaces. The reasons for this are discussed next.

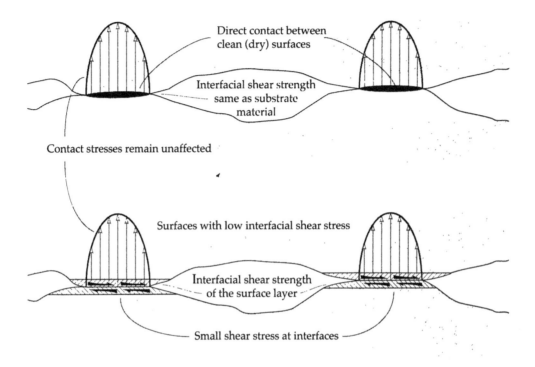

FIGURE 8.3   Lubrication by a low shear strength layer formed at asperity peaks.

Model of Adsorption on Sliding Surfaces

Organic polar molecules such as fatty acids and alcohols adsorb on to metallic surfaces and are not easily removed. Speculation about the role of these substances in lubrication has a long history [3]. Effective adsorption is the reason why a metallic surface still feels greasy or slippery after being wetted by a fatty substance and will remain greasy even after vigorous wiping of the surface by a dry cloth. Adsorption on a metallic surface of organic polar molecules produces a low friction, mono-molecular layer on the surface as shown in Figure 8.4. The polarity of the adsorbate is essential to the lubrication mechanism. Polarity means that a molecule is asymmetrical with a different chemical affinity at either end of the molecule. For example, one end of a molecule which is the carboxyl group of a fatty acid, '-**COOH**', is strongly attracted to the metallic surface while the other end which is an alkyl group, '-**CH₃**', is repellant to almost any other substance.

Strong adsorption ensures that almost every available surface site is occupied by the fatty acid to produce a dense and robust film. The repulsion or weak bonding between the contacting alkyl groups ensures that the shear strength of the interface is relatively low. The ratio of τ/p_y and therefore the friction coefficient is low compared to bare metallic surfaces in contact. This is the adsorption model of lubrication first postulated by Hardy and Doubleday [4,5] and later developed by Bowden and Tabor [6]. The fatty acids are particularly effective because of their strong polarity, but other organic compounds such as alcohols and amines have sufficient polarity to be of practical use.

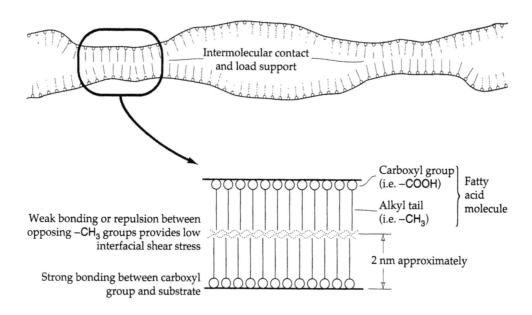

FIGURE 8.4   Low friction mono-molecular layer of adsorbed organic polar molecules on metallic surfaces.

From the viewpoint of lubrication, adsorption can be divided into two basic categories: '**physisorption**' and '**chemisorption**'. The latter generally occurs at higher temperatures than the former and is consequently more useful as a lubrication mechanism in practical applications.

· *Physisorption*

Physisorption or 'physical adsorption' is the classical form of adsorption. Molecules of adsorbate may attach or detach from a surface without any irreversible changes to the surface or the adsorbate. Most liquids and gases physisorb to most solid surfaces, but there is almost always an upper temperature limit to this process. In physisorption van der Waals or dispersion forces provide the bonding between substrate and adsorbate as illustrated in Figure 8.5.

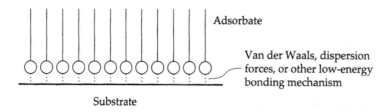

FIGURE 8.5   Schematic illustration of physisorption.

Physisorption is effective in reducing friction provided that temperatures do not rise much above ambient temperature. This effect is illustrated in Figure 8.6 where the results from friction experiments with a ball traversing a platinum surface covered with solid paraffin (docosane), fatty acid (stearic) and copper laurate, or a copper surface covered with 1% lauric acid in paraffin oil are shown [6]. Docosane is a straight hydrocarbon without any carboxyl groups, while lauric acid is a fatty acid similar to stearic acid but has a shorter chain length. The results obtained show that the boundary lubricating properties of fatty acids depend on the type of metallic surface they lubricate. Reactive metals such as copper are well lubricated by fatty acids while unreactive metals such as platinum are not.

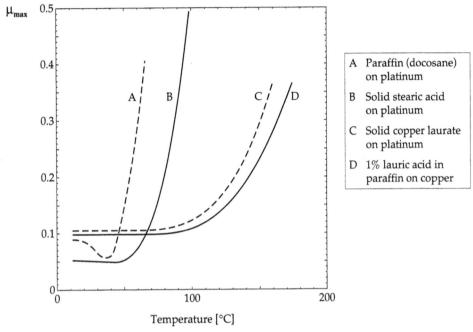

FIGURE 8.6   Effect of temperature on friction of platinum and copper surfaces lubricated by docosane, fatty acids and copper laurate [6].

It can be seen from Figure 8.6 that there is a sharp rise in friction at some 'transition temperature' which is about 45°C for the platinum surfaces lubricated with docosane and about 70°C for the platinum surfaces lubricated with stearic acid. A similar sharp increase in friction is observed for copper laurate applied to platinum and lauric acid applied to copper but the transition temperatures are much higher, about 100°C. The difference in performance between copper and platinum is that the former has sufficient reactivity to induce chemisorption and to produce a mono-molecular layer of copper laurate. Similar friction characteristics are achieved when solid copper laurate is directly applied to platinum. The friction transition temperatures manifested by sharp rises in friction for docosane and stearic acid lubricating platinum occurred at temperatures close to the melting points of docosane (44°C) and stearic acid (69°C). This proximity is not coincidental and relates to the phenomenon of surface melting of the monomolecular layer of adsorbate [7]. An ordered layer of adsorbate is critical to the effectiveness of adsorption lubrication. When the melting point is exceeded, this order is lost and the adsorbed film ceases to function as a lubricating layer, as illustrated schematically in Figure 8.7.

Where physisorbed films are well established, there is strong evidence that the adsorbate molecules form a close-packed normally aligned layer. Early studies revealed that the average area per molecule of n-octadecylamine on platinum was 0.3 [nm²] [8]. In a later more accurate study using radio-actively labelled stearic acid, a packing density of 0.189 [nm²] per molecule was found [9]. This is very close to the theoretical maximum packing density of 0.185 [nm²] per molecule.

Breakdown in the structure of a physisorbed film with increasing temperature was also studied by X-ray diffraction methods [10-13]. The X-ray diffraction pattern from the adsorbed film was observed only at low temperatures. As the temperature increased, this pattern gradually faded to reveal the pattern of the underlying metal. In a separate study it was also demonstrated that breakdown of the low temperature crystalline structure of fatty acids occurred on platinum at a temperature close to the melting point of each acid [14].

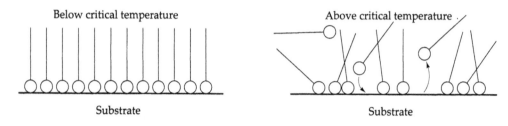

Ordered film analogous to solid   Disordered film analogous to liquid

FIGURE 8.7   Surface melting of adsorbed film.

· *Chemisorption*

Chemisorption or 'chemical adsorption' is an irreversible or partially irreversible form of adsorption which involves some degree of chemical bonding between adsorbate and substrate as illustrated schematically in Figure 8.8.

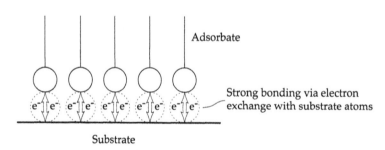

FIGURE 8.8   Mechanism of chemisorption.

Since most common metals, e.g., iron, are reactive, this form of adsorption has practical significance. The strength of chemical bonding between the adsorbate and the substrate which affects the friction transition temperature depends on the reactivity of the substrate material as shown in Table 8.2 [15]. All the materials listed in Table 8.2 were lubricated with lauric acid and the friction transition temperature was determined. The fraction of lauric acid reacting with the metallic surface expressed in terms of the percentage of a retained monolayer after washing was also measured by radio-tracers.

It can be seen from Table 8.2 that zinc, cadmium, copper and magnesium are comparatively well lubricated by lauric acid with low coefficients of friction and high transition temperatures. The common characteristic of all these metals is that some of the lauric acid has been irreversibly adsorbed onto the metallic surface, even though magnesium shows anomalously low reactivity. On the other hand, the inert metals such as platinum and silver show high friction coefficients and low transition temperatures. Glass, which is virtually inert, is also poorly lubricated by lauric acid. No permanent retention of lauric acid is found on platinum, silver or glass. Other metals such as nickel, aluminium and chromium also conform to the pattern of high friction coefficient and poor retention of lauric acid. The reasons for the low reactivity of these metals are, however, still unclear.

TABLE 8.2   Frictional data for lauric acid lubricating metals of varying reactivity [15].

Material	Coefficient of friction at 20°C	Transition temperature [°C]	% acid* reacting	Type of sliding at 20°C
Zinc	0.04	94	10.0	Smooth
Cadmium	0.05	103	9.3	Smooth
Copper	0.08	97	4.6	Smooth
Magnesium	0.08	80	Trace	Smooth
Platinum	0.25	20	0.0	Intermittent
Nickel	0.28	20	0.0	Intermittent
Aluminium	0.30	20	0.0	Intermittent
Chromium	0.34	20	Trace	Intermittent
Glass	0.3 – 0.4	20	0.0	Intermittent (irregular)
Silver	0.55	20	0.0	Intermittent (marked)

* Estimated amount of acid involved in the reaction assuming formation of a normal salt.

A major difference between chemisorption and physisorption is that chemisorbed films are, at least in part, irreversibly bound to the substrate surface. Even washing by strong solvents, which removes physisorbed films, does not remove chemisorbed films. It was found experimentally, for example, that while some stearic acid was removed by strong solvents a certain minimum quantity equal to 38% of a close packed monolayer always remained [9]. However, not all adsorbates are sufficiently reactive to initiate chemisorption. A long-chain alcohol did not show any retention even on the base metals [16] while stearic acid showed permanent retention on base metals such as zinc and cadmium but not on noble metals such as platinum and gold.

Although chemisorption has the irreversible characteristics of a chemical reaction it does not generally proceed to the stage where the original molecule is destroyed. In some environments, however, e.g., vacuum, the complete destruction of a molecule can take place. If, for example, fatty acids were applied as a dilute vapour in a vacuum to a clean iron surface then they would completely decompose to simple gases such as methane, carbon monoxide and hydrogen [17]. This process is an example of catalytic decomposition of organic compounds on clean metallic surfaces. The process may take place in practical wear situations, for example, when a nascent surface is produced under severe load. The nascent surfaces produced during the wear process are usually hot and very catalytic during their short lifetime. Since a strongly adsorbed monolayer is not formed a high coefficient of friction results.

· *Influence of the Molecular Structure of the Lubricant on Adsorption Lubrication*

The molecular structure or shape of the adsorbate has a very strong influence on the effectiveness of lubrication. In addition to the basic requirement that the adsorbing molecules be polar, preferably with an acidic end group for attraction to a metallic surface, the shape of the molecule must also facilitate the formation of close packed monolayers. This latter requirement virtually ensures that only linear molecules are suitable for this purpose. Although the molecules can be of different sizes as shown in Figure 8.9, the size of the molecule is critical. It was found, for example, that the friction transition temperature for fatty acids increased when their molecular weight was raised [6]. More importantly, there is a critical minimum chain length of fatty acids required in order to provide effective lubrication.

It was found that the minimum chain length for effective lubrication is **n = 9** (pelargonic acid) [6]. An increase in 'n' from 9 to 18 (stearic acid) raises the friction transition temperature by about 40°C. Short chain fatty acids with **n ≤ 8** do not show any useful lubricating properties.

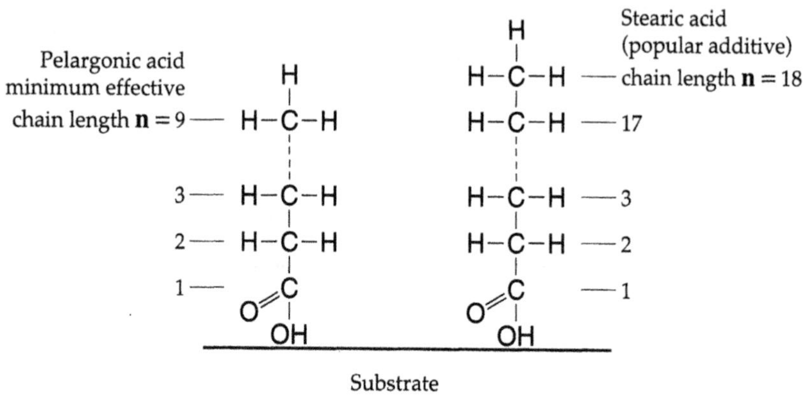

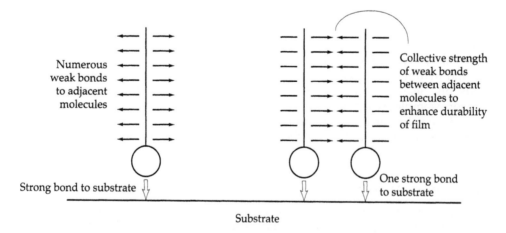

FIGURE 8.9   Chain length of a fatty acid.

The effect of chain length on lubrication may be explained in terms of the relatively weak bonding between **CH$_2$** groups of adjacent fatty acid molecules compared to the bonding at the base of the film as illustrated in Figure 8.10. It seems that a sufficiently large number of paired **CH$_2$** is required to ensure the strength of the adsorbed monolayer.

The effect of chain length is quite strong. For example, **n = 18** alcohol provides a lower coefficient of friction when used with steel than **n = 12** fatty acid, despite the far stronger attraction fatty acids have to metals [18].

FIGURE 8.10   Bonding between fatty acid molecules to ensure the strength of the adsorbed monolayer.

Deviations from the ideal linear molecular shape can severely degrade the lubricating properties of an adsorbate. For example, the differences in friction characteristics become clearly visible for various isomers of octadecanol which include linear and branched molecular configurations [19]. This effect is illustrated in Figure 8.11 where the friction coefficients of a steel ball on a steel plate lubricated by varying concentrations of stearic and isostearic acid in paraffinic oil are shown [20].

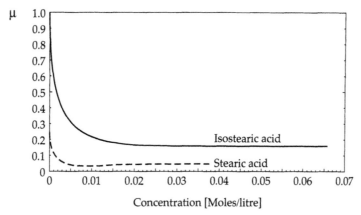

FIGURE 8.11   Effect of varying concentrations of stearic and isostearic acid in paraffinic oil on the coefficient of friction [20].

The difference between the molecular shape of stearic and isostearic acid is that in the latter there are 17 main chain carbon atoms with one branching to the side as opposed to 18 main chain carbon atoms in the former. As can be seen from Figure 8.11 this small difference causes the coefficient of friction between steel surfaces to almost triple. The possible effect of the branched isomerism is illustrated in Figure 8.12.

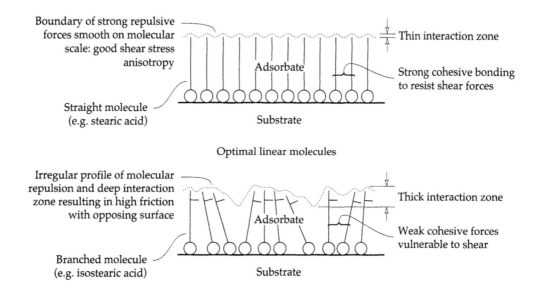

FIGURE 8.12   Disruption of adsorbate film structure by branched molecule.

The branched molecular shape results in two detrimental effects:

· complete surface coverage is difficult to achieve so that the probability of metallic contact is increased,

· there is a deeper interaction zone between opposing adsorbate surfaces, allowing stronger bonding between adsorbate films, resulting in higher coefficients of friction.

Although it is often assumed that fatty acids are among the most effective adsorption lubricants available, other organic compounds such as amines are also fairly effective and are used as lubricant additives to reduce friction. However, the range of commonly cited compounds is fairly narrow.

Hydrocarbons containing silicon and oxygen groups have also been tried. These compounds are generally referred to in the literature as 'silanes' (note that this can easily be confused with an entirely different group of compounds). It was found that under repeated sliding the durability of monolayers formed by silanes was far superior to other adsorption lubricants [21]. The structure of a monolayer of the typical silane compound is shown in Figure 8.13.

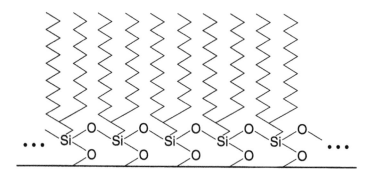

FIGURE 8.13 Structure of the monolayer of a silane compound, an adsorption lubricant with durability superior to fatty acids [21].

The critical difference between the silane and the fatty acid monolayers is the lateral anchoring between silane molecules caused by bonding between adjacent oxygen and silicon atoms. Removal of individual molecules, which creates holes in the film, is effectively prevented by the strong lateral bonds so the monolayer can sustain at least **10,000** cycles of sliding without any increase in friction. In contrast, a monolayer of stearic acid fails after **100** cycles under the same conditions.

It has also been found that an additional hydroxyl group on the fatty acid chain enables cross polymerization of the adsorbate film at high additive concentrations, resulting in a significant reduction in friction [22]. An adsorbate with its modified structure is shown in Figure 8.14.

Much of the knowledge of adsorbate films is still provisional. Although effective forms of adsorption are known, these can always be superseded by newly developed adsorbates.

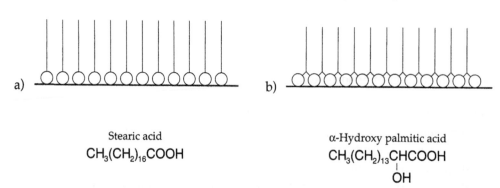

a)

Stearic acid

$CH_3(CH_2)_{16}COOH$

b)

α-Hydroxy palmitic acid

$CH_3(CH_2)_{13}CHCOOH$

OH

FIGURE 8.14 Diagram of a fatty acid (a) and a polymerized derivative (b) [22].

· *Influence of Oxygen and Water*

Atmospheric oxygen and water are always present in lubricated systems unless actively excluded. These two substances are found to have a strong influence on adsorption lubrication since chemically active metals such as iron react with oxygen and water. A surface film of oxide is formed on the metallic surface immediately after contact is made with oxygen. This oxide film is later hydrated by water. Unless the conditions of wear are severe, the oxide film usually survives sliding damage and forms a substrate for adsorbates. The removal of these oxide films by severe wear, however, can result in the failure of adsorption lubrication.

An early study of this phenomenon performed by Tingle involved the temporary removal of the oxide film from a metallic surface by a cutting tool as illustrated in Figure 8.15 [23].

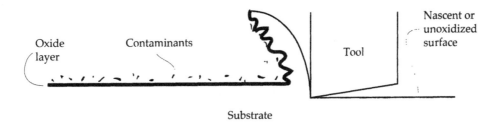

FIGURE 8.15   Removal of oxide films from metallic surfaces by a cutting tool.

A surface layer of the material of thickness about 50 - 100 [μm] is removed. This ensures the complete removal of the surface oxide film which is in fact less than 1 [μm] thick. Unless a high vacuum is maintained, the oxide film rapidly reforms [24]. On the other hand, if the surface is covered by a lubricant, then a virtually unoxidized surface may persist for perhaps as long as a few seconds [25]. Placing the slider directly behind the tool on the steel surface covered with oil enables the measurement of the frictional characteristics of a virtually unoxidized surface. Some information about the effectiveness of adsorption lubrication under severe conditions when metal oxide films become disrupted by wear can be obtained in this manner. The schematic diagram of the apparatus is shown in Figure 8.16 [23].

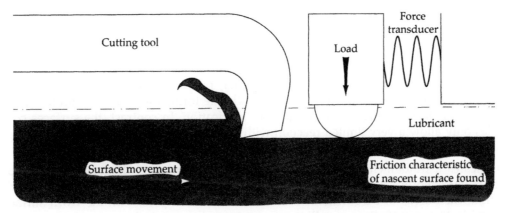

FIGURE 8.16   Schematic diagram of an apparatus for the evaluation of lubricant frictional characteristics with clean metallic surfaces (adapted from [23]).

The frictional characteristics of some chemically active and noble metals are shown in Figure 8.17. The friction tests were performed on uncut metallic surfaces which have previously been polished and cleaned by abrasion underwater, on cut metallic surfaces under a layer of lubricant and on surfaces previously cut in air and washed with water. In this manner the effects of an aged oxide film, nascent surface and recently formed oxide film on friction were assessed. The lubricant used was a solution of lauric acid (a fatty acid) in purified paraffinic oil [23].

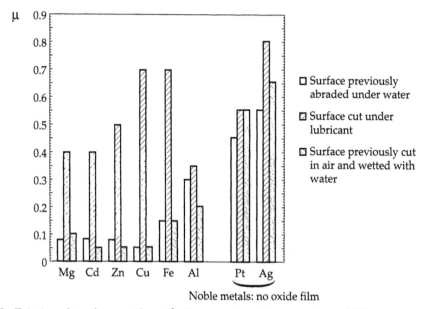

FIGURE 8.17 Friction data for metals with clean and oxidized surfaces [23].

It can be seen from Figure 8.17 that all the metals except platinum and silver exhibit a significant rise in coefficient of friction when the surface is unoxidized. Platinum and other noble metals are lubricated by the mechanism of physisorption which is insensitive to substrate chemistry and therefore unaffected by the presence of any oxide or contaminant films. For other metals, some of which are commonly used as bearing materials (e.g., iron, copper and zinc), the unoxidized surface has a very strong effect on the coefficient of friction. This means that if the oxide film covering these metals is removed, e.g., by severe wear, then a lubrication functioning by adsorption will fail. The similarity in friction data obtained for the 'aged' and 'freshly formed' oxide film indicates that the ageing or maturing of an oxide film is not particularly important.

The reason why nascent surfaces do not allow the establishment of adsorbed films may be due to their extreme reactivity. As discussed earlier, fatty acids decompose to form gaseous hydrocarbons in the presence of clean surfaces [17]. The formation of gas in minute quantities on a surface is clearly entirely unfavourable to lubrication. In contrast, the surfaces covered by oxides only allow a very limited reaction with the fatty acids in the form of chemisorption, which is in fact fundamental to lubrication [26,27].

This indicates a fundamental weakness of adsorption lubrication. If during wear, asperity contact is sufficiently severe to remove not only adsorbed layers but also the underlying oxide film, then areas of bare metallic surface can form and persist on the worn surfaces as illustrated in Figure 8.18. Bare metallic surfaces are prone to seizure or severe wear and this problem is discussed further in the chapter on 'Adhesion and Adhesive Wear'.

Despite the importance of this work, Tingle's contribution has largely been ignored in the literature and his experiments have never been repeated. The effect of ambient oxygen and water on friction and wear has also been studied. It has been found that lubricating oil is ineffective in preventing severe wear in a steel-on-steel sliding contact without oxygen and water [28]. It has also been found that oxygen alone gives more favourable results than water without oxygen but the combination of oxygen and water provides the lowest friction and wear. Studies of a range of lubricants and lubricant additives revealed that ambient oxygen and water enhance the functioning of most lubricants except certain phosphorous additives for which water has a harmful effect [29-31].

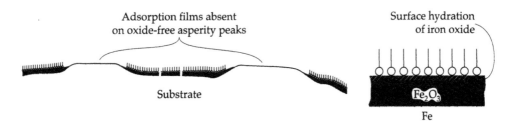

FIGURE 8.18   Formation of bare metallic surface, unfavourable for adsorption lubrication, by removal of oxide films.

· *Dynamic Nature of Adsorption under Sliding Conditions*

Almost all of the fundamental research on adsorption lubrication is devoted to mono-molecular films of adsorbate which have been allowed to reach chemical and thermal equilibrium. As more recently observed, however, it is extremely unlikely that adsorbate films can equilibriate under sliding conditions [32]. Most of the data available on adsorption films has been obtained under rigorously controlled conditions during which cleaning of the specimen surfaces and deposition of the lubricant films take many hours. The friction tests themselves are performed at extremely slow sliding speeds with several minutes between successive sliding contacts. In modern equipment, however, such as high speed gears, the repetition rate of frictional contacts may reach several hundred cycles per minute. The differences between a dynamic form of adsorption lubrication and the classic equilibrium model remain poorly understood.

An experiment designed to find whether friction is determined by the balance between adsorption and removal of the monolayer by friction under dynamic conditions was conducted on a steel-on-steel contact based on a ball and cylinder apparatus [33]. The coefficient of friction was measured as a function of concentration of various surfactants (adsorbing agents) in pure hexadecane (inert neutral carrier fluid). Surfactants studied were fatty acids, i.e., lauric, myristic, palmitic, stearic and behenic of varying chain length 'n' from 11 to 21. A transition concentration was found for all lubricants tested where the friction coefficient declined sharply from a value characteristic of hexadecane to a level dictated by the additive or surfactant as shown in Figure 8.19.

The 'transition concentration', where the rate of decline of friction coefficient with increasing concentration of fatty acid decelerates, has a value close to 0.5 [mol/m^3] for most of the fatty acids. This concentration is modelled as the minimum concentration where the fatty acids will replenish a friction-damaged adsorbate film under the conditions of repetitive sliding contact. It is also assumed that adsorption lubrication is not effective unless the adsorbate film is in near-perfect condition, i.e., has very few holes or vacant sites in it. In the case where equilibrium adsorption prevails, a simple linear dependence of friction with fatty acid concentration would be expected. The rate-limiting step in the formation of an adsorbate film under sliding conditions is believed to be re-adsorption and a minimum concentration of fatty acid is required for this process to occur within the time available between successive sliding contacts. This model is illustrated schematically in Figure 8.20.

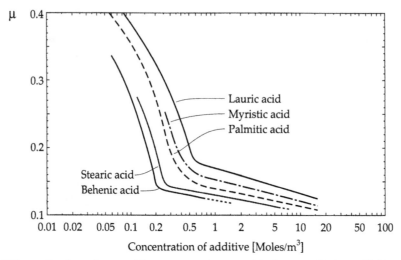

FIGURE 8.19   Effect of solute fatty acid concentration on friction coefficients [33].

Some models of adsorption under dynamic conditions suggest that the rapidity of processes in high speed wearing contacts can in fact provide favourable as opposed to destructive effects on lubrication. A phenomenon known as the '**Borsoff effect**', where the apparent limiting temperature for adsorption lubrication is raised by increasing the speed of a rotary sliding mechanism, has been modelled in terms of the suppression of desorption of an adsorbed film [34,35]. In this model, it is hypothesized that if the pulses of frictional heat become shorter than the average residence time of an adsorbed molecule, then the activation energy of desorption becomes significant. The adsorbed film would have a greater chance of survival at high sliding speeds where pulses of frictional heat become very short and also usually more intense. This idea, however, did not persist in the literature and it has to be concluded that dynamic effects comprise another poorly understood aspect of lubrication by adsorbates.

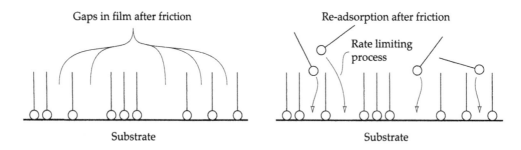

FIGURE 8.20   Model of adsorption kinetics under sliding conditions.

· *Mixed Lubrication and Scuffing*

Most sliding contacts of practical importance, e.g., high speed gearing, are not lubricated by either purely hydrodynamic, elastohydrodynamic or by classical adsorption lubrication. Usually two lubrication mechanisms act simultaneously and both are essential for lowering friction and wear in the contact. In many cases most of the applied load is supported by hydrodynamic or EHL lubrication. However, some additional lubrication mechanism is required to reduce friction and wear in contacts between large asperities from opposing surfaces. Even if the fraction of load supported by non-hydrodynamic means is small, severe wear and perhaps seizure can occur if this additional component of lubrication is not available. This particular lubrication regime where several mechanisms act simultaneously is termed '**mixed lubrication**'. The current model of this lubrication regime is illustrated schematically in Figure 8.21.

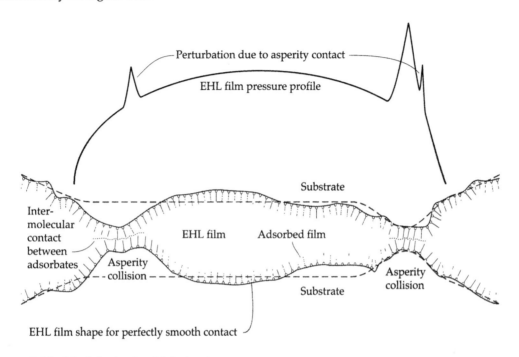

FIGURE 8.21   Model of mixed lubrication.

Mixed lubrication allows much smaller film thicknesses than pure hydrodynamic lubrication or EHL. Reduced film thickness coincides with increased load and contact pressure, if other factors remain unchanged, and this characteristic is the basic reason for the importance of 'mixed lubrication'.

Although in most cases when this lubrication regime is active the collisions between asperities are prevented from inducing any severe forms of wear, a sudden and severe mode of lubrication failure known as '**scuffing**' or '**scoring**' in the U.S.A can occur. This can cause serious industrial problems since scuffing can occur precipitately in an apparently well lubricated contact. Scuffing often takes place in heavily loaded gears.

An example of changes in the oil film thickness for the root and pitch line of the gear teeth versus applied load is shown in Figure 8.22. The film thickness is shown as the voltage drop across the contact. It can be seen that the film is thicker at the pitch line where there is almost pure rolling between contacting surfaces while at the root of the gear tooth where a significant amount of sliding is present the film is thinner.

At a certain load level a rapid collapse in oil film thickness at the root of the gear tooth occurred. This was also manifested by a sharp rise in friction and destruction of the wearing surfaces. It can be seen from Figure 8.22 that there is no gradual decline in film thickness to a zero value and there is no pre-indicator of film collapse. These characteristics constitute a major limitation in the application of lubricated gears.

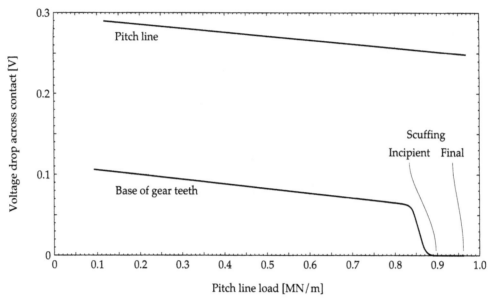

FIGURE 8.22 Experimental observation of oil film collapse and initiation of scuffing in heavily loaded gears [36].

The rapidity of scuffing and the destruction of the original surfaces greatly impede any investigations into the original causes of scuffing. In fact, in cases of severe scuffing, the oil may burn and the steel teeth may sustain metallurgical modification as well as plastic deformation. Scuffing may result in the altered microstructure of a gear surface. During scuffing of steel gears, the surface temperature of the teeth may become very high. It is also probable that the rate of temperature increase is extremely high. These conditions favour the formation of austenite, which is found in concentrations of up to 60% austenite fraction on the surface of scuffed gear teeth to a depth of 5 [μm] [134]. For example, in one study scuffing failure was induced in steel samples with tempered martensite microstructure [134]. The examination of the scuffed surfaces revealed the presence of austenite, up to 60% by volume in some of the samples. A plausible explanation for the austenite formation during scuffing requires very high temperature and/or high heating rates to be reached on the surfaces [134]. A generalized view of events leading to severe scuffing is illustrated in Figure 8.23.

It is possible for systems to recover from a mild scuffing which further demonstrates the complexity of the problem. A comprehensive review of scuffing models can be found elsewhere [37,118].

In simple terms, it can be reasoned that the desorption of an adsorbed film sets in motion a train of events leading to the complete destruction of a mechanical component. This view, however attractive, is only part of the description of scuffing and there are many other influences occurring which make scuffing an almost intractable problem. The best known theory of scuffing, particularly in gears, is the Blok limiting frictional temperature theory [38,39]. It was postulated by Blok that when a critical temperature is reached on the sliding surfaces, scuffing will be initiated. The temperature on a sliding surface is the sum of ambient temperature, steady state frictional heating and transient friction temperature which is a function of load and sliding speed. The critical temperature was observed to be in the region of **150°C**. The theory unfortunately lacks a specific explanation as to why there should be a critical temperature. It has often been assumed that this temperature relates to the desorption temperature but experimental studies suggest that this is only a crude approximation. The concept of steady state and transient temperatures in a mixed lubrication sliding contact is illustrated in Figure 8.24.

The concept of temperature under mixed lubrication shown in Figure 8.24 is far from simple. Exactly which temperatures contribute to scuffing has rarely been discussed in detail partly because of the difficulty in measuring transient temperatures only occurring on the surface of a moving object and attained for about **10** [ms]. Blok's critical temperature criterion is not particularly reliable since, although a critical temperature does exist, it varies in a complex and unpredictable manner [40] and the criterion usually underestimates the scuffing resistance of a sliding contact, occasionally giving an overestimate of permissible loads and sliding speeds [41].

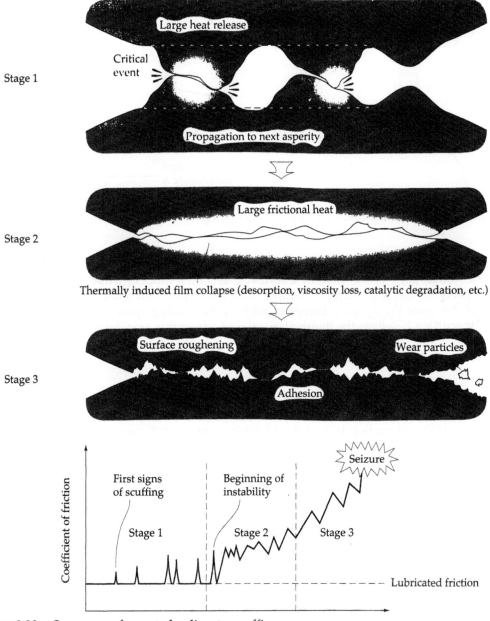

FIGURE 8.23   Sequence of events leading to scuffing.

For these reasons, one reliable means of avoiding scuffing is to apply relatively conservative gear design standards. Also, as will be discussed later, EP lubricants can prevent scuffing when the limit of adsorption lubrication is reached. There is, however, no quantitative measure available for estimating, for example, how much extra load can be applied if special anti-scuffing lubricants are used.

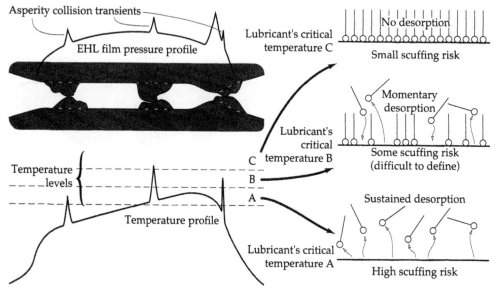

FIGURE 8.24 Model of transient and steady state temperatures in a mixed lubrication sliding contact.

Interpretation of scuffing in terms of friction transition temperatures has been the object of extensive studies [e.g., 42-44]. Initially a series of tests on a model friction apparatus with low sliding speed to suppress frictional temperature rises was conducted [42], followed by the testing of the same theory on tribometers with higher sliding speeds [43,44]. The following model for scuffing based on the thermodynamics of adsorption was developed [44]:

$$\ln C = -E_a \, / \, (RT_t) + \text{constant} \tag{8.3}$$

where:

C	is the concentration of the additive in the solvent base stock [%wt];
E_a	is the adsorption heat of the additive on the metallic surface [kJ/kmol];
R	is the universal gas constant [kJ/kmolK];
T_t	is the friction transition temperature [K].

According to this model, when the transition temperature is exceeded, damage to the adsorbate film is more rapid than film repair so that the adsorption film is progressively removed. High friction and wear are then inevitable.

For a narrow range of experimental conditions, agreement between the model (8.3) and experimental data was obtained. In Figure 8.25 the relationship between the concentration of fatty acids of varying chain length dissolved in purified inert mineral oil and transition temperature is shown [42].

The data provide a linear plot which is in agreement with theory. The gradient of the graph, which is a measure of the heat of adsorption, is also approximately the same as the heat of adsorption determined by more exact tests. On the other hand, it has been found that a 1% solution of oleic acid improved the lubrication capabilities of white oil even though the transition temperature for oleic acid was clearly exceeded [45]. In other words, this indicates that fatty acids do not function merely by adsorption lubrication and some new theories are necessary.

An attempt to estimate the critical temperature in EHL contacts where scuffing or film failure is likely to occur based on heat of adsorption, sliding speed and melting point of lubricant has been made [35,113,114]. The practical applications of the expression found are extremely limited because the data referring to one of the variables, i.e., melting temperature of the lubricant, is only available for pure compounds, not for mixtures, which commercial oils are.

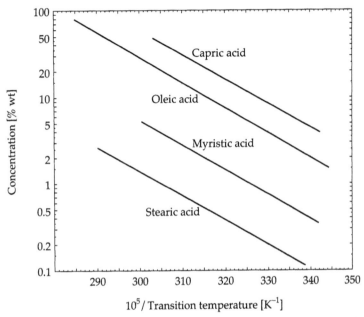

FIGURE 8.25   Relationship between friction transition temperature and concentration of some adsorption additives [42].

It should be realized that an EHL pressure field also affects both the concentration of additives within the EHL contact and the mechanism of adsorption lubrication by raising the critical temperature for desorption [115,125]. It has been found that under EHL pressure the concentration of additives in plain mineral oil (but not certain synthetic oils) tends to decline to less than half the bulk oil concentration [125]. Although the causes of this effect remain unclear, the implications for additive function within the EHL contact and its consequences on scuffing appear very significant. According to Langmuir's theory of adsorption, elevated pressure increases the fraction of the surface covered by adsorbate for any given temperature. The critical temperature for scuffing to occur is modelled as the temperature where the fractional surface coverage by adsorbate declines to less than half of the available atomic sites on the surface [116]. The critical temperature is approximately 150°C in slow speed sliding experiments but is between 300°C and 400°C in full scale scuffing tests where a substantial EHL pressure field is present [115]. Thus the equations which attempt to predict the critical temperature of scuffing according to Blok's theory but which do not allow for the pressure dependence might give incorrect results.

The increase in critical desorption temperature with pressure suggests a mechanism of combined instability in a lubricated contact. Consider an experiment conducted on a 'two disc' apparatus where the discs are subjected to a progressive increase in load until scuffing occurs. Assume that initially the EHL pressure is low and the desorption temperature is close to that obtained in low speed tests. As the load is increased the contact temperatures, frequency of asperity contact and hydrodynamic pressure increase. The critical desorption temperature also increases so that effective adsorption lubrication is maintained. At some level of load, however, a limiting EHL pressure is reached. The limiting EHL pressure may either be due to lubricant characteristics or may be determined by the hardness of the disc materials. Further increases in load merely tend to increase the contact area or shift a greater proportion of the load onto asperity contacts. At this stage, two events may occur: either there is direct desorption of the adsorbate lubricating films caused by excessive contact temperature or there is a progressive collapse in EHL pressure caused by the asperity interference. When pressure declines the adsorbed films become unstable. The collapse in pressure can be limited to asperity contacts only while the bulk pressure field remains unaffected. If there is a localized reduction in hydrodynamic pressure then the critical desorption temperature will precipitately decline to allow local desorption of the adsorbate film. Scuffing will then be initiated from localized adhesive contacts between asperity peaks denuded of adsorbate film. This mechanism of combined instability is illustrated in Figure 8.26.

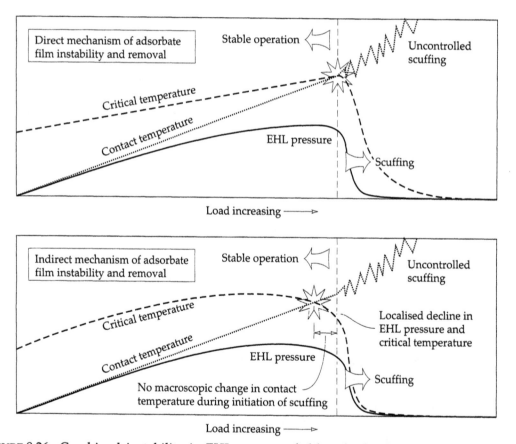

FIGURE 8.26   Combined instability in EHL pressure field and adsorbate lubricating films as cause of scuffing.

As may be deduced from Figure 8.26, it is still unclear whether any collapse in EHL or micro-EHL induces desorption and scuffing or whether desorption occurs first and the resultant surface damage causes the cessation of EHL.

A phenomenon of catalytic oil decomposition is believed to contribute to scuffing too. At high temperatures found in a heavily loaded EHL film, whenever asperity contact occurs, i.e., when the EHL film thickness becomes comparable to the combined surface roughness of the contacting surfaces, the exposure of nascent surfaces worn by asperity interaction may directly affect the base oil of the lubricant which may have severe consequences for the lubricant film. It is known that a major feature of nascent surface is its elevated catalytic activity compared to quiescent, oxidized metal [17,119-121]. Nascent surface typically catalyzes decomposition reactions of organic compounds found in oil to release low molecular weight products that are often gaseous [119]. Such catalysis can have a destructive effect on the lubricating capacity of an oil. It has been suggested that scuffing occurs when there is sufficient nascent surface exposed by mechanical wear to cause the rate of chemical degradation of oil inside the contact to exceed the rate at which it can be replenished [121]. When a critical rate of degradation is reached, contact closure by partial failure of lubrication further reduces the supply rate of oil. This causes a sharp transition to unstable lubrication and scuffing from stable lubrication below a critical load. The catalysed decomposition products are unlikely to facilitate lubrication since they do not possess the physical capacity to sustain the extremely high shear rates prevailing in the EHL contacts. These products tend to accumulate between asperity contacts, and when their concentration becomes high enough, lack of lubrication occurs, leading to scuffing [115]. In addition, the decomposition products most probably surround the contact excluding fresh lubricating oil or else under the influence of extreme frictional heating they react and chemically bind both sliding surfaces. This last effect could cause a catastrophic rise in friction levels. The schematic illustration of this '**catalytic model of scuffing**' is shown in Figure 8.27.

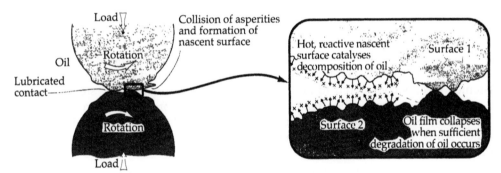

Figure 8.27   Schematic illustration of catalytic model of scuffing of metal surfaces lubricated by an oil.

An important method of controlling scuffing would appear to be the prevention of nascent metal surface by covering the sliding surfaces with coatings of non-metallic materials such as ceramics. Steel gears coated with titanium nitride and carbide have been found to offer good scuffing resistance in gearbox tests compared to uncoated steel gears [122]. Selection of a stable lubricant is also important since the decomposition of perfluoroalkyether lubricating oils has been found to initiate scuffing and wear of metal surfaces [123,124].

In other more mechanically orientated studies, it was also found that operational parameters such as loading history and run-in procedure have a strong influence on scuffing and measured critical temperatures [46-48]. The critical temperature appears therefore to be a function of many parameters not just pressure, heat of adsorption and sliding speed. Despite the poor understanding of scuffing, research in this area has become scant in recent years. This may be due to the fact that scuffing belongs to 'industrial tribology' which has a relatively low priority compared to other aspects of tribology [49].

· *Metallurgical Effects*

The effect of alloying and heat treatment to produce a specific microstructure also exerts a major influence on whether a low coefficient of friction can be obtained by oil-based lubrication. Frictional characteristics of steel-on-steel contacts versus temperature for two steels, a martensitic plain carbon steel and an austenitic stainless steel, are shown in Figure 8.28 [50]. Both steels are lubricated by mineral oil.

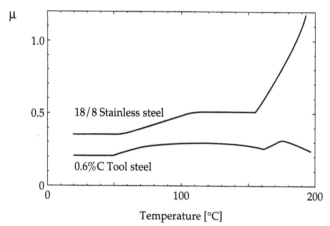

FIGURE 8.28   Frictional characteristics of plain carbon and stainless steels versus temperature under mineral oil lubrication [50].

It can be seen from Figure 8.28 that the coefficient of friction of the austenitic stainless steel rises sharply at 160°C, reaching values greater than unity by 200°C. In favourable contrast, the coefficient of friction of the plain carbon martensitic steel remains moderate in the range of 0.2 - 0.3. The difference between these two steels can be explained in terms of reactivity since the austenitic steel is considered to be less reactive than the martensitic steel because of the latter's greater lattice strain. The greater reactivity causes more rapid formation or repair of oxide films and re-adsorption of surfactant films under conditions of repeated sliding contact. In another study [52] it was found that austenitic steels have lower friction transition temperatures than martensitic steels.

Similar tests conducted with additive enriched oils revealed that low alloy steels exhibit lower coefficients of friction than high alloy steels [51]. It appears that both the phase of the steel and the alloying content are the controlling factors in lubricant performance. For example, chromium was found to raise the scuffing load for austenitic steels [53] while the contrary effect was found in other cases [54-56] where martensitic and ferritic steels were tested. In a comprehensive study where the effect of different alloying elements on scuffing resistance was tested, it was found that irrespective of the alloying elements, the microstructure has a controlling effect on scuffing load. For example, ferrite gives the highest scuffing loads and since martensite and cementite are less 'reactive', they lower the scuffing load [57].

Austenite is the most unsuitable phase and gives very low scuffing loads. The failure load for austenites is less than one-tenth of the failure load for ferritic steels [57]. Thus hardening of steels does not provide increased protection against scuffing since this induces martensite with a corresponding reduction in ferrite.

·   *Interaction between Surfactant and Carrier Fluid*

In the model of adsorption lubrication discussed so far, the fatty acid or surfactant was either applied neat to the test surface or as a solution in an inert fluid. In practice, the 'carrier fluid' or 'base stock' can also influence the lubrication mechanism. It was found that the heat of adsorption of stearic and palmitic acid on iron powder was up to 50% greater with hexadecane as the carrier fluid than with heptane [58]. The heat of adsorption dictates the friction transition temperature. For example, if hexadecane is used as a carrier fluid in preference to heptane, a higher friction transition temperature can be expected.

This aspect of adsorption lubrication has also been relatively neglected, partly because of the difficulty in manipulating mineral oil as a carrier fluid. With the adoption of synthetic oils which offer a much wider freedom of chemical specification, systematic optimization of the heat of adsorption may eventually become practicable.

8.4   HIGH TEMPERATURE - MEDIUM LOAD LUBRICATION MECHANISMS

There has always been much interest in oil based lubrication mechanisms which were effective at high temperatures.

The primary difficulty associated with lubrication is temperature, whether this is the result of process heat, e.g., a piston ring, or due to frictional energy dissipation, e.g., a high speed gear. Once the temperature limitations of adsorption lubrication were recognized the search began for 'high temperature mechanisms'. Although these mechanisms have remained elusive, some interesting phenomena have been discovered.

Two basic mechanisms involved in high temperature lubrication at medium loads have been found: chain matching and formation of thick films of soapy or amorphous material. Chain matching is the modification of liquid properties close to a sliding surface in a manner similar to the 'low temperature - low load' mechanism but effective at far higher temperatures and contact pressures, and dependent on the type of additive used. The thick colloidal or greasy films are deposits of material formed in the sliding contact by chemical reaction. They separate the opposing surfaces by a combination of very high viscosity and entrapment in the contact.

Chain Matching

Chain matching refers to the improvement of lubricant properties which occurs when the chain lengths of the solute fatty acid and the solvent hydrocarbon are equal. This is a concept which is not modelled in detail but which has periodically been invoked to explain some unusual properties of oil-based lubricants.

In a series of 'four-ball' tests the scuffing load was found to increase considerably when the dissolved fatty acid had the same chain length as the carrier fluid lubricant [43]. An example of scuffing load data versus chain length of various fatty acids is shown in Figure 8.29. Three carrier fluids (solvents) were used in the experiments, hexadecane, tetradecane and decane of chain lengths of **16**, **14** and **10**, respectively.

The maximum in scuffing load occurred at a fatty acid chain length of **10** for decane, **14** for tetradecane and **16** for hexadecane. To explain this effect, it was hypothesized that a coherent viscous layer forms on the surface when chain matching occurred. This is similar to the 'low temperature - low load' mechanism discussed previously except that much higher contact stresses, > 1 [GPa], and higher temperatures, > 100°C, are involved and furthermore the mechanism is dependent on the type of additive used. It was suggested that when chain matching occurs, a thin layer with an ordered structure forms on the metallic surface. The additive, since it usually contains polar groups, may even act by bonding this layer to the surface. If the chain lengths do not match then a coherent surface structure cannot form and the properties of the surface-proximal liquid remain similar to those of the disordered state of bulk fluid as shown in Figure 8.30.

To support this argument, the near surface viscosity under hydrodynamic squeeze conditions was measured and a large viscosity was found when chain matching was present [43]. The relationship between the viscosity calculated from squeeze rates versus distance from the surface for pure hexadecane and hexadecane plus fatty acids of varying chain length is shown in Figure 8.31.

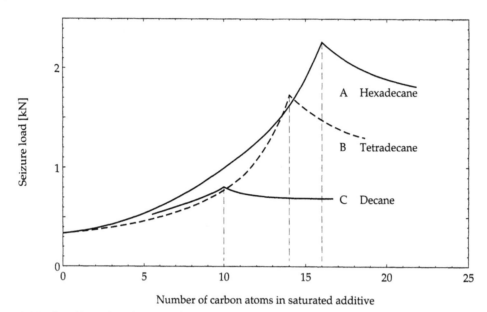

FIGURE 8.29 Scuffing loads as a function of fatty acid chain length for various aliphatic hydrocarbon carrier oils [43].

Although chain matching has been confirmed in other studies [59,60], many researchers have failed to detect this effect and still remain sceptical [33]. Recently, however, an influence of fatty acids on EHL film thickness was also detected [61]. Film thickness or separation distance versus rolling speed under EHL lubrication by pure hexadecane and hexadecane with stearic acid present as a saturated solution is shown in Figure 8.32.

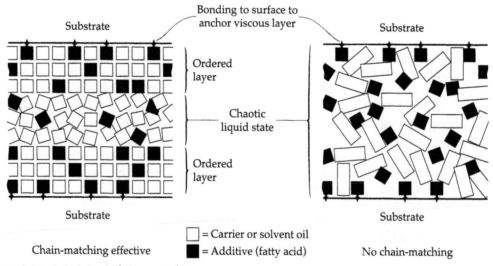

FIGURE 8.30 Model of chain matching.

It can be seen from Figure 8.32 that EHL film thicknesses for pure hexadecane and a hexadecane solution of stearic acid diverge significantly. At very low speeds hexadecane gives no residual film on the surface while the stearic acid/hexadecane solution gives separation of about 2 [nm]. This effect can be attributed to an adsorbed layer of stearic acid. As speed increases and an EHL film is generated the film thickness for both lubricating liquids becomes the same and the effect of stearic acid is diminished.

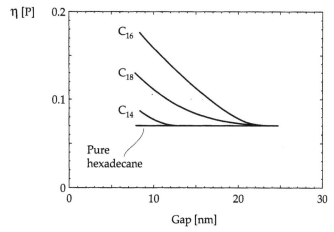

FIGURE 8.31 Viscosity versus distance between squeezing surfaces for pure hexadecane and hexadecane with dissolved fatty acids of chain lengths 14, 16 and 18 [43].

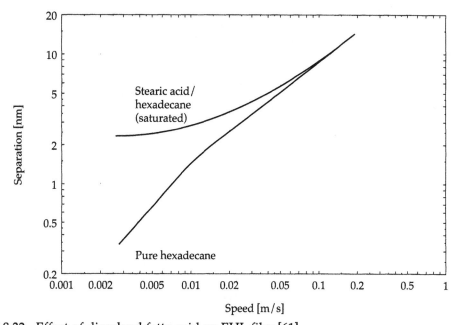

FIGURE 8.32 Effect of dissolved fatty acid on EHL film [61].

The effects of various fatty acids on friction, i.e., lauric, palmitic and stearic acid added to hexadecane, were tested under heavily loaded conditions between sliding steel surfaces [62]. At low friction a layer of adsorbate, thicker than a monolayer, was detected by contact resistance measurements. After the friction transition temperature was exceeded and the friction coefficient rose, this layer seemed to decline to negligible values. However, the highest friction transition temperature of about 240°C was recorded when the chain length of the fatty acid matched that of the hexadecane, i.e., at **16** which corresponds to palmitic acid. For the other acids, the friction transition temperature was much lower, between 120°C and 160°C.

Thick Films of Soapy or Amorphous Material

Almost all additives used to control friction and wear can react chemically with the worn metallic surface. This means that in addition to adsorbate films and viscous surface layers, a layer of reaction product can also form on the sliding contact surface. It is virtually impossible to control this process once the additive is present in the oil. Until recently this aspect of additive interaction was hardly considered since the reaction products were usually assumed to be extraneous debris having little effect on film thicknesses, friction and wear. Recently, however, the idea of films thicker than a mono-molecular adsorbate layer but thinner than the typical EHL film thickness has been developed [62,63,67-69]. The thickness of this film is estimated to be in the range of **100 - 1000** [nm] and the limitations of desorption at high frictional temperatures have been avoided. The consistency or rheology of these films varies from soapy, which implies a quasi-liquid, to a powder or amorphous solid.

· *Soap Layers*

Soap layers are formed by the reaction between a metal hydroxide and a fatty acid which results in soap plus water. If reaction conditions are favourable, there is also a possibility of soap formation between the iron oxide of a steel surface and the stearic acid which is routinely added to lubricating oils. The iron oxide is less reactive than alkali hydroxides but, on the other hand, the quantity of 'soap' required to form a lubricating film is very small. Soap formation promoted by the heat and mechanical agitation of sliding contact was proposed to model the frictional characteristics of stearic acid [62,63]. In the theory of adsorption lubrication, it was assumed that only a monolayer of soap would form by chemisorption between the fatty acid and underlying metal oxide, e.g., copper oxide and lauric acid to form copper laurate. No fundamental reason was given as to why the reaction would be limited to a monolayer.

The soap formed by the reaction between a fatty acid and metal is believed to lubricate by providing a surface layer much more viscous than the carrier oil as shown schematically in Figure 8.33 [62].

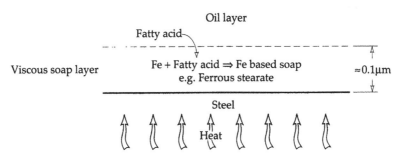

FIGURE 8.33   Formation of a viscous soap layer on steel by a reaction between iron and a fatty acid in lubricating oil.

The presence of a viscous layer functioning by the mechanism of hydrodynamic lubrication was deduced from electrical contact resistance measurements [62]. When there was a measurable and significant contact resistance, the thick viscous layer was assumed to be present. Dependence on hydrodynamic lubrication was tested by applying the Stribeck law. According to the Stribeck law, the following relationship applies at the limit of hydrodynamic lubrication:

$$\log U + \log \upsilon - \log W = \text{constant} \tag{8.4}$$

where:

U   is the sliding velocity [m/s];

υ is the kinematic viscosity [m^2/s];

W   is the load [N].

The apparatus used to measure friction was a reciprocating steel ball on a steel plate, oscillating at short amplitude and high frequency as shown schematically in Figure 8.34. The value of the constant in equation (8.4) was found by measuring the loads and velocities where oil film collapse, manifested by a sharp increase in temperature (Figure 8.34), occurred during lubrication by plain mineral oil. Assuming that the constant is only a function of film geometry and independent of the lubricant it is possible to calculate the viscosity of the soap film. An example of the experimental results obtained with 0.3% stearic acid in hexadecane is shown in Figure 8.35 [62].

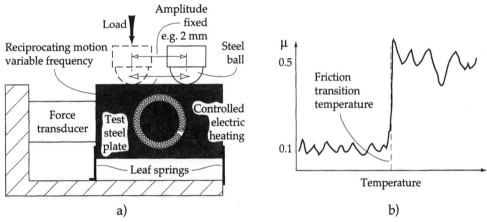

FIGURE 8.34 Experimental principles involved in detecting viscous soap layers during reciprocating sliding: a) schematic diagram of the test apparatus and b) sharp increase in friction temperature indicating collapse of lubricating film (adapted from [62]).

It can be seen from Figure 8.35 that the calculated viscosity is in the range between **200** and **2000** [cS] which is similar to the viscosity of a soap under the same temperature.

The limitation associated with this mode of lubrication is that like chemisorption, reaction with an oxidized metallic substrate is a pre-requisite. Steels and other active metals such as copper or zinc would probably form soap layers whereas noble metals and non-oxide ceramics are unlikely to do so.

· *Amorphous Layers*

It is known from common experience that the process of sliding involves grinding which can reduce the thickness of any interposed object. Lumps of solid can be ground into fine powders and, at the extreme, a crystal lattice can be dismantled into an amorphous assembly of atoms and molecules. This process is particularly effective for brittle or friable substances. As discussed already, many lubricant additives function by reacting with a substrate to form a deposit or film of reacted material which is inevitably subjected to the process of comminution imposed by sliding. This material, finely divided (i.e., as very fine particles) or with an amorphous molecular structure, can have some useful load carrying properties and can also act as a lubricant.

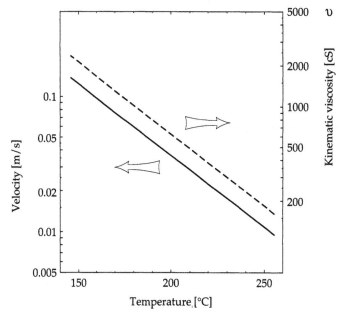

FIGURE 8.35   Relation between temperature, sliding speed and viscosity of the soap layer formed in sliding contact during lubrication by stearic acid in hexadecane [62].

The process of amorphization of interposed material can be illustrated by a bubble raft analogue of a crystal lattice. Each bubble is analogous to an atom and, when closely packed, the bubbles resemble a crystal lattice if regular and an amorphous distribution if irregularly arranged [64]. An example of a bubble raft model of a sliding interface is shown in Figure 8.36.

Material close to the sliding surfaces tends to be crystalline because of the tendency to align with a plane surface. The bulk of the material, however, is amorphous because the shearing caused by sliding does not follow exact planes parallel to the sliding direction. Instead, transient ripples of shear waves completely disrupt any pre-existing crystal structure as shown in Figure 8.37.

Amorphous layers of phosphates containing iron and zinc have been found in steel sliding contacts when zinc dialkyldithiophosphate (ZnDDP) was used as a lubricant additive [65,66]. The formation of these amorphous layers is associated with anti-wear action by the ZnDDP for reasons still unclear.

Finely divided matter as small as the colloidal range of particles has been shown to be capable of exerting a large pressure of separation between metallic surfaces [67]. Very little pressure is required to compress a spherical powder particle to a lozenge shape, but when this lozenge shaped particle is further deformed to a lamina, the contact pressure rises almost exponentially. This can be visualized by considering the indentation of a layer of powder supported by a hard surface using a hemispherical punch. Initial indentation requires little force but it is very hard to penetrate the powder completely. The deformation process of a soft spherical powder particle is illustrated schematically in Figure 8.38.

When the compression force is sufficiently large, the soft material is entrapped within the harder surface as illustrated in Figure 8.38. The resultant strain in the hard material may cause permanent deformation which could be manifested by scratching and gouging [67]. The compression tests reported were performed without simultaneous sliding. The films deposited by ZnDDP presumably have the ability to roll and shear within the sliding contact while individual 'lumps' of material are not further divided into smaller pieces.

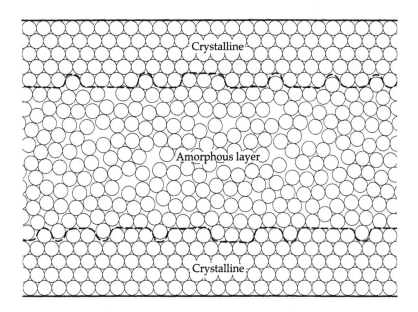

FIGURE 8.36   Bubble raft analogy of crystal/amorphous structure of the material separating sliding surfaces [64].

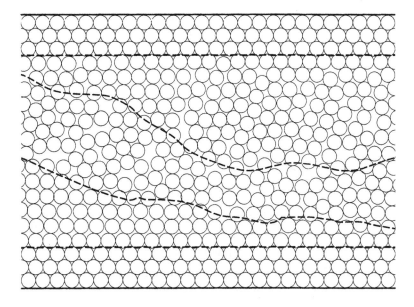

FIGURE 8.37   Rippling shear fronts under sliding using the bubble raft analogy as a mechanism of destruction of the crystal lattice [64].

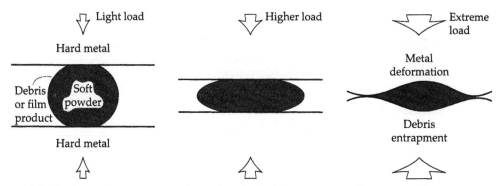

FIGURE 8.38 Deformation process of a soft material between two harder surfaces.

These deposited films of solid powder or amorphous material on metallic surfaces would not suffer from the drawback of desorption at a limiting temperature or viscosity loss with increasing temperature as adsorbed films do. The mechanism which is involved in formation of these films evidently forms the basis of the high temperature lubricating properties of ZnDDP as compared to fatty acids. This research, however, is very recent and the current models will most probably be revised in the future.

A further intriguing aspect of phosphorus-based additives is the spontaneous formation of deposits under a full elastohydrodynamic film [68,69]. Both pure phosphonate esters and solutions of phosphonate esters in paraffin tested by optical interferometry showed an increase in the EHL film thickness from about 200 [nm] to 400 [nm] over 2 [hours] of testing at 100°C. An example of this increase in EHL film thickness versus rolling time for a 3% solution of didodecyl phosphonate in purified mineral oil is shown in Figure 8.39.

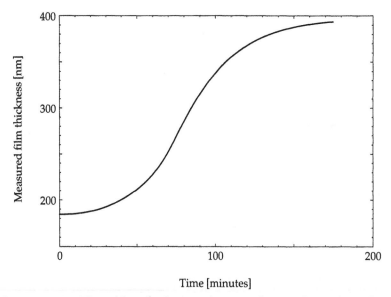

FIGURE 8.39 Increase in EHL film thickness due to chemical reaction between the iron substrate and the phosphonate additive [68].

Surface analysis of the rolling track of the EHL contact revealed that the deposited layer was composed of a polymerized network of iron phosphate with some organic groups included. This indicated that the iron had reacted with the phosphorous additive to form a thick layer on the surface. The layer was extremely viscous and waxy in consistency and almost insoluble in organic solvents. Polymerization by cross-linkage between phosphate and iron atoms was also detected. It therefore seemed possible that an irregular network of repeating phosphate and organic groups and iron atoms formed to create an amorphous structure. This particular process may be the only confirmed observation of a so called '**friction polymer**' [70]. Friction polymer in general terms refers to the polymerization of hydrocarbon lubricants in contact with metal whose oxide film has been removed by friction. A clean metallic surface, particularly of steel, is believed to have strong powers of catalysis which can induce the formation of hydrocarbon polymer films on the worn surface. These films are believed to reduce friction and wear.

8.5   HIGH TEMPERATURE - HIGH LOAD LUBRICATION MECHANISMS

A lubrication mechanism acting at high temperature and high load is generally known as lubrication by sacrificial reaction films or '**Extreme Pressure lubrication**' (often abbreviated to EP lubrication). This mechanism takes place in lubricated contacts in which loads and speeds are high enough to result in high transient friction temperatures sufficient to cause desorption of available adsorption lubricants. When desorption of adsorbed lubricants occurs, another lubrication mechanism based on sacrificial films is usually the most effective means available of preventing seizure or scuffing. The significance of temperature has led to the suggestion that this mode of lubrication be termed 'Extreme Temperature Lubrication' [6] but this term has never gained wide acceptance. It seems that the term 'Extreme Temperature' is too ambiguous for practical use since oils are never used at extreme temperatures and furthermore the contact stresses under which EP lubricants are effective and commonly used considerably exceed the limiting contact stresses of many high temperature lubricants. For example, sulphur-based EP additives used in pin-on-disc sliding tests ensure moderate wear rates up to 2 [GPa] while methyl laurate (an adsorption lubricant) fails at about 1.3 [GPa] and allows scuffing. Plain mineral oil shows an even lower ability to operate under high contact stress, resulting in excessive wear rates at contact pressures below 1 [GPa] [71].

Model of Lubrication by Sacrificial Films

Current understanding of EP lubrication is based on the concept of a sacrificial film. The existence of this film is difficult to demonstrate as it is thought to be continuously destroyed and reformed during the wear process. However, a great deal of indirect evidence has been compiled to support the existence of such sacrificial films. The model of lubrication by a sacrificial film between two discs is illustrated in Figure 8.40.

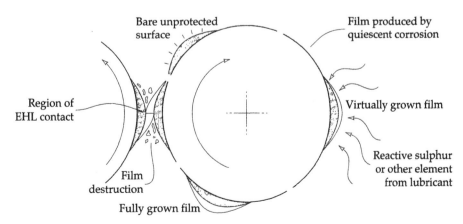

FIGURE 8.40   Model of lubrication by a sacrificial film.

The main effect of severe contact loads is to remove the oxide film from asperity peaks during contact with opposing surfaces. As already mentioned, the oxide-free surface of most metals is extremely reactive. If a lubricant additive containing sulphur, chlorine or phosphorus is present then a sulphide, chloride, phosphide or phosphate film rapidly forms on the exposed or 'nascent' surface. The adhesion between opposing asperities covered with these films is much less than for nascent metallic surfaces and this forms the basis of the lubricating effect. The asperities are able to slide past each other with minimum of damage and wear while the film material is destroyed by the shearing that inevitably occurs. If this mechanism fails, asperity adhesion and severe wear occur as described in the section on 'Mixed Lubrication and Scuffing'. In general terms the lubrication mechanism by sacrificial films depends on rapid film formation by a reactive EP additive and on sufficient time and temperature for the reaction films to form.

The evidence for the formation of sacrificial films has gradually been gathered over time. It was originally observed that when wear tracks and contacts were lubricated by oils containing sulphur, the sulphur accumulated in the heavily loaded regions [72-74]. The concept of an iron sulphide film was then proposed [75] and later confirmed when surface analysis was sufficiently advanced to detect traces of iron sulphide on the steel surface [76]. Films of iron sulphide were then produced on steel surfaces to test their lubricating effect and it was found that their survival time in a sliding contact was very short [77].

When the temperature of rubbing steel specimens was deliberately lowered below the 'EP start temperature' (i.e., the minimum temperature for which the EP reaction becomes effective, producing a lubricating effect), any lubricating effect rapidly disappeared even though it has been well developed at a higher temperature [78]. In a more elaborate test, the friction characteristics of a carbon steel pin sliding against a stainless steel ball were compared with those of a stainless steel pin sliding against a carbon steel ball while lubricated by an EP oil. The load capacity of the stainless steel pin sliding against a carbon steel ball was higher than when the friction pair materials were exchanged. Static corrosion tests of the sulphur additive outside of the sliding contact showed that stainless steel did not react or corrode as rapidly as plain carbon steel [78]. Since the pin is subjected to a very intense wearing contact, a sacrificial film is unlikely to form on its surface whatever the material. It is therefore preferable for the ball to be made of reactive material, i.e., carbon steel, to allow a sacrificial film to form on the surface outside the sliding contact, giving higher load capacity and better wear resistance. A model of this film formation is schematically illustrated in Figure 8.41.

Additive Reactivity and Its Effect on Lubrication

In order for an EP additive to effectively form sacrificial films it must be chemically active and react with worn metallic surfaces [75,79]. An '**active**' EP additive gives a higher seizure load than a '**mild**' EP additive [76]. The seizure load is the load sufficient to cause seizure of the balls in a 'four-ball' test. In this test one ball is rotated under load against three stationary balls until seizure occurs. This test is commonly used in characterizing lubricating oils. Exact comparisons of additive chemical reactivity and EP performance are rather rare in the literature because of the limited practical need for them and experimental difficulties involved. One test measuring corrosion by EP additives and load carrying capacity was conducted on a '**hot wire**' and '**four-ball**' test rigs simultaneously [80]. The operating principle of a hot wire corrosion apparatus is shown in Figure 8.42. A wire submerged in a bath of the test oil is heated by electric current to induce corrosion of the wire. Since the corrosion product, e.g., iron sulphide, usually has a much higher resistivity than the metal wire, the increase in resistance provides a measure of the depth of the corrosion.

The temperature of the wire, which also affects its resistance, is held constant during the test. There is a short period of time required for the wire to reach a steady temperature. The

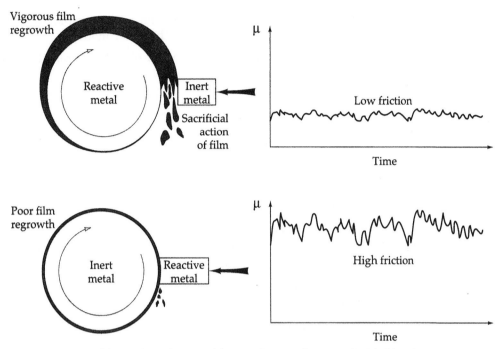

FIGURE 8.41   Favourable and unfavourable conditions for sacrificial film formation.

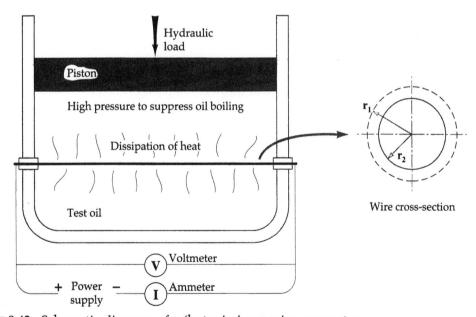

FIGURE 8.42   Schematic diagram of a 'hot wire' corrosion apparatus.

corrosion rates are usually sufficiently slow so that it can be assumed that no corrosion occurs during this period. The 'hot wire' method has certain drawbacks since it is possible for non-uniform corrosion of the wire to occur which may corrupt the experimental data. Unless a high pressure is maintained in a test chamber a vapour jacket will surround the wire. The vapour has a different chemical characteristic from the liquid oil and additive reactivity becomes a function of hydraulic pressure [81]. The 'hot wire' method is an adequate and effective method for demonstrating a general relationship between chemical activity and lubricating effect for a wide range of additives. However, an exact comparison between similar additives also requires an independent confirmation by other methods.

To specifically evaluate both reactivity and lubricating effect of oils, two parameters 'K' and 'K$_1$' are introduced, i.e.:

$$K = logW_{test\ oil} / logW_{white\ oil} \qquad (8.5)$$

and

$$K_1 = log\kappa_{test\ oil} / log\kappa_{white\ oil} \qquad (8.6)$$

where:

K	is the relative load capacity of the EP lubricant;
K$_1$	is the relative corrosivity of the EP lubricant;
W$_{test\ oil}$	is the mean Hertzian load where gross wear begins in a four-ball test with test oil. The mean of several measurements is usually taken [N];
W$_{white\ oil}$	is the mean Hertzian load where gross wear begins in a four-ball test with white oil. The mean of several measurements is usually taken [N];
$\kappa_{test\ oil}$	is the corrosion constant of the test oil [m^2/s];
$\kappa_{white\ oil}$	is the corrosion constant of the white oil [m^2/s].

The corrosion constant 'κ' can be deduced from the Wagner parabolic law of high temperature corrosion, i.e.:

$$d^2 = \kappa t \qquad (8.7)$$

where:

d	is the average depth of corrosion [m];
t	is the corrosion time [s].

An example of the plot of 'K' versus 'K$_1$' for various sulphur, chlorine and phosphorus based EP lubricants is shown in Figure 8.43.

It is evident from Figure 8.43 that the performance of a lubricant is proportional to its corrosivity or film formation rate. It is also clear that there are fundamental differences among the additives depending on the active element. Sulphur is the most effective element, i.e., it provides the greatest lubricating effect for the least corrosion, followed by phosphorus and chlorine. The reasons for this variation are still unknown.

A large range of compounds have been tested as EP additives and it was found that effective additives have a weakly bonded active element. A frequently studied additive is dibenzyldisulphide (DBDS), the molecular structure of which is shown in Chapter 3, Figure 3.18. In this molecule, two sulphur atoms are arranged in a chain linking two benzyl groups. This molecular structure is relatively weak and the sulphur is easily released to react with contacting metal. It is also possible to increase the sulphur chain length, e.g., have four sulphur atoms in one molecule, and this further increases the instability or reactivity of the compound. Pure elemental sulphur without any inhibiting organic radical is highly reactive with steel and other metals and hence can be very effective as an EP additive for severe conditions. More information on the formulation of the molecular structure of EP additives can be found elsewhere [82].

For any given application, the reactivity of the EP additive is usually adjusted to provide just sufficient lubricating effect without causing excessive corrosion. If the additive is too reactive,

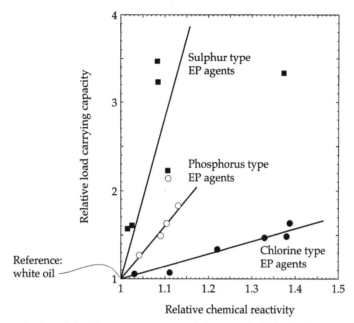

FIGURE 8.43   The relationship between corrosivity and lubricating effect of some EP lubricants [80].

e.g., elemental sulphur, the attendant corrosion can cause almost as much wear and damage as the scuffing that would have otherwise occurred. Wear induced by EP additives is discussed further in the chapter on 'Corrosive and Oxidative Wear'.

Nascent Metallic Surfaces and Accelerated Film Formation

Sliding between metals under high contact pressures is generally believed to disrupt naturally occurring oxide layers and expose a '**nascent surface**'. As discussed earlier, this phenomenon may impede adsorption lubrication. For sacrificial film lubrication, the exposure of the nascent surface can accelerate the formation of these films and promote a lubricating effect. It is very difficult to demonstrate experimentally the transient removal of oxide films during sliding contact but some evidence of this was provided by applying the technique of ellipsometry [83]. Ellipsometry is an optical technique that detects oxide films by the interference between fractions of light reflected from the base and outer surface of an oxide film. Enhancement of the experimental method with a laser and light polarizer revealed that very small patches of surface, typically 10 - 30 [μm] in diameter, with no detectable oxide film formed on the wearing surface of hard martensitic steel during high stress (0.1 [GPa]) sliding contact [83].

A nascent surface is far more reactive than an oxidized surface because (i) there is no oxide barrier between the metal and reactants, (ii) the surface atoms release electrons known as '**Kramer electrons**' to initiate reactions [84] and (iii) the nascent surface formed by sliding contains numerous defects which provide catalytic sites for reactions [85]. These characteristics of nascent surfaces are schematically illustrated in Figure 8.44 and are discussed in detail elsewhere [86].

The release of electrons is critical to the initiation of a reaction between the EP additive and the metal. Low energy electrons emitted by the surface ionize molecules of the additive and then these ionic radicals (transformed additive molecules) adsorb onto positive points on the surface [87]. The electron emission is associated with initial oxidation of the surface by

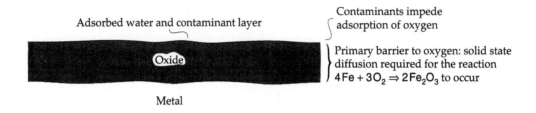

Adsorbed water and contaminant layer

Contaminants impede adsorption of oxygen

Oxide

Primary barrier to oxygen: solid state diffusion required for the reaction $4Fe + 3O_2 \Rightarrow 2Fe_2O_3$ to occur

Metal

Normal oxidized surface

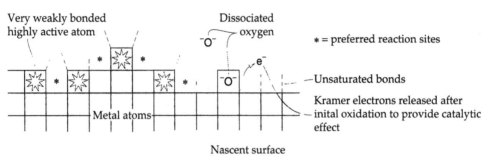

Very weakly bonded highly active atom

Dissociated oxygen

$* =$ preferred reaction sites

—Unsaturated bonds

Metal atoms

Kramer electrons released after inital oxidation to provide catalytic effect

Nascent surface

FIGURE 8.44 Characteristics of nascent surfaces.

atmospheric oxygen [84,88] or possibly by initial sulphidization. The positive points or electron vacancies have only a minute lifetime. On the surface of a typically conductive metal this is only about 10^{-13} [s] but this is sufficient to initiate chemical reactions. Surface adsorption reactions are therefore activated by the low energy electrons and may progress very rapidly. The model of an ionic reaction mechanism between EP additives and a metallic surface is shown in Figure 8.45.

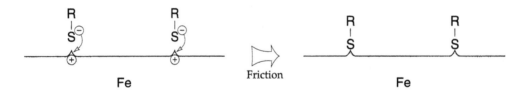

Fe

Friction

Fe

FIGURE 8.45 Ionic model of reaction between an additive and a worn surface [87].

Measurements of reaction rates between sulphur and a nascent steel surface [89] show that the rate of sulphidization by elemental sulphur is about **1000** times more rapid for a nascent surface than for an oxide surface. This speed of reaction ensures that a thin film of sulphidized material, perhaps **5** [nm] thick, would form in a few milliseconds instead of over several seconds. A sacrificial film based on this thin film could be sustained even at high contact rates, i.e., when the angular speed is several thousand revolutions per minute [25]. It was also found that the presence of a nascent steel surface lowered the temperature at which a measurable reaction rate occurred between steel and EP additives [90].

To summarize, the nascent surfaces play a significant role in lubrication by sacrificial films through (i) raising the film formation rates to a level sufficient to sustain lubrication by the sacrificial film mechanism in practical high speed contacts and (ii) confining the corrosive attack by the EP additive, particularly at low temperatures, to asperity peaks, i.e., to locations where a nascent surface is most often found. These characteristics are illustrated schematically in Figure 8.46.

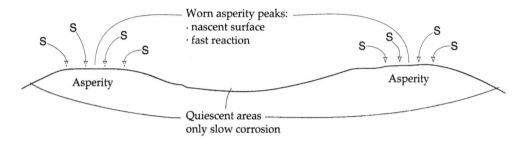

FIGURE 8.46   Preferential formation of sacrificial films at localized areas of nascent metallic surface.

Influence of Oxygen and Water on the Lubrication Mechanism by Sacrificial Films

Oxygen and water or atmospheric moisture exert a strong influence on EP lubrication as well as on adsorption lubrication. Oxygen is a competing reagent to the EP active element, e.g., sulphur, for the nascent steel surface. The chemistry of oxidation is fundamentally similar to sulphidization, chloridization and even phosphidization. Oxygen is present in air and is not excluded from most sliding contacts. Pure films of sulphide or of other EP active elements are rarely found in practical systems. It is found, for example, that when EP lubrication is effective the wear debris is mostly iron oxides [91]. High concentrations of sulphur on worn surfaces are found to coincide with scuffed regions while smooth surfaces are covered with oxygen rich layers [92]. More recently, however, it was found that, provided scuffing did not occur, oxygen is more densely concentrated on worn surfaces than sulphur when lubricated by EP additives [93]. Films rich in sulphur were found only if oxygen was deliberately excluded from the surrounding environment by imposing an atmosphere of pure nitrogen [93]. The requirements of thermodynamic equilibrium ensure that sulphides are eventually oxidized to sulphates and later oxides in the presence of oxygen. In a study of oxygen and sulphur interactions with clean metallic surfaces under high vacuum, it was found that a monoatomic layer of sulphur, probably bonded as sulphide to the iron surface, was gradually converted to iron oxide when oxygen was admitted to the vacuum [94].

These characteristics of an EP lubricated system can be interpreted in terms of a chemical system subject to mechanical intervention. At the asperity peaks where a nascent surface is repeatedly formed, sulphidization occurs because this is the far more rapid process. Any sulphide debris from the sacrificial films collects in the oil where it is oxidized by dissolved oxygen. In between the asperity peaks, a slower form of corrosion or chemical attack occurs which causes the compounds closest to thermodynamic equilibrium to form, i.e., oxides or oxidation products of sulphides. This duplex structure and formation of the EP film is illustrated schematically in Figure 8.47.

The lubrication mechanisms under EP conditions, i.e., high temperature and high load, are only indirectly controlled by the presence of an additive. The lubrication mechanism therefore functions largely according to the Le Chatelier principle. When conditions are moderate the system responds by producing the compounds which are closest to thermodynamic equilibrium. On the other hand, under severe conditions, i.e., onset of scuffing, chemical reactions respond to the most immediate disturbance, which is a large quantity of unreacted metallic surface. In that case, the most rapid means of neutralizing the metallic surface become the primary response.

A further complication in the elucidation of the role of oxygen is the difficulty in precisely defining the structure of the EP film. As discussed above, chemical reactions and film

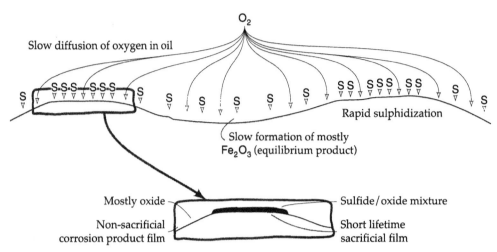

FIGURE 8.47 Formation and structure of the EP film in the presence of atmospheric oxygen.

formation at asperity peaks may be very different from events occurring in the grooves and spaces between asperities. Surface analysis is most effective in providing an average chemical composition of a surface. However, the establishment of a surface 'map' of, for example, sulphides is a very laborious if not impossible task requiring the application of modern surface analysis techniques. In other words it is relatively simple to determine the composition of the bulk of the surface while overlooking the small but critical areas of asperity peaks. Most work has shown that EP films are at least **1** [μm] thick simply because that thickness of sulphide films has been observed on worn surfaces [76,95]. Although much thinner films between **10** and **50** [nm] have also been suggested [78,96], they are still too thick to be sacrificial films.

The presence of these 'tough' thick films was originally used as evidence for EP lubrication. The rapid destruction of thick sulphide films by sliding contact precludes their survival in EP systems apart from the relatively unworn spaces between asperities. In a study of sulphide films on steel surfaces, it was found that the product of several hours corrosion by sulphur on steel, i.e., a thick sulphide film, was destroyed by less than 20 passes of a steel slider [97] which in a high speed sliding contacts would last for about one second or less. An EP film where there is an accumulation of corrosion product between asperities with short-lifetime or transient films present on asperity peaks is suggested as the most probable [25] and is illustrated in Figure 8.48.

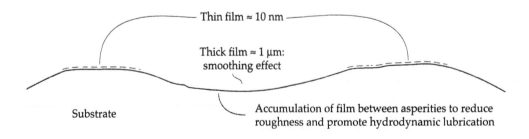

FIGURE 8.48 Probable structure of the EP film.

Oxygen is not only a strong chemical reagent which actively modifies the chemistry of a sacrificial film, but it has also been found to improve the lubricating effect. The heat of adsorption of fatty acids on iron sulphide is raised considerably by a small amount of oxidation [97]. The heat of adsorption of unoxidized iron sulphide is similar to iron oxides. This means that the role of oxygen may be to extend the temperature limits of adsorption lubrication when surfactants and EP additives are simultaneously employed as additives. Oxygen may also suppress excessive corrosion by EP additives since it was found that removal of oxygen from a lubricated system caused a more severe sulphidization without any increase in maximum load capacity [98]. A mixed oxide/sulphide film was also found to have a much higher load capacity than a pure sulphide film under certain conditions of load and sliding [99]. A graph of critical load versus measured ratios of surface sulphur to oxygen concentrations from the wear scars of a four-ball test is shown in Figure 8.49. The critical load is defined as the maximum load which permits smooth sliding up to 200°C.

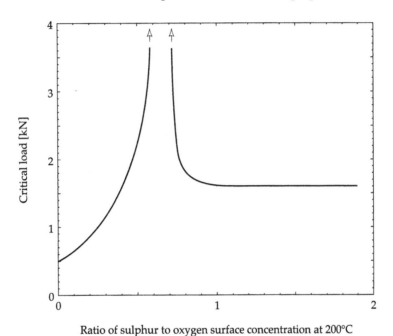

Ratio of sulphur to oxygen surface concentration at 200°C

FIGURE 8.49   Influence of sulphur to oxygen ratio in the wear scar film on the critical load [99].

When only a film of oxides is present then the critical load is very low, less than what is provided by a predominantly sulphidized film. By far the highest level of critical load is found when the sulphur and oxygen are present in approximately equal quantities. These data may, however, relate more to relative distributions of oxides and sulphides on and between asperity peaks than to film composition.

Differences between EP additives in terms of their chemical composition are also revealed by the differences in the oxide, sulphide and sulphate composition of wear scar films [100]. Films on metallic surfaces lubricated by sulphur-based EP additives consisted of a mixture of oxides, sulphides and occasionally sulphates. Each additive shows a characteristic composition of sulphides and oxides for a given set of sliding conditions. The relative oxide/sulphide/sulphate composition may depend on the rates of sulphidization and oxidation during film formation [25]. When elemental sulphur is present, sulphide films form very rapidly and remain as almost pure sulphide until their removal by wearing contact. When a milder additive is used, the sulphide forms slowly, if at all, and is usually heavily contaminated by oxide and sulphate.

In contrast to oxygen, the influence of water or moisture on sacrificial film lubrication has scarcely been studied. One investigation found that water can interfere with the functioning of EP additives but the reason for this effect has not been suggested [28].

Mechanism of Lubrication by Milder EP Additives

Measurements of sulphidization rates on nascent steel surfaces revealed that not all EP additives show rapid reaction rates [89]. The sacrificial film mechanism of EP lubrication is probably not the only mechanism that can occur and there are differing views on this subject. Most of the existing theories suggest that EP additives also function by a modified form of adsorption lubrication. The sulphidization reaction is not necessarily considered to be spontaneous, and EP additives are thought to be initially adsorbed onto the surface [76,101,102]. This adsorption provides a useful lubricating effect or wear reducing effect at moderate loads and is called the '**anti-wear**' effect which is often abbreviated to '**AW**'. An increase in loads, sliding speeds or operating temperatures causes the adsorbed additive to decompose on the worn surface, leaving the sulphur atom (or any other active element) to react with the iron of the worn metal. This mechanism is illustrated in Figure 8.50.

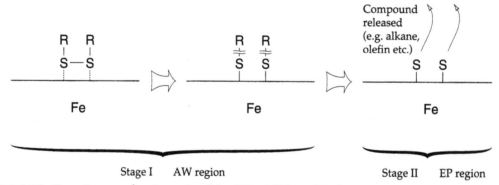

FIGURE 8.50   Reaction mechanism of milder EP additives [101].

When the additive finally decomposes to produce a sulphide film, organic residue molecules such as alkanes and olefins are released. Identification of these molecules in order to confirm this model has proved difficult because of their extreme dilution caused by the lubricating oil. It has also been suggested that the conversion of an adsorbed film of additive to a sulphide film occurs mainly in the wearing contact where the temperatures are highest. Outside of the wearing contact, adsorption of the additive is the predominant process. This mechanism appears to be more appropriate for milder additives such as DBDS than for elemental sulphur, since these additives are not observed to form sulphide films rapidly at the operating temperatures of lubricating oil, i.e., **100 - 180°C** [89]. The model of this mechanism is illustrated in Figure 8.51.

The unresolved aspect of this model is how the adsorbed film simultaneously protects against strong adhesion between opposing asperities and decomposes to a sulphide film. As will be discussed in the chapter on 'Adhesion and Adhesive Wear', more than a monolayer of film material is required to prevent adhesion between asperities.

Function of Active Elements Other Than Sulphur

Phosphorus and chlorine are generally believed to provide a lubricating effect similar to sulphur. When a phosphorous compound containing a phosphate radical is reactive to a metallic surface, a metallic phosphate film is formed on the worn surface, increasing the load capacity characteristic of EP lubrication [30,103]. Tricresylphosphate (TCP) and zinc dialkyldithiophosphate (ZnDDP) are usually used for this purpose although, as discussed in Chapter 3, the zinc and sulphur present in ZnDDP considerably complicate the film formation by this additive [104]. The lubricating effect of phosphorous additives depends on the presence of oxygen but a clear pattern of interaction has not yet been established. Although in early tests, oxygen was found to promote the lubricating effect of TCP [30,103], in other studies it was observed that ZnDDP, but not other phosphorous compounds, gave enhanced lubrication in the presence of oxygen [105].

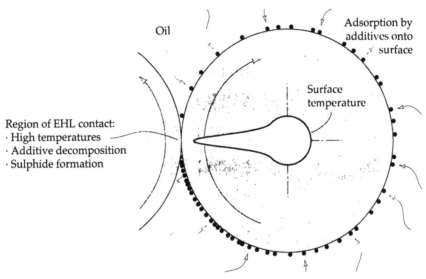

FIGURE 8.51   Sulphide film formation in high contact temperatures as a model for mild EP additives.

Chlorine is useful as an active element in EP additives. A chlorinated fatty acid ester was found to give a higher load capacity in a four-ball test than either sulphur or phosphorus compounds [106]. Sliding experiments in vacuum have shown that a film thickness of only 1 [nm] of ferrous chloride is sufficient to minimize friction on a clean metallic surface. A much thicker film, however, of about 50 [nm], is required to suppress subsurface deformation [135]. The mechanism of lubrication by chlorine based additives is believed to involve sacrificial films in the same manner as sulphur [75]. One limitation of chlorine additives is the low melting temperature of iron chlorides. When the transient temperature in a sliding contact reaches 680°C, melting of the iron chloride causes failure of the sacrificial film mechanism, resulting in seizure [107]. Most chlorine compounds used as additives are toxic, e.g., chlorinated paraffins, or else decompose to release hydrochloric acid in the presence of water. Interest in these additives is therefore limited.

EP lubrication is not only restricted to sulphur, phosphorus and chlorine. Any element capable of reacting with a metallic surface to form a sacrificial film can be suitable. Additives based on tin, which reacts with iron oxides to form an organometallic complex containing both iron and tin on the surface, have also been proposed [108]. Research on these complexes is only in its initial stage.

Lubrication with Two Active Elements

Practical experience has revealed that the combination of two active elements, e.g., sulphur and phosphorus or sulphur and chlorine, gives a much stronger lubricating effect than one element alone. The sulphur-phosphorus system is most widely used because of the instability of chlorine compounds.

The effect of combining additives is demonstrated in Figure 8.52, which shows data from Timken tests. The mineral oils used in these tests were enriched with the additives dibenzyl disulphide and dilauryl hydrogen phosphate. By using these additives separately and together, the effect of phosphorus, sulphur and phosphorus-sulphur on seizure load was found [109].

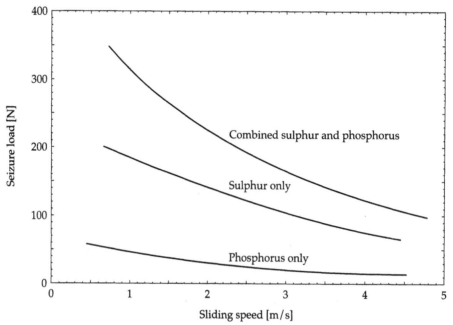

FIGURE 8.52 Comparison of seizure loads for sulphur, phosphorus and sulphur-phosphorus enriched lubricants [109].

It can be seen that although the phosphorous additive by itself is ineffective as compared to the sulphur additive, the combination of phosphorus and sulphur is significantly better than either additive acting in isolation. Unfortunately the Timken test imposes severe sliding conditions on the lubricant which may not be representative of typical operating conditions of practical machinery. The IAE (Institute of Automotive Engineers) and IP (Institute of Petroleum) 166 gear tests conducted with the same additives revealed that the phosphorus based additive allowed the same seizure or failure loads for a much smaller concentration than the sulphur based additive [109]. The critical difference between the Timken test and the gear tests is the slide/roll ratio. The Timken test involves pure sliding while the gear tests only impose sliding combined with rolling. It appears that sulphur originated surface films are more resistant to the shearing of pure sliding than films formed from phosphorous additives.

The chemistry of steel surfaces after lubrication by sulphur-phosphorus oils was also studied [109,110]. Films found on wear scars formed under severe conditions, e.g., the Timken test, consisted mostly of sulphur. However, under milder load and lower slide/roll ratios, which are characteristic for general machinery, it was found that phosphorus predominates in the wear scar films. This pattern of film chemistry versus sliding severity is illustrated schematically in Figure 8.53.

It can be seen from Figure 8.53 that a sulphur-phosphorus based lubricant provides considerable versatility in lubricating performance. The sulphur is essential to prevent seizure under abnormally high loads and speeds while phosphorus maintains low friction and wear rates under normal operating conditions.

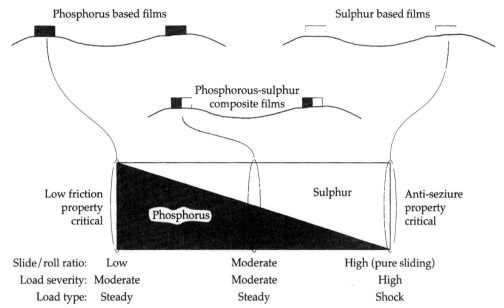

FIGURE 8.53 Dependence of sulphur-phosphorus wear scar film chemistry on severity of sliding conditions.

The relative benefits of sulphur versus phosphorus can also be discussed in terms of their ability to provide effective lubrication under shock loading. It was found that sulphur based additives tend to provide better lubrication, i.e., maintain a moderate coefficient of friction, during a precipitate increase in load than phosphorus based additives [109]. Phosphorus based additives are characterized by a progressive decline in friction and accumulation of phosphorus on the worn surface. It appears that in these cases mechanisms other than sacrificial film lubrication may be involved. The most probable mechanism seems to be lubrication by an amorphous layer, which was discussed previously.

Temperature Distress

Temperature distress is a term used to describe high friction occurring over a relatively narrow band of intermediate temperature in lubrication by an oil. An example of this effect is shown in Figure 8.54 which illustrates the friction coefficient versus temperature results from a four-ball test where the lubricant tested is white oil with tributylphosphate [111].

The tests were conducted at a relatively high contact stress of approximately 2 [GPa] and, to ensure negligible frictional transient temperatures, at a very low sliding speed of 0.2 [mm/s]. Friction, initially moderate at room temperature, rises to a peak between 100 and 150°C followed by a sharp decline at higher temperatures. This phenomenon is the result of a significant difference between the desorption temperature of surfactants from the steel surface and the lowest temperature where rapid sacrificial film formation can occur. In this test, the surfactants were relatively scarce, consisting only of impurities or oxidation products in the white oil. In practical oil formulations, however, surfactants are carefully chosen so that the desorption temperature is higher than the 'start temperature' of sacrificial film lubrication. The concept of wide temperature range lubrication, which is achieved by employing in tandem adsorption and sacrificial film lubrication, is illustrated in Figure 8.55.

It can be seen that when only the fatty acid is applied, the coefficient of friction is quite low below a critical temperature and then sharply rises. Conversely, when the EP additive (in an EP lubricant) is acting alone, the coefficient of friction remains high below a critical

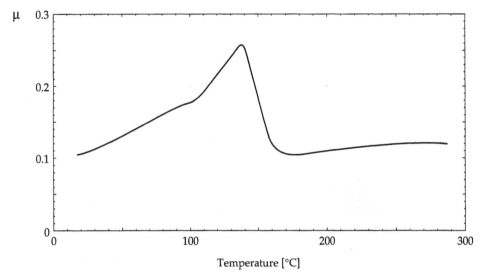

FIGURE 8.54 Experimental friction characteristic of a phosphate EP lubricant versus temperature [111].

temperature and then there is a sharp drop. Effective lubrication, i.e., a low coefficient of friction over a wide range of temperatures, is obtained when these two additive types are combined. This model of temperature distress assumes that the mechanisms of adsorption and sacrificial film lubrication are entirely independent. The formation of partially oxidized sulphide films can influence the desorption temperature so that the range of temperature distress is not necessarily the exact temperature difference between desorption and sacrificial film formation acting in isolation.

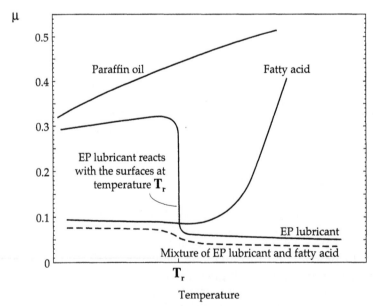

FIGURE 8.55 Co-application of adsorption and sacrificial film lubrication to ensure a wide temperature range of lubrication function [6].

Speed Limitations of Sacrificial Film Mechanism

As discussed in this chapter, sacrificial films formed on severely loaded surfaces require some finite period of time to reform between successive sliding contacts. In most research it is assumed that the formation time is so short that it does not exert a significant limitation on lubricant performance. It was found, for example, that EP additives were effective in raising the maximum load before scuffing only at low sliding speeds [112]. In low speed tests performed under pure sliding using a pin-on-ring machine, when an EP additive was present, the scuffing load was increased by a factor of **2** compared to that of plain oil. At higher speeds the EP additives had almost no effect on the scuffing load. It is speculated that at high speeds the sacrificial films did not form and as a result the EP additives were ineffective.

Tribo-emission from Worn Surfaces

Tribo-emission is a term describing the emission of electrons, ions and photons as a response to friction and wear processes. The mechanisms involved in tribo-emission are complex and not known in detail [130]. However, it is speculated that triboemission precedes and is even necessary for tribochemical reactions to occur in the tribocontact. The best researched is the emission of already mentioned low energy electrons (Figures 8.44 and 8.45), also called exoelectrons. One of the mechanisms proposed, involving tribo-emission of electrons, is described below.

During wear surface cracks are generated as a result of severe deformation of the worn surface. In general, when a crack forms there is an imbalance of electrons on opposite faces of the crack [e.g., 126-128]. This imbalance is particularly evident in ionic solids which are composed of alternating layers of anions and cations. For example, when a crack develops in aluminium oxide, one side of the crack will contain oxide anions while the opposite side will contain aluminium cations. The narrow gap between opposing faces of a crack causes formation of a large electric field gradient (electric field gradient is controlled by the distance between opposite electric charges). This electric field is sufficient to cause electron escape from the anions [128]. It is believed that not all the electrons which escape from the anions are collected by the cations on the opposing crack face. This results in tribo-emission or the release of electrons into the wider environment under the action of sliding. The phenomenon is schematically illustrated in Figure 8.56.

In dry sliding tests under vacuum, ceramics exhibit a strong tribo-emission of electrons because of their ionic crystalline structure while metals reveal a lesser tendency since the high electron mobility in a metal tends to equalize electron distribution on either side of the crack. Tribo-emission also occurs during sliding in air or under a lubricant but the electrons are not easily detected as their path length in air is much shorter than that in vacuum. Water and possibly other gases or liquids may influence tribo-emission of electrons by chemisorption on the exposed surfaces of the crack about to release electrons. Irradiation by high energy radiation such as gamma rays appears to activate worn surfaces to significantly raise the level of tribo-emission; the detailed physical causes of this phenomenon are still poorly understood [129].

Tribo-emission of positive and negative ions, as well photons, has been detected during wear of ceramics in n-butane of various pressure [127]. In this case the wear mechanism was explained in terms of gas discharge due to high electric field generated on the wear surface when charges are separated. The ionized gas molecules may then recombine generating molecules different from the original gas. A completely different mechanism of tribo-emission was also suggested for a similar ceramic-diamond abrasive contact [130]. Tribo-emission from MgO scratched by diamond was attributed to excited defects created by abrasion in the solid phase.

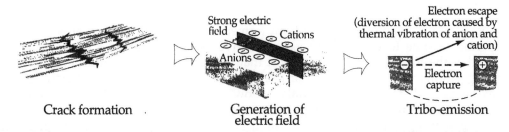

Figure 8.56   Schematic illustration of the mechanism of crack-induced tribo-emission.

The tribo-emission accelerates chemical reactions such as oxidation or polymerization of the lubricant under boundary lubrication conditions [127] and is an example of mechanical activation. The tribo-emission is beneficial if it promotes formation of wear and friction reducing surface films but is harmful if these films or a lubricant are degraded to produce a sludge or other forms of debris. Therefore it is important to know whether the tribo-emission triggers the tribochemical reactions and whether these reaction products influence wear and friction characteristics.

8.6   BOUNDARY AND EP LUBRICATION OF NON-METALLIC SURFACES

Most of the discussion on boundary and EP lubrication in this chapter refers to lubrication of metallic surfaces. Increased interest in the tribological applications of ceramics has resulted in more research into boundary lubrication of ceramics, especially at elevated temperatures. Both EP [131] and detergent-type additives [132] were found to form boundary lubricating layers on silicon nitride in the 'four-ball' tester. EDX analysis revealed, however, that the tribochemical reactions on silicon nitride were different from those found on steel surfaces when the same detergent-type additives were used. Since ceramics are less reactive than metals the effectiveness of typical adsorption and antiwear additives in many cases appears to be lower for ceramic-ceramic contacts than for ceramic-metal contacts [133]. Although a sacrificial iron phosphate film was detected on the silicon nitride surface when it was slid against steel with vapour phase lubrication of oleic acid and TCP, the triboreaction took place on the steel surface [133]. When self-mated silicon nitride was lubricated by the same vapour phase, much higher wear was recorded.

Some ceramics are capable of initiating tribochemical reactions and ionic ceramics appear to be more reactive than covalent ceramics. For example, the refrigerant gas, HFC-134a (chemical formula CF_3CH_2), studied during the friction tests in vacuum resulted in a reduced friction coefficient of alumina and zirconia sliding against an alumina ball from about 0.8 in high vacuum to about 0.2 in 0.1 atmospheric pressure of HFC-134a [136]. A smaller reduction in the friction coefficient, from about 0.8 to about 0.4, was also observed for silicon nitride and silicon carbide sliding against alumina [136]. This reduction in friction coefficient was associated with the decomposition of HFC-134a to form fluoride films on the worn ceramic surfaces [136]. The surface chemistry of these ceramics is not yet fully understood, but the formation of surface films, for example, of aluminium fluoride, may be instrumental.

On the other hand, boundary lubrication by sacrificial films of oxides and hydroxides is much more effective for ceramics than for metals [117]. For example, silicon nitride can be lubricated by thin layers of silicon oxide and alumina by alumina hydroxide formed in the tribocontact. In contrast with EP sacrificial films on metal surfaces, ceramic oxides and hydroxides do not require high temperatures to be generated.

More information on lubrication of ceramics can be found in Chapter 16.

8.7   SUMMARY

Lubrication by chemical and physical interaction between an oil-based lubricant and a surface (usually metal) is essential to the operation of most practical machinery. Four basic forms of this lubrication are identified: (i) the formation of an ultra-viscous layer close to the worn surface, (ii) the shielding of an oxidized metal surface by a mono-molecular layer of adsorbed linear surfactants, (iii) the separation of contacting surfaces by entrapped layers of finely divided and perhaps amorphous debris and (iv) the suppression of metal to metal contact at extreme pressures by the temperature dependent formation of sacrificial films of corrosion product on worn metallic surfaces. Each lubrication mechanism has certain merits and disadvantages but they all contribute to the reduction of wear and friction under conditions where other lubrication mechanisms such as hydrodynamic and elastohydrodynamic lubrication are ineffective. This is achieved by the addition of some relatively cheap and simple chemicals to the oil. It is possible to describe fairly precisely how a particular additive functions in terms of friction and wear control. However, the prediction of lubricant performance from chemical specification is still not possible and this constrains research to testing for specific applications. This task remains a future challenge for research.

REVISION QUESTIONS

8.1   What is the film thickness or separation distance between two contacting surfaces in boundary lubrication?

8.2.   Are liquid properties such as viscosity relevant to the mechanism of boundary lubrication?

8.3   Under what conditions of load and speed is boundary lubrication most likely to occur?

8.4   How do boundary lubricants such as fatty acids function?

8.5   What is the significance of molecular polarity?

8.6   What parameter determines a fundamental limit to boundary lubrication?

8.7   In thermodynamic terms what is the difference between chemisorption and physisorption?

8.8   Which of the above mentioned two forms of adsorption has the greater capacity to sustain higher loads and speeds?

8.9   Can there be chemisorption on gold?

8.10   Give examples of boundary lubricating films formed by physical and chemical adsorption.

8.11   Describe an alternative theory of boundary lubrication that differs from Hardy's adsorption theory.

8.12   What is the difference between boundary and EP lubricating films?

8.13   What are the active chemical elements present in most EP additives?

8.14   How do the chemically reactive elements present in EP additives function?

8.15   Why should EP additives react (at a measurable rate) with a worn surface but not with an unworn surface even if the temperature rise due to friction has been compensated for?

8.16   When sulphur-based EP lubricants are specified, what precaution must be taken in the design of surrounding equipment?

8.17   What is the cause of 'temperature distress'?

8.18   In a metal rolling mill, what was the physical parameter that determined whether heat streaking occurred?

8.19   What was the probable solution to the heat streaking problem?

8.20   A roving explorer presents a plant specimen to the Natural Products Laboratory of the University of the Tropics with claims about its superior lubricating powers. The plant, even in a half-dried state, emits a repulsive odour, especially from its globular fruits. Discuss what is the likely source of its lubricating powers and how one would test for them. Suggest how one could obtain a useful extract from the plant. Clue: A non-metallic, stable element in the same column of the periodic table as oxygen is well known for its noxious and malodorous compounds.

REFERENCES

1 C.M. Allen and E. Drauglis, Boundary Layer Lubrication: Monolayer or Multilayer, *Wear*, Vol. 14, 1969, pp. 363-384.

2 G.J. Fuks, The Properties of Solutions of Organic Acids in Liquid Hydrocarbons at Solid Surfaces, Research in Surface Forces, editor B.V. Derjagin, Moscow (Trans-Consultants Bureau, New York, 1962), pp. 79-88.

3 Lord Rayleigh (J.W. Strutt), On the Lubricating and Other Properties of Thin Oily Films, *Phil. Mag. J. Science*, 6th Series, Vol. 35, 1918, pp. 157-163.

4 W.B. Hardy and I. Doubleday, Boundary Lubrication - The Paraffin Series, *Proc. Roy. Soc., London*, Series A, Vol. 100, 1921, pp. 550-574.

5 W.B. Hardy and I. Doubleday, Boundary Lubrication - The Temperature Coefficient, *Proc. Roy. Soc., London*, Series A, Vol. 101, 1922, pp. 487-492.

6 F.P. Bowden and D. Tabor, The Friction and Lubrication of Solids, Part 1, Clarendon Press, Oxford, 1950.

7 D. Tabor, Desorption or 'Surface Melting' of Lubricant Films, *Nature*, Vol. 147, 1941, pp. 609- 610.

8 W.C. Bigelow, D.L. Pickett and W.A. Zisman, Oleophobic Monolayers, I. Films Adsorbed from Solution in Non-Polar Liquids, *Journal of Colloid Science*, Vol. 1, 1946, pp. 513-538.

9 C.O. Timmons, R.L. Patterson and L.B. Lockhart, Adsorption of Carbon-14 Labelled Stearic Acid on Iron, *Journal of Colloid and Interface Science*, Vol. 26, 1968, pp. 120-127.

10 K. Tanaka, Molecular Arrangement in Thin Films of Some Paraffins, *Mem. Coll. Sic. Kyoto Imp. Univ.*, Vol. A23, 1941, pp. 195-205.

11 J.M. Cowley, Electron Diffraction by Fatty Acid Layers on Metal Surfaces, *Trans. Faraday Soc.*, Vol. 44, 1948, pp. 60-68.

12 K.G. Brummage, An Electron-Diffraction Study of the Structure of Thin Films of Normal Paraffins, *Proc. Roy. Soc., London*, Series A, Vol. 188, 1947, pp. 414-426.

13 K.G. Brummage, An Electron-Diffraction Study of the Heating of Straight Chain Organic Films and its Application to Lubrication, *Proc. Roy. Soc., London*, Series A, Vol. 191, 1947, pp. 243-252.

14 J.W. Menter, Effect of Temperature on Lateral Spacing of Fatty Acid Monolayers on Metals - Electron Diffraction Study by Transmission, *Research (London)*, Vol. 3, 1950, pp. 381-382.

15 F.P. Bowden, J.N. Gregory and D. Tabor, Lubrication of Metal Surfaces by Fatty Acids, *Nature*, Vol. 156, 1945, pp. 97-98.

16 F.P. Bowden and A.C. Moore, Physical and Chemical Adsorption of Long Chain Polar Compounds, *Research (London)*, Vol. 2, 1949, pp. 585-586.

17 D.W. Morecroft, Reactions of Octadecane and Decoic Acid with Clean Iron Surfaces, *Wear*, Vol. 18, 1971, pp. 333-339.

18 M. Beltzer and S. Jahanmir, Role of Dispersion Interactions Between Hydrocarbon Chains in Boundary Lubrication, *ASLE Transactions*, Vol. 30, 1987, pp. 47-54.

19 P. Studt, The Influence of the Structure of Isometric Octadecanols on their Adsorption from Solution on Iron and Their Lubricating Properties, *Wear*, Vol. 70, 1981, pp. 329-334.

20 S. Jahanmir and M. Beltzer, An Adsorption Model for Friction in Boundary Lubrication, *ASLE Transactions*, Vol. 29, 1986, pp. 423-430.

21 E. Ando, Y. Goto, K. Morimoto, K. Ariga and Y. Okahata, Frictional Properties of Monomolecular Layers of Silane Compounds, *Thin Solid Films*, Vol. 180, 1989, pp. 287-291.

22 H. Okabe, T. Ohmori and M. Masuko, A Study on Friction-Polymer Type Additives, Proc. JSLE. Int. Tribology Conf., 8-10 July 1985, Tokyo, Publ. Elsevier, pp. 691-696.

23 E.D. Tingle, The Importance of Surface Oxide Films in the Friction and Lubrication of Metals, Part 2, The Formation of Lubrication Films on Metal Surfaces, *Trans. Faraday Soc.*, Vol. 326, 1950, pp. 97-102.

24 F.P. Fehlner and N.F. Mott, Low Temperature Oxidation, *Oxidation of Metals*, Vol. 2, 1970, pp. 59-99.

25 A.W. Batchelor and G.W. Stachowiak, Some Kinetic Aspects of Extreme Pressure Lubrication, *Wear*, Vol. 108, 1986, pp. 185-199.

26 R. Dubrisay, Alteration of Metals by Organic Acids Dissolved in Non-Aqueous Liquids, *Comptes Rendus, Academie des Sciences*, Vol. 210, 1940, pp. 533-534.

27 C.F. Prutton, D.R. Frey, D. Turnbull and G. Dlouhy, Corrosion of Metals by Organic Acids in Hydrocarbon Solution, *Industrial Engineering Chemistry*, Vol. 37, 1945, pp. 90-100.

28   I.B. Goldman, J.K. Appeldoorn and F.F. Tao, Scuffing as Influenced by Oxygen and Moisture, *ASLE Transactions*, Vol. 13, 1970, pp. 29-38.

29   R.O. Daniels and A.C. West, The Influence of Moisture on the Friction and Surface Damage of Clean Metals, *Lubrication Engineering*, Vol. 11, 1955, pp. 261-266.

30   D. Godfrey, The Lubrication Mechanism of Tricresyl Phosphate on Steel, *ASLE Transactions*, Vol. 8, 1965, pp. 1-11.

31   I.L. Goldblatt and J.K. Appeldoorn, The Antiwear Behaviour of Tricresylphosphate (TCP) in Different Atmospheres and Different Base Stocks, *ASLE Transactions*, Vol. 13, 1970, pp. 203-214.

32   Y. Kimura and H. Okabe, Tribology, An Introduction, Publ. Youkandou, Tokyo, Japan, 1982.

33   H. Okabe, M. Masuko and K. Sakurai, Dynamic Behaviour of Surface-Adsorbed Molecules Under Boundary Lubrication, *ASLE Transactions*, Vol. 24, 1981, pp. 467-473.

34   E.P. Kingsbury, Some Aspects of the Thermal Desorption of a Boundary Lubricant, *Journal of Applied Physics*, Vol. 29, 1958, pp. 888-891.

35   C.N. Rowe, Some Aspects of the Heat of Adsorption in the Function of a Boundary Lubricant, *ASLE Transactions*, Vol. 9, 1966, pp. 101-111.

36   M. Ibrahim and A. Cameron, Oil Film Thickness and the Mechanism of Scuffing in Gear Teeth, Proc. Lubrication and Wear Convention, 1963, Inst. Mech. Engrs. Publ., London, 1963, pp. 70-80.

37   A. Dyson, Scuffing, A Review, *Tribology International*, Vol. 8, 1975, Part 1: pp. 77-87, Part 2: pp. 117-122.

38   H. Blok, Les Temperatures de Surface dans des Conditions de Graissage sous Pression Extreme, Proc. Second World Petroleum Congress, Paris, Vol. 4, 1937, pp. 151-82.

39   H. Blok, The Flash Temperature Concept, *Wear*, Vol. 6, 1963, pp. 483-494.

40   A. Dorinson and K.C. Ludema, Mechanics and Chemistry in Lubrication, Elsevier Amsterdam, 1985, pp. 457-469.

41   T.E. Tallian, Discussion to 'The Effects of Temperature and Metal Pairs on Scuffing', M.W. Bailey and A. Cameron, *ASLE Transactions*, Vol. 16, 1973, pp. 121-131.

42   J.J. Frewing, The Heat of Adsorption of Long-Chain Compounds and their Effect on Boundary Lubrication, *Proc. Roy. Soc., London*, Series A, Vol. 182, 1944, pp. 270-285.

43   T.C. Askwith, A. Cameron and R.F. Crouch, Chain Length of Additives in Relation to Lubricants in Thin Film and Boundary Lubrication, *Proc. Roy. Soc., London*, Series A, Vol. 291, 1966, pp. 500-519.

44   W.J.S. Grew and A. Cameron, Thermodynamics of Boundary Lubrication and Scuffing, *Proc. Roy. Soc., London*, Series A, Vol. 327, 1972, pp. 47-59.

45   O. Beeck, J.W. Givens and A.E. Smith, On the Mechanism of Boundary Lubrication: I. The Action of Long-Chain Polar Compounds, *Proc. Roy. Soc., London*, Series A, Vol. 177, 1940, pp. 90-102.

46   R.S. Fein, Operating Procedure Effects on Critical Temperatures, *ASLE Transactions*, Vol. 10, 1967, pp. 373-385.

47   H.J. Carper, P.M. Ku and E.L. Anderson, Effect of Some Material and Operating Variables on Scuffing, *Mechanism and Machine Theory*, Vol. 8, 1973, pp. 209-225.

48   J.C. Bell, A. Dyson and J.W. Hadley, The Effect of Rolling and Sliding Speed on the Scuffing of Lubricated Steel Discs, *ASLE Transactions*, Vol. 18, 1975, pp. 62-73.

49   G.H. Kitchen, The Economics of Tribology, Proc. Int. Tribology Conference, Melbourne, The Institution of Engineers, Australia, National Conference Publication No. 87/18, December, 1987, pp. 424-427.

50   W.J.S. Grew and A. Cameron, Role of Austenite and Mineral Oil in Lubricant Failure, *Nature*, Vol. 217, 1968, pp. 481-482.

51   F.G. Rounds, The Influence of Steel Composition on Additive Performance, *ASLE Transactions*, Vol. 15, 1972, pp. 54-66.

52   M.W. Bailey and A. Cameron, The Effects of Temperature and Metal Pairs on Scuffing, *ASLE Transactions*, Vol. 16, 1973, pp. 121-131.

53   R.M. Matveevsky, V.M. Sinaisky and I.A. Buyanovsky, Contribution to the Influence of Retained Austenite Content in Steels on the Temperature Stability of Boundary Lubricant Layers in Friction, *Transactions ASME, Journal of Tribology*, Vol. 97, 1975, pp. 512-515.

54   H. Diergarten, J. Stocker and H. Werner, Erfahrungen mit dem Vierkugelapparat zur Beurteilung von Schmierstoffen, *Erdol und Kohle*, Vol. 8, 1955, pp. 312-318.

55   K.H. Kloos, Werkstoffpaarung und Gleitreibungsverhalten in Fertigung und Konstruktion, *Fortschrittberichte der VDI Zeitschriften*, Reihe 2, 1972, pp. 1-91.

56   A. Begelinger, A.W.J. de Gee and G. Salomon, Failure of Thin Film Lubrication - Function-Orientated Characterization of Additives and Steels, *ASLE Transactions*, Vol. 23, 1980, pp. 23-24.

57   K.-H. Habig, P. Feinle, Failure of Steel Couples Under Boundary Lubrication: Influence of Steel Composition, Microstructure and Hardness, *Transactions ASME, Journal of Tribology*, Vol. 109, 1987, pp. 569-576.

58   A.J. Groszek, Heats of Preferential Adsorption of Boundary Additives at Iron Oxide/Liquid Hydrocarbon Interfaces, *ASLE Transactions*, Vol. 13, 1970, pp. 278-287.

59   F. Hirano and T. Sakai, The Chain Matching Effect on Performance of Mechanical Seals, Proc. 9th Int. Conf. Fluid Sealing, BHRA, 1981, pp. 429-444.

60   F. Hirano, N. Kuwano and N. Ohno, Observation of Solidification of Oils under High Pressure, Proc. JSLE Int. Tribology Conf., 8-10 July, 1985, Tokyo, Japan, Elsevier, pp. 841-846.

61   G.J. Johnston, R. Wayte and H.A. Spikes, The Measurement and Study of Very Thin Lubricant Films in Concentrated Contacts, *Tribology Transactions*, Vol. 34, 1991, pp. 187-194.

62   A. Cameron and T.N. Mills, Basic Studies on Boundary, E.P. and Piston-Ring Lubrication Using a Special Apparatus, *ASLE Transactions*, Vol. 25, 1982, pp. 117-124.

63   P. Cann, H.A. Spikes and A. Cameron, Thick Film Formation by Zinc Dialkyldithiophosphate, *ASLE Transactions*, Vol. 26, 1986, pp. 48-52.

64   D. Mazuyer, J.M. Georges and B. Cambou, Shear Behaviour of an Amorphous Film with Bubbles Soap Raft Model, Proc. 14th Leeds-Lyon Symp. on Tribology, Interface Dynamics, Sept. 1987, editors: D. Dowson, C.M. Taylor, M. Godet and D. Berthe, Elsevier, 1988, pp. 3-9.

65   M. Belin, J.M. Martin, J.L. Mansot, Role of Iron in the Amorphization Process in Friction-Induced Phosphate Glasses, *Tribology Transactions*, Vol. 32, 1989, pp. 410-413.

66   J.M. Martin, M. Belin, J.L. Mansot, H. Dexpert and P. Lagarde, Friction-Induced Amorphization with ZDDP - an EXAFS Study, *ASLE Transactions*, Vol. 29, 1986, pp. 523-531.

67   J. Dimnet and J.M. Georges, Some Aspects of the Mechanical Behaviour of Films in Boundary Lubrication, *ASLE Transactions*, Vol. 25, 1982, pp. 456-464.

68   I.N. Lacey, G.H. Kelsall and H.A. Spikes, Thick Antiwear Films in Elastohydrodynamic Contacts. Part 1: Film Growth in Rolling/Sliding EHD Contacts, *ASLE Transactions*, Vol. 29, 1986, pp. 299-305.

69   I.N. Lacey, G.H. Kelsall and H.A. Spikes, Thick Antiwear Films in Elastohydrodynamic Contacts. Part 2: Chemical Nature of the Deposited Films, *ASLE Transactions*, Vol. 29, 1986, pp. 306-311.

70   M.J. Furey, The Formation of Polymeric Films Directly on Rubbing Surfaces to Reduce Wear, *Wear*, Vol. 26, 1973, pp. 369-392.

71   A. Dorinson, Influence of Chemical Structures in Sulfurized Fats on Anti-Wear Behaviour, *ASLE Transactions*, Vol. 14, 1971, pp. 124-134.

72   E.H. Loeser, R.C. Wiquist and S.B. Twist, Cam and Tappet Lubrication, IV, Radio-Active Study of Sulphur in the E.P. film, *ASLE Transactions*, Vol. 2, 1959-1960, pp. 199-207.

73   V.N. Borsoff and C.D. Wagner, Studies of Formation and Behaviour of an Extreme Pressure Film, *Lubrication Engineering*, Vol. 13, 1975, pp. 91-99.

74   R.B. Campbell, The Study of Hypoid Gear Lubrication Using Radio-Active Tracers, Proc. Lubrication and Wear Convention, Inst. Mech. Engrs. Publ., London, 1963, pp. 286-290.

75   C.F. Prutton, D. Turnbull and G. Dlouhy, Mechanism of Action of Organic Chlorine and Sulphur Compounds in Extreme-Pressure Lubrication, *J. Inst. Petroleum*, Vol. 32, 1946, pp. 90-118.

76   K.G. Allum and E.S. Forbes, The Load Carrying Mechanism of Some Organic Sulphur Compounds - An Application of Electron Microprobe Analysis, *ASLE Transactions*, Vol. 11, 1968, pp. 162-175.

77   B.A. Baldwin, Relationship Between Surface Composition and Wear, An X-Ray Photo-Electron Spectroscopic Study of Surfaces Tested with Organo-Sulphur Compounds, *ASLE Transactions*, Vol. 19, 1976, pp. 335-344.

78   H.A. Spikes and A. Cameron, Additive Interference in Dibenzyl Disulphide Extreme Pressure Lubrication, *ASLE Transactions*, Vol. 17, 1974, pp. 283-289.

79   W. Davey, Some Observations on the Mechanism of the Development of Extreme Pressure Lubricating Properties by Reactive Sulfur in Mineral Oil, *J. Inst. Petroleum*, Vol. 31, 1945, pp. 154-158.

80   T. Sakurai and K. Sato, Study of Corrosivity and Correlation Between Chemical Reactivity and Load Carrying Capacity of Oils Containing Extreme Pressure Agent, *ASLE Transactions*, Vol. 9, 1966, pp. 77-87.

81 M. Masuko, N. Naganuma and H. Okabe, Acceleration of the Thermal Reaction of Sulfur with Steel Surface Under Increased Pressure, *Tribology Transactions*, Vol. 33, 1990, pp. 76-84.

82 A. Dorinson and K.T. Ludema, Mechanics and Chemistry in Lubrication, Elsevier, Amsterdam, 1985, pp. 255-307.

83 J.L. Lauer, N. Marxer and W.R. Jones, Ellipsometric Surface Analysis of Wear Tracks Produced by Different Lubricants, *ASLE Transactions*, Vol. 29, 1986, pp. 457-466.

84 J. Ferrante, Exoelectron Emission from a Clean, Annealed Magnesium Single Crystal During Oxygen Adsorption, *ASLE Transactions*, Vol. 20, 1977, pp. 328-332.

85 E.A. Gulbransen, The Role of Minor Elements in the Oxidation of Metals, *Corrosion*, Vol. 12, 1956, pp. 61-67.

86 K. Meyer, Physikalich-Chemische Kristallographie, Gutenberg Buchdruckerei, 1977, German Democratic Republic.

87 Cz. Kajdas, On a Negative-Ion Concept of EP Action of Organo-Sulfur Compounds, *ASLE Transactions*, Vol. 28, 1985, pp. 21-30.

88 T.F. Gesell, E.T. Arakawa and T.A. Callcott, Exoelectron Emission During Oxygen and Water Chemisorption on Fresh Magnesium Surface, *Surface Science*, Vol. 20, 1970, pp. 174-178.

89 A.W. Batchelor, A. Cameron and H. Okabe, An Apparatus to Investigate Sulfur Reactions on Nascent Steel Surfaces, *ASLE Transactions*, Vol. 28, 1985, pp. 467-474.

90 K. Meyer, H. Berndt and B. Essiger, Interacting Mechanisms of Organic Sulphides with Metallic Surfaces and their Importance for Problems of Friction and Lubrication, *Applications of Surface Science*, Vol. 4, 1980, pp. 154-161.

91 O.D. Faut and D.R. Wheeler, On the Mechanism of Lubrication by Tricresylphosphate (TCP) - The Coefficient of Friction as a Function of Temperature for TCP on M-50 Steel, Vol. 26, 1983, pp. 344-350.

92 R.O. Bjerk, Oxygen, An "Extreme-Pressure Agent", *ASLE Transactions*, Vol. 16, 1973, pp. 97-106.

93 M. Masuko, Y. Ito, K. Akatsuka, K. Tagami and H. Okabe, Influence of Sulphur-base Extreme Pressure Additives on Wear Under Combined Sliding and Rolling Contact, Proc. Kyushu Conference of JSLE, Oct., 1983, pp. 273-276 (in Japanese).

94 D.H. Buckley, Oxygen and Sulfur Interactions with a Clean Iron Surface and the Effect of Rubbing Contact in These Interactions, *ASLE Transactions*, Vol. 17, 1974, pp. 201-212.

95 E.P. Greenhill, The Lubrication of Metals by Compounds Containing Sulphur, *J. Inst. Petroleum*, Vol. 34, 1948, pp. 659-669.

96 J.J. McCarroll, R.W. Mould, H.B. Silver and M.C. Sims, Auger Electron Spectroscopy of Wear Surfaces, *Nature (London)*, Vol. 266, 1977, pp. 518-519.

97 K. Date, Adsorption and Lubrication of Steel with Oiliness Additives, Ph.D. thesis, London University, 1981.

98 M. Tomaru, S. Hironaka and T. Sakurai, Effects of Some Oxygen on the Load-Carrying Action of Some Additives, *Wear*, Vol. 41, 1977, pp. 117-140.

99 T. Sakai, T. Murakami and Y. Yamamoto, Optimum Composition of Sulfur and Oxygen of Surface Film Formed in Sliding Contact, Proc. JSLE. Int. Tribology Conf., July 8-10, Tokyo, Japan, Elsevier, 1985, pp. 655-660.

100 B.A. Baldwin, Wear Mitigation by Anti-Wear Additives in Simulated Valve Train Wear, *ASLE Transactions*, Vol. 26, 1983, pp. 37-47.

101 E.S. Forbes, The Load Carrying Action of Organic Sulfur Compounds, a Review, *Wear*, Vol. 15, 1970, pp. 87-96.

102 E.S. Forbes and A.J.D. Reid, Liquid Phase Adsorption/Reaction Studies of Organo-Sulfur Compounds and their Load Carrying Mechanism, *ASLE Transactions*, Vol. 16, 1973, pp. 50-60.

103 D. Godfrey, The Lubrication Mechanism of Tricresylphosphate on Steel, *ASLE Transactions*, Vol. 8, 1965, pp. 1-11.

104 E.H. Loeser, R.C. Wiquist and S.B. Twist, Cam and Tappet Lubrication, Part III, Radio-Active Study of Phosphorus in the E.P. Film, *ASLE Transactions*, Vol. 1, 1958, pp. 329-335.

105 P.A. Willermet, S.K. Kandah, W.O. Siegl and R.E. Chase, The Influence of Molecular Oxygen on Wear Protection by Surface-Active Compounds, *ASLE Transactions*, Vol. 26, 1983, pp. 523-531.

106 M. Kawamura, K. Fujita and K. Ninomiya, Lubrication Properties of Surface Films Under Dry Conditions, *Journal of JSLE.*, International Edition, No. 2, 1981, pp. 157-162.

107 P.V. Kotvis, L. Huezo, W.S. Millman and W.T. Tysoe, The Surface Decomposition and Extreme-Pressure Tribological Properties of Highly Chlorinated Methanes and Ethanes on Ferrous Surfaces, *Wear*, Vol. 147, 1991, pp. 401-419.

108 D. Ozimina and C. Kajdas, Tribological Properties and Action Mechanism of Complex Compounds of Sn(II) and Sn(IV) in Lubrication of Steel, *ASLE Transactions*, Vol. 30, 1987, pp. 508-519.

109 K. Kubo, Y. Shimakawa and M. Kibukawa, Study on the Load Carrying Mechanism of Sulphur-Phosphorus Type Lubricants, Proc. JSLE. Int. Tribology Conf., 8-10 July, 1985, Tokyo, Japan, Elsevier, pp. 661-666.

110 A. Masuko, M. Hirata and H. Watanabe, Electron Probe Microanalysis of Wear Scars of Timken Test Blocks on Sulfur-Phosphorus Type Industrial Gear Oils, *ASLE Transactions*, Vol. 20, 1977, pp. 304-308.

111 R.M. Matveevsky, Temperature of the Tribochemical Reaction Between Extreme-Pressure (E.P.) Additives and Metals, *Tribology International*, Vol. 4, 1971, pp. 97-98.

112 G. Bollani, Failure Criteria in Thin Film Lubrication With E.P. Additives, *Wear*, Vol. 36, 1976, pp. 19-23.

113 T.A. Stolarski, A Contribution to the Theory of Lubricated Wear, *Wear*, Vol. 59, 1980, pp. 309-322.

114 C.N. Rowe, Role of Additive Adsorption in the Mitigation of Wear, *ASLE Transactions*, Vol. 13, 1970, pp. 179-188.

115 S.C. Lee and H.S. Cheng, Scuffing Theory Modelling and Experimental Correlations, *Transactions ASME, Journal of Tribology*, Vol. 113, 1991, pp. 327-334.

116 H.A. Spikes and A. Cameron, A Comparison of Adsorption and Boundary Lubricant Failure, *Proc. Roy. Soc., London*, Series A, Vol. 336, 1974, pp. 407-419.

117 S.M. Hsu, Boundary Lubrication: Current Understanding, *Tribology Letters*, Vol. 3, 1997, pp 1-11.

118 W.F. Bowman and G.W. Stachowiak, A Review of Scuffing Models, *Tribology Letters*, Vol. 2, No. 2, 1996, pp. 113-131.

119 K. Meyer, H. Berndt and B. Essiger, Interacting Mechanisms of Organic Sulphides With Metallic Surfaces and Their Importance for Problems of Friction and Lubrication, *Applications of Surface Science*, Vol. 4, 1980, pp. 154-161.

120 S. Mori and Y. Shitara, Chemically Active Surface of Gold Formed by Scratching, *Applied Surface Science*, Vol. 68, 1993, pp. 605-607.

121 A.W. Batchelor and G.W. Stachowiak, Model of Scuffing Based on the Vulnerability of an Elastohydrodynamic Oil Film to Chemical Degradation Catalyzed by the Contacting Surfaces, *Tribology Letters*, Vol. 1, No. 4, 1995, pp. 349-365.

122 A. Douglas, E.D. Doyle and B.M. Jenkins, Surface Modification for Gear Wear, Proc. Int. Tribology Conference, Melbourne, The Institution of Engineers Australia, National Conference Publication No. 87/18, December, 1987, pp. 52-58.

123 M.A. Keller and C.S. Saba, Catalytic Degradation of a Perfluoroalkylether in a Thermogravimetric Analyzer, *Tribology Transactions*, 1998, Vol. 41, pp. 519-524.

124 D.J. Carre, Perfluoroalkylether Oil Degradation: Inference of FeF_3 Formation on Steel Surfaces Under Boundary Conditions, *ASLE Transactions*, Vol. 29, 1986, pp. 121-125.

125 Y. Hoshi, N. Shimotamai, M. Sato and S. Mori, Change of Concentration of Additives Under EHL Condition - Observation by Micro-FTIR, *The Tribologist, Proc. Japan Society of Tribologists*, Vol. 44, No. 9, 1999, pp. 736-743.

126 B. Rosenblum, P. Braunlich and L. Himmel, Spontaneous Emission of Charged Particles and Photons During Tensile Deformation of Oxide-Covered Metals Under Ultrahigh Vacuum Conditions, *Journal of Applied Physics*, Vol. 48, 1997, pp. 5263-5273.

127 K. Nakayama and H. Hashimoto, Triboemission, Tribochemical Reaction and Friction and Wear in Ceramics Under Various N-Butane Gas Pressures, *Tribology International*, Vol. 29, 1996, pp. 385-393.

128 C. Kajdas, Physics and Chemistry of Tribological Wear, Proceedings of the 10th International Tribology Colloquium, Technische Akademie Esslingen, Ostfildern, Germany, 9-11 January, 1996, Volume I (editor: Wilfried J. Bartz), publ. Technische Akademie Esslingen, 1996, pp. 37-62.

129 Y. Enomoto, H. Ohuchi and S. Mori, Electron Emission and Electrification of Ceramics during Sliding, Proceedings of the First Asia International Conference on Tribology, ASIATRIB'98, Beijing, publ. Tsinghua University Press, 1998, pp. 669-672.

130 J.T. Dickinson, L. Scudiero, K. Yasuda, M-W. Kim and S.C. Langford, Dynamic Tribological Probes: Particle Emission and Transient Electrical Measurements, *Tribology Letters*, Vol. 3, 1997, pp. 53-67.

131 R.S. Gates and S.M. Hsu, Silicon Nitride Boundary Lubrication: Effect of Phosphorus-Containing Organic Compounds, *Tribological Transactions*, Vol. 39, 1996, pp. 795-802.

132 R.S. Gates and S.M. Hsu, Silicon Nitride Boundary Lubrication: Effect of Sulfonate, Phenate and Salicylate Compounds, *Tribology Transactions*, Vol. 43, 2000, pp. 269-274.

133 W. Liu, E.E. Klaus and J.L. Duda, Wear Behaviour of Steel-on-Si_3N_4 and Systems with Vapor Phase Lubrication of Oleic Acid and TCP, *Wear* , Vol. 214, 1998, pp. 207-211.

134 J. Hersberger, O.O. Ajayi, J. Zhang, H. Yoon and G.R. Fenske, Formation of Austenite during Scuffing Failure of SAE 4340 Steel, *Wear*, Vol. 256, 2004, pp. 159-167.

135 F. Gao, P.V. Kotvis and W.T. Tysoe, The Friction, Mobility and Transfer of Tribological Films: Potassium Chloride and Ferrous Chloride on Iron, *Wear*, Vol. 256, 2004, pp. 1005-1017.

136 P. Cong, T. Li and S. Mori, Friction-Wear Behavior and Tribochemical Reactions of Different Ceramics in HFC-134a Gas, *Wear*, Vol. 252, 2002, pp. 662-667.

SOLID LUBRICATION
AND
SURFACE TREATMENTS

9.1   INTRODUCTION

Solid lubricants have many attractive features compared to oil lubricants, and one of the obvious advantages is their superior cleanliness. Solid lubricants can also provide lubrication at extremes of temperature, under vacuum conditions or in the presence of strong radioactivity. Oil usually cannot be used under these conditions. Solid lubrication is not new; the use of graphite as a forging lubricant is a traditional practice. The scope of solid lubrication has, however, been greatly extended by new technologies for depositing the solid film onto the wearing surface. The lubricant deposition method is critical to the efficiency of the lubricating medium, since even the most powerful lubricant will be easily scraped off a wearing surface if the mode of deposition is incorrect.

Specialized solid substances can also be used to confer extremely high wear resistance on machine parts. The economics of manufacture are already being transformed by the greater lifetimes of cutting tools, forming moulds, dies, etc. The wear resistant substances may be extremely expensive in bulk, but when applied as a thin film they provide an economical and effective means of minimizing wear problems. The questions of practical importance are: what are the commonly used solid lubricants? What distinguishes a solid lubricant from other solid materials? What is the mechanism involved in their functioning? What are the methods of application of solid lubricants? What are the wear resistant coatings and methods for their deposition? In this chapter the characteristic features of solid lubricants and basic surface treatments are discussed.

9.2   LUBRICATION BY SOLIDS

In the absence of lubrication provided by liquids or gases, most forms of solid contact involve considerable adhesion between the respective surfaces. Strong adhesion between contacting surfaces nearly always causes a large coefficient of friction because most materials resist shear parallel to the contact surface as effectively as they resist compression normal to the contact face. However, some materials exhibit anisotropy of mechanical properties, i.e., failure occurs at low shear stresses, resulting in a low coefficient of friction at the interface. Anisotropy of mechanical properties or, in simple terms, planes of weakness are characteristic of lamellar solids. If these lamellae are able to slide over one another at relatively low shear stresses then the lamellar solid becomes self-lubricating. This mechanism is schematically illustrated in Figure 9.1.

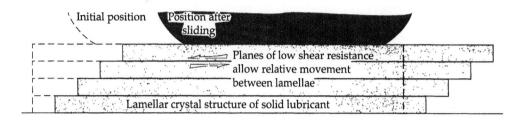

FIGURE 9.1   Mechanism of lubrication by lamellar solids.

This intuitive model of solid lubrication, which still has not been unequivocally demonstrated, was formally stated by Bragg in 1928 [1] to explain the lubricating properties of graphite, which is a classic example of a lamellar solid with lubricating properties. Unfortunately very few of the lamellar solids known offer useful lubricating properties.

Some non-lamellar solids, e.g., silver, can also reduce friction and wear when applied as a thin film to the wearing surfaces. Therefore a second mechanism of solid lubrication referring to films of soft metals on a hard substrate has been suggested by Bowden and Tabor [2]. Since the soft metallic film is thin, the hard substrate determines the contact area and no matter how thin the soft metallic layer is, the shear strength of asperities in contact is determined by the softer and weaker metal. Consequently the product of asperity shear strength and contact area, which determines frictional force, becomes quite low under such conditions. This principle is illustrated schematically in Figure 9.2.

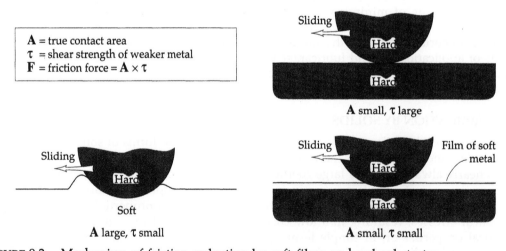

FIGURE 9.2   Mechanism of friction reduction by soft films on hard substrates.

Lamellar solids and soft films provide the two fundamental modes of solid lubrication currently employed. Since there may be other modes, for example, the low friction characteristic of melting wear, which has scarcely been examined, it is thought that the basic concepts of solid lubrication may sustain considerable revision in the future.

9.2.1   LUBRICATION BY LAMELLAR SOLIDS

Lamellar solids with useful lubricating properties exhibit three essential characteristics:

·   the lamellar structure deforms at very low shear stress levels;

·   the lamellar solid adheres strongly to the worn surface;

· there is no decomposition or other form of chemical degradation at the operating temperature and in the environment.

All these conditions unfortunately create significant limitations on the usefulness of solid lubricants.

As already mentioned, not all lamellar solids are capable of interlamellar sliding at low shear stresses. For example, mica and talc, although very similar chemically and crystallographically, exhibit a large difference in the level of adhesion between lamellae. Freshly cleaved mica sheets show very strong adhesion force, preventing continuous smooth sliding between the sheets [3,4]. The values of coefficient of friction measured were found to exceed **100** at low loads and even at the highest load attainable without fracturing of the mica, the coefficient of friction measured was **35** or more. The results obtained for talc were in total contrast to those of mica with very low values of both friction and adhesion. The reason for this large discrepancy is believed to lie in the nature of bonding between the mica lamellae and the talc lamellae. When mica is cleaved, positive potassium ions and negative oxygen ions are exposed so that if two mica surfaces are brought into contact there is strong electrostatic attraction between corresponding oxygen and potassium ions. Since this feature is absent with talc, there is only a much weaker van der Waals bonding acting between lamellae. The difference between these mechanisms is illustrated schematically in Figure 9.3.

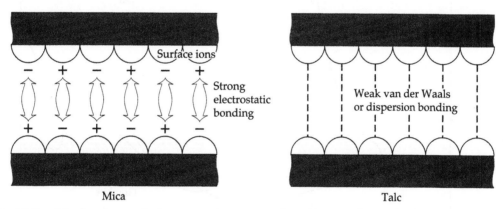

FIGURE 9.3   Mechanism of electrostatic strong bonding and weak dispersion bonding between lamellae.

Good solid lubricants therefore exhibit only weak bonding between lamellae. Although adhesion between lamellae is highly undesirable, adhesion of lamellae to the worn surface is essential. In general, material that does not adhere to a worn surface is quickly removed by the sweeping action of sliding surfaces. This is schematically illustrated in Figure 9.4.

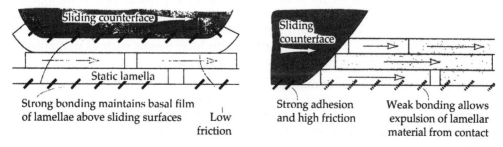

FIGURE 9.4   Effect of adhesion strength of the solid lubricant lamellae to the worn surface on friction.

The frictional properties of graphite, molybdenum disulphide and talc applied as powders to sliding steel surfaces are found to be quite different although all three substances have the required lamellar crystal structure [5]. Molybdenum disulphide and graphite exhibit low coefficients of friction between **0.1** and **0.3** for temperatures ranging from room temperature up to 400°C. On the other hand, as shown in Figure 9.5, talc shows a high value of the coefficient of friction of about 0.9 - 1 once the temperature exceeds 200°C.

After sliding, transferred layers of graphite and molybdenum disulphide were found on the worn surfaces. The crystal structure of these transferred layers showed orientation of the lamellae parallel to the worn surface. It was found that talc was transferred in much smaller quantities than graphite and molybdenum disulphide, with negligible orientation of lamellae parallel to the worn surface. To explain the poor performance of talc, it has been suggested that talc, unlike graphite and molybdenum disulphide, is too soft to be mechanically embedded in the surface [5]. Other studies, however, have emphasized that adhesion plays the decisive role in the mechanism of solid lubrication and this is a widely accepted view, although the supporting evidence is still incomplete [6-8].

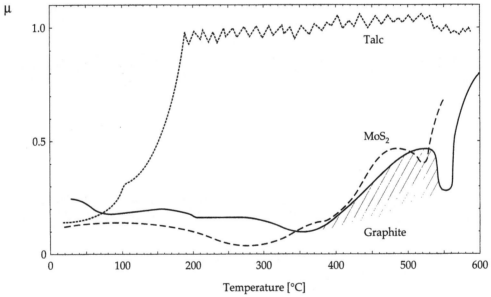

FIGURE 9.5   Relationship between coefficient of friction and temperature for graphite, talc and molybdenum disulphide [5].

The mechanism of bonding between a solid lubricant and a worn surface has not been investigated thoroughly at the atomic scale. In the case of steel surfaces it is thought that the sulphur ions in the molybdenum disulphide bond with the iron in a steel surface [9], but the lubricating effect of molybdenum disulphide is not only limited to steel or other reactive metals. For example, it shows a lubricating effect with inert metals such as platinum [5]. The mechanism by which graphite bonds to the surface is also unclear.

There are also chemical and environmental limitations imposed on the functioning of solid lubricants which are quite important since these lubricants are often required to function under extreme conditions. The basic problems are associated with oxidation or decomposition at high temperatures and contamination by water and are discussed in the next section.

Friction and Wear Characteristics of Lamellar Solids

The tribological characteristics of a large number of inorganic substances with lamellar crystal structures have been examined. A lamellar crystal structure with planes of weakness in shear is found in some metal dichalcogenicides and a few metal halides. Typical examples of dichalcogenicides with useful tribological characteristics are molybdenum disulphide, tungsten disulphide and molybdenum ditelluride. Halides with the required crystal structure include cadmium iodide and nickel iodide. On the other hand, graphite constitutes a unique class of solid lubricant since it is an uncombined chemical element. For reasons not yet fully understood, out of the many existing lamellar solids, only graphite and molybdenum disulphide offer a superior lubrication performance, and therefore these two substances have been much more extensively investigated than the others and are commonly used as solid lubricants.

· Graphite and Molybdenum Disulphide

The tribological characteristics of lubricating films of graphite and molybdenum disulphide are very similar. This is partly because of their considerable similarity in crystal structure. The unit cells of crystal structure of molybdenum disulphide and graphite are shown in Figure 9.6.

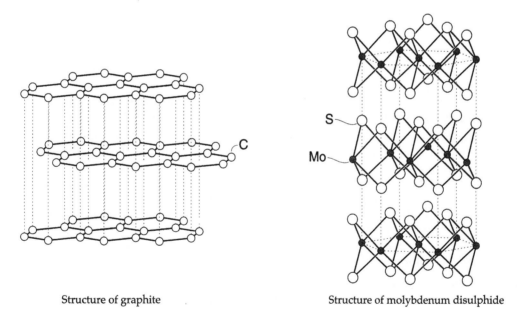

Structure of graphite   Structure of molybdenum disulphide

FIGURE 9.6   Crystal structures of molybdenum disulphide and graphite.

It can be seen from Figure 9.6 that in both materials strong chemical bonds between associated atoms form planes of high strength, while in a direction normal to these planes atoms are far apart and bonding is relatively weak. This bond strength anisotropy is far more acute in the case of molybdenum disulphide than in graphite. In graphite the surface energy along cleavage planes is relatively high and sliding between these planes is facilitated by the presence of small amounts of oxygen and water [10]. It is thought that oxygen and water adsorb on the surface of graphite lamellae and suppress bonding between lamellae.

Apart from having a lamellar crystal structure, a layer structure is also present in both molybdenum disulphide and graphite. The layered structure of molybdenum disulphide is shown in Figure 9.7 where layers about 1 [μm] thick are clearly visible.

5μm

FIGURE 9.7   Layered structure of molybdenum disulphide [11].

The layers of molybdenum disulphide are quite flexible and can slide over each other repeatedly without damage [11]. It was found that under repeated sliding, films of molybdenum disulphide can move significant distances over the worn surface [12]. The lubrication mechanism of graphite and molybdenum disulphide is believed to be a result of the relatively free movement of adjacent layers in these substances.

There are some clear distinctions in performance between graphite and molybdenum disulphide films on steel under atmospheric conditions [13]. Both graphite and molybdenum disulphide exhibit a decrease in scuffing failure load with increased sliding speed. In general, however, graphite films fail at lower loads and exhibit shorter lifetimes than molybdenum disulphide films. The limiting contact stresses for graphite are a little over half that of molybdenum disulphide. It has been found that the coefficient of friction in the presence of molybdenum disulphide declines asymptotically with increasing load to a value of about 0.05 [14]. Although sliding speed does not seem to have an influence on the coefficient of friction [14], the life of solid lubricant films is greatly reduced by increases in sliding speed [15].

The failure mode of molybdenum disulphide films involves a relatively specialized mechanism [16]. The smooth, continuous films formed by a solid lubricant over the wearing surfaces fail by '**blistering**' rather than by directly wearing out. The mechanism of blistering is illustrated schematically in Figure 9.8.

During the 'blistering' process, circular patches of the lubricant with a diameter of 0.1 to 1 [mm] detach from the substrate. These blisters originate from 'micro-blisters' about 1 [μm] in diameter which are formed very soon after the beginning of sliding. The blisters, when formed, are not immediately destroyed by the wearing contact and can be pressed back onto the wearing surface many times. The number of blisters per unit area of worn surface increases with duration, although their diameter slightly decreases [16]. When the blisters are sufficiently numerous, failure of the lubricating film begins with large scale detachment of the solid lubricant from the substrate. It has also been shown that graphite films can fail by blister formation [17].

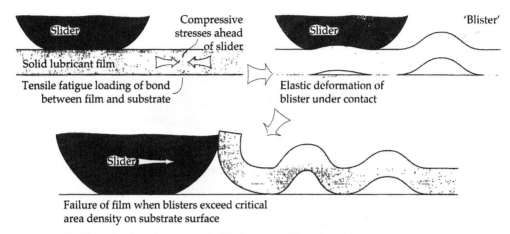

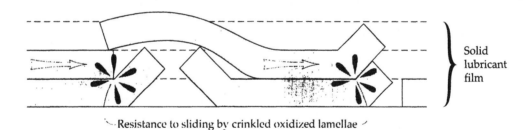

FIGURE 9.8   Failure mechanism of solid lubricant films by 'blistering'.

The blistering process can be greatly accelerated by atmospheric oxygen. The lifetime of molybdenum disulphide films in a vacuum is more than 10 times higher than that obtained in air, and even quite small traces of oxygen can significantly reduce the life of solid lubricant films [16]. To explain this it has been hypothesized that atmospheric oxygen causes oxidation of the edges of the molybdenum disulphide lamellae [11]. When unoxidized, the lamellae edges are thought to be relatively smooth, facilitating mutual sliding. After oxidation has taken place, crinkling and pitting of the edges occur which hinders sliding. Consequently, blistering is a result of buckling under compressive stress caused by the hindrance of lamella movement due to oxidized edges. This process of oxidative crinkling of molybdenum disulphide lamellae and the associated failure of the lubricating film is shown schematically in Figure 9.9.

FIGURE 9.9   Schematic illustration of the oxidation induced failure of solid lubrication by molybdenum disulphide.

The solution to this problem is in blending of graphite and molybdenum disulphide which results in a superior lubricant. Adding antimony-thioantimonate ($Sb(SbS_4)$) gives further improvement in lubricant's performance [15]. Although this effect has long been known in the trade [16], it has only recently been scientifically investigated [18]. The optimum combination of molybdenum disulphide, graphite and antimony-thioantimonate contains only a surprisingly small fraction of molybdenum disulphide, i.e., the precise composition is 18.75% molybdenum disulphide, 56.25% graphite and 25% antimony-thioantimonate [15].

The effect has been explained in terms of graphite forming a layered structure with the molybdenum disulphide [11]. Since the graphite lamellae are considered to be more resistant to distortion by oxidation than the molybdenum disulphide lamellae, a mix of distorted and undistorted lamellae is less prone to blistering than distorted lamellae acting alone. The model of the beneficial effect of graphite on the durability of solid lubricating films is illustrated schematically in Figure 9.10.

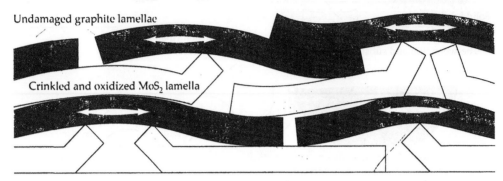

Undamaged graphite lamellae

Crinkled and oxidized MoS_2 lamella

Uncrinkled graphite lamellae maintain smooth movement

FIGURE 9.10 Schematic illustration of graphite suppressing the effect of oxidative crinkling of molybdenum disulphide.

The role of antimony-thioantimonate is not clearly understood. It has been suggested, however, that this compound acts as a sacrificial anti-oxidant [12]. The oxidation product of antimony-thioantimonate, antimony trioxide (Sb_2O_3), has also been found to improve the life of formulations based on molybdenum disulphide [19,20]. It was hypothesized that antimony trioxide acts as a soft plastic 'lubricant' in critical high temperature asperity contacts [21].

The functioning of solid lubricants can be severely affected by the environmental conditions, i.e., high temperatures and the presence of oxygen. The two commonly used solid lubricants, graphite and molybdenum disulphide, are not exempt from these problems. It was found, for example, that graphite fails to function in air at temperatures greater than 500°C due to rapid oxidation [5]. Oxidation or 'burning' also causes large friction increases with molybdenum disulphide over about the same range of temperatures as with graphite, as illustrated in Figure 9.5.

Removal or absence of air (as in space flight) also has a profound effect on the performance of graphite. For example, the coefficient of friction of graphite in a vacuum at temperatures below 800°C is about 0.4, much higher than that observed in air, and declines slightly at higher temperatures, as illustrated in Figure 9.11.

On the other hand, molybdenum disulphide retains a low friction coefficient in the absence of air, providing a vital lubricant for space vehicles. The coefficient of friction of molybdenum disulphide measured in a vacuum exhibits a constant value of about 0.2 over the temperature range from room temperature to about 800°C, where decomposition of the molybdenum disulphide to molybdenum metal and gaseous sulphur occurs [22].

The effect of water on the frictional performance of molybdenum disulphide is only slight. For example, it was found that when dry nitrogen was replaced by moist nitrogen the coefficient of friction increased from 0.1 to 0.2 [23]. In air, trace amounts of water caused an increase in friction similar to that of moist nitrogen [24].

One of the practical aspects of solid lubrication is the initial thickness of the lubricating film. The effect of the initial solid lubricant film thickness on friction and wear characteristics has been investigated with a view to achieving long film life [15]. It was found that only a thin layer of solid lubricant is needed to achieve effective lubrication. The thickness of the solid lubricant film declines rapidly from the initial value to a steady state value between **2** and **4** [μm] and is maintained until failure of the film by the blistering process occurs.

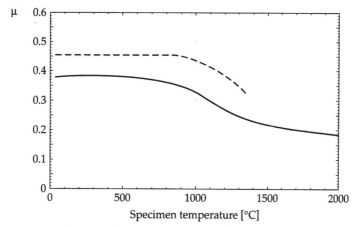

FIGURE 9.11 Friction coefficient of two types of graphite in a vacuum versus temperature [22].

· *Carbon-Based Materials Other Than Graphite*

Apart from graphite, other compounds of carbon have useful lubricating properties. Excluding the organic polymers, such as PTFE, which are discussed in Chapter 16, two substances, phthalocyanine and graphite fluoride, have shown potential usefulness as lubricants.

Phthalocyanine is a generic term for a series of organic compounds which are useful as dye pigments and the detailed description of the compound can be found in any standard chemical text. Phthalocyanine consists of several interconnected cyclic carbon groups linked together by nitrogen atoms. A metal such as iron may also be incorporated at the centre of the molecule. The structure of phthalocyanine with and without a central metal atom is shown in Figure 9.12. Since phthalocyanines often exhibit a lamellar crystal structure, this has generated speculation about their lubricating capacity.

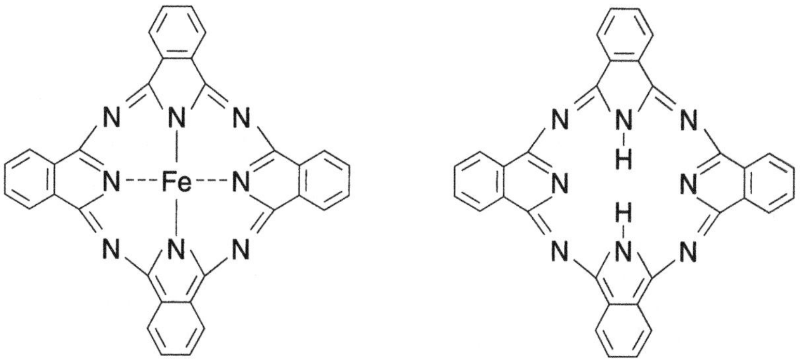

FIGURE 9.12 Molecular structure of phthalocyanine, with and without a central metal atom.

Phthalocyanine shows a good lubrication performance under high contact stresses and sliding speeds when sprayed directly onto the surface while molybdenum disulphide requires a more careful method of deposition [25]. It was found that the lubricating performance of phthalocyanine in sliding steel contacts is quite similar to graphite but inferior to molybdenum disulphide [26]. The load carrying mechanism of phthalocyanine depends on a visible film of material deposited on the surface, in a manner similar to graphite and molybdenum disulphide [26]. On the other hand, it has also been found that for a copper on sapphire contact phthalocyanine provides superior lubricating performance to molybdenum disulphide [27]. Since the main research efforts in solid lubricants are directed to finding a compound with a lubricating performance better than that of molybdenum disulphide, interest in phthalocyanine has waned.

Graphite fluoride was introduced as a solid lubricant in the 1960's [28]. It must be synthesized artificially by an intricate method which has limited its use. Graphite fluoride is a complex substance whose lubrication performance surpasses molybdenum disulphide. Graphite fluoride is an impure substance described by the chemical formula $(CF_x)_n$ where 'x' may vary from 0.7 to 1.12 but is usually close to 1. The crystal structure of graphite fluoride is believed to be lamellar, resembling graphite or molybdenum disulphide. It was found that under certain conditions of load and speed, a lubricating film of graphite fluoride offers a much greater durability than a film of molybdenum disulphide [28]. This is demonstrated in Figure 9.13 where the relationships between wear life and coefficient of friction under moderate loads for graphite fluoride and molybdenum disulphide are shown.

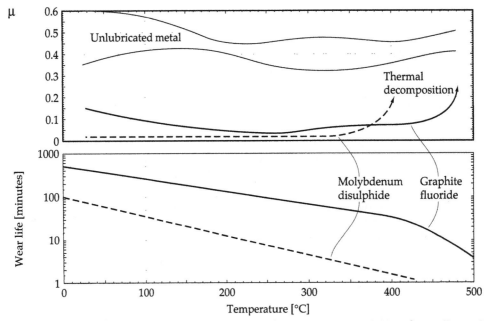

FIGURE 9.13 Comparison of friction and wear characteristics of graphite fluoride and molybdenum disulphide [28].

It can be seen from Figure 9.13 that graphite fluoride films are more durable than molybdenum disulphide films under moderate loads. Friction coefficients and upper temperature limits of these two substances are very close to each other. Furthermore, graphite fluoride has an advantage of providing more durable films in moist air than in dry air [28]. The performance of graphite fluoride as a solid lubricant, however, is strongly influenced by the technique of film deposition (e.g., burnishing) and this is probably the most important factor in controlling its performance [29].

· *Minor Solid Lubricants*

Substances such as tungsten disulphide and non-stoichiometric niobium sulphide, although exhibiting promising friction and wear characteristics, have not, for various reasons, attained general acceptance by lubricant users. It has been noted that the planar hexagonal crystal structure considered as optimal for lubrication is not confined only to molybdenum disulphide and it is shared by many other metal dichalcogenicides [30,31].

The sulphides, selenides and tellurides of metals such as tungsten, niobium and tantalum form planar hexagonal or trigonal structures (with a few exceptions) and exhibit coefficients of friction even lower than molybdenum disulphide. For example, compounds such as tungsten selenide, niobium sulphide, tantalum sulphide and selenide offer coefficients of friction less than 0.1 while the coefficient of friction of molybdenum disulphide measured at the same conditions is 0.18 [30]. Although these substances exhibit a lubricating capacity exceeding that of molybdenum disulphide, they unfortunately have certain disadvantages which are their high cost and scarcity. Tantalum and niobium are exotic metals produced only in small quantities.

Synthetic non-stoichiometric compounds also exhibit useful lubricating properties. For example, it was found that synthetic non-stoichiometric niobium disulphide (Nb_1+xS_2) can sustain a higher seizure load than molybdenum disulphide in a Falex test and exhibits a lower friction coefficient in a four-ball test [32]. The synthesis technique of niobium disulphide is critical to the performance of this compound and is reflected in its high price. In contrast, molybdenum disulphide occurs naturally in the appropriate crystalline form as molybdenite and is therefore relatively cheap and plentiful.

In some applications, apart from cost, an important consideration is the oxidation resistance of solid lubricants. For example, the oxidation resistance of tungsten is about 100°C higher than that of molybdenum disulphide [33]. Other compounds such as tantalum disulphide and diselenide and vanadium diselenide also display better oxidation resistance at high temperatures than either graphite or molybdenum disulphide [30].

Apart from the dichalcogenicides only very few inorganic compounds can be considered as solid lubricants. Although cadmium iodide and cadmium bromide can lubricate copper, the coefficient of friction in a vacuum is higher than when molybdenum disulphide is used [34]. Both cadmium bromide and iodide have the layer-lattice crystal structure. Unfortunately, these compounds are toxic and soluble in water.

At extreme temperatures in a corrosive environment, very few lubricants except the fluorides of reactive metals are stable. The lubricating properties of mixtures of calcium and barium fluorides were studied at high temperatures [35]. Although these lubricants give satisfactory performance at temperatures above 260°C with a coefficient of friction about 0.2, at temperatures below 260°C they exhibit a high coefficient of friction at low sliding velocities and poor adhesion to the substrate.

9.2.2 REDUCTION OF FRICTION BY SOFT METALLIC FILMS

Soft plastic metals such as gold, silver, indium and lead have often been used as solid lubricants by applying them as a thin surface layer to a hard substrate, e.g., carbon steel. The application of these metallic layers can result in a significant reduction in the coefficient of friction as shown in Figure 9.14.

Indium is a soft metal resembling lead in mechanical properties. It can be seen from Figure 9.14 that a usefully low friction coefficient only occurs at high loads, where it is thought that the contact area is controlled by the hard substrate [36,37].

Lubrication by thin metallic films is particularly useful in high vacuum applications where, in the absence of oxygen, particles from the metallic film can be repeatedly transferred between the sliding surfaces [38]. At low temperatures, however, brittleness of the soft metal may become a problem, with the soft metal film prone to flake off the worn surface [36]. In general, thin metallic films do not offer equal or superior lubrication to molybdenum disulphide and for this reason interest has been limited, although there are some exceptions [e.g., 39].

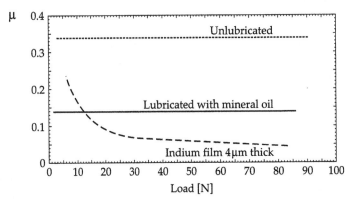

FIGURE 9.14 Effect of indium surface film on the frictional characteristics of steel [2].

Noble metal coatings show also some useful lubricating characteristics. For example, ultra-thin silver coatings offer promisingly low friction coefficients in vacuum. Tests conducted in ultra-high vacuum (UHV) of diamond pins sliding against uncoated silicon and Ag-coated silicon surfaces revealed that a single atomic mono-layer coating of silver reduced the friction coefficient from just about 1 to 0.01 or less [100]. This effect cannot only be attributed to a classical Bowden and Tabor's thin film model of low friction, as shown in Figure 9.2, but also to a high d-bond characteristic of the outer electron shells of silver. A high d-bond characteristic of coating material seems to correlate well with low adhesion and friction coefficient in vacuum [101]. A change in crystalline structure of the material when the film thickness is reduced to 1 [nm] or less might be a controlling factor [102].

Soft plastic metals can also be useful as solid lubricant additives to a mechanically stronger material. For example, chromium carbide coatings enriched with silver, barium fluoride and calcium fluoride have been investigated with the aim of developing a surface film suitable for high temperature sliding contacts [40,41]. The chromium carbide forms the substrate, silver provides lubrication from room temperature to 400°C, and above 400°C lubrication is provided by barium and calcium fluorides which are high temperature solid lubricants. This specially formulated material, when deposited on a sliding surface, forms a lubricating and wear resistant coating with a coefficient of friction of about 0.2 or less.

Reduction of Friction by Metal Oxides at High Temperatures

At high temperatures, metal oxides may become relatively ductile and begin to act as solid lubricants. Although, in general, metal oxides exhibit a lubricating effect at high temperatures, not all of them show a reduction in the coefficient of friction in the range of temperatures which are useful for practical applications. Yellow lead oxide, PbO, is probably the most useful of the metal oxides and can provide good lubrication at high temperatures. The steel sliding tests conducted at temperatures of approximately 600°C revealed that among the many oxides tested, only lead oxide (PbO) and molybdenum trioxide (MoO_3) offered a substantial reduction in the coefficient of friction compared to the unlubricated case [42].

9.2.3 DEPOSITION METHODS OF SOLID LUBRICANTS

The durability of a solid lubricant film depends critically on the method of deposition of the solid lubricant on the substrate. Although it is relatively easy to devise a solid lubricant film that provides a low friction coefficient, it is much more difficult to ensure that this film will last for **1** million or more cycles of wearing contact. Firm adhesion between the film and the substrate is a pre-requisite for prolonged survival of the film and, as discussed in previous sections, the durability of a film is controlled by environmental factors as well as load and speed. The method of deposition dictates the level of adhesion between the film and the substrate. There are two modes of solid lubricant deposition:

- the traditional methods, which involve either spraying or painting the lubricant on the surface followed by burnishing of the film or frictional transfer of lubricant;

- the modern methods, which depend on the properties of plasma in a moderate vacuum to produce a lubricant film of high quality.

These two modes of application of solid lubricant have their particular merits and demerits, i.e., the traditional methods are easy to perform and do not require specialized equipment; the modern methods are more specialized and intricate but give better lubricating films with superior performance.

Traditional Methods of Solid Lubricant Deposition

A widely used method of applying solid lubricants such as molybdenum disulphide and graphite is to paint or spray a mixture of the solid lubricant and a 'binder' on to the surface. The binder is a substance which hardens on exposure to air or after heating in an oven and is used to bond the solid lubricant to the surface. Examples of binders that harden on exposure to air are acrylic and alkyd resins [43]. Binders requiring heating in order to harden are termed 'thermosets' and include phenolic and epoxy resins. The hardening temperature for the thermosetting resins is usually close to **200**°C. Another class of binders intended for high temperatures or exposure to nuclear radiation are based on inorganic materials such as low melting point glass. There are a wide range of binders available and the relative merits of different products are described in detail [e.g., 29,33]. A small quantity of a volatile 'carrier fluid' may also be present to liquefy the lubricant. Binders are usually present in the lubricant blend as **33**% by volume and usually a hard binder is preferable to a soft binder. Thermosetting binders are usually harder than air-setting binders. The exact type of binder has a considerable effect on the performance of the lubricant film in terms of the coefficient of friction and film lifetime. Phenolic binders give some of the best results [33]. The required thickness of solid lubricant film is between **3** and **10** [μm] and the control of thickness during application of the lubricant is one of the more exacting requirements of the entire process. Intricate or convolute shape of the component, for example, gear teeth, heightens the difficulty involved. Prior to coating the component must be carefully degreased to ensure good film adhesion. Sand-blasting with fine particles such as 220-mesh alumina provides the optimum surface roughness, which is between **0.4** and **2** [μm] RMS.

After the film deposition process is completed a careful running-in procedure is applied. Loads and speeds are increased gradually to the required level to allow conditioning of the film [33]. If the solid lubricant film is properly applied, it should give a low coefficient of friction lasting one million cycles or more [33].

Another method of solid lubricant replenishment is based on frictional transfer of material. A rod or cylinder of solid lubricant is placed in the same wear track as the load bearing contact. The rod sustains wear to produce a transfer film of solid lubricant on the wear track [90]. This method is effective for a variety of solid lubricants and worn substrates. For example, a soft metal alloy based on lead and tin was deposited on steel to provide lubrication in a vacuum [91] while graphite was deposited on stellite alloys in air [92]. Stellite was effectively lubricated by this method to a temperature of about 400°C. Further increase in temperature caused the bulk oxidation of the graphite rod, resulting in the reduction in the diameter of the rod at the wearing surface which prevented the effective lubrication over the whole contact width [92].

Modern Methods of Solid Lubricant Deposition

The demanding requirements of space technology for solid lubrication led to the development of vacuum-based solid lubricant deposition techniques. In recent years, these techniques have been adopted by almost every industry. The use of a vacuum during a coating process has some important advantages over coating in air, i.e., contaminants are excluded and the solid lubricant can be applied as a plasma to the substrate. The significance of the plasma is that solid lubricant is deposited on the surface as individual atoms and ions of high energy. The adhesion and crystal structure of the film are improved by this effect and a longer lifetime of the film can be obtained. It has been found, for example, that a 200 [nm] film of sputtered molybdenum disulphide lasted more than five times longer than a much thicker 13 [μm] resin bonded molybdenum disulphide film [44]. This illustrates the potential advantages of vacuum-coating technology, i.e., a better friction and wear characteristic with less solid lubricant used; the thinner films enable conservation of the expensive solid lubricant and allow close tolerances to be maintained on precision machinery.

Strong bonding of a lubricant film depends on solid state adhesion between the film and the substrate [45]. As discussed in Chapter 12, strong solid state adhesion occurs when there are no intervening contaminants and the surfaces are in close contact. A big advantage of vacuum coating is that major sources of contaminants such as oxygen and water are excluded. Under vacuum, special cleaning processes can be carried out that remove residual contaminants from the surface of the substrate. Typically, the substrate surface is subjected to argon ion bombardment, which dislodges oxide films and water without destroying the microstructure of the substrate by over-heating. Argon ion bombardment is performed by admitting argon gas at a pressure of approximately 1 [Pa] and raising the substrate material to a large negative potential.

These two measures by themselves, i.e., the removal of contaminant sources and the cleaning of the surface, are insufficient to ensure a long-lasting solid lubricant film. It has been found experimentally that the solid lubricant must be projected at the substrate with considerable energy before a high performance lubricating film is obtained. In the early days of developing the deposition techniques a vapour of coating material was admitted to the vacuum containing the substrate, but this did not give the desirable results. It was found that a difference in electric potential between the source of coating material and the substrate is essential in obtaining a high performance lubricating film. If the coating material is deposited on the substrate largely as ions as opposed to atoms, the greater mobility of deposited ions favours the development of a superior crystal structure of the lubricant which determines its performance.

Two processes, ion plating and sputtering, have been found to be a very effective means of applying solid lubricants under vacuum. There are also other processes that are being developed or which already exist, but these are basically refinements of the same process. Incidentally, these processes are the same as for wear resistant coatings and therefore are discussed in detail later in this chapter.

The efficiency of the lubricant film is almost totally dependent on the coating method used. This is illustrated in Figure 9.15 where the durability of a gold lubricating film produced by different coating methods is shown. The annealed steel substrate was coated with gold film and fretted against the glass [46]. There are clear differences obtained by the application of different coating techniques. Therefore the subject of coating technology is intensively researched while, by comparison, the search for superior solid lubricants is pursued by only a few specialized groups.

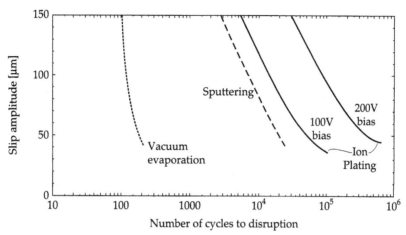

FIGURE 9.15 Comparison of the durability of a gold lubricant film produced by different coating techniques under fretting conditions [46].

Solid Lubricants as Additives to Oils and Polymers

Solid lubricants can be added to oils and polymers to improve their friction and wear properties. The solid lubricant most commonly added to oils is molybdenum disulphide. Finely ground molybdenum disulphide is added to oils in concentrations around **1%** by weight to form a colloidal dispersion in the oil. If the molybdenum disulphide is merely stirred into the oil it will rapidly precipitate and an improvement in lubrication is not achieved. A **1%** colloidal dispersion of molybdenum disulphide in uncompounded gear oil can raise the mechanical efficiency of worm gear drives by **1 - 3%** depending on the base oil [47]. A **1%** dispersion of molybdenum disulphide in mineral oil reduced the wear rate by a factor of **2** in a series of ball-on-cylinder wear tests [48]. Seizure loads and scuffing loads in various test machines were also increased by the addition of suspended molybdenum disulphide [49,50]. Although in most cases **1%** concentration by weight of molybdenum disulphide in oil is sufficient, improvements were still obtained at higher concentrations reaching **5%** [49].

However, an increase in wear when molybdenum disulphide is added to oil has also been reported [51]. Under moderate conditions of sliding speed and load where molybdenum disulphide is not expected to improve lubrication, abrasive impurities in the solid lubricant can cause rapid wear [51]. Silica in particular accentuates wear when in concentrations above **0.01%**, and pyrites (iron sulphide) are also destructive [51]. The quality, i.e., cleanliness, of the solid lubricant added to oil is therefore critical. Although solid lubricant additives are suitable for extremes of loads and speeds, they are not suitable for reducing wear under moderate conditions. Molybdenum disulphide suspensions provide a limited reduction in friction and wear when added to an oil containing sulphur based additives or zinc dialkyldithiophosphate. On the other hand, the presence of detergents or dispersants in the oil, such as calcium sulphonate, inhibits the lubricating action of molybdenum disulphide [48,50].

The mechanism of lubrication by molybdenum disulphide dispersed in oil has unfortunately received very little attention. It is widely believed, however, that molybdenum disulphide provides a complementary role to surfactants. Where there is a worn surface devoid of surfactant, it is hypothesized that molybdenum disulphide particles adhere to form a lubricating film. A conceptual model of solid lubrication by molybdenum disulphide, which occurs only when there are no surfactants to block adhesion by lamellae of solid lubricant to the worn surface, is illustrated schematically in Figure 9.16.

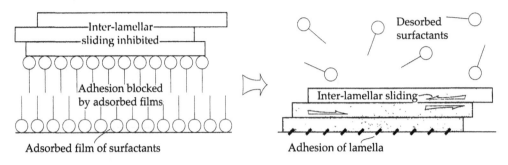

FIGURE 9.16 Conceptual model of the mechanism of lubrication by molybdenum disulphide suspended in oil.

It has been found that molybdenum disulphide lubricates by film formation on a worn surface at high temperatures where all surfactants, both natural and artificial, are unlikely to adsorb on worn surfaces [52]. However, evidence which confirms that molybdenum disulphide is only effective beyond the desorption temperature of the specific surfactants is absent from the published literature.

Solid lubricants are also used to improve the frictional characteristics of polymers [33]. In general they do offer some improvement but the effectiveness of solid lubricants added to polymers depends on the type of polymer used. The greatest improvements in polymer friction and wear characteristics are achieved with polymers of moderate lubricity such as nylon and polyimide [53]. For example, the addition of graphite to nylon results in a reduction of the coefficient of friction from 0.25 to 0.18 and a small reduction in wear [53]. On the other hand, it has also been shown that molybdenum disulphide when added to nylon oxidizes during wear and does not develop an effective transfer film [54]. Under these conditions, the friction performance of the nylon/molybdenum disulphide blend was found to be inferior to plain nylon [54].

In polyimides the addition of the same amount of graphite reduced the coefficient of friction to less than half of pure polyimide and significantly reduced wear. Although molybdenum disulphide showed the same reduction of coefficient of friction as the graphite/polyimide blend, its reduction in wear rate was inferior to that of the graphite/polyimide blend [53].

Improvements achieved by adding molybdenum disulphide and graphite to polytetrafluoroethylene (PTFE) are very limited [55,56]. The coefficients of friction for PTFE filled with graphite and molybdenum disulphide are very similar to that of unfilled PTFE and slightly lower than those obtained with most other fillers [55].

Interest in graphite has recently been extended by the incorporation of carbon fibres into polymers. Carbon fibres offer a unique combination of mechanical reinforcement and lubricity [57]. It has been shown that a carefully formulated polyimide/carbon fibre composite can sustain high contact loads and maintain a friction coefficient close to 0.2 at temperatures reaching 300°C with very low wear rates [58,59].

9.3 WEAR RESISTANT COATINGS AND SURFACE TREATMENTS

Wear resistant coatings consist of carefully applied layers of usually hard materials which are intended to give prolonged protection against wear. Abrasive wear, adhesive wear and fretting are often reduced by wear resistant coatings. There are numerous methods of applying hard materials. For example, sputtering and ion plating are used in a similar manner as in the deposition of solid lubricants to generate thin coatings. Other methods are used to deposit very thick layers of hard material. Applications of wear resistant coatings are found in every industry and, for example, include mining excavator shovels and crushers [60], cutting and forming tools in the manufacturing industries [61], rolling bearings in liquefied natural gas pumps [62], etc. In most of these applications, wear rather than friction is the critical problem. Another benefit of hard-coating technology is that a cheap substrate material can be improved by a coating of an exotic, high performance material. Most engineering items are made of steel and it is often found that some material other than steel is needed to fulfil the wear and friction requirements. Many wear resistant materials are brittle or expensive and can only be used as a coating, so improved coating technology has extended the control of wear to many previously unprotected engineering components.

9.3.1 TECHNIQUES OF PRODUCING WEAR RESISTANT COATINGS

There are many different methods of applying wear resistant or hard coatings to a metal substrate currently in use [e.g., 63-65]. New techniques continue to appear as every available technology is adapted to deposit a wear resistant coating more efficiently. The wear resistance of a surface can also be improved by localized heat treatment, i.e., thermal hardening, or by introducing alloying elements, e.g., nitriding or carburizing. Many of these methods have been in use for many years but unfortunately suffer from the disadvantage that the substrate needs to be heated to a high temperature. Carburizing, nitriding and carbonitriding in particular suffer from this problem. Various coating techniques available with their principal merits and demerits are listed in Table 9.1.

TABLE 9.1   Available techniques for modifying the surface to improve its tribological characteristics.

Physical and chemical vapour deposition	Thin discrete coating; no limitations on materials
Ion implantation	Thin diffuse coating; mixing with substrate inevitable
Surface welding	Suitable for very thick coatings only; limited to materials stable at high temperatures; coated surfaces may need further preparation
Thermal spraying	Very thick coatings possible but control of coating purity is difficult
Laser glazing and alloying	Thick coatings; coating material must be able to melt
Friction surfacing	Simple technology but limited to planar surfaces; produces thick metal coating
Explosive cladding	Rapid coating of large areas possible and bonding to substrate is good. Can give a tougher and thicker coating than many other methods
Electroplating	Wide range of coating thicknesses, but adhesion to substrate is poor and only certain materials can be coated by this technique

The thinner coatings are usually suitable for precision components while the thicker coatings are appropriate for large clearance components.

Coating Techniques Dependent on Vacuum or Gas at Very Low Pressure

Plasma based coating methods are used to generate high quality coatings without any limitation on the coating or substrate material. The basic types of coating processes currently in use are physical vapour deposition (PVD), chemical vapour deposition (CVD) and ion implantation. These coating technologies are suitable for thin coatings for precision components. The thickness of these coatings usually varies between **0.1** and **10** [μm]. These processes require enclosure in a vacuum or a low pressure gas from which atmospheric oxygen and water have been removed. As mentioned already the use of a vacuum during a coating process has some important advantages over coating in air. The exclusion of contaminants results in strong adhesion between the applied coating and substrate and greatly improves the durability of the coating.

· *Physical Vapour Deposition*

This process is used to apply coatings by condensation of vapours in a vacuum. The extremely clean conditions created by vacuum and glow discharge result in near perfect adhesion between the atoms of coating material and the atoms of the substrate. Porosity is also suppressed by the absence of dirt inclusions. PVD technology is extremely versatile. Virtually any metal, ceramic, intermetallic or other compounds that do not undergo dissociation can be easily deposited onto substrates of virtually any material, i.e., metals, ceramics, plastics or even paper. Therefore the applications of this technology range from the decorative to microelectronics, over a significant segment of the engineering, chemical, nuclear and related industries. In recent years, a number of specialized PVD techniques have been developed and extensively used. Each of these techniques has its own advantages and range of preferred applications. Physical vapour deposition consists of three major techniques: evaporation, ion plating and sputtering.

Evaporation is one of the oldest and most commonly used vacuum deposition techniques. This is a relatively simple and cheap process and is used to deposit coatings up to **1** [mm] thick. During the process of evaporation the coating material is vaporized by heating to a temperature of about **1000 - 2000**°C in a vacuum typically **10⁻⁶** to **1** [Pa] [64]. The source material can be heated by electrical resistance, eddy currents, electron beam, laser beam or arc discharge. Electric resistance heating usually applies to metallic materials having a low melting point while materials with a high melting point, e.g., refractory materials, need higher power density methods, e.g., electron beam heating. Since the coating material is in the electrically neutral state it is expelled from the surface of the source. The substrate is also pre-heated to a temperature of about **200 - 1600**°C [64]. Atoms in the form of vapour travel in straight lines from the coating source towards the substrate where condensation takes place. The collisions between the source material atoms and the ambient gas atoms reduce their kinetic energy. To minimize these collisions the source to substrate distance is adjusted so that it is less than the free path of gas atoms, e.g., about **0.15 - 0.45** [m]. Because of the low kinetic energy of the vapour the coatings produced during the evaporation exhibit low adhesion and therefore are less desirable for tribological applications compared to other vacuum based deposition processes. Furthermore, because the atoms of vapour travel in straight lines to the substrate, this results in a 'shadowing effect' for surfaces that do not directly face the coating source and common engineering components such as spheres, gears, moulds and valve bodies are difficult to coat uniformly. The evaporation process is schematically illustrated in Figure 9.17.

Ion plating is a process in which a phenomenon known as 'glow discharge' is utilized. If an electric potential is applied between two electrodes immersed in gas at reduced pressure, a stable passage of current is possible. The gas between the electrodes becomes luminescent, hence the term 'glow discharge'. When sufficient voltage is applied the coating material can be transferred from the 'source' electrode to the 'target' electrode which contains the substrate. The process of ion plating therefore involves thermal evaporation of the coating material in a manner similar to that used in the evaporation process and ionization of the vapour due to the presence of a strong electric field and previously ionized low pressure gas, usually argon. The argon and metal vapour ions are rapidly accelerated towards the substrate surface, impacting it with a considerable energy. Under these conditions, the coating material becomes embedded in the substrate with no clear boundary between film and substrate. Usually prior to ion plating the substrate is subjected to high-energy inert gas (argon) ion bombardment, causing a removal of surface impurities which is beneficial since it results in better adhesion. The actual coating process takes place after the surface of the substrate has been cleaned. However, the inert gas ion bombardment is continued without interruptions. This causes an undesirable effect of decreasing deposition rates since some of the deposited material is removed in the process. Therefore for the coating to form the deposition rate must exceed the sputtering rate. The heating of the substrate by intense gas bombardment may also cause some problems. The most important aspect of ion plating which distinguishes this process from the others is modification of the microstructure and composition of the deposit caused by ion bombardment [65]. Ion plating processes can be classified into two general categories: glow discharge (plasma) ion plating conducted in a low vacuum of **0.5** to **10** [Pa] and ion beam ion plating (using an external ionization source) performed in a high vacuum of 10^{-5} to 10^{-2} [Pa] [64]. The ion plating process is schematically illustrated in Figure 9.18.

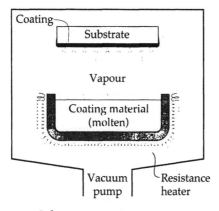

FIGURE 9.17 Schematic diagram of the evaporation process.

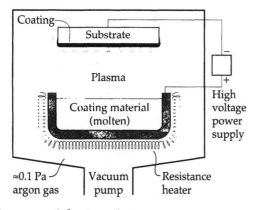

FIGURE 9.18 Schematic diagram of the ion plating process.

<u>Sputtering</u> is based on dislodging and ejecting the atoms from the coating material by bombardment of high-energy ions of heavy inert or reactive gases, usually argon. In sputtering the coating material is not evaporated and instead ionized argon gas is used to dislodge individual atoms of the coating substance. For example, in glow-discharge sputtering a coating material is placed in a vacuum chamber, which is evacuated to 10^{-5} to 10^{-3}

[Pa] and then back-filled with a working gas, e.g., argon, to a pressure of **0.5** to **10** [Pa] [64]. The substrate is positioned in front of the target so that it intercepts the flux of dislodged atoms. Therefore the coating material arrives at the substrate with far less energy than in ion plating so that a distinct boundary between film and substrate is formed. When atoms reach the substrate, a process of very rapid condensation occurs. The condensation process is critical to coating quality and unless optimized by the appropriate selection of coating rate, argon gas pressure and bias voltage, it may result in a porous crystal structure with poor wear resistance.

The most characteristic feature of the sputtering process is its universality. Since the coating material is transformed into the vapour phase by mechanical (momentum exchange) rather than a chemical or thermal process, virtually any material can be coated. Therefore the main advantage of sputtering is that substances which decompose at elevated temperatures can be sputtered and substrate heating during the coating process is usually negligible. Although ion plating produces an extremely well bonded film, it is limited to metals and thus compounds such as molybdenum disulphide which dissociate at high temperatures cannot be ionplated. Sputtering is further subdivided into direct current sputtering, which is only applicable to conductors, and radio-frequency sputtering, which permits coating of non-conducting materials, for example, electrical insulators. In the latter case, a high frequency alternating electric potential is applied to the substrate and to the 'source' material. The sputtering process is schematically illustrated in Figure 9.19.

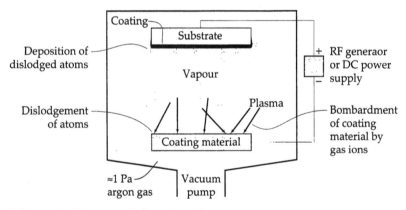

FIGURE 9.19 Schematic diagram of the sputtering process.

· *Chemical Vapour Deposition*

In this process the coating material, if not already in the vapour state, is formed by volatilization from either a liquid or a solid feed. The vapour is forced to flow by a pressure difference or the action of the carrier gas towards the substrate surface. Frequently reactant gas or other material in vapour phase is added to produce a metallic compound coating. For example, if nitrogen is introduced during titanium evaporation, then a titanium nitride coating is produced. The coating is obtained by either thermal decomposition or chemical reaction (with gas or vapour) near the atmospheric pressure. The chemical reactions usually take place in the temperature range between **150** and **2200**°C at pressures ranging from **50** [Pa] to atmospheric pressure [64]. Since the vapour will condense on any relatively cool surface that it contacts, all parts of the deposition system must be at least as hot as the vapour source. The reaction portion of the system is generally much hotter than the vapour source but considerably below the melting temperature of the coating. The substrate is usually heated by electric resistance, inductance or infrared heating. During the process the coating material is deposited, atom by atom, on the hot substrate. Although CVD coatings usually exhibit excellent adhesion, the requirements of high substrate temperature limit their applications to substrates which can withstand these high temperatures. The CVD process at low pressure allows the deposition of coatings with superior quality and uniformity over a large substrate area at high deposition rates [64]. The CVD process is schematically illustrated in Figure 9.20.

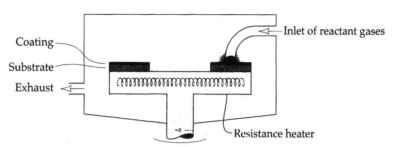

FIGURE 9.20 Schematic diagram of the CVD process.

· *Physical-Chemical Vapour Deposition*

This is a hybrid process which utilizes glow discharge to activate the CVD process. It is broadly referred to as 'plasma enhanced chemical vapour deposition' (PECVD) or 'plasma assisted chemical vapour deposition' (PACVD). In this process the techniques of forming solid deposits by initiating chemical reactions in a gas with an electrical discharge are utilized. Many of the phenomena characteristic to conventional high temperature CVD are employed in this process. Similarly the same principles that apply to glow discharge plasma in sputtering apply to CVD. In this process the coating can be applied at significantly lower substrate temperatures, of about **100 - 600**°C, because of the ability of high-energy electrons produced by glow discharge, at pressures ranging from **1** to **500** [Pa], to break chemical bonds and thus promote chemical reactions. Virtually any gas or vapour, including polymers, can be used as source material [64]. For example, during this process a diamond coating can be produced from carbon in methane or in acetylene [88]. Amorphous diamond-like coatings in vacuum can attain a coefficient of friction as low as 0.006 [96]. Although contamination by air and moisture tends to raise this coefficient of friction to about 0.02-0.07, the diamond-like coating still offers useful wear resistance under these conditions [97-99]. The mechanism responsible for such low friction is still not fully understood. The PECVD process is schematically illustrated in Figure 9.21.

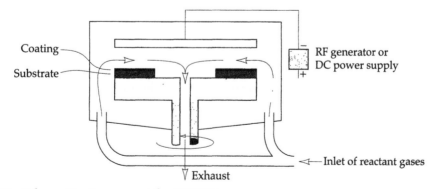

FIGURE 9.21 Schematic diagram of the PECVD process.

· *Ion Implantation*

The energy of ions in a plasma can be raised to much higher levels than is achieved in either ion plating or sputtering. If sufficient electrical potential is applied then the plasma can be converted to a directed beam which is aimed at the material to be coated, allowing the controlled introduction of the coating material into the surface of the substrate. This process is known as ion implantation. During the process of ion implantation, ions of elements, e.g., nitrogen, carbon or boron, are propelled with high energy at the specimen surface and penetrate the surface of the substrate. This is done by means of high-energy ion beams containing the coating material in a vacuum typically in the range 10^{-3} to 10^{-4} [Pa]. A specialized non-equilibrium microstructure results which is very often amorphous as the original crystal structure is destroyed by the implanted ions [66]. The modified near-surface layer consists of the remnants of a crystal structure and interstitial implanted atoms. The mass of implanted ions is limited by time, therefore compared to other surfaces, the layers of ion-implanted surfaces are very shallow, about **0.01** to **0.5** [μm]. The thickness limitation of the implanted layer is the major disadvantage of this method. The coatings generated by ion implantation are only useful in lightly loaded contacts. The technique allows for the implantation of metallic and non-metallic coating materials into metals, cermets, ceramics or even polymers. The ion implantation is carried out at low temperatures. Despite the thinness of the modified layer, a long-lasting reduction in friction and wear can be obtained, for example, when nitrogen is implanted into steel. The main advantage of the ion implantation process is that the treatment is very clean and the deposited layers very thin, hence the tolerances are maintained and the precision of the component is not distorted. Ion implantation is an expensive process since the cost of the equipment and running costs are high [64]. The ion implantation process is schematically illustrated in Figure 9.22.

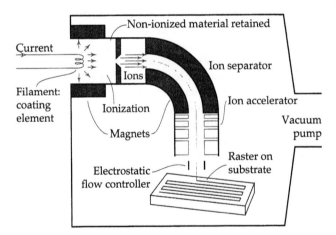

FIGURE 9.22 Schematic diagram of the ion implantation process.

More detailed information about surface coating techniques can be found elsewhere [45,64,65].

Coating Processes Requiring Localized Sources of Intense Heat

A localized intense source of heat, e.g., a flame, can provide a very convenient means of depositing coating material or producing a surface layer of altered microstructure. Coating methods in common use that apply this principle are surface welding, thermal spraying and laser hardening or surface melting.

· *Surface Welding*

In this technique the coating is deposited by melting of the coating material onto the substrate by a gas flame, plasma arc or electric arc welding process. A large variety of materials that can be melted and cast can be deposited by this technique. During the welding process a portion of the substrate surface is melted and mixed together with the coating material in the fusion zone, resulting in good bonding of the coating to the substrate. Welding is used in a variety of industrial applications requiring relatively thick, wear resistant coatings ranging from about **750** [μm] to a few millimetres [64]. Welding processes can be easily automated and are capable of depositing coatings on both small components of intricate shape and large flat surfaces.

There is a variety of specialized welding processes, e.g., oxyfuel gas welding (OGW), shielded metal arc welding (SMAW), submerged arc welding (SAW), gas metal arc welding (GMAW), gas tungsten arc welding (GTAW), etc., which are described in detail elsewhere [e.g., 64]. A schematic diagram of the typical welding process is shown in Figure 9.23.

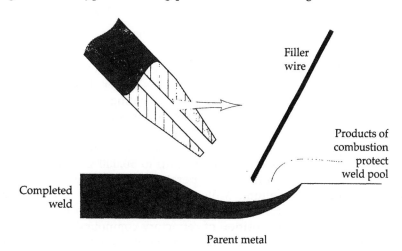

FIGURE 9.23 Schematic diagram of the welding process.

· *Thermal Spraying*

This is the most versatile process of deposition of coating materials. During this process the coating material is fed to a heating zone where it becomes molten and then is propelled to the pre-heated substrate. Coating material can be supplied in the form of rod, wire or powder (most commonly used). The distance from the spraying gun to the substrate is in the range of **0.15** to **0.3** [m] [64]. The molten particles accelerated towards the substrate are cooled to a semimolten condition. They splatter on the substrate surface and are instantly bonded primarily by mechanical interlocking [64]. Since during the process a substantial amount of heat is transmitted to the substrate it is therefore water cooled. There are a number of techniques used to melt and propel the coating material and the most commonly applied are flame spraying, plasma spraying, detonation-gun spraying, electric arc spraying and others.

<u>Flame Spraying</u> utilizes the flame produced from combustion gases, e.g., oxyacetylene and oxyhydrogen, to melt the coating material. Coating material is fed at a controlled rate into the flame where it melts. The flame temperature is in the range of **3000** to **3500**°C. Compressed air is fed through the annulus around the outside of the nozzle and accelerates the molten or semimolten particles onto the substrate. The process is relatively cheap and is characterized by high deposition rates and efficiency. The flame sprayed coatings, in general, exhibit lower

bond strength and higher porosity than the other thermally sprayed coatings. The process is widely used in industry, i.e., for corrosion resistant coatings. A schematic diagram of this process is shown in Figure 9.24.

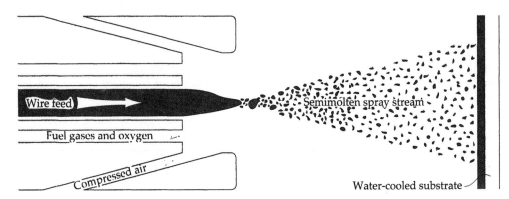

FIGURE 9.24 Schematic diagram of the flame spraying process.

Plasma Spraying is different from the plasma-based coating methods described previously since the coating metal is deposited as molten droplets rather than as individual atoms or ions. The technique utilizes an electric arc to melt the coating material and to propel it as a high-velocity spray onto the substrate. In this process gases passing through the nozzle are ionized by an electric arc, producing a high temperature stream of plasma. The coating material is fed to the plasma flame where it melts and is propelled to the substrate. The temperature of the plasma flame is very high, e.g., up to **30,000°C** and can melt any coating material, e.g., ceramics [89]. The highest temperatures are achieved with a monoatomic carrier gas such as argon and helium. Molecular gases such as hydrogen and nitrogen produce lower plasma temperatures because of their higher heat capacity. Therefore plasma spraying is suitable for the rapid deposition of refractory compounds which are usually hard in order to form thick hard surface coatings. The very high particle velocity in plasma spraying compared to flame spraying results in very good adhesion of the coating to the substrate and a high coating density. The application of an inert gas in plasma spraying gives high purity, oxide free deposits. Although it is possible to plasma spray in open air the oxidation of the heated metal powder is appreciable and the application of inert gas atmosphere is advantageous. The quality of coating is critical to the wear resistance of the coating, i.e., adhesion of the coating to the substrate and cohesion or bonding between powder particles in the coating must be strong. These conditions often remain unfulfilled when the coating material is deposited as partially molten particles or where the shrinkage stress on cooling is allowed to become excessive [67]. Plasma spraying is commonly used in applications requiring wear and corrosion resistant surfaces, i.e., bearings, valve seats, aircraft engines, mining machinery and farm equipment. A schematic diagram of the plasma spraying process is shown in Figure 9.25.

Detonation Gun Spraying is similar in some respects to flame spraying. The mixture of a metered amount of coating material in a powder form with a controlled amount of oxygen and acetylene is injected into the chamber where it is ignited. The powder particles are heated and accelerated at extremely high velocities towards the substrate where they impinge. The process is repeated several times per second. The coatings produced by this method exhibit higher hardness, density and adhesion (bonding strength) than can be achieved with conventional plasma or flame spraying processes. The coating porosity is also very fine. Unfortunately, very hard materials cannot be coated by this process because the high velocity gas can cause surface erosion. Wear and corrosion resistant coatings capable of operating at elevated temperatures are produced by this method. They are used in applications where close tolerances must be maintained, i.e., valve components, pump plungers, compressor rods, etc. A schematic diagram of this process is shown in Figure 9.26.

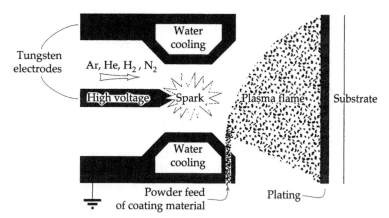

FIGURE 9.25 Schematic diagram of the plasma spraying process.

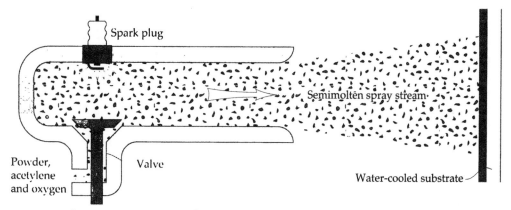

FIGURE 9.26 Schematic diagram of the detonation gun spraying process.

Electric Arc Spraying differs from the other thermal spraying processes since there is no external heat source such as a gas flame or electrically induced plasma [64]. In this process an electric arc is produced by two converging wire electrodes. Melting of the wires occurs at the high arc temperature and molten particles are atomized and accelerated onto the substrate by the compressed air. The use of an inert atomizing gas might result in improved characteristics of some coatings by inhibiting oxidation. The wires are continuously fed to balance the sprayed material. Since there is no flame touching the substrate like in the other thermal spraying processes, the substrate heating is lower. The adhesion achieved during this process is higher than that of flame sprayed coatings under comparable conditions. During this process coatings of mixed metals, e.g., copper and stainless steel, can be produced. A schematic diagram of this process is shown in Figure 9.27.

· *Laser Surface Hardening and Alloying*

Laser hardening is a form of thermal hardening where a high power laser beam, such as from a carbon dioxide laser (with the beam power up to **15** [kW]), is scanned over a surface to cause melting to a limited depth. Rapid cooling of the surface by the unheated substrate results in a hard quenched microstructure with a fine grain size formed on re-solidification [68,69]. Surface alloying is also possible if the surface of the substrate is pre-coated with the alloying element or the alloying element is fed into the path of the laser beam. This process is also known as laser cladding. The coating material is mixed together with the melted top layer of the substrate and subsequently solidifies. Because of the very large temperature gradients, mixing of the molten material is intense. A strong bond between the modified layer and the substrate is formed since the substrate is never exposed to any atmospheric contaminants. For example, a stainless steel layer on a steel substrate can be produced by pre-coating steel with chromium and then melting the surface with the laser beam. To produce a **500** [μm] thick layer of **1%** stainless steel, a pre-coating of **5** [μm] thick chromium is required. Although laser treatment can be performed in the open air the oxidation rate, e.g., of steel, can be high and destructive. Therefore it is often preferable to apply this process in an inert gas atmosphere. The process is particularly useful in applications where access to the surface to be treated is more easily achieved by the laser than any other method, e.g., a torch. The area coverage by this process is relatively slow and the overlap areas between successive laser passes have inferior properties and microstructure [89]. A schematic diagram of laser surface alloying is shown in Figure 9.28.

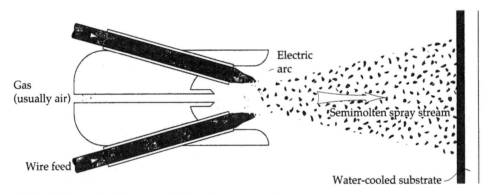

FIGURE 9.27 Schematic diagram of electric arc spraying process.

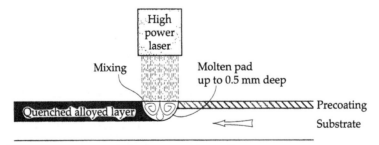

FIGURE 9.28 Schematic diagram of the laser surface alloying process.

The application of laser surface modification technology is not only limited to metallic components. For example, the wear and corrosion resistance of concrete can significantly be improved by the application of laser surface melting. A high power diode laser directed at the surface of the concrete causes surface melting and the formation of a glaze of about 0.75 [mm] thick. This glaze is much harder than the untreated concrete, i.e., Mohs hardness of 6 compared to 2 for plain concrete. The formation of the amorphous and hard glaze leads to a much greater wear and corrosion resistance of concrete [103].

Coating Processes Based on Deposition in the Solid State

It would be very convenient to directly join the coating material and substrate without intermediate processes such as plasma-based coating. Under certain circumstances this is possible, although there are some comparatively severe limitations on the utility of such methods. Two basic methods of direct joining or bonding are explosive bonding and friction surfacing. These two methods do not require a carefully controlled environment or a localized heat source and can be performed in the open air.

Friction Surfacing is an adaptation of friction welding where a material from a rod is bonded to a flat surface by a combination of rotation and high contact force. It was discovered that if the flat surface was moved while the rod was pressed against it and simultaneously rotated then a layer of transferred material was deposited on the flat surface. This constituted a relatively simple way of rapidly depositing a thick layer of metal [70]. Friction surfacing has been studied as a simple and robust way of re-surfacing worn military and agricultural equipment in remote areas such as the interior of Australia [70]. A major simplification of this coating technology compared to other coating methods is that there is no necessity for the exclusion of atmospheric oxygen during the coating process. However, the provision of an inert gas atmosphere does improve adhesion or bonding between the coating and the substrate [70]. Shape limitations of the substrate, i.e., that friction surfacing is only practicable for plane surfaces or objects with axial symmetry, e.g., metal extrusions, as opposed to complex surfaces, e.g., gear teeth, restricts the application of this otherwise promising and simple coating technology. A schematic diagram of friction surfacing process is shown in Figure 9.29a.

Explosive Cladding, also known as explosive bonding or explosive welding, is essentially a solid-phase welding process during which bonding is produced by high velocity collision between the substrate and the coating material. The high velocity is achieved by a controlled explosion. In most cases, the coating material in the form of a sheet is placed at a small angle of incidence to the substrate. A protective buffer, usually in the form of rubber sheet, is placed on top of the coating material. When the explosives in the form of sheet or slurry are detonated behind the buffer, contact between the sheet of coating material and the substrate spreads out from the end of the sheet closest the substrate. A front forms at the edge of the contact where the sheet is momentarily folded. Strong bonding of the cladding material is facilitated by the expulsion of contaminants and oxide layers as a jet of fragmented or molten material in front of the impacting metal surfaces. The removal of contaminants and oxides is caused by the extremely high impact speed of the opposing surfaces during explosive bonding. At the apex of the front, substrate and coating material melt. Since the metal flow around the collision point is unstable and oscillating it often produces a rippled or wavy interface between the substrate and the coating material. No external heat is required in this process. Virtually any combination of metals and alloys, which otherwise cannot be bonded, e.g., aluminium and steel, can be bonded by this process. Very high pressures of about **3** [GPa] generated during the process restrict the thickness of the coatings to the layers thicker than **0.3** [mm] as thinner layers could rupture. The process is used in manufacturing, e.g., corrosion resistant coatings for chemical, marine and petrochemical industries. The inconvenience of explosives, the limitation of this method to large flat surfaces and the requirement for the coating material to be tough does, however, severely curtail the usefulness of this technique. A schematic diagram of the explosive bonding process is shown in Figure 9.29b.

Miscellaneous Coating Processes

There is a wide range of coating processes which are extensively used for applications requiring resistance to corrosion and mild wear. These coating processes are very much simpler and cheaper than the processes already described. For example, coatings can be deposited by dipping the substrate in a coating material, spraying the coating material in an atomized liquid form, e.g., the technique commonly used for paint applications, by utilizing brush pad or roller, by chemical deposition or electroplating. Although the adhesion of these coatings is sometimes not adequate for severe tribological applications, they can be used as corrosion resistant coatings and as soft, low shear strength solid lubricant and metallic coatings for sliding wear applications.

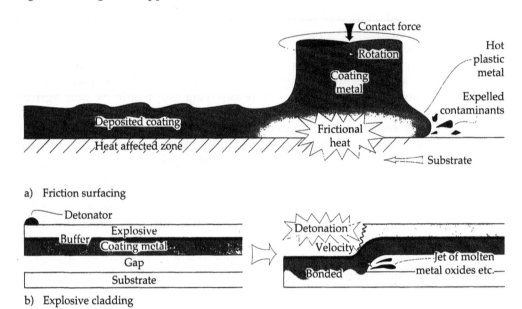

a) Friction surfacing

b) Explosive cladding

FIGURE 9.29 Schematic diagram of the coating processes based on deposition in the solid state.

Electroplating is a well established process with proven benefits in controlling corrosion and wear resistance. This process is a convenient way of applying coatings of metals with high melting points such as chromium, nickel, copper, silver, gold, platinum, etc., onto the substrate. The electroplating system consists of an electrolytic bath, two electrodes and a DC power source. A conducting solution which contains a salt or other compound of the metals to be deposited is placed in the bath. When an electrical potential is applied to the electrodes, i.e., one is the material to be coated and the other is the donor electrode, the metal is deposited on the substrate by electrochemical dissolution from the donor electrode as schematically illustrated in Figure 9.30.

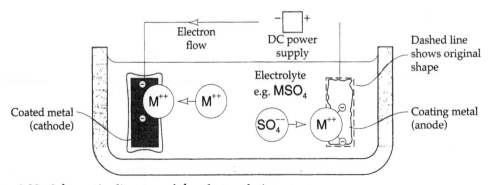

FIGURE 9.30 Schematic diagram of the electroplating process.

Since, in general, the process is conducted under atmospheric conditions and material is deposited with low energy, the coating-substrate adhesion is poor. Coatings can be applied by this method to most metallic surfaces.

9.3.2 APPLICATION OF COATINGS AND SURFACE TREATMENTS IN WEAR AND FRICTION CONTROL

There is a wide range of coating techniques, and careful selection of the appropriate coating material and method is a pre-requisite for an effective coating. Prior to selecting the coating material and method, the first question to be asked is whether wear or friction is of greater concern. If the prime objective is to reduce friction, then a solid lubricant coating should be selected and the coating method will, in most cases, be either sputtering or a combination of painting and baking.

To suppress wear by the application of coatings, it is first necessary to determine the mechanism of wear occurring, e.g., whether abrasive wear or some other form of wear is present. Although most coatings can suppress several forms of wear, each type of coating is most effective at preventing a few specific wear mechanisms. Therefore during the selection process of the most effective coating to suppress wear in a particular situation, i.e., coating optimization, the prevailing wear mechanism must first be recognized and assessed. The basic characteristics of the coatings which can be achieved by the methods described in the previous section in terms of wear control are summarized in Figure 9.31.

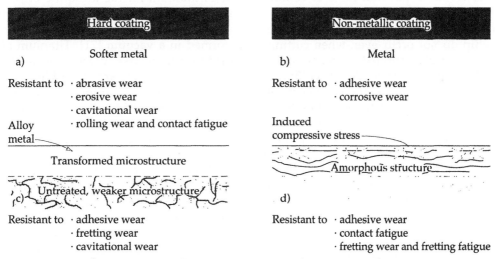

FIGURE 9.31 Basic characteristics of coatings in terms of wear control.

It can be seen from Figure 9.31 that while the optimization of a coating to resist abrasive wear is relatively simple, i.e., it is sufficient to produce a thick hard surface layer with toughness high enough to prevent coating fracture, other wear mechanisms require much greater care in coating optimization.

Characteristics of Wear Resistant Coatings

Studies of wear resistant coatings reveal that hard coatings are most effective in suppressing abrasive wear. An example of this finding is illustrated in Figure 9.32, which shows the wear rate of a pump rotor as a function of the hardness of the coating applied to the surface. It can be seen from Figure 9.32 that the abrasive wear rate declines to a negligible value once a PVD coating of titanium nitride, which is characterized by extremely high hardness, is employed.

In this example, abrasive wear was caused by very fine contaminants present in the pumped fluid and the size of the abrasives was sufficiently small for a thin PVD coating to be effective. In other applications where the abrasive particles are much larger, thicker coatings are more appropriate.

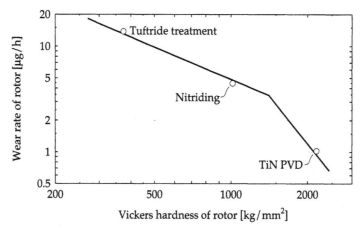

FIGURE 9.32   Example of the resistance of a hard coating, TiN, to abrasion [60].

It was also found that thin films of ceramics such as titanium nitride are quite effective in suppressing adhesive wear in poorly lubricated and high stress contacts. For example, when a cutting tool is coated with titanium nitride, adhesion and seizure between the tool and the metal chip do not occur even when cutting is performed in a vacuum [71]. Titanium nitride coatings were also applied to gears, and the scuffing tests on coated and uncoated gears revealed that the critical load and scuffing resistance for coated gears are much higher [71,72]. This coating also reduces the coefficient of friction in unlubricated sliding as well as wear rates, e.g., coefficients of friction close to 0.1 between titanium nitride and zirconium nitride coatings on hardened bearing steel have been observed [73]. Unfortunately, titanium nitride coatings do not provide corrosion resistance [74]. Since zirconium and hafnium belong to the same IVB group of the periodic table of chemical elements as titanium, some similarity in wear properties of their compounds can be expected. In fact, hafnium nitride was found to give the best wear resistance performance in tests on cutting tools [75]. Zirconium nitride is also extremely useful as a coating [73,76]. It should also be mentioned that for hard coatings to be effective, an adequate substrate hardness is essential [63]. Therefore hardened steels and materials such as stellite are generally used as a substrate for this type of coating.

Fretting wear can be mitigated by the use of hard coatings, e.g., carbides, especially at small amplitudes of fretting movement [77]. However, at higher fretting amplitudes, spalling of the carbide coatings renders them ineffective.

Coatings produced by ion implantation, in certain applications, can also provide large reductions in wear. Since the coatings produced by this technique are very thin, they are only effective in reducing wear at low load levels as illustrated in Figure 9.33.

It can be seen from Figure 9.33 that in dry unlubricated sliding of stainless steel, nitrogen ion implantation reduces the wear rate by a factor of 10 or more at light loads. Nitrogen ion implantation was found to be very efficient in reducing the wear and friction of titanium and titanium alloys [79]. Titanium and its alloys are notorious for their susceptibility to seizure in dry sliding, and implantation by nitrogen ions reduces the coefficients of friction in dry sliding to a value as low as 0.15. It has also been found that nitrogen ion implantation is effective in reducing the fretting wear and surface damage in stainless steel [80].

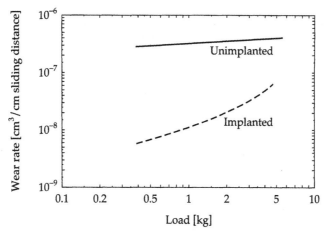

FIGURE 9.33 Effect of nitrogen ion implantation on wear rates of stainless steel in unlubricated sliding [78].

It is, however, difficult to give general rules for the applicability of the ion implantation technique since the results are only specific to a particular combination of substrate and implanted material. Since there are about 10 substrate metals in common use, e.g., steel, cast iron, aluminium, copper, titanium, etc., and theoretically the entire periodic table of elements is available for implantation, the number of combinations of test materials far exceeds available resources and time. Currently the most commonly implanted elements are nitrogen, carbon and boron, and the conclusions about the usefulness of ion implantation may change as 'new' implantation elements are discovered, e.g., combined yttrium and nitrogen implantation [81].

In some cases the incorrect choice of implantation material for coating may actually result in increased wear rates. It has been shown that implantation of stainless steel by argon increased the coefficient of friction from 0.8 to 1.0 in dry unlubricated sliding at room temperature while implantation by boron reduced the coefficient of friction to about 0.15 [82]. The causes of wear reduction by ion implantation are largely unknown.

It has been hypothesized that ion implantation causes surface hardening, passivation and loss of adhesion [79]. Although the hardening is effective for pure metals, hardened alloys such as martensitic steel are not hardened by ion implantation [79].

An interesting feature of nitrogen ion implantation is that the effect of implantation persists after wear has exceeded the depth of implantation, in some cases by a factor of 10 or more [79]. It was observed that nitrogen migrated inwards with the wear surface [83]. As with other characteristics of ion implantation, this phenomenon is also poorly understood.

Surface hardening by high power lasers also results in reductions in wear for a wide range of applications. For example, the flank and rake faces of a cutting tool made of high speed steel showed less wear after laser hardening [84]. However, the reductions in wear achieved by the application of laser hardening are not as dramatic as those obtained by nitride coatings. On the other hand, laser hardening has been found to be very beneficial in the unlubricated sliding of cast irons [85,86]. This effect is illustrated in Figure 9.34 where laser hardened cast iron exhibits superior wear behaviour to untreated cast iron.

It can be seen from Figure 9.34 that the transition from mild to severe wear is suppressed by laser hardening and the mitigation of wear at high loads is clear.

Laser surface alloying has also been found to effectively reduce wear under fretting. For example, a zirconium alloyed layer formed on the surface of carbon steel caused a reduction in the volume of fretting wear by at least a factor of 5 [87].

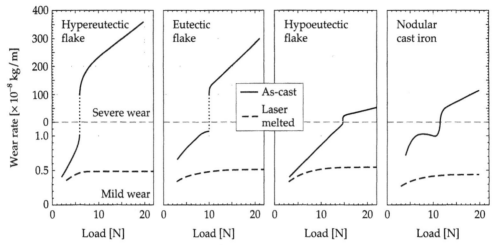

FIGURE 9.34 Effect of laser hardening on the wear rates of various cast irons [85].

The performance of non-metallic coatings such as tungsten carbide used for rolling elements is related to the operating conditions. For example, it was found that **100** to **200** [μm] thick plasma-sprayed coatings on steel and ceramic balls fail by surface wear when lubrication is poor or by sub-surface delamination when lubrication is effective [93].

Wear-resistant coatings can be as vulnerable to oxidative wear as monolithic metal substrates. For example, copper causes rapid wear of cutting tools coated with titanium nitride, titanium carbide or a combination of both compounds. It was found that the primary cause of rapid wear of the titanium nitride and carbide coatings is a catalytic effect of copper on the oxidation of the nitride and carbide to titanium oxide, which is then rapidly worn away. In contrast, the oxidation of chromium nitride in air is much slower than titanium nitride [94], thus permitting the chromium nitride to effectively protect machining tools from wear by copper [95].

New Trends In Coating Technology

There is a large variety of different types of coatings used across the industry. New technologies continue to emerge and new coatings are continuously being developed to meet the needs and the requirements of the modern industry. Some recent developments in coating technology are briefly described in this section.

· Diamond-Like Carbon Coatings

Diamond-like carbon coatings, or DLC films as they are widely known, offer both extremely low friction coefficients and considerable protection against wear at high contact stresses. DLC coatings are usually deposited using plasma-assisted CVD or various PVD techniques. DLC films, in general, are effective in vacuum and inert atmospheres. Hydrogenated DLC films exhibit the lowest friction in vacuum. The introduction of hydrogen or oxygen molecules to the vacuum chamber increases the friction slightly [105]. The effect of water molecules depends on film composition. For some compositions, such as hydrogen free DLCs, water molecules can drastically reduce friction compared to that in vacuum, while for other compositions, such as hydrogenated DLCs, friction can be slightly increased [105]. The worst performance in water was observed for amorphous hydrogenated carbon (a-C:H) coating [104] and the coating was rapidly worn through. The improvement was achieved with the hydrogenated DLC coatings with a multilayered structure and doped with, for example, Si, W or Cr. These coatings provided low wear and low friction coefficient of about 0.1 when slid in water against alumina balls [104].

Closely related to DLC films but with a different crystalline structure is tetrahedral amorphous carbon, i.e., ta-Carbon, which provides coatings with useful low friction and wear properties on steel substrates. Ta-Carbon not only resists detrimental effects of water and air but is found to depend on the presence of water molecules for its low coefficient of friction of about 0.07 [104,105]. It is thought that the crystalline structure of ta-Carbon has only three out of the four carbon bonds fully saturated, i.e., the fourth bond remains dangling or extending out of the surface. In high vacuum these dangling bonds interact with the counterface, resulting in a high friction until water, oxygen or hydrogen molecules are introduced to the chamber and passivate these bonds [105].

DLC coatings also provide good rolling contact fatigue resistance. It was found that at high contact stresses, exceeding 5 [GPa], thin coatings, of slightly less than 1 [μm], of either DLC or titanium nitride (TiN) significantly improved the rolling contact fatigue resistance [106].

· *Carbide and Nitride Coatings*

Carbon nitride coatings provide low friction coefficients and are extremely hard, i.e., similar to diamond. When deposited as 0.1 [μm] films on a silicon substrate they provide a coefficient of friction as low as 0.05 when slid against the silicon nitride. These coatings are sensitive to water and oxygen, but a moderate flow of nitrogen gas around the sliding contact reduces the friction coefficient to as low as 0.01, although 0.05 is more common. The coating is believed to wear out by a process of low-cycle fatigue against hard asperities from the counterface [107]. The wear of these coatings is sensitive to a contact load, with no observed wear occurring below a critical load level for a contacting spherical diamond. Progressive load increase results in 'feathery' and then plate-like wear debris [108].

Titanium nitride coatings, known for their excellent wear resistance, have been used by the industry for many years. Some examples are described in the previous section. New coating materials, emerging now, are combined nitride-carbides or carbonitrides and contain other metals apart from titanium. For example, titanium carbonitride (TiC$_x$N$_{(1-x)}$, where 'x' denotes a non-stoichiometric composition) coatings, deposited by PVD and by filtered arc deposition, seem to offer better wear resistance than pure TiN coatings [109].

While titanium nitrides and carbides are widely cited in the literature and used as valuable wear resistant materials, other carbides of transition metals may also be useful. Tantalum and molybdenum carbides have bulk hardnesses of 16.6 and 15.5 [GPa] (respectively), which are softer than titanium nitride with a hardness in excess of 20 [GPa]. Tantalum and molybdenum carbide coatings on steel substrates with a supporting layer of titanium nitride exhibit low friction coefficients of approximately 0.1 to 0.2 during dry sliding against alumina balls [110] while a higher friction coefficient, of approximately 0.4, was observed during sliding against steel balls. Tantalum and molybdenum carbides are used as protective coatings for steel moulds during the casting of aluminium and aluminium alloys [111].

In some cases the type of surrounding gas can affect the coatings performance. For example, steel balls coated with TiN sliding against uncoated steel discs showed significantly lower friction coefficients in a nitrogen atmosphere than in air [112]. It was observed that in air an iron oxide layer rapidly formed on the worn surface of the discs and controlled the wear process. The replacement of air with nitrogen resulted in the removal of the oxide layer and reversion of the friction coefficient to its lower value [112].

Chromium nitride is also viewed as a potentially useful substitute for titanium nitride due to its higher thermal stability, lower deposition temperature (which reduces damage to the substrate), better wear and corrosion resistance [113]. In addition, favourable interactions with lubricating oil are achieved through the oil entrapment in between the grain boundaries on the chromium nitride films [113].

The addition of boron to TiN coatings results in the formation of titanium-boron-nitride coatings. This was found to provide useful reductions in friction and wear coefficients. The crystalline structure of the films is, however, quite complex and is usually described as a nano-structured composite of TiB_2 and TiN grains [114].

· *Thick Coatings*

Thick coatings, usually produced by plasma spraying, are widely used in the mineral processing industry. The mineral processing industry contains innumerable examples of high stress abrasion where a softer metal substrate requires protection from indentation and gouging by hard grits. For this application, a thick, hard coating is the appropriate protective measure and plasma spraying is suitable for the deposition of such coatings of ceramics or refractory, hard metals.

Chromic oxide (Cr_2O_3) is widely used as a wear resistant coating for high temperature conditions. However, as the friction coefficient at high temperatures is rather high, solid lubricants such as calcium fluoride and silver oxide can be blended into the plasma-sprayed film to achieve a desirable effect, i.e., reduction in high temperature friction [115]. With the addition of these solid lubricants there is, however, an increase in wear rate. These solid lubricants appear to be more effective with plasma-sprayed zirconia coatings. A useful reduction in friction and wear was achieved for the temperature range of 200 to 800°C when both calcium fluoride and silver oxide were blended into the zirconia. These solid lubricants are, however, ineffective at room temperature and at temperatures exceeding 800°C because of possible melting [116].

In common with much thinner coatings, boron doping at a concentration of 0.2% by weight is found to reduce the friction and wear of nickel aluminide-chromium carbide ($Ni_3Al-Cr_7C_3$) composite coatings formed by self-propagating high temperature casting on carbon steel. These coatings have shown good sliding wear resistance at moderate to high temperatures [117].

· *Nano-Engineered Coatings*

With advances in coating technology, it is now possible to develop structured coatings to produce superior engineering properties such as hardness and durability. The typical structure of such coatings is a multi-layer configuration, and for thin coatings of about 50 [nm] thickness, 'nano-engineering' is involved. These nano-engineered coatings have become the subject of rapidly progressing research and development.

In one form of nano-engineered coatings, the adhesion and stiffness of successive layers of the coating were optimized to produce a durable lubricating film. For example, a nano-engineered coatings composed of the polymer trilayer (sandwiched) film of 20 to 30 [nm] thick deposited on silicon provided a very low friction coefficient when slid against steel at contact stresses exceeding 1 [GPa] [118].

Very fine multi-layer coatings of iron nitride (FeN) and TiN, where the thickness of each layer of FeN is about 5 [nm] and the thickness of the TiN layer is about 2 [nm] for a total coating thickness of several hundred nanometres, appear to offer a better scratch resistance in nano-indentation tests than simple iron nitride coatings [119]. It seems that the TiN layers effectively reinforce the FeN layers to give a lower friction coefficient during the scratch test. The FeN coatings are used for the improved data storage density in computer hard disk drives.

· *Other Coatings*

Most coating projects relate to steel or ferrous alloys as substrates because these are by far the most widely used materials. However, non-ferrous metals or alloys also require special protective coatings. For these applications, non-metallic materials are often used since they can be effective yet relatively cheap. For example, thin organic coatings of acrylic, polyurethane or polyester resins are often applied to the surfaces of hot-dipped Al-Zn coated steel sheets. The purpose of such organic coatings is to reduce adhesive wear and galling between the steel sheet and the processing rollers. Without the organic coating the hot-dipped Al-Zn alloy coating has poor galling resistance, which leads to frequent product spoilage [120].

It is known that copper is prone to adhesive and abrasive wear due to its low hardness and high ductility. Coatings, approximately 20 [μm] thick, of titanium-copper inter-metallic compounds forming on copper discs by the pack-cementation process showed useful reductions in the friction and wear coefficients during the sliding tests with steel pins [121]. Cementation process is based on the solid state diffusion when the material to be enriched is placed in close contact with the additive material and held there for long periods of time at high temperatures.

9.4   SUMMARY

Solid lubricants and surface treatments have rapidly evolved in recent decades from simple and traditional methods to extremely sophisticated technologies. These developments are part of an effort to eliminate the limitations imposed by oil-based lubrication and, in the process, are changing the general perception of the limits of wearing contacts. Knowledge of the mechanisms behind these improvements in lubrication and wear resistance is, in most cases, very limited. The methods employed in most studies on surface coatings are empirical and there is relatively little information available on which surface treatment is the most suitable for a particular application. This is a new area subjected to extensive research and the number of new surface treatments and coating technologies available to control friction and wear is rapidly increasing.

REVISION QUESTIONS

9.1   Name the two most widely used solid lubricants?

9.2   What is a characteristic material property of solids used as solid lubricants?

9.3   What two criteria distinguish a solid lubricant from a material with no useful lubricating properties?

9.4   Name a lamellar solid which is not effective as a solid lubricant.

9.5   Why is talc ineffective as a solid lubricant, despite its lamellar crystalline structure and friable nature?

9.6   Is graphite intrinsically a self-lubricating solid?

9.7   What is the cause of 'dusting' in the commutators of aviation electric motors or generators?

9.8   What is the most important advantage of molybdenum disulphide (MoS_2) over graphite?

9.9   In air, which of the two lubricants, MoS_2 or graphite, has the higher temperature limit?

9.10   How do graphite and MoS_2 degrade at high temperatures in air?

9.11   Apart from applying MoS_2 as a spray, paint, powder or as a suspension in liquid, name a method of depositing a film of MoS_2 to give a high quality low wear - low friction surface.

9.12 MoS_2 is often added to oils and greases to produce a superior lubricant but additives in the oil often prevent the MoS_2 from being effective. What is the cause of this?

9.13   Name another lamellar solid that has been considered as a solid lubricant.

9.14   Are there any non-lamellar solids used as solid lubricants?

9.15   Describe the two main purposes of tribological surface coatings.

9.16   How can surface treatments reduce surface fatigue (cracking) problems?

9.17   What is the main disadvantage of traditional processes such as carburizing and nitriding?

9.18   Name an application of coating technology that is of great importance to manufacturing.

9.19   Name two materials widely used as ingredients of hard coatings.

9.20   What happens to the crystalline structure of a metal when subjected to an intense flux of nitrogen ions, e.g., as in ion implantation?

9.21   What are the key differences between the friction characteristics of a thin discrete film of solid lubricant and an integral diffuse coating deposited by, for example, ion plating? Clue: Consider the record of friction versus sliding distance or cycle number in a friction test.

9.22   How might the operating environment affect the usefulness of solid lubricant and wear resistant films? Clue: Consider the role of water and oxygen as chemical reagents.

9.23   Are laser surface treatments exclusively for metals and metal-like materials such as engineering ceramics? Provide a counter-example if the answer is negative.

REFERENCES

1   W.L. Bragg, Introduction to Crystal Analysis, G. Bell and Sons, London, 1928.

2   F.P. Bowden and D. Tabor, Friction and Lubrication of Solids, Part 1, Oxford University Press, 1954.

3   A.I. Bailey and J.S. Courtney-Pratt, Real Contact and Shear Strength of Monomolecular Layers, *Proc. Roy. Soc., London*, Series A, Vol. 227, 1955, pp. 500-515.

4   F.P. Bowden and D. Tabor, Friction and Lubrication of Solids, Part 2, Oxford University Press, 1964.

5   R.F. Deacon and J.F. Goodman, Lubrication by Lamellar Solids, *Proc. Roy. Soc., London*, Series A, Vol. 243, 1958, pp. 464-482.

6   E.R. Braithwaite and J. Hickman, Dry Film Lubrication of Metals, *Metal Industries*, Vol. 104, 1964, pp. 190-192.

7   H.E. Sliney, Dynamics of Solid Lubrication as Observed by Optical Microscopy, *ASLE Transactions*, Vol. 21, 1978, pp. 109-117.

8   Y. Tsuya, Lubrication with Molybdenum Disulfide Film Under Various Conditions, *ASLE Transactions*, Vol. 15, 1972, pp. 225-232.

9   A.I. Brudnyi and A.F. Karmadonov, Structure of Molybdenum Disulphide Lubricant Film, *Wear*, Vol. 33, 1975, pp. 243-249.

10   P.J. Bryant, P.L. Gutshall and L.H. Taylor, A Study of Mechanisms of Graphite Friction and Wear, *Wear*, Vol. 7, 1964, pp. 118-126.

11   M.N. Gardos, The Synergistic Effects of Graphite on the Friction and Wear of Molybdenum Disulphide Films in Air, *Tribology Transactions*, Vol. 31, 1988, pp. 214-227.

12   J.W. McCain, A Theory and Tester Measurement Correlation About Molybdenum Disulphide Dry Film Lubricant Wear, *Journal of Society of Aerospace, Material and Process Engineers*, Vol. 6, 1970, pp. 17-28.

13   J.K. Lancaster, Lubrication by Transferred Films of Solid Lubricants, *ASLE Transactions*, Vol. 8, 1965, pp. 146-155.

14   A.J. Haltner and C.S. Oliver, Frictional Properties of Some Solid Lubricant Films Under High Load, *Journal of Chemical Engineering Data*, Vol. 6, 1961, pp. 128-130.

15   W.J. Bartz and J. Xu, Wear Behaviour and Failure Mechanism of Bonded Solid Lubricants, *Lubrication Engineering*, Vol. 43, 1987, pp. 514-521.

16   G. Salomon, A.W.J. de Gee and J.H. Zaat, Mechano-Chemical Factors in Molybdenum Disulphide Film Lubrication, *Wear*, Vol. 7, 1964, pp. 87-101.

17   J.W. Midgley and D.G. Teer, An Investigation of the Mechanism of the Friction and Wear of Carbon, *Transactions ASME, Journal of Basic Engineering*, Vol. 85, 1963, pp. 488-494.

18   W.J. Bartz, Tribological Behaviour of Three Component Bonded Solid Lubricant Films, Proc. JSLE Int. Tribology Conference, 8-10 July, 1985, Tokyo, Japan, Elsevier, pp. 419-424.

19   M. Campbell and V. Hopkins, Development of Polyimide Bonded Solid Lubricants, *Lubrication Engineering*, Vol. 23, 1967, pp. 288-294.

20   R.D. Hubbel and B.D. McConnell, Wear Behaviour of Polybenzimidazole Bonded Solid-Film Lubricants, *Transactions ASME, Journal of Lubrication Technology,* Vol. 92, 1970, pp. 252-257.

21   P.W. Centers, The Role of Oxide and Sulfide Additions in Solid Lubricant Compacts, *Tribology Transactions,* Vol. 31, 1988, pp. 149-156.

22   G.W. Rowe, Some Observations on the Frictional Behaviour of Boron Nitride and Graphite, *Wear,* Vol. 3, 1960, pp. 274-285.

23   A.J. Haltner and C.S. Oliver, Effect of Water Vapour on the Friction of Molybdenum Disulphide, *Ind. Eng. Chem. Fundamentals,* Vol. 5, 1966, pp. 348-355.

24   J. Gansheimer, Neue Erkenntnisse uber die Wirkungsweise von Molybdansulfid als Schmierstoff, *Schmiertechnik,* Vol. 11, 1964, pp. 271-280.

25   H.H. Krause, S.L. Cosgrave and C.M. Allen, Phthalocyanines as High Temperature Lubricants, *Journal of Chemical Engineering Data,* Vol. 6, 1961, pp. 112-118.

26   G. Salomon, A. Begelinger and A.W.J. De Gee, Friction Properties of Phthalocyanine Pigments, *Wear,* Vol. 10, 1967, pp. 383-396.

27   P.A. Grattan and J.K. Lancaster, Abrasion by Lamellar Solid Lubricants, *Wear,* Vol. 10, 1967, pp. 453-468.

28   R.L. Fusaro and H.E Sliney, Graphite Fluoride, a New Solid Lubricant, *ASLE Transactions,* Vol. 13, 1970, pp. 56-65.

29   R.L. Fusaro, Graphite Fluoride Lubrication: the Effect of Fluorine Content, Atmosphere and Burnishing Technique, *ASLE Transactions,* Vol. 22, 1977, pp. 15-24.

30   P.M. Magie, A Review of the Properties and Potentials of the New Heavy Metal Derivative Solid Lubricants, *Lubrication Engineering,* Vol. 22, 1966, pp. 262-269.

31   W.E. Jamison and S.L. Cosgrove, Friction Characteristics of Transition-Metal Disulfides and Di-Selenides, *ASLE Transactions,* Vol. 14, 1971, pp. 62-72.

32   S. Hironaka, M. Wakihara, H. Einode, M. Taniguchi, T. Moriuchi and T. Hanzawa, Lubricity of Synthetic Niobium Sulfides with Layer Structure, Proc. JSLE International Tribology Conf., 8-10 July 1985, Tokyo, Japan, Elsevier, pp. 389-394.

33   F.J. Clauss, Solid Lubricants and Self-Lubricating Solids, Academic Press, New York, 1972.

34   A.J. Haltner, Sliding Behavior of Some Layer Lattice Compounds in Ultrahigh Vacuum, *ASLE Transactions,* Vol. 9, 1966, pp. 136-148.

35   H.E. Sliney, T.N. Strom and G.P. Allen, Fluoride Solid Lubricants for Extreme Temperatures and Corrosive Environments, *ASLE Transactions,* Vol. 8, 1965, pp. 307-322.

36   R.A. Burton and J.A. Russell, Forces and Deformations of Lead Films in Frictional Processes, *Lubrication Engineering,* Vol. 21, 1965, pp. 227-233.

37   Y. Tsuya and R. Takagi, Lubricating Properties of Lead Films on Copper, *Wear,* Vol. 7, 1964, pp. 131-143.

38   S. Miyake and S. Takahashi, Small-Angle Oscillatory Performance of Solid-Lubricant Film-Coated Ball Bearings for Vacuum Applications, *ASLE Transactions,* Vol. 30, 1987, pp. 248-253.

39   H. Kondo, K. Maeda and N. Tsushima, Performance of Bearings with Various Solid Lubricants in High Vacuum and High Speed Conditions, Proc. JSLE, International Tribology Conf., 8-10 July 1985, Tokyo, Japan, Elsevier, pp. 787-792.

40   H.E.Sliney, The Use of Silver in Self-Lubricating Coatings for Extreme Temperatures, *ASLE Transactions,* Vol. 29, 1986, pp. 370-376.

41   C. Della Corte and H.E. Sliney, Composition Optimization of Self-Lubricating Chromium-Carbide Based Composite Coatings for Use to 760°C, *ASLE Transactions,* Vol. 30, 1987, pp. 77-83.

42   M.B. Petersen, S.F. Murray and J.J. Florek, Consideration of Lubricants for Temperatures above 1000°F, *ASLE Transactions,* Vol. 2, 1960, pp. 225-234.

43   R.M. Gresham, Solid-Film Lubricants: Unique Products for Unique Lubrication, *Lubrication Engineering,* Vol. 44, 1988, pp. 143-145.

44   T. Spalvins, Friction Characteristics of Sputtered Solid Film Lubricants, NASA TM X-52819, 1970.

45   D.H. Buckley, Surface Effects in Adhesion, Friction, Wear and Lubrication, Elsevier, Amsterdam 1981.

46   N. Ohmae, T. Tsukizoe and T. Nakai, Ion-Plated Thin Films for Anti-Wear Applications, *Transactions ASME, Journal of Lubrication Technology,* 1978, Vol. 100, pp. 129-135.

47   P.J. Pacholke and K.M. Marshek, Improved Worm Gear Performance with Colloidal Molybdenum Disulfide Containing Lubricants, *Lubrication Engineering*, Vol. 43, 1987, pp. 623-628.

48   W.J. Bartz and J. Oppelt, Lubricating Effectiveness of Oil-Soluble Additives and Molybdenum Disulfide Dispersed in Mineral Oil, *Lubrication Engineering*, Vol. 36, 1980, pp. 579-585.

49   J. Gansheimer, Influence of Certain Vapours and Liquids on the Frictional Properties of Molybdenum Disulphide, *ASLE Transactions*, Vol. 10, 1967, pp. 390-399.

50   W.J. Bartz, Influence of Extreme-Pressure and Detergent-Dispersant Additives on the Lubricating Effectiveness of Molybdenum Disulphide, *Lubrication Engineering*, Vol. 33, 1977, pp. 207-215.

51   J.P. Giltrow, Abrasion by Impurities in MoS_2 , Part 2 - Practical Tests and the Development of a Quality Control, *Tribology International*, Vol. 7, 1974, pp. 161-168.

52   R.J. Rolek , C. Cusano and H.E. Sliney, The Influence of Temperature on the Lubricating Effectiveness of MoS_2 Dispersed in Mineral Oils, *ASLE Transactions*, Vol. 28, 1985, pp. 493-502.

53   J.K. Lancaster, Dry Bearings: a Survey of Materials and Factors Affecting Their Performance, *Tribology*, Vol. 6, 1973, pp. 219-251.

54   W. Liu, C. Huang, L. Gao, J. Wang and H. Dang, Study of the Friction and Wear Properties of MoS_2-Filled Nylon 6, *Wear*, Vol. 151, 1991, pp. 111-118.

55   K. Tanaka and S. Kawakami, Effect of Various Fillers on the Friction and Wear of Polytetrafluoroethylene-Based Composites, *Wear*, Vol. 79, 1982, pp. 221-234.

56   M.B. Low, The Effect of the Transfer Film on the Friction and Wear of Dry Bearing Materials for a Power Plant Application, *Wear*, Vol. 52, 1979, pp. 347-363.

57   J.P. Giltrow and J.K. Lancaster, Friction and Wear of Polymers Reinforced with Carbon Fibres, *Nature (London)*, Vol. 214, 1967, pp. 1106-1107.

58   R.L. Fusaro, Polyimides Formulated from a Partially Fluorinated Diamine for Aerospace Tribological Applications, NASA, TM-83339, 1983.

59   R.L. Fusaro and W.F. Hady, Tribological Properties of Graphite-Fiber-Reinforced Partially Fluorinated Polyimide Composites, *ASLE Transactions*, Vol. 29, 1986, pp. 214-222.

60   S. Asanabe, Applications of Ceramics for Tribological Components, *Tribology International*, Vol. 20, 1987, pp. 355-364.

61   E. Bergmann, J. Vogel and R. Brink, Criteria for the Choice of a PVD Treatment for the Solution of Wear Problems, Proc. Int. Tribology Conference, Melbourne, The Institution of Engineers, Australia, National Conference Publication No. 87/18, December, 1987, pp. 65-74.

62   M. Tomaru, Application of Surface Treatment on Rolling Bearings, *Transactions JSLE*, Vol. 31, 1986, pp. 593-594.

63   W.H. Roberts, Surface Engineering, Proc. Int. Tribology Conference, Melbourne, The Institution of Engineers, Australia, National Conference Publication No. 87/18, December, 1987, pp. 438-451.

64   B. Bhushan and B.K. Gupta, Handbook of Tribology, Materials, Coatings and Surface Treatments, McGraw-Hill, Inc., 1991.

65   R.F. Bunshah (editor), Deposition Technologies for Films and Coatings, Developments and Applications, Noyes Publications, 1982.

66   A.E. Berkowitz, W.G. Johnston, A. Mogro-Campero, J.L. Walter and H. Bakhru-Suny, Structure and Properties Changes During Ion Bombardment of Crystalline $Fe_{75}B_{25}$, Metastable Materials Formation by Ion Implantation, Proceedings of the Materials Research Society, Annual Meeting, November 1981, Boston, U.S.A., editors: S.T. Picraux and W.J. Choyke, Elsevier Science Publ. Co. New York, 1982, pp. 195-202.

67   M.M. Mayuram and R. Krishnamurthy, Tribological Characteristics of Sprayed Surfaces, Proc. Int. Tribology Conference, Melbourne, The Institution of Engineers, Australia, National Conference Publication No. 87/18, December, 1987, pp. 203-207.

68   S.T. Picraux and E.L. Pope, Tailored Surface Modification by Ion Implantation and Laser Treatment, *Science*, Vol. 224, 1984, pp. 615-622.

69   C.T. Walters, A.H. Clauer and B.P. Fairand, Pulsed Laser Surface Melting of Fe-Base Alloys, Proc. 2nd Int. Conf. on Rapid Solidification Processing: Principles and Technologies, 23-26 March 1980, editors: R. Mehrabian, B.H. Kear and M. Cohen, Reston, Virginia, U.S.A., Baton Rouge, Claitor's Publ. Division, L.A., 1980, pp. 241-245.

70   E.M. Jenkins and E.D. Doyle, Advances in Friction Deposition - Low Pressure Friction Surfacing, Proc. Int. Tribology Conference, Melbourne, The Institution of Engineers, Australia, National Conference Publication No. 87/18, December, 1987, pp. 87-94.

71   A. Douglas, E.D. Doyle and B.M. Jenkins, Surface Modification for Gear Wear, Proc. Int. Tribology Conference, Melbourne, The Institution of Engineers, Australia, National Conference Publication No. 87/18, December, pp. 52-58.

72   Y. Terauchi, H. Nadano, M. Kohno and Y. Nakamoto, Scoring Resistance of TiC and TiN-Coated Gears, *Tribology International*, Vol. 20, 1987, pp. 248-254.

73   S. Ramalingam and S. Kim, Tribological Characteristics of Arc Coated Hard Compound Films, Proc. Int. Tribology Conference, Melbourne, The Institution of Engineers, Australia, National Conference Publication No. 87/18, December, pp. 403-408.

74   Y.W. Lee, Adhesion and Corrosion Properties of Ion-Plated TiN, Proc. 6th Int. Conf. on Ion and Plasma Assisted Techniques, Brighton, U.K., C.E.P. Consultants, Edinburgh, 1987, pp. 249-251.

75   J.J. Oakes, A Comparative Evaluation of HfN, Al_2O_3, TiC and TiN Coatings on Cemented Carbide Tools, *Thin Solid Films*, Vol. 107, 1983, pp. 159-165.

76   R.G. Duckworth, Hot Zirconium Cathode Sputtered Layers for Useful Surface Modification, First International Conference on Surface Engineering, Brighton U.K., 25-28 June 1985, Paper 42, pp. 167-177, Publ. Welding Institute, Abingdon, U.K.

77   R.C. Bill, Fretting Wear and Fretting Fatigue, How Are They Related?, *Transactions ASME, Journal of Lubrication Technology*, Vol. 105, 1983, pp. 230-238.

78   G. Dearnaley, The Ion-implantation of Metals and Engineering Materials, *Transactions Institute of Metal Finishing*, Vol. 56, 1978, pp. 25-31.

79   C.J. McHargeu, Ion Implantation in Metals and Ceramics, *International Metals Reviews*, Vol. 31, 1986, pp. 49-76.

80   J.P. Hirvonen and J.W. Mayer, Fretting Wear of Nitrogen-Implanted AISI 304 Stainless Steel, *Materials Letters*, 1986, Vol. 4, pp. 404-408.

81   G. Dearnaley, Adhesive, Abrasive and Oxidative Wear in Ion-Implanted Metals, *Materials Science and Engineering*, Vol. 69, 1985, pp. 139-147.

82   M. Hirano and S. Miyake, Boron and Argon Ion Implantation Effect on the Tribological Characteristics of Stainless Steel, Proc. JSLE. Int. Tribology Conference, 8-10 July, 1985, Tokyo, Japan, Elsevier, pp. 245-250.

83   E. Lo Russo, P. Mazzoldi, I. Scotoni, C. Tosello and S. Tosto, Effect of Nitrogen-Ion Implantation on the Unlubricated Sliding Wear of Steel, *Applied Physics Letters*, Vol. 34, 1979, pp. 627-629.

84   Ming-Jen Hsu and P.A. Mollian, Cutting Tool Wear of Laser-Surface-Melted High Speed Steels, *Wear*, Vol. 127, 1988, pp. 253-268.

85   P.W. Leech, Comparison of the Sliding Wear Process of Various Cast Irons in the Laser-Surface-Melted and as-Cast Forms, *Wear*, Vol. 113, 1986, pp. 233-245.

86   W.J. Tomlinson, R.F. O'Connor and T.A. Spedding, Running-in Wear of a Grey Cast Iron and the Effect of Laser Transformation Hardening, *Tribology International*, Vol. 21, 1988, pp. 302-308.

87   A.W. Batchelor, G.W. Stachowiak, G.B. Stachowiak, P.W. Leech and O. Reinhold, Control of Fretting Friction and Wear of Roping Wire by Laser Surface Alloying and Physical Vapour Deposition Coatings, *Wear*, Vol. 152, 1992, pp. 127-150.

88   J.C. Angus, Diamond and Diamond-Like Films, *Thin Solid Films*, Vol. 216, 1992, pp. 126-133.

89   B.C. Oberlander and E. Lugscheider, Comparison of Properties of Coatings Produced by Laser Cladding and Conventional Methods, *Materials Science and Technology*, Vol. 8, 1992, pp. 657-665.

90   J.K. Lancaster, Transfer Lubrication for High Temperatures, A Review, *Trans. ASME. Journal of Tribology*, Vol. 107, 1985, pp. 437-443.

91   T. Kayaba, K. Kato and H. Ohsaki, The Lubricating Properties of Friction-Coating Films of Pb-Sn Alloys in High Vacuum, Proc. JSLE International Tribology Conference, Tokyo, 8-10, July 1985, Vol. 1, Japan Society of Lubrication Engineers, Tokyo, 1985, pp. 209-214.

92   A.W. Batchelor, N.L. Loh and M. Chandrasekaran, Lubrication of Stellite at Ambient and Elevated Temperatures by Transfer Films from a Graphite Slider, *Wear*, Vol. 198, 1996, pp. 208-215.

93   R. Ahmed and M. Hadfield, Rolling Contact Performance of Plasma Sprayed Coatings, *Wear*, Vol. 220, 1998, pp. 80-91.

94 H. Ichimura and A. Kawana, High Temperature Oxidation of Ion-Plated CRN Films, *Journal of Materials Research*, Vol. 9 (1), 1994, pp. 151-155.

95 T. Sato, T. Besshi, D. Sato and K. Inouchi, Evaluation of Wear and Tribological Properties of Coatings Rubbing Against Copper, *Wear*, Vol. 220, 1998, pp. 154-160.

96 C. Donnett, M. Belin, J.C. Auge, J.M. Martin, A. Grill and V. Patel, Tribochemistry of Diamond-Like Carbon Coatings in Various Environments, *Surface and Coatings Techn.*, Vol. 68/69, 1994, pp. 626-631.

97 K.-H. Habig, Fundamentals of Tribological Behaviour of Diamond, Diamond-Like Carbon and Cubic Boron Nitride Coatings, *Surface and Coatings Techn.*, Vol. 76/77, 1995, pp. 540-547.

98 E.I. Meletis, A. Erdemir and G.R. Fenske, Tribological Characteristics of DLC Films and Duplex Plasma Nitriding/DLC Coating Treatment, *Surface and Coatings Techn.*, Vol. 73, 1995, pp. 39-45.

99 K. Holmberg, J. Koskinen, H. Ronkainen, J. Vihersalo, J.-P. Hirvonen and J. Likonen, Tribological Characteristics of Hydrogenated and Hydrogen-Free Diamond-Like Carbon Coatings, *Diamond Films and Techn.*, Vol. 4, 1994, pp. 113-129.

100 M. Goto, F. Honda and M. Uemura, Extremely Low Coefficient of Friction of Diamond Sliding Against Ag Thin Films on Si (111) Surface under Ultra-High Vacuum Condition, *Wear*, Vol. 252, 2002, pp. 777-786.

101 K. Miyoshi and D.H. Buckley, Adhesion and Friction of Single-Crystal Diamond in Contact with Transition Metals, *Applied Surface Science*, Vol. 6, 1980, pp. 161-172.

102 M. Goto and F. Honda, Film-Thickness Effect of Ag Lubricant Layer in the Nano-Region, *Wear*, Vol. 256, 2004, pp. 1062-1071.

103 J. Lawrence and L. Li, The Wear Characteristics of High Power Diode Laser Generated Glaze on the Ordinary Portland Cement Surface of Concrete, *Wear*, Vol. 246, 2000, pp. 91-97.

104 H. Ronkainen, S. Varjus and K. Holmberg, Tribological Performance of Different DLC Coatings in Water-Lubricated Conditions, *Wear*, Vol. 249, 2001, pp. 267-271.

105 J. Andersson, R.A. Erck and A. Erdemir, Friction of Diamond-Like Carbon Films in Different Atmospheres, *Wear*, Vol. 254, 2003, pp. 1070-1075.

106 S. Stewart and R. Ahmed, Rolling Contact Fatigue of Surface Coatings – A Review, *Wear*, Vol. 253, 2002, pp. 1132-1144.

107 K. Kato, N. Umehara and K. Adachi, Friction, Wear and N_2-Lubrication of Carbon Nitride Coatings: A Review, *Wear*, Vol. 254, 2003, pp. 1062-1069.

108 D.F. Wang and K. Kato, In Situ Examination of Wear Particle Generation in Carbon Nitride Coatings by Repeated Sliding Contact Against a Spherical Diamond, *Wear*, Vol. 253, 2002, pp. 519-526.

109 S.W. Huang, M.W. Ng, M. Samandi and M. Brandt, Tribological Behaviour and Microstructure of $TiC_xN_{(1-X)}$ Coatings Deposited by Filtered Arc, *Wear*, Vol. 252, 2002, pp. 566-579.

110 E. Martinez, U. Wiklund, J. Esteve, F. Montala and L.L. Carreras, Tribological Performance of TiN Supported Molybdenum and Tantalum Carbide Coatings in Abrasion and Sliding Contact, *Wear*, Vol. 253, 2002, pp. 1182-1187.

111 J. Esteve, E. Martinez, A. Lousa, F. Montala, L.L. Carreras, Microtribological Characterization of Group V And VI Metal-Carbide Resistant Coatings Effective in the Metal Casting Industry, *Surface Coatings Technology*, Vol. 133-134, 2000, pp. 314-318.

112 C.W. Choo and Y.Z. Lee, Effects of Oxide Layer on the Friction Characteristics between TiN Coated Ball and Steel Disk in Dry Sliding, *Wear*, Vol. 254, 2003, pp. 383-390.

113 S. Ortmann, A. Savan, Y. Gerbig and H. Haefke, In-Process Structuring of CrN Coatings, and Its Influence on Friction in Dry and Lubricated Sliding, *Wear*, Vol. 254, 2003, pp. 1099-1105.

114 J.L. He, S. Miyake, Y. Setsuhara, I. Shimizu, M. Suzuki, K. Numata and H. Saito, Improved Anti-Wear Performance of Nanostructured Titanium Boron Nitride Coatings, *Wear*, Vol. 249, 2001, pp. 498-502.

115 J.H. Ouyang and S. Sasaki, Effects of Different Additives on Microstructure and High-Temperature Tribological Properties of Plasma-Sprayed Cr_2O_3 Ceramic Coatings, *Wear*, Vol. 249, 2001, pp. 56-67.

116 J.H. Ouyang, S. Sasaki and K. Umeda, Microstructure and Tribological Properties of Low-Pressure Plasma-Sprayed ZrO_2-CaF_2-Ag_2O Composite Coating at Elevated Temperature, *Wear*, Vol. 249, 2001, pp. 440-451.

117 P. La, Q. Xue and W. Liu, Effects of Boron Doping on Tribological Properties of Ni_3Al-Cr_7C_3 Coatings Under Dry sliding, *Wear*, Vol. 249, 2001, pp. 94-100.

118 A. Sidorenko, H.-S. Ahn, D.-I. Kim, H. Yang and V.V. Tsukruk, Wear Stability of Polymer Nanocomposite Coatings With Trilayer Architecture, *Wear*, Vol. 252, 2002, pp. 946-955.

119 X.C. Lu, B. Shi, L.K.Y. Li, J. Luo and J.I. Mou, Nanoindentation and Nanotribological Behavior of Fe-N/Ti-N Multilayers with Different Thickness of Fe-N Layers, *Wear*, Vol. 247, 2001, pp. 15-23.

120 P. Carlsson, U. Bexell and M. Olsson, Friction and Wear Mechanisms of Thin Organic Permanent Coatings Deposited on Hot-Dip Coated Steel, *Wear*, Vol. 247, 2001, pp. 88-99.

121 M.R. Bateni, F. Ashrafizadeh, J.A. Szpunar and R.A.L. Drew, Improving the Tribological Behavior of Copper through Novel Ti-Cu Inter-Metallic Coatings, *Wear*, Vol. 253, 2002, pp. 626-639.

10 FUNDAMENTALS OF CONTACT BETWEEN SOLIDS

10.1 INTRODUCTION

Surfaces of solids represent a very complex form of matter, far more complicated than a mere plane. There is a variety of defects and distortions present on any real surface. These surface features, ranging from bulk distortions of the surface to local microscopic irregularities, exert a strong influence on friction and wear. The imperfections and features of a real surface influence the chemical reactions which occur with contacting liquids or lubricants while the visible roughness of most surfaces controls the mechanics of contact between the solids and the resulting wear. The study of surfaces is relatively recent and the discoveries so far give rise to a wide range of questions for the technologist or tribologist, such as: what is the optimum surface? Is there a particular type of optimum surface for any specific application? Why are sliding surfaces so prone to thermal damage? How can wear particles be formed by plastic deformation when the operating loads between contacting surfaces are relatively very low? Although some of these questions can be answered with the current level of knowledge, the others remain as fundamental research topics. The characteristics of friction are also of profound importance to engineering practice. Seemingly mundane phenomena, such as the difference between static and kinetic friction, are still not properly understood and their control to prevent technical problems remains imperfect. The basic question: what is the mechanism of 'stick-slip'?, i.e., the vibration of sliding elements caused by a large difference between static and kinetic friction, has yet to be answered. In this chapter, the nature of solid surfaces, contact between solids and its effects on wear and friction are discussed.

10.2 SURFACES OF SOLIDS

At all scales of size, surfaces of solids contain characteristic features which influence friction, wear and lubrication in a manner independent of the underlying material. There are two fundamental types of features of special relevance to wear and friction:

·   atomic-scale defects in a nominally plain surface which provide a catalytic effect for lubricant reactions with the worn surface;

·   the surface roughness which confines contact between solids to a very small fraction of the nominally available contact area.

Surfaces at a Nano Scale

Any surface is composed of atoms arranged in some two-dimensional configuration. This configuration approximates to a plane in most cases but there are nearly always significant deviations from a true plane. The atoms of the solid body can be visualized as hard spheres packed together with no loose space. To form an exact plane or perfectly flat exterior surface, the indices of the crystal planes should be orientated to allow a layer of atoms to lie parallel to the surface. Since this is rarely the case the atom layers usually lie inclined to the surface. As a result a series of terraces is formed on the surface generating a quasi-planar surface [1]. The terraces between atom layers are also subjected to imperfections, i.e., the axis of the terrace may deviate from a straight path and some atoms might be missing from the edge of the terrace. Smaller features such as single atoms missing from the surface or an additional isolated atom present on the surface commonly occur. This model of the surface is known in the literature as the 'terrace ledge kink' (TLK) model. It has been suggested that close contact between surface atoms of opposing surfaces is hindered by this form of surface morphology. Consequently wear and friction are believed to be reduced in severity by the lack of interfacial atomic contact [2]. The TLK surface model and contact between two opposing real surfaces are shown schematically in Figure 10.1.

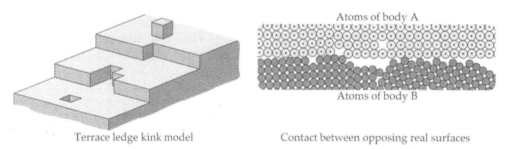

Terrace ledge kink model   Contact between opposing real surfaces

FIGURE 10.1 TLK surface model and contact between two opposing real surfaces (adapted from [2]).

TLK surface features such as terraces, ledges, kinks, missing atoms and 'ad-atoms' provide a large number of weakly bonded atoms. Atoms present on the surface have a lower bonding strength than interior atoms because they have a lower number of adjacent atoms. It has been observed that without all of these imperfections surfaces would probably be virtually inert to all chemical reactants [3]. These surface features facilitate chemical reactions between the surface and the lubricant. The reaction between lubricant and surface often produces a surface layer or 'film' which reduces friction and wear. Furthermore, the substrate material may be deformed plastically, which increases the number of dislocations reaching the surface. Dislocations form strong catalytic sites for chemical reactions and this effect is known as 'mechanical activation' [4]. An intense plastic deformation at a worn surface is quite common during wear and friction and the consequent mechanical activation can exert a strong influence on the formation of a lubricating film.

The composition of surface atoms may be quite different from the nominal or bulk composition since the alloying elements and impurities in a material tend to segregate at the surface. For example, carbon, sulphur and silicon tend to segregate in steel, while aluminium will segregate in copper [5]. Most materials, e.g., steel or copper, are not manufactured to a condition of thermodynamic or chemical equilibrium. Materials tend to be manufactured at a high temperature where impurities are relatively soluble, and then they are cooled rapidly to ambient temperature. Therefore most engineering materials contain a supersaturated solution of impurities which tend to be gradually released from the solvent material. Surface heating and chemical attack by lubricants during sliding contact also contribute to accentuation of surface segregation of contaminants and secondary moieties [6]. Another factor which influences surface segregation of impurities is plastic deformation below the annealing temperature of the worn metal [7]. Intense deformation of the material below the worn surface takes place in unlubricated sliding contacts and the resulting increased dislocation density is believed to provide a dense network of crystal lattice defects which facilitate the diffusion of impurity atoms. Quite large effects on friction and wear with relatively small alloying additions to pure metals have been observed and surface analysis revealed that significant changes in the friction and wear coefficients are usually accompanied by surface segregation [5].

In microscopic, clean contacts where direct contact between crystalline lattices is possible, crystalline anisotropy becomes significant. For crystalline silicon in dry sliding, the friction coefficient was observed to vary between 0.2 and 0.4 as the lattice orientation to the contact changed with rotation [108].

It should be mentioned that apart from, for example, the prescient work of Landheer et al. [2], contact between surfaces is generally modelled in terms of continuum mechanics. The first specific mention of the atomic nature of contact can be found in the work by Buckley [5] and Ferrante and Smith [102] where electron transfer and the Jellium model of adhesive contact is invoked. New models are developed involving 'atomistic' solid contacts, i.e., involving interactions between individual atoms on both contacting solids. More information on 'atomistic' solid contacts can be found in Chapter 17.

Surface Topography

Surface imperfections at an atomic level are matched by macroscopic deviations from flatness. Almost every known surfaces, apart from the cleaved faces of mica [8], are rough. Roughness means that most parts of a surface are not flat but form either a peak or a valley. The typical amplitude between peaks and valleys for engineering surfaces is about 1 [μm]. The profile of a rough surface is almost always random unless some regular features have been deliberately introduced. The random components of the surface profiles look very much the same whatever their source, irrespectively of the absolute scale of size involved [9]. This is illustrated in Figure 10.2 where a series of surface roughness profiles extracted from machined surfaces and from the surface of the earth and the moon (on a large scale) are shown.

Another unique property of surface roughness is that, if repeatedly magnified, increasing details of surface features are observed down to the nanoscales. Also the appearance of the surface profiles is the same regardless of the magnification [9,10]. This self-similarity of surface profiles is illustrated in Figure 10.3.

It has been observed that surface roughness profiles resemble electrical recordings of white noise and therefore similar statistical methods have been employed in their analysis. The introduction of statistical methods to the analysis of surface topography was probably due to Abbott and Firestone in 1933 [11] when they proposed a bearing area curve as a means of profile representation [9]. This curve representing the real contact area, also known as the Abbott curve, is obtained from the surface profile. It is compiled by considering the fraction of surface profile intersected by an infinitesimally thin plane positioned above a datum plane. The intersect length with material along the plane is measured, summed together and plotted as a proportion of the total length. The procedure is repeated through a number of slices. The proportion of this sum to the total length of bearing line is considered to represent the proportion of the true area to the nominal area [9]. Although it can be disputed that this procedure gives the bearing length along a profile, it has been shown that for a random surface the bearing length and bearing area fractions are identical [12,9]. The obtained curve is in fact an integral of the height probability density function '$p(z)$' and if the height distribution is Gaussian, then this curve is nothing else than the cumulative probability function '$P(z)$' of classical statistics. The height distribution is constructed by plotting the number or proportion of surface heights lying between two specific heights as a function of the height [9]. It is a means of representing all surface heights. The method of obtaining the bearing area curve is illustrated schematically in Figure 10.4. It can be seen from Figure 10.4 that the percentage of bearing area lying above a certain height can easily be assessed.

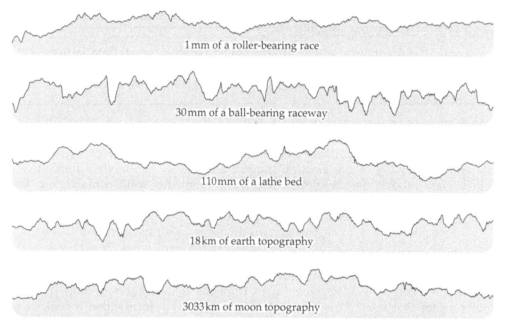

FIGURE 10.2 Similarities between random profiles of rough surfaces whether natural or artificial (adapted from [9]).

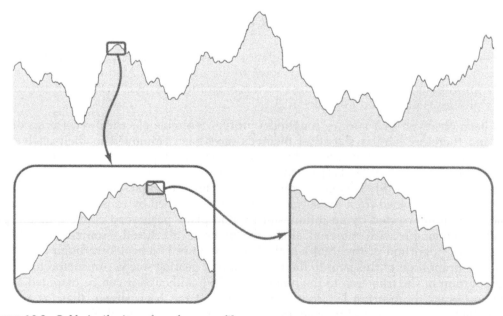

FIGURE 10.3 Self-similarity of surface profiles.

Although in general it is assumed that most surfaces exhibit Gaussian height distributions, this is not always true. For example, it has been shown that machining processes such as grinding, honing and lapping produce negatively skewed height distributions [13] while some milling and turning operations can produce positively skewed height distributions [9]. In practice, however, many surfaces exhibit symmetrical Gaussian height distributions.

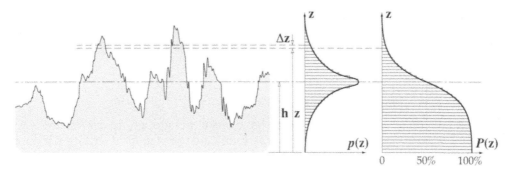

FIGURE 10.4   Determination of a bearing area curve of a rough surface; **z** is the distance perpendicular to the plane of the surface, **Δz** is the interval between two heights, **h** is the mean plane separation, **p(z)** is the height probability density function and **P(z)** is the cumulative probability function [9].

Plotting the deviation of surface height from a mean datum on a Gaussian cumulative distribution diagram usually gives a linear relationship [14,15]. A classic example of Gaussian profile distribution observed on a bead-blasted surface is shown in Figure 10.5. The scale of the diagram is arranged to give a straight line if a Gaussian distribution is present.

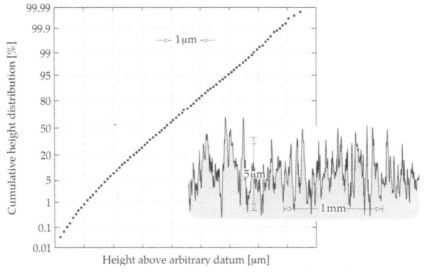

FIGURE 10.5   Experimental example of a Gaussian surface profile on a rough surface (adapted from [14]).

During mild wear the peaks of the surface asperities are truncated, resulting in a surface profile consisting of plateaux and sharp grooves. In such profiles the asperity heights are distributed according to not one but two Gaussian constants, i.e., Gaussian surface profile exhibits bi-modal behaviour [40]. Truncation of surface asperities is found to be closely related to a 'running-in' process where a freshly machined surface is worn at light loads in order to be able to carry a high load during service.

It should also be realized that most real engineering surfaces consist of a blend of random and non-random features. The series of grooves formed by a shaper on a metal surface are a prime example of non-random topographical characteristics. On the other hand, bead-blasted surfaces consist almost entirely of random features because of the random nature of this process. The shaped surface also contains a high degree of random surface features which gives its rough texture. In general, non-random features do not significantly affect the contact area and contact stress provided that random roughness is superimposed on the non-random features.

Characterization of Surface Topography

A number of techniques and parameters have been developed to characterize surface topography. The most widely used surface descriptors are the statistical surface parameters. A new development in this area involves surface characterization by fractals.

· *Characterization of Surface Topography by Statistical Parameters*

Real surfaces are difficult to define. In order to describe the surface at least two parameters are needed, one describing the variation in height (i.e., height parameter) and the other describing how height varies in the plane of the surface (i.e., spatial parameter) [9]. The deviation of a surface from its mean plane is assumed to be a random process which can be described using a number of statistical parameters.

<u>Height characteristics</u> are commonly described by parameters such as the centre-line-average or roughness average (**CLA** or 'R_a'), root mean square roughness (**RMS** or 'R_q'), mean value of the maximum peak-to-valley height ('R_{tm}'), ten-point height ('R_z') and many others. In engineering practice, however, the most commonly used parameter is the roughness average. Some of the height parameters are defined in Table 10.1.

The 'R_a' represents the average roughness over the sampling length. The effect of a single spurious, non-typical peak or valley (e.g., a scratch) is averaged out and has only a small effect on the final value. Therefore, because of the averaging employed, one of the main disadvantages of this parameter is that it can give identical values for surfaces with totally different characteristics. Since the 'R_a' value is directly related to the area enclosed by the surface profile about the mean line, any redistribution of material has no effect on its value. The problem is illustrated in Figure 10.6 where the material from the peaks of a 'bad' bearing surface is redistributed to form a 'good' bearing surface without any change in the 'R_a' value [9].

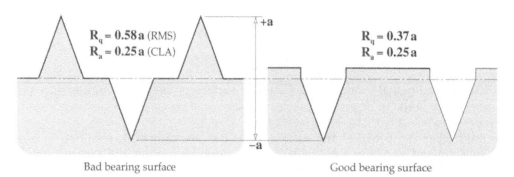

FIGURE 10.6   Effect of averaging on 'R_a' value [9].

TABLE 10.1   Commonly used height parameters.

Roughness average (CLA or $\mathbf{R_a}$)	$$R_a = \frac{1}{L}\int_0^L |z|\, dx$$	
Root mean square roughness (RMS or $\mathbf{R_q}$)	$$R_q = \sqrt{\frac{1}{L}\int_0^L z^2\, dx}$$	
Maximum peak-to-valley height ($\mathbf{R_t}$)	Largest single peak-to-valley height in five adjoining sample lengths $$R_t = \frac{1}{5}\sum_{i=1}^{5} R_{max_i}$$	
Ten-point height ($\mathbf{R_z}$)	Average separation of the five highest peaks and the five lowest valleys within the sampling length $$R_z = \frac{p_1 + ... + p_5 + v_1 + ... + v_5}{5}$$	

where:

   L   is the sampling length [m];

   z   is the height of the profile along '**x**' [m].

The 'good' bearing surface illustrated schematically in Figure 10.6 in fact approximates to most worn surfaces where lubrication is effective. Such surfaces tend to exhibit the favourable surface profile, i.e., quasi-planar plateaux separated by randomly spaced narrow grooves.

The problem associated with the averaging effect can be rectified by the application of the RMS parameter since, because it is weighted by the square of the heights, it is more sensitive than '$\mathbf{R_a}$' to deviations from the mean line.

Spatial characteristics of real surfaces can be described by a number of statistical functions. Some of the commonly used functions are shown in Table 10.2. Although two surfaces can have the same height parameters their spatial arrangement and hence their wear and frictional behaviour can be very different. To describe spatial arrangement of a surface the autocovariance function (**ACVF**) or its normalized form the autocorrelation function (**ACF**), the structure function (**SF**) or the power spectral density function (**PSDF**) are commonly used

[9]. The autocovariance function or the autocorrelation function are most popular in representing spatial variation. These functions are used to discriminate between the differing spatial surface characteristics by examining their decaying properties. Their limitation, however, is that they are not sensitive enough to be used to study changes in surface topography during wear. Wear usually occurs over almost all wavelengths and therefore changes in the surface topography are hidden by ensemble averaging and the autocorrelation functions for worn and unworn surfaces can look very similar as shown in Table 10.2 [9].

This problem can be avoided by the application of a structure function [9,16]. Although this function contains the same amount of information as the autocorrelation function, it allows a much more accurate description of surface characteristics. The power spectral density function as a spatial representation of surface characteristics seems to be of little value. Although Fourier surface representation is mathematically valid the very complex nature of the surfaces means that even a very simple structure needs a very broad spectrum to be well represented [9].

A detailed description of the height and spatial surface parameters can be found elsewhere [e.g., 9,41].

Multi-Scale Characterization of Surface Topography

A characteristic feature of the engineering surfaces is that they exhibit topographical details over a wide range of scales; from nano- to micro-scales. It has been shown that surface topography is a nonstationary random process for which the variance of height distribution (RMS^2) depends on the sampling length (i.e., the length over which the measurement is taken) [17]. Therefore the same surface can exhibit different values of the statistical parameters when a different sampling length or an instrument with a different resolution is used. This leads to certain inconsistencies in surface characterization [18]. The main problem is associated with the discrepancy between the large number of length scales that a rough surface contains and the small number of particular length scales, i.e., sampling length and instrument resolution, that are used to define the surface parameters. Therefore traditional methods used in 3-D surface topography characterization provide functions or parameters that strongly depend on the scale at which they are calculated. This means that these parameters are not unique for a particular surface [e.g., 42-44]. Since this 'one-scale' characterization provided by statistical functions and parameters is in conflict with the multi-scale nature of tribological surfaces, new 'multi-scale' characterization methods still need to be developed. Recent developments in this area have been concentrated on four different approaches:

- Fourier transform methods;
- wavelet transformation methods;
- fractal methods and
- hybrid fractal-wavelet method.

For the characterization of surfaces by wavelet and fractal methods the 3-D surface topography data is presented in the form of range images [45,46]. In these images the surface elevation data is encoded into a pixel brightness value, i.e., the brightest pixel, depicted by the grey level of '255', represents the highest elevation point on the surface, while the darkest pixel, depicted by the grey level of '0', represents the lowest elevation point on the surface [46].

TABLE 10.2   Statistical functions used to describe spatial characteristics of the real surfaces (adapted from [9]).

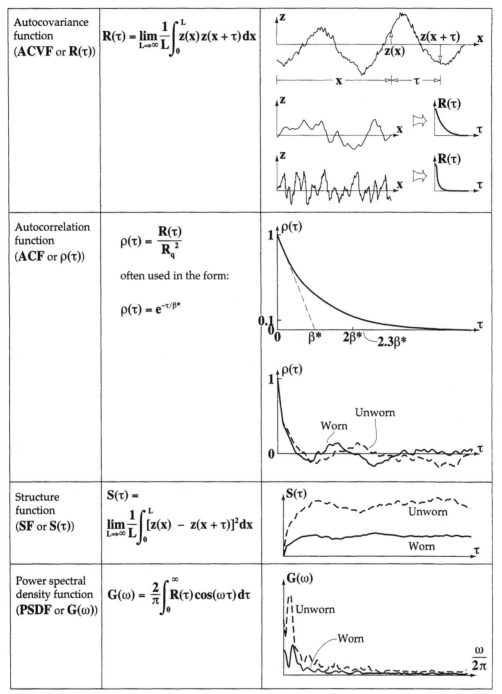

Autocovariance function (**ACVF** or **R(τ)**)	$R(\tau) = \lim_{L \Rightarrow \infty} \frac{1}{L} \int_0^L z(x)\,z(x+\tau)\,dx$	
Autocorrelation function (**ACF** or ρ(τ))	$\rho(\tau) = \dfrac{R(\tau)}{R_q^{\,2}}$ often used in the form: $\rho(\tau) = e^{-\tau/\beta^*}$	
Structure function (**SF** or **S(τ)**)	$S(\tau) =$ $\lim_{L \Rightarrow \infty} \frac{1}{L} \int_0^L [z(x) - z(x+\tau)]^2 dx$	
Power spectral density function (**PSDF** or **G(ω)**)	$G(\omega) = \dfrac{2}{\pi} \int_0^\infty R(\tau)\cos(\omega\tau)\,d\tau$	

where:

τ   is the spatial distance [m];

β*   is the decay constant of the exponential autocorrelation function [m];

ω   is the radial frequency [m⁻¹], i.e., $\omega = 2\pi/\lambda$, where 'λ' is the wavelength [m].

· *Characterization of Surface Topography by Fourier Transform*

Fourier transform methods allow to decompose the surface data into complex exponential functions of different frequencies. The Fourier methods were used to calculate the power spectrum and the autocorrelation function in order to obtain the surface topography parameters [e.g., 41,47-49]. However, the problem with the application of these methods to surfaces is that they provide results which strongly depend on the scale at which they are calculated, and hence they are not unique for a particular surface. This is because the Fourier transformation provides only the information whether a certain frequency component exists or not. As the result, the surface parameters calculated do not provide information about the scale at which the particular frequency component appears.

· *Characterization of Surface Topography by Wavelets*

Wavelet methods allow to decompose the surface data into different frequency components and then to characterize it at each individual scale. The wavelet methods were used to decompose the topography of a grinding wheel surface into long and small wavelengths [50], to analyse 3-D surface topography of orthopaedic joint prostheses [51] and others. When applying wavelets the surfaces are usually first decomposed into roughness, waviness and form, and then the changes in surface peaks, pits and scratches together with their locations are obtained at different scales. However, there are still major difficulties in extracting the appropriate surface texture parameters from wavelets [44].

An example of the application of a wavelet transform to decompose a titanium alloy surface image at two different levels is shown in Figure 10.7. Each decomposition level contains a low resolution image and three images containing the vertical, horizontal and diagonal details of the original image at a particular scale. The low resolution images and the detail images were obtained by applying a combination of low-pass and/or high-pass filters along the rows and columns of the original image and a downsampling operator. The original image can be reconstructed back from these images obtained by using mirror filters and an upsampling operator [52].

· *Characterization of Surface Topography by Fractals*

Fractal methods allow to characterize surface data in a scale-invariant manner. Usually fractal dimensions, since they are both 'scale-invariant' and closely related to self-similarity, are employed to characterize rough surfaces [19]. The basic difference between the characterization of real surfaces by statistical methods and fractals is that the statistical methods are used to characterize the disorder of the surface roughness while the fractals are used to characterize the order behind this apparent disorder [10].

The variation in height 'z' above a mean position with respect to the distance along the axis 'x' of the surface profile obtained by stylus or optical measurements, Figure 10.8, can be characterized by the Weierstrass-Mandelbrot function which has the fractal dimension 'D' and is given in the following form [19]:

$$z(x) = G^{(D-1)} \sum_{n=n_1}^{\infty} \frac{\cos 2\pi \gamma^n x}{\gamma^{(2-D)n}} \qquad \text{for } 1 < D < 2 \text{ and } \gamma > 1 \qquad (10.1)$$

where:

$z(x)$ is the function describing the variation of surface heights along 'x';

G is the characteristic length scale of a surface [m]. It depends on the degree of surface finish. For example, 'G' for lapped surfaces was found to be in the range of 1×10^{-9} to about 12.5×10^{-9} [m], for ground surfaces about 0.1×10^{-9} - 10×10^{-9} [m] while for shape turned surfaces '**G**' is about 7.6×10^{-9} [m] [20];

$\mathbf{n_1}$ is the lowest frequency of the profile, i.e., the cut-off frequency, which depends on the sampling length '**L**', i.e., $\gamma^{n_1} = 1/L$ [m^{-1}] [17,19];

γ is the parameter which determines the density of the spectrum and the relative phase difference between the spectral modes. Usually $\gamma = 1.5$ [20];

γ^n are the frequency modes corresponding to the reciprocal of roughness wavelength, i.e., $\gamma^n = 1/\lambda_n$ [m^{-1}] [19];

$\mathbf{D}$ is the fractal dimension which is between **1** and **2**. It depends on the degree of surface finish. For example, '**D**' for lapped surfaces was found to be in the range of **1.7** to about **1.9**, for ground surfaces about **1.6** while for shape turned surface '**D**' is about **1.8** [20].

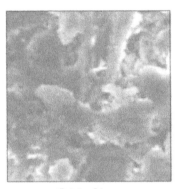

Original image

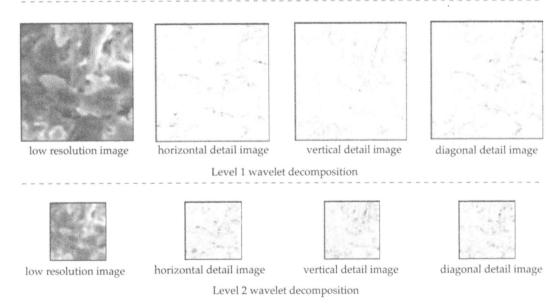

low resolution image horizontal detail image vertical detail image diagonal detail image

Level 1 wavelet decomposition

low resolution image horizontal detail image vertical detail image diagonal detail image

Level 2 wavelet decomposition

Figure 10.7 Example of application of wavelet transform to titanium alloy surface image.

The Weierstrass-Mandelbrot function has the properties of generating a profile that does not appear to change regardless of the magnification at which it is viewed. As the magnification is increased, more fine details become visible and so the profile generated by this function closely resembles the real surfaces. In analytical terms, the Weierstrass-Mandelbrot function is non-differentiable because it is impossible to obtain a true tangent to any value of the function. The fractal dimension and other parameters included in the Weierstrass-Mandelbrot function provide more consistent indicators of surface roughness than conventional parameters such as the standard deviation about a mean plane. This is because the fractal dimension is independent of the sampling length and the resolution of the instrument which otherwise directly affect the measured roughness [17].

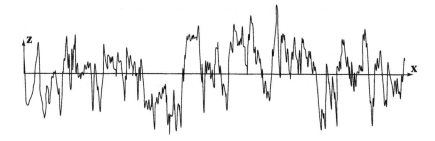

FIGURE 10.8 Example of a surface profile obtained by stylus or optical measurements.

Although the Weierstrass-Mandelbrot function appears to be very similar to a Fourier series, there is a basic difference. The frequencies in a Fourier series increase in an arithmetic progression as multiples of a basic frequency, while in a Weierstrass-Mandelbrot function they increase in a geometric progression [19]. In Fourier series the phases of some frequencies coincide at certain nodes which make the function appear non-random. With the application of a Weierstrass-Mandelbrot function this problem is avoided by choosing a non-integer 'γ', and taking its powers to form a geometric series. It was found that $\gamma = 1.5$ provides both phase randomization and high spectral density [20].

The parameters '**G**' and '**D**' can be found from the power spectrum of the Weierstrass-Mandelbrot function (10.1) which is in the form [19,21]:

$$S(\omega) = \frac{G^{2(D-1)}}{2\ln\gamma} \frac{1}{\omega^{(5-2D)}}$$

(10.2)

where:

$S(\omega)$ is the power spectrum [m^3];

ω is the frequency, i.e., the reciprocal of the wavelength of roughness, [m^{-1}], i.e., the low frequency limit corresponds to the sampling length while the high frequency limit corresponds to the Nyquist frequency which is related to the resolution of the instrument [19].

The fractal dimension '**D**' is obtained from the slope '**m**' of the log-log plot of '$S(\omega)$' versus 'ω', i.e.:

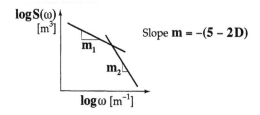

The parameter 'G', which determines the location of the spectrum along the power axis and is a characteristic length scale of a surface, is obtained by equating the experimental variance of the profile to that of the Weierstrass-Mandelbrot function [19,20].

The constants 'D', 'G' and 'n_1' of the Weierstrass-Mandelbrot function form a complete set of scale independent parameters which characterize an isotropic rough surface [20]. When they are known then the surface roughness at any length scale can be determined from the Weierstrass-Mandelbrot function [20].

It should also be mentioned that in the fractal model of roughness, as developed so far, the scale of roughness is imagined to be unlimited. For example, if a sufficient sampling distance is selected, then macroscopic surface features, i.e., ridges and craters would be observed. In practice, engineering surfaces contain a limit to roughness, i.e., the surfaces are machined 'smooth' and in this respect, the fractal model diverges from reality.

There have been various techniques developed to evaluate the fractal dimension from a profile, e.g., horizontal structuring element method (HSEM) [53], correlation integral [54,55] and fast Fourier transform (FFT) [e.g., 56], modified 1-D Richardson method [57] and others. However, it was found that fractal dimensions calculated from surface profiles exhibit some fundamental limitations, especially when applied to the characterization of worn surfaces [56,58,59]. For example, it was shown that the fractal dimensions calculated fail to distinguish between the two worn surfaces [60]. Also tests conducted on artificially generated profiles demonstrated that the problem of choosing any particular algorithm for the calculation of fractal dimension from a profile is not a simple one since there is no way of knowing the 'true' or even 'nominal' fractal dimension of the surface profile under consideration [59].

Attempts have also been made to apply fractal methods to characterization of 3-D surface topographies. For example, it has been shown that surface fractal dimensions could be used to characterize surfaces exhibiting fractal nature [56], surface profiles produced by turning, electrical discharge and grinding [61], isotropic sandblasted surfaces and anisotropic ground surfaces [62], engineering surfaces measured with different resolutions [63], etc. The most popular methods used to calculate surface fractal dimension are the ε-blanket [64], box-counting [65], two-dimensional Hurst analysis [56], triangular prism area surface [66] and variation method [67], generalized fractal analysis based on a Ganti-Bhushan model [63] and the patchwork method [68]. The basic limitation of these methods is that they work well only with isotropic surfaces, i.e., with surfaces which exhibit the same statistical characteristics in all directions [45]. Majority of surfaces, however, are anisotropic, i.e., they exhibit different surface patterns along different directions.

In order to overcome this limitation and characterize the surface in all directions, a modified Hurst Orientation Transform (HOT) method was developed [69]. The HOT method allows calculation of Hurst coefficients (H), which are directly related to surface fractal dimensions, i.e., D = 3-H, in all possible directions. These coefficients, when plotted as a function of orientation, reveal surface anisotropy [45,69].

The problem is that none of the methods mentioned provide a full description of surface topography since they were designed to characterize only particular morphological surface features such as surface roughness and surface directionality. Even though a modified HOT method allows for a characterization of surface anisotropy, it still does not provide a full description of the surface topography. It seems that fractal methods currently used only work well with surfaces that conform to a fractional Brownian motion (FBM) model and are self-similar with uniform scaling.

Recently a new approach, called a Partitioned Iterated Function System (PIFS), has been tried. This approach is based on the idea that since most of the complex structures observed in nature can be described and modelled by a combination of simple mathematical rules [e.g.,

70,71], it is reasonable to assume that, in principle, it should be possible to describe a surface by a set of such rules.

It can be observed that any surface image, containing 3-D surface topography data, exhibits a certain degree of 'self-transformability', i.e., one part of the image can be transformed into another part of the image reproducing itself almost exactly [72]. In other words, a surface image is composed of image parts which can be converted to fit approximately other parts located elsewhere in the image [45]. This is illustrated in Figure 10.9 which shows a mild steel surface with the 'self-transformable' parts marked by the squares.

The PIFS method is based on these affine transformations and allows to encapsulate the whole information about the surface in a set of mathematical formulae [44,45]. These formulae when iteratively applied into any initial image result in a sequence of images which converge to the original surface image. This is illustrated in Figure 10.10 where the sets of rules found for a surface image shown in Figure 10.9 were applied to some starting image, i.e., black square.

Figure 10.9   Range image of a mild steel surface with marked 'self-transformable' parts.

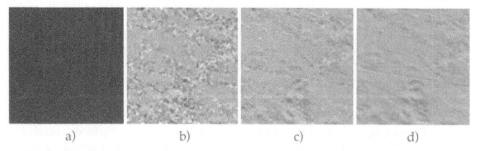

a)   b)   c)   d)

Figure 10.10 The application of the PIFS data obtained for the mild steel surface image shown in Figure 10.9: a) initial image, b) 1 iteration, c) 4 iterations and d) 12 iterations.

It can be seen from Figure 10.10 that an almost exact replica of the original image has been obtained from the 'black' image only after 12 iterations of PIFS data. Since a relatively accurate description of the whole surface is obtained the PISF method may also be used to classify the surfaces into specific groups.

· *Characterization of Surface Topography by Combination of Wavelets and Fractals*

As mentioned already most of the tribological surfaces exhibit multiscale and nonstationary characteristics, i.e., they exhibit topographical features over a wide range of scales from nano-

(at atomic or molecular level) to micro-scales (up to hundreds of micrometers) and these features are superimposed on each other and located at different positions on the surface. Parameters or functions currently used in surface characterization strongly depend on the scale at which they are calculated, i.e., they are not unique for a particular surface and hence they do not provide sufficient information about these objects. The drawback of fractals is that they characterize surfaces at all scales while the wavelets provide a description at any particular scale. A possible solution to these problems is to use a combination of fractals and wavelets. Recently, a hybrid fractal-wavelet technique, based on the combination of fractal and wavelet methods, has been developed allowing for the 3-D characterization of often complex tribological surfaces with a unique precision and accuracy, without the need for any parameters. First, the surface topography features are broken down into individual scale components by wavelets and then fractals are applied to provide a surface topography description over the finest achievable range of scales [109-111].

Optimum Surface Roughness

In practical engineering applications the surface roughness of components is critical as it determines the ability of surfaces to support load [22]. It has been found that at high or very low values of 'R_q' only light loads can be supported while the intermediate 'R_q' values allow for much higher loads. This is illustrated schematically in Figure 10.11 where the optimum operating region under conditions of boundary lubrication is determined in terms of the height and spatial surface characteristics. If surfaces are too rough then excessive wear and eventual seizure might occur. On the other hand, if surfaces are too smooth, i.e., when $\beta^* < 2$ [μm], then immediate surface failure occurs even at very light loads [22].

It is found that most worn surfaces, where lubrication is effective, tend to exhibit a favourable surface profile, i.e., quasi-planar plateaux separated by randomly spaced narrow grooves. In this case, the profile is still random which allows the same analysis of contact between rough surfaces as described below, but a skewed Gaussian profile results.

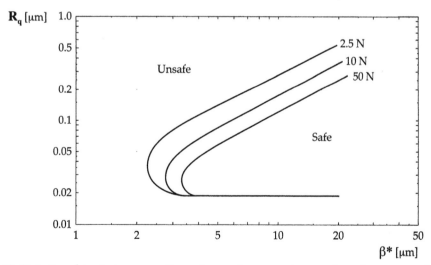

FIGURE 10.11 Relationship between safe and unsafe operating regions in terms of the height and spatial surface characteristics [22].

10.3   CONTACT BETWEEN SOLIDS

Surface roughness limits the contact between solid bodies to a very small portion of the apparent contact area. The true contact over most of the apparent contact area is only found at extremely high contact stresses which occur between rocks at considerable depths below the surface of the earth and between a metal-forming tool and its workpiece. Contact between solid bodies at normal operating loads is limited to small areas of true contact between the high spots of either surface. In brake blocks momentary points of true contact are confined to contact plateaux that form a fraction of the total contact area [103]. The random nature of roughness prevents any interlocking or meshing of surfaces. True contact area is therefore distributed between a number of micro-contact areas. If the load is raised, the number of contact areas rather than the 'average' individual size of contact area is increased, i.e., an increase in load is balanced by newly formed small contact areas. A representation of contact between solids is shown schematically in Figure 10.12.

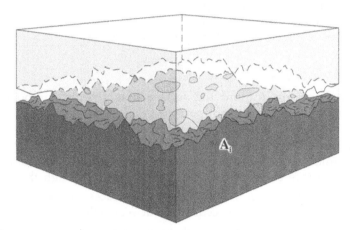

Figure 10.12   Real contact area of rough surfaces in contact; A_r is the true area of contact, i.e.,

$$A_r = \sum_{i=1}^{n} A_i,\ n \text{ is the number of asperities [23].}$$

The real contact area is a result of deformation of the high points of the contacting surfaces which are generally referred to as asperities. Contact stresses between asperities are large, as shown in Figure 10.13, and in some cases localized plastic deformation may result.

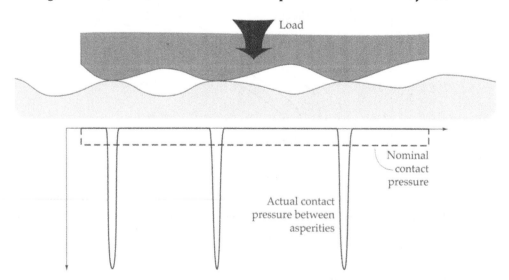

Figure 10.13 Contact stresses between the asperities.

Although in the early theories of surface contact it was assumed that the true contact area arose from the plastic deformation of asperities [24], it was found later that a large proportion of the contact between the asperities is entirely elastic [25,26]. The relationship between the true area of contact and the load is critically important since it affects the law of friction and wear.

Model of Contact between Solids Based on Statistical Parameters of Rough Surfaces

Contact between an idealized rough surface and a perfectly smooth surface was first analyzed by assuming that a rough surface is approximated by a series of hierarchically superimposed spherical asperities as shown in Figure 10.14 [27].

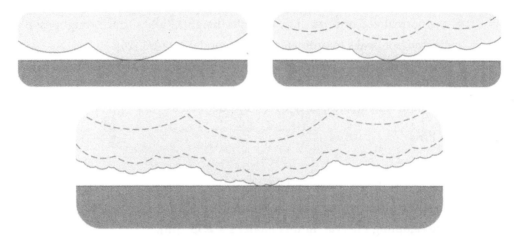

Figure 10.14 Contact between idealized rough surfaces of varying levels of detail and a smooth plane (adapted from [27]).

As can be seen from Figure 10.14 the surface is modelled by spherical asperities of differing scales of size. It was found that as the complexity of the model is increased by superimposing spherical asperities of a new order of magnitude on existing ones the true area of contact is proportional to load at a power close to unity. The relationships between true contact area 'A_r' and load 'W' for the three geometries shown in Figure 10.14 were found to be the following: 1st order $A_r \, \alpha \, W^{4/5}$; 2nd order $A_r \, \alpha \, W^{14/15}$; 3rd order $A_r \, \alpha \, W^{44/45}$. Therefore, it has been deduced that since real surfaces are even more complex than these idealized surfaces, then the true area of the multiple elastic contact between the asperities should be directly proportional to load.

Although the model of a surface composed of a series of hemispheres is highly idealized, more sophisticated analyses have shown that a random surface profile also contains a wide spectrum of asperity curvature to give the same relationship between contact area and load [14,25]. The range of curvatures found on a real, complex surface ensures a near linear dependence between true contact area and load. Therefore, in more exact analysis, statistical functions which describe the random nature of rough surfaces have been employed.

One of the first models of contact between two real surfaces incorporating their random statistical nature was introduced by Greenwood and Williamson [14]. This was followed by the work of Whitehouse and Archard [25], Onions and Archard [26], Pullen and Williamson [28], Nayak [29] and others [e.g., 30-32]. In these models statistical methods are applied to describe the complex nature of the contact between two rough surfaces. For example, in the Onions and Archard model [26], which is based on the Whitehouse and Archard statistical model [25], the true contact area is given by the following expression:

$$A_r = n\pi A (2.3\beta^*)^2 \int_d^\infty (z^* - d) \int_0^\infty \frac{f^*(z^*,C)}{NC} dC\,dz^*$$ (10.3)

where:

A_r	is the true area of contact [m²];
n	is the number of asperities per unit area of apparent contact;
A	is the apparent contact area [m²];
β^*	is the correlation distance obtained from the exponential autocorrelation function of a surface profile [m];
z^*	is the normalized ordinate, i.e., $z^* = z/\sigma$ (height/RMS surface roughness);
N	is the ratio of peaks to ordinates. In this model $N = 1/3$ [26];
d	is the normalized separation between the datum planes of either surface, i.e., $d = h/\sigma$;
C	is the dimensionless asperity curvature defined as:

$$C = l^2/r\sigma$$

where:

h	is the mean plane separation [m];
σ	is the RMS surface roughness [m];
l	is the sampling interval. In this model $l = 2.3\beta^*$ [m];
f^*	is the probability density function of peak heights and curvatures;
r	is the mean asperity radius [m] defined as:

$$r = \frac{2\pi^{0.5}(2.3\beta^*)^2}{9\sigma}$$

The expression for a total load is given in a form [26]:

$$W = \frac{4}{3} A\sigma E' (2.3\beta^*) \int_d^\infty (z^* - d)^{1.5} \int_0^\infty \frac{f^*(z^*,C)}{N\sqrt{C}} dC\,dz^*$$ (10.4)

where:

W	is the total load [N];
E'	is the composite Young's modulus (eq. 7.35) [Pa].

The ratio of load to real contact area, i.e., the mean contact pressure, is therefore given by:

$$p_{mean} = \frac{W}{A_r} = \frac{4\sigma E'}{3\pi n (2.3\beta^*)} \frac{\displaystyle\int_d^\infty (z^* - d)^{1.5} \int_0^\infty \frac{f^*(z^*,C)}{\sqrt{C}} dC\,dz^*}{\displaystyle\int_d^\infty (z^* - d) \int_0^\infty \frac{f^*(z^*,C)}{C} dC\,dz^*}$$ (10.5)

It can clearly be seen from equation (10.5) that the ratio of load to true contact area depends only on the material properties defined by the Young's modulus and the asperity geometry. The apparent contact area is eliminated from the equation. Therefore there is a proportionality between contact load and true contact area for most rough surfaces. However, there is no definite proof yet that the contact load and tangential friction force should be proportional. There seems to be only one general argument which is that since friction force is determined by events occurring on the atomic scale, a proportionality of friction force to true contact area will extend even down to patches of contact area a few micrometres in diameter, the size typical of contact areas between rough surfaces.

Although there is a certain degree of variation between the models based on statistical methods which are used to describe the chaotic nature of real surfaces, in essence they are quite similar.

The asperities of surfaces in contact may also sustain localized plastic deformation. It has been shown, however, that even if the asperities are deformed plastically, the true contact area is still linearly proportional to load at moderate values [28]. At large loads, the true contact area reaches a limiting value which is close to but less than the apparent contact area. Even at extreme levels of contact stress, deep grooves and depressions in the surface remain virtually intact. As mentioned already in Chapter 7 the probability of plastic deformation depends on the surface topography and material properties and is defined by the plasticity index. Over the years the definition of the plasticity index evolved as the contact models became more precise. The plasticity indices for three major contact models are shown in Table 10.3.

Table 10.3.   Plasticity indices for three surface contact models.

Greenwood and Williamson (G-W) model [14]	Whitehouse and Archard (W-A) model [25,26]	Bower and Johnson (B-J) model [73,74,76]
$\psi = \left(\dfrac{E'}{H}\right) \times \left(\dfrac{\sigma^*}{r}\right)^{0.5}$	$\psi^* = \left(\dfrac{E'}{H}\right) \times \left(\dfrac{\sigma}{\beta^*}\right)$	$\psi_s = \left(\dfrac{E'}{p_s}\right) \times \left(\sigma\kappa\right)^{0.5}$

where:

E' is the composite Young's modulus (eq. 7.35) [Pa];

H is the hardness of the deforming surface [Pa];

σ^* is the standard deviation of the surface peak height distribution [m];

r is the asperity radius, constant in the G-W model [m];

σ is the RMS surface roughness, 'σ' refers to the harder surface in the B-J model [m];

β^* is the correlation distance [m];

κ is the asperity tip curvature of the harder surface [m^{-1}];

p_s is the shakedown pressure of the softer surface [Pa]. Note that 'p_s' is a function of friction coefficient, i.e., it decreases with the increase of 'μ' [74];

ψ_s is the plasticity index for repeated sliding [73,74,76].

In G-W and W-A models for ψ and $\psi^* < 0.6$ elastic deformation dominates and if ψ and $\psi^* > 1$ a large portion of contact will involve plastic deformation. Plastic deformation causes the surface topography to sustain considerable permanent change, i.e., flattening of asperities. Protective films may also fracture and allow severe wear to occur. When 'ψ' or 'ψ^*' is in the range of **0.6 - 1** the mode of deformation is in doubt. In the B-J model for values of $\psi_s < 1$ the wear rate is negligible, and as ψ_s increases from **1.0** to **3.5** the wear coefficients increase by several orders of the magnitude [74].

Plasticity index 'ψ_s' for repeated sliding is similar to plasticity index 'ψ' for static normal contact; the only difference is that the shakedown pressure 'p_s' replaces the indentation hardness '**H**'. The shakedown pressure is the limiting pressure dictating the type of asperity deformation, i.e., below the shakedown pressure elastic deformation dominates while above it plastic flow occurs at every load cycle [73]. It has been found that the shakedown pressure decreases with increasing 'ψ_s' and roughness 'σ' of the harder surface [75].

Model of Contact between Solids Based on the Fractal Geometry of Rough Surfaces

Instead of modelling a rough surface as a series of euclidean shapes approximating the asperities, e.g., spheres, an attempt has been made to apply fractal geometry or the non-euclidean geometry of chaotic shapes to analyse the contact between rough surfaces [10]. Fractals, since they provide a quantitative measure of surface texture incorporating its multi-scale nature, have been introduced to the model. The Majumdar-Bhushan fractal model has been developed for elastic-plastic contacts between rough surfaces and is based on the spectrum of a surface profile defined by equation (10.2). The contact load for $D \neq 1.5$ is calculated from the following expression [10]:

$$\frac{W}{A_a E'} = \frac{4\sqrt{\pi}}{3} G^{*(D-1)} g_1(D) A_r^{*D/2} \left[\left(\frac{(2-D)A_r^*}{D} \right)^{(3-2D)/2} - a_c^{*(3-2D)/2} \right]$$
$$+ K\phi g_2(D) A_r^{*D/2} a_c^{*(2-D)/2} \qquad (10.6)$$

and for $D = 1.5$ from the relation [10]:

$$\frac{W}{A_a E'} = \sqrt{\pi} G^{*1/2} \left(\frac{A_r^*}{3} \right)^{3/4} \ln\left(\frac{A_r^*}{3a_c^*} \right) + \frac{3K\phi}{4} \left(\frac{A_r^*}{3} \right)^{3/4} a_c^{*1/4} \qquad (10.7)$$

where:

W	is the total load [N];
E'	is the composite Young's modulus (eq. 7.35) [Pa];
D	is the fractal dimension, i.e., $1 < D < 2$;
K	is the factor relating hardness '**H**' to the yield strength 'σ_y' of the material, i.e., $H = K\sigma_y$ and '**K**' is in the range **0.5 - 2** [10,33];
g_1, g_2	are parameters expressed in terms of the fractal dimension '**D**', i.e.:

$$g_1(D) = \frac{D}{(3-2D)} \left(\frac{2-D}{D} \right)^{D/2} \qquad \text{and} \qquad g_2(D) = \left(\frac{D}{2-D} \right)^{(2-D)/2}$$

G*	is the roughness parameter, i.e., $G^* = G/\sqrt{A_a}$;
G	is the characteristic length scale of a surface (fractal roughness) [m];
L	is the sampling length [m];
A_a	is the apparent contact area, i.e., $A_a = L^2 [m^2]$;

A_r^* is the nondimensional real contact area, i.e., A_r / A_a;

A_r is the real contact area [m^2], i.e.:

$$A_r = \frac{D}{2-D} a_l$$

a_l is the contact area of the largest spot [m^2];

a_c is the critical contact area demarcating elastic and plastic regimes [m^2], i.e.:

$$a_c = \frac{G^2}{\left(\frac{\pi K \phi}{2}\right)^{2/(D-1)}}$$

ϕ is a material property parameter, i.e., $\phi = \sigma_y / E'$.

The first part of equations (10.6) and (10.7) represents the total 'elastic' load while the second term represents the 'plastic' load. When the largest spot 'a_l' is less than the critical area for plastic deformation, $a_l < a_c$, only plastic deformation will take place and the load is given by [10]:

$$\frac{W}{A_a E'} = K \phi A_r^* \tag{10.8}$$

An important feature of this model is the assumption that the radius of curvature of a contact spot is a function of the area of the spot. This is in contrast to the G-W model which assumed that the radius of curvature was the same for all asperities.

Summarizing, it can be seen that the Onions-Archard model provides a single value for the ratio of load to real contact area while the fractal model allows for variation between surfaces which depends on the fractal dimension 'D'. The variation in fractal dimension is usually small, since the theoretical limits of 'D' are $1 < D < 2$, so that all surfaces approximate to a simple proportionality between contact load and true contact area.

There is a fundamental difference between earlier models, e.g., Greenwood and Williamson [14], and the fractal model. The early models show that the load exponent for the load to contact area ratio is fixed and is approximately equal to one, i.e., $A_r \propto W$. Archard, when studying the contacts between the surfaces modelled by successive hierarchies of hemispheres, suggested that as the surface became more complex this exponent would change in value [27]. At that time, however, the full implications of these findings were never realized. On the other hand, the fractal model implies that the load exponent for the load to contact area ratio is not a constant but instead varies within a narrow range which is dictated by the limits of the fractal dimension, i.e., $A_r \propto W^{2/(3-D)}$ [10] where $1 < D < 2$. In other words, the model implies that all rough surfaces exhibit a weak non-linear proportionality between load and true contact area as the exact load-area relationship is affected by the surface texture which is in turn described by the fractal dimension. It should also be mentioned that the current fractal model applies for contacts under light loads where the asperity interaction is negligible. Therefore extrapolation to higher levels of load may give unreliable results. An interesting observation could be made when relating this model to scuffing. During scuffing and seizure one would expect the fractal dimension to rise. If this happens then the real contact area is likely to rise and this would result in an increase in the frictional force.

A detailed review of contact models between stationary surfaces can be found elsewhere [77,78].

Effect of Sliding on Contact between Solid Surfaces

Almost all analyses of contact between solids are based on stationary contacts where no sliding occurs between the surfaces. Parameters such as the real contact area and the average contact stress under sliding are of critical importance to the interpretation of wear and friction so that analysis of solid to solid contact under sliding is a major objective for future research.

A qualitative description of some characteristic features of contacts between asperities during sliding has been obtained from studies of hard asperities indenting a soft material [34]. Three distinct stages of contact were observed: static contact, i.e., where the tangential force is small, the stage just before the gross movement of the asperity, i.e., when the tangential force is at its maximum level, and unrestricted movement of the asperity. Tangential force is equivalent to frictional force in real sliding contacts. These three stages of asperity contact are illustrated schematically in Figure 10.15.

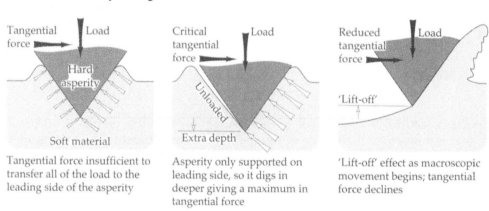

FIGURE 10.15 Schematic illustration of the transition from static contact to sliding contact for a hard asperity on a soft surface.

As can be seen from Figure 10.15, at low levels of tangential force, the hard asperity is supported on both flanks by deformed material. When a critical level of tangential force is reached, one flank on which the force is acting becomes unloaded and, at the same time, the asperity sinks deeper into the softer material, providing a compensating increase in real contact area. Once the asperity begins to move or slide across the soft material, an accumulation of deformed material provides sufficient support for the asperity to rise above the level of static contact. As a result the tangential force declines since the support to the asperity is provided by material which has a relatively 'short dimension' in the direction of sliding. The short dimension of accumulated material reduces the amount of material required to be sheared as compared to the earlier stages of sliding.

Contact between asperities is therefore fundamentally affected by sliding and a prime effect of sliding is to cause the separation of surfaces by a small distance. Real contact is then confined to a much smaller number of asperities than under stationary conditions. Wear particles tend to reduce the number of asperity contacts which under certain conditions can rapidly form large wear particles. This effect of asperity separation may contribute to the characteristic of some worn surfaces which exhibit a topography dominated by a number of large grooves or lumps. The process of reduction of asperity contact induced by sliding is illustrated schematically in Figure 10.16.

As sliding and wear proceed, trapped debris tend to modify the nature of contact between the sliding surfaces. Debris entrapment, which is most common between prominent asperities, causes the progressive growth of a lump that becomes sufficiently large to displace the asperities as the site of true contact. The average diameter and depth of these lumps also increase with increasing frictional power [104]. This effect, illustrated schematically in Figure 10.17, is most likely to occur at high contact loads, where the space between asperities is flattened [103].

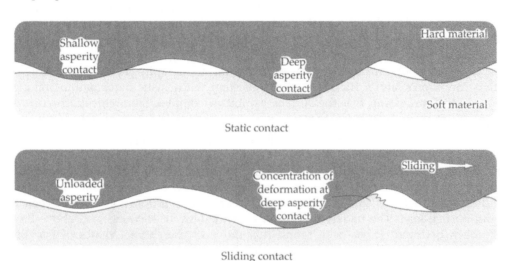

FIGURE 10.16 Reduction in asperity contact under sliding as compared to static conditions.

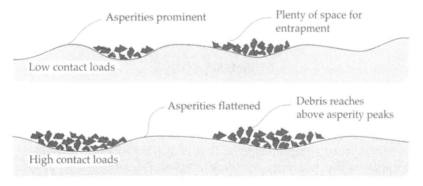

Figure 10.17 Schematic illustration of the debris entrapment effect between contacting asperities under low and high contact loads.

10.4   FRICTION AND WEAR

Friction is the dissipation of energy between sliding bodies. Four basic empirical laws of friction have been known for centuries since the work of da Vinci and Amonton:

- there is a proportionality between the maximum tangential force before sliding and the normal force when a static body is subjected to increasing tangential load;

- the tangential friction force is proportional to the normal force in sliding;

- friction force is independent of the apparent contact area;

- friction force is independent of the sliding speed.

In the early studies of contacts between the real surfaces it was assumed that since the contact stresses between asperities are very high, the asperities must deform plastically [24]. This assumption was consistent with Amonton's law of friction, which states that the friction force is proportional to the applied load, providing that this force is also proportional to the real contact area. However, it was later shown that the contacting asperities, after an initial plastic deformation, attain a certain shape where the deformation is elastic [27]. It has been demonstrated on a model surface made up of large irregularities approximated by spheres with a superimposed smaller set of spheres, which were supporting an even smaller set (as shown in Figure 10.14), that the relationship between load and contact area is almost linear despite the contact being elastic [27]. It was found that a nonlinear increase in area with load at an individual contact is compensated by the increasing number of contacts. A similar tendency was also found for real surfaces with random topography [14,26]. It therefore became clear that Amonton's law of friction is also consistent with elastic deformations taking place at the asperities, providing that the surface exhibits a complex hierarchical structure so that several scales of microcontact can occur.

The proportionality between friction force and normal load has led to the definition of **'kinetic'** and **'static'** coefficients of friction. In many reference books, coefficients of friction are quoted as 'properties' of certain combinations of materials. This approach, however, is very simplistic since the coefficients of friction are dependent on parameters such as temperature and sliding speed and, in some instances, there is no exact proportionality between friction force and normal load. The underlying reasons for the laws of friction listed above have only recently been deduced. It has been found that much of the characteristics of friction are a result of the properties of rough surfaces in contact.

Apart from the dissipation of energy between sliding bodies, friction results in the generation of noise. In most applications frictional noise is a nuisance that must be controlled. Frictionally generated vibrations associated with noise emission can additionally be harmful. Noise generation is usually controlled by lubrication to provide smooth, silent sliding as well as to suppress friction and wear [81].

Onset of Sliding and Mechanism of Stick-Slip

The difference between the 'static' and 'kinetic' coefficients of friction has been known for a considerable period of time. In the analysis of dynamic systems a certain discontinuity between static and kinetic friction forces is usually assumed. Detailed investigation of forces and movements at the onset of sliding has revealed that whenever there is a friction force, sliding must occur even to a minute extent. This is also a property of contact between rough surfaces and explains the necessity in some mechanical analyses to assume that limited sliding occurs below the static friction load. A prime example of this is the creep of railway wheels during acceleration and braking or steering around corners. It has been found that the revolved distance on the contacting surface of the railway wheels never exactly corresponds to the distance travelled. When a railway wagon proceeds along a curved track, the wheels closest to the centre of curvature will skid unless frictional creep takes place, allowing for a difference between the rolling speed of the wheel and the linear speed along the rail. In almost all railway wagons, wheels are rigidly fixed to the axle, so that both wheels must rotate at the same speed irrespective of differing linear speed along the respective rails.

An oscillation between static and kinetic levels of friction can also occur and this is known as **'stick-slip'**. Stick-slip is a phenomenon where the instantaneous sliding speed of an object does not remain close to the average sliding speed. Instead, the sliding speed continuously varies between almost stationary periods and moments of very high speed. Stick-slip depends on the variation in the friction coefficient at low sliding speeds and on the vibrational characteristics of the system. In many cases the suppression of stick-slip can be as important as reducing the overall coefficient of friction because of the destructive nature of the vibration caused.

The measurements of small sliding movements between the solids revealed a continuity between any level of friction force. For example, the coefficients of friction of indium and lead blocks on a steel surface increase gradually with sliding speed for both metals, until a certain limiting level of frictional force is reached [35]. Once this level of force is reached, the coefficient of friction maintains a steady value over the range of velocities, declining gradually after reaching a critical velocity level as shown in Figure 10.18.

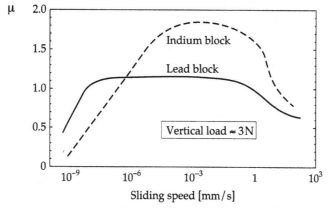

FIGURE 10.18 Variation of friction coefficient for indium and lead blocks sliding on a steel surface [35].

It has also been shown that a lightly loaded steel on steel contact under varying frictional load sustains a reversible displacement [36]. This is believed to be the result of elastic movements of the surface asperities. The scale of the movement is approximately 1 [μm]. Evidence of this phenomenon is illustrated in Figure 10.19.

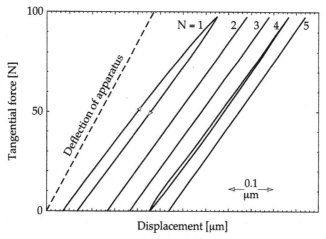

FIGURE 10.19 Reversible tangential movement of steel surfaces in contact under a tangential friction force; **N** is the number of loading cycles, e.g., **N = 1** is the first cycle [36].

It was also found that the rate at which friction force is applied has a considerable influence on frictional characteristics [36]. This effect is illustrated in Figure 10.20 where the friction force versus displacement for a low rate of frictional force application (20 [N/s]) and a high rate (20,000 [N/s]) is shown.

It can be seen from Figure 10.20 that low rates of friction force application correspond to the classical model of friction, i.e., at a critical friction load sliding is initiated and friction force declines discontinuously. However, at the high rate of friction force application, there is no discontinuity in friction force and sliding movement. After sliding is initiated the friction force further continues to increase reaching a maximum. Although there is no model of the underlying mechanism, some inferences based on 'stick-slip' phenomenon have been suggested. The occurrence of 'stick-slip' depends on the stiffness of the system in which the sliding contact takes place. The applied friction force depends on the stiffness of the support system and also the displacement or stretching of the system. The rate at which the support system is displaced corresponds to the rate of load application. At moderate rates of friction change, the rate or speed of the support system displacement follows the rate at which the friction rises or falls. If there is a very rapid change in friction force, then the support structure adjacent to the sliding contact moves at a linear speed determined by its resonant frequency. When the support structure is able to resonate, then a severe 'stick-slip' motion may occur. If the stiffness of the system is too low, then the rate of load application is also low and a discontinuity in friction occurs. For a stiffer system, however, this discontinuity can be suppressed and smooth motion is possible. Although this view of 'stick-slip' motion is only hypothetical, it serves to illustrate the complex nature of the phenomenon in the absence of more detailed research.

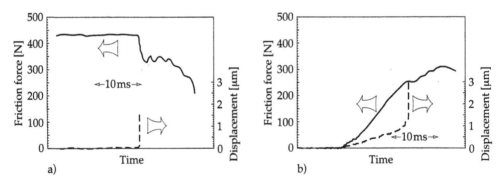

FIGURE 10.20 Effect of the rate of friction force application on friction characteristics: a) low rate and b) high rate [36].

'Stick-slip' is also a function of the wear and friction mechanisms. A large difference in friction coefficient between apparently static conditions and gross sliding implies that smooth sliding is impossible. A fundamental impediment to smooth sliding occurs when the sliding surfaces tend to adhere to each other. Under this condition, true smooth sliding is impossible and the opposing surfaces are forced to move against each other by a series of small jumps between successive adhesive contacts.

Structural Differences between Static and Sliding Contacts

The distinction between a static and a sliding contact is generally understood to be important, with fundamental differences in structure and physical processes believed to occur. The difficulties in observing a sliding contact through opaque bodies largely prevented tribologists from resolving the physical phenomena involved. Recently, however, direct observations of sliding contact by real-time radiography (X-ray microscopy) revealed much of its hidden detail [79,80]. It was found that while a static contact can be described in terms of a random distribution of point contacts, in accordance with the Greenwood-Williamson model, this model is not applicable to a sliding contact. A basic feature of sliding contact is that it is distributed over a lesser number of larger contact areas rather than a large number of contact points. It appears that contact between dry sliding surfaces, e.g., during oxidative wear or fretting, is controlled by a series of lamellar bodies (compacted wear particles) which are formed from material from both sliding surfaces. These areas do not have a fixed location inside the contact but instead move slowly across the surface as sliding progresses [80]. The frictional interaction appears to be controlled by mechanical inter-locking between 'lumps' on opposing surfaces as schematically illustrated in Figure 10.21.

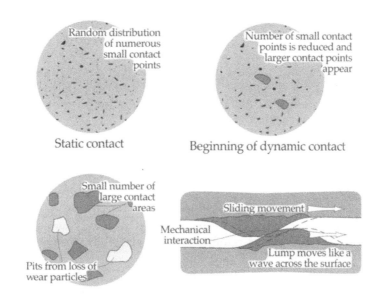

FIGURE 10.21 A comparison between static and dynamic contact.

The question is: why should the number of true contact points between opposing surfaces be greatly diminished by sliding motion? One reason may be that sliding does result in a greater degree of separation between the opposing surfaces [34]. When the opposing surfaces move apart, only the larger asperities on the surface can remain in contact. Another cause may be a more subtle process, which requires sufficient sliding distance before its effect is observed, i.e., the irreversible trend towards expulsion of wear particles once they leave the wearing contact. It can be imagined that at the stage when the wear particles are still inside the wearing contact, but are no longer bonded to a specific site, they can move. This movement is strongly affected by the sliding. In statistical terms it is possible for the wear particle to move outside the sliding contact and then to return. However, as can be readily appreciated, once the wear particles exit the sliding contact, they are very unlikely to return. This concept is schematically illustrated in Figure 10.22. Irreversible departures of wear particles from the boundaries of the sliding contact cause changes in the statistical distribution of wear particles within the sliding contact. It appears that these few wear particles, which by pure chance happen to remain in the sliding contact, survive to form lumps or related structures. Work-hardening of these lumps and compressive forces may cause them to become embedded in the sliding surfaces. Sliding movement may also force the lumps to move slowly across the wearing surface while simultaneously accumulating more material.

Irreversibility of wear particle expulsion from the contact zone persists even at the very small amplitudes of micro-sliding. The most common example of micro-sliding is fretting wear, which is described in Chapter 15. Statistical bias in the survival of wear particles appears to be the cause of the formation of distinctively non-uniform contact structure of segregated debris layers inside a fretted wear scar [82].

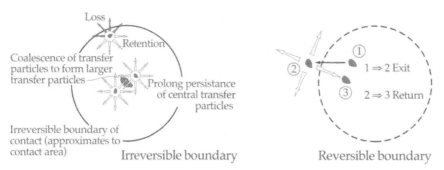

FIGURE 10.22 Schematic illustration of the loss of transfer particles at the edge of contact.

In some cases of fretting wear, e.g., ball-on-plate, the variation in contact pressure can exert the following additional effects [82]:

· close to the centre of the contact where contact pressure is at a maximum and the gradient of pressure, with respect to distance, is at a minimum wear particles tend to move the least (the lack of movement would render the wear particles more likely to coalesce) and

· at the edge of the contact where the gradient of pressure, with respect to distance, is at a maximum wear particles are more likely to move towards the boundaries of the contact zone. This effect reinforces the irreversibility of wear particle expulsion from the contact zone.

Surface temperature also exerts a strong influence on the mechanics of sliding contact. Elevated sliding speeds result in high surface temperatures, which may lead to softening and possibly even melting of the surface layers. Lump formation is observed to decline at high sliding speeds [79,80], apparently because of this softening and melting of the surface layers and particles. A smaller number of interlocking lumps at high sliding speeds may be the reason why dry friction coefficients decline with rise in sliding speed for most combinations of sliding materials.

Friction and Other Contact Phenomena in Rolling

When a cylindrical or spherical object rolls across a smooth surface of sufficient hardness to support its weight, it is generally observed that there is only a small amount of friction to oppose the rolling motion. This is the operating principle behind many vital mechanical components such as roller or ball bearings, railway wheels and rubber tyres. Rolling motion not only entails friction but also other phenomena such as corrugation and traction. Corrugation is the formation of a wave-shaped profile on the rolled surface by repeated rolling contact. Traction is the ability of a roller or sphere to sustain a tangential contact force while continuing to roll with negligible resistance to motion.

A coefficient of rolling friction is defined as the force required to maintain steady rolling, divided by the load carried by the roller. For hard smooth rollers such as those made of steel or other hard metals, the coefficient of rolling friction is very low with typical values ranging between **0.01** and **0.001**. Rolling friction coefficients are not necessarily minute as it is possible to select systems, with physical parameters differing widely from hard metal rollers, that display much higher values of rolling friction coefficients. For example, a roller or sphere made of soft material that adheres to the underlying surface would generate a higher level of rolling friction. A physical representation of this might be a coalescence of several wear particles that had been rolled into a single sphere or roller while retaining strong adhesion to the underlying surface.

Traction is the phenomenon which enables all wheeled vehicles to accelerate or decelerate and ascend or descend hills. Traction is a form of friction but is distinguished from it because of its usefulness in propulsion of vehicles. The maximum amount of traction available to any rolling contact is equal to the product of the normal contact load and the coefficient of adhesion of the rolling contact. The coefficient of adhesion is defined as the ratio of the maximum tangential force that can be sustained at the rolling contact and the normal contact force. In other words the coefficient of adhesion defines the resistance to skidding by the rolling element when a braking torque is imposed such as occurs during braking of a railway wheel or car tyre. The coefficient of adhesion typically has a value ranging from **0.1** to **1.0** and is distinct from the coefficient of rolling friction. For most mechanical systems, the coefficient of adhesion should be as high as possible while the coefficient of rolling friction should be as low as possible. For example, a low rolling friction coefficient enables a railway train to minimize energy consumption when travelling while a high coefficient of adhesion allows the train to stop if required. The difference between the coefficient of rolling friction and the coefficient of adhesion and the role of hardness and adhesion in controlling the coefficient of rolling friction is schematically illustrated in Figure 10.23.

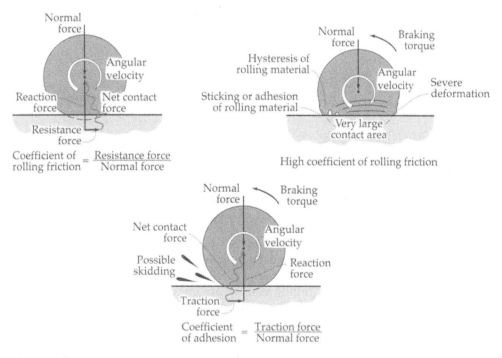

FIGURE 10.23 The difference between coefficient of rolling friction and coefficient of adhesion and the role of hardness and adhesion in controlling the coefficient of rolling friction.

It was observed that the fundamental cause of rolling friction is micro-slip between the contacting surfaces [83]. Carter [93] has also found that a wheel cannot generate traction without additional micro-slip in response to the tangential forces at the rolling contact. Micro-slip is a very limited amount of tangential movement that occurs in regions within the rolling contact without gross sliding occurring over the whole of the contact area. Classical kinematic theory of rolling by rigid bodies predicts that only normal motion occurs at a point contact for a sphere or a line contact for a roller. As all known materials, even hard metals and ceramics, are not perfectly rigid, they elastically deform to produce, depending on the geometry of contacting bodies, a circular, rectangular or elliptical contacts. This elastic deformation reduces the radius of rolling by a minute amount and so causes the surface of rolling body to fail to move sufficiently fast inside the rolling contact. In order to compensate for this lack of speed, creeping movement is initiated between the rolling surfaces [84-86]. This creeping movement is generated by micro-slip at the margins of the rolling contact. The phenomenon of micro-slip and creeping movement is illustrated schematically in Figure 10.24.

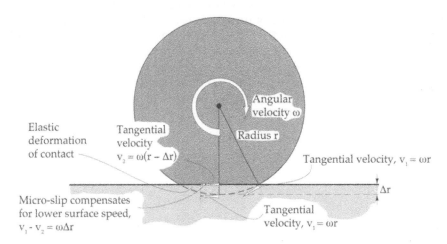

FIGURE 10.24 Schematic illustration of micro-slip and creeping movement in a rolling contact.

When the roller or sphere sustains traction, the micro-slip increases in level and extent over the rolling contact area [94,95]. When micro-slip prevails over the entire rolling contact area, gross sliding or skidding of the roller (or sphere) will commence.

A fundamental difference between rolling and sliding friction is that other energy dissipation mechanisms, which are negligible for sliding friction, become significant for rolling because of the very low friction level. Major sources of energy dissipation, which are not discussed further here, are aerodynamic drag of the rapidly rotating roller and repetitive compression of air inside a pneumatic tyre. Another important source of energy dissipation is hysteresis in the mechanical response of the rolling material. Hysteresis means that the compressive stresses ahead of the centre of the rolling contact are greater than the compressive stresses behind the rolling contact. Ahead is defined as not yet reached by the centre of the rolling contact while behind is defined as already rolled on by the centre of the rolling contact. The resulting asymmetry in compressive stresses generates reaction forces that oppose the rolling motion. For example, hysteresis is found to be the principal component of rolling friction in polymers [87]. An effect similar to mechanical hysteresis may also be generated by adhesion between the roller and the rolled surface [96]. Adhesion behind the rolling contact causes the compressive forces behind the rolling contact to be less than the compressive forces ahead of the rolling contact. Adhesive effects are significant for rubbers [96] where the adhesion is generated by van der Waals bonding between atoms of the opposing surfaces [97].

Rolling is nearly always associated with high levels of contact stress, which can be sufficient at high contact loads to cause plastic deformation in the rolling contact. Plastic deformation not only causes the surface layers of the roller and rolled surface to accumulate plastic strain, but may also cause corrugation to occur. Corrugation is the transformation of a smooth, flat surface into a surface covered by a wave-form like profile aligned so that the troughs and valleys of the wave-form profile lie perpendicular to the direction of rolling. The wavelength of corrugations varies from 0.3 [mm] on the discs of Amsler test machines to 40 ~ 80 [mm] on railway tracks [98]. Another term used to describe corrugations, especially longer wavelength corrugations, is facets. Although the causes of corrugation are unclear, there is evidence that vibration of the rolling wheel and metallurgical factors exhibit a strong influence [99,100]. It was found that the peaks of the corrugations on steel surfaces were significantly harder than the troughs between the corrugations [101]. It is believed that corrugation occurs when a lump of plastically deformed material is formed at the leading edge of the rolling contact. This lump periodically grows to a maximum size before being released behind the rolling contact to form a corrugation [88,89].

According to theoretical models of the deformation and slip involved in rolling friction, it appears that there is a linear relationship between contact force and the drag force opposing rolling [84]. The geometry of the rolling contact has a strong influence on rolling friction, and the coefficient of rolling friction is inversely related to the rolling radius. At low loads where elastic deformation dominates, the coefficient of rolling friction is inversely proportional to the square root of the rolling radius; at higher contact loads where plastic deformation is significant, the coefficient of rolling friction is inversely proportional to the rolling radius [90]. Basic materials parameters also exert an effect, the coefficient of rolling friction is inversely related to the Young's modulus of the rolling material [90]. Temperature exerts a strong effect on the coefficient of rolling friction of polymers since the mechanical hysteresis of the polymer is controlled by temperature [92]. The coefficient of adhesion in a rolling steel contact was found to decline with speed in the range from 20 to 500 [km/hr] [91].

Concentration of Frictional Heat at the Asperity Contacts

The inevitable result of friction is the release of heat and, especially at high sliding speeds, a considerable amount of energy is dissipated in this manner. The released heat can have a controlling influence on friction and wear levels due to its effect on the lubrication and wear processes. Almost all of the frictional heat generated during dry contact between bodies is conducted away through the asperities in contact [24]. Since the true contact area between opposing asperities is always considerably smaller than the apparent contact area, the frictional energy and resulting heat at these contacts becomes highly concentrated with a correspondingly large temperature rise as illustrated schematically in Figure 10.25.

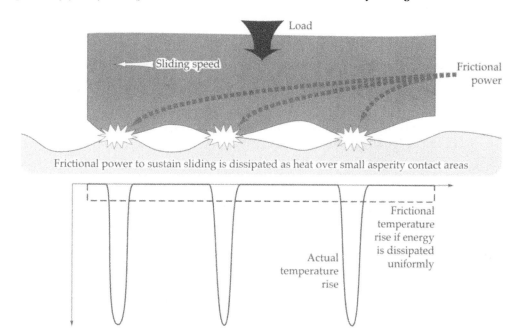

FIGURE 10.25 Concentration of frictional energy at the asperity contacts.

This concentration of frictional energy over small localized areas has a significant influence on friction and wear. Local temperatures can rise to very high values even with a relatively small input of frictional energy. For example, a frictional temperature rise was exploited by paleolithic man to ignite fires by rotating a stick against a piece of wood.

Surface heating from frictional energy dissipation also causes the surface layers of a material to expand. Where such heating is localized, a small area of surface becomes elevated from the rest of the surface which has not sustained thermal expansion. This effect is known as a 'thermal mound' since the shape of this temperature-induced structure resembles a gently sloping hill or mound. When the wearing surface is flat, the distribution of thermal mounds tends to be random along with the distribution of frictional energy dissipation. When the contact of the wearing surface is controlled by asperities, the asperities which sustain the greatest amount of frictional energy dissipation will expand the most and lift apart the remaining asperities. The effect of thermal mound formation results in the concentration of frictional energy dissipation and mechanical load on a few asperities only. This effect is transient and once the source of frictional energy is removed, i.e., by stopping the moving surfaces, the thermal mounds disappear.

Shear rates between contacting solids can also be extremely high as often only a thin layer of material accommodates the sliding velocity difference. The determination of surface temperature as well as the observation of wear is difficult as the processes are hindered by the contacting surfaces. In the majority of sliding contacts, the extremes of temperature, stress and strain can only be assessed indirectly by their effect on wear particles and worn surfaces.

The frictional temperatures can, for example, be measured by employing the 'dynamic thermocouple method' [24]. The method involves letting two dissimilar metals slide against each other. Frictional temperature rises at the sliding interface cause a thermo-electric potential to develop which can be measured. For example, significant temperature rises were detected by this method when a constantan alloy was slid under unlubricated conditions against steel at a velocity of 3 [m/s] [24]. Momentary temperature rises reaching 800°C but only lasting for approximately 0.1 [ms] occurring on a random basis were observed. It is speculated that these temperature rises are the result of intense localized metal deformation between asperities in contact.

Thermoelastic Instability and Transient Hump Formation

During sliding an originally flat surface can deform by a localized concentration of frictional heat, which causes differential thermal expansion. Due to this expansion, a shallow hump is formed on the surface, leading to a further concentration of frictional contact at the original site. A loop with a positive feedback is formed and the phenomenon is known as thermoelastic instability (TEI), which is schematically illustrated in Figure 10.26. A recent review and analysis of TEI can be found elsewhere [105].

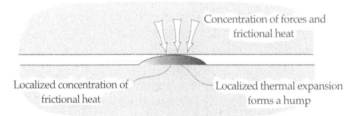

Figure 10.26 Model of the formation of thermoelastic instabilities.

Tribo-Electrification of Sliding Contacts

It is common in sliding contacts for a voltage difference of a few microvolts to be generated between metal surfaces. A possible source of this potential difference is the Seebeck effect, which is a form of thermo-electricity and only occurs between dissimilar metals. For similar or identical metals in sliding, different mechanisms prevail. It is thought that the formation of wear particles or transfer particles leads to electron loss or their transfer from one sliding surface to the other [106]. In dry open-air sliding, the magnitude of the sliding induced voltage is proportional to the product of electrical resistivity and hardness divided by the surface energy [107].

Wear between Surfaces of Solids

As already discussed the contact between surfaces of solids at moderate pressures is limited to contacts between asperities of opposing surfaces. Most forms of wear are the result of events occurring at asperity contacts. There could, however, be some exceptions to this rule, e.g., erosive wear which involves hard particles colliding with a surface.

It has been postulated by Archard that the total wear volume is proportional to the real contact area times the sliding distance [37]. A coefficient 'K' which is the proportionality constant between real contact area, sliding distance and the wear volume has been introduced, i.e.:

$$V = K\, A_r l = K\, l\, \frac{W}{H}$$

(10.9)

where:

V	is the wear volume [m³];
K	is the proportionality constant;
A_r	is the real area of the contact [m²];
W	is the load [N];
H	is the Vickers hardness of the softer surface [Pa];
l	is the sliding distance [m].

The 'K' coefficient, also known as the '**Archard coefficient**', '**wear coefficient**' or sometimes the '**wear constant**', is widely used as an index of wear severity [112]. The coefficient can also be imagined as the proportion of asperity contacts resulting in wear. The value of 'K' is never supposed to exceed unity and in practice 'K' has a value of **0.001** or less for all but the most severe forms of wear. The low value of 'K' indicates that wear is caused by only a very small proportion of asperity contacts. In almost all cases, asperities slide over each other with little difficulty and only a minute proportion of asperity contacts results in the formation of wear particles.

It has to be mentioned, however, that the same term 'wear coefficient' is often used to describe the '**specific wear rate**' '**k**' defined by the following expression [112]:

$$k = \frac{V}{W \times L}$$

(10.10)

where:

k	is the specific wear rate (sometimes called the '**wear factor**') [m³/Nm];
V	is the wear volume [m³];

W   is the normal load [N];

L   is the sliding distance [m].

The use of the same term describing two different parameters can be confusing. Therefore when consulting the publications containing wear data, it is best to verify first the meaning of wear coefficient from its dimensions, i.e., '**K**' is dimensionless while '**k**' has the dimension of $[m^3/Nm]$. Materials with specific wear rate of, for example, 10^{-14} $[m^3/Nm]$ or higher would be classified as not particularly wear resistant. Materials with good wear resistance would exhibit specific wear rate about 10^{-16} $[m^3/Nm]$ or lower. Materials with specific wear rate as low as 10^{-17} or 10^{-18} $[m^3/Nm]$ have also been developed. It should also be mentioned that some wear data published in the literature is presented as a 'wear rate' defined usually as the volume loss per time or per sliding distance.

It has also been suggested that wear particles are the result of a cumulative process of many interactions between randomly selected opposing asperities [38]. The combination of opposing asperities during sliding at any one moment can easily be imagined as continuously changing. A gradual or incremental mode of wear particle formation allows for extensive freedom for variation or instability in the process. Statistical analysis of wear data reveals that there is a short term 'memory' inherent in wear processes, i.e., any sample of a wear rate is related to the immediately preceding wear rates, although there seems to be no correlation with much earlier wear rates [39]. Therefore wear prediction is extremely difficult.

10.5   SUMMARY

Real surfaces are composed of surface features ranging in size from individual atoms to visible grooves and ridges. Most surface features affect wear and friction. Since almost all surfaces are rough, in terms of solid contact they cannot be approximated by a flat plane. The basic laws of friction are a result of the control of solid contact by rough surfaces. The topography of the contacting surfaces therefore has a decisive effect on wear and friction. Rough surfaces have very small areas of real contact with the opposing surface and this causes wear and friction to be determined by high contact stresses and extreme concentrations of frictional energy even though the nominal contact stress and total frictional energy can be small.

Friction has traditionally been divided into static and kinetic friction. Exact measurements of microscopic sliding movements reveal that as the friction force acting on a contact is progressively increased, microscopic sliding movement occurs for all levels of friction force and the maximum friction force occurs at some specific sliding speed. The basic difference between gross sliding and sliding movements at small levels of friction force is that these latter movements are reversible. A major consequence of the difference between static and kinetic coefficients of friction is 'stick-slip' or discontinuous sliding. Stick-slip is often present when the supporting structure of the sliding contact has insufficient stiffness to follow the rapid changes in frictional force that can occur.

Wear results from direct contact between the individual asperities at sliding interfaces and, in almost all situations, many asperity interactions are required before wear occurs.

REVISION QUESTIONS

10.1   Would one expect there to be more points of true contact between two stationary objects or between two moving objects? How would wear and transfer particles affect the number of contact points?

10.2   In engineering, contact pressure is calculated by dividing normal load by contact area. Such pressure is called 'nominal pressure'. Why?

10.3   What happens to alloying and impurity elements in a metal subjected to the heat and mechanical working of sliding contact and what is the tribological significance of this?

10.4   Give examples of how the surface roughness can affect a tribological contact.

10.5   What surface finish is used in engine cylinder liners to optimize wear and friction between the liners and the piston rings?

10.6   According to the Hertzian theory of elastic contact (Chapter 7) the relationship between contact area '**A**' and contact load '**W**' can be written as $A \propto W^{2/3}$. Since the contact area between real surfaces is due to the elastic deformations of the asperities, is there the same relationship between '**A**' and '**W**'?

10.7   If some sales literature claimed a 3-fold decrease in friction rate, would you regard it with interest or with scepticism?

10.8   Is the concept of 'static friction' an exact explanation of events occurring between contacting surfaces?

10.9   The term 'friction' is most commonly used to describe the resistance to movement between two solid bodies and it is called a 'solid friction'. What other forms of friction exist?

10.10   In many engineering handbooks there are tables showing dry coefficients of friction for various coating pairs of materials. Does it mean that the dry coefficient of friction is constant for a given pair?

10.11   Give examples of tribological contact where friction is often proportional to normal load and where it is not.

10.12   Why should fluctuations in low level friction coefficients be of critical importance in the operation of high power gear units? Suggest a design measure to allow for these fluctuations.

10.13   What natural phenomenon was probably critical to the invention of a wheel for transport applications?

10.14   Where is the flange of a railway wheel most likely to touch the rails?

10.15   In the absence of flange lubrication, what is the tribological term used to describe the sliding conditions at flange-rail contact?

10.16   What is the cause of the 'squealy' noise during flange rail contact?

10.17   How does the flange lubricant suppress the stick-slip?

10.18   Why are dry plain bearings so prone to frictional seizure, especially after some wear has occurred?

10.19   Wear prediction using mathematical equations is still unsuccessful in many tribological situations. Explain why.

10.20   Give examples where friction is beneficial.

REFERENCES

1   J. Benard (editor), Adsorption on Metal Surfaces, Elsevier, Amsterdam, 1983.

2   D. Landheer, A.J.G. Dackus and J.A. Klostermann, Fundamental Aspects and Technological Implications of the Solubility Concept for the Prediction of Running Properties, *Wear*, Vol. 62, 1980, pp. 255-286.

3   E.A. Gulbransen, The Role of Minor Elements in the Oxidation of Metals, *Corrosion*, Vol. 12, 1956, pp. 61-67.

4   K. Meyer, Physikalisch-Chemische Kristallographie, Copyright VEB Deutscher Verlag fur Grundstoffindustrie, Leipzig, Gutenberg Buchdruckerei, Weimar, 1988.

5   D.H. Buckley, Surface Effects in Adhesion, Friction, Wear and Lubrication, Elsevier, Amsterdam, 1981.

6   D. Godfrey, Chemical Changes in Steel Surfaces during Extreme-Pressure Lubrication, *ASLE Transactions*, Vol. 5, 1962, pp. 51-66.

7   R. Kothari and R.W. Vook, The Effect of Cold Work on Surface Segregation of Sulphur on Oxygen-Free High Conductivity Copper, *Wear*, Vol. 157, 1992, pp. 65-79.

8   J. Van Alsten and S. Granick, Friction Measured with a Surface Forces Apparatus, *Tribology Transactions*, Vol. 32, 1989, pp. 246-250.

9   T.R. Thomas (editor), Rough Surfaces, Longman Group Limited, 1982.

10   A. Majumdar and B. Bhushan, Fractal Model of Elastic-Plastic Contact between Rough Surfaces, *Transactions ASME, Journal of Tribology*, Vol. 113, 1991, pp. 1-11.

11   E.J. Abbott and F.A. Firestone, Specifying Surface Quality, *Mechanical Engineering*, Vol. 55, 1933, pp. 569-572.

12   E.F. Finklin, The Bearing Area of Surfaces, *Transactions ASME, Journal of Lubrication Technology*, Vol. 90, 1968, pp. 329-330.

13   R.S. Sayles and T.R. Thomas, A Stochastic Explanation of Some Structural Properties of a Ground Surface, *Int. Journal of Production Research*, Vol. 14, 1976, pp. 641-655.

14   J.A. Greenwood and J.B.P. Williamson, Contact of Nominally Flat Surfaces, *Proc. Roy. Soc., London*, Series A, Vol. 295, 1966, pp. 300-319.

15   J.B.P. Williamson, The Microtopography of Surfaces, *Proc. Inst. of Mech. Engrs.*, Vol. 182, Pt. 3K, 1967-1968, pp. 21-30.

16   R.S. Sayles and T.R. Thomas, The Spatial Representation of Surface Roughness by Means of the Structure Function: a Practical Alternative to Correlation, *Wear*, Vol. 42, 1977, pp. 263-276.

17   R.S. Sayles and T.R. Thomas, Surface Topography as a Non-Stationary Random Process, *Nature*, Vol. 271, 1978, pp. 431-434.

18   J.I. McCool, Relating Profile Instrument Measurements to the Functional Performances of Rough Surfaces, *Transactions ASME, Journal of Tribology*, Vol. 109, 1987, pp. 264-270.

19   A. Majumdar and B. Bhushan, Role of Fractal Geometry in Roughness Characterization and Contact Mechanics of Surfaces, *Transactions ASME, Journal of Tribology*, Vol. 112, 1990, pp. 205-216.

20   A. Majumdar and C.L. Tien, Fractal Characterization and Simulation of Rough Surfaces, *Wear*, Vol. 136, 1990, pp. 313-327.

21   M.V. Berry and Z.V. Lewis, On the Weierstrass-Mandelbrot Fractal Function, *Proc. Roy. Soc., London*, Series A, Vol. 370, 1980, pp. 459-484.

22   W. Hirst and A.E. Hollander, Surface Finish and Damage in Sliding, *Proc. Roy. Soc., London*, Series A, Vol. 337, 1974, pp. 379-394.

23   H. Czichos, Tribology; A System Approach to the Science and Technology of Friction, Lubrication and Wear, Elsevier, Amsterdam, 1978.

24   F.P. Bowden and D. Tabor, The Friction and Lubrication of Solids, Part I, Clarendon Press, Oxford, 1954.

25   D.J. Whitehouse and J.F. Archard, The Properties of Random Surfaces of Significance in Their Contact, *Proc. Roy. Soc. London*, Series A, Vol. 316, 1970, pp. 97-121.

26   R.A. Onions and J.F. Archard, The Contact of Surfaces Having a Random Structure, *Journal of Physics*, Series D: Appl. Phys., Vol. 6, 1973, pp. 289-304.

27   J.F. Archard, Elastic Deformation and the Laws of Friction, *Proc. Roy. Soc., London*, Series A, Vol. 243, 1957, pp. 190-205.

28   J. Pullen and J.B.P. Williamson, On the Plastic Contact of Rough Surfaces, *Proc. Roy. Soc., London*, Series A, Vol. 327, 1972, pp. 159-173.

29   P.R. Nayak, Random Process Model of Rough Surfaces, *Transactions ASME, Journal of Lubrication Technology*, Vol. 93, 1971, pp. 398-407.

30   A.W. Bush, R.D. Gibson and T. R Thomas, The Elastic Contact of a Rough Surface, *Wear*, Vol. 35, 1975, pp. 87-111.

31   P.K. Gupta and N.H. Cook, Junction Deformation Models for Asperities in Sliding Interactions, *Wear*, Vol. 20, 1972, pp. 73-87.

32   B. Bhushan, Tribology of Mechanics of Magnetic Storage Devices, Sprigler-Verlag, 1990.

33   B. Bhushan, Analysis of the Real Area of Contact Between a Polymeric Magnetic Medium and a Rigid Surface, *Transactions ASME, Journal of Tribology*, Vol. 106, 1984, pp. 26-34.

34   J.M. Challen, L.J. MacLean and P.L.B. Oxley, Plastic Deformation of a Metal Surface in Sliding Contact With a Hard Wedge: Its Relation to Friction and Wear, *Proc. Roy. Soc., London*, Series A, Vol. 394, 1984, pp. 161-181.

35   J.T. Burwell and E. Rabinowicz, The Nature of the Coefficient of Friction, *Journal of Applied Physics*, Vol. 24, 1953, pp. 136-139.

36   M. Eguchi and T. Yamamoto, Dynamic Behaviour of a Slider under Various Tangential Loading Conditions, Proc. JSLE. Int. Tribology Conference, 8-10 July 1985, Tokyo, Japan, Elsevier, 1986, pp. 1047-1052.

37   J.F. Archard, Single Contacts and Multiple Encounters, *Journal of Applied Physics*, Vol. 32, 1961, pp. 1420-1425.

38   Y. Kimura and H. Okabe, Review of Tribology, Youkandou Press, Tokyo, (in Japanese), 1982.

39   S.C. Lim, C.J. Goh and L.C. Tang, The Interdependence of Wear Events during Slow Sliding - a Statistical Viewpoint, *Wear*, Vol. 137, 1990, pp. 99-105.

40 K. Naoi, K. Sasjima and T. Tsukuda, A Quantitative Evaluation of Truncation Wear Based on Three-Dimensional Surface Asperity Changes, *Proc. JAST*, Vol. 4, 1999, pp. 452-459.

41 D.J. Whitehouse, Handbook of Surface Metrology, Bristol; Philadelphia: Institute of Physics Pub., 1994.

42 C.Y. Poon, B. Bhushan, Comparison of Surface Roughness Measurements by Stylus Profiler, AFM and Non-Contact Optical Profiler, *Wear*, Vol. 190, 1995, pp. 76-88.

43 H. Zahouani, R. Vargiolu, Ph. Kapsa, J.L. Loubat, T.G. Mathia, Effect of Lateral Resolution on Topographical Images and Three-Dimensional Functional Parameters, *Wear*, Vol. 219, 1998, pp. 114-123.

44 P. Podsiadlo and G.W. Stachowiak, Scale-Invariant Analysis Tribological Surfaces, Proceedings of the International Leeds-Lyon Tribology Symposium, 'Lubrication at the Frontier', September 1999, Elsevier, 2000.

45 G.W. Stachowiak and P. Podsiadlo, Surface Characterization of Wear Particles, *Wear*, Vol. 225-229, 1999, pp. 1171-1185.

46 P. Podsiadlo and G.W. Stachowiak, 3-D Imaging of Wear Particles Found in Synovial Joints, *Wear*, Vol. 230, 1999, pp. 184-193.

47 W.P. Dong, P.J. Sullivan and K.J. Stout, Comprehensive Study of Parameters for Characterising Three-Dimensional Topography. IV: Parameters for Characterising Spatial and Hybrid Properties, *Wear*, Vol. 178, 1994, pp. 45-60.

48 Z. Peng and T.B. Kirk, Two-Dimensional Fast Fourier Transform and Power Spectrum for Wear Particle Analysis, *Tribology International*, Vol. 30, 1997, pp. 583-590.

49 D.M. Tsai and C.F. Tseng, Surface Roughness Classification for Castings, *Pattern Recognition*, Vol. 32, 1999, pp. 389-405.

50 Y. Wang, K. S. Moon, A methodology for the multi-resolution simulation of grinding wheel surface, *Wear*, Vol. 211, 1997, pp. 218-225.

51 X.Q. Jiang, L. Blunt, K.J. Stout, Three-Dimensional Surface Characterization for Orthopaedic Joint Prostheses, *Proceedings of Institute of Mechanical Engineers, Part H*, Vol. 213, 1999, pp. 49-68.

52 J.-L. Starck, F. Murtagh, A. Bijaoui, Image Processing and Data Analysis: The Multiscale Approach, New York, Cambridge University Press, 1998.

53 G. Borgerfors, Distance Transforms in Arbitrary Dimensions, *Comp. Vision, Graphics Image Proc.*, Vol. 27, 1984, pp. 321-345.

54 P. Grassberger and I. Procaccia, Characterisation of Strange Attractors, *Phys. Rev. Letters*, Vol. 50, 1983, pp. 346-349.

55 K. Judd, An Improved Estimator of Dimension and Comments on Providing Confidence Intervals, *Phys. D.*, Vol. 56, 1992, pp. 216-228.

56 J.C. Russ, Fractal Surfaces, Plenum Press, New York, 1994.

57 M.G. Hamblin and G.W. Stachowiak, Application of the Richardson Technique to the Analysis of Surface Profiles and Particle Boundaries, *Tribology Letters*, Vol. 1, 1995, pp. 95-108.

58 W. P. Dong, P. J. Sullivan and K. J. Stout, Comprehensive Study of Parameters for Characterising Three-Dimensional Surface Topography, II: Statistical Properties of Parameter Variation, *Wear*, Vol. 167, 1993, pp. 9-21.

59 M.G. Hamblin and G.W. Stachowiak, Measurement of Fractal Surface Profiles Obtained from Scanning Electron and Laser Scanning Microscope Images and Contact Profile Meter, *Journal of Computer Assisted Microscopy*, Vol. 6, No. 4, 1994, pp. 181-194.

60 C. Tricot, P. Ferland and G. Baran, Fractal Analysis of Worn Surfaces, *Wear*, Vol. 172, 1994, pp. 127-133.

61 M. Hasegawa, J. Liu, K. Okuda, M. Nunobiki, Calculation of the Fractal Dimensions of Machined Surface Profiles, *Wear*, Vol. 192, 1996, pp. 40-45.

62 J. Lopez, G. Hansali, H. Zahouani, J.C. Le Bosse, T. Mathia, 3D Fractal-Based Characterisation for Engineered Surface Topography, *International Journal of Machine Tools and Manufacture*, Vol. 35, 1995, pp. 211-217.

63 S. Ganti, B. Bhushan, Generalized Fractal Analysis and Its Applications to Engineering Surfaces, *Wear*, Vol. 180, 1995, pp. 17-34.

64 S. Peleg, J. Naor, R. Harley and D. Avnir, Multiresolution texture analysis and classification, *IEEE Transactions on Pattern Analysis Machine Intelligence*, Vol. 4 , 1984, pp. 518-523.

65 J.J. Gangepain and C. Roques-Carmes, Fractal Approach to Two Dimensional and Three Dimensional Surface Roughness, *Wear*, Vol. 109, 1986, pp. 119-126.

66   K.C. Clarke, Computation of the Fractal Dimension of Topographic Surfaces Using the Triangular Prism Surface Area Method, *Computers and Geosciences*, Vol. 12, 1986, pp. 713-722.

67   B. Dubuc, S.W. Zucker, C. Tricot, J-F. Quiniou and D. Wehbi, Evaluating the Fractal Dimension of Surfaces, *Proc. Roy. Soc. London*, Series A425, 1989, pp. 113-127.

68   C.A. Brown, P.D. Charles, W.A. Johnsen and S. Chesters, Fractal Analysis of Topographic Data by The Patchwork Method, *Wear*, Vol. 161, 1993, pp. 61-67.

69   P. Podsiadlo and G.W. Stachowiak, The Development of Modified Hurst Orientation Transform for the Characterization of Surface Topography of Wear Particles, *Tribology Letters*, Vol. 4, 1998, pp. 215-229.

70   P. Prusinkiewicz and A. Lindenmayer, The Algorithmic Beauty of Plants, Springer-Verlag, New York, 1990.

71   M.F. Barsney and L.P. Hurd, Fractals Everywhere, Academic Press, San Diego, 1988.

72   Y. Fisher (editor), Fractal Image Compression. Theory and Application, Springer-Verlag, New York, 1995.

73   K.L. Johnson, Contact Mechanics and the Wear of Metals, *Wear*, Vol. 190, 1995, pp. 162-170.

74   A. Kapoor, K.L. Johnson and J.A. Williams, A Model for the Mild Ratchetting Wear of Metals, *Wear*, Vol. 200, 1996, pp. 38-44.

75   A. Kapoor, J.A. Williams and K.L. Johnson, The Steady State Sliding of Rough Surfaces, *Wear*, Vol. 175, 1995, pp. 81-92.

76   A.F. Bower and K.L. Johnson, The Influence of Strain Hardening on Cumulative Plastic Deformation in Rolling and Sliding Line Contact, *Journal of Mech. Phys. Solids*, Vol. 37, 1989, pp. 471-493.

77   B. Bhushan, Contact Mechanics of Rough Surfaces in Tribology: Single Asperity Contact, *Appl. Mech. Rev.*, Vol. 49, 1996, pp. 275-298.

78   G. Liu, Q. Wang and C. Lin, A Survey of Current Models for Simulating the Contact between Rough Surfaces, *Tribology Transactions*, Vol. 42, 1999, pp. 581-591.

79   M. Chandrasekaran, A.W. Batchelor and N.L. Loh, Direct Observation of Frictional Seizure of Mild Steel Sliding on Aluminium by X-ray Imaging, Part 1, Methods, *Journal of Materials Science*, Vol. 35, 2000, pp. 1589-1596.

80   M. Chandrasekaran, A.W. Batchelor and N.L. Loh, Direct Observation of Frictional Seizure of Mild Steel Sliding on Aluminium by X-ray Imaging, Part 2, Mechanisms, *Journal of Materials Science*, Vol. 35, 2000, pp. 1597-1602.

81   A.A. Seireg, Friction and Lubrication in Mechanical Design, Marcel Dekker Inc., New York, 1998.

82   Y. Fu, A.W. Batchelor and N.L. Loh, Study on Fretting Wear Behavior of Laser Treated Coatings by X-ray Imaging, *Wear*, Vol. 218, 1998, pp. 250-260.

83   D. Dowson, History of Tribology, Longmans Group, 1979, page 25.

84   J.J. Kalker, Three-Dimensional Elastic Bodies in Rolling Contact, Kluwer Academic Publishers, Dordrecht, 1990.

85   J.J. Kalker, A Fast Algorithm for the Simplified Theory of Rolling Contact, *Vehicle System Dynamics*, Vol. 11, 1982, pp. 1-13.

86   J.J. Kalker, The Computation of Three-Dimensional Rolling Contact With Dry Friction, *Int. Journal for Numerical Methods in Engineering*, Vol. 14, 1979, pp. 1293-1307.

87   D. Tabor, The Mechanism of Rolling Friction; II The Elastic Range, *Proc. Roy. Soc., London*, Series A, Vol. 229, 1955, pp. 198-220.

88   W.R. Tyfour, J.H. Breynon and A. Kapoor, The Steady State Behaviour of Pearlitic Rail Steel under Dry Rolling Sliding Contact Conditions, *Wear*, Vol. 180, 1995, pp. 79-89.

89   A. Kapoor, Wear by Plastic Ratchetting, *Wear*, Vol. 212, 1997, pp. 119-130.

90   Y. Uchiyama, Control of Rolling Friction, *The Tribologist, Journal of Japanese Society of Tribologists*, Vol. 44, 1999, pp. 487-492.

91   K. Ohno, Rolling Friction and Control between Wheel and Rail, *The Tribologist, Journal of Japanese Society of Tribologists*, Vol. 44, 1999, pp. 506-511.

92   I. Sekiguchi, Rolling Friction and Control of Polymeric Materials, *The Tribologist, Journal of Japanese Society of Tribologists*, Vol. 44, 1999, pp. 493-499.

93   F.W. Carter, On the Action of a Locomotive Driving Wheel, *Proc. Roy. Soc., London, Series A*, Vol. 112, 1926, pp. 151-157.

94   J.J. Kalker, Wheel Rail Rolling Contact Theory, *Wear*, Vol. 144, 1991, pp. 243-261.

95 K.L. Johnson, Contact Mechanics, Cambridge University Press, Cambridge, 1985.

96 M. Barquins, Adherence, Friction and Wear of Rubber-Like Materials, *Wear*, Vol. 158, 1992, pp. 87-117.

97 K.L. Johnson, K. Kendall and A.D. Roberts, Surface Energy and Contact of Elastic Solids, *Proc. Roy. Soc., London, Series A*, Vol. 324, 1971, pp. 301-313.

98 D. Pupaza and J.H. Beynon, The Use of Vibration Monitoring in Detecting the Initiation and Prediction of Corrugations in Rolling-Sliding Contact Wear, *Wear*, Vol. 177, 1994, pp. 175-183.

99 Y. Suda, Effects of Vibration System and Rolling Conditions on the Development of Corrugations, *Wear*, Vol. 144, 1991, pp. 227-242.

100 E. Tassilly and N. Vincent, Rail Corrugations, Analytical Model and Field Tests, *Wear*, Vol. 144, 1991, pp. 163-178.

101 H.G. Feller and K. Walf, Surface Analysis of Corrugated Rail Treads, *Wear*, Vol. 144, 1991, pp. 153-161.

102 J. Ferrante and J.R. Smith, A Theory of Adhesion at a Bimetallic Interface: Overlap Effects, *Surface Science*, Vol. 38, 1973, pp. 77-92.

103 M. Eriksson, F. Bergman and S. Jacobson, On the Nature of Tribological Contact in Automotive Brakes, *Wear*, Vol. 252, 2002, pp. 26-36.

104 D. Bettege and J. Starcevic, Topographic Properties of the Contact Zones of Wear Surfaces in Disc Brakes, *Wear*, Vol. 254, 2003, pp. 195-202.

105 L. Aferante, M. Ciavarella, P. Decuzzi and G. Demelio, Transient Analysis of Frictionally Excited Thermoelastic Instability in Multi-Disk Clutches and Brakes, *Wear*, Vol. 254, 2003, pp. 136-146.

106 Y.C. Chiou, Y.P. Chang, R.T. Lee, Tribo-Electrification Mechanism for Self-Mated Metals in Dry Severe Wear Process, Part I Pure Hard Metals, *Wear*, Vol. 254, 2003, pp. 606-615.

107 Y.C. Chiou, Y.P. Chang, R.T. Lee, Tribo-Electrification Mechanism for Self-Mated Metals in Dry Severe Wear Process, Part II Pure Soft Metals, *Wear*, Vol. 254, 2003, pp. 616-624.

108 H.H. Gatzen and M. Beck, Investigations on the Friction Force Anisotropy of the Silicon Lattice, *Wear*, Vol. 254, 2003, pp. 1122-1126.

109 P. Podsiadlo and G.W. Stachowiak, Hybrid Fractal-Wavelet Method for Characterization of Tribological Surfaces - a Preliminary Study, *Tribology Letters*, Vol. 13 (4), 2002, pp. 241-250.

110 P. Podsiadlo and G.W. Stachowiak, Multi-scale Representation of Tribological Surfaces, *Proc. Inst. Mech. Engrs., Part J: Journal of Engineering Tribology*, Vol. 216, No. J6, 2002, pp. 463-479.

111 P. Podsiadlo, and G.W. Stachowiak, Fractal-Wavelet Based Classification of Tribological Surfaces, *Wear*, Vol. 254, 2003, pp. 1189-1198.

112 Glossary of Terms, ASM Handbook, Vol., 18, Friction, Lubrication and Wear Technology, ASM International, USA, 1992, p. 21.

11 ABRASIVE, EROSIVE AND CAVITATION WEAR

11.1 INTRODUCTION

Wear by abrasion and erosion are forms of wear caused by contact between a particle and solid material. Abrasive wear is the loss of material by the passage of hard particles over a surface [1]. Erosive wear is caused by the impact of particles against a solid surface. Cavitation is caused by the localized impact of fluid against a surface during the collapse of bubbles. Abrasion and erosion in particular are rapid and severe forms of wear and can result in significant costs if not adequately controlled [2]. Although all three forms of wear share some common features, there are also some fundamental differences, e.g., a particle of liquid can cause erosion but cannot abrade. These differences extend to the practical consideration of materials selection for wear resistance due to the different microscopic mechanisms of wear occurring in abrasion, erosion or cavitation. The questions are: where are abrasive, erosive or cavitation wear likely to occur? When do these forms of wear occur and how can they be recognized? What are the differences and similarities between them? Will the same protective measures, e.g., material reinforcement, be suitable for all these forms of wear? What is the effect of temperature on these wear mechanisms? Will the use of hard materials suppress all or only some of these forms of wear? The practising engineer needs answers to all these questions and more. The fundamental mechanisms involved in these three forms of wear and the protective measures that can be taken against them are discussed in this chapter.

11.2 ABRASIVE WEAR

Abrasive wear occurs whenever a solid object is loaded against particles of a material that have equal or greater hardness. A common example of this problem is the wear of shovels on earth-moving machinery. The extent of abrasive wear is far greater than may be realized. Any material, even if the bulk of it is very soft, may cause abrasive wear if hard particles are present. For example, an organic material, such as sugarcane, is associated with abrasive wear of cane cutters and shredders because of the small fraction of silica present in the plant fibres [3]. A major difficulty in the prevention and control of abrasive wear is that the term 'abrasive wear' does not precisely describe the wear mechanisms involved. There are, in fact, almost always several different mechanisms of wear acting in concert, all of which have different characteristics. The mechanisms of abrasive wear are described next, followed by a review of the various methods of their control.

Mechanisms of Abrasive Wear

It was originally thought that abrasive wear by grits or hard asperities closely resembled cutting by a series of machine tools or a file. However, microscopic examination has revealed that the cutting process is only approximated by the sharpest of grits and many other more indirect mechanisms are involved. The particles or grits may remove material by microcutting, microfracture, pull-out of individual grains [4] or accelerated fatigue by repeated deformations as illustrated in Figure 11.1.

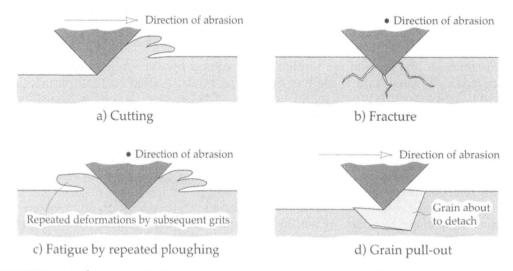

FIGURE 11.1 Mechanisms of abrasive wear: microcutting, fracture, fatigue and grain pull-out.

The first mechanism illustrated in Figure 11.1a, cutting, represents the classic model where a sharp grit or hard asperity cuts the softer surface. The material that is cut is removed as wear debris. When the abraded material is brittle, e.g., ceramic, fracture of the worn surface may occur (Figure 11.1b). In this instance wear debris is the result of crack convergence. When a ductile material is abraded by a blunt grit, then cutting is unlikely and the worn surface is repeatedly deformed (Figure 11.1c). In this case wear debris is the result of metal fatigue. The last mechanism illustrated (Figure 11.1d) represents grain detachment or grain pull-out. This mechanism applies mainly to ceramics where the boundary between grains is relatively weak. In this mechanism the entire grain is lost as wear debris.

Cutting

Much of this more complex view of abrasive wear is relatively new since, like all forms of wear, the mechanisms of abrasive wear are hidden from view by the materials themselves. Until recently, direct demonstrations of abrasive wear mechanisms were virtually non-existent. The development of the Scanning Electron Microscope (SEM) has provided a means of looking at some aspects of abrasive wear in closer detail. In one study [5] a rounded stylus was made to traverse a surface while under observation by SEM. In another study [6] a pin on disc wear rig was constructed to operate inside the SEM, to allow direct observations of wear. Two basic mechanisms were revealed: a cutting mechanism and a wedge build up mechanism with flake-like debris [5]. This latter mechanism, called '**ploughing**', was found to be a less efficient mode of metal removal than '**microcutting**'. In a separate study with a similar apparatus it was found that random plate-like debris were formed by a stylus scratching cast iron [7]. It is probable that in an actual wear situation the effect of cutting alone is relatively small since much more material is lost by a process that has characteristics of both cutting and fatigue.

The presence of a lubricant is also an important factor since it can encourage cutting by abrasive particles [5]. When a lubricant is present, cutting occurs for a smaller ratio of grit penetration to grit diameter than in the unlubricated case. This implies that if a grit is rigidly held, e.g., embedded in a soft metal, and is drawn under load against a harder metal in the presence of a lubricant, then a rapid microcutting form of abrasive wear is more likely to occur than when no lubricant is present.

The geometry of the grit also affects the mechanism of abrasive wear. It has been observed that a stylus with a fractured surface containing many '**microcutting edges**' removes far more material than unfractured pyramidal or spheroidal styluses [8]. Similarly, a grit originating from freshly fractured material has many more microcutting edges than a worn grit which has only rounded edges.

Beneath the surface of the abraded material, considerable plastic deformation occurs [9,10]. This process is illustrated in Figure 11.2.

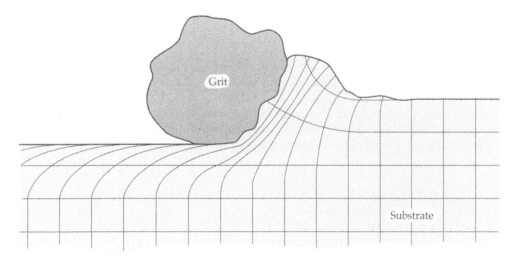

FIGURE 11.2   Subsurface deformation during passage of a grit.

As a result of this subsurface deformation, strain hardening can take place in the material which usually results in a reduction of abrasive wear.

Fracture

Visual evidence of abrasive wear by brittle fracture was found by studying the subsurface crack generation caused by a sharp indenter on a brittle transparent solid [12] as illustrated in Figure 11.3.

Three modes of cracking were found [12]: vent cracks propagating at 30° to the surface, localized fragmentation, and a deep median crack. When grits move successively across the surface, the accumulation of cracks can result in the release of large quantities of material. Brittle fracture is favoured by high loads acting on each grit and sharp edges on the grit, as well as brittleness of the substrate [13]. Since in most cases material hardening has the disadvantage of reducing toughness, it may be possible that a hardened material which resists abrasive wear caused by lightly loaded blunt grits will suddenly wear very rapidly when sharp heavily loaded grits are substituted. Hence a material which is wear resistant against moving, well worn grits (e.g., river sand) might be totally unsuitable in applications which involve sharp edged particles, such as the crushing of freshly fractured quartz.

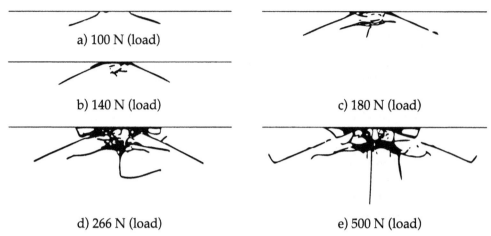

a) 100 N (load)

b) 140 N (load)

c) 180 N (load)

d) 266 N (load)

e) 500 N (load)

FIGURE 11.3   Generation of cracks under an indenter in brittle solids (adapted from [12]).

Fatigue

The repeated strain caused by grits deforming the area on the surface of a material can also cause metal fatigue. Detailed evidence for sideways displacement of material and the subsequent fracture has been found [11]. An example of the sideways material displacement mechanism is given in Figure 11.4 which shows a transverse section of an abrasion groove. Wear by repeated sideways displacement of material would also be a relatively mild or slow form of abrasive wear since repeated deformation is necessary to produce a wear particle.

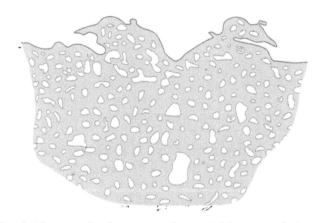

FIGURE 11.4   Example of sideways displacement of material by a grit (adapted from [11]).

Grain Pull-Out

Grain detachment or pull-out is a relatively rare form of wear which is mainly found in ceramics. This mechanism of wear can become extremely rapid when inter-grain bonding is weak and grain size is large.

Modes of Abrasive Wear

The way the grits pass over the worn surface determines the nature of abrasive wear. The literature denotes two basic modes of abrasive wear:

- · two-body and
- · three-body abrasive wear.

Two-body abrasive wear is exemplified by the action of sandpaper on a surface. Hard asperities or rigidly held grits pass over the surface like a cutting tool. In three-body abrasive wear the grits are free to roll as well as slide over the surface, since they are not held rigidly. The two- and three-body modes of abrasive wear are illustrated schematically in Figure 11.5.

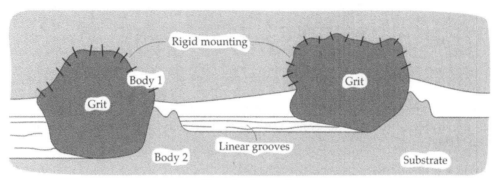

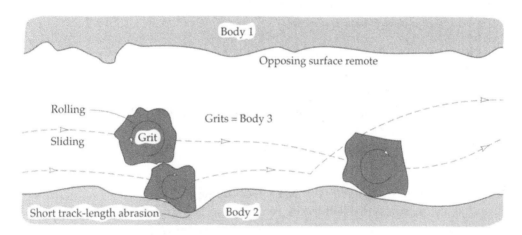

FIGURE 11.5   Two and three-body modes of abrasive wear.

Until recently these two modes of abrasive wear were thought to be very similar; however, some significant differences between them have been revealed [14]. It was found that three-body abrasive wear is ten times slower than two-body wear since it has to compete with other mechanisms such as adhesive wear [15]. Properties such as hardness of the 'backing wheel', which forces the grits onto a particular surface, were found to be important for three-body but not for two-body abrasive wear. Two-body abrasive wear corresponds closely to the 'cutting tool' model of material removal whereas three-body abrasive wear involves slower mechanisms of material removal, though very little is known about the mechanisms involved [16]. It appears that the worn material is not removed by a series of scratches as is the case with two-body abrasive wear. Instead, the worn surface displays a random topography suggesting gradual removal of surface layers by the successive contact of grits [17].

Analytical Models of Abrasive Wear

In one of the simplest and oldest models of abrasive wear a rigidly held grit is modelled by a cone indenting a surface and being traversed along the surface as shown in Figure 11.6. In this model it is assumed that all the material displaced by the cone is lost as wear debris. Although this is a simplistic and inaccurate assumption it is still used because of its analytical convenience.

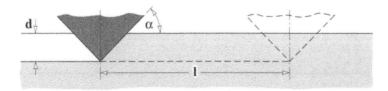

FIGURE 11.6   Model of abrasive wear by a single grit.

In this model of abrasive wear the individual load on the grit is the product of the projected area of the indentation by the cone and the material's yield stress under indentation (hardness) [18], i.e.:

$$W_g = 0.5\pi(d\cot\alpha)^2 H \tag{11.1}$$

where:

    W_g is the individual load on the grit [N];

    d is the depth of indentation [m];

    α is the slope angle of the cone (Figure 11.6);

    H is the material's yield stress under indentation (hardness) [Pa].

The approximate volume of the material removed by the cone is the product of the cross-sectional area of the indentation '$d^2\cot\alpha$' and the traversed distance 'l', i.e.:

$$V_g = ld^2\cot\alpha \tag{11.2}$$

where:

    V_g is the volume of material removed by the cone [m³];

    l is the distance travelled by the cone (Figure 11.6) [m].

Substituting for 'd' from equation (11.1) into equation (11.2) results in an expression for the worn volume of material in terms of the load on the grit, the shape of the grit, and the sliding distance, i.e.:

$$V_g = \frac{2l\tan\alpha}{\pi H} \times W_g \tag{11.3}$$

The total wear is the sum of the individual grit worn volumes of the material:

$$V_{tot} = \Sigma V_g = \frac{2l\tan\alpha}{\pi H} \times \Sigma W_g$$

or:

$$V_{tot} = \frac{2l\tan\alpha}{\pi H} \times W_{tot} \tag{11.4}$$

where:

    V_{tot} is the total wear [m³];

    W_{tot} is the total load [N].

Equation (11.4) assumes that all the material displaced by the cone in a single pass is removed as wear particles. This assumption is dubious since it is the mechanism of abrasive wear which determines the proportion of material removed from the surface. However, equation (11.4) has been used as a measure of the efficiency of abrasion by calculating the ratio of real wear to the wear computed from equation (11.4) [19].

A more elaborate and exact model of two-body abrasive wear has recently been developed [21]. In this model it is recognized that during abrasive wear the material does not simply disappear from the groove gouged in the surface by a grit. Instead, a large proportion of the gouged or abraded material is envisaged as being displaced to the sides of the grit path. If the material is ductile, this displaced portion remains as a pair of walls to the edges of the abrasion groove. An idealized cross section of an abrasion groove in ductile abrasive wear is shown in Figure 11.7.

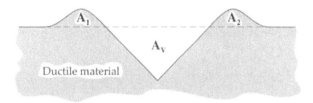

FIGURE 11.7   Model of material removal and displacement in ductile abrasive wear.

A new parameter 'f_{ab}', defined as the ratio of the amount of material removed from the surface by the passage of a grit to the volume of the wear groove is introduced, i.e.:

$$f_{ab} = 1 - (A_1 + A_2) / A_v \tag{11.5}$$

where:

f_{ab} is the ratio of the amount of material removed by the passage of a grit to the volume of the wear groove; $f_{ab} = 1$ for ideal microcutting, $f_{ab} = 0$ for ideal microploughing and $f_{ab} > 1$ for microcracking;

A_v is the cross-sectional area of the wear groove [m²];

$(A_1 + A_2)$ is the cross-sectional area of the material displaced at the edges of the groove (Figure 11.7) when the material is ductile [m²].

The volumetric wear loss 'ΔV_l' in terms of the sliding distance 'l' is given by:

$$\Delta V_l = \Delta V / l = f_{ab} A_v \tag{11.6}$$

where:

ΔV_l is the volumetric wear loss in terms of sliding distance [m²].

The linear wear rate or depth of wear per sliding distance 'l' in the ductile mode is expressed as:

$$\Delta V_{d,ductile} = \Delta V / lA = f_{ab} A_v / A \tag{11.7}$$

where:

ΔV_d is the linear wear rate or depth of wear per sliding distance;

A is the apparent grit contact area [m²]. For example, the apparent contact area in pin-on-disc experiments is the contact area of the pin with the disc.

The ratio of the worn area in true contact with the abrading grits to the apparent area is given by [36]:

$$A_v/A = \phi_1 p / H_{def} \qquad (11.8)$$

where:

ϕ_1 is a factor depending on the shape of the abrasive particles, e.g., the experimentally determined value for particles of pyramidal shape is **0.1**;

p   is the externally applied surface pressure. The pressure is assumed to have a uniform value, e.g., uniformly loaded sandpaper [Pa];

H_{def} is the hardness of the material when highly deformed [Pa].

For ductile materials, a relationship for 'f_{ab}' in terms of the effective deformation on the wearing surface and the limiting deformation of the same material in a particular abrasion system was derived from the principles of plasticity [36], i.e.:

$$f_{ab} = 1 - (\varphi_{lim} / \varphi_s)^{2/\beta} \qquad (11.9)$$

where:

φ_{lim} is the limiting plastic strain of the material in the abrasion system. A value of $\varphi_{lim} \approx 2$ is typical;

φ_s is the effective plastic strain on the wearing surface;

β is a term describing the decline in strain or deformation with depth below the surface. This quantity is mainly influenced by the work-hardening behaviour of the abraded material. Typically $\beta = 1$.

It can be seen from equation (11.9) that the value of the parameter 'f_{ab}' is closely related to material properties but is also dependent on the characteristics of abrasion, e.g., grit sharpness.

For the modelling of abrasive wear of brittle materials the parameter 'f_{ab}' is modified to allow for the tendency of the abraded material to spall at the sides of grooves as shown in Figure 11.8.

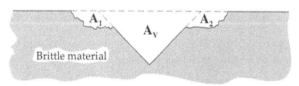

FIGURE 11.8   Model of material removal in brittle abrasive wear.

In this case, the areas 'A_1' and 'A_2' are negative because the brittle material does not pile up at the sides as with ductile material but instead fractures to further widen the groove and the expression for 'f_{ab}' becomes:

$$f_{ab} = 1 + |A_1 + A_2| / A_v \qquad (11.10)$$

The expression for linear wear rate in the brittle mode is given by the expression [21]:

$$\Delta V_{d,brittle} = \phi_1 p / H_{def} + \phi_3 A_f D_{ab} p^{1.5} H^{0.5} \mu^2 \Omega / K_{IC}^2 \qquad (11.11)$$

where:

ϕ_3 is a factor depending on the shape of cracking (additional fracture to the formed grooves, Figure 11.8) during abrasive wear. For pyramidal shape particles $\phi_3 \approx 0.12$;

A_f is the area fraction of material flaws such as brittle lamellae;

D_{ab} is the effective size of the abrasive particles [m]. Typical values are between **30** and **100** [μm];

H_{def} is the hardness of the deformed abraded material [Pa];

H is the hardness of the undeformed abraded material [Pa];

μ is the coefficient of friction at the leading face of the abrasive particles. For the unlubricated condition $\mu = 0.1 - 0.5$ [10];

K_{IC} is the fracture toughness under tension [$m^{0.5}$ Pa];

Ω is a parameter defined as:

$$\Omega = 1 - \exp(- (p/p_{crit})^{0.5}) \qquad (11.12)$$

where:

p_{crit} is the critical surface pressure for any material containing cracks or lamellae of very brittle material [Pa].

In situations where:

$$p \leq p_{crit} \quad \text{then} \quad \Omega = 0$$

The critical surface pressure is given in the form:

$$p_{crit} = \phi_2 \lambda K_{IIC}^2 / (D_{ab}^2 H \mu^2) \qquad (11.13)$$

where:

ϕ_2 is a geometrical factor relating to the effectiveness of the shape of the abrasive particle on abrasive wear. A typical value for a pyramidal shape particle is $\phi_2 \approx 1$;

λ is the mean free path between brittle defects [m], e.g., for martensitic steels $\lambda = 40 - 120$ [μm] is typical;

K_{IIC} is the fracture toughness of the abraded material under shear [$m^{0.5}$Pa]. For example, for tool steel K_{IIC} is between **10** and **20** [$m^{0.5}$MPa] and for nodular cast iron between **30** and **50** [$m^{0.5}$ MPa] [36].

Theoretically the total amount of abrasive wear is equal to the sum of ductile and brittle wear. In most applications, however, either ductile or brittle wear takes place.

From the presented model the limitations of applying hard but brittle materials as abrasion resistant materials are clear. The generally recognized hardness of the material is not the only factor critical for its abrasive wear resistance. The material's toughness is also critical. It can be seen from equation (11.11) that if 'K_{IC}' is small then very large wear rates may result.

In practice, it cannot be assumed that any grit will abrade a surface, i.e., remove material. If the grit is sufficiently blunt then the surface material will deform without generation of wear debris as illustrated in Figure 11.9.

The deformation of a soft surface by hard wedge-shaped asperities has been described by three different models depending on the friction and wear regimes [10].

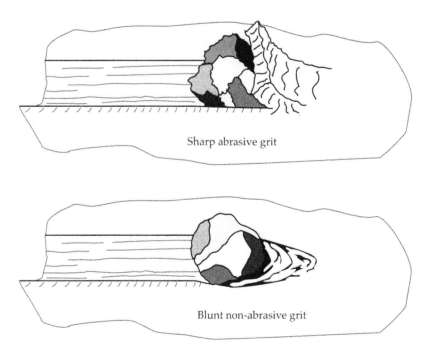

FIGURE 11.9   Cessation of abrasion with increasing grit bluntness.

· <u>Wave formation model (Rubbing model)</u>

In this model, characterized by low friction, a soft surface is plastically deformed, forming a wave which is pushed away by a hard asperity. Wear debris may eventually be formed by fatigue processes. The model applies to smooth surfaces with weak interface between the asperities.

The coefficient of friction in this model is given in the following form:

$$\mu = \frac{A\sin\alpha + \cos(\text{arc cos} f - \alpha)}{A\cos\alpha + \sin(\text{arc cos} f - \alpha)} \tag{11.14}$$

where:

μ is the coefficient of friction ($0 \leq \mu < 1$);

α is the slope angle of the asperity (Figure 11.6);

f   is the coefficient of interfacial adhesion between the asperity and the worn surface. For a dry contact in air 'f' is in the range **0.1 - 0.6** [10];

A   is the coefficient defined as:

$$A = 1 + 0.5\pi + \text{arc cos} f - 2\alpha - 2\text{arc sin}[(1 - f)^{-0.5}\sin\alpha]$$

Equation (11.14) clearly illustrates that the degree of lubrication, which is represented by the 'coefficient of interfacial adhesion', can affect the coefficient of friction.

· <u>Wave removal model (Wear model)</u>

In this model a wave of plastically deformed material is removed from the surface producing wear particles. The process is characterized by high friction and high wear rates. The model applies to smooth surfaces with strong interface between the asperities.

The coefficient of friction associated with this model is given by:

$$\mu = \frac{[1 - 2\sin\beta + (1 - f^2)^{0.5}]\sin\alpha + f\cos\alpha}{[1 - 2\sin\beta + (1 - f^2)^{0.5}]\cos\alpha - f\sin\alpha} \qquad (11.15)$$

where:

β is the coefficient defined as:

$$\beta = \alpha - 0.25\pi - 0.5\text{arc }\cos f + \text{arc }\sin[(1 - f)^{-0.5}\sin\alpha]$$

Chip formation model (Cutting model)

The deformation of a soft material proceeds by a microcutting mechanism and a layer of material is removed as a chip. The model applies to rough surfaces.

The coefficient of friction for this model is in the following form:

$$\mu = \tan(\alpha - 0.25\pi + 0.5\text{arc }\cos f) \qquad (11.16)$$

Calculated values of coefficient of friction for these three models plotted as a function of the slope angle of the asperity 'α' and the coefficient of interfacial adhesion between the asperity and the worn surface '**f**' are shown in Figure 11.10.

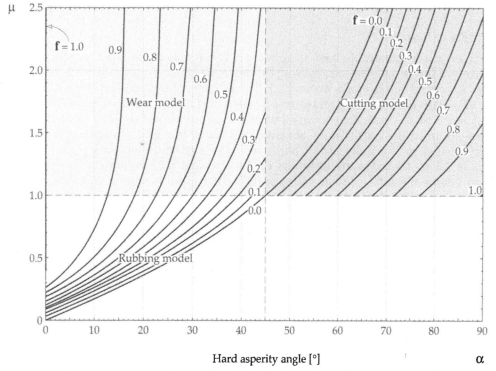

FIGURE 11.10 Variation of coefficient of friction in three models of soft surface deformation by hard wedge-shaped asperities [10].

It can be seen from Figure 11.10 that for a fixed value of the coefficient of interfacial adhesion '**f**' the friction increases with the increasing surface roughness expressed in terms of the asperity slope angle 'α' while for a fixed value of 'α' an increase in '**f**' results in increased friction 'μ' in the rubbing model and decreased friction in the cutting model. This may explain why lubrication (defined by the '**f**' value) under different conditions may inhibit or accelerate abrasive wear. The models presented predict that lubrication inhibits wear for smooth surfaces (low asperity slope angle 'α') and promotes wear for rough surfaces (high asperity slope angle).

The above models indicate that there is no absolute value of 'asperity sharpness' determining abrasion, instead the effect of asperity sharpness in the form of the 'asperity slope angle' is coupled to the coefficient of interfacial adhesion. This means that an asperity which is relatively benign in a lubricating medium may become much more abrasive in a non-lubricated contact.

An attempt was also made to model the brittle mode of abrasive wear [13,20] and some limited agreement with wear data was obtained. The equations developed are highly specialized and show a non-linear dependence of wear rate on grit load, fracture toughness and hardness of the abraded material. A fundamental weakness of this model is that no distinction is made between abrading and non-abrading grits. In essence, the classic assumption of two-body abrasive wear is made, i.e., that all grits are equally sharp and are uniformly loaded against the wearing surface.

As may be surmised, none of the expressions listed in the above models is entirely suitable for the practical prediction of abrasive wear rates. Even in the elaborate model, only the highly controlled situation of ideal two-body abrasive wear by a single grit is analyzed. Modelling of wear rates under complex conditions like three-body abrasive wear, which is one of the most important industrial problems, still remains unattempted.

Abrasivity of Particles

A particle or grit is usually defined as abrasive when it can cause rapid or efficient abrasive wear. In most instances, the hardness of the material must be less than **0.8** of the particle hardness for rapid abrasion to occur [22]. It has been observed, however, that a limited amount of abrasive wear and damage to a surface (e.g., bearing surfaces) still occurs unless the yield stress of the material exceeds that of the abrasive particle [22]. Very slow abrasive wear persists until the hardness of abrasive and worn material are equal. Some materials with soft phases or not fully strain hardened may sustain some wear until the material hardness is 1.2 to 1.4 times the hardness of the abrasive [22]. A conceptual graph of wear resistance versus the ratio of material to abrasive hardness is shown in Figure 11.11. Wear resistance is usually defined as the reciprocal of wear rates and relative wear resistance is defined as the reciprocal of wear rate divided by the reciprocal wear rate of a control material.

Natural minerals vary considerably in hardness and abrasivity. The Vickers hardness of minerals used to define the Mohs scale of hardness have been measured by Tabor [23] and Mott [24]. The hardness of typical minerals given in Mohs and Vickers is listed in Table 11.1 [23-25].

A major problem in deciding whether a mixture of natural minerals such as rock is abrasive or not is the uncertainty in the value of its hardness. There is a natural variability in hardness of rock, especially where a soft matrix contains hard particles. It is also difficult to measure the hardness of rock with standard tests such as Brinell and Rockwell because of the brittleness of the rock. Low load Knoop and Vickers tests seem to give the most reliable results in these cases [114].

Silicon carbide, which is an artificial mineral, has a hardness of 3000 [VHN] (Vickers Hardness Number) or 30 [GPa]. Quartz (1100 [VHN]) and harder minerals are the main cause of abrasive wear problems of tough alloy steels which have a maximum hardness of 800 [VHN]. Quartz is particularly widespread in the form of sand and is perhaps the most common agent of abrasion. The abrasivity of coal is not usually caused by the carbonaceous minerals such as vitrinite which are relatively soft but by contaminant minerals such as pyrites and hematite [25]. Identification of the mineral in the grits which causes the excessive abrasive wear is an important step in the diagnosis and remedy of this phenomenon. On the other hand, minerals which are too soft to abrade, e.g., calcite, may still wear a material, but the mechanisms involved are different, e.g., thermal fatigue [26].

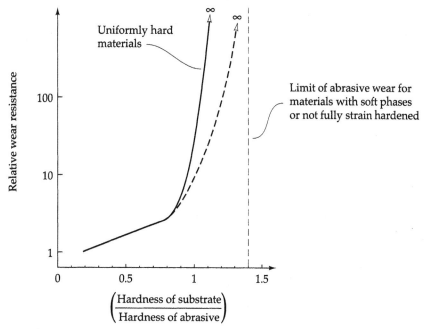

FIGURE 11.11 Relative abrasive wear resistance versus hardness ratio of worn to abrasive material.

TABLE 11.1   Hardness of typical minerals.

Substance	Mohs' scale	Hardness (VHN)
Talc	1	2 – 3
Gypsum	2	36 – 76
Calcite	3	109 – 172
Fluorite	4	190 – 250
Vitrinite (coal constituent)	4 – 5	294
Apatite	5	566 – 850
Orthoclase	6	714 – 795
Hematite	6 – 7	1038
Quartz	7	1103 – 1260
Pyrite (iron sulphide, cubic form)	7 – 8	1500
Marcasite (iron sulphide, orthorhombic form)	7 – 8	1600
Topaz or garnet	8	1200 – 1648
Corundum	9	2060 – 2720
Diamond	10	8000 – 10 000

A more complex constraint is the brittleness of the abrasive. If the grits are too brittle then they may break up into fine particles, thus minimizing wear [2]. If the abrasive is too tough then the grits may not fracture to provide the new cutting faces necessary to cause rapid wear [2,7,8]. The sharp faces of the grits will gradually round-up and the grits will become less efficient abrasive agents than angular particles [27] as illustrated in Figure 11.12.

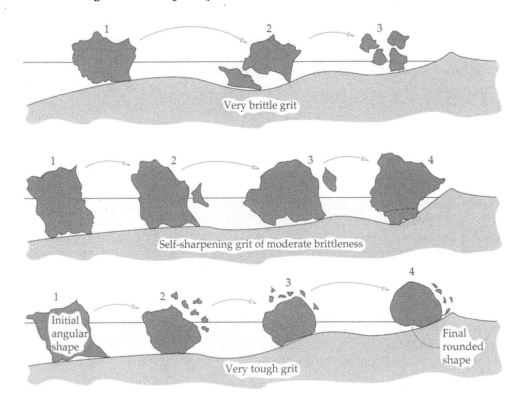

FIGURE 11.12 Effect of grit brittleness and toughness on its efficiency to abrade.

Another factor controlling the abrasivity of a particle is the size and geometry of a grit. The size of a grit is usually defined as the minimum size of a sphere which encloses the entire particle. This quantity can be measured relatively easily by sieving a mineral powder through holes of a known diameter. The geometry of a grit is important in defining how the shape of the particle differs from an ideal sphere and how many edges or corners are present on the grit. The non-sphericity of most particles can be described by a series of radii beginning with the minimum enclosing radius and extending to describe the particle in progressively more detail as shown in Figure 11.13.

Three parameters are identified as significant in grit description: overall grit size or the minimum enclosing diameter, the radance, and the roughness of a particle [27]. The radance is described as the second moment of the radius vector '$R(\theta)$' about the mean radius based on overall cross sectional area. The roughness is defined as the sum of the squares of higher order radii above the fourth order of a corresponding Fourier series divided by the mean radius squared [27]. In other work common abrasives such as SiC, Al_2O_3 and SiO_2 have been characterized using aspect ratio (width/length) and perimeter^2/area shape parameters [63]. It was found that the erosion rate increased with increasing P^2/A and decreasing W/L for these three types of abrasive particles [63].

Recently two new numerical parameters describing the angularity of particles have been introduced [108-110]. One of the parameters, called '**spike parameter - linear fit**' (SP), is based

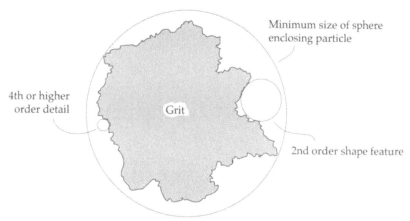

FIGURE 11.13 Method of defining grit geometry by a series of radii.

on representing the particle boundary by a set of triangles constructed at different scales and is calculated in the following manner [108]. A particle boundary is 'walked' around at a fixed step size in a similar manner as used in calculating the boundary fractal dimension [111-113]. The start and the end point at each step is represented by a 'triangle' as illustrated in Figure 11.14a [108,109]. It has been assumed that the sharpness and size of these triangles are directly related to particle abrasivity, i.e., the sharper (smaller apex angle) and larger (perpendicular height) the triangles are the more abrasive is the particle. The sharpness and size of these triangles has been described by a numerical parameter called the 'spike value', i.e.,

$sv = \cos\left(\frac{\theta}{2}\right)h$ (where: 'h' is the perpendicular height of the triangle while 'θ' is the apex angle as shown in Figure 11.14a). For each step around the particle boundary the spike values are calculated for the largest and sharpest triangles. From the spike values obtained a 'spike parameter - linear fit' is calculated according to the following formula [108,109]:

$$SP = \frac{\Sigma\left[\Sigma(sv_{max}/h_{max})\right]/m}{n}$$

(11.17)

where:

sv_{max} is $\max\left[\cos\left(\frac{\theta}{2}\right)h\right]$ for a given step size;

h_{max} is the height at 'sv_{max}';

m   is the number of valid 'sv' for a given step size;

n   is the number of different step sizes used.

The other parameter, called '**spike parameter - quadratic fit**' (SPQ), is based on locating a particle boundary centroid 'O' and the average radius circle [110], as illustrated in Figure 11.14b. The areas outside the circle, 'spikes', are deemed to be the areas of interest while the areas inside the circle are omitted. For each protrusion outside the circle, i.e., 'spike', the local maximum radius is found and this point is treated as the spike's apex [110]. The sides of the 'spike', which are between the points 's-m' and 'm-e', Figure 11.14b, are then represented by fitting quadratic polynomial functions. Differentiating the polynomials at the 'm' point yields the apex angle 'θ' and the spike value 'sv', i.e., $sv=\cos\theta/2$. From the spike values 'spike parameter - quadratic fit' is then calculated according to the formula [110]:

$$SPQ = sv_{average}$$

(11.18)

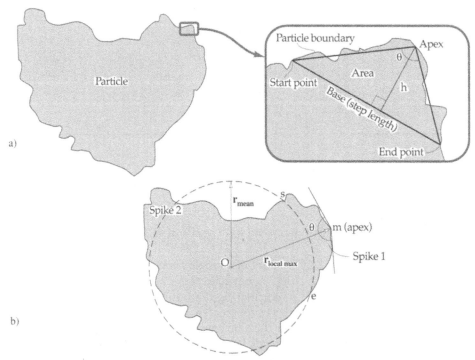

Figure 11.14 Schematic illustration of particle angularity calculation methods of a) 'spike parameter - linear fit' (SP) and b) 'spike parameter - quadratic fit' (SPQ) (adapted from [108 and 110]).

One of the advantages of SPQ over SP is that it considers only the boundary features, i.e., protrusions, which are likely to come in contact with the opposing surface.

It was found that both SP and SPQ correlate well with abrasive wear rates, i.e., two-body, three-body abrasive and erosive wear [109,110,113]. This is illustrated in Figure 11.15 where the abrasive wear rates obtained with chalk counter-samples are plotted against the angularity parameters.

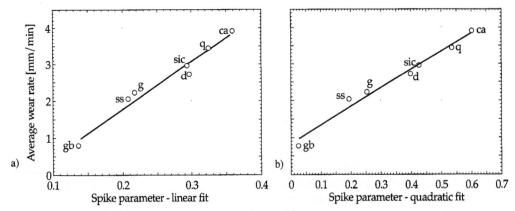

Figure 11.15 Relationship between wear rates and particle angularity described by a) 'spike parameter - linear fit' and b) 'spike parameter - quadratic fit' (SPQ) for different abrasive grit types, i.e., 'gb' - glass beads, 'ss' - silica sand, 'g' - garnet, 'd' - natural industrial diamonds, 'sic' - silicon carbide, 'q' - crushed quartz and 'ca' - crushed sintered alumina (adapted from [108 and 109]).

It has been found that below 10 [μm] diameter the grits are too small to abrade under certain conditions [15,19]. The wear rate of an abrasive for constant contact pressure and other conditions increases non-linearly with grit diameter up to about 50 [μm] and reaches a limiting value with grit diameter of about 100 [μm] for most metals [28]. For polymers at high contact pressures, the wear rate is found to increase with grit diameter up to at least 300 [μm] [28]. Experimental data of these trends are shown in Figure 11.16.

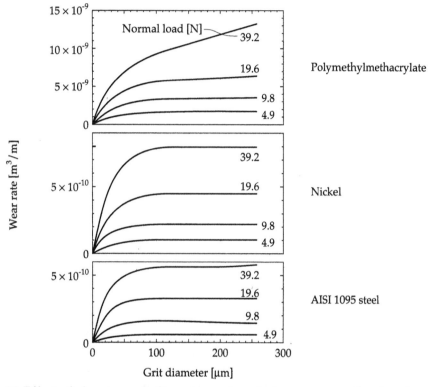

FIGURE 11.16 Effect of abrasive grit diameter and contact pressure on the abrasive wear rate of a polymer (polymethylmethacrylate, PMMA), nickel and AISI 1095 steel [28].

A fundamental limit to the abrasiveness of particles at extremely small grit diameters is the surface energy of the abraded material. As grit size decreases, the proportion of frictional energy used for the creation of a new surface increases. For grits within the typical size range of 5 to 300 [μm], the formation of a new surface consumes less than 0.1% of the energy absorbed by plastic deformation. With extremely fine grits the formation of a new surface would absorb a much larger fraction of the available energy [29].

Abrasive Wear Resistance of Materials

The basis of abrasive wear resistance of materials is hardness and it is generally recognized that hard materials allow slower abrasive wear rates than softer materials. This is supported by experimental data, an example of which is shown in Figure 11.17. The relative abrasive wear resistance for a variety of pure metals and alloys after heat treatment is plotted against the corresponding hardness of the undeformed metal [30-32]. Relative abrasive wear resistance is defined as wear rate of control material/wear rate of test material. A typical control material is EN24 steel [e.g., 30-32]. The abrasive material used in these tests was carborundum with a hardness of 2300 [VHN] and a grit size of 80 [μm]. The tests were conducted in the two-body mode of abrasive wear with a metallic pin worn against a carborundum abrasive paper.

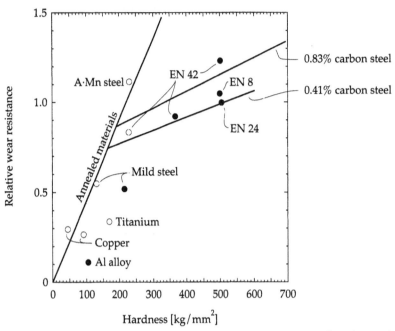

FIGURE 11.17 Relative abrasive wear resistance versus undeformed hardness for pure metals and alloys [adapted from 30-32].

The high hardness of the abrasive ensured that all metals were subjected to rapid abrasive wear with no metal exceeding **0.8** of the abrasive hardness. The mechanism of abrasive wear was mostly microcutting with chip formation clearly observed in most tests [30,31]. Abrasive wear resistance of typical steel alloys was tested in the laboratory on specially designed test rigs and also in field trials. The obtained results, shown in Table 11.2, demonstrate a correlation between the relative wear resistance recorded in the field and that determined by the laboratory tests [32,34]. As a reference EN24 steel of hardness 5100 [MPa] was used.

A proportionality between relative wear resistance and hardness is observed for plastics but the proportionality constant, i.e., relative wear resistance = constant × hardness, is about **3.2** times higher than for metals [35]. The loss of proportionality between hardness and the relative wear rate for hardened metals is the result of defining the wear resistance in terms of the undeformed hardness of the metal. In abrasion, severe subsurface deformation is inevitable and the hardness at high strains is a controlling property. The other factor controlling wear rates is the tendency for material to be displaced rather than removed as wear debris. It was found that the reciprocal of abrasive wear rate 'V_d' when plotted against wear debris hardness 'H_{deb}' divided by the volume loss to the volume of wear groove ratio 'f_{ab}', in the ductile wear mode (eq. 11.5), confirms an approximate linear relationship between the reciprocal of wear rate and H_{deb}/f_{ab} ratio for a variety of ferrous and non-ferrous alloys and pure metals, as shown in Figure 11.18 [21]. The significance of this finding is that the abrasive wear of all metallic materials conforms to the same general relationship between wear rate and material parameters. There seems to be no fundamental distinction between alloyed hardened metals and pure annealed metals.

In assessing the resistance of a material to abrasive wear, it is clearly necessary to consider its hardness at large strains, not the conventional hardness measured at relatively low plastic strains.

TABLE 11.2   Relative wear resistance recorded in the field and in laboratory tests [32,34].

		H [kg/mm²]	Laboratory results			Field results		
			Corundum cloth 180 grit	Quartz paper 40 grit	Quartz paper 180 grit	Light soil ironstone	Light soil flint	Quartz sand soil (Kenya)
Carbon and low-alloy steels								
EN42	0.74%C steel	500	1.13		1.14	1.20	1.14	
	0.74%C steel	650	1.22	1.23	1.58	1.42	1.37	
	0.74%C steel	820	1.53	1.80	2.06	1.76	1.95	2.32
EN8	0.43%C steel	500	1.00			1.05	1.02	
	0.43%C steel	600	1.11	1.17	1.26	1.34	1.34	1.45
EN24	0.37%C, NiCrMo steel	350	0.97		0.94		0.72	0.86
	0.37%C, NiCrMo steel reference	500	1.00	1.00	1.00	1.00	1.00	1.00
Alloy steels								
A·Mn	austenitic manganese steel	220	1.38	1.27	1.60	1.09	1.08	1.39
KE275	0.40%C, 10%W, 3%Cr hot die steel	600	1.39	1.37	2.89	1.66	1.67	2.48
C·Cr	2%C, 14%Cr die steel	700	1.75	1.78	11.7	1.94	2.07	
C·Cr	2%C, 14%Cr die steel	860/900	2.04	3.50	32.6	2.93	3.34	
Cast hard-facing alloys								
Delcrome	3%C, 30%Cr, Fe base	610	2.12	2.25	129	2.28	3.32	9.60
Stellite 1	2.5%C, 33%Cr, 13%W, Co base	630	1.71	2.29	26.9	2.49	4.26	10.3
White cast irons								
NiHard	3%C, 1.7%Cr, 3%Ni	700	1.52	1.50	5.95	1.71	2.50	
W.I.	3.6%C	700	1.53	1.59	4.32	2.32	3.81	

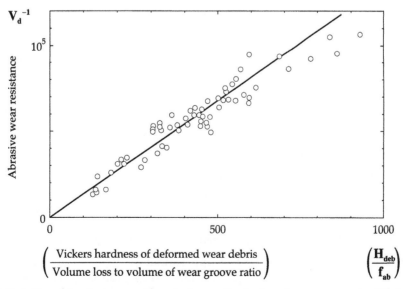

FIGURE 11.18 Relationship between the reciprocal of abrasive wear rate 'V_d' versus wear debris hardness 'H_{deb}' divided by the volume of material loss to the volume of wear groove ratio 'f_{ab}' [21].

Another parameter determining resistance to abrasive wear is the brittleness of a material. This is a major limitation with ceramics [2,4]. If the material cracks during abrasion then rapid wear by fracture of surface layers occurs. Given that brittleness usually increases with hardness, there is a limit to the improvement in abrasive wear resistance that can be achieved by raising hardness.

Compressive residual stresses have also been observed to influence the resistance to abrasive wear. A study of plasma sprayed aluminium oxide coatings with a sealant of aluminium phosphate showed that compressive residual stress correlated well with increased resistance to wear by quartz sand during a rubber wheel abrasion test (ASTM G65) [115].

· *Abrasive Wear Resistance of Steels*

The abrasive wear resistance of steels can be considerably enhanced by judicious selection of hardness and metallurgical phase. Selection of a steel depends on the hardness of the abrasive. For example, if the abrasive is relatively soft, i.e., hardness is less than 1000 [VHN], then it is possible to select a steel of hardness that would be greater than **0.8 × hardness of the abrasive** and quenched martensite with a hardness of approximately 800 [VHN] would be suitable. Unfortunately this approach finds relatively few applications because most abrasives present in natural minerals are harder than 1000 [VHN] and it is often necessary to choose the metallurgical phase which exhibits the greatest resistance to wear by a 'hard' abrasive. So in this case, abrasive wear resistance is not quite synonymous with hardness. The abrasive wear resistance of a steel to a hard abrasive is determined by the relative proportions of austenite, bainite, martensite, pearlite and ferrite and by the presence of cementite. A general result of many different tests is that austenite and bainite, which are softer than martensite, are more resistant to abrasive wear by a hard abrasive [36]. This superior wear resistance is believed to be a result of the greater ductility and toughness of austenite and bainite which suppress the more rapid forms of abrasive wear, such as microcutting and brittle fracture [36]. Austenitic steels function by forming a tough work-hardened layer under conditions of heavy abrasion [32,34,37,38] which can only be removed from the surface with difficulty.

For low-alloy plain carbon steels, the influence of metallurgical phase depends on whether a hyper-eutectoid or hypo-eutectoid steel is selected. For hypo-eutectoid steels, bainite is the most abrasion resistant phase, with tempered martensite and ferrite/pearlite offering successively less wear resistance [39]. For hyper-eutectoid steels the presence and morphology of cementite (iron carbide) inclusions have the dominant influence. With the higher carbon content, the annealed microstructure is superior to hardened (martensitic) hyper-eutectoid steels [39]. The cause of this reversal of wear resistance is the inhibition of abrasion grooves by hard carbide inclusions. The morphology of the carbide inclusions is critical to abrasive wear resistance. The most wear resistant microstructure contains lamellar cementite inclusions of the pearlitic form. When the cementite is present as spherical inclusions, there is less improvement in wear resistance because the spheres do not provide rigid barriers to plastic deformation. This distinction is illustrated in Figure 11.19.

Spherical inclusions of iron carbide can, however, improve the abrasive wear resistance of a steel by raising the yield stress of the steel according to the Hall-Petch effect [40]. If the size of the grits is small compared to the carbide inclusions, there is an additional improvement in wear resistance provided by the direct blockage of abrasion grooves by hard inclusions as illustrated in Figure 11.20. This process is known as the 'stand-out effect'.

The stand-out effect becomes significant at approximately 10% volume fraction of pearlite content [41]. For coarse abrasives, the wear resistance rises gradually with increasing pearlite content while the wear resistance for smaller grits increases sharply at 10% volume fraction. A schematic representation of wear resistance versus volume fraction of pearlite is shown in Figure 11.21.

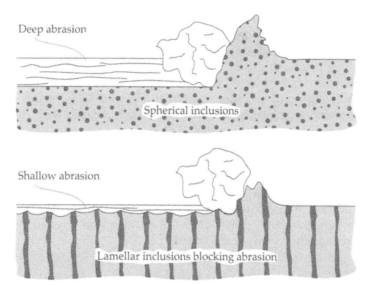

FIGURE 11.19 Influence of carbide inclusion morphology on the abrasion process.

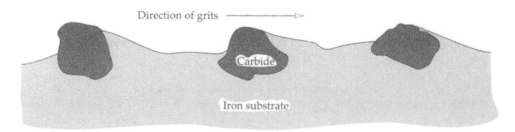

FIGURE 11.20 'Stand-out effect' or inhibition by large carbide inclusions of abrasion by small grits.

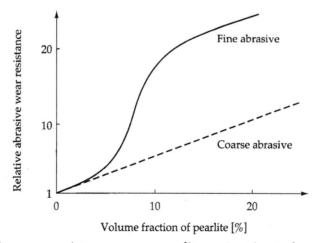

FIGURE 11.21 Abrasive wear resistance versus pearlite content in steels.

Alloying of steels with elements such as chromium, manganese and nickel results in considerable improvements in abrasive wear resistance. The classic abrasion resistant steel is **'Hadfield's'** manganese steel with a composition of 12% Mn and 1.2% C [42]. The high manganese content allows for a virtually pure austenitic steel with a trace of martensite to form [43]. Hadfield's steel is tough as well as abrasion resistant and therefore is particularly suitable for situations where rocks as well as grits impact on the wearing surface. Other abrasion resistant compositions are 0.55 - 0.65% C, 0.8 - 1.5% Cr known as '**1% chromium steel**' and '**NiHard**' which is 0.5% Si, 3 - 4% C, 2 - 4% Ni and 1 - 2% Cr [44]. Where the carbon content of an alloy is high and carbides have been allowed to form during heat treatment, additional abrasive wear resistance is provided by carbide inclusions by the same mechanism as pearlite. Alloying elements useful for this purpose are chromium and molybdenum since the carbides obtained are extremely hard. The hardness of chromium carbide is about 1300 [VHN] and that of molybdenum carbide is about 1500 [VHN]. On the other hand, elements which form relatively soft carbides, e.g., nickel and manganese, should be avoided as these can accentuate abrasive wear [2]. A steel containing carbides can possess up to four times the abrasive wear resistance of the corresponding carbide free steel [2]. The peak of abrasive resistance occurs at approximately 30% volume of carbide and beyond this level brittleness appears to cause the reduction in wear resistance.

· *Abrasive Wear Resistance of Polymers and Rubbers*

As mentioned already polymers, although soft, can have a surprisingly high degree of resistance to abrasive wear compared with a metal of the same hardness [35]. The relative abrasive wear resistance of plastics and soft metals is shown in Figure 11.22 [35]. The superior durability of polymers can generally be attributed to their very high resistance to abrasion by blunt grits [45] as compared to metals, and their inability to fracture grits to produce fresh sharp edges.

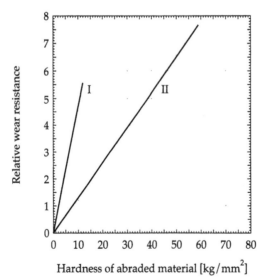

FIGURE 11.22 Abrasive wear resistances for plastics and metals of similar hardness; (I) plastics (e.g., L54, L68, nylon 6, low-pressure polyethylene, high-pressure polyethylene, polyfluoroethylene), (II) metals (e.g., silver, zinc, cadmium, lead) [35].

Abrasive wear properties of plastics can be strongly affected by additives such as fillers and plasticizers. Usually an optimum level of a filler compound is found which gives the minimum wear as demonstrated in Figure 11.23 [35].

Plasticizer was found to have a detrimental effect on the abrasive wear resistance of PVC [35] since it softens the polymer. The abrasive wear resistance of polymers, however, is radically altered by the presence of glass fibres, etc., which form composites and this will be discussed in the chapter on 'Wear of Non-Metallic Materials'.

Pure abrasive wear of plastics is generally thought to be less common than with metals [7,45]. The grits tend to wear plastics by indentation fatigue which is a much slower wear process.

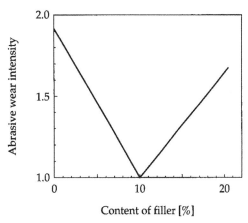

FIGURE 11.23 Effect of filler (titanium dioxide) on the abrasive wear intensity of a plastic (polypropylene) [35].

Two-body and three-body modes of abrasive wear are also very different for plastics. With two-body abrasion, when sandpaper is used, the wear rate is linearly proportional to load, but with three-body abrasion the wear of plastics has a non-linear dependence on load [35]. The reasons for this variation are still unclear.

With rubber there is a further mechanism of wear occurring during abrasion which involves molecular degradation [46]. Rubber sustains very large strains during abrasion and this can cause chain-scission of the polymer molecules inside the rubber. The broken ends of the molecule are highly reactive since they have become chemical radicals. They rapidly combine with oxygen to form oxidation products. The interaction between these radicals and oxygen is similar to the degradation of mineral oil discussed in Chapter 3. The degradation products form a fine 'oily' dust which is a characteristic feature of rubber abrasion. This strain-induced degradation mechanism in rubber is illustrated in Figure 11.24.

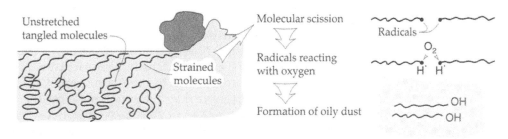

FIGURE 11.24 Formation of powdery rubber degradation products by abrasion-induced strain in rubber.

· Abrasive Wear Resistance of Ceramics

Ceramic materials are in general extremely hard and therefore should possess good abrasive wear resistance. An example of a hard ceramic mineral is alumina which reaches a Mohs hardness of **9** in the form of corundum. Field tests on agricultural ploughs fitted with alumina surfaces demonstrated greatly reduced wear in comparison to conventional steel ploughs [47]. Chipping by impacting stones, however, resulted in accelerated wear close to the cutting edge of the alumina plough. This implied that a tougher ceramic would be more useful for this application. Brittleness is a limiting factor in the abrasive wear of alumina and the dominant mechanism of abrasion is by grain detachment as shown in Figure 11.1d [4]. The development of ceramics resistant to abrasive wear involves the application of composite ceramics with enhanced toughness. For example, alumina blended with zirconia shows an increase in toughness and this can result in increased wear resistance [48]. Ceramic-matrix composites containing metal fibres to raise toughness have also been found to have superior abrasive wear resistance to the pure ceramic [49].

Effect of Temperature on Abrasive Wear

The effect of temperature on abrasive wear can be divided into:

- the influence of ambient temperature,
- the role of temperature rises induced by plastic deformation of the worn material on contact with grits.

The effects caused by these forms of heating are not similar. The influence of elevated ambient temperature on abrasive wear has scarcely been studied, probably due to experimental difficulties. Some limited tests of the abrasive wear of copper and aluminium showed only a small increase in wear at temperatures up to 400°C for copper and no effect for aluminium [50]. With the temperature increase there is a corresponding decline in the hardness of both the worn material and the abrasive grit. This trend was recorded in experiments conducted up to temperatures of 2000°C where most metals and metallic carbides and nitrides showed the same proportional decline in hardness with temperature. It was found that when a temperature of about '**0.8 × melting point**' was reached the hardness of most materials was negligible, although non-metallic minerals such as silicon nitride and silicon carbide maintain their hardness until very close to the melting point [51]. When considering the effect of temperature on the abrasive wear of steel by, for example, silica (quartz) and alumina (corundum), the melting points of these materials, steel ~1500°C, quartz 1710°C and alumina 2045°C, become relevant and must be considered. As temperature is raised, the ratio of abrasive hardness to steel hardness increases more sharply for alumina than for quartz. Alumina is therefore expected to cause more severe high temperature abrasive wear of steel than quartz. This prediction, however, still remains to be tested experimentally.

The temperature increase caused by plastic deformation during abrasion is associated with high grit speeds [52]. Dynamic thermocouple measurements with an electrically conductive abrasive reveal that temperatures as high as 1000°C can be reached during abrasion [53]. The critical difference between the effects of a temperature rise in the worn metal imposed by high grit speeds and changes in ambient temperature is that the grits remain relatively cool due to the transient nature of abrasion. Contact between a grit and the worn surface would be particularly short in the three-body abrasive wear mode, so that any heat generated in the deformed material would not diffuse into the grit. It is possible then that transient thermal softening occurs only in the deformed material while the grit remains with its hardness virtually unaltered. The localization of deformation heat during high speed abrasion is illustrated in Figure 11.25.

If the grit remains relatively cool during abrasion it also maintains its hardness while the worn material effectively softens. Thus at high grit speeds, soft minerals begin to wear hard materials significantly. An example of this phenomenon is the wear of steel by coal free of hard contaminants [52]. The speed dependent softening effect is reduced at high temperatures because of the reduced strain energy of deformation [50]. The effect of high temperature is to soften a material so that there is less local heating of the deformed material for a given amount of deformation.

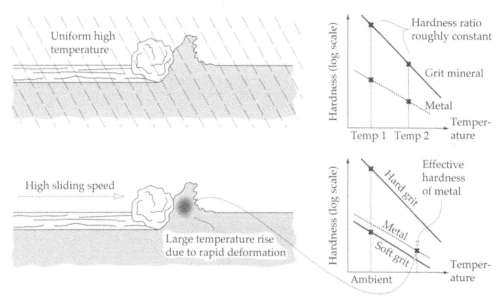

FIGURE 11.25 Temperature effects on abrasion under uniformly hot conditions and under high rates of frictional energy released during rapid abrasion.

Another effect of high temperatures is to cause a form of wear which depends on the combined action of oxidation and removal of oxide layers by abrasion. The oxidation of steels in air is much more rapid at 600°C than at 20°C [54], and as temperature rises, the removal of steel as oxide becomes more significant. The detailed mechanism of oxidative-abrasive wear is discussed in the chapter on 'Corrosive and Oxidative Wear'.

Effect of Moisture on Abrasive Wear

Moisture has a strong influence on abrasive wear rates. Usually abrasive wear rates increase with moisture content in the atmosphere but there are occasions when a contrary effect occurs [2]. Prediction of the moisture effect for any particular case is difficult. The grit may either be just sufficiently weakened by moisture to produce a larger number of new cutting edges, or severe grit weakening may occur causing disintegration of the grits into non-abrasive, fine particles. The worn material may also be weakened by moisture, e.g., glass [55]. For the same abrasive and worn material, two-body abrasive wear may increase with humidity while the three-body abrasive wear rate may either increase or decrease. The data summarizing the effects of water and humidity on the wear of some selected materials is shown in Table 11.3 [2].

Water may also introduce corrosive agents into the abrasive wear system, e.g., dilute acids. This causes a combined corrosive-abrasive wear [56] which has certain fundamental similarities with oxidative-abrasive wear mentioned previously.

Non-aqueous fluids such as lubricants can also affect abrasive wear. When stearic acid is applied as a lubricant to a three-body abrasive wear system, the abrasion of the harder of the two metal surfaces is increased [14]. The mechanism responsible for this may be that the abrasive is preferentially embedded in the softer material and wears the harder material by microcutting when lubrication is effective. When lubrication is absent, the slower ploughing form of abrasion predominates.

TABLE 11.3   Water and humidity effects on abrasive wear (RH - relative humidity) [2].

Material and wear conditions		Percentage change with respect to 'dry' wear		
		50% RH	100% RH	Wet
Aluminium alloy 6063-T6:	Fixed SiC abrasive	+ 20	+ 10	− 10
1040 steel, pearlitic:	Fixed SiC abrasive Loose SiC abrasive, 3-body	+ 5 + 20	+ 10 + 175	0
Pyrex glass:	Fixed SiC abrasive Loose SiC abrasive, 3-body	+ 15 − 12	+ 30 + 220	0
Sintered alumina:	Fixed SiC abrasive Loose Al_2O_3 abrasive, 2-body			− 99 + 300
WC/6-10%Co:	Loose Al_2O_3 abrasive, 2-body Cutting sandstone			+ 54 − 36
Flexible PVC:	Loose SiO_2 abrasive, 2-body			+ 200
Nylon (polyamide) 6.6:	Loose SiO_2 abrasive, 2-body			− 58

Control of Abrasive Wear

Since abrasive wear is the most rapid form of wear and causes the largest costs to industry, several methods have been developed to minimize the losses incurred. The basic method of abrasive wear control or suppression is to raise the hardness of the worn surface until its value is at least **0.8** of the grit hardness. No other form of wear allows such a simple rationale for its prevention. There are of course complications such as the prevention of brittleness while raising the hardness which can be overcome only to a certain degree.

Abrasive wear is usually suppressed by the application of a hard material or hard coating. Most of the hard materials are more expensive than the customary materials so the first question to be answered is, what is the nature of the problem caused by abrasive wear? If the issue is survival of the worn part against gross wear, e.g., soles of shoes, then the choice of abrasion resistant material is determined by the cost of the replacement. With industrial machinery, however, small amounts of abrasive can severely affect its overall performance, e.g., in hydraulic systems. The assessment of performance losses imposed by abrasive wear can often be impossible to quantify or may require very elaborate testing. An example of this problem is the gradual wear of sugarcane shredder hammers by silica from the sugarcane [57]. Sugarcane millers observed that small amounts of wear caused the hammers to become rounded and prevented the cane from being properly 'shredded' before 'crushing' to extract sugar. In other words, the wear of the hammers caused the sugarcane millers to lose a certain amount of sugar. The problem of wear was solved by replacing the hardened steel hammers with tungsten carbide. Since the hardness of tungsten carbide is about 1100 [VHN] (or 11 [GPa]) it effectively resisted abrasive wear by the prevailing silica which has a hardness of about 1150 [VHN] (11.5 [GPa]). The extended maintenance free period of the shredder hammers and the improved cane preparation quality justified the extra expense of using hard tungsten carbide, which is five times the cost of steel.

It has to be mentioned, however, that with the adoption of tungsten carbide sugarcane shredding hammers, magnetic separators had to be employed to remove 'tramp iron' (worn cane cutting blades and other iron-ware thrown into the sugarcane by apathetic farmers). Before this protection was introduced, tungsten carbide hammers were immediately shattered at the first ingress of contaminant metal and their life expectancy was low [57,58].

Hard surface coatings are becoming more widely used as a convenient means of suppressing abrasive wear. The thin layers of coating can be deposited onto any steel component and this allows economy in the use of expensive materials. The fabrication of wear-resistant components is also simplified since materials such as tungsten carbide are hard to machine or weld and also have some other undesirable features such as brittleness, i.e., they shatter on impact. Wear resistant coatings are discussed in more detail in Chapter 9 on 'Solid Lubricants and Surface Treatments'.

11.3   EROSIVE WEAR

Erosive wear is caused by the impact of particles of solid or liquid against the surface of an object. Erosive wear occurs in a wide variety of machinery and typical examples are the damage to gas turbine blades when an aircraft flies through dust clouds and the wear of pump impellers in mineral slurry processing systems. In common with other forms of wear, mechanical strength does not guarantee wear resistance and a detailed study of material characteristics is required for wear minimization. The properties of the eroding particle are also significant and are increasingly being recognized as a relevant parameter in the control of this type of wear.

Mechanisms of Erosive Wear

Erosive wear involves several wear mechanisms which are largely controlled by the particle material, the angle of impingement, the impact velocity and the particle size. If the particle is hard and solid then it is possible that a process similar to abrasive wear will occur. Where liquid particles are the erodent, abrasion does not take place and the wear mechanisms involved are the result of repetitive stresses on impact.

The term 'erosive wear' refers to an unspecified number of wear mechanisms which occur when relatively small particles impact against mechanical components. This definition is empirical by nature and relates more to practical considerations than to any fundamental understanding of wear. The known mechanisms of erosive wear are illustrated in Figure 11.26.

The angle of impingement is the angle between the eroded surface and the trajectory of the particle immediately before impact as shown in Figure 11.27. A low angle of impingement favours wear processes similar to abrasion because the particles tend to track across the worn surface after impact. A high angle of impingement causes wear mechanisms which are typical of erosion.

The speed of the erosive particle has a very strong effect on the wear process. If the speed is very low then stresses at impact are insufficient for plastic deformation to occur and wear proceeds by surface fatigue. When the speed is increased to, for example, 20 [m/s] it is possible for the eroded material to deform plastically on particle impact. In this regime, which is quite common for many engineering components, wear may occur by repetitive plastic deformation. If the eroding particles are blunt or spherical then thin plates of worn material form on the worn surface as a result of extreme plastic deformation. If the particles are sharp then cutting or brittle fragmentation is more likely. Brittle materials, on the other hand, wear by subsurface cracking. At very high particle speeds melting of the impacted surface might even occur [59].

The size of the particle is also of considerable relevance and most of the erosive wear problems involve particles between **5** and **500** [μm] in size, although there is no fundamental

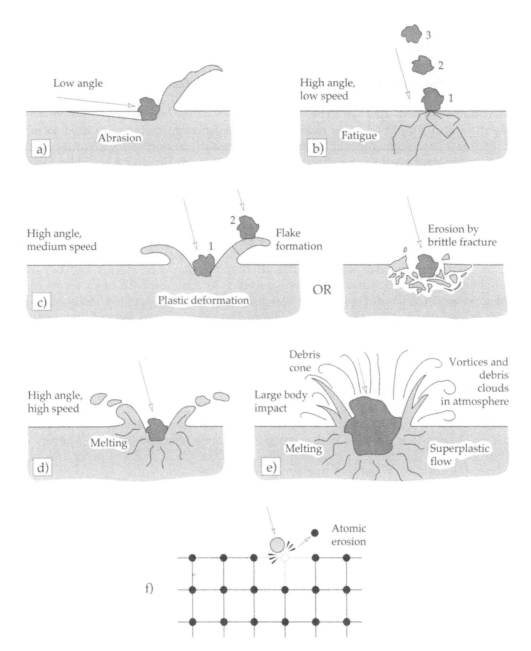

FIGURE 11.26 Possible mechanisms of erosion; a) abrasion at low impact angles, b) surface fatigue during low speed, high impingement angle impact, c) brittle fracture or multiple plastic deformation during medium speed, large impingement angle impact, d) surface melting at high impact speeds, e) macroscopic erosion with secondary effects, f) crystal lattice degradation from impact by atoms.

reason why eroding particles should be limited to this size range. A low earth orbit (LEO) satellite provides an example of erosive wear by minute particles. The satellite is subject to erosion by impacting oxygen and nitrogen atoms from the outer atmosphere [60] and this eventually causes degradation of the satellite casing. In space, there are also innumerable meteorites which 'erode' any larger asteroid or moon [61]. For both material degradation in the LEO satellites and planetary meteorite bombardment, impact speeds of eroding particles are very high and the specific wear mechanism is different from what is usually understood by erosive wear. During impact by atmospheric atoms, the crystal lattice of the bombarded material is degraded to form an eroded structure. In erosion by meteorites, the large size and speed result in a macroscopic damage process where effects such as the eddying of the atmosphere around the impact site are also significant.

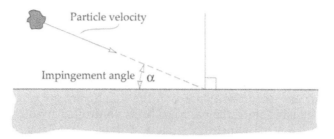

FIGURE 11.27 Impingement angle of a particle causing erosion of surface.

Effect of Impingement Angle and Impact Speed on Erosive Wear Rate

Impingement angles can range from 0° to 90°. At zero impingement angle there is negligible wear because the eroding particles do not impact the surface, although even at relatively small impingement angles of about 20°, severe wear may occur if the particles are hard and the surface is soft. Wear similar to abrasive wear prevails under these conditions. If the surface is brittle then severe wear by fragmentation of the surface may occur, reaching its maximum rate at impact angles close to 90°. The relationship between wear rate and impingement angle for ductile and brittle materials is shown in Figure 11.28

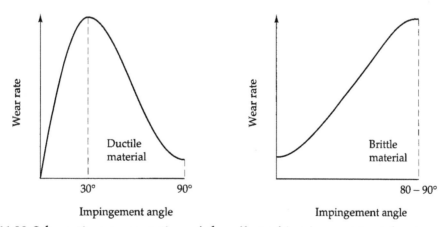

FIGURE 11.28 Schematic representation of the effect of impingement angle on wear rates of ductile and brittle materials.

In cases when erosion shows a maximum at low impingement angles, it is concluded that the **'ductile mode of erosive wear'** prevails. Conversely if the maximum is found at high impingement angles then the **'brittle mode'** is assumed.

The impact speed of the particle has a very strong effect on wear rate. There is often a threshold velocity below which wear is negligibly small. For medium to high speeds covering most practical problems, the relationship between wear rate and impact velocity can be described by a power law, i.e.:

$$-dm / dt = kv^n \qquad\qquad (11.19)$$

where:

m	is the mass of the worn specimen (negative, since wear involves mass loss) [kg];
t	is the duration of the process [s];
k	is an empirical constant;
v	is the impact velocity [m/s];
n	is a velocity exponent.

The value of the exponent 'n' is usually in the range between **2** and **3** for solid particles which is slightly in excess of any prediction based on the kinetic energy of the particles. Equation (11.19) is not comprehensive since the value of 'k' is controlled by other parameters such as particle density and shape for which no analytical data is available. It is one of the early equations used to demonstrate the effect of velocity on wear rate, e.g., as particle speed increases **10** times the wear rate can increase between **100** and **1000** times.

Effect of Particle Shape, Hardness, Size and Flux Rates on Erosive Wear Rate

Particle characteristics are an important but relatively poorly researched aspect of the erosion problem. It is known that hard particles cause higher wear rates than soft particles [62]. The sharpness of the particle has also been recognized as accelerating erosive wear [63,113]. Both of these parameters have been included in numerical models of erosive wear [64,103]. The ratio of particle hardness to substrate hardness seems to be a controlling parameter [64]. The significance of particle hardness becomes apparent when the hardness of some erosives, e.g., alumina, are compared to that of standard materials such as mild steel. In this instance the ratio of particle to substrate hardness is about **10**. The effect of particle hardness on wear depends on the particular mode of erosive wear taking place, e.g., ductile or brittle. In the brittle mode the effect of particle hardness is much more pronounced than in the ductile mode [64].

It is impossible to isolate hardness completely from other features of the particle such as its shape. Even if the particle is hard but relatively blunt then it is unlikely to cause severe erosive wear. A blunt particle has a mostly curved surface approximating to a spherical shape while a sharp particle consists of flat areas joined by corners with small radii which are critical to the process of wear.

Variations in particle size in the range typical of engineering applications can cause fundamental changes in the erosion mechanism. A series of erosion tests on glass, steel, graphite and ceramics revealed that as particle size was increased from 8.75 [μm] to 127 [μm] in diameter the mode of erosion changed from ductile to brittle. This caused the erosive wear peak to move from about a 30° to about an 80° impingement angle and even more significantly resulted in a dramatic increase in erosive wear rates as shown in Figure 11.29 [65]. In both cases silicon carbide impinging at a speed of 152 [m/s] was used as the erosive agent.

It can also be seen from Figure 11.29 that particle size not only affects the wear rate but drastically alters the ranking of materials in terms of wear resistance. When the small particles were used as the erosive agent the materials ranked according to their wear resistance are in the following order: high density alumina > annealed aluminium > plate glass > high density magnesia > graphite and hardened steel. In this case, apart from the annealed aluminium, erosive wear rate depends on the hardness of the material. Work hardening of the aluminium could be significant in this instance. On the other hand, when the large particles were used as the erosive agent, the order changes to annealed aluminium

> hardened steel > high density alumina > high density magnesia > plate glass > graphite. So in this case toughness of the material is important. Materials which are neither tough nor hard, e.g., graphite, show inferior erosion resistance.

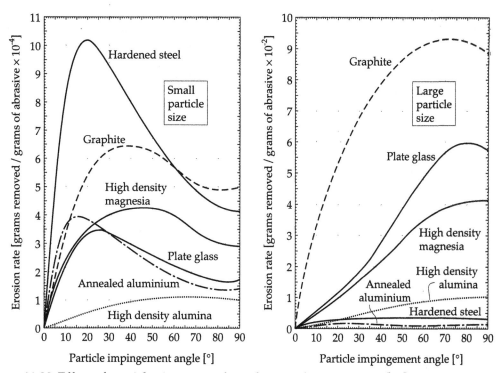

FIGURE 11.29 Effect of particle size on mode and rates of erosive wear [65].

The change in wear modes is believed to be a consequence of the average spacing of defects, e.g., holes or cracks in a solid. If the impinging particles are very small then only a minority of the impingement sites will coincide with a defect. The impingement site is a zone of highly stressed material directly beneath the particle on impact and similar in size to the particle. Plastic deformation is encouraged by an absence of defects and is the predominant mode of metal removal for small particles. Since repeated plastic deformation is required to remove material, this form of wear is relatively slow. For larger eroding particles, a defect is almost always present in the impingement site and material removal by brittle processes is therefore favoured. Since crack formation is rapid the brittle mode of erosion can be a very destructive form of wear.

The particle flux rate, or the mass of impacting material per unit area and time, is another controlling parameter of erosive wear rates. Erosive wear rate is proportional to the flux rate up to a certain limiting value of wear. This limit has been observed in many studies and is believed to be the result of interference between rebounding particles and arriving particles. The limiting particle flux rate is highly variable, ranging from as low as **100** [kg/m²s] for elastomers to as high as **10,000** [kg/m²s] for erosion against metals by large and fast particles [66]. It is also possible for wear rates to decrease marginally when the limiting flux is exceeded.

The incubation period of erosive wear refers to the period of time from the start of erosion to the onset of measurable positive wear. During the incubation period, wear may either be negligible or may appear to be negative. This latter characteristic is caused by eroding particles becoming trapped in the worn material. The incubation period is generally believed to relate to the accumulation of subsurface damage, e.g., cracks or strained material which are the precursors of wear particle release. Once the incubation period has passed, wear usually proceeds at a constant rate.

Erosive Wear by Liquid

Liquid can cause as much erosion damage as solids provided that impact velocities are sufficiently high. A prime example of this problem is damage to aeroplanes flying through clouds or turbine blades in wet steam. A series of elegant experiments conducted by Bowden and Brunton [67] revealed the basic mechanism of liquid erosion. In these experiments, cylindrical droplets of water were propelled with very high velocity at a target. High speed photography enabled observations of events at impact to reveal the transient formation of shock waves within the liquid projectile. The shock waves allow for release of the impact pressure. A high impact pressure is sustained until the shock or pressure relief waves have passed through the liquid. In Figure 11.30 a conceptual diagram of the fluid particle (cylindrical in shape) impacting the surface and the resulting impact force-time history is shown.

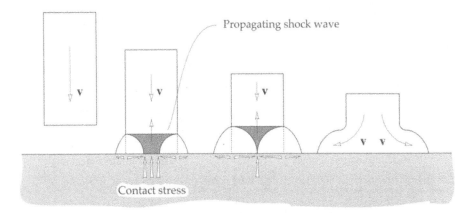

FIGURE 11.30 Erosion mechanism by liquid particles on a solid surface.

The contact pressure on impact can be estimated from the following formula:

$$\mathbf{p} = \rho \mathbf{v_s v} \qquad\qquad (11.20)$$

where:

  $\mathbf{p}$ is the contact pressure on impact [Pa];

  ρ is the density of the fluid [kg/m³];

  $\mathbf{v_s}$ is the speed of sound in fluid [m/s];

  $\mathbf{v}$ is the fluid velocity at impact [m/s].

The contact pressures generated by impacting droplets of fluid can be quite high. For example, for a water droplet impinging at a velocity of **250** [m/s] the impact pressure is:

$$\mathbf{p} = 1000 \times 1500 \times 250 = 375 \ [\mathbf{MPa}]$$

For a water droplet impinging at **1000** [m/s] the estimated pressure rise reaches the extremely high value of **1.5** [GPa].

The duration of the impact pressure is determined by the speed at which pressure release waves reach the centre of the droplet. These pressure waves move at the speed of sound, and for a **3** [mm] diameter water droplet the duration of the impact is about **1** [μs] (the speed of sound in water is **1500** [m/s]).

Wear is a result of a series of transient contact stress pulses in the impacted material. The mechanism of wear depends on the liquid velocity. At low velocities, the worn material is firstly roughened uniformly, with the subsequent formation of random craters. Lips form at the edge of the craters which may then be removed by later impacts. At high velocities holes or pits are formed in the worn material by impacting droplets. If a brittle material is involved, wear by fracture may occur.

Erosive wear by water drops is a major cause of wear in steam-turbine blades and the fuselage of supersonic aircraft when the impingement velocity exceeds 250 [m/s] and the droplet diameter is greater than 200 [μm]. Impact by droplets larger than 800 [μm] in diameter results in large impact forces. On uncoated steels, a surface layer of austenite without carbides is found to provide good wear resistance to liquid erosion. The subsequent transformation of austenite to martensite at the surface, which generates a surface compressive residual stress, also helps increase the wear resistance [116]. High Velocity Oxygen Fuel (HVOF) sprayed coatings are useful at lower values of impact velocity while at high impact velocities the brittleness of HVOF coatings becomes a limiting factor [116].

Chlorinated water widely used by the industry may affect the erosive wear rates. For example HVOF coatings, developed for corrosion-erosion resistance against chloride-contaminated steam and water droplets, showed significantly accelerated wear due to the oxidization of carbides. It was found that nickel based, chromium alloyed HVOF coatings were the most effective [117]. The rate of erosion-corrosion in boiler equipment working with chloride contaminated water can be so high that the boiler temperature has to be reduced to ensure the adequate lifetime of boiler components [117].

Early studies have shown that erosive wear resistance is proportional to material toughness, hence UHMWPE erodes more slowly than polyester resin [68]. As can be seen from equation (11.19) the dependence of wear rate on impact velocity is extreme. The value of the exponent '**n**' in equation (11.19) for erosive wear by liquid particles is typically between **4** and **6** for metals and polymers but reaches **12** for glass. An incubation period of wear may also occur during which material loss is negligible. The length of the incubation period (which is never very long) is inversely proportional to impact velocity. Most studies conducted are related to erosion by water and there has been a limited amount of work on other fluids. A high density of fluid is believed to promote wear. An example is tetrachloromethane (CCl_4) which has a density of approximately 1700 [kg/m^3] and causes more rapid erosive wear than water [69].

Effect of Temperature on Erosive Wear

The rate and mechanism of erosive wear are influenced by temperature. The primary effect of temperature is to soften the eroded material and increase wear rates. The effects of temperature on erosion of stainless steel are shown in Figure 11.31 [70]. The erosive agent is silicon carbide impinging stainless steel at a speed of 30 [m/s] in a nitrogen atmosphere.

It is not until temperatures higher than 600°C are reached that the erosion rate shows significant increase. This temperature coincides with the softening point of the steel. There is a strong correlation between the mechanical properties of the material at the temperature of erosion and wear rate as shown in Figure 11.32 [71].

When high temperature erosion of metals occurs in an oxidizing medium, corrosion can take place and further accelerate wear. Material is removed from the eroding surface as a relatively brittle oxide and this process of wear can be far more rapid than the erosion of ductile metal. At sufficiently high temperatures, however, the underlying metal does not come into contact with the impinging particles because of the thick oxide layer present [72] and then oxidation rates, not mechanical properties, control the erosive wear. Corrosion and oxidation accelerated wear is further discussed in the chapter on 'Corrosive and Oxidative Wear'.

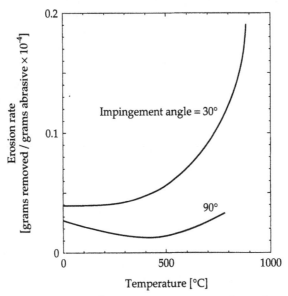

FIGURE 11.31 Effect of temperature on the erosive wear rate of stainless steel [70].

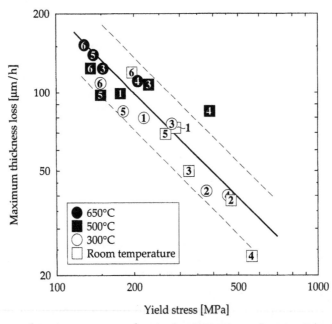

FIGURE 11.32 Relationship between mechanical properties of materials and erosion rate at elevated temperatures: 1) carbon steel, 2) 1.25Cr-1Mo-V steel, 3) 2.25Cr-1Mo steel, 4) 12Cr-1Mo-V steel, 5) 304 steel and 6) alloy 800 [71].

Effect of Erosion Media on Erosive Wear

Most erosive agents are conveyed by a medium, e.g., water or air. A mixture of erosive particles and liquid medium is known as a slurry. The characteristics of the medium have a surprisingly strong effect on the final wear rate. Controlling factors relate to the bulk properties of the medium, i.e., viscosity, density and turbulence, and to its microscopic properties such as corrosivity and lubrication capacity. It has been shown that small additions of lubricants to erosive slurries can significantly reduce wear [73,74]. The ability of the liquid medium to provide cooling during particle impingement is also important [73,74].

In terms of bulk properties, the drag forces imposed by a viscous slurry on the erosive particles can affect wear by altering the impingement angle. This is demonstrated schematically in Figure 11.33 [75].

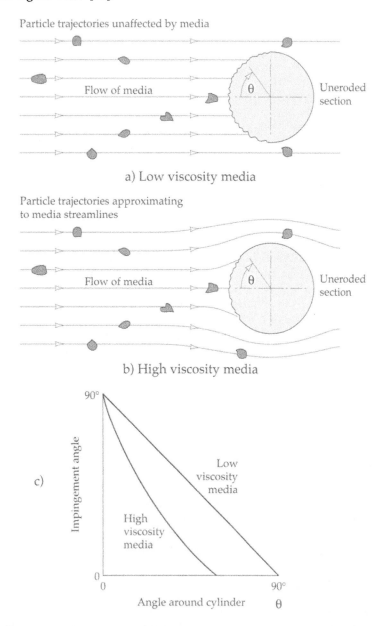

FIGURE 11.33 Effect of medium on impingement angle by erosive particles [75].

It can be seen that the increased particle drag forces imposed by the more viscous medium shift particle impingement to the sides of the eroding cylinder. The effect of the medium is to alter the location and form of wear since the impingement angle is reduced by the shift to the cylinder sides. The medium-induced reduction in impingement angle causes an increase in abrasion-type mechanisms of erosive wear. If an estimation of wear rates in a real machine is required then a comprehensive analysis of particle trajectories is essential. For example, an analysis performed for the inlet blades of a gas turbine gave an excellent agreement between predicted and actual location of wear spots [76]. An example of erosive particle trajectories between gas turbine blades is shown in Figure 11.34 [76].

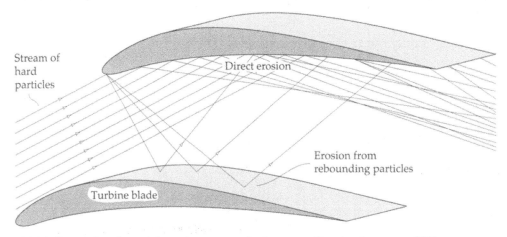

FIGURE 11.34 Example of particle trajectory analysis to predict erosive wear [76].

The effect of a medium is assessed in terms of the 'collision efficiency' which is the ratio of particles that actually hit a wearing surface to the theoretical number of particle impacts in the absence of any medium. It was found that the collision efficiency declines from a limiting value of **1** for large particles, e.g., **750** [μm] size, to less than **0.1** for small particles of **75 - 90** [μm] size at medium viscosities of **0.005** [Pas] [77]. The reduction in collision efficiency is due to the viscous medium sweeping the particles past the wearing surface as shown in Figure 11.35. The erosive wear rate was found to closely follow the same trend as collision efficiency which indicates that the primary effect of a liquid medium is to divert particles from the wearing surface. Increasing particle velocity reduces the influence of medium, so that at high slurry velocities, only large particles are affected by the medium's viscosity [77].

Turbulence of the medium accelerates erosive wear as particle impingement is more likely to occur in turbulent flow than in laminar flow where the medium tends to draw the particles parallel to the surface [78]. The difference between particle behaviour in laminar and turbulent flow of the medium is illustrated in Figure 11.35.

An exception to this rule is where the laminar flow is directed normally to the surface which is the case when a jet of fluid impinges against a surface. In this case, wear is concentrated directly beneath the jet and a relatively unworn annular area surrounds the wear scar. This phenomenon is known as the 'halo effect'. The effect of increasing turbulence with distance from the jet is outweighed by the concentration of erosion directly beneath the jet [78].

Erosive Wear Resistance of Materials

Material characteristics exert a strong effect on erosive wear and have been extensively studied. In a similar manner to abrasive wear, it is found that improvements in mechanical

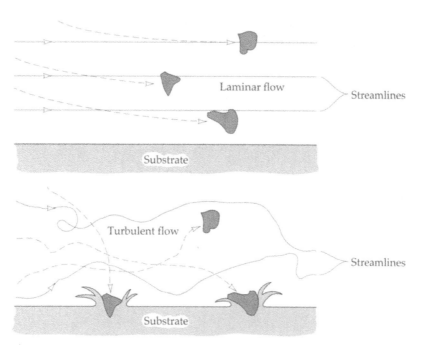

FIGURE 11.35 Effect of flow on erosive wear.

properties do not always coincide with superior erosive wear resistance. For example, erosive wear rates may increase when a material is deliberately hardened. The difficulty with materials optimization for wear reduction is that the characteristics of erosive wear as well as the material characteristics control the wear rate. An illustration of this rule is provided by the comparison of the relative erosion resistance of metals as a function of impingement angle. When the impingement angle is shallow, a hardened steel shows lower wear than a soft steel; the converse is true at high impingement angles. This is illustrated in Figure 11.36 where the erosive wear rate, at two different impingement angles of 15° and 90°, is shown as a function of material hardness for various metals and grades of steel hardness [36,79]. The abrasive used was silicon carbide of diameter about 1 [mm] impinging at a velocity of 30 [m/s].

At the shallow impingement angle, it is evident that the hardness and work-hardening ability of materials suppress a quasi-abrasive process of wear. In this case, materials can be rated according to the hardness of the pure metal. It can be seen from Figure 11.36 that at an impingement angle of 15° the most wear resistant metal is cobalt while the second worst is copper. When the impingement angle is 90° the ranking of materials changes significantly, and copper has the second best while cobalt has the third worst wear resistance. Heat treatment of steel to increase hardness improves erosive wear resistance at low impact angles but lessens the erosive wear resistance at high impact angles. To summarize, the effects of small differences in, for example, hardness or alloy content between similar materials cannot be viewed in isolation from the overall system characteristics of erosive wear. In order to define a material's erosive wear resistance it is only useful to consider broad classes of materials, e.g., polymers, ceramics and metals, where distinctive differences are present and are not obscured by the effects of variables such as velocity or impingement angle. There is no general recipe for a high level of erosive wear resistance. Because of the two different erosive wear protection mechanisms that can take place, high wear resistance can be achieved by more than one type of material. In some cases the material can be extremely hard and tough so that the impacting particle is unable to make any impression on the surface. This is the approach adapted when developing metallic or ceramic erosion resistant materials. Alternatively, the material can be tough but with an extremely low elastic modulus so that the kinetic energy of the particles is harmlessly dissipated. These contrasting wear protection mechanisms are illustrated in Figure 11.37.

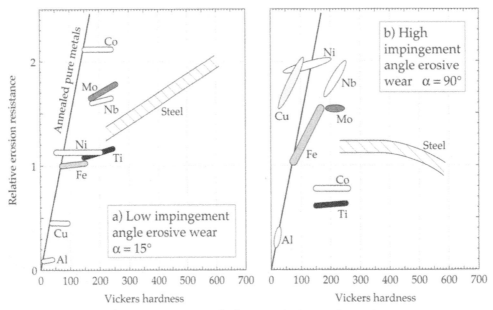

FIGURE 11.36 Effects of primary material characteristics and erosion parameters on erosive wear rate [36,79].

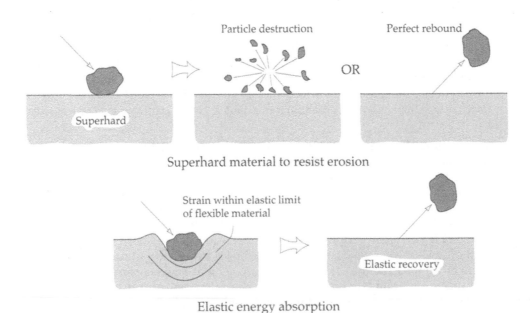

FIGURE 11.37 Comparison of the high and low elastic modulus modes of erosive wear protection.

Rubber is generally believed to provide good erosion resistance by elastic absorption of particle energy although this has not been demonstrated experimentally. It has been shown that the first particle impact causes no visible damage to a rubber surface and that wear depends on slow fatigue processes [80]. Unfilled rubber shows good erosive wear resistance but surprisingly is not resistant to abrasive wear [80].

The choice of erosion resistant material may also be compromised by other considerations such as operating temperature or material transparency. Clearly, temperatures in excess of 200°C preclude polymers from service, but if a transparent material is required for a specific application then metals are not particularly useful. For example, materials for aircraft windscreens, apart from being transparent, are required to be resistant to high speed erosion by sand, dust and rain [81]. It was found that polymethylmethacrylate was the best candidate for this application since it is both tough and shows a minimum of transparency loss by erosion damage.

The relative merits and demerits of metals, polymers and ceramics as erosive wear resistant materials are summarized in Table 11.4.

· Erosive Wear Resistance of Steels

The literature available on the effect of steel microstructure on erosive wear rates suggests that a ductile steel is the most wear resistant. Hardening of steel to form martensite offers little improvement except at very low impingement angles, and the formation of massive or lamellar carbides reduces erosive wear resistance. The selection of steel for erosive wear minimization is therefore different from the case of abrasive wear. For low alloy carbon steels, the ferritic phase with sufficient spheroidal carbide inclusions to induce strengthening is very effective against erosive wear [82]. Pearlitic steels show inferior wear resistance to spheroidized steels. It was found that the erosive wear of steels shows the classical ductile erosion characteristic, i.e., a maximum wear rate at a low impingement angle of 30°, with subsurface and surface cracking [82]. This suggests that the erosive wear resistance of steels is limited by a lack of ductility.

TABLE 11.4   Relative qualities of erosive wear resistant materials.

Material	Relative qualities regarding erosive wear resistance
Metals	Large range of toughness and hardness to suit any particle or impingement angle. Prone to high temperature corrosion and softening effects; corrosive media also harmful.
Ceramics	Very hard and increasingly tougher grades available. Resistant to high temperatures and corrosive media. Poor erosive wear resistance when brittle mode prevails.
Polymers	Tough polymers and rubbers provide good erosion resistance even in corrosive media. Usage is restricted however by a relatively low temperature limit.

For very soft erosive particles such as coal, the inclusion of carbides promotes wear resistance slightly [25]. Alloying of steel or cast iron to obtain a microstructure containing a significant amount of retained austenite is an effective means of reducing erosive wear [83,84]. Adding about 2.5 wt% of silicon to 0.7 wt% carbon steel or about 0.45 wt% of silicon to 2.54 wt% cast iron results in good erosive wear resistance [83,84].

The optimum heat treatment of this steel or cast iron includes a relatively long austempering time where all the martensite is removed and only retained austenite and bainitic ferrite are present. As a general rule, however, ductility rather than hardness should be enhanced in steels for improved erosive wear resistance.

· *Erosive Wear Resistance of Polymers*

Polymers are gaining importance as erosive wear resistant materials for applications where metals are unsuitable, e.g., where transparency to visible light or other radiation is required. The erosive wear resistance of polymers is generally poorer than that of steel. In Figure 11.38 erosive wear rates of reinforced polymers such as chopped graphite fibre in a thermoplastic polyphenylene sulphide (PPS), woven aramid fibre in reinforced epoxy laminate (Kevlar 49/epoxy), graphite fibre (T-300) in bismaleamide polyamide resin and carbon steel (AISI 1018) at an impingement velocity of 31 [m/s] are shown [85].

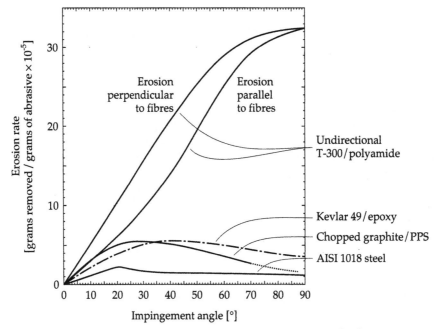

FIGURE 11.38 Erosive wear of reinforced polymers and carbon steel [85].

It can be seen from Figure 11.38 that the polymers showing the brittle mode of erosive wear characterized by high wear rates at high impingement angles are considerably inferior to steel. On the other hand, the erosive wear resistance of polymers eroding in a ductile mode is comparable to that of steel. There is, however, no consistent correlation between ductility and erosion resistance for polymers. For example, nylon erodes in the ductile mode but has poor erosive wear resistance [86]. The ranking of commonly used polymers in terms of their erosive wear resistance is as follows: polyurethane > fluorocarbon > polycarbonate > polymethylmethacrylate > nylon [86,87].

Apart from the direct effect of wear which is the formation of a wear scar, lateral displacement and rippling of a polymer are also possible. This effect is particularly pronounced at low impingement angles around 30°. The mechanism of rippling and lateral displacement of an eroded polymer is illustrated in Figure 11.39.

Another erosive wear characteristic of polymers is that a long wear 'incubation period' is typical where even a weight gain may be recorded. This is due to eroding particles becoming embedded in the much softer polymer [86].

The erosion of certain polymers, in particular elastomers, may be accelerated by oxidation and other forms of chemical degradation [66]. Water and gases are present on the surfaces of hydrophilic particles. Many common minerals, e.g., silica or sand, are hydrophilic. During impact of these particles with rubber, the water or oxygen on their surface will react with the rubber. Chemical reaction is facilitated by the temperature rise which occurs on impact which causes the formation of a mechanically weak surface layer on the rubber. This process of chemical degradation is further enhanced if there are relatively long periods of time between successive impacts at the same position, i.e., low levels of erosive particle flux. In such cases the average reaction time for surface degradation is longer as the temperature rise on impact persists for some time afterwards. With increasing erosive particle flux the ratio of wear mass to eroding particle mass decreases. This decline in erosive wear intensity is noticeable even at low levels of erosive particle flux of approximately 1 [kg/m²s] [66]. The mechanism of chemical degradation in rubber during the process of erosive wear is illustrated in Figure 11.40.

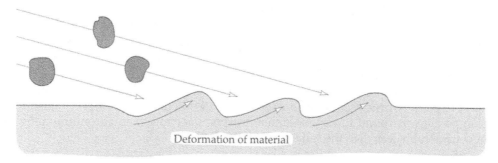

FIGURE 11.39 Rippling and lateral displacement of a polymer during erosive wear at low impingement angles.

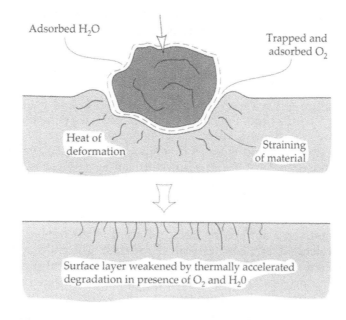

FIGURE 11.40 Chemical degradation and formation of a weakened surface layer (on rubber) induced by the impact of an eroding particle.

· *Erosive Wear of Ceramics and Cermets*

Ceramics are finding use as erosive wear resistant materials particularly at high temperatures where common metals either fail or show inferior wear resistance. The principal disadvantage of ceramic materials for this application is their brittleness which may result in accelerated wear in certain cases. Ceramics such as alumina, partially stabilized zirconia, zirconia toughened alumina, silicon nitride and silicon carbide have been studied for their erosive wear resistance. It was found that oxide ceramics such as alumina, zirconia and zirconia toughened alumina appear to have the higher erosive wear resistance compared to silicon nitride and carbide [88]. Partially stabilized zirconia, however, does not have a significantly higher erosive wear resistance in comparison to alumina ceramics despite its higher bulk toughness [89]. Cermets consisting of tungsten carbide grains in a cobalt binder matrix are also used for erosive wear resistance. In these materials, preferential wear of the cobalt binder appears to be the rate determining factor whereas the tungsten carbide grains are relatively durable against erosive wear [90]. Unlike abrasive wear, during erosive wear the harder carbide grains do not shield the softer cobalt matrix from impacting particles [90].

An important application of ceramics and ceramic composites as erosive wear resistant materials is their use at high temperatures. Metallic materials such as steel are often more wear resistant than ceramics at ambient temperatures but are inferior at high temperatures. At elevated temperatures, metals become excessively soft while ceramic become more ductile which suppresses the brittle mode of erosive wear. A silicon carbide fibre - silicon carbide matrix composite was found to have a higher erosive wear rate than chromium alloy steel at 25°C but considerably less than the same steel at 850°C [91].

11.4 CAVITATION WEAR

Cavitation wear is known to damage equipment such as propellers or turbine blades operating in wet steam, and valve seats. Wear progresses by the formation of a series of holes or pits in the surface exposed to cavitation. The entire machine component can be destroyed by this process. Operation of equipment, e.g., propellers, is often limited by severe vibration caused by cavitation damage.

Mechanism of Cavitation Wear

The characteristic feature of cavitation is the cyclic formation and collapse of bubbles on a solid surface in contact with a fluid. Bubble formation is caused by the release of dissolved gas from the liquid where it sustains a near-zero or negative pressure. Negative pressures are likely to occur when flow of liquid enters a diverging geometry, i.e., emerging from a small diameter pipe to a large diameter pipe. The down-stream face of a sharp sided object moving in liquids, e.g., ship propeller, is particularly prone to cavitation. The ideal method of preventing cavitation is to avoid negative pressures close to surfaces, but in practice this is usually impossible.

When a bubble collapses on a surface the liquid adjacent to the bubble is at first accelerated and then sharply decelerated as it collides with the surface. The collision between liquid and solid generates large stresses which can damage the solid. Transient pressures as high as **1.5** [GPa] are possible. The process of bubble collapse together with experimental evidence of a hole formed in a metal surface by bubble collapse are shown in Figure 11.41 [92].

The cavitation crater, shown in Figure 11.41, was produced on the surface of indium which is soft. Harder materials such as ceramics are unlikely to form a deep hole under the same conditions. Cracking and spallation are the predominant modes of wear for hard brittle materials. Almost all materials suffer some kind of subsurface damage by cavitation and accumulated work-hardening and crack formation are commonly observed [93]. In some cases when the cavitation is intense, the density of holes may be sufficient to reduce the worn material to a porous matrix or 'sponge'. Although cavitation involves a similar process of collision between a liquid and a solid as occurs in erosion by liquids there are some significant differences. Cavitation wear is a much milder process than erosive wear. In cavitation wear particles are detached per millions of cavitations whereas only a few thousand impacts by droplets are enough to cause erosive wear [94]. Cavitation wear has an 'incubation period' like erosive wear but the weight gain found in erosive wear is not possible unless the cavitated material absorbs liquid.

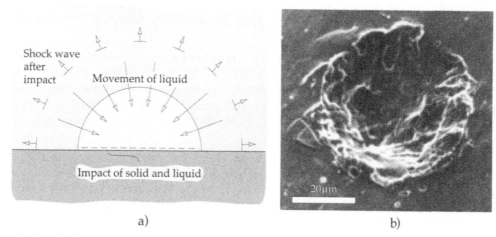

Shock wave after impact

Movement of liquid

Impact of solid and liquid

20μm

a)   b)

FIGURE 11.41 Mechanism of cavitation wear: a) mechanism of bubble collapse and b) experimental evidence of damage by cavitation to a metallic (indium) surface [92].

Cavitation wear is strongly controlled by surface tension of liquid and its reduction may significantly reduce the rate of wear [118]. The comparatively high surface tension of water promotes large bubble formation, a higher impact energy during the collapse of a bubble and less likelihood of a large bubble collapsing into smaller bubbles. When the surface tension of water is reduced from 70 to 30 [N/m] by the addition of wetting agents, the cavitation wear rate sharply declines [118].

When cavitation occurs in corrosive media, stress corrosion cracking may accentuate the wear process. An example of this can be found in the difference in cavitation wear rates between fresh and salt water [95].

Cavitation wear can be accelerated by the simultaneous occurrence of erosive wear, in other words synergistic interaction between these two wear mechanisms is possible. If the cavitating fluid contains erosive particles, then the collapsing bubbles cause the particles to hit the worn surface at high velocity. The rate of wear is higher than either cavitation or erosion alone. For example, this phenomenon takes place in hydraulic turbines operating in sandy water [96].

Cavitation wear is not always entirely undesirable as it finds some unique applications in medical treatment. Kidney stones were traditionally removed by surgery which is inevitably painful and has a certain risk of post operative complications. Extracorporeal shock wave lithotripsy (ESWL) allows destruction and removal of the stones without the need for surgical intervention [104]. ESWL involves strong ultrasonic vibrations which cause intense cavitation around the stones (which are immersed in urine). Wear of the kidney stones proceeds by the repeated impact of high-speed microjets which occur during the collapse of cavitation bubbles near the surface of the stone [105,106]. After a sufficient period of treatment, the stones are reduced to a fine powder which can be excreted by the patient. The destruction rate of kidney stones depends on the material of which they are composed. For example, stones composed of calcium apatite, magnesium ammonium phosphate and calcium oxalate are brittle and rapidly fracture under cavitation [107]. On the other hand, cavitation wear of stones composed of cystine is much slower.

Cavitation Wear Resistance of Materials

A basic determinant in the choice of material for protection against cavitation wear is the physical scale of the device where the cavitation takes place. Cavitation can occur in components ranging from propellers to dam spillways. For large-scale structures, concrete based materials are often used, e.g., concrete reinforced with chopped steel fibres, polymer impregnated concrete or concrete coated with epoxy resin. More information on these materials can be found elsewhere [97].

A material with good cavitation wear resistance is rubber since its low modulus of elasticity allows the bubble collapse energy to be dissipated harmlessly. However, rubber loses its effectiveness at extremes of cavitation intensity. There are applications where rubber is unsuitable for other reasons, e.g., high temperatures. Epoxy resins are used as a coating for components vulnerable to cavitation but these are also ineffective at high intensity cavitation.

A basic feature of cavitation is its preferential attack on the weakest phase of a material. An example of this is found in the significance of graphite inclusions on the cavitation wear of cast iron. The graphite inclusions provide the required crack initiation centres for rapid wear by brittle fracture [98]. A similar process affects cermets which often contain a hard material such as tungsten carbide particles surrounded by a softer metallic matrix. Cavitation can dislodge the tungsten carbide by gradual removal of the surrounding matrix. Thus the improvement in wear resistance is dependent on the properties of the binder rather than the tungsten carbide [99]. Materials which protect against cavitation usually have a uniform microstructure with an absence of large mechanical differences between phases. The mechanism of cavitation wear in multi-phase materials is schematically illustrated in Figure 11.42.

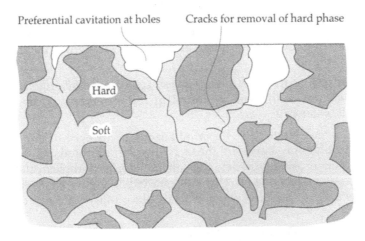

Preferential cavitation at holes   Cracks for removal of hard phase

Hard

Soft

FIGURE 11.42 Preferential attack by cavitation of the weaker phase in a microstructure.

Of the ferrous metals, stainless steels are more resistant to cavitation than cast iron. Plain carbon steels are not often considered as materials providing protection against cavitation because most cavitation problems occur in water. With stainless steels, the ferrite phase is inferior to the austenite phase and the martensitic phase has the best resistance. Austenitic chromium-manganese-nitrogen (Cr-Mn-N) stainless steels have been found to offer better resistance to cavitation than conventional chromium-nickel-molybdenum stainless steels. The additional benefit is that the substitution of manganese and nitrogen as alloying elements reduces the cost of the stainless steel. It is believed that the pronounced work-hardening characteristics of the austenitic (but not ferritic) Cr-Mn-N stainless steels promotes the resistance to cavitatation wear [119]. Hadfield's steel or the manganese steels provide the best cavitation resistance of the austenitic steels. Of the more recently developed materials, high-nitrogen austenitic steels (HNAS) offer attractively high levels of resistance to cavitation wear. HNAS are characterized by low carbon contents and nitrogen content exceeding 0.4%, while mechanical properties combine high yield strength with good ductility and toughness. Vigorous strain hardening of these steels under bubble impact was attributed as the cause for the high resistance to cavitation wear by HNAS [120].

Where cast iron must be used, for example in cylinder liners, the level of free carbon and to a lesser extent free carbide should be minimized. Most ceramic materials appear to lack sufficient toughness and do not show particularly good cavitation resistance characteristics despite their high hardness [100]. Of the bearing metals, the descending order of cavitation resistance is leaded bronze, tin-based white metal, Cu-Pb alloy and lead-based white metal [101]. This particular order is pre-determined by the presence of lead as a matrix material. Because of its softness lead has very inferior cavitation resistance. There has only been quite limited research work conducted on the cavitation resistance of non-ferrous metals. Corrosion resistant titanium alloys have a cavitation resistance similar to that of stainless steels [100]. For severe cavitation problems, cobalt alloys such as 'Stellite' are particularly useful. It should be mentioned that cobalt is more wear resistant to cavitation than to erosion [100]. In ductile materials plastic deformation occurs under cavitation and key characteristics of plastic deformation are dynamic recovery and work-hardening. Materials with high Stacking Fault Energy (SFE) display rapid dynamic recovery and poor work-hardening characteristics. For alpha aluminium bronzes, at least, SFE is inversely related to cavitation wear resistance [121]. This indicates that for alpha aluminium bronzes, plastic deformation under bubble collapse is significant and that the energy of bubble collapse is not entirely absorbed by plastic deformation.

One of the fundamental characteristics of cavitation wear is a fatigue-type damage process which allows some useful comparisons of the relative wear resistances of metals based on metal fatigue theory [102]. It is found that the cavitation wear rates of a range of pure metals correlate well with a fatigue strength parameter which is the product of the nominal fatigue failure stress at zero cycles and the exponent of stress increase during cyclic plastic strain. The nominal fatigue stress at zero cycles is found by extrapolation from experimentally observed fatigue data and detailed derivations of these parameters are given elsewhere [102]. The relationship, obtained from experimental data, between the cavitation wear resistance expressed in terms of maximum thickness loss and modified fatigue failure stress is shown in Figure 11.43.

It can be seen from Figure 11.43 that metals with poor fatigue resistance properties have in general poorer cavitation wear resistance but this not always means that improving the fatigue resistance of a metal will necessarily diminish the cavitation wear rate.

11.5   SUMMARY

The various forms of wear caused by contact between a particle and a surface have been described in this chapter. The three basic forms of particle-surface interaction, i.e., abrasion, erosion and cavitation, are shown to consist of many specific wear mechanisms. Some wear mechanisms may occur in more than one form of particle-surface interaction. Abrasive and erosive wear in particular were initially thought to consist of one or two relatively simple mechanisms but it is now realized that many processes are involved and some of them are not yet well understood. Despite the basic similarities of these three forms of wear there are also fundamental differences between them which require different methods to be applied in the practical control of wear. Abrasive, erosive and cavitation wear are particularly amenable to control by careful materials selection and many wear resistant materials have been developed for this purpose. However, a material which is resistant to, for example, abrasive wear may fail under erosive or cavitation wear so that materials optimization for a specific application is essential.

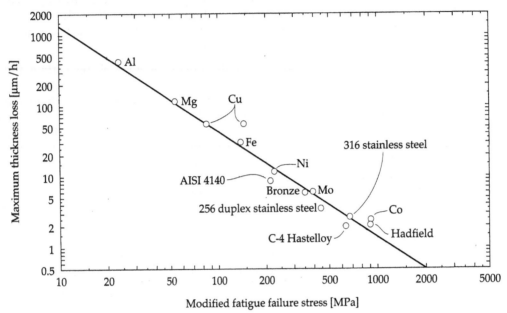

FIGURE 11.43 Relationship between the cavitation wear resistance expressed in terms of maximum thickness loss and modified fatigue failure stress [102].

REVISION QUESTIONS

11.1   Is abrasive wear a slow and therefore benign form of wear?

11.2   Abrasive wear has been described as 'wear by a series of miniature cutting tools'. Is this an oversimplification and if so indicate why?

11.3   What is the difference between two-body and three-body abrasive wear and why is this difference important?

11.4   Does heat treatment of metallic alloys to gain increases in hardness offer large improvements in wear resistance to a hard abrasive?

11.5   What is the ratio of material hardness to grit hardness, which distinguishes a hard abrasive from a soft abrasive?

11.6   How can abrasive wear due to soil and plant silica be suppressed (in general terms)?

11.7   Are polymers abrasion resistant despite having a low hardness?

11.8   What coating properties are needed in order to resist abrasive wear?

11.9   What properties of abrasive particles (grits) affect abrasive and erosive wear?

11.10   When does an abrasive with sufficient hardness fail to abrade a material?

11.11   What is the main difference between abrasive and erosive wear?

11.12   Wear surface morphology in erosive wear shows many similarities to abrasive wear. Under what particular condition does it differ fundamentally?

11.13   Give an example where abrasive wear is similar to erosive wear.

11.14   How can liquid droplets erode a hard material?

11.15   Provide examples of liquid droplet erosion.

11.16   What are the two modes of erosive wear?

11.17   To prolong the life of a mechanical device exposed to grits impinging at near 90° impingement angles, what material property would you optimize?

11.18   How does grit size affect the erosion process?

11.19   Name an erosive wear mechanism exclusively involving a medium that does not initiate the abrasive wear by itself.

11.20   Cutting tools and slurry pumps are typical examples of applications where abrasive and erosive wear is found. What materials are generally selected for these applications in order to combat wear?

11.21   Give an example where erosive wear by solids and liquids is used in engineering to our advantage.

11.22   It has been found that in some applications rubber components resist erosive wear better than hard and tough metallic alloys. What is the reason for this?

11.23   Is cavitation wear an entirely destructive process?

11.24   What is the controlling material property or service characteristic for cavitation wear under typical conditions?

REFERENCES

1   Research Group on Wear of Engineering Materials, Glossary of Terms and Definitions in the Field of Friction, Wear and Lubrication, Tribology O.E.S.D. Publications, Paris 1969.

2   M.A. Moore, Abrasive Wear, ASM Materials Science Seminar on Fundamentals of Friction and Wear of Materials, 4-5 October 1980, Pittsburgh, Pennsylvania, editor: D.A. Rigney, Metals Park, Ohio, Publ. ASM, 1981, pp. 73-118.

3   R.B. Sharp, Plant Silica: An Abrasive Constituent of Plant Matter, *J. Agricultural Engineering Research*, Vol. 7, 1962, pp. 214-220.

4   M.V. Swain, Microscopic Observations of Abrasive Wear of Polycrystalline Alumina, *Wear*, Vol. 35, 1975, pp. 185-189.

5   T. Kayaba, The Latest Investigations of Wear by the Microscopic Observations, *JSLE Transactions*, Vol. 29, 1984, pp. 9-14.

6   S.C. Lim and J.H. Brunton, A Dynamic Wear Rig for the Scanning Electron Microscope, *Wear*, Vol. 101, 1985, pp. 81-91.

7   T.R. Jr. Bates, K.C. Ludema and W.A. Brainard, A Rheological Mechanism of Penetrative Wear, *Wear*, Vol. 30, 1974, pp. 365-375.

8   S.K. Dean and E.D. Doyle, Significance of Grit Morphology in Fine Abrasion, *Wear*, Vol. 35, 1975, pp. 123-129.

9   M.A. Moore and R.M. Douthwaite, Plastic Deformation below Worn Surfaces, *Metallurgical Transactions*, Vol. 7A, 1978, pp. 1833-1839.

10   J.M. Challen and P.L.B. Oxley, An Explanation of the Different Regimes of Friction and Wear Using Asperity Deformation Models, *Wear*, Vol. 53, 1979, pp. 229-243.

11   O. Vingsbo and S. Hogmark, Wear of Steels, ASM Materials Science Seminar on Fundamentals of Friction and Wear of Materials, 4-5 October 1980, Pittsburg, Pennsylvania, editor: D.A. Rigney, Metals Park, Ohio, Publ. ASM, 1981, pp. 373-408.

12   K. Phillips, Study of the Free Abrasive Grinding of Glass and Fused Silica, Ph.D. Thesis, University of Sussex, United Kingdom, 1975.

13   M.A. Moore and F.S. King, Abrasive Wear of Brittle Solids, *Wear*, Vol. 60, 1980, pp. 123-140.

14   N. Emori, T. Sasada and M. Oike, Effect of Material Combination in Rubbing Parts on Three Body Abrasive Wear, *JSLE Transactions*, Vol. 30, 1985, pp. 53-59.

15   T. Sasada, M. Oike and N. Emori, The Effects of Abrasive Grain Size on the Transition between Abrasive and Adhesive Wear, *Wear*, Vol. 97, 1984, pp. 291-302.

16   R.W. Johnson, The Use of the Scanning Electron Microscope to Study the Deterioration of Abrasive Papers, *Wear*, Vol. 12, 1968, pp. 213-216.

17   A. Misra and I. Finnie, A Classification of Three-Body Abrasive Wear and Design of a New Tester, ASTM Int. Conf. on Wear of Materials, 1979 Dearborn, Michigan, USA, editors: K.C. Ludema, W.A. Glaeser and S.K. Rhee, pp. 313-318.

18   E. Rabinowicz, Friction and Wear of Materials, John Wiley and Sons, 1965.

19   J. Larsen-Basse, Influence of Grit Diameter and Specimen Size on Wear during Sliding Abrasion, *Wear*, Vol. 12, 1968, pp. 35-53.

20   A.G. Evans and T.R. Wilshaw, Quasi-Static Particle Damage in Brittle Solids - I: Observations, Analysis and Implications, *Acta Metallurgica*, Vol. 24, 1976, pp. 939-956.

21   K-H. Zum Gahr, Modelling of Two-Body Abrasive Wear, *Wear*, Vol. 124, 1988, pp. 87-103.

22   R.C.D. Richardson, Wear of Metals by Relatively Soft Abrasives, *Wear*, Vol. 11, 1968, pp. 245-275.

23   D. Tabor, Mohs's Hardness Scale - A Physical Interpretation, *Proc. Phys. Soc.*, Vol. 67, Part 3, 1954, pp. 294-257.

24   B.W. Mott, Micro Indentation Hardness Testing, Butterworths, London, 1956.

25   G.A. Sargent and D. Saigal, Erosion of Low-Carbon Steel by Coal Particles, *ASLE Transactions*, Vol. 29, 1986, pp. 256-266.

26   J. Larsen-Basse, Wear of Hard Metals in Rock Drilling: A Survey of the Literature, *Powder Metallurgy*, Vol. 16, 1973, pp. 1-32.

27   P.A. Swanson, A.F. Vetter, The Measurement of Abrasive Particle Shape and Its Effect on Wear, *ASLE Transactions*, Vol. 28, 1985, pp. 225-230.

28   H. Sin, N. Saka and N.P. Suh, Abrasive Wear Mechanisms and the Grit Size Effect, *Wear*, Vol. 55, 1979, pp. 163-170.

29   M.A. Moore, Energy Dissipation in Abrasive Wear, ASTM Int. Conf. on Wear of Materials, 1979 Dearborn, Michigan, USA, editors: K.C. Ludema, W.A. Glaeser and S.K. Rhee, pp. 636-638.

30   M.M. Khruschov, Resistance of Metal to Wear by Abrasion as Related to Hardness, Proc. Conf. on Lubrication and Wear, Inst. Mech. Engrs. Publ., London, 1967, pp. 635-639.

31   M.M. Khruschov, M.A. Babichev, Investigation of the Wear of Metals and Alloys by Rubbing on Abrasive Surface, *Friction and Wear in Machines (Trenie i Iznos v Mash.)*, Inst. of Machines, Acad. Sci. U.S.S.R., Moscow, 1956, pp. 351-358.

32   R.C.D. Richardson, The Abrasive Wear of Metals and Alloys, *Proc. Inst. Mech. Engrs.*, Vol. 182, 1967, Pt. 3A, pp. 410-414.

33   F.T. Barwell, Bearing Systems, Principles and Practice, Oxford University Press, 1979.

34   R.C.D. Richardson, Laboratory Simulation of Abrasive Wear Such as that Imposed by Soil, *Proc. Inst. Mech. Engrs.*, Vol. 182, Pt. 3A, 1967, pp. 29-31.

35   G.M. Bartenev and V.V. Laurentev, Friction and Wear of Polymers, Elsevier, Amsterdam, 1981.

36   K-H. Zum Gahr, Microstructure and Wear of Materials, Elsevier, Amsterdam, 1987.

37   K-H. Zum Gahr, The Influence of Thermal Treatments on Abrasive Wear Resistance of Tool Steels, *Z. Metallkde*, Vol. 68, 1977, pp. 783-792.

38   H.S. Avery, Work Hardening in Relation to Abrasion Resistance, Materials for the Mining Industry Symposium, editor: R.Q. Barr, Greenwich, Conn., USA, 1974, Climax Molybdenum Co., 1974, pp. 43-77.

39   L. Xu and N.F. Kennon, A Study of the Abrasive Wear of Carbon Steels, *Wear*, Vol. 148, 1991, pp. 101-112.

40   J. Larsen-Basse and K.G. Matthew, Influence of Structure on the Abrasion Resistance of A1040 Steel, *Wear*, Vol. 14, 1969, pp. 199-206.

41   M.A. Moore, The Relationship between the Abrasive Wear Resistance, Hardness and Microstructure of Ferritic Materials, *Wear*, Vol. 28, 1974, pp. 59-68.

42   British Standard 3100-1457.

43   R.W.K. Honeycombe, Steels, Microstructure and Properties, Edward Arnold (Publishers) Ltd., 1981.

44   British Standard 3100-1956.

45   J.K. Lancaster, Abrasive Wear of Polymers, *Wear*, Vol. 14, 1969, pp. 223-229.

46   A.G. Veith, The Most Complex Tire-Pavement Interaction: Tire Wear, The Tire Pavement Interface, ASTM STP 929, editors: M.G. Pottinger and T.J. Yager, ASTM, Philadelphia, 1986, pp. 125-158.

47   A.G. Foley, C.J. Chisholm and V.A. McLees, Wear of Ceramic-Protected Agricultural Subsoilers, *Tribology International*, Vol. 21, 1988, pp. 97-103.

48   A. Krell and P. Blank, On Abrasive Wear of Zirconia-Toughened Alumina, *Wear*, Vol. 124, 1988, pp. 327-330.

49 D. Holtz, R. Janssen, K. Friedrich and N. Claussen, Abrasive Wear of Ceramic-Matrix Composites, *J. European Ceramic Society*, Vol. 5, 1989, pp. 229-232.

50 S. Soemantri, A.C. McGee and I. Finnie, Some Aspects of Abrasive Wear at Elevated Temperatures, *Wear*, Vol. 104, 1985, pp. 77-91.

51 A.G. Atkins and D. Tabor, Hardness and Deformation Properties of Solids at Very High Temperatures, *Proc. Roy. Soc.*, Series A, Vol. 292, 1966, pp. 441-459.

52 N. Wing, The Transformation of Soft-Abrasive Wear into Hard-Abrasive Wear under the Effect of Frictional Heat, *Tribology Transactions*, Vol. 32, 1989, pp. 85-90.

53 M.A. Moore, A Preliminary Investigation of Frictional Heating during Wear, *Wear*, Vol. 17, 1971, pp. 51-58.

54 O. Kubaschewski and B.E. Hopkins, Oxidation of Metals and Alloys, Butterworths, London, 1967.

55 J. Larsen-Basse, Influence of Atmospheric Humidity on Abrasive Wear - II: 2-body Abrasion, *Wear*, Vol. 32, 1975, pp. 9-14.

56 G.R. Hoey and J.S. Bednar, Erosion-Corrosion of Selected Metals in Coal Washing Plants Environments, *Materials Performance*, Vol. 22, 1983, pp. 9-14.

57 C.M. Perrott, Materials and Design to Resist Wear, Private Communication, Source ref. V. Mason, Implications of Recent Investigations on Shredder Hammer Tip Materials, 44th Conf. Queensland Society of Sugar Cane Technologists, 1977, pp. 255-259.

58 K.F. Dolman, Alloy Development; Shredder Hammer Tips, Proceedings of Australian Society of Sugar Cane Technologists, 5th Conference, 1983, pp. 281-287.

59 C.S. Yust and R.S. Crouse, Melting at Particle Impact Sites during Erosion of Ceramics, *Wear*, Vol. 51, 1978, pp. 193-196.

60 A. Garton, W.T.K. Stevenson and P.D. McLean, The Stability of Polymers in Low Earth Orbit, *Materials and Design*, Vol. 7, 1986, pp. 319-323.

61 P.H. Schultz and D.E. Gault, Atmospheric Effects on Martian Ejecta Emplacement, *Journal of Geophysical Research*, 1979, Vol. 84, pp. 7669-7687.

62 J.E. Goodwin, W. Sage and G.P. Tilly, Study of Erosion by Solid Particles, *Proc. Inst. Mech. Engrs.*, Vol. 184, 1969-1970, pp. 279-289.

63 S. Bahadur and R. Badruddin, Erodent Particle Characterization and the Effect of Particle Size and Shape on Erosion, *Wear*, Vol. 138, 1990, pp. 189-208.

64 G.W. Stachowiak and A.W. Batchelor, Dimensional Analysis Modelling Tribological Data, Proc. Int. Tribology Conference, Melbourne, The Institution of Engineers, Australia, National Conference Publication No. 87/18, December, 1987, pp. 255-259.

65 G.L. Sheldon and I. Finnie, On the Ductile Behaviour of Nominally Brittle Materials during Erosive Cutting, *Transactions ASME*, Vol. 88B, 1966, pp. 387-392.

66 J.C. Arnold and I.M. Hutchings, Flux Rate Effects in the Erosive Wear of Elastomers, *Journal of Materials Science*, Vol. 24, 1989, pp. 833-839.

67 F.P. Bowden and J.H. Brunton, The Deformation of Solids by Liquid Impact at Supersonic Speeds, *Proc. Roy. Soc.*, Series A, Vol. 263, 1961, pp. 433-450.

68 H. Busch, G. Hoff and G. Langben, Rain Erosion Properties of Materials, *Phil. Trans. Roy. Soc.*, Series A, Vol. 260, 1966, pp. 168-178.

69 N.L. Hancox and J.H. Brunton, The Erosion of Solids by the Repeated Impact of Liquid Drops, *Phil. Trans. Roy. Soc.*, Series A, Vol. 266, 1966, pp. 121-139.

70 A.V. Levy and Y-F. Man, Surface Degradation of Ductile Materials in Elevated Temperature Gas-Particle Streams, *Wear*, Vol. 111, 1986, pp. 173-186.

71 Y. Shida and H. Fujikawa, Particle Erosion Behaviour of Boiler Tube Materials at Elevated Temperature, *Wear*, Vol. 103, 1985, pp. 281-296.

72 D.J. Stephenson, J.R. Nicholls and P. Hancock, Particle-Surface Interactions during the Erosion of a Gas Turbine Material (MarM002) by Pyrolytic Carbon Particles, *Wear*, Vol. 111, 1986, pp. 15-29.

73 A.V. Levy, N. Jee and P. Yau, Erosion of Steels in Coal-Solvent Slurries, *Wear*, Vol. 117, 1987, pp. 115-127.

74 A.V. Levy and G. Hickey, Liquid-Solid Particle Slurry Erosion of Steels, *Wear*, Vol. 117, 1987, pp. 129-158.

75 H. Hojo, K. Tsuda and T. Yabu, Erosion Damage of Polymeric Material by Slurry, *Wear*, Vol. 112, 1986, pp. 17-28.

76   W. Tabakoff, Study of Single-Stage Axial Flow Compressor Performance Deterioration, *Wear*, Vol. 119, 1987, pp. 51-61.

77   H. McI. Clark, On the Impact Rate and Impact Energy of Particles in a Slurry Pot Erosion Tester, *Wear*, Vol. 147, 1991, pp. 165-183.

78   S. Dosanjh and J.A.C. Humphrey, The Influence of Turbulence on Erosion by a Particle-Laden Fluid Jet, *Wear*, Vol. 102, 1985, pp. 309-330.

79   I. Kleis, Grundlagen der Werkstoffauswahl bei der Bekampfung des Strahlverschleisses, *Zeitschrift fur Werkstofftech.*, Vol. 15, 1984, pp. 49-58.

80   J.C. Arnold and I.M. Hutchings, The Mechanisms of Erosion of Unfilled Elastomers by Solid Particle Impact, *Wear*, Vol. 138, 1990, pp. 33-46.

81   P.V. Rao and D.H. Buckley, Angular Particle Impingement Studies of Thermoplastic Materials at Normal Incidence, *ASLE Transactions*, Vol. 29, 1986, pp. 283-298.

82   A.V. Levy, The Solid Particle Erosion Behaviour of Steel as a Function of Microstructure, *Wear*, Vol. 68, 1981, pp. 269-287.

83   S.M. Shah, J.D. Verhoeven and S. Bahadur, Erosion Behaviour of High Silicon Bainitic Structures, I: Austempered Ductile Cast Iron, *Wear*, Vol. 113, 1986, pp. 267-278.

84   S.M. Shah, S. Bahadur and J.D. Verhoeven, Erosion Behaviour of High Silicon Bainitic Structures, II: High Silicon Steels, *Wear*, Vol. 113, 1986, pp. 279-290.

85   K.V. Pool, C.K.H. Dharan and I. Finnie, Erosive Wear of Composite Materials, *Wear*, Vol. 107, 1986, pp. 1-12.

86   S.M. Walley, J.E. Field and P. Yennadhiou, Single Solid Particle Impact Erosion Damage on Polypropylene, *Wear*, Vol. 100, 1984, pp. 263-280.

87   J. Zahavi and G.F. Schmitt, Jr., Solid Particle Erosion of Polymer Coatings, *Wear*, Vol. 71, 1981, pp. 191-210.

88   S. Srinivasan and R.O. Scattergood, R Curve Effects in Solid Particle Erosion of Ceramics, *Wear*, Vol. 142, 1991, pp. 115-133.

89   S. Srinivasan and R.O. Scattergood, Erosion of Transformation Toughening Zirconia by Solid Particle Impact, *Adv. Ceram. Mater.*, Vol. 3, 1988, pp. 345-52.

90   S.F. Wayne, J.G. Baldoni and S.T. Buljan, Abrasion and Erosion of WC-Co with Controlled Microstructures, *Tribology Transactions*, Vol. 33, 1990, pp. 611-617.

91   A.V. Levy and P. Clark, The Erosion Properties of Alloys for the Chemical Industry, *Wear*, Vol. 151, 1991, pp. 337-350.

92   A. Karimi and F. Avellan, Comparison of Erosion Mechanisms in Different Types of Cavitation, *Wear*, Vol. 113, 1986, pp. 305-322.

93   K.R. Trethewey, T.J. Haley and C.C. Clark, Effect of Ultrasonically Induced Cavitation on Corrosion Behaviour of a Copper-Manganese-Aluminium Alloy, *British Journal of Corrosion*, Vol. 23, 1988, pp. 55-60.

94   C.R. Preece and J.H. Brunton, A Comparison of Liquid Impact Erosion and Cavitation Erosion, *Wear*, Vol. 60, 1980, pp. 269-284.

95   W.J. Tomlinson and M.G. Talks, Cavitation Erosion of Laser Surface Melted Phosphoric Grey Irons, *Wear*, Vol. 129, 1989, pp. 215-222.

96   H. Jin, F. Zheng, S. Li and C. Hang, The Role of Sand Particles on the Rapid Destruction of the Cavitation Zone of Hydraulic Turbines, *Wear*, Vol. 112, 1986, pp. 199-205.

97   P. Veerabhadra Rao, Evaluation of Epoxy Resins in Flow Cavitation Erosion, *Wear*, Vol. 122, 1988, pp. 77-95.

98   T. Okada, Y. Iwai and A. Yamamoto, A Study of Cavitation Erosion of Cast Iron, *Wear*, Vol. 84, 1983, pp. 297-312.

99   C.J. Heathcock, A. Ball and B.E. Protheroe, Cavitation Erosion of Cobalt-Based Stellite Alloys, Cemented Carbides and Surface-Treated Low Alloy Steels, *Wear*, Vol. 74, 1981-1982, pp. 11-26.

100   A. Karimi and J.L. Martin, Cavitation Erosion of Materials, *International Metals Reviews*, Vol. 31, 1986, pp. 1-26.

101   T. Okada, Y. Iwai and Y. Hosokawa, Resistance to Wear and Cavitation Erosion of Bearing Alloys, *Wear*, Vol. 110, 1986, pp. 331-343.

102   R.H. Richman and W.P. McNaughton, Correlation of Cavitation Erosion Behaviour with Mechanical Properties of Metals, *Wear*, Vol. 140, 1990, pp. 63-82.

103   W.J. Head, M.E. Harr, The Development of a Model to Predict the Erosion of Materials by Natural Contaminants, *Wear*, Vol. 15, 1970, pp. 1-46.

104   C. Chaussy, First Clinical Experience With Extracorporeally Induced Destruction of Kidney Stones by Shock Waves, *Journal of Urology*, Vol. 249, 1982, pp. 417-420.

105   M. Delius, W. Brendel and G. Heine, A Mechanism of Gallstone Destruction by Extracorporeal Shock Waves, *Naturwissenschaften*, Vol. 75, 1988, pp. 200-201.

106   C.J. Chuong, P. Zhong, H.J. Arnott and G.M. Preminger, Stone Damage Modes During Piezo-Electric Shock Wave Delivery, in Shock Wave Lithotripsy 2: Urinary and Biliary Lithotripsy, editors: J.E. Lingeman and D.M. Newman, Plenum Press, New York, 1989.

107   P. Zhong, C.J. Chuong, R.D. Goolsby and G.M. Preminger, Microhardness Measurements of Renal Calculi: Regional Differences and Effects of Microstructure, *Journal of Biomedical Materials Research*, Vol. 26, 1992, pp. 1117-1130.

108   M.G. Hamblin and G.W. Stachowiak, A Multi-Scale Measure of Particle Abrasivity, *Wear*, Vol. 185, 1995, pp. 225-233.

109   M.G. Hamblin and G.W. Stachowiak, A Multi-Scale Measure of Particle Abrasivity and its Relation to Two Body Abrasive Wear, *Wear*, Vol. 190, 1995, pp. 190-196.

110   M.G. Hamblin and G.W. Stachowiak, Description of Abrasive Particle Shape and its Relation to Two-Body Abrasive Wear, *Tribology Transactions*, Vol. 39, 1996, pp. 803-810.

111   M.G. Hamblin and G.W. Stachowiak, Comparison of Boundary Fractal Dimension from Projected and Sectioned Particle Images, Part I - Technique Evaluation, *Journal of Computer Assisted Microscopy*, Vol. 5, 1993, pp. 291-300.

112   M.G. Hamblin and G.W. Stachowiak, Comparison of Boundary Fractal Dimension from Projected and Sectioned Particle Images, Part II - Dimension Changes, *Journal of Computer Assisted Microscopy*, Vol. 5, 1993, pp. 301-308.

113   G.W. Stachowiak, Particle Angularity and Its Relationship to Abrasive and Erosive Wear, *Wear*, Vol. 241, 2000, pp. 214-219.

114   U. Beste and S. Jacobson, Micro-Scale Hardness Distribution of Rock Types Related to Drill Wear, *Wear*, Vol. 254, 2003, pp. 1147-1154.

115   S. Ahmaniemi, M. Vippola, P. Vuoristo, T. Mantyla, M. Buchmann and R. Gadow, Residual Stresses in Aluminium Phosphate Sealed Plasma Sprayed Oxide Coatings and Their Effect on Abrasive Wear, *Wear*, Vol. 252, 2002, pp. 614-623.

116   B.S. Mann and V. Arya, HVOF Coating and Surface Treatment for Enhancing Droplet Erosion Resistance of Steam Turbine Blades, *Wear*, Vol. 254, 2003, pp. 652-667.

117   M.A. Uusitalo, P.M.J. Vuoristo and T.A. Mantyla, Elevated Temperature Erosion-Corrosion of Thermal Sprayed Coatings in Chlorine Containing Environments, *Wear*, Vol. 252, 2002, pp. 586-594.

118   Y. Iwai and S. Li, Cavitation Erosion in Waters Having Different Surface Tensions, *Wear*, Vol. 254, 2003, pp. 1-9.

119   W. Liu, Y.G. Zheng, C.S. Liu, Z.M. Yao and W. Ke, Cavitation Erosion Behavior of Cr-Mn-N Stainless Steels in Comparison With 0Cr13Ni5Mo Stainless Steel, *Wear*, Vol. 254, 2003, pp. 713-722.

120   W. Fu, Y. Zheng and X. He, Resistance of a High Nitrogen Austenitic Steel to Cavitation Erosion, *Wear*, Vol. 249, 2001, pp. 788-791.

121   X.F. Zhang and L. Fang, The Effect of Stacking Fault Energy on the Cavitation Erosion Resistance of Alpha-Phase Aluminium Bronzes, *Wear*, Vol. 253, 2002, pp. 1105-1110.

A D H E S I O N
A N D
A D H E S I V E   W E A R

12.1   INTRODUCTION

Adhesive wear is a very serious form of wear characterized by high wear rates and a large unstable friction coefficient. Sliding contacts can rapidly be destroyed by adhesive wear and, in extreme cases, sliding motion may be prevented by very large coefficients of friction or seizure. Metals are particularly prone to adhesive wear hence its practical significance. Most lubricant failures in sliding metal contacts result in adhesive wear since this relates to a breakdown in the lubricant's basic function of providing some degree of separation between the sliding surfaces. If sliding surfaces are not separated, adhesion and subsequent wear are almost inevitable. The questions of practical importance are: which metals are most prone to adhesion and adhesive wear? How can adhesive wear be recognized and controlled? In this chapter the process of adhesion between surfaces is described together with the resulting wear mechanism.

12.2   MECHANISM OF ADHESION

Most solids will adhere on contact with another solid to some extent provided certain conditions are satisfied. Adhesion between two objects casually placed together is not observed because intervening contaminant layers of oxygen, water and oil are generally present. The earth's atmosphere and terrestrial organic matter provide layers of surface contaminant on objects which suppress very effectively any adhesion between solids. Adhesion is also reduced with increasing surface roughness or hardness of the contacting bodies. Actual observation of adhesion became possible after the development of high vacuum systems which allowed surfaces free of contaminants to be prepared. Adhesion and sliding experiments performed under high vacuum showed a totally different tribological behaviour of many common materials from that observed in open air. Metallic surfaces free of oxide films under high vacuum exhibited the most dramatic changes and partly for this reason have been widely studied.

Metal-Metal Adhesion

Apart from noble metals such as gold and platinum any other metal is always covered by an oxide film when present in unreacted form in an oxidizing atmosphere. The oxide film is often so thin as to be invisible and the metal appears shiny and pure. This film, which may be only a few nanometres thick, prevents true contact between metals and hinders severe wear unless deliberately removed [1].

It has been found in experiments conducted in vacuum that as the degree of surface contamination is reduced, adhesion between metallic surfaces becomes very large [1]. In these experiments the metal was first heated to melt off the oxide film. A schematic diagram of the apparatus to measure the adhesion of clean surfaces under vacuum is shown in Figure 12.1.

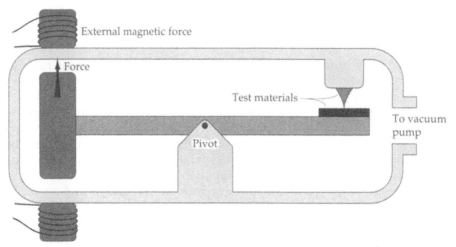

FIGURE 12.1 Schematic diagram of the apparatus for measurements of adhesion between metals [2].

As can be seen from Figure 12.1 a stylus is loaded against a flat surface and the strength of adhesion is determined by measuring the force needed to pull the two surfaces apart. Adhesion force data for various metals against iron measured at **0.2** [mN] of a contact load and **10⁻¹⁰** [Torr] of a chamber pressure are shown in Table 12.1 [2].

It is evident from Table 12.1 that in all cases the adhesion or separation force is greater than the contact force. The tendency to adhere does not discriminate between metals on the basis of their mutual solubility or relative atomic size. The greatest adhesion occurs for a combination of like materials, i.e., iron to iron, but many other combinations of unlike metals also show quite high adhesions. The ratio of adhesion force to contact force can be very high, about **20** or more in some cases. The bonding process is almost instantaneous and can occur at moderate or low temperatures [2].

Numerous tests on a wide variety of metal combinations have shown that when there is strong adhesion, transfer of the weaker metal to the stronger occurs as illustrated schematically in Figure 12.2.

FIGURE 12.2 Process of metal transfer due to adhesion.

TABLE 12.1   Adhesion force of various metals against iron in vacuum [2].

Metal	Solubility in iron [atomic %]	Adhesion force to iron [mN]
Iron		> 4.0
Cobalt	35	1.2
Nickel	9.5	1.6
Copper	< 0.25	1.3
Silver	0.13	0.6
Gold	< 1.5	0.5
Platinum	20	1.0
Aluminium	22	2.5
Lead	Insoluble	1.4
Tantalum	0.20	2.3

The strong adhesion observed between metals can be explained by electron transfer between contacting surfaces. Numerous free electrons are present in metals and on contact electrons may be exchanged between the two solids to establish bonding. The 'Jellium model' [3] is used to describe this effect. The electrons are not bound by a rigid structure and providing that the distance between two bodies in contact is sufficiently small, i.e., < 1 [nm], they can move from one body to another. As a result the electrons can bond two solids despite their differing atomic structures. It has been found that the calculated values of the strength of adhesion between two metals [4,5] are considerably in excess of experimental values [7]. This is attributed to the difficulty in determining a true value of the contact area between atoms of opposing surfaces.

It is theorized that when different metals are in contact, the metal with a higher electron density donates electrons to the other metal as illustrated in Figure 12.3.

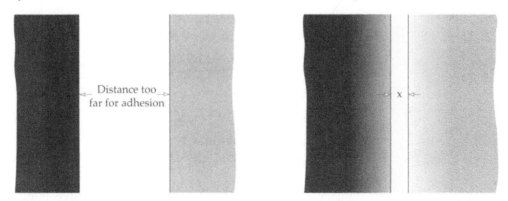

FIGURE 12.3   Jellium electron exchange model of adhesive contact between metals [4]; x is equivalent to atomic dimensions, i.e., less than 1 [nm].

All metals show a strong tendency to adhere on contact with another solid but there are significant differences between particular elements. Metals mainly exist in four principal types of crystal structure: face-centred cubic, body-centred cubic, hexagonal close packed and tetragonal. It has been found experimentally that metals with hexagonal close packed structure show much less adhesion than other crystal structures [2,7]. High hardness, large elastic moduli and surface energy of the metal also suppress adhesion [7]. The graph of the coefficient of adhesion versus hardness for a number of pure metals is shown in Figure 12.4 [7], where the coefficient of adhesion is defined as the ratio of rupture force to contact force. It can be seen from Figure 12.4 that for metals with similar hardness but different crystal structure, e.g., aluminium and zinc or lead and tin, there are significant differences in adhesion.

The reason for the difference in adhesion between metals of similar hardness is believed to lie in the necessity for some degree of plastic deformation between asperities before a true contact can be established. Hexagonal close packed metals have far fewer slip systems and are therefore less ductile than face-centred and body-centred metals, which results in their lower adhesion.

Adhesion between metals is also influenced by the 'chemical reactivity' or electropositivity of the individual metals [2,8]. Chemically active metals, such as aluminium, bond more readily and therefore show stronger adhesion than noble metals. This suggests that face-centred cubic crystal lattice metal with a high level of chemical activity would show a particularly strong adhesion. Such metals are usually unsuitable for unlubricated sliding contacts.

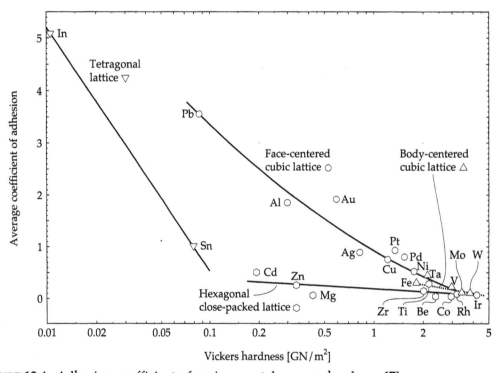

FIGURE 12.4 Adhesion coefficient of various metals versus hardness [7].

Summarizing, the electron transfer between metals allows a strong adhesive bond to be formed between two identical or different metallic elements. A limiting factor in adhesion is the minimum load which causes plastic flow and therefore the establishment of a true contact between surfaces.

Metal-Polymer Adhesion

In the extensive series of experiments conducted into metal-polymer contact and adhesion under high vacuum it has been revealed that metals and polymers can also show a high degree of adhesion [2]. Adhesion observed between a tungsten surface and polymers such as polytetrafluoroethylene and polyimide is strong enough to cause polymer to transfer to the metallic surface when the two materials are separated. The strength of adhesion is found to be related to the presence of reactive non-metals, such as fluorine, in the polymer [2]. Surface atoms of the polymer are believed to bond with surface atoms of the metal and this can occur irrespective of the inertness of the polymer in bulk.

Most polymers adhere to other materials by van der Waals forces. In most wear situations, this form of adhesion is not strong enough for lumps of material to be torn out on rupture of the contact [9,10].

Strong adhesion between a metal and a polymer based on chemical interaction forms the basis for the mechanism of polymer on metal wear. Van der Waals forces, although they do not directly cause adhesive wear, provide a significant component of frictional resistance for elastomers such as rubber.

Metal-Ceramic Adhesion

Under suitable conditions, quite strong adhesion between metals and ceramics can occur [2,11-13]. The common factor in adhesion between various ceramics and metals is their chemical affinity. It has been found that only metals which do not form stable oxides exhibit low friction coefficients against ceramics [11]. In contacts with ceramics, metals such as copper, aluminium and nickel show high friction coefficients while the coefficients of friction of gold and silver (unstable oxides) are low.

Although the oxygen ions present on the surface of aluminium oxide are already bonded to aluminium, an additional interaction with contacting metal atoms is possible according to the laws of quantum chemistry [11]. The trend for contact between other metals and ceramics is similar [13].

Metals usually have a cohesive strength lower than most engineering ceramics so that on rupture of the adhesive contact, fragments of metal are often left adhering to the ceramic to form a transfer film. Adhesion of ceramics to metals is greatly reduced by surface contamination in a manner similar to metal-metal contacts. These issues will be discussed in more detail in Chapter 16 on 'Wear of Non-Metallic Materials'.

Polymer-Polymer and Ceramic-Ceramic Adhesion

Adhesion based on electron transfer is less likely to take place in contacts lacking a metal counterface. Very little is known about the mechanisms of adhesion between non-metallic materials [14]. It is known that there exists a weak to moderate level of adhesion as a result of van der Waals forces acting between almost all contacting materials [9,10,14]. Attractive forces have been found between quartz surfaces [14] and mica surfaces [10]. Rubber was also found to adhere to glass and polymer [15]. In all of these cases van der Waals forces were clearly the largest component contributing to adhesion.

The markedly different mechanical properties of polymers and ceramics illustrate well the difference between inter-atom attraction and bulk adhesion. Polymers have one of the lowest elastic moduli of commonly used engineering materials whereas ceramics have one of the highest. Most of the surfaces are rough and for the contacting surfaces to reach a proximity similar to the size of an atom or less, the deformation of the surface asperities must take place. Forces required to deform the asperities act in opposition to the adhesion forces and reduce the overall net adhesion force. The adhesion therefore is strongly influenced by the size of the asperities. A relationship illustrating the dependence of adhesion on surface roughness for elastic solids has been developed by Fuller and Tabor [16], i.e.:

$$\mathbf{K} = E\sigma^{3/2}/(r^{1/2}\Delta\gamma) \tag{12.1}$$

where:

   K   is the coefficient of reduction in adhesion by asperity deformation forces;

   E   is the Young's modulus [Pa];

   σ is the standard deviation of the asperity height distribution (RMS) [m];

   r   is the average radius of curvature of individual asperities [m];

   $\Delta\gamma$ is the change in surface energy on contact between the two surfaces [J/m^2].

It is assumed that for **K < 10** strong adhesion occurs, while for **K > 10** asperity deformation forces cause the net adhesion force for elastic materials to be small. Expression (12.1) clearly shows that adhesion is more sensitive to surface roughness for materials with high Young's modulus, i.e., for soft materials the range of surface roughness, over which adhesion occurs, is much wider than for hard materials. For example, in elastomers weak adhesion takes place when the surface roughness 'R_q' is above **1** [μm], while in hard materials, i.e., ceramics, weak adhesion occurs at much lower values of surface roughness when 'R_q' is greater than **5** [nm]. Below these transition values of surface roughness strong adhesion occurs. Therefore adhesion between hard elastic solids can be significantly reduced by even very small surface irregularities. If plastic deformation between contacting asperities takes place then adhesion is enhanced.

Summarizing, polymers and ceramics in contact show a similar adhesion mechanism caused by van der Waals forces. The net adhesion force for ceramics contacting ceramics is greatly reduced, however, due to their high hardness.

Effects of Adhesion between Wearing Surfaces

Strong adhesion between the asperities of wearing surfaces has two effects: a large component of frictional force is generated and the asperities may be removed from the surface to form wear particles or transfer layers.

· *Friction Due to Adhesion*

An adhesive theory of friction was developed almost half a century ago by Bowden and Tabor [17]. As discussed already in the chapter on 'Boundary and Extreme Pressure Lubrication', in simple terms, the coefficient of friction is defined as:

$$\mu = \tau/p_y \qquad\qquad (12.2)$$

where:

   τ is the effective shear stress of the material [Pa];

   p_y is the plastic flow stress (yield pressure) of the material [Pa].

It is argued that the effective shear stress acting on the surface should be close to the bulk value which is about **0.2** of the yield stress giving μ = **0.2**. If the materials of contacting bodies differ then the yield stress of the softer material and the shear stress of the weaker material or the interface shear stress, whichever is the least, are used in equation (12.2).

The highest adhesion occurs between identical metals, whereas bimetallic combinations exhibit weaker adhesion and therefore lower friction. Heterogeneous materials such as steels and cast irons often show moderate adhesion because of the interference by inclusions and non-metallic phases present in their microstructure. Corresponding coefficients of friction are also lower for these materials.

There are, however, doubts about this adhesive theory of friction and some controversy is extant in the literature. It has been found that most of the frictional resistance in lubricated or atmospheric conditions is due to the deformation of asperities rather than the fracture of adhesive bonds [18]. Frictional forces due to adhesion are dominant when there is a total absence of lubrication and such circumstances correspond to the original experiments performed in vacuum [17]. The friction theory in the simple form presented so far implies that the limiting values of friction are less than unity. In practice much higher values of the coefficient of friction are observed and the reasons for this are explained in the next section.

· *Junction Growth between Contacting Asperities as a Cause of Extreme Friction*

The implication of the adhesion experiments conducted in high vacuum [2] is that as the sources of surface contamination are progressively removed, the levels of adhesion and therefore friction rise precipitately. Conversely, when gas or contaminants are introduced to clean surfaces, friction levels decline to the moderate values typically found under atmospheric conditions.

The coefficient of friction of iron against iron as a function of surrounding gas pressure was measured in a specially designed apparatus [17,19,20]. A schematic diagram of the apparatus is shown in Figure 12.5. An iron sphere is driven against an iron surface by a solenoid. The friction force is measured by the deflection of a silica spring. The test chamber is connected to a vacuum pump and heating coils are supplied for thermal cleaning of the iron surfaces.

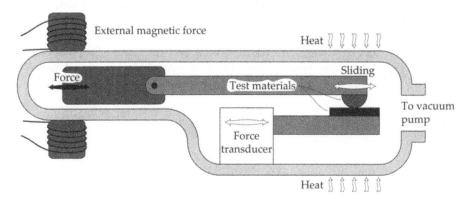

FIGURE 12.5 Schematic diagram of the apparatus for measurements of friction in a vacuum [20].

The relationship between the friction coefficient of iron sliding against iron in the presence of various amounts of oxygen is shown in Figure 12.6.

In high vacuum a total seizure between the contacting samples occurs. As oxygen is supplied to the iron surface, a film of iron oxide forms, resulting in a reduced coefficient of friction. When this film reaches a certain thickness the strong adhesion between metallic iron is replaced by a weaker adhesion between iron oxide which is probably controlled by van der Waals forces.

The coefficient of friction between clean iron surfaces is very high, up to $\mu = 3$. The simple theory of adhesion, described in the previous section, fails to predict such high values of friction coefficient, and in order to explain this phenomenon the process of '**asperity junction growth**' is considered [21]. In the plastically deforming adhesion junction both normal and tangential stresses are involved.

To explain the 'asperity junction growth' process assume that initially there is a normal load acting on the asperity which is high enough for the asperity to plastically yield. Since the contact is in the 'plastic state', i.e., material flows, the contact area will easily be increased when the tangential stress is introduced. The increase in the contact area will result in a

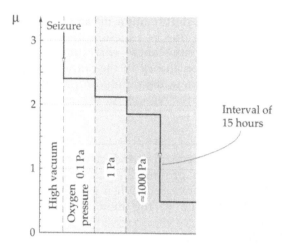

FIGURE 12.6 Effect of oxygen on the friction of clean iron [20].

reduction in the normal pressure (i.e., the same load is now carried by an increased area), as illustrated schematically in Figure 12.7. The increased contact area will also enable a larger tangential force to be sustained. The tangential force and the contact area will grow until the maximum (yield) shear stress of the material is reached (it is implicitly assumed here that under sliding conditions each asperity contact is loaded to a maximum stress prior to rupture). As a result the coefficient of friction will also increase. Since the loop with a positive feedback is created, the system may become unstable. The onset of instability is followed by a rapid increase in the coefficient of friction which eventually leads to seizure of the operating parts.

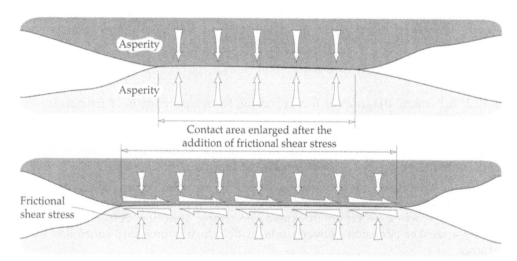

FIGURE 12.7 Schematic diagram of asperity junction growth under frictional force.

In precise terms the mechanism of junction growth can be described by considering the von Mises yield criterion. According to this criterion a material will plastically deform when:

$$p^2 + 3\tau^2 = p_y^2 \qquad\qquad (12.3)$$

where:

  p is the normal contact stress (pressure) [Pa];

τ is the effective shear stress in the contact [Pa];

p_y is the plastic flow stress of the material [Pa].

Since the plastic yielding of a junction is controlled by the combined effect of the normal stress 'p' and tangential stress 'τ' a similar relation to describe its behaviour was proposed [21], i.e.:

$$p^2 + C\tau^2 = p_0^2 \tag{12.4}$$

where:

C   is an arbitrary constant assumed to have a value close to **10**;

p_0 is the plastic flow stress of the material in the absence of tangential (frictional) force [Pa].

The other variables are as already defined.

It can be seen from equation (12.4) that when a normal load only is acting on the asperity, i.e., $\tau = 0$ then:

$$p = p_0$$

The stresses 'p', 'τ' and 'p_0' can be expressed as follows:

$$p = W/A_r$$

$$\tau = F/A_r$$

$$p_0 = W/A_{ro}$$

where:

F   is the friction force [N];

W   is the normal force (load) [N];

A_r is the real area of contact with tangential force present [m^2];

A_{ro} is the real area of contact in the absence of tangential force [m^2].

Substituting for these expressions into (12.4) gives:

$$(W/A_r)^2 + C(F/A_r)^2 = (W/A_{ro})^2 \tag{12.5}$$

Rearranging (12.5) yields a relationship between the increase in real contact area and the tangential force, i.e.:

$$\frac{A_r}{A_{ro}} = \left[1 + C\frac{F^2}{W^2}\right]^{0.5} \tag{12.6}$$

It can be seen from equation (12.6) that increasing the tangential force causes the adhesion to increase since the real area of contact grows. For example, if **C = 10** and the ratio of tangential force to normal force is **0.3** then the contact area is enlarged by a factor of **1.4**. The enlargement of the real contact area is particularly marked at the high values of '**F**' which are observed for clean surfaces. Increased tangential force is accommodated by the increase in real contact area until the yield shear stress is reached at the interface between asperities and the macroslip takes place.

For high values of yield shear stress this condition is hard to reach because the increase in the contact area is almost matched by the increase in the tangential force (as 'F' exceeds 'W', A_r/A_{ro} tends to $C^{0.5}F/W$). In contrast it can be seen from equation (12.6) that for small values of tangential force and a limiting shear stress, the increase in 'A_r' is negligible so that the relationship $\mu = \tau/p_y$ is approximately true. In cases of extremely high adhesion and limiting interfacial asperity shear stress, the rate of increase in the real contact area with tangential force is sufficient to maintain an approximately constant asperity interface shear stress. This is because the ratio of tangential force to contact area does not change significantly so complete seizure of the sliding members can occur.

It should also be mentioned that the plastic flow which occurs at the asperities is accompanied by work hardening in most metals. Since the strength of the welded junction is often higher than that of the softer metal the shear occurs along a plane which is different from that defined by the localized welding. The overall effect of work hardening on the coefficient of friction, however, is quite small compared to the effect of junction growth.

· Seizure and Scuffing

Very high friction coefficients found on clean surfaces under a vacuum can also occur in practical mechanical contacts when there is a breakdown or absence of lubrication. Plain bearings and gear teeth are susceptible to this problem. Figure 12.8 shows the typical appearance of scuffed gear teeth. It can be seen that the smooth as machined surface of the teeth is completely disrupted and displays signs of strong adhesion and adhesive fracture.

FIGURE 12.8   Adhesion between gear teeth resulting in scuffing.

Under these conditions normal operation of the gear is impossible and considerable damage to the unit from overheating as well as adhesion will result. In most sliding contacts, such as bearings, gears, chains and cams, the cause of rapid and sometimes catastrophic failures is adhesion and adhesive wear. Although other wear mechanisms can also cause problems, in general, these problems are of a milder form.

· Asperity Deformation and Formation of Wear Particles

The combined action of adhesion between asperities and sliding motion causes severe plastic deformation of the asperities. To observe and study the events that are likely to occur between sliding and adhering asperities in an actual wearing contact is virtually impossible. To facilitate such studies the contact between two asperities was simulated by two pointed plates as illustrated in Figure 12.9 [22]. The plates are forced together by a vertical load and moved against each other by a hydraulic ram. The entire system can be fitted into a scanning electron microscope for observation.

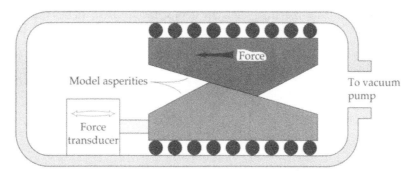

FIGURE 12.9 Schematic diagram of the experimental apparatus to study adhesive wear processes [22].

Since the position of the plates is maintained at a constant level by the slideway, a close representation of a wearing contact where individual asperities move along a horizontal plane and sustain transient loads when in contact with opposing asperities is obtained. The mechanism of shearing and cracking to form a transfer particle in the adhesive contact between asperities is illustrated schematically in Figure 12.10 [22].

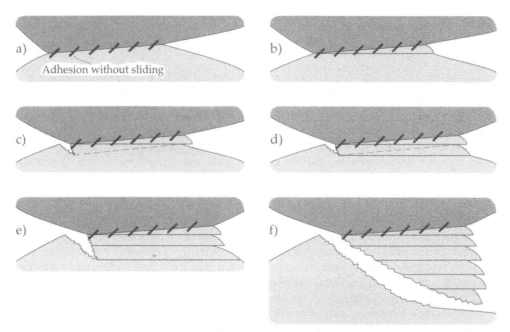

FIGURE 12.10 Schematic diagram of the formation of an adhesive transfer particle [22].

Material in the softer or sharper asperity deforms in a series of shear bands to accommodate the relative movement, i.e., there is no sliding along the asperity contact line. When each shear band reaches a certain limit, a crack is initiated or an existing crack progresses till a new shear band is formed. The crack extends across the asperity and eventually a particle detaches from the deformed asperity.

It has been found that asperities with large slope angles, i.e., 'sharper asperities', tend to lose material to asperities with small slope angles [22]. Material properties have a strong influence on asperity deformation and the severity of adhesive wear. Experiments conducted on model asperities [17] revealed that the contacting asperities of brittle materials tend to break away cleanly with little deformation and produce fewer wear particles compared to ductile materials [17]. It appears that ductility has an undesirable effect of accentuating adhesive wear.

In the contacts between asperities which do not produce wear particles, there may still be extensive plastic deformation as illustrated in Figure 12.11 [23].

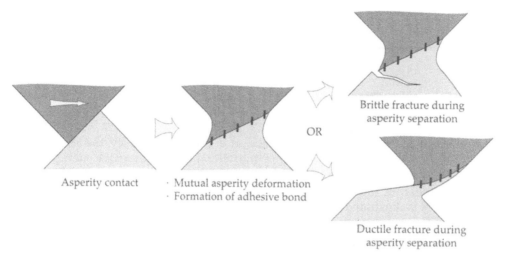

Brittle fracture during
asperity separation

OR

Asperity contact   · Mutual asperity deformation
               · Formation of adhesive bond

Ductile fracture during
asperity separation

FIGURE 12.11 Alternative model of deformation in adhesive asperity contact [23].

The evidence of such severe plastic deformation and/or surface cracking producing a sharply skewed worn asperity profile has been confirmed by scanning electron microscopy studies [23].

The particle of metal detached from one of the asperities, i.e., as shown in Figure 12.10, remains attached to the other surface. Depending on conditions it may subsequently be removed by further asperity contact to form a true wear particle or it will remain on the surface to form a '**transfer film**'.

· *Transfer Films*

The formation of transfer films is a characteristic feature of adhesive wear where material is transferred from one surface to another before being released as a wear particle. It distinguishes adhesive wear from most other wear mechanisms. In the early studies of this phenomenon it was found that brass rubbed against steel leaves a film of transferred brass on the steel surface which eventually covers the wear track [24]. The transferred brass was found to be highly work-hardened and probably capable of wearing the brass sample itself. This observation of inter-metallic transfer was confirmed later by tests on a variety of combinations of metals in sliding [25,26]. Examples of metallic transfer film are shown in Figure 12.12.

The formation of a transfer film or transfer particles can have a dramatic effect on the wear rate [27]. The process of transfer particle formation and removal is illustrated schematically in Figure 12.13. It can be seen from Figure 12.13 that the transfer particle can lift the pin away from the opposing surface and this causes an apparently 'negative wear rate'. The evidence of this phenomenon is illustrated in Figure 12.14, where the wear depth incurred when a zinc pin is slid against a zinc disc is shown. Periods of apparently negative wear are followed by a step-form of positive wear as transfer particles are formed and released.

Tests with other metals such as iron, molybdenum, nickel, copper, silver and aluminium for possible combinations of sliding partners show that wherever there is mutual solubility, e.g., copper and silver, the same pattern occurs, but if the two metals are insoluble, e.g., iron and silver, then lumpy transfer does not occur [27].

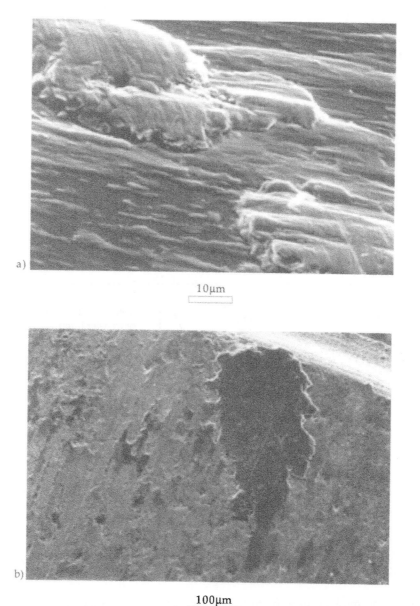

a)

10μm

b)

100μm

FIGURE 12.12 Examples of metallic film transfer: a) brass film transfer on alumina and b) Al-Si alloy transfer film onto a piston ring.

When different metals are slid on each other, a form of mechanical alloying occurs and the transfer particle consists of lamella of the two metals [27]. At the beginning transfer particles accumulate material from both surfaces in small bits. As the transfer particle grows bigger it becomes flattened between the sliding surfaces, producing a lamellar structure. The possible mechanism involved in this process and an example of such a particle are shown schematically in Figure 12.15.

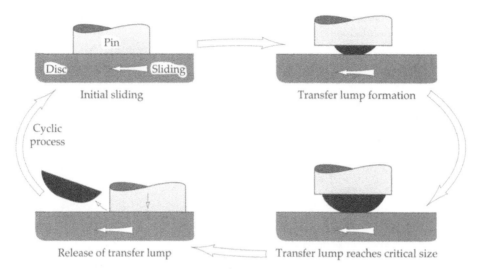

FIGURE 12.13 Formation and removal of a transfer particle (adapted from [27]).

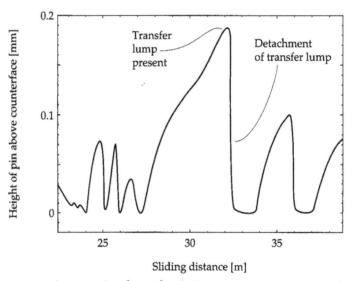

FIGURE 12.14 Variation of wear depth with sliding distance for zinc sliding against zinc (adapted from [27]).

From the tests performed on various test rigs, e.g., pin on disc machines, it has been found that transfer films have the following specific characteristics which clearly distinguish them from other films on worn surfaces [24,28]:

· the equilibrium rate of transfer to the disc or ring is equal to the rate of wear of the pin;

· wear particles are formed entirely from the transfer layer on the disc or ring with no direct wear of the pin;

· patches of transfer layer are usually larger than individual transfer particles so agglomeration of the particles occurs on the disc or ring surface;

· wear particles are generally larger than transfer particles;

· transfer particles are usually the same size as the area of real contact which suggests that just a few transfer particles are carrying the load;

· although the number of transfer particles and the area covered by transfer film increase with load, the thickness of the transfer film remains approximately constant;

· transfer particles are generally harder than the substrate material due to severe work hardening and are capable of producing grooves in the surface.

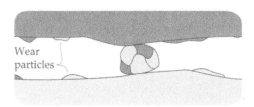

1) Early growth stage of transfer particle

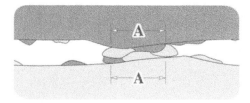

2) Depressed transfer particle contacting with area **A** determined by the flow pressure

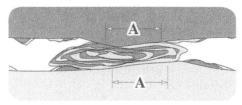

3) Press-slide flattening

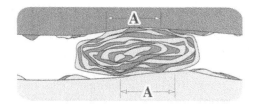

4) Grown transfer particle just before removal

5) Transfer particle

FIGURE 12.15 Formation of lamellar structure transfer particles (adapted from [27]).

The mechanism of groove formation involves ploughing of the softer substrate material by work-hardened transfer particles [29]. The ploughing is a very inefficient form of cutting which can lead to crack formation on the worn surface as a result of high tensile stresses. The mechanism of ploughing by transfer particles is illustrated in Figure 12.16.

The formation of such coarse grooves on worn surfaces is frequently observed when adhesive wear occurs. These grooves are usually formed on the sliding member with larger wear track area, e.g., on the disc or ring in pin on disc/ring machines.

Summarizing, transfer films can greatly modify the sliding characteristics of materials. When a transfer film is present as thick lumps, smooth sliding is impossible and the load is carried by a few or just one transfer particle. When a transfer particle is released there is an abrupt movement of the sliding surfaces to compensate for this. In extreme cases the transfer particles can fail to detach and grow to cause total seizure of the sliding interfaces. However, not all transfer films are undesirable. The wear of polymers, in particular, depends on very thin transfer films which allow for low friction. Solid lubricants also function by forming thin layers on the wearing surfaces. These layers can be transferred from one wearing surface to another which is useful when inaccessible contacts have to be lubricated.

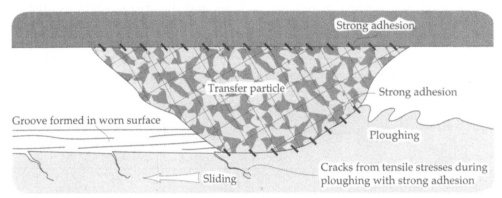

FIGURE 12.16 Mechanism of groove formation on worn surfaces by work-hardened transfer particles.

12.3 CONTROL OF THE ADHESIVE WEAR

If adhesive wear is allowed to proceed uncontrolled various undesirable consequences can follow. High friction with the possibility of seizure and the growth of transfer particles can result. In some cases transfer particles can jam the sliding contact, e.g., if it is annular. A further problem caused by adhesive wear is an extremely high wear rate and severe surface damage as illustrated in Figure 12.17.

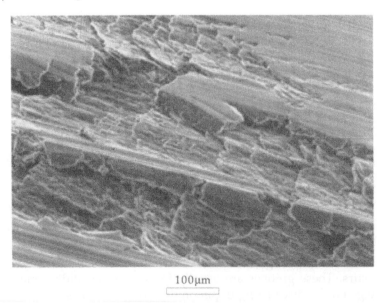

100μm

Figure 12.17 Al-Si alloy surface worn by adhesive wear. Note the formation of wear particles.

Experimental results have shown a much higher probability of wear particle generation due to asperity contacts during adhesive wear compared to, for example, abrasive wear [30]. The probability ranged from 10^{-2} for a mild steel/mild steel contact to 10^{-7} for a tungsten carbide/tungsten carbide contact [30]. For example, in cases of brass sliding on tool steel, where the adhesion is severe, about 0.2% of total asperity contacts results in wear particle formation. On the other hand, for adhesive wear of stellite on tool steel the percentage of the total asperity contacts producing particles is about 0.02% [31]. Successful operation of machinery, however, relies on far lower ratios of wear particles to asperity contacts, e.g., a ratio of $1/10^6$ is typical of mild wear [30]. Rapid wear is therefore a main reason why adhesive wear must usually be carefully controlled and suppressed.

Fortunately it is a relatively simple matter to reduce or even eliminate adhesion between solids. Contaminant layers of surface oxides and material impurities contribute to the reduction of adhesive wear. Adhesion can also be controlled by the application of specially formulated lubricants and careful selection of sliding materials.

Contaminant Layers Formed Due to Surface Oxidation and Bulk Impurities

Oxidation of metal surfaces can lower adhesion to acceptable levels. Almost all metals, when exposed to air, form very rapidly an oxide film of about 5 [nm] in thickness [32]. A 5 [nm] film is so thin that it is transparent and the metal remains shiny, but it radically changes the surface properties of the metal. The oxidized surface can be considered 'contaminated' as non-metal atoms are present on the surface. Oxygen, but not passive gases such as nitrogen or argon, is very useful as a universal 'lubricant'. Moisture [17] accentuates the effect of oxygen but the reasons for this are not yet clear. For example, adhesive wear of ceramics is notably reduced by adsorbed moisture [11,13].

Surface oxidation of metals occurs rapidly even at low temperatures and the rate at which oxidation proceeds is often limited only by the supply of oxygen until the critical 5 [nm] thickness is reached [33]. This means that under atmospheric pressure, oxide films, if removed, are reformed in just a few microseconds. This rapid re-growth means that a protective oxide film in a wearing contact can be sustained indefinitely. It is only under a vacuum or under a thick layer of oil that oxygen starvation may cause problems [34].

However, it has been shown that a small amount of contamination is necessary for strong adhesion [35]. Oxygen atoms, in particular, can bond both surfaces of contacting metals when present in very small quantities. The occurrence of strong adhesion between weakly contaminated surfaces agrees with high friction and seizure often taking place under atmospheric conditions. When air is present, even if oxide films are destroyed by wear, some contamination of nascent surfaces exposed by wear occurs. The fact that these surfaces can seize confirms the model of adhesion being promoted by small amounts of contamination but suppressed by heavy contamination.

Bulk material impurities reduce adhesive wear to a lesser but still useful extent compared to surface oxidation. The high temperatures associated with friction and wear promote the migration of impurities to the surface. Studies of the enrichment of exposed surfaces by the minor constituents of a material reveal that the adhesive wear resistance of iron is significantly improved by small quantities of carbon and sulphur [2]. Surface analysis of the worn carbon-rich irons shows a carbon layer covering the worn surface. Hence for this reason alloys and composite materials are usually superior to pure materials in terms of adhesive wear resistance.

Lubricants

The primary purpose of a lubricant is to suppress adhesive wear by providing a superior form of surface 'contamination'. Fatty acids and other polar organic substances are usually blended into lubricating oils. If these fatty acids adsorb on the top of the existing oxide layers, further reductions in friction and wear are obtained [17]. Lubrication mostly involves the application of this basic principle which has already been discussed in detail in Chapter 8.

Certain oils also contain additives rich in sulphur, phosphorus or chlorine. Metal surfaces can form sulphide, phosphide or chloride films just as readily as oxide films [2]. These films are intended to form when oxide films break down to provide a protective surface film against very high friction or seizure.

The importance of minimizing adhesive wear is such that lubricants are specially formulated to control it even at the cost of promoting other forms of wear or surface damage, e.g., corrosive wear, which will be discussed in the next chapter.

Favourable Combinations of Sliding Materials

The likelihood of severe adhesive wear occurring varies significantly between different combinations of sliding materials. A careful choice of materials can yield benefits of minimized wear and friction. An example of this is the combination of steel and bronze used for the shaft and bush in journal bearings The general rule is to avoid sliding similar or identical materials against each other [17].

The effects of mutual solubility of metals on their adhesion are still far from clear. Strong arguments have been made that metals which are mutually soluble should not be slid against each other [36]. However, closer examination of the data published [e.g., 37,38] shows that of the metals slid against iron, aluminium, which is only of a limited solubility in iron, causes higher friction than chromium, which is completely soluble. This can be explained by Buckley's hypothesis [2] which suggests that chemically active metals are electropositive, i.e., electron donors, and show much stronger adhesion to iron than passive or inert metals. Since aluminium is more electropositive than chromium its adhesion to iron is stronger than that of chromium.

Lead, tin, copper and silver, which are widely used in plain bearings to reduce friction, are metals of low chemical activity. Despite the cost, the noble metals, silver and gold, are used as bearing materials in demanding applications [2], e.g., silver coated worm gears or piston rings. Impure materials are also less reactive than pure materials and therefore wear less. For example, the performance of steel against steel is better than that of pure iron against pure iron [17].

Some polymers can also be used in sliding combinations with hard metals and the friction behaviour depends on the polymer characteristics. Ceramics sliding against themselves or against metals show quite high friction and adhesive wear rates once the temperature is high enough to desorb a lubricating moisture film [11]. These topics are discussed further in Chapter 16 on 'Wear of Non-Metallic Materials'.

12.4 SUMMARY

A well-disguised tendency for all materials to mutually adhere when brought into a close contact is the basic cause of adhesive wear. Although atmospheric contaminants and lubricants provide effective means of preventing adhesive wear they can never entirely eliminate it. Adhesion results in high coefficients of friction and serious damage to the contacting surfaces. In extreme cases, when adhesive wear is fully established, the friction and wear rate can be so high that it may be impossible for the contacting surfaces to continue sliding. Adhesive wear is the fundamental cause of failure of most metal sliding contacts and therefore its effective prevention is essential to proper functioning of engineering machinery.

REVISION QUESTIONS

12.1   What is the cause of adhesive wear?

12.2   Why does significant adhesion not occur when two metal bodies are pressed together under terrestrial conditions?

12.3   Under what conditions can a clean metal surface be observed?

12.4   What is the component of air that strongly affects adhesion?

12.5   If there is good adhesion between two surfaces, how much greater is the separation force than the original contact force?

12.6   What is the cause of adhesive film transfer between two contacting bodies of dissimilar mechanical strengths?

12.7   Is a large quantity of contaminant needed to significantly reduce adhesion between two contacting bodies?

12.8   Which class of materials adhere the most strongly with like and unlike substances?

12.9   What material properties promote adhesion in metals? Give examples of metals that should be avoided when adhesion is likely to occur.

12.10   In dry sliding of metallic alloys what steps can be undertaken to minimize adhesive wear?

12.11   Name two important mechanisms of adhesion.

12.12   The first stage in the mechanism of adhesive wear is adhesion between the contacting asperities. What, in general terms, is the second stage in ductile materials?

12.13   Why is rolling contact so much less likely to initiate adhesive wear than sliding contact?

12.14   Give an example of a beneficial effect of adhesive wear.

12.15   Can adhesive wear occur in lubricated contacts?

12.16   What is an effective way of controlling adhesive wear in lubricated contacts?

REFERENCES

1   F.P. Bowden and G.W. Rowe, The Adhesion of Clean Metals, *Proc. Roy. Soc., London*, Series A, Vol. 233, 1956, pp. 429-442.

2   D.H. Buckley, Surface Effects in Adhesion, Friction, Wear and Lubrication, Elsevier, 1981.

3   J.M. Ziman, Electrons in Metals - A Short Guide to the Fermi Surface, Taylor and Francis, London, 1963.

4   J. Ferrante and J.R. Smith, A Theory of Adhesion at a Bimetallic Interface: Overlap Effects, *Surface Science*, Vol. 38, 1973, pp. 77-92.

5   J. Ferrante and J.R. Smith, Metal Interfaces: Adhesive Energies and Electronic Barriers, *Solid State Communications*, Vol. 20, 1976, pp. 393-396.

6   H. Czichos, Tribology - a System Approach to the Science and Technology of Friction, Lubrication and Wear, Elsevier, 1978.

7   M.E. Sikorski, Correlation of the Coefficient of Adhesion With Various Physical and Mechanical Properties of Metals, *Transactions ASME, Series D - Journal of Basic Engineering*, Vol. 85, 1963, pp. 279-285.

8   D.H. Buckley, The Influence of the Atomic Nature of Crystalline Materials in Friction, *ASLE Transactions*, Vol. 11, 1968, pp. 89-100.

9   W.C. Wake, Adhesion and the Formulation of Adhesives, Applied Science Publishers, London, 1982, 2nd edition.

10   B.J. Briscoe and D. Tabor, Surface Forces in Friction and Adhesion, Faraday Special Discussions, Solid-Solid Interfaces, *Chem. Soc. London*, No. 2, 1972, pp. 7-17.

11   D.H. Buckley and K. Miyoshi, Tribological Properties of Structural Ceramics, NASA Technical Memorandum 87105, Lewis Research Centre, Cleveland, Ohio, 1985.

12   Y. Tsuya, Tribology of Ceramics, Proc. JSLE. Int. Tribology Conference, 8-10 July 1985, Tokyo, Japan, Elsevier, 1986, pp. 641-646.

13   D.H. Buckley and K. Miyoshi, Friction and Wear of Ceramics, *Wear*, Vol. 100, 1984, pp. 333-353.

14   W. Black, J.G.V. de Jongh, J.Th. Overbeek and M.J. Sparnaay, Measurements of Retarded van der Waals Forces, *Trans. Faraday Soc.*, 1960, Vol. 56, pp. 1597-1608.

15   A.D. Roberts, Surface Charge Contribution in Rubber Adhesion and Friction, *Journal of Physics, Series D: Applied Physics*, Vol. 10, 1977, pp. 1801-1819.

16   K.N.G. Fuller and D. Tabor, The Effect of Surface Roughness on the Adhesion of Elastic Solids, *Proc. Roy. Soc., London,* Series A, Vol. 345, 1975, pp. 327-342.

17   F.P. Bowden and D. Tabor, The Friction and Lubrication of Solids, Part 2, Oxford University Press, 1964.

18   J.M. Challen and P.L.B. Oxley, Plastic Deformation of a Metal Surface in Sliding Contact with a Hard Wedge: Its Relation to Friction and Wear, *Proc. Roy. Soc., London,* Series A, Vol. 394, 1984, pp. 161-181.

19   F.P. Bowden and T.P. Hughes, Friction of Clean Metals and the Influence of Adsorbed Gases. The Temperature Coefficient of Friction, *Proc. Roy. Soc., London,* Series A, Vol. 172, 1939, pp. 263-279.

20   F.P. Bowden and J.E. Young, Friction of Clean Metals and the Influence of Adsorbed Films, *Proc. Roy. Soc., London,* Series A, Vol. 208, 1951, pp. 311-325.

21   J.S. McFarlane and D. Tabor, Relation between Friction and Adhesion, *Proc. Roy. Soc., London,* Series A, Vol. 202, 1950, pp. 244-253.

22   T. Kayaba and K. Kato, The Analysis of Adhesive Wear Mechanism by Successive Observations of the Wear Process in SEM, Proc. Int. Conf. on Wear of Materials, Dearborn, Michigan, 16-18 April 1979, editors: K.C. Ludema, W.A. Glaeser and S.K. Rhee, Publ. American Society of Mechanical Engineers, New York, 1979, pp. 45-56.

23   O. Vingsbo, Wear and Wear Mechanisms, Proc. Int. Conf. on Wear of Materials, Dearborn, Michigan, 16-18 April 1979, editors: K.C. Ludema, W.A. Glaeser and S.K. Rhee, Publ. American Society of Mechanical Engineers, New York, 1979, pp. 620-635.

24   M. Kerridge and J.K. Lancaster, The Stages in a Process of Severe Metallic Wear, *Proc. Roy. Soc., London,* Series A, Vol. 236, 1956, pp. 250-264.

25   M. Cocks, Interaction of Sliding Metal Surfaces, *Journal of Applied Physics,* Vol. 33, 1962, pp. 2152-2161.

26   M. Antler, Processes of Metal Transfer and Wear, *Wear,* Vol. 7, 1964, pp. 181-204.

27   T. Sasada, S. Norose and H. Mishina, The Behaviour of Adhered Fragments Interposed between Sliding Surfaces and the Formation Process of Wear Particles, Proc. Int. Conf. on Wear of Materials, Dearborn, Michigan, 16-18 April 1979, editors: K.C. Ludema, W.A. Glaeser and S.K. Rhee, Publ. American Society of Mechanical Engineers, New York, 1979, pp. 72-80.

28   D.A. Rigney, L.H. Chen, M.G.S. Naylor and A.R. Rosenfeld, Wear Process in Sliding Systems, *Wear,* Vol. 100, 1984, pp. 195-219.

29   K. Komvopoulos, N. Saka and N.P. Suh, The Mechanism of Friction in Boundary Lubrication, *Transactions ASME, Journal of Tribology,* Vol. 107, 1985, pp. 452-463.

30   J.F. Archard, Single Contacts and Multiple Encounters, *Journal of Applied Physics,* Vol. 32, 1961, pp. 1420-1425.

31   J.F. Archard, Contact and Rubbing of Flat Surfaces, *Journal of Applied Physics,* Vol. 24, 1953, pp. 981-988.

32   N.D. Tomashov, Theory of Corrosion and Protection of Metals, MacMillan, New York, 1966.

33   F.P. Fehlner and N.F. Mott, Low Temperature Oxidation, *Oxidation of Metals,* Vol. 2, 1970, pp. 59-99.

34   A.W. Batchelor and G.W. Stachowiak, Some Kinetic Aspects of Extreme Pressure Lubrication, *Wear,* Vol. 108, 1986, pp. 185-199.

35   W. Hartweck and H.J. Grabke, Effect of Adsorbed Atoms on the Adhesion of Iron Surfaces, *Surface Science,* Vol. 89, 1979, pp. 174-181.

36   E. Rabinowicz, The Influence of Compatibility on Different Tribological Phenomena, *ASLE Transactions,* Vol. 14, 1971, pp. 206-212.

37   A.E. Roach, C.L. Goodzeit and R.P. Hunnicutt, Scoring Characteristics of 38 Different Elemental Metals in High-Speed Sliding Contact With Steel, *Transactions ASME,* Vol. 78, 1956, pp. 1659-1667.

38   C.L. Goodzeit, R.P. Hunnicutt and A.E. Roach, Frictional Characteristics and Surface Damage of 39 Different Elemental Metals in Sliding Contact with Iron, *Transactions ASME,* Vol. 78, 1956, pp. 1669-1676.

C O R R O S I V E
A N D
O X I D A T I V E   W E A R

13.1   INTRODUCTION

Corrosive and oxidative wear occur in a wide variety of situations both lubricated and unlubricated. The fundamental cause of these forms of wear is a chemical reaction between the worn material and a corroding medium which can be a chemical reagent, reactive lubricant or even air. Corrosive wear is a general term relating to any form of wear dependent on a chemical or corrosive process whereas oxidative wear refers to wear caused by atmospheric oxygen. Both these forms of wear share the surprising characteristic that a rapid wear rate is usually accompanied by a diminished coefficient of friction. This divergence between friction and wear is a very useful identifier of these wear processes. The questions are: how can corrosive and oxidative wear be controlled? To what extent do the chemical reactions and corrosion processes dictate the mechanism and rate of wear? Are corrosive and oxidative wear relatively benign or can they cause immediate failure of machinery? The answers to these questions are in the understanding of the mechanisms of corrosive and oxidative wear which are discussed in this chapter.

13.2   CORROSIVE WEAR

The surface chemical reactions which are beneficial in preventing adhesive wear will, if unchecked, lead to a considerable loss of the underlying material. If a material (metal) is corroded to produce a film on its surface while it is simultaneously subjected to a sliding contact then one of the four following processes may occur [1]:

·   a durable lubricating film which inhibits both corrosion and wear may be formed;

·   a weak film which has a short life-time under sliding contact may be produced and a high rate of wear may occur due to regular formation and destruction of the films. The friction coefficient may or may not be low in this instance;

·   the protective surface films may be worn (e.g., by pitting) and a galvanic coupling between the remaining films and the underlying substrate may result in rapid corrosion of the worn area on the surface;

·   the corrosive and wear processes may act independently to cause a material loss which is simply the sum of these two processes added together.

These hypothetical models of corrosive wear are illustrated schematically in Figure 13.1.

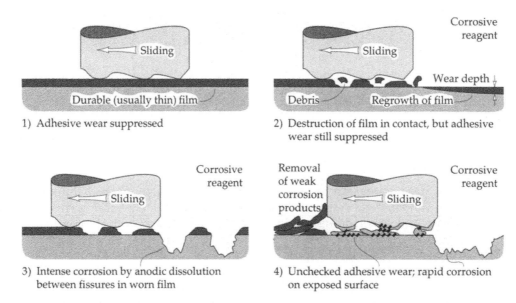

1) Adhesive wear suppressed

2) Destruction of film in contact, but adhesive wear still suppressed

3) Intense corrosion by anodic dissolution between fissures in worn film

4) Unchecked adhesive wear; rapid corrosion on exposed surface

FIGURE 13.1 Models of interaction between a corrosive agent and a worn surface.

The first process is dominated by the formation of durable lubricating films. If such films prevail then the worn contacts are well lubricated and corrosive wear does not occur. Unfortunately very few corrosion product films are durable so that this category of film formation is rarely seen in practice. The second process is related to the formation of a sacrificial or short life-time corrosion product film under sliding contacts. This is the most common form of corrosive wear since most corrosion films consist of brittle oxides or other ionic compounds. For example, the oxides of iron are extremely brittle at all but very high temperatures [2]. During this process corrosive wear occurs on the 'active areas' of the worn surface [45]. Active areas, which are equivalent to the true contact area, corrode actively compared to the surrounding passivated surface. The concept of active areas was used to explain most aspects of the corrosive wear of steel in dilute sulphuric acid [45]. At moderate loads, wear is entirely a result of rapid corrosion at these 'active areas' which are regenerated by mechanical film removal during sliding contact. At higher loads, an additional subsurface mechanism of wear may cause a rapid increase in wear [45]. The third process relates to wear in highly corrosive media while the fourth process is effectively limited to extremely corrosive media where the corrosion products are very weak and are probably soluble in the liquid media. It is very unlikely that wear and corrosion, if occurring in the same system, can proceed entirely independently since the heat and mechanical agitation of a sliding contact would almost inevitably accelerate corrosion.

The formation and subsequent loss of sacrificial or short life-time corrosion films is the most common form of corrosive wear. This form of wear can be modelled as a process of gradual buildup of a surface film followed by a near instantaneous loss of the film after a critical period of time or number of sliding contacts is reached. Since most corrosion films passivate or cease to grow beyond a certain thickness, this is a much more rapid material loss than static corrosion alone. The process of corrosive wear by repeated removal of passivating films is shown schematically in Figure 13.2 [3].

The model implies that a smooth worn surface together with corrosion products as wear debris are produced during the wear process which is well confirmed in practice [4]. Figure 13.2 suggests that static corrosion data could be applied to find the wear rate which would obviously be extremely useful. Unfortunately frictional temperature rises and mechanical activation during the process of tribocorrosion prevent this [5].

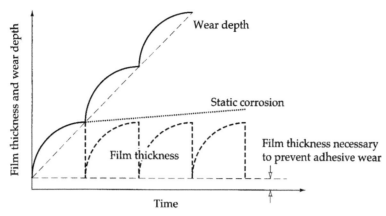

FIGURE 13.2   Model of corrosive wear by repeated removal of passivating films [3].

Recently, a model of corrosive wear occurring in hard-coated metallic alloys has been proposed. When such a coating is present, a sub-surface controlled mechanism of corrosive wear may become significant. Defects in the hard coating allow the corrosive fluid to enter cavities lying below it. Corrosion, accentuated by a galvanic coupling between the coating and the substrate, then proceeds to form voluminous corrosion products beneath the external hard coating. The accumulation of corrosion products eventually forces the outer coating to rise, forming a blister, which is then fractured and fragmented by sliding contact, forming a large pit in the surface [46], as schematically illustrated in Figure 13.3.

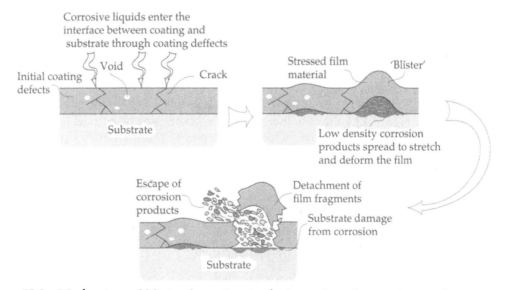

Figure 13.3   Mechanism of blister formation in thick coating of corrosion products.

Typical examples of corrosive wear can be found in situations when overly reactive EP additives are used in oil (condition sometimes dubbed as 'lubricated wear' [6]) or when methanol, used as a fuel in engines, is contaminated with water and the engine experiences a rapid wear [7].

Another example of corrosive wear, extensively studied in laboratory conditions, is that of cast iron in the presence of sulphuric acid [8]. The corrosivity of sulphuric acid is very sensitive to the water content and increases with acid strength until there is less water than acid. Pure or almost pure acid is only weakly corrosive and has been used as a lubricant for chlorine compressors where oils might cause an explosion [9]. Typical wear and frictional characteristics of cast iron at different concentrations of sulphuric acid are shown in Figure 13.4.

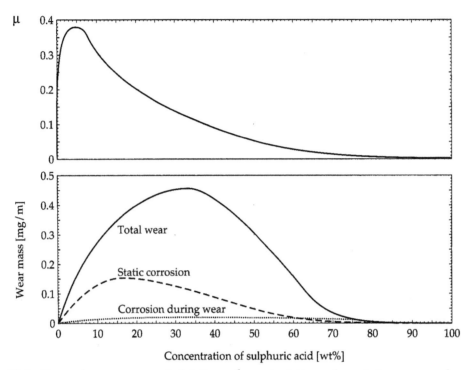

FIGURE 13.4 Corrosion, wear and friction characteristics of cast iron as a function of concentration of sulphuric acid in water at a sliding speed of 0.13 [m/s] and an apparent contact stress of 1.7 [MPa] [8].

It can be seen from Figure 13.4 that the increase in the coefficient of friction is not followed immediately by the increase in wear rate which is the specific characteristic of corrosive wear. The concentration of sulphuric acid significantly affects the wear rate which is also reflected by the changes in the wear scar morphology as shown in Figure 13.5.

FIGURE 13.5 Worn surface of cast iron at various concentrations of sulphuric acid [8].

It can be seen from Figure 13.5 that at low concentrations of sulphuric acid, up to 30% strength, a severe surface attack with pitting prevails. These concentrations of acid are so corrosive that a significant amount of material is lost as pitting corrosion of the nonload bearing areas of the surface and this represents the wear mechanism at extreme levels of corrosivity. At 65% acid strength a sacrificial film mechanism of wear becomes dominant while at 96% acidity a thin long-life film is established which results in a significant reduction of wear and friction.

A common instance of corrosive wear is that due to oxygen present in oil. It has been shown that for steel contacts lubricated by oil, most of the wear debris consists of iron oxide and that removal of oxygen from the oil virtually eliminates wear [4]. When oxygen is present the worn surfaces become smooth with no pits which indicates wear by periodic removal of a sacrificial film. This form of wear, however, is usually classified as oxidative wear and is discussed in more detail further on in this chapter.

The mechanism of corrosive wear and smoothing of the worn surface for oil covered surfaces in contact with air is illustrated in Figure 13.6.

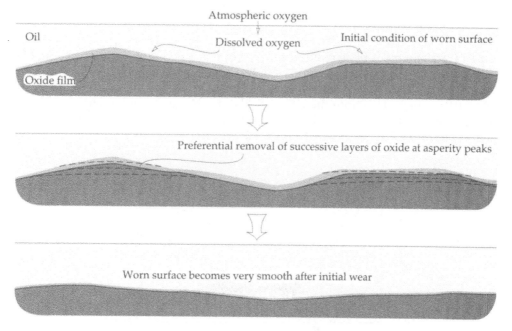

FIGURE 13.6 Mechanism of smoothing of the worn surface by mild corrosive wear resulting from atmospheric oxygen dissolved in the lubricating oil.

The adoption of ethanol as a fuel for vehicles creates many problems of corrosive wear in engines. The difficulty with using ethanol or a solution of ethanol in gasoline (known as 'gasohol') is that ethanol is hygroscopic and causes water contamination of the fuel. The absorbed water initiates a form of corrosive wear on engine surfaces [7]. Concentrations of water as low as 1% can cause significant increases in wear [7] and strict precautions against water contamination of fuel must be undertaken. The problem of corrosive wear is even more acute when methanol is substituted for ethanol. Absorption of water causes a methanol-gasoline solution to divide into two phases, a methanol-gasoline phase and a methanol-water phase and when this happens, damage to the engine can be very severe.

Corrosive wear is accentuated at elevated temperatures. For example, a 20°C increase in temperature may double the corrosive wear rate [8]. Therefore cooling of the operating surfaces is necessary to suppress many corrosive wear problems.

Wear of hard materials by much softer materials is always difficult to explain. It seems that the corrosive wear may provide the explanation of the wear of steel by rubber in sliding contact. It is suspected that the frictional heat and mechanical activation are sufficient to induce a surface chemical reaction between rubber and steel. A metallo-organic compound, consisting of iron and part of the molecular chain of the original elastomer (rubber) molecule, is believed to be formed, together with oxidized organic iron compound [47]. These compounds are probably mechanically weak and easily detach from the worn surface to complete the cycle of the corrosive wear mechanism.

Metals and alloys immersed in liquid metals are also subjected to corrosion accompanied by the formation of intermetallic compounds. Under sliding in liquid metals, a form of corrosive wear develops with periodic removal of an outer layer of corrosion product on the worn surface. Many intermetallic compounds formed are harder than the parent metals and may lead to abrasion of the parent metals by particles of the newly formed inter-metallic compounds [48]. There is significant variation in the wear resistance of metallic alloys, such as cobalt and iron-based superalloys, immersed in liquid zinc [48].

Transition between Corrosive and Adhesive Wear

As the corrosivity of a medium is reduced it may become a good lubricant at a certain level of load and sliding speed. However, an excessive reduction of corrosivity or reactivity of a lubricant may result in severe adhesive wear because of the insufficient generation of protective surface films. Therefore the composition of lubricants can be optimized to achieve a balance between corrosive and adhesive wear which gives the minimum wear rate, as illustrated in Figure 13.7.

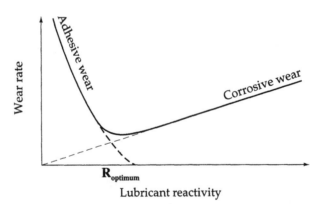

FIGURE 13.7   Balance between corrosive and adhesive wear.

It can be seen from Figure 13.7 that if the lubricant reactivity is too low then adhesive wear dominates which can be severe. On the other hand, if the lubricant reactivity is too high, then corrosive wear becomes excessive. Thus there is an optimal lubricant reactivity for particular operating conditions.

The transition between sufficient corrosion to generate protective films and adhesive wear is present in most systems [10]. The transition is dependent on load since as the load is increased, more asperities from opposing surfaces make contact at any given moment so that the average time between successive contacts for any single asperity is reduced. Consequently a higher additive or media reactivity is required when load is increased. Load dependence on lubricant reactivity is illustrated schematically in Figure 13.8.

The classic evidence of the transition between corrosive and adhesive wear is provided by experiments involving steel contacts lubricated by oil containing various concentrations of dissolved oxygen [10]. In the tests conducted in a 'four-ball' machine it was found that the wear rate was reduced as oxygen concentration was lowered until a critical point was reached

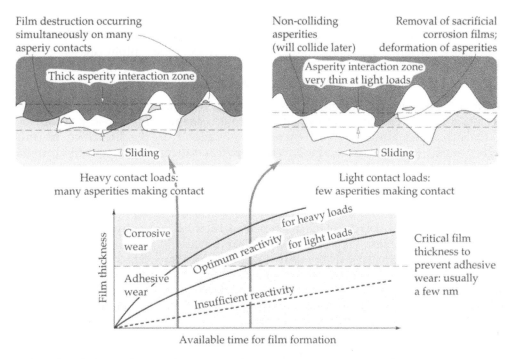

FIGURE 13.8   Load dependence of transition between corrosive and adhesive wear.

at which the wear rate sharply increased. There was an optimum oxygen concentration in the oil which gave a minimum wear rate of the steel for any specific level of load. The optimum oxygen concentration also rose with increasing load as shown in Figure 13.9.

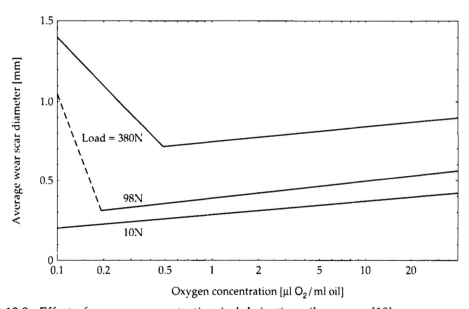

FIGURE 13.9   Effect of oxygen concentration in lubricating oil on wear [10].

A similar balance between corrosive and adhesive wear was also observed when sulphur was used as the corrosive agent [11]. In tests where copper pins were slid against a hard steel disc lubricated by pure hexadecane with various concentrations of sulphur, a clear minimum in wear rate at 0.01 [wt%] of sulphur was obtained.

There is, however, no universal value of corrosive agent concentration that gives a minimum wear rate since it varies with load and temperature and has to be found experimentally for any particular sliding contact. For example, sulphur-based EP additives function by controlled corrosion and the choice of the EP additive (i.e., mild or strong) is based on finding the minimum wear rate for specific conditions.

Although the presence of the minimum wear rate has been conclusively demonstrated, there is still a lack of evidence as to whether there is a change in the wear mechanism at this point. Only a small amount of information is available in the literature on this topic although it seems that detailed studies of wear debris and worn surfaces could provide some answers.

Synergism between Corrosive and Abrasive Wear

Abrasion can accelerate corrosion by the repeated removal of passivating films and a very rapid form of material loss may result. This wear process is particularly significant in the mineral processing industries where slurries containing corrosive chemicals and abrasive grits must be pumped, transported and stirred. The first report of the process which was well disguised by the fact that although corrosion was occurring the surface remained untarnished due to the prompt removal of corrosion products was provided by Zelders [12]. The generally accepted model of corrosive-abrasive wear is shown in Figure 13.10. The model is based on cyclic film formation and removal by corrosive and abrasive action, respectively. Material is lost in a pattern similar to the model of corrosive wear illustrated in Figure 13.2.

This mechanism of wear prevails when the rate of mechanical abrasion under dry conditions is less than the corrosion rate without abrasive wear [13]. When mechanical abrasion is more intense, corrosive effects become insignificant [13]. It is probable that when corrosion is slow compared to abrasive wear the grits remove underlying metal with little interference from the corrosion film. Resistance of materials under corrosion-abrasion depends on their resistance to corrosion. It has been shown that a soft but non-corrodible organic polymer can be more long-lasting as a lining of a slurry pipe than a hard but corrodible steel [14].

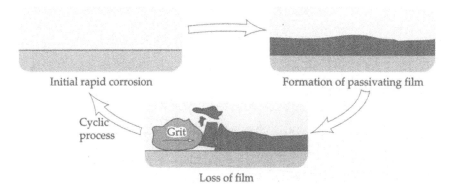

FIGURE 13.10 Cyclic removal of corrosion product films by abrasion (adapted from [13]).

Although most of the studies on corrosive-abrasive wear have concentrated on aqueous systems, lubricating oils contaminated by grits may also cause wear by the same process. It has been found that the contamination of lubricating oils by silica (sand) or iron oxides at levels as low as 0.01% by weight can so dramatically accelerate the wear of rotary compressors that the unit fails after a few hours of operation [15]. In such cases it appears that corrosive-abrasive wear dominates since large quantities of iron-oxide debris as well as metallic debris are found.

When the oxide formed by corrosion is much harder than the parent metal, there might be a significant acceleration of wear, even when sliding against a much softer counterface. A soft counterface can provide a resilient bed for abrasive particles of oxide, leading to rapid wear of the metallic surface [49]. This mode of corrosive-abrasive wear is of significance to biomedical implants such as orthopaedic prostheses. Corrosion-abrasion of the titanium-aluminium-vanadium alloys used as implant materials leads to their rapid wear even in contact with a soft UHMWPE counterface. Perhaps for this reason, these alloys are not usually used for articulating surfaces in combination with UHMWPE in artificial hip replacements.

Tribochemical Polishing

It needs to be realized that the corrosive wear is not always an entirely destructive process. For example, corrosive wear can be utilized to produce very smooth surfaces where the component is polished in a moderately active chemical reagent or water. This technique is known as tribochemical or chemo-mechanical polishing and can be used to polish a range of ceramics and metals. The principle of tribochemical polishing is similar to that illustrated in Figure 13.6. The thin corrosion film is cyclically formed and removed from asperities by a rubbing solid and the worn surface is gradually levelled attaining at the end a very high level of polish. It appears that the only common metal that cannot be effectively polished in this manner is aluminium [39]. Aluminium is covered with a film of aluminium oxide which is chemically unreactive compared to the underlying aluminium. This means that the aluminium oxide tends to remain intact while an intense corrosion of aluminium, exposed by small holes in the oxide layer, occurs resulting in rough and pitted surface.

In some cases of tribochemical polishing the depth of indentation is comparable to the diameter of an atom [50]. In such cases, it is more useful to consider the removal of individual atoms rather than cyclic destruction of a corrosion film. It is hypothesized that the initial stage of tribochemical polishing is the chemical reaction between surface atoms of the processed surface and reactants in the polishing fluid. As a consequence of this reaction or reactions the surface atoms lose their strong bonding to the subsurface atoms and may be detached by mechanical contact [50].

Tribopolishing is not only limited to metals. In hard ceramics, sliding contact between boron carbide coating and steel leads to polishing of the boron carbide. Boron carbide is much harder than steel and the polishing is caused by mild corrosive wear in which boron carbide oxidizes to form softer boric oxides and acids (in humid conditions) [51].

Tribochemical polishing has been used to generate very smooth surfaces on silicon nitride, silicon and silicon carbide [39]. When silicon nitride slides in water the tribocorrosion products formed at the contacting asperities dissolve and this subsequently results in very flat surfaces with extremely low surface roughness. For example, surface roughness as low as Ra = 0.5 [nm] at 50 [μm] cut-off and 4 [nm] at 8 [mm] cut-off has been reported [39]. This method of surface polishing does not produce surface defects such as scratches and pits often found on traditionally polished surfaces with fine abrasive particles.

Tribochemical polishing is effective only when the following conditions are satisfied:

· friction at the asperities is sufficient to generate tribocorrosion products,

· mechanical stresses are low to avoid any surface deformation or fracture, and

· corrosive medium is not too aggressive to cause corrosion outside the frictional contact [39].

13.3   OXIDATIVE WEAR

Oxidative wear is the wear of dry unlubricated metals in the presence of air or oxygen. Atmospheric oxygen radically changes the friction coefficients and wear rates of dry sliding metals and there are several different mechanisms involved in the process. Oxidative wear was postulated when changes in the chemical composition of wear debris generated in dry sliding of steels under different levels of load and sliding speed were observed [16]. It was found that when the load and sliding speed were high enough to increase the frictional contact temperature to several hundred degrees Celsius, the wear debris changed from metallic iron to iron oxides. It was hypothesized later that when thick oxide films were formed on the worn surfaces then 'mild wear' prevailed [17] and if the thick oxide films were absent or broken down then 'severe wear', which is a form of adhesive wear, was inevitable [17]. Oxidative or mild wear shows a moderate and stable coefficient of friction of about **0.3 - 0.6** compared to much larger fluctuating values for severe wear. The characteristic features of oxidative wear are smooth wear surfaces and small oxidized wear debris.

Instances of oxidative wear can be found in cases when a high process temperature causes rapid oxidation and the formation of thick oxide films. Examples are found in some metal working operations such as hot rolling and drawing of steels. The hole piercer used in the hot drawing of tubes, Figure 13.11, provides a particularly good example of oxidative wear [18]. A multilayer 'cap' of scale and deformed metal accumulates at the tip of the piercer and the thickness of the scale (thick oxide film) can be as high as **0.1** [mm].

A more mundane example of oxidative wear takes place in train wheels when a cast iron brake block is applied to a rotating steel tyre.

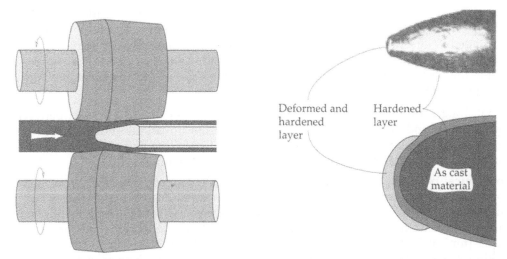

FIGURE 13.11 Thick oxidative wear scales formation on piercing tools (adapted from [18]).

Kinetics of Oxide Film Growth on Metals at High and Low Temperatures

Oxide films are present on almost all metals and will form on any clean metal surface exposed to oxygen even at cryogenic temperatures. The oxidation rate of metals is dependent on temperature as is expected of a chemical reaction. The kinetics of metallic surface oxidation has a controlling influence on oxidative wear. At low or ambient temperatures, e.g., 20°C, the oxidation of metal is initially rapid and is immediately followed by the passivation of the surface which limits the oxide film thickness [19]. The limiting film thickness can be as low as **2** [nm] (about 5 atom layers) for steels when the temperature is below 200°C [20]. If the temperature of steel is increased to, for example, 500°C, almost unlimited oxidation occurs which results in a very thick oxide film, e.g., in the range of **1 - 10** [μm]. The distinction between these two forms of oxidation is illustrated in Figure 13.12.

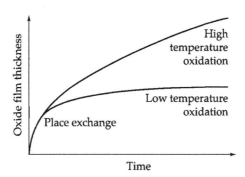

FIGURE 13.12 Kinetics of metal oxidation at high and low temperatures.

The difference in oxidation kinetics results from much more rapid movement of oxygen or metal ions across the oxide film at high temperatures, when solid state diffusion is sufficient to ensure adequate transport of the metal or oxide ions necessary for a continuous film growth [19]. At low temperatures, an electric field exerted by the difference in electrochemical potential across the film [20] or an activated mechanism of 'place exchange' where atoms of oxygen and iron exchange positions in the oxide film crystal lattice [21] is necessary to ensure oxidation. These latter mechanisms are limited, however, to very thin oxide films which is the reason for the effective termination of oxidation at low temperatures once a critical oxide film thickness is reached [20,21].

At low temperatures, the oxide films are extremely beneficial since they form rapidly and effectively suppress adhesive wear. If a system operates under mild oxidational conditions wear is greatly reduced. The oxide layers formed are supported by the strain-hardened substrate layers which are generated due to plastic deformation. At high temperatures, however, oxidation resembles corrosion in its high rate of reaction and can become a direct cause of increased wear. This rapid oxidation at high temperatures forms the base of oxidative wear. The high temperatures can either be imposed externally or can be due to high frictional heating at high speeds and loads as illustrated in Figure 13.13.

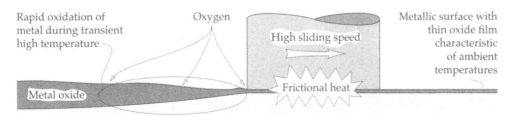

FIGURE 13.13 Rapid oxidation of metallic surfaces at high contact temperatures.

· *Oxidative Wear at High Sliding Speeds*

At sliding speeds above **1** [m/s] the surface flash temperatures can be as high as several hundred degrees Celsius and if the load is low enough to permit mild wear, oxide films several micrometres thick can buildup on the worn surface [22]. Under these conditions the oxidation proceeds very rapidly, especially at the high contact spots. Because the oxide layers formed are thick enough to physically separate the wearing surfaces, it is reasoned that the oxidative wear which occurs must be due to the formation and removal of these thick oxide layers. A fairly accurate model to predict wear rates from basic parameters such as load, speed, and static oxidation characteristics has been developed [23]. This model, however, involves the theory of oxide film growth which, although well established, is very complex [19] and too specialized to be described in detail in this book.

At the onset of sliding contact, at high speed, the thin oxide films present on unworn steel surfaces are rapidly destroyed and the friction and wear rates increase, initiating a period of severe wear. Then by some poorly understood processes the worn surface recovers and a state of mild wear is reached. The thick oxide layers are established and the wear rate declines markedly. When each oxide layer reaches a critical thickness, it becomes too weak to withstand the load and frictional shear stress and is removed during the sliding.

An alternative mechanism of oxide layer removal is due to a fatigue process which is initiated after a certain number of contacts with the opposing surface is reached [24,25]. The sequence of events associated with the formation and removal of oxide layers is illustrated schematically in Figure 13.14 [26].

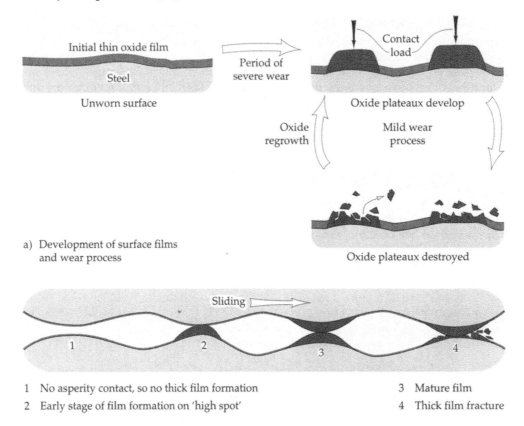

a) Development of surface films and wear process

1   No asperity contact, so no thick film formation
2   Early stage of film formation on 'high spot'
3   Mature film
4   Thick film fracture

b) Film morphology under steady-state wear

FIGURE 13.14 Mechanism of oxidative wear at high sliding speeds (adapted from [26]).

It should be realised that at high sliding speeds any asperity of a surface is subjected to a random sequence of short periods of high temperature oxidation when contact is made with asperities of an opposing surface as shown schematically in Figure 13.15. The kinetics of oxidation is governed by the temperature level at the asperity contacts, i.e., 'high spots'.

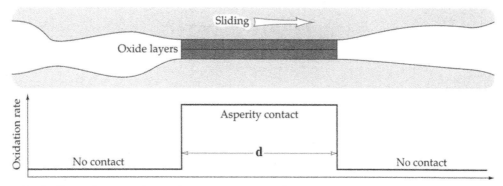

FIGURE 13.15 Periodic rapid oxidation between the asperities in dry high-speed sliding contact; **d** is average diameter of thick oxide patches [26].

The average frequency and duration of the high temperature oxidation periods and the oxidation temperature, as well as the average lifetime of the oxide layers, determine the wear rate under these conditions [23]. The oxidative wear model was developed based on the classical theories of high temperature oxidation [23]. This, however, seems to be a highly controversial assumption since there are distinct differences between oxidation under 'static conditions' and 'tribo-oxidation' [26]. For example, in static corrosion, a uniform temperature across the oxide film is almost certain. In contrast, in tribo-oxidation a temperature variation across the oxide film is probable [27] and this will almost certainly affect the oxidation rate. The contact frequency, between asperities of opposite surfaces in the sliding contact, seems to affect the oxidative wear rate. However, it is not easy to discriminate between the effect of contact frequency and the effect of increasing frictional temperatures with the increase in sliding speed [52].

· *Oxidative Wear at Low Sliding Speeds*

When a steel surface is exposed to air and subjected to low speed sliding wear the initial thin (about 2 [nm]) films are rapidly worn away and a period of severe or adhesive wear results. At low sliding speeds below **1** [m/s] frictional temperature rises are not high enough to cause rapid oxidation at the asperity tips. Although thick oxide films still form on the worn surface, they are the result of wear debris accretion, not direct oxidation. The fractured oxides and oxidized metallic wear particles compact to form oxide 'islands' on the worn surface. The area of these 'islands' increases with the sliding distance. The development of 'islands' is accompanied by a progressive reduction in the coefficient of friction [28]. The top surface of the 'islands' is smooth and consists of plastically deformed fine oxide debris. Directly underneath this top layer there is mixture of much larger oxide and oxidized particles. This sequence of events taking place during the process of oxidative wear at low sliding speeds is schematically illustrated in Figure 13.16.

The process of wear at low sliding speeds is particularly effective in forming debris consisting of a finely divided mixture of oxides and metal. Wear particles are formed and successively deformed, a process which creates a continuous supply of nascent metallic surface for oxidation by atmospheric oxygen. For example, oxidative wear of magnesium alloy sliding against hardened steel counterfaces was observed to be associated with mild wear of the magnesium alloy. Wear debris consisted of a mixture of magnesium metal, magnesium oxides, iron and iron oxides. The non-oxide fraction of the wear debris increased with the rise in contact load [53]. The mechanism of formation of such particles involving wear debris oxidation and oxide-metal blending is illustrated schematically in Figure 13.17.

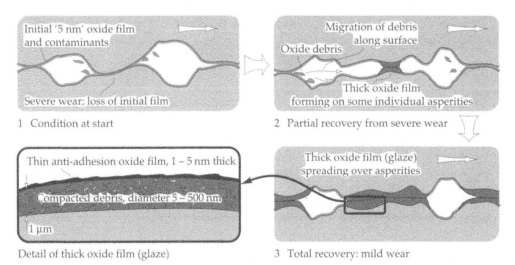

1 Condition at start   2 Partial recovery from severe wear

Detail of thick oxide film (glaze)   3 Total recovery: mild wear

FIGURE 13.16 Mechanism of oxidative wear at low sliding speeds (adapted from [26]).

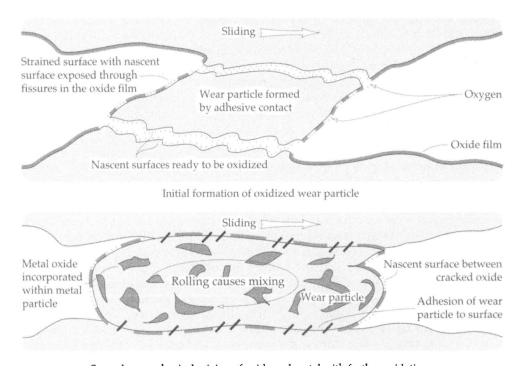

Initial formation of oxidized wear particle

Secondary mechanical mixing of oxide and metal with further oxidation

FIGURE 13.17 Formation of debris consisting of a mixture of oxide and metal.

The variation in friction coefficient with sliding distance or more precisely with the number of sliding cycles depends on the spread of thick oxidized layers over the surface. It declines from an initially high value to a more moderate value as the thick oxidized layers attain almost complete coverage of the worn surface [28].

· *Oxidative Wear at High Temperature and Stress*

When the temperature is progressively increased from close to ambient to several hundred degrees Celsius, oxidative wear of a metal becomes more intense. The time necessary for the development of the wear protective, compacted oxide layers, is reduced [29] and the quantity of oxidized wear particles and the thickness of the oxide film are dramatically increased. This is associated with the increase in oxidation rate at higher temperatures. Usually the fractured fine metallic debris which remains on the worn surface is oxidized and compacted into a 'glaze'. As the glaze spreads over the worn surface the wear process becomes 'mild'. A practical example of this form of wear is found in gas turbine components where thermal cycling causes slow periodic movements between contacting surfaces [29].

During a high temperature oxidative wear, a 'glaze' commonly forms on ferrous alloys and hard metal alloys based on cobalt and nickel (e.g., Stellite). The glaze resembles a glassy phase and lies above a layer of oxidized metal with deformed metal below as illustrated in Figure 13.18.

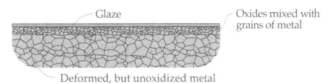

Figure 13.18 Schematic illustration of glaze formation on the metallic surfaces during a high temperature oxidative wear.

In microstructural terms, the glaze can be classified as 'nano-structured' where extreme fragmentation and deformation of the metal and oxide particles have occurred [54].

A similar oxide-debris glaze formation was also observed in nickel-based alloys, such as Nimonic 80A [55]. After an initial period of high sliding friction, a gradual decline to a long-term value of sliding friction was observed. Once the glaze is established, wear rates decline to negligible values. A minimum temperature close to 250°C is required for the glaze to form. Below this temperature friction and wear remain comparatively high. Glaze formation is sensitive to partial pressure of oxygen, i.e., glaze formation is inhibited and wear accelerated for partial pressures lower than 50 [kPa] (~0.5 atmospheres) [55].

Nickel-based alloys such as Inconel are structurally useful to temperatures as high as 800°C. It has been found that the wear rates of these materials in open air tend to increase at temperatures above 400°C and 800°C while the friction coefficient exhibits a contrary trend [40]. This decline in friction coefficient is attributed to increased plasticity of the oxide film at elevated temperatures. Stellite, which is a cobalt based alloy, also exhibits a similar transition from mechanical wear, characterised by severe plastic deformation, to oxidative wear at temperatures exceeding 400°C. However, the rise in wear rate with temperature is much lower than that for nickel based alloys.

At high operating temperatures, the undisturbed oxidization rate of metals becomes very high and if erosive wear occurs, the eroding particles are unlikely to reach the underlying metal. Tests on iron and nickel based alloys at 550°C revealed that erosion accelerated the oxidization of the alloy by mechanical damage to the protective oxide layer. This was caused by cracking and lamellar detachment (flaking) of the oxide layer which enhanced oxygen access to the metal alloy [56].

At high contact stresses, e.g., at the contact between rolled steel and its hot rolling work roll, a different mode of oxidative wear can occur. In this case, the contact stresses are sufficient to cause plastic deformation of the roll surface and fracture of the oxide film. A phenomenon known as 'peeling' occurs where oxide penetrates the worn surface to enclose a quantity of surface material [41]. The brittleness of the oxide (mostly iron oxide) facilitates detachment of lumps of metal by sudden fracture of the oxide. Loss of large lumps of steel from an apparently undamaged roll then occurs, hence the name 'peeling'. Peeling can cause rapid damage to the roll and should be avoided wherever possible.

· *Oxidative Wear at Low Temperature Applications*

Oxidative wear persists even at very low temperature applications, e.g., in rocket engine turbopumps [42-44]. Rocket turbopumps contain rolling bearings which operate under immersion in liquid oxygen or liquefied hydrocarbon gases. As these bearings must remain functional for the duration of the rocket flight, careful control of wear in the hostile environment of liquid oxygen is essential. A comparative laboratory wear tests of ball bearings in liquid oxygen and liquid hydrogen revealed more rapid wear by liquid oxygen [44]. The examination of worn bearings from space shuttle turbopumps indicated that excessive damage was due to adhesive wear/shear peeling following the breakdown of the oxide scale formed on the balls and rings [42].

· *Transition between Oxidative and Adhesive Wear*

Sharp transitions, referred to as 'T_1' and 'T_2' [30], between oxidative (mild) and adhesive (severe) wear are observed in metal-to-metal dry sliding contacts. The transition loads depend on the material properties of the sliding surfaces and their relative velocity. The relationship between wear rates and applied load for sliding steel contacts is schematically illustrated in Figure 13.19.

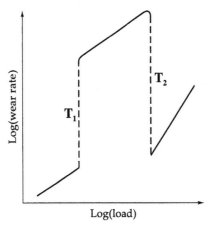

FIGURE 13.19 Transition between oxidative and adhesive wear (adapted from [30]).

Below the transition 'T_1' the surfaces are separated by oxidized layers and the wear debris consists of small oxidized particles. Between 'T_1' and 'T_2' higher contact loads cause the breakdown of the protective oxide layers and metallic particles appear in the wear debris. Above the 'T_2' transition the oxide layers are established again and are protected by a hardened subsurface layer (white layer). Below 'T_1' and above 'T_2' the wear is classified as mild (oxidative) and between 'T_1' and 'T_2' as severe (adhesive).

The mild to severe wear transition in metals is generally associated with a transition from oxidative wear to adhesive wear, in which oxide films are presumed to be largely absent. However, another form of mild to severe wear transition is observed for ferritic stainless steels where mild wear is associated with the formation of protective oxides such as ferrosoferric oxide (Fe_3O_4) and $FeCrO_4$ while severe wear is associated with the formation of brittle oxides such as FeO and Cr_2O_3 [57].

· *Oxidative Wear under Lubricated Conditions*

Under lubricated conditions, surface temperatures are moderate since friction coefficients are usually low and consequently oxidative wear is relatively slow. Even when there is a temperature peak occurring in the contact area, the movement of oxygen to the reaction surface is inhibited by extremely low solute diffusivity coefficients in oil or other liquids at high pressures [31]. The extremely high oil pressures encountered in elastohydrodynamic contacts cause the viscosity of the oil to rise and solute diffusivity coefficients to fall. It is a general law for many fluids that with varying pressure, the product of viscosity and diffusivity is nearly constant [32]. It is quite probable that unless the oxygen is already adsorbed onto the surface outside the contact, it will not react with the surface [31].

Oxidation processes play an important role in boundary lubrication, where asperity contacts frequently occur. Oxidation of the asperities prevents direct metal-to-metal contact. It was found quite early on that oxide films on the sliding surfaces are necessary for the boundary films formed by lubricant additives to be effective [33,34]. Increased wear has been observed when oxygen is removed from the lubricant [35]. As shown in Figure 13.9, the wear rate in boundary lubrication depends on the oxygen concentration and the optimum concentration for minimum wear is often found [10].

The wear debris of oil-lubricated steel contacts was found to be extremely fine oxide particles and an oxide film of thickness approximately 100 [nm] was observed to build up on the worn surface [4]. It appears that wear in such circumstances is by the periodic loss of thin oxide films which form slowly at relatively low temperatures.

Means of Controlling Corrosive and Oxidative Wear

The addition of corrosion inhibitors to the lubricating oil or process fluid can be an effective means of controlling corrosive wear [36]. However, in lubricated systems there is a significant risk of interfering with vital lubrication additives so that ad hoc addition of corrosion inhibitors to lubricating oils is unwise. The corrosion inhibitors can function by the adsorption or the formation of a passivating layer on the operating surface.

Many corrosion inhibitors work by forming a strongly adsorbed monomolecular layer on the protected surface. This layer acts as a barrier which prevents oxygen and water from reaching the surface. The corrosion inhibitor may, however, displace adsorbed layers of lubricants and promote adhesive wear. EP lubrication, which is a form of controlled corrosion, may also be disrupted by corrosion inhibitors. Most oils are used to lubricate a number of contacts, all of which usually operate under different loads and sliding speeds. Consequently if, for example, a corrosion inhibitor is added to suppress corrosive wear in one contact, much more severe adhesive wear may be initiated in an adjacent contact.

Where there is no danger of disrupting lubrication, e.g., in a contact immersed in the process fluid, then the addition of corrosion inhibitors to the wetting fluid can produce a significant reduction in the wear rate [36]. The severity of corrosion and wear determines the selection of an optimum corrosion inhibitor. When corrosion is severe but wear is mild, then a corrosion inhibitor, which forms a passivating film, is the most suitable. When loads or wear are severe but corrosion is relatively mild, then an inhibitor that functions by adsorption to produce a lubricating layer is the most suitable. In this case, even a weak corrosion inhibitor may be effective. When both corrosion and wear are severe, an effective corrosion inhibitor which adsorbs strongly to the worn surface is essential [36].

Since most forms of corrosive wear involve electrochemical reactions, it may be possible to suppress the wear by imposing a cathodic potential on the wearing surface. Tests on a titanium alloy worn in the presence of sulphuric acid revealed that although the application of cathodic protection could suppress wear it could not completely prevent it [37]. This deficiency of cathodic protection resulted from the evolution of hydrogen on the wearing surface leading to hydrogen embrittlement [37,38]. It appears that a wearing surface, probably due to the high level of sustained stress, is more sensitive to hydrogen embrittlement than a surface subjected to static corrosion.

There is only limited research data published on the control of oxidative wear since this form of wear is relatively harmless when compared to other forms of wear, e.g., adhesive wear. Under certain conditions oxygen and the formation of oxides can substantially reduce wear, and then the oxidative wear can be regarded as beneficial. In other cases, e.g., when very hard ion implanted surfaces are used, the excessive oxidative wear has to be controlled. The most direct method of suppressing oxidative wear is to remove the source of oxygen which, in almost all cases, is air. Exclusion of air could be achieved by providing a flow of nitrogen to the wearing contacts, but care must be taken to prevent a complete exclusion of oxygen as this may result in severe adhesive wear. This method of nitrogen ventilation, however, is not practical. It might be possible to modify the metallic surface by coatings in order to produce an oxidation resistant surface, but this method still needs to be confirmed experimentally.

13.4 SUMMARY

Whenever the two independent processes of corrosion and wear occur simultaneously, it is almost certain that there will be a strong mutual interaction. Except in cases where a limited degree of corrosion, or more exactly, surface film forming reaction, is essential to prevent adhesive wear, this interaction causes an undesirable acceleration of material loss rate. Practical remedies to this problem include the removal of corrosive agents or the substitution of an inert material for the sliding contact. The reduction of temperature at the sliding contact slows down the corrosion rate and also helps to minimize the overall wear damage. Corrosion inhibitors based on strong adsorption to the corroding surface can be effective in controlling corrosive wear; however, they may interfere with the adsorption lubrication of the sliding contact of immediate concern or else with an adjacent contact lubricated by the same oil or process fluid. Other types of inhibitors, e.g., those which involve the formation of a passive layer, are usually ineffective.

Oxidative wear occurs when oxygen can access a hot sliding metallic contact. This form of wear is similar to corrosive wear except for the tendency of metallic oxides to mix with the worn metal and form debris layers of mixed metal and oxide. It may be possible to suppress this form of wear in a similar way as corrosive wear is suppressed, i.e., by chemical inhibition, but this has not apparently been widely tested. It is also not always desirable to suppress oxidative wear. Oxidative wear is characterized by reduced wear and friction compared to adhesive wear and therefore it is often deliberately induced, especially in dry sliding conditions.

REVISION QUESTIONS

13.1   What lubricant characteristic is required for corrosive wear to occur?

13.2   What connection does adhesive wear have with corrosive wear?

13.3   Is it a universal rule that friction coefficients and wear rates rise or fall in parallel?

13.4   What distinguishes corrosive films that minimize wear from those that accelerate wear?

13.5   A mechanical property of most metal oxides, in particular iron oxide, contributes to the mechanism of corrosive wear. What is it?

13.6   Under conditions of corrosive wear does the friction coefficient increase with wear rate?

13.7   What kind of wear debris would you expect to be formed by corrosive wear?

13.8   What must happen on the worn surface for mild corrosive wear to take place? Give an example of such wear.

13.9   The balance between adhesive and corrosive wear often determines the optimum lubricant. What is the deciding criterion for the optimum lubricant (i.e., gives minimum wear rate)?

13.10   Could oxidative wear occur under an atmosphere of pure nitrogen?

13.11 Very smooth and flat (planar) surfaces are critical in the manufacture of integrated circuits. Suggest a method, based on tribological principles, to produce such surfaces.

13.12 Approximately how thick are the oxide films that build up on a worn surface under conditions of oxidative wear?

13.13 How does a lubricant suppress oxidative wear?

13.14 What kind of debris is formed by severe wear? Is it chemically combined and small in size?

13.15 How does load determine the transition from initial severe wear to mild wear?

13.16 How many theories of oxidative wear are currently accepted and what is the difference between them?

13.17 Are oxide films entirely absent from a surface suffering severe wear and does the oxygen have any influence on the wear process?

13.18 Is corrosive wear accelerated by lack of grain boundaries on a worn surface?

13.19 Most examples of corrosive wear involve metals. Is it possible for corrosive wear to occur with non-metals, e.g., polymers? Give reasons and provide examples.

13.20 How would one control oxidative wear? Would one recommend thin hard coatings of, for example, titanium nitride?

REFERENCES

1 G.W. Rengstorff, K. Miyoshi and D.H. Buckley, Interaction of Sulfuric Acid Corrosion and Mechanical Wear of Iron, *ASLE Transactions*, Vol. 29, 1986, pp. 43-51.

2 D.R. Holmes and R.T. Pascoe, Strain Oxidation Interaction in Steels and Model Alloys, *Werkstoff und Korrosion*, Vol. 23, 1972, pp. 859-870.

3 F.F. Tao, A Study of Oxidation Phenomena in Corrosive Wear, *ASLE Transactions*, Vol. 12, 1969, pp. 97- 105.

4 M. Masuko, Y. Itoh, K. Akatsuka, K. Tagami and H. Okabe, The Influence of Sulphur-Based Extreme Pressure Additives on Wear under Combined Rolling and Sliding, 1983 Annual Conference, Japan Society of Lubrication Engineers, Nagasaki, Publ. JSLE, Tokyo, 1983, pp. 273-276.

5 S.M. Hsu and E.E. Klaus, Estimation of the Molecular Junction Temperature in Four-Ball Contacts by Chemical Reaction Rate Studies, *ASLE Transactions*, Vol. 21, 1978, pp. 201- 210.

6 G.H. Benedict, Correlation of Disk Machines and Gear Tests, *Lubrication Engineering*, Vol. 4, 1968, pp. 591-596.

7 Y. Yahagi and Y. Mizutani, Corrosive Wear of Steel in Gasoline-Ethanol-Water Mixtures, *Wear*, Vol. 97, 1984, pp. 17-26.

8 Y. Yahagi and Y. Mizutani, Corrosive Wear of Cast Iron (1) - Influence of Sulphuric Acid, *JSLE Transactions*, Vol. 31, 1986, pp. 883-888.

9 A. Beerbower, Discussion to reference No. 1 (by G.W. Rengstorff, K. Miyoshi and D.H. Buckley, Interaction of Sulfuric Acid Corrosion and Mechanical Wear of Iron, *ASLE Transactions*, Vol. 29, 1986, pp. 43-51), *ASLE Transactions*, Vol. 29, 1986, pp. 51.

10 E.E. Klaus and H.E. Bieber, Effect of Some Physical and Chemical Properties of Lubricants on Boundary Lubrication, *ASLE Transactions*, Vol. 7, 1964, pp. 1-10.

11 T. Sakurai, Status and Future Research Directions of Liquid Lubricant Additives, *JSLE Transactions*, Vol. 29, 1984, pp. 79-86.

12 H.G. Zelders, La Corrosion Superficielle dans le Circuit de Lavage des Charbonnages des Mines de l'Etat Neerlandais, *Metaux et Corrosion*, Vol. 65, 1949, pp. 25-76.

13 A.W. Batchelor and G.W. Stachowiak, Predicting Synergism between Corrosion and Abrasive Wear, *Wear*, Vol. 123, 1988, pp. 281-291.

14 H. Hocke and H.N. Wilkinson, Testing Abrasion Resistance of Slurry Pipeline Materials, *Tribology International*, Vol. 11, 1978, pp. 289-294.

15 F.F. Tao and J.K. Appledoorn, An Experimental Study of the Wear Caused by Loose Abrasive Particles in Oil, *ASLE Transactions*, Vol. 13, 1970, pp. 169-178.

16 R. Mailander and K. Dies, Contributions to Investigation in the Process Taking Place during Wear, *Arch. Eisenhuettenwissenschaft*, Vol. 16, 1943, pp. 385-398.

17 J.F. Archard and W. Hirst, The Wear of Metals under Unlubricated Conditions, *Proc. Roy. Soc., London*, Series A, Vol. 236, 1956, pp. 397-410.

18 A. Ohnuki, Deformation Processing, *JSLE Transactions*, Vol. 28, 1983, pp. 53-56.

19 A.T. Fromhold, Theory of Metal Oxidation, Volume 1, Fundamentals, Elsevier, Amsterdam, 1976.

20 F.P. Fehlner and N.F. Mott, Low Temperature Oxidation, *Oxidation of Metals*, Vol. 2, 1970, pp. 56-99.

21 D.D. Eley and P.R. Wilkinson, Adsorption and Oxide Formation on Aluminium Films, *Proc. Roy. Soc., London*, Series A, Vol. 254, 1960, pp. 327-342.

22 T.F.J. Quinn, The Dry Wear of Steel as Revealed by Electron Microscopy and X-ray Diffraction, *Proc. Inst. Mech. Engrs.*, London, Vol. 182, 1967-1968, Pt. 3N, pp. 201-213.

23 T.F.J. Quinn, J.L. Sullivan and D.M. Rawson, New Developments in the Oxidational Theory of the Mild Wear of Steels, Proc. Int. Conf. on Wear of Materials, Dearborn, Michigan, 16-18 April 1979, editors K.C. Ludema, W.A. Glaeser and S.K. Rhee, Publ. American Society of Mechanical Engineers, New York, 1979, pp. 1-11.

24 A. Ohnuki, Friction and Wear of Metals at High Temperatures, *JSLE Transactions*, Vol. 30, 1985, pp. 329-334.

25 V. Aronov, Kinetic Characteristics of the Transformation and Failure of the Surface Layers Under Dry Friction, *Wear*, Vol. 41, 1977, pp. 205-212.

26 A.W. Batchelor, G.W. Stachowiak and A. Cameron, The Relationship between Oxide Films and the Wear of Steels, *Wear*, Vol. 113, 1986, pp. 203-223.

27 G.A. Berry and J.R. Barber, The Division of Frictional Heat - a Guide to the Nature of Sliding Contact, *Transactions ASME, Journal of Tribology*, Vol. 106, 1984, pp. 405-415.

28 J.E. Wilson, F.H. Stott and G.C. Wood, The Development of Wear Protective Oxides and Their Influence on Sliding Friction, *Proc. Roy. Soc., London*, Series A, Vol. 369, 1980, pp. 557-574.

29 J. Glascott, G.C. Wood and F.H. Stott, The Influence of Experimental Variables on the Development and Maintenance of Wear-Protective Oxides during Sliding of High-Temperature Iron-Base Alloys, *Proc. Inst. Mech. Engrs.*, London, Vol. 199, Pt. C, 1985, pp. 35-41.

30 N.C. Welsh, The Dry Wear of Steels, Part I - General Pattern of Behaviour, *Phil. Trans. Roy. Soc.*, Vol. 257A, 1964, pp. 31-50.

31 B.A. Baldwin, Wear Mitigation by Anti-Wear Additives in Simulated Valve Train Wear, *ASLE Transactions*, Vol. 26, 1983, pp. 37-47.

32 N.S. Isaacs, Liquid Phase High Pressure Chemistry, John Wiley, New York, 1981, pp. 181-351.

33 E.D. Tingle, Influence of Water on the Lubrication of Metals, *Nature*, Vol. 160, 1947, pp. 710-711.

34 F.B. Bowden and J.E. Young, Friction of Clean Metals and the Influence of Adsorbed Films, *Proc. Roy. Soc., London*, Series A, Vol. 208, 1951, pp. 311-325.

35 I.M. Feng and H. Chalk, Effects of Gases and Liquids in Lubricating Fluids on Lubrication and Surface Damage, *Wear*, Vol. 4, 1961, pp. 257-268.

36 X. Jiang, Q. Sun, S. Li and J. Zhang, Effect of Additives on Corrosive Wear of Carbon Steel, *Wear*, Vol. 142, 1991, pp. 31-41.

37 X. Jiang, S. Li, C. Duan and M. Li, A Study of the Corrosive Wear of Ti-6Al-4V in Acidic Medium, *Wear*, Vol. 129, 1989, pp. 293-301.

38 X. Jiang, S. Li, C. Duan and M. Li, The Effect of Hydrogen on Wear Resistance of a Titanium Alloy in Corrosive Medium, *Lubrication Engineering*, Vol. 46, 1990, pp. 529-532.

39 T.E. Fischer, Tribochemistry of Ceramics: Science and Applications, New Directions in Tribology, editor I.M. Hutchings, Plenary and Invited Papers from the First World Tribology Congress, London, MEP Publications Ltd., 1997, pp. 211-215.

40 M. Chandrasekaran, I.K.C. Chin, A.W. Batchelor and N.L. Loh, In-situ Surface Modification of Inconel 625 under the Influence of External Heating during Tribological Testing, Proc. SMT-10, 10th Surface Modification Technologies Conference, Singapore, Nov. 1996, pp. 881-890, publ. Institute of Materials, London, 1997.

41 R. Colas, J. Ramirez, I. Sandoval, J.C. Morales, L.A. Leduc, Damage in Hot Rolling Work Rolls, *Wear*, Vol. 230, 1999, pp. 56-60.

42 T.J. Chase, Wear Modes Active in Angular Contact Ball Bearings Operating in Liquid Oxygen Environment of the Space Shuttle Turbopumps, *Lubrication Engineering, Journal of STLE*, Vol. 49, 1993, pp. 313-322.

43 M. Nosaka, M. Oike, M. Kikuchi, K. Kamijo and M. Tajiri, Self-Lubricating Performance and Durability of Ball Bearings for the LE-7 Liquid Oxygen Rocket-Turbopump, *Lubrication Engineering, Journal of STLE*, Vol. 49, 1993, pp. 677-688.

44 M. Nosaka, M. Oike, M. Kikuchi, R. Nagao and T. Mayumi, Evaluation of Durability for Cryogenic High-Speed Ball Bearings of LE-7 Rocket Pumps, *Lubrication Engineering, Journal of STLE*, Vol. 52, 1996, pp. 221-232.

45 I. Garcia, D. Drees and J.P. Celis, Corrosion-Wear of Passivating Materials in Sliding Contacts Based on a Concept of Active Wear Track Area, *Wear*, Vol. 249, 2001, pp. 452-460.

46 P.A. Dearnley and G. Aldrich-Smith, Corrosion-Wear Mechanisms of Hard Coated Austenitic 316L Stainless Steels, *Wear*, Vol. 256, 2004, pp. 491-499.

47 S.W. Zhang, H. Liu and R. He, Mechanisms of Wear of Steel by Natural Rubber in Water Medium, *Wear*, Vol. 256, 2004, pp. 226-232.

48 K. Zhang and L. Battiston, Friction and Wear Characterization of Some Cobalt- and Iron-Based Superalloys in Zinc Alloy Baths, *Wear*, Vol. 252, 2002, pp. 332-344.

49 I. Serre, N. Celati, R.M. Pradeilles-Duval, Tribological and Corrosion Wear of Graphite Ring Against a Ti6Al4V Disk in Artificial Sea Water, *Wear*, Vol. 252, 2002, pp. 711-718.

50 Y. Zhao, L. Chang and S.H. Kim, A Mathematical Model for Chemical-Mechanical Polishing Based on Formation and Removal of Weakly Bonded Molecular Species, *Wear*, Vol. 254, 2003, pp. 332-339.

51 S.J. Harris, G.G. Kraus, S.J. Simko, R.J. Baird, S.A. Gebremariam and G. Doll, Abrasion and Chemical-Mechanical Polishing between Steel and a Sputtered Boron Carbide Coating, *Wear*, Vol. 252, 2002, pp. 161-169.

52 I. Garcia, A. Ramil and J.P. Celis, A Mild Oxidation Model Valid for Discontinuous Contacts in Sliding Wear Tests: Role of Contact Frequency, *Wear*, Vol. 254, 2003, pp. 429-440.

53 H. Chen and A.T. Alpas, Sliding Wear Map of Magnesium Alloy Mg-9Al-0.9Zn (AZ91), *Wear*, Vol. 246, 2000, pp. 106-116.

54 I.A. Inman, S. Datta, H.L. Du, J.S. Burnell-Gray and Q. Luo, Microscopy of Glazed Layers Formed during High Temperature Sliding at 750°C, *Wear*, Vol. 254, 2003, pp. 461-467.

55 J. Jiang, F.H. Stott and M.M. Stack, A Generic Model for Dry Sliding Wear of Metals at Elevated Temperatures, *Wear*, Vol. 256, 2004, pp. 973-985.

56 R. Norling and I. Olefjord, Erosion-Corrosion of Fe- and Ni-Based Alloys at 550°C, *Wear*, Vol. 254, 20003, pp. 173-184.

57 M. Aksoy, O. Yilmaz and M.H. Korkut, The Effect of Strong Carbide-Forming Elements on the Adhesive Wear Resistance of Ferritic Stainless Steel, *Wear*, Vol. 249, 2001, pp. 639-646.

14

FATIGUE WEAR

14.1 INTRODUCTION

In many well-lubricated contacts, adhesion between the two surfaces is negligible, yet there is still a significant rate of wear. This wear is caused by deformations sustained by the asperities and surface layers when the asperities of opposing surfaces make contact. Contacts between asperities accompanied by very high local stresses are repeated a large number of times in the course of sliding or rolling, and wear particles are generated by fatigue propagated cracks, hence the term 'fatigue wear'. Wear under these conditions is determined by the mechanics of crack initiation, crack growth and fracture. Worn surfaces contain very high levels of plastic strain compared to unworn surfaces. This strain and the consequent modification of the material's microstructure have a strong effect on the wear processes.

The term 'contact fatigue' or 'surface fatigue' commonly used in the literature is technical jargon for surface damage caused by a repeated rolling contact. It refers to the initial damage on a smooth surface and is most often used in the context of rolling bearings. Rolling bearings rely on smooth undamaged contacting surfaces for reliable functioning. A certain number of rolling contact cycles must elapse before surface defects are formed, and their formation is termed 'contact fatigue'. Once the rolling surfaces of a bearing are pitted, its further use is prevented due to excessive vibration caused by pits passing through the rolling contact. Bearing failure caused by contact fatigue is usually sudden and is highly undesirable especially when the bearing which is critical to the proper functioning of the machinery, e.g., in a jet engine, is involved. It is common experience that when a rolling bearing fails, e.g., in a car axle, much labour-intensive dismantling and re-assembly are usually required. For these reasons, contact fatigue, particularly of lubricated metal contacts, has been the subject of most of the research programs related to wear under rolling contact.

Practical questions arise, such as what are the characteristic features of fatigue wear? What is the mechanism involved in wear particle formation in this wear mode? How can fatigue wear be recognized and controlled? Can lubrication be effective in controlling fatigue wear? Fatigue wear can cause severe problems which prevent the effective functioning of essential equipment, and an engineer should know the answers to these questions and many more. The fundamental characteristics of fatigue wear and ways of controlling it are discussed in this chapter.

14.2 FATIGUE WEAR DURING SLIDING

Examination of worn surfaces in cross section reveals intense deformation of the material directly below the worn surface [e.g., 1]. For example, it has been shown that under conditions of severe sliding with a coefficient of friction close to unity, material within 0.1 [mm] of the surface is shifted in the direction of sliding due to deformation caused by the frictional force. Also, close to the surface the grain structure is drawn out and orientated parallel to the wearing surface [1]. Obviously under the lower coefficients of friction which prevail in lubricated systems, this surface deformation is less or may even be absent.

Strains caused by shearing in sliding are present some depth below the surface reaching the extreme values at the surface. The strain levels in a deformed surface layer are illustrated schematically in Figure 14.1 [2].

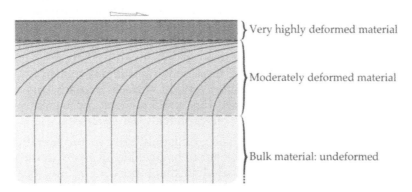

FIGURE 14.1 Strain levels in a deformed surface [2].

The strain induced by sliding eventually breaks down the original grain structure at the surface to form dislocation cells. These cells can be described as submicron regions, relatively free from dislocations, which are separated by regions (walls) of highly tangled dislocations [3,4]. A cellular structure directly beneath worn surfaces of metallic materials has been observed in transmission electron microscopy studies [e.g., 3-5]. The new structure is found to be similar, if not identical, to the structure occurring in heavily worked metals [3].

Materials vary greatly in their tendency to form dislocation cells which, according to general metallurgical theory, depends on stacking fault energy, i.e., high stacking fault energy promotes cell formation. For example, aluminium, copper and iron have a high stacking fault energy and therefore readily form dislocation cells. At the interface the cells are elongated in the direction of sliding and are relatively thin, resembling layers of flat 'tiles'. The high energy cell boundaries are probable regions for void formation and crack nucleation [3]. The formation of a wear particle can be initiated at the cell walls which are orientated perpendicular to the direction of sliding [4] since the crack can propagate along the cell boundary. Alternatively, the crack can be initiated at a weak point below the surface and subsequently propagate to the surface resulting in the release of a wear particle.

Plastic deformation of the surface layer under sliding was simulated by moving a hard blunt wedge against a soft flat surface. It was found that during the process material piles up in front of the moving wedge without detaching from the surface. During repetitive sliding, the piled up material does not move with the wedge and instead the wedge passes continuously through the protuberance of deformed surface. This movement resembles a wave and therefore the concept of waves of material being driven across a surface by hard asperities has been suggested [6]. To accommodate the 'wave', very high strains are sustained, leading to the cracking of the material in the wave. However, how closely these experimental findings correspond to strain processes occurring between the asperities in wearing contacts still needs to be investigated. The example of such a 'wave' of deformed material is shown in Figure 14.2.

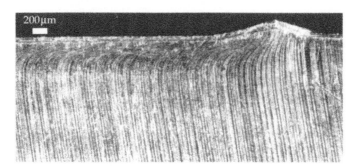

FIGURE 14.2   Accumulation of material on the surface due to the passage of a blunt wedge and the resulting plastic deformation [6].

Surface Crack Initiated Fatigue Wear

Cracks and fissures have frequently been observed on micrographs of worn surfaces. The mechanism of surface crack initiated fatigue wear is illustrated schematically in Figure 14.3. A primary crack originates at the surface at some weak point and propagates downward along weak planes such as slip planes or dislocation cell boundaries. A secondary crack can develop from the primary crack or alternatively the primary crack can connect with an existing subsurface crack. When the developing crack reaches the surface again a wear particle is released [7].

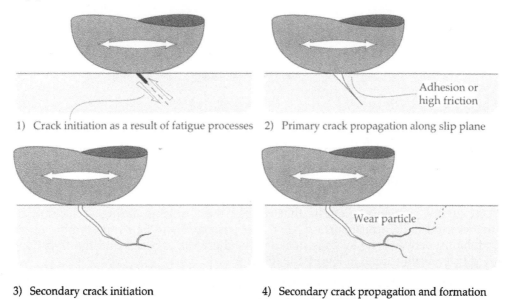

FIGURE 14.3   Schematic illustration of the process of surface crack initiation and propagation.

Therefore it has been found that during unlubricated sliding, in particular reciprocal sliding, wear particles can form due to the growth of surface initiated cracks [1,8-10]. During sliding, planes of weakness in the material become orientated parallel to the surface by the already discussed deformation processes, and laminar wear particles are formed by a surface crack, reaching a plane of weakness as illustrated in Figure 14.4.

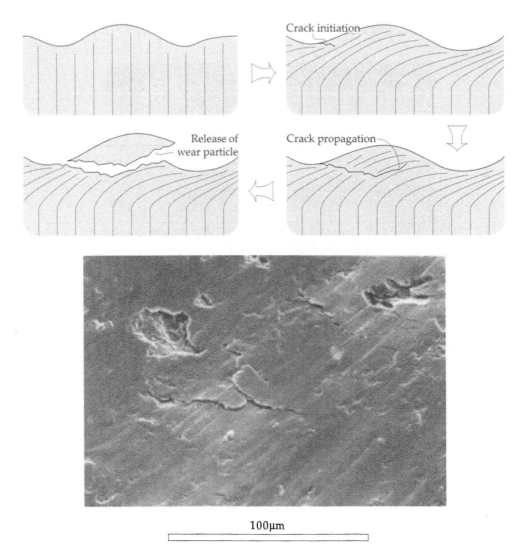

FIGURE 14.4 Schematic illustration of mechanism of wear particles formation due to growth of surface initiated cracks (adapted from [11]) and an example of fatigue wear particle formation on cast iron.

The accumulated evidence suggests that fatigue wear during sliding is a result of crack development in the deformed surface layer. This also seems to be supported by an observed proportionality between the average thickness of the wear particles and the thickness of the deformed layer as illustrated in Figure 14.5 [9].

It has been demonstrated in wear and fatigue experiments conducted with chemically active and noble metals under various pressures that this type of wear is strongly affected by the presence of oxygen [10]. For example, wear rates and the reciprocal of fatigue lives as a function of atmospheric pressure for chemically active metals, e.g., nickel, and noble metals, e.g., gold, are shown in Figures 14.6 and 14.7, respectively.

It can be seen from Figure 14.6 that for the chemically active metals there is a distinct increase in wear rate and decrease in fatigue life with increased pressure, which is a function of oxygen concentration. On the other hand, it can be seen from Figure 14.7 that for noble metals, i.e., gold, wear rate and fatigue life are independent of pressure.

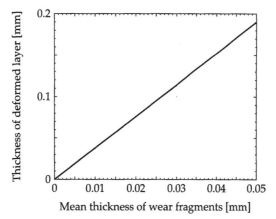

FIGURE 14.5 Relationship between mean wear particle thickness and the thickness of the deformed layer [9].

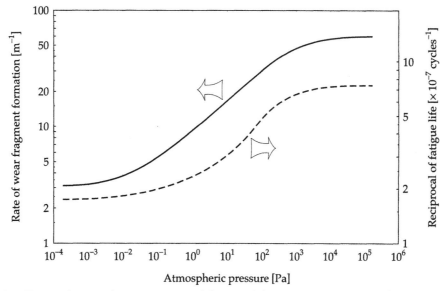

FIGURE 14.6 Dependence of wear rates and fatigue life of nickel on atmospheric pressure [10].

This phenomenon can be explained in terms of the ability of metals to form oxide films in the vicinity of the crack tips, as schematically illustrated in Figure 14.8. Surfaces of nickel and copper rapidly form films of oxide when exposed to air, whereas gold forms, at most, an adsorbed layer of oxygen. If an oxide film can form on the freshly exposed surface at a crack root, then healing of the crack by adhesion of the fracture faces cannot take place. Lowering the pressure of air (and oxygen) slows down the oxide film formation to the point where crack healing may occur. Therefore fatigue life can considerably be extended when oxygen and oxide films are absent. This effect does not take place in metals that do not form oxide films.

Subsurface Crack Initiated Fatigue Wear

During sliding contact between two bodies, much of the damage done to the material of each body occurs beneath the sliding surfaces. A worn surface can remain quite smooth and lacking in obvious damage, while a few micrometres below the surface processes leading to the formation of a wear particle are taking place [12].

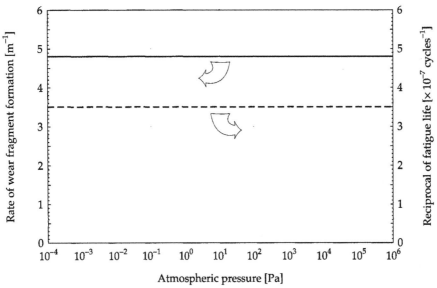

FIGURE 14.7 Dependence of wear rates and fatigue life of gold on atmospheric pressure [10].

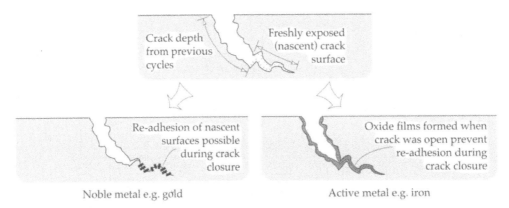

FIGURE 14.8 Effects of oxide films on surface crack development.

When two surfaces are brought into sliding contact, the random asperity topography on both surfaces is soon replaced by a smooth surface [13] or a series of grooves aligned in the direction of sliding [14]. If the two sliding materials differ in hardness, the softer material loses its asperities first but those on the harder surface also eventually disappear. Cyclic plastic deformation occurs over the whole area of the worn surface as the loaded asperities pass over the surface and the frictional traction forces intensify the plastic deformation [15]. Strain in the material immediately beneath the worn surface may reach extremely high levels [16] but does not contribute directly to crack growth since a triaxial compressive stress field occurs directly beneath a contacting asperity [13]. If a crack cannot form at the surface it will form some distance below the surface where the stress field is still sufficiently intense for significant crack growth.

Most engineering materials contain inclusions and other imperfections which act as nuclei for void formation under plastic deformation. These voids form a plentiful supply of initiators for crack growth as illustrated schematically in Figure 14.9.

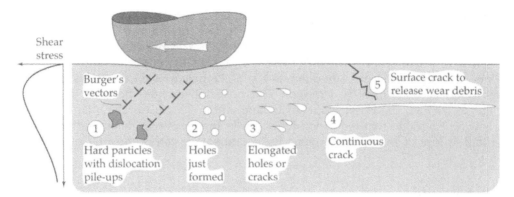

FIGURE 14.9   Illustration of a process of subsurface crack formation by growth and link up of voids (adapted from [12]).

The development of voids by plastic deformation is a result of dislocation pile up at hard inclusions. These voids enlarge with further deformation since they act as traps for dislocations. Crack growth does not proceed very near the surface because of the large plastic zone around its tip, but is confined to a narrow range of depth where hydrostatic or triaxial stress is small but shear stresses are still large [16]. All these factors favour the growth of a crack parallel to but beneath the surface. At some unspecified point the crack finally turns upwards to the surface and a long thin laminar particle is released. The name of this theory, introduced by Suh in 1973, relates to the wear particle shape and it is known as 'the delamination theory of wear' [12]. The theory has been confirmed by a number of microscopy studies and experimental data [e.g., 15,17].

The hypothesis that void nucleation is a necessary step in the formation of a wear particle suggests that very clean materials with no inclusions will exhibit very low wear rates. This prediction has been confirmed experimentally by measuring the wear rates of a range of metals sliding unlubricated against steel [17]. Pure iron, several different steels and pure copper were tested. It was found that pure copper (99.96% purity) slid against steel gives a wear rate ten times lower than any other material despite exhibiting the highest coefficient of friction of all the materials tested. On the other hand, a steel rich in carbide particles shows a low coefficient of friction and gives one of the highest wear rates. The wear rate was found to increase with inclusion density in the material, while friction was determined by adhesion factors so that complex impure materials exhibited the lowest friction coefficient [17].

Unfortunately the current delamination theory of wear does not include the effect of surface temperature rises on wear, which are inevitable in sliding contacts. It seems that at the surface, where temperatures are highest, the material is continually recrystallizing or stress relieving, unlike the material below the surface. Although the recrystallization of surface layers has been observed [5], the effect of temperature on delamination wear is yet to be investigated.

Effect of Lubrication on Fatigue Wear during Sliding

The effect of lubrication or friction reduction on the wear of steel in terms of the delamination model of wear has been studied [18]. A hardened martensitic steel ball (Vickers hardness 840) was slid against a normalized steel shaft of Vickers hardness 215 [18]. Pure hexadecane with or without various lubricant additives was used as a means of varying the friction while keeping other factors that might affect the wear rate as unchanged as possible.

The application of a lubricant reduces friction, and variation in friction levels was found to affect the delamination type wear mechanism [18]. It was found that the delamination type wear mechanism in boundary lubricated sliding is predominant at medium levels of friction. Although delamination may also occur at high levels of friction, it is usually overshadowed by adhesive wear and material transfer. At low levels of friction, subsurface shear forces are insufficient to initiate crack growth. In tests conducted in lubricated sliding contacts the occurrence of delamination was predominant at a coefficient of friction between **0.2** and **0.4** and was often accompanied by adhesion and material transfer. Between these values of coefficient of friction, the classic signs of delamination, i.e., subsurface planar cracks, were found. It was also found that reducing friction slows the process of delamination wear so that an eventual failure of the sliding contact by delamination can occur after a long period of operation.

Plastic Ratchetting

A new model of progressive plastic deformation of surfaces during repeated sliding has recently been proposed to describe metallic wear [75]. The model, known as 'plastic ratchetting', explains the generation of plate-like metallic wear particles, often observed in boundary lubricated contacts, in terms of the accumulation of plastic strains. Ratchetting takes place when contact pressure exceeds the elastic shakedown limit [75]. The word 'ratchetting' describes the process where large plastic strains are gradually accumulated by the superposition of small, unidirectional shear strains that are generated at each load cycle. The accumulated strain is known as ratchetting strain. The deformed material raptures when the accumulated strain exceeds the critical value of plastic strain. This mode of wear is different from both high-cycle fatigue wear due to elastic strains and low-cycle fatigue wear due to large cyclic plastic strains.

In lubricated sliding (low friction) ratchetting results in thin metallic films being extruded from the contact that then form 'filmy wear' particles by breaking off [76]. The generation of thin particles when a hard ball is slid on a soft rough surface is illustrated schematically in Figure 14.10 [76].

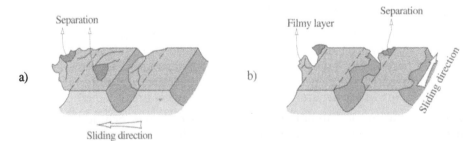

Figure 14.10 Schematic illustration of the mechanism of 'filmy wear' particles generation due to plastic ratchetting (adapted from [76]).

It can be noticed from Figure 14.10 that when the asperities (shown as machining grooves) are perpendicular to the sliding direction, the material is extruded from one side of the grooves only. When the asperities are parallel to the sliding direction, the material is extruded from both sides. The plastic ratchetting model is largely restricted to cases when one surface is much harder than the other and plastic deformation occurs on the softer surface. However, the case when both contacting surfaces are deformed (often the more realistic case) is not yet well defined.

14.3   FATIGUE WEAR DURING ROLLING

During rolling the local contact stresses are very high, concentrated over a small area and repetitive, and wear mechanisms are determined mostly by material characteristics and operating conditions as illustrated schematically in Figure 14.11.

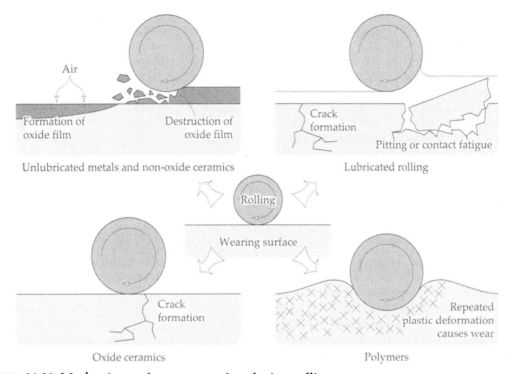

FIGURE 14.11 Mechanisms of wear occurring during rolling.

In the absence of lubrication, wear of metals or other oxidizable solids proceeds by the cyclic fragmentation of oxide films and their subsequent reformation. In other words, a form of oxidative wear occurs. The most common example of this dry oxidative rolling wear is found on railway wheels which develop smooth worn surfaces and gradually wear out over several years of service [19]. Oxidizable ceramics, e.g., the nitrides, display a similar form of wear, although the oxides, particularly silicon dioxide formed from silicon nitride or silicon carbide, can form durable lubricant films which can suppress wear quite significantly.

When lubrication is applied to metals and an effective EHL film is formed, solid to solid contact is prevented and the repeated destruction of surface oxide films is suppressed or terminated. Any wear or surface destruction that occurs is entirely a result of cyclic stress variation from the repeated rolling contact. Wear or contact fatigue proceeds by the formation of cracks in the surface that eventually allow a wear particle to detach from the surface. It should be noted that under dry conditions, wear by crack formation or pitting and spalling can also occur when contact stresses are excessively high [19,20]. The latter is especially notable in brittle solids such as oxide ceramics which tend to wear by pitting or crack formation under rolling contacts.

When a hard body, e.g., steel, is rolled against a much softer polymer, deformation of the polymer is almost inevitable. Wear occurring in this case is the result of cyclic deformation of the polymer. Rapid wear may occur until changes in contact geometry caused by wear reduce contact stresses sufficiently to allow only mild wear.

Causes of Contact Fatigue

In very broad terms the causes of contact fatigue can be summarized as due to rolling material limitations, lubrication or operating conditions. Material for rolling contacts must be of very high quality since any imperfections present can act as initiation sites for developing cracks. The surface finish must also be of high quality since the cracks can originate from surface imperfections and irregularities. The presence of a lubricant can have a significant effect on contact fatigue by preventing a true contact between the rolling bodies. Even the highest quality rolling bearing made of the finest steel will only provide limited duty if lubrication is neglected. Since most contact fatigue problems occur in rolling bearings the prevailing form of lubrication where contact fatigue is of practical importance is elastohydrodynamic lubrication (EHL). The characteristics of EHL therefore have a significant influence on contact fatigue and an understanding of the fundamentals of EHL is essential to any interpretation of contact fatigue phenomena. The size and quantity of wear debris and contaminants present in the lubricant are very important. Wear debris and contaminants can affect the contact fatigue by denting and scratching the contacting surfaces when passing through the EHL contact and thus creating new sites for crack development. The operating conditions such as the stress level and the amount of slip present in the rolling contact can also significantly affect contact fatigue.

· *Asperity Contact during EHL and the Role of Debris in the Lubricant in Contact Fatigue*

As was emphasized in previous chapters, it is either difficult or uneconomical to attain perfect lubrication, i.e., where solid to solid contact is completely prevented. When some solid contact occurs, even if it is very occasional, sufficient surface damage to initiate contact fatigue can take place. Solid to solid contact can occur when asperities from the opposing surfaces interact or when debris passes through the elastohydrodynamic contact. Contact fatigue can originate from the subsequent damage done to the surface, e.g., scratches and dents, as shown schematically in Figure 14.12.

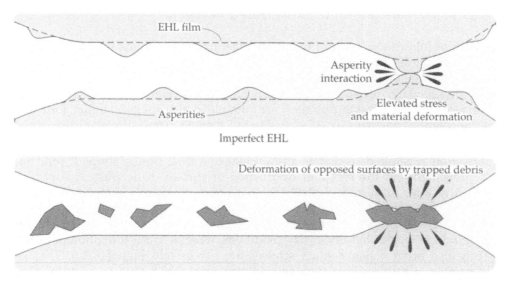

FIGURE 14.12 Contact between asperities and oversized debris entrainment as a cause of contact fatigue.

Asperity interaction usually occurs when the minimum film thickness becomes unusually low, when the surface roughness is high or when the bearing is overloaded. The film thickness may become unusually low when lubricant viscosity is reduced by elevated temperature or by high shear rates. As described in Chapter 7 on 'Elastohydrodynamic Lubrication' the minimum film thickness should be greater than 4 times the composite surface roughness to ensure full separation of the surfaces by an EHL film. For poorly finished surfaces this condition might not be fulfilled and contact may be established between the asperities during rolling. Interaction between the asperities can also occur when the bearing is overloaded. It is therefore desirable to select the lubricant, surface finish and operating conditions so that interaction is minimized.

When the minimum dimension of the debris (e.g., thickness of planar debris) is greater than the minimum film thickness, damage to the contacting surfaces is inevitable [e.g., 21-23]. For example, it was shown that careful filtration of lubricating oil can significantly extend the life of rolling bearings [e.g., 22,24]. There is usually a number of large wear particles present in engineering equipment and therefore rolling bearing surfaces can be subjected to damage by oversized debris entrainment. In practice all precautions should be taken to ensure lubricant cleanliness.

· *Material Imperfections*

Material imperfections such as inclusions, weak grain boundaries and zones of high residual stress are an important source of initiation sites for the development of cracks and formation of wear particles. These can all cause contact fatigue. Inclusions are particularly detrimental to contact fatigue resistance, as shown schematically in Figure 14.13, and should be avoided if at all possible. Steel cleanliness is therefore critical to rolling bearing durability [25].

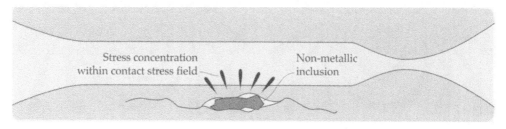

Material imperfections

FIGURE 14.13 Material imperfection as the cause for contact fatigue.

· *Plastic Deformation in Wheel-Rail Contacts*

Classical studies of rolling contact fatigue mainly focus on bearings and gears where the rolling materials, e.g., hardened steel, do not sustain bulk plastic deformation. However, in softer materials, such as steel rails, there is some plastic deformation occurring in rolling contacts. Repeated rolling contacts result in a lateral deformation of the surface layers of rolled material in the direction of rolling, caused by wheel sliding and high tangential stresses. Where the lateral deformation is sufficiently severe to obliterate the original grain structure of the metal and replace it with aligned lamellar grains, cracks are likely to propagate along the deformed grain boundaries [77], as schematically illustrated in Figure 14.14. Such extreme plastic deformation and resultant crack growth plays a significant role in the fatigue of the rolled surface. The cracks, called 'head checks', are often found on rail heads on curved parts of the rail track. Special surface coatings, produced by a laser-cladding technique, with higher hardness than that of typical pearlite steel rails, tend to improve the contact fatigue of rails [77].

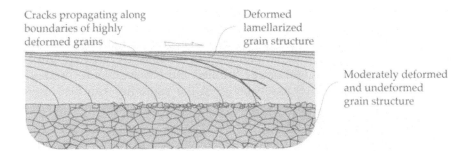

Cracks propagating along
boundaries of highly
deformed grains

Deformed
lamellarized
grain structure

Moderately deformed
and undeformed
grain structure

Figure 14.14 Cracks development caused by severe lateral surface plastic deformation induced by a rolling contact.

Self-Propagating Nature of Contact Fatigue Cracks

Once a crack is initiated in the rolling surface, further growth of the crack is facilitated by changes in the Hertzian contact stress field [26]. A photo-elastic stress field of a Hertzian contact between two cylinders with an artificially introduced crack is shown in Figure 14.15.

FIGURE 14.15 Photo-elastic stress field of a Hertzian contact between two cylinders with an introduced crack in one of the cylinders [26].

It can be seen from Figure 14.15 that the presence of a crack disrupts the normal Hertzian stress field of unflawed solids and introduces a large stress concentration around the crack tip. The photo-elastic stress micrograph displays shear stress which is the stress promoting crack growth so that it is evident that the presence of a small crack introduces conditions extremely favourable to further crack growth [26]. Once the crack length becomes comparable in size to the Hertzian contact diameter or depth at which the maximum shear stress occurs, it may be assumed that rapid crack extension begins.

Subsurface and Surface Modes of Contact Fatigue

Contact fatigue failure can develop from either surface or subsurface defects. Therefore there are two modes of contact fatigue, surface and subsurface, which produce pits of distinctly different shape [27]. The mechanisms of surface and subsurface failure modes are illustrated schematically in Figure 14.16.

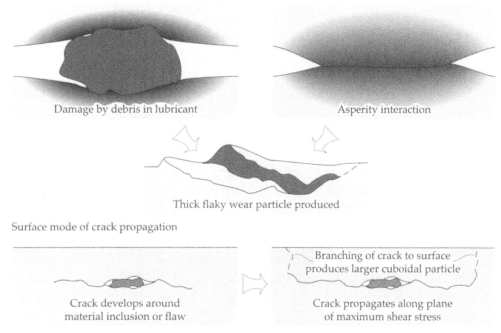

Damage by debris in lubricant

Asperity interaction

Thick flaky wear particle produced

Surface mode of crack propagation

Crack develops around
material inclusion or flaw

Branching of crack to surface
produces larger cuboidal particle

Crack propagates along plane
of maximum shear stress

Subsurface mode of crack propagation

FIGURE 14.16 Schematic illustration of the surface and subsurface modes of contact fatigue.

Examples of surface and subsurface initiated spalls are shown in Figures 14.17 and 14.18. It can be seen from Figure 14.17 that the surface mode of contact fatigue is characterized by a shallow pit with multiple cracking or flaking at one side of the pit. On the other hand, the subsurface mode of contact fatigue, shown in Figure 14.18, produces a much deeper pit with clearly defined edges and limited multiple cracking. These two modes of pit morphology are associated with different causes of contact fatigue.

The surface mode of contact fatigue is associated with failure caused by insufficient film thickness, excessive surface roughness or oversized debris present in the lubricant. Insufficient film thickness or excessive surface roughness affects the contact fatigue through the asperity interactions [e.g., 28-30]. An EHL film, of sufficient thickness to prevent interaction between the asperities, results in a significant reduction of surface-originated spalls [31,32]. Oversized debris passing through the EHL contact can scratch and dent the surface. The surface dents and scratches act as points of stress concentration from which rapidly developing cracks can originate [e.g., 23,24,28,33,34].

The subsurface mode of failure is usually caused by cracks propagating from material imperfections situated close to the plane of maximum shear stress within the Hertzian contact [e.g., 27,33]. With improvements in the cleanliness of steels, subsurface-originated contact fatigue has become a rarity in rolling contact bearings compared to surface-originated contact fatigue. However, in gears the subsurface mode of fatigue wear still appears to be significant with the cause believed to be crack initiation by immobile dislocations [80].

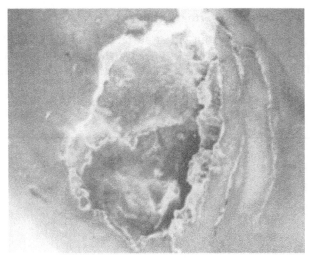

1 mm

FIGURE 14.17 Surface initiated spall.

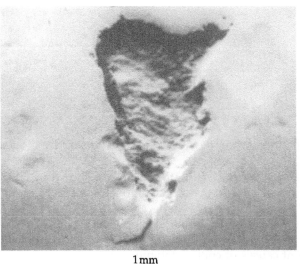

1 mm

FIGURE 14.18 Subsurface initiated spall.

The surface-originated mode of contact fatigue is often known as flaking failure because of the flake-like debris generated. However, thin flake-like debris can also be generated in the subsurface mode of contact fatigue. In rolling contacts, during the initial stages of failure, small cracks can form parallel to the surface, typically about 20 [μm] below the bearing surface. These cracks grow until they reach a length of about 100 to 300 [μm] before initiating the growth of a pair of descending cracks that reach deeper into the bearing material [78,79]. The flake debris is mostly formed from the release of material lying above the original parallel crack. This mechanism is illustrated schematically in Figure 14.19.

Microscopic examination of components can therefore provide much information on the cause of their failure.

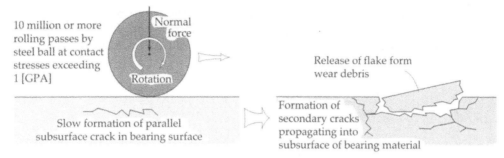

Figure 14.19 Mechanism of crack and debris formation in surface-originated contact fatigue.

Since the interfacial thermal resistance across the crack is much higher than the equivalent passage through a solid metal, a developing crack can block the flow of frictional heat from the surface layers to the interior of the substrate. In particular, if a crack is oriented parallel to the surface but at some small depth below it, then surface heating above the crack is more extreme than for the rest of the worn surface. An elevated degree of physical softening or even melting and enhanced chemical reactivity may then occur in the surface layers directly above the crack. These abnormal characteristics are caused by higher frictional temperature rises that would not otherwise occur without the crack. The extra heating of these surface layers tends to promote the thermal mound formation and this effect may also be a contributing factor to fatigue wear.

When cracks reach a critical size, spalling occurs and usually large quantities of debris are released. The debris is usually chunky and large in size so that it will cause further damage to the contacting surfaces as it passes through the contact. Often contact fatigue failure is also accompanied by the release of 'spherical particles' as shown in Figure 14.20.

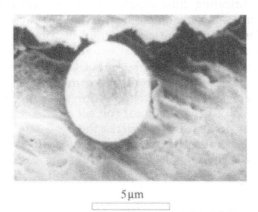

5 μm

FIGURE 14.20 Example of a spherical particle formed during contact fatigue.

Although the origin of precursor to spherical particles is still unknown it is generally believed that the relative motion between opposite crack faces causes some material to detach from one side of the crack and form the precursor to the spherical particle [35-37]. The movement of the crack faces continues to roll the particle and simultaneously deform it. A spherical particle consisting of a mixture of metal and oxide eventually forms after several million cycles of rolling movement. The mechanism of spherical particle formation is illustrated schematically in Figure 14.21.

As suggested in Figure 14.21 a sharp bend in the crack can form the precursor to the spherical particle. However, the formation of the precursor can also invoke adjacent inclusions as initiation points of cracks to release the particles which either contribute to normal debris or are rolled up to form spherical particles [38]. Spherical particles produced by contact fatigue are very distinctive and can provide an early warning of contact fatigue occurring in bearing systems.

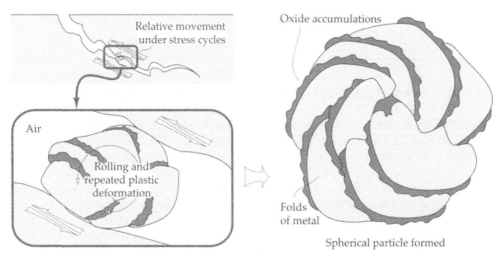

FIGURE 14.21 Mechanism of spherical particle formation during contact fatigue.

Effect of Lubricant on Contact Fatigue

The lubricant has a critical role in the nature of rolling wear and contact fatigue. Apart from its capacity to form a lubricating film separating the interacting surfaces, the chemical components of the lubricant can exert some influence on wear and fatigue in rolling contacts. EHL films modify the Hertzian pressure and traction distributions. They also influence the number and severity of asperity interactions and affect the stress concentrations in the vicinity of surface defects [39]. However, the question is whether EHL films are entirely beneficial in preventing contact fatigue or if they exhibit some significant limitations. As discussed already in Chapter 7, there is a pressure spike present in the EHL pressure distribution. The pressure spike generates local contact stresses which are much greater than the maximum contact stress predicted by Hertzian theory. There has been much speculation whether this pressure spike induces contact fatigue by a stress overload and reduces the fatigue life of a bearing [40]. The role of micro-EHL in suppressing or sometimes promoting contact fatigue is another aspect of contact fatigue that is still poorly understood.

Hydraulic Pressure Crack Propagation

A known demerit of EHL films is the cyclic changes in traction and contact pressure which occur during rolling contact at any point on a rolling surface. The crack present on the surface can be enhanced in EHL contacts by the mechanism known as 'hydraulic pressure crack propagation' illustrated schematically in Figure 14.22. It is suggested that the process occurs in three stages: an initial crack opening phase caused by traction forces ahead of the rolling contact, the filling of the crack with lubricant and its subsequent pressurization when traction forces and contact stresses close the crack [41]. The hydraulic crack propagation mechanism is thought to be particularly significant in the surface mode of contact fatigue [42].

It can be seen from Figure 14.22 that the mechanism of hydraulic crack propagation promotes rapid crack growth after the initiation stage in lubricated rolling contacts. Once a small crack is formed, the combined action of stress concentration at the crack tip and extreme lubricant pressure within the crack force it to extend rapidly. Hydraulic crack propagation might be suppressed by selecting a lubricant of high viscosity and compressibility [43]. This combination of properties would introduce a pressure loss due to lubricant flow down the restricted crack space to the crack tip [41].

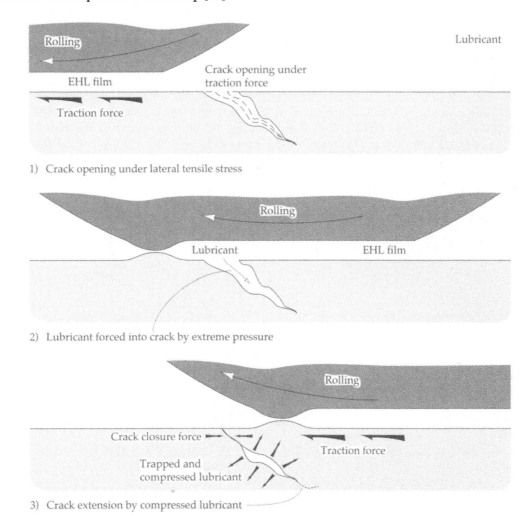

FIGURE 14.22 Schematic illustration of the mechanism of hydraulic pressure crack propagation.

Direct observations of the passage of a large dent (cavity) through an EHL contact, using optical interferometry, confirm the early theories postulated by Way [42]. It was found that under negative slide-to-roll ratio, i.e., the roller is rotating too slowly to ensure pure rolling contact at the contact zone, fluid appears to be trapped inside the dent and then released as a pulse of fluid as soon as the dent emerges from the EHL contact [81].

Chemical Effects of Lubricant Additives, Oxygen and Water on Contact Fatigue

Lubricants can react chemically with the rolling surfaces and convey many other reactive substances such as lubricant additives, oxygen and water. The significance of chemical interaction to sliding wear has already been described and a similar degree of chemical interaction has been found for rolling wear. Some lubricant additives have been found to suppress contact fatigue [e.g., 44-46] but some were found to promote contact fatigue [e.g., 44,46]. The effect was found to be proportional to the chemical reactivity of additives, their quantity and the type of material [e.g., 44,46,47].

Of greatest practical concern is the deleterious effect of water on contact fatigue resistance of steels [e.g., 48-51]. It has been shown that as little as **10** [ppm] of water reduces the fatigue life by about **10%** [50]. The increase in water concentration in oil progressively reduces the fatigue life. For example, **0.01%** of water reduces fatigue life by about **32 - 48%** [52] while with **6%** of water concentration fatigue life is reduced by about **70%** [53].

It is thought that a form of hydrogen embrittlement occurs at the crack tip where unoxidized metallic (steel) surface is in contact with water dissolved in the lubricant. Water dissociates in contact with unoxidized metal to release hydrogen ions which then permeate the metal. Interstitial hydrogen ions cause dislocation pinning and embrittlement of the metal [54]. A major function of lubricant additives in rolling contacts is to provide an adsorbed film at the crack tip which prevents adsorption or dissociation of water. A simplified model of the acceleration of crack extension by water present in the lubricant and suppression of detrimental water effect by a lubricant additive are illustrated in Figure 14.23.

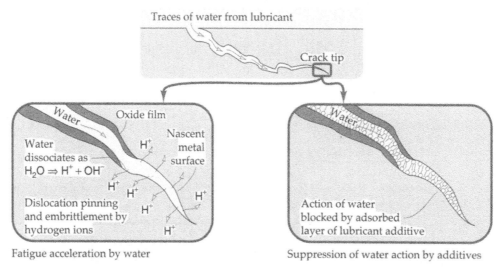

FIGURE 14.23 Simplified model of the acceleration of crack extension by water present in the lubricant and the suppression of the detrimental water effect by a lubricant additive.

To completely eliminate water from lubricating oils is very difficult and often impractical. Water can be introduced to the lubricant, for example, through the seals if the system is operating in an aqueous environment, or even through contact with moist air [50]. Therefore the effects of various oil additives on reducing the influence of water on contact fatigue have been studied. It was found that additives such as isoamyl alcohol, imidazoline [55,56] and a nonstoichiometric inorganic glassy compound consisting of oxides of '**B**', '**P**', '**Mg**' and '**K**' (known as '**nsic-bp1**') [53] are useful for the suppression of detrimental water effects on contact fatigue.

The formation of a hydrophobic surface film, shown in Figure 14.23, is most likely to occur with long chain hydrocarbon additives [57]. Three other mechanisms by which oil additives suppress the deleterious effect of water on fatigue life have also been suggested. These include a chemical reaction between water and the additive [53], proton neutralization [57] and water sequestration (by increasing water solubility in the oil and preventing its adsorption on the surface) [57].

Materials Effect on Contact Fatigue

Material properties play an important role in contact fatigue. The basic requirements of a material for rolling bearings is that it is sufficiently hard to withstand the Hertzian contact stresses and is suitable for manufacturing of high precision balls, rollers and rings. High carbon steel is the most widely used material in rolling contacts as it is relatively cheap and its hardness is high compared to most other metals. Steel rolling bearings are limited to operating temperatures below 320°C and rely on the continuous lubrication by oil. Apart from hardened steels, cast irons, such as ductile cast irons, have also found applications in machine components in rolling contacts.

A typical composition of austempered ductile iron (ADI) is 3.45% C, 2.71% Si, 1.26% Cu, 0.86% Ni and lesser amounts of manganese and chromium [82]. Partially chilled ductile iron (PCDI) has a similar composition except for the omission of copper and nickel. It has been observed that a major cause for crack nucleation were the graphite nodules (provided that there were no other major defects in the crystalline phases). A small nodule the size of about 15 [μm] in diameter and a large nodule density appears to exhibit a stronger effect than fewer but larger nodules. It was found that the rolling contact fatigue resistance of PCDI is superior to ADI [82].

Other metals have scarcely been used for this purpose because they are either relatively soft or more expensive. Titanium was suggested as a possible material for dry rolling contacts, i.e., railway wheels [58]. Since this metal is less dense than steel it could offer weight reductions, but this idea was not apparently developed further.

With increasing demands for rolling bearings capable of operating at very high temperatures, ceramic materials are now being applied and tested as rolling bearing materials. For example, silicon nitride rolling bearings are able to operate at temperatures reaching 1000°C in either dry conditions or in non-lubricating fluids such as cryogenic liquids (i.e., liquefied atmospheric gases), tap water or even saltwater [83].

The failure mode in ceramics under rolling is similar to that of metals. Ceramics tend to wear by pitting or crack formation even under dry conditions since chemical attack by air is limited to hydrolysis by moisture. Crack formation is, however, significantly modified by the microstructure of the ceramic. For example, in silicon nitride the grain boundaries formed preferred paths for crack propagation, resulting in grain pull-out as a dominating wear mode under rolling [59].

In silicon nitride rolling element bearings, pre-existing surface cracks, known as 'line cracks', appear to be the main initiator of spalling. The fatigue cracks propagate from the initial defect forming round conical pits with stepped surfaces. The brittleness of silicon nitrides leads to propagation of cracks in either direction relative to the direction of rolling. This is different from fatigue crack propagation in steels [84]. Useful rolling service life of silicon nitride is achieved at contact stress of about 5 to 6 [GPa] [84].

A critical property or characteristic of any steel selected for rolling contacts is its cleanliness or lack of non-metallic inclusions. Any type of inclusion acts as a stress raiser and can promote contact fatigue. Vacuum-melting or vacuum-deoxidizing of steels significantly reduces the number of inclusions and improves resistance to contact fatigue [e.g., 60-64].

It has been speculated that residual compressive stresses present in the material can have some effect on reducing the maximum shear stress inside the Hertzian contact field and thus delaying the onset of crack growth [e.g., 65-69] but the experimental evidence is not conclusive [70].

Influence of Operating Conditions on Rolling Wear and Contact Fatigue

The most direct influence of operating conditions on rolling wear and contact fatigue is whether rolling speeds and loads allow an elastohydrodynamic lubricating film to form. If the lubricant film thickness is relatively small compared to the surface roughness and some sliding as well as rolling occurs, then a form of lubricated oxidative wear takes place, as described in the chapter on 'Corrosive and Oxidative Wear'. This form of wear is very similar to dry rolling wear where the effect of increased load is to introduce wear by spalling which only begins after an initial period of purely oxidative wear on the rolling metal surfaces [19,71]. When the film thickness is sufficient to fully separate the rolling surfaces then a genuine form of contact fatigue occurs with phenomena such as hydraulic crack propagation taking part.

Traction or the generation of frictional stresses in the contact can significantly affect fatigue wear in both rolling and sliding contacts. As discussed already in the chapter on 'Elastohydrodynamic Lubrication', traction is the positive use of frictional forces to transmit mechanical energy. In traction there is always a speed difference, i.e., sliding, between two rolling bodies (a small speed difference is a pre-requisite for traction). It is generally well known that contact fatigue is very sensitive to sliding. Even a small amount of slip introduced to the rolling contact causes a decline in fatigue life. The interesting effect of traction is that a faster moving body has a longer fatigue life than the slower body which experiences sliding [72,73]. This characteristic of traction is illustrated schematically in Figure 14.24.

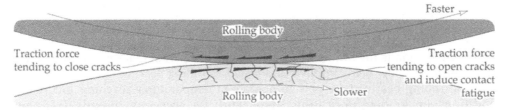

FIGURE 14.24 Effect of traction on contact fatigue lives of two rolling bodies.

It is reasoned that this effect is related to the mechanism of hydraulic crack propagation [42,74]. For the faster moving body, traction forces tend to maintain closed cracks in the surface which impedes lubricant pressurization inside the crack. The converse effect occurs for the slower body where traction force tends to open the cracks, causing inflow of the lubricant just prior to entering the Hertzian contact region and hydraulic crack propagation becomes relatively rapid.

The other mechanism suggested is due to the difference in plastic flow patterns occurring in the faster and in the slower moving bodies [73]. In the slower moving body, plastic deformation of the surface layer is significant and the layer moves in the same direction as rolling [73]. In the faster moving body the plastic deformation of the surface layer is much less significant and the layer moves in the direction opposite to the direction of rolling. This leads to differences in crack propagation for the faster and the slower moving bodies. In slower moving bodies, microcracks tend to propagate in alignment with the texture of the material, i.e., along the streamline of the deformed material, while in the faster moving body they tend to propagate into the substrate, crossing the texture of the deformed material, hence meeting more resistance. This leads to the difference in fatigue life in the faster and slower moving bodies. In order to optimize the fatigue life under traction it has been suggested that the surfaces of slower moving bodies should be made harder than the surfaces of the faster moving bodies [73].

14.4   MEANS OF CONTROLLING FATIGUE WEAR

The most effective method of preventing fatigue-based wear is to lower the coefficient of friction between two interacting bodies so that surface traction forces are insufficient for delamination in sliding or contact fatigue in rolling to occur. The other very important aspect in controlling fatigue wear in both sliding and rolling is the material's 'cleanliness'. Clean materials with minimum imperfections or inclusions should be selected for sliding and rolling contacts. Enhancement of material properties, i.e., hardness, to reduce crack growth can also be beneficial in some cases, but this method is limited by the increased brittleness of hard materials. Care needs to be taken when selecting combination of sliding materials. In particular, sliding contacts between identical materials should be avoided as this may provoke adhesive wear.

14.5   SUMMARY

During sliding a consistent pattern of events relating to subsurface plastic deformation, crack formation and subsequent release of wear debris is evident. The role of material properties in determining wear rates involves factors influencing crack initiation and propagation. A material with the minimum of microscopic flaws and inclusions will usually give low fatigue wear rates. The lack of relative motion between contacting asperities in rolling contacts ensures that wear during rolling is relatively slow compared to sliding wear. Wear under dry rolling is sufficiently slow to allow many mechanical components to operate without lubrication or other forms of wear protection for a certain limited period of time. The application of lubrication reduces the level of wear during rolling still further so that a considerable period of rolling must elapse before the first wear particle is produced. Rolling bearings, however, rely on near perfect contact surfaces for reliable functioning and the release of even one wear particle can terminate the useful life of the bearing. The slow release of a single wear particle associated with a pit in the rolling surface is known as contact fatigue. The study of contact fatigue has been developed to a very specialized level in order to predict the useful life of rolling bearings, but reliable prediction is still far from perfect. Fatigue-based wear is inevitable for all sliding and rolling contacts so that the onset of delamination or spalling could be considered as an acceptable limit to the working life of the component provided that only gradual failure occurs.

REVISION QUESTIONS

14.1   Under what conditions does fatigue wear occur?

14.2   Is strong adhesion between contacting asperities needed for this form of wear to occur?

14.3   Explain briefly how a wear particle is formed by fatigue wear.

14.4   How could oxygen or a lubricant affect this form of wear?

14.5   Is it reasonable to assume that fatigue wear is explained by one mechanism and therefore can be alleviated by one set of remedies? Give reasons in support of your answer.

14.6   Why should the fatigue wear rate of gold not show the same sensitivity to partial pressure of oxygen as does that of nickel? On the basis of the information given, what other metal might exhibit similar fatigue wear characteristics to gold?

14.7   Are dislocations present at the surface of a deformed body? Of the dislocations created close to the surface, what happens to them subsequently?

14.8   What happens to the subsurface dislocations after repeated plastic deformations by opposing harder asperities?

14.9   Why should lamellar wear debris be formed?

14.10   What are the size of lamellar wear particles?

14.11   What feature of a metal favours subsurface fatigue and delamination wear?

14.12   Would you expect fatigue processes in the worn surface to directly affect the friction coefficient?

14.13   What is the basic cause of material loss in a lubricated rolling contact?

14.14   Is direct contact between asperities a prerequisite for material loss by rolling contact fatigue?

14.15   How might the presence of oil in a rolling contact influence the growth of surface cracks? (That is to say, apart from the previously mentioned suppression of wear.)

14.16   What is the mechanical process involved in the generation of lamellar wear particles?

14.17   What is a major cause of pitting in rolling bearings or gears?

14.18   Why is a material's purity so important in the control of contact fatigue or delamination wear?

14.19   Name the type of material impurity most likely to promote delamination wear and describe how it acts.

REFERENCES

1   N. Soda, Y. Kimura and A. Tanaka, Wear of Some F.C.C. Metals during Unlubricated Sliding Part I: Effects of Load, Velocity and Atmospheric Pressure, *Wear*, Vol. 33, 1975, pp. 1-16.

2   D.A. Rigney and J.P. Hirth, Plastic Deformation and Sliding Friction of Metals, *Wear*, Vol. 53, 1979, pp. 345-370.

3   D.A. Rigney and W.A. Glaeser, The Significance of Near Surface Microstructure in the Wear Process, *Wear*, Vol. 46, 1978, pp. 241-250.

4   I.I. Garbar and J.V. Skorinin, Metal Surface Layer Structure Formation under Sliding Friction, *Wear*, Vol. 51, 1978, pp. 327-336.

5   R.C. Bill and D. Wisander, Recrystallization as a Controlling Process in the Wear of Some F.C.C. Metals, *Wear*, Vol. 41, 1977, pp. 351-363.

6   J.M. Challen, L.J. McLean and P.L.B. Oxley, Plastic Deformation of a Metal Surface in Sliding Contact with a Hard Wedge: Its Relation to Friction and Wear, *Proc. Roy. Soc., London*, Series A, Vol. 394, 1984, pp. 161-181.

7   D.H. Buckley, Surface Effects in Adhesion, Friction, Wear and Lubrication, Elsevier, Amsterdam, 1981.

8   N. Soda, Y. Kimura and A. Tanaka, Wear of Some F.C.C. Metals during Unlubricated Sliding Part II: Effects of Normal Load, Sliding Velocity and Atmospheric Pressure on Wear Fragments, *Wear*, Vol. 35, 1975, pp. 331-343.

9   N. Soda, Y. Kimura and A. Tanaka, Wear of Some F.C.C. Metals during Unlubricated Sliding Part III: a Mechanical Aspect of Wear, *Wear*, Vol. 40, 1976, pp. 23-35.

10   N. Soda, Y. Kimura and A. Tanaka, Wear of Some F.C.C. Metals during Unlubricated Sliding Part IV: Effects of Atmospheric Pressure on Wear, *Wear*, Vol. 43, 1977, pp. 165-174.

11   Y. Kimura, Mechanisms of Wear - the Present State of Our Understanding, *Transactions JSLE*, Vol. 28, 1983, pp. 709-714.

12   N.P. Suh, The Delamination Theory of Wear, *Wear*, Vol. 25, 1973, pp. 111-124.

13   N.P. Suh and H.C. Sin, On Prediction of Wear Coefficients in Sliding Wear, *ASLE Transactions*, Vol. 26, 1983, pp. 360-366.

14   N.P. Suh and H.C. Sin, The Genesis of Friction, *Wear*, Vol. 69, 1981, pp. 91-114.

15   N.P. Suh, S. Jahanmir, E.P. Abrahamson, A.P.L. Turner, Further Investigation of the Delamination Theory of Wear, *Transactions ASME, Journal of Lubrication Technology*, Vol. 96, 1974, pp. 631-637.

16   S. Jahanmir and N.P. Suh, Mechanics of Subsurface Void Nucleation in Delamination Wear, *Wear*, Vol. 44, 1977, pp. 17-38.

17   S. Jahanmir, N.P. Suh and E.P. Abrahamson, Microscopic Observations of the Wear Sheet Formation by Delamination, *Wear*, Vol. 28, 1974, pp. 235-249.

18   S. Jahanmir, The Relationship of Tangential Stress to Wear Particle Formation Mechanisms, *Wear*, Vol. 103, 1985, pp. 233-252.

19   V. Aronov and S. Kalpakjian, Wear Kinetics of Rail and Wheel Steels in the Dry Friction Condition, *Wear*, Vol. 61, 1980, pp. 101-110.

20   H. Krause and G. Poll, Wear of Wheel-Rail Surfaces, *Wear*, Vol. 113, 1986, pp. 103-122.

21   J.A. Perrotto, R.R. Riano and S.F. Murray, Effect of Abrasive Contamination on Ball Bearing Performance, *Lubrication Engineering*, Vol. 35, 1979, pp. 698-705.

22   R.S. Sayles and P.B. MacPherson, The Influence of Wear Debris on Rolling Contact Fatigue, Proc. Symposium on Rolling Contact Fatigue Testing of Bearing Steels sponsored by ASTM Committee A-1 on Steel, Stainless Steel, and Related Alloys, Phoenix, 12-14 May 1981, editor: J.J.C. Hoo, 1981, pp. 255-274.

23   T.E. Tallian, Prediction of Rolling Contact Fatigue Life in Contaminated Lubricant, Part II: Experimental, *Transactions ASME, Journal of Lubrication Technology*, Vol. 98, 1976, pp. 384-392.

24   S.H. Loewenthal and D.W. Moyer, Filtration Effects on Ball Bearing Life and Condition in a Contaminated Lubricant, *Transactions ASME, Journal of Lubrication Technology*, Vol. 101, 1979, pp. 171-179.

25   A.B. Jones, Metallographic Observations of Ball Bearing Fatigue Phenomena, Symposium on Testing of Bearings, 49th annual meeting of American Society for Testing Materials, Buffalo, N.Y., 24-28 June, 1946, preprint No. 45, 1946, 14p.

26   N. Outsuku and T. Muragami, Photoelastic Experiments on the Influence of Oil Films and Cracks on the Contact Stress Field during Sliding, Proc. Japan Society of Lubrication Engineers Annual Conference, Kyushu, October, 1983, pp. 369-372.

27   W.E. Littmann and R.L. Widner, Propagation of Contact Fatigue from Surface and Subsurface Origins, *Transactions ASME, Journal of Basic Engineering*, Vol. 88, 1966, pp. 624-636.

28   T.E. Tallian, On Competing Failure Modes in Rolling Contact, *ASLE Transactions*, Vol. 10, 1967, pp. 418-439.

29   T.E. Tallian, J.I. McCool and L.B. Sibley, Partial Elastohydrodynamic Lubrication in Rolling Contact, Proc. Symposium on Elastohydrodynamic Lubrication, Leeds, September, 1965, Inst. Mech. Engrs. Publ., London, Vol. 180, Pt. 3B, 1965-1966, pp. 169-184.

30   P.H. Dawson, Rolling Contact Fatigue Crack Initiation in a 0.3% Carbon Steel, *Proc. Inst. Mech. Engrs.*, London, Vol. 183, Pt. 4, 1968-1969, pp. 75-83.

31   P.H. Dawson, Further Experiments on the Effect of Metallic Contact on the Pitting of Lubricated Rolling Surfaces, *Proc. Inst. Mech. Engrs.*, London, Vol. 180, pt. 3B, 1965-1966, pp. 95-100.

32   C.A. Foord, C.G. Hingley and A. Cameron, Pitting on Steel Under Varying Speeds and Combined Stresses, *Transactions ASME, Journal of Lubrication Technology*, Vol. 91, 1969, pp. 282-290.

33   J.A. Martin and A.D. Eberhardt, Identification of Potential Failure Nuclei in Rolling Contact Fatigue, *Transactions ASME, Journal of Basic Engineering*, Vol. 89, 1967, pp. 932-942.

34   S. Borgese, Electron Fractographic Study of Spalls Formed in Rolling Contact, *Transactions ASME, Journal of Basic Engineering*, Vol. 89, 1967, pp. 943-948.

35   D. Scott and G.H. Mills, Spherical Particles in Rolling Contact Fatigue, *Nature*, Vol. 241, 1973, pp. 115-116.

36   B. Loy and R. McCallum, Mode of Formation of Spherical Particles in Rolling Contact Fatigue, *Wear*, Vol. 24, 1973, pp. 219-228.

37   D. Scott and G.H. Mills, Spherical Debris - Its Occurrence, Formation and Significance in Rolling Contact Fatigue, *Wear*, Vol. 24, 1973, pp. 235-242.

38   A.W. Ruff, Metallurgical Analysis of Wear Particles and Wearing Surfaces, National Bureau of Standards Report No. NBS1R 74-474, Washington, 1974.

39   T.E. Tallian, Elastohydrodynamic Effects in Rolling Contact Fatigue, Proc. 5th Leeds-Lyon Symp. on Tribology, Elastohydrodynamics and Related Topics, editors: D. Dowson, C.M. Taylor, M. Godet and D. Berthe, September 1978, Inst. Mech. Engrs. Publ., London, 1979, pp. 253-281.

40   L. Houpert, E. Ioannides, J.C. Kuypers and J. Tripp, The Effect of the EHD Pressure Spike on Rolling Bearing Fatigue, *Transactions ASME, Journal of Tribology*, Vol. 109, 1987, pp. 444-451.

41   M. Kaneta and Y. Murakami, Effect of Oil Hydraulic Pressure on Surface Crack Growth in Rolling/Sliding Contact, *Tribology International*, Vol. 20, 1987, pp. 210-217.

42   S. Way, Pitting Due to Rolling Contact, *Transactions ASME, Journal of Applied Mechanics*, Vol. 2, 1935, pp. A49-A58.

43   F.G. Rounds, Effects of Base Oil Viscosity and Type on Bearing Ball Fatigue, *ASLE Transactions*, Vol. 5, 1962, pp. 172-182.

44   D. Scott, Study of the Effect of Lubricant on Pitting Failure of Balls, Proc. Conf. on Lubrication and Wear, Inst. Mech. Engrs. Publ., London, 1957, pp. 463-468.

45   L. Arizmendi, A. Rincon and J.M. Bernardo, The Effect of a Solid Additive on Rolling Fatigue Life, *Tribology International*, Vol. 18, 1985, pp. 17-20.

46   F.G. Rounds, Influence of Steel Composition on Additive Performance, *ASLE Transaction*, Vol. 15, 1972, pp. 54-56.

47   F.G. Rounds, Some Aspects of Additives on Rolling Contact Fatigue, *ASLE Transaction*, Vol. 10, 1967, pp. 243-255.

48   L. Grunberg and D. Scott, The Acceleration of Pitting Failure by Water in the Lubricant, *Journal of the Institute of Petroleum*, Vol. 44, 1958, pp. 406-410.

49   P. Schatzberg and I.M. Felsen, Influence of Water on Fatigue Failure Location and Surface Alteration during Rolling Contact Lubrication, *Transactions ASME, Journal of Lubrication Technology*, Vol. 91, 1969, pp. 301-307.

50   P. Schatzberg and I.M. Felsen, Effects of Water and Oxygen during Rolling Contact Lubrication, *Wear*, Vol. 12, 1968, pp. 331-342.

51   R.E. Cantley, The Effect of Water in Lubricating Oil on Bearing Fatigue Life, *ASLE Transactions*, Vol. 20, 1977, pp. 244-248.

52   I.M. Felsen, R.W. McQuaid and J.A. Marzani, Effect of Seawater on the Fatigue Life and Failure Distribution of Hydraulic Fluid Flood Lubricated Angular Contact Ball Bearings, *ASLE Transactions*, Vol. 15, 1972, pp. 8-17.

53   L. Arizmendi, A. Rincon and J.M. Bernardo, The Effect of a Solid Additive on Rolling Fatigue Life, Part 2: Behaviour in a Water Accelerated Test, *Tribology International*, Vol. 18, 1985, pp. 282-284.

54   L. Grunberg, D.T. Jamieson and D. Scott, Hydrogen Penetration in Water-Accelerated Fatigue of Rolling Surfaces, *Philosophical Magazine*, Vol. 8, 1963, pp. 1553-1568.

55   L. Grunberg and D. Scott, The Effect of Additives on the Water Induced Pitting of Ball Bearings, *Journal of the Institute of Petroleum*, Vol. 46, 1960, pp. 259-266.

56   D. Scott, Further Data on the Effect of Additives on the Water Induced Pitting of Ball Bearings, *Journal of the Institute of Petroleum*, Vol. 48, 1962, pp. 24-25.

57   W.R. Murphy, C.J. Polk and C.N. Rowe, Effect of Lubricant Additives on Water-Accelerated Fatigue, *ASLE Transactions*, Vol. 21, 1978, pp. 63-70.

58   H. Krause and J. Scholten, Wear of Titanium and Titanium Alloys under Conditions of Rolling Stress, *Transactions ASME, Journal of Lubrication Technology*, Vol. 100, 1978, pp. 199-207.

59   J.F. Braza, H.S. Cheng and M.E. Fine, Silicon Nitride Wear Mechanisms: Rolling and Sliding Contact, *Tribology Transactions*, Vol. 32, 1989, pp. 439-446.

60   D. Scott, The Effect of Steel Making, Vacuum Melting and Casting Techniques on the Life of Rolling Bearings, *Vacuum*, Vol. 19, 1969, pp. 167-168.

61   R.F. Johnson and J.F. Sewel, The Bearing Properties of 1% Steel as Influenced by Steel Making Practice, *Journal of Iron and Steel Inst.*, Vol. 196, 1960, pp. 414-444.

62   H. Styri, Fatigue Strength of Ball Bearing Races and Heat Treated 52100 Steel Specimens, *Proc. ASTM*, Vol. 51, 1951, pp. 682-697.

63   T.W. Morrison, T. Tallian, H.O. Walp and G.H. Baile, The Effect of Material Variables on the Fatigue Life of AISI 52100 Steel Ball Bearings, *ASLE Transactions*, Vol. 5, 1962, pp. 347-364.

64   R.L. Widner, An Initial Appraisal of the Contact Fatigue Strength of Electron Beam Melted Bearing Steel, *Transactions ASME, Journal of Lubrication Technology*, Vol. 94, 1972, pp. 174-178.

65   J.O. Almen, Effect of Residual Stress on Rolling Bodies, Proc. Symposium on Rolling Contact Phenomena, Warren, Michigan, October 1960, editor: J.B. Bidwel, Elsevier, Amsterdam, 1962, pp. 400-424.

66   R.L. Scott, R.K. Kepple and M.H. Miller, The Effect of Processing Induced Near Surface Residual Stress on Ball Bearing Fatigue, Proc. Symposium on Rolling Contact Phenomena, Warren, Michigan, October 1960, editor: J.B. Bidwel, Elsevier, 1962, 301-316.

67   A.J. Gentile and A.D. Martin, The Effects of Prior Metallurgically Induced Compressive Residual Stress on Metallurgical and Endurance Properties of Overload Tested Ball Bearings, American Society of Mechanical Engineers, Paper 65-WA/CF-7, 1965.

68   E.V. Zaretsky, R.J. Parker and W.J. Anderson, A Study of Residual Stress Induced during Rolling, *Transactions ASME, Journal of Lubrication Technology*, Vol. 91, 1969, pp. 314-319.

69   H. Muro and N. Tsushima, Microstructural, Microhardness and Residual Stress Changes Due to Rolling Contact, *Wear*, Vol. 15, 1970, pp. 309-330.

70   R.K. Kepple and R.L. Mattson, Rolling Element Fatigue and Macroresidual Stress, *Transactions ASME, Journal of Lubrication Technology*, Vol. 92, 1970, pp. 76-82.

71   P.J. Bolton, P. Clayton and I.J. McEwen, Wear of Rail and Tire Steels under Rolling/Sliding Conditions, *ASLE Transactions*, Vol. 25, 1982, pp. 17-24.

72   B.W. Kelley, Lubrication of Concentrated Contacts, Interdisciplinary Approach to the Lubrication of Concentrated Contacts, Troy, New York, NASA SP-237, 1969, pp. 1-26.

73   N. Soda and T. Yamamoto, Effect of Tangential Traction and Roughness on Crack Initiation/Propagation during Rolling Contact, *ASLE Transactions*, Vol. 25, 1982, pp. 198-205.

74   W.E. Littmann, Discussion to reference [73] (by N. Soda and T. Yamamoto Effect of Tangential Traction and Roughness on Crack Initiation/Propagation during Rolling Contact, *ASLE Transactions*, Vol. 25, 1882, pp. 198-205), *ASLE Transactions*, Vol. 25, 1982, pp. 206-207.

75   K.L. Johnson, Contact Mechanics and the Wear of Metals, *Wear*, Vol. 190, 1995, pp. 162-170.

76   T. Akagaki and K. Kato, Plastic Flow Processes in Flow Wear under Boundary Lubricated Conditions, *Wear*, Vol. 117, 1987, pp. 179-186.

77   F.J. Franklin, G.-J. Weeda, A. Kapoor and E.J.M. Hiensch, Rolling Contact Fatigue and Wear Behaviour of the Infrastar Two-Material Rail, *Wear*, Vol. 258, 2005, pp. 1048-1054.

78   Y. Fujii and K. Maeda, Flaking Failure in Rolling Contact Fatigue Caused by Indentations on Mating Surface, (I) Reproduction of Flaking Failure Accompanied by Cracks Extending Bi-Directionally Relative to the Load Movement, *Wear*, Vol. 252, 2002, pp. 787-798.

79   Y. Fujii and K. Maeda, Flaking Failure in Rolling Contact Fatigue Caused by Indentations on Mating Surface, (II) Formation Process of Flaking Failure Accompanied by Cracks Extending Bi-Directionally Relative to the Load Movement, *Wear*, Vol. 252, 2002, pp. 799-810.

80   Y. Ding and N.F. Rieger, Spalling Formation Mechanism for Gears, *Wear*, Vol. 254, 2003, pp. 1307-1317.

81   A.V. Olver, L.K. Tiew, S. Medina and J.W. Choo, Direct Observations of a Micropit in an Elastohydrodynamic Contact, *Wear*, Vol. 256, 2004, pp. 168-175.

82   R.C. Dommarco and J.D. Salvande, Contact Fatigue Resistance of Austempered and Partially Chilled Ductile Irons, *Wear*, Vol. 254, 2003, pp. 230-236.

83   L. Wang, R.W. Snidle and L. Gu, Rolling Contact Silicon Nitride Bearing Technology: A Review of Recent Research, *Wear*, Vol. 246, 2000, pp. 159-173.

84   Y. Wang and M. Hadfield, A Study of Line Defect Fatigue Failure of Ceramic Rolling Elements in Rolling Contact, *Wear*, Vol. 253, 2002, pp. 975-985.

15 FRETTING AND MINOR WEAR MECHANISMS

15.1 INTRODUCTION

Fretting occurs wherever short amplitude reciprocating sliding between contacting surfaces is sustained for a large number of cycles. It results in two forms of damage: surface wear and deterioration of fatigue life. The extent of wear and surface damage is much greater than suggested by the magnitude of sliding distance. Reciprocating movements as short as 0.1 [μm] in amplitude can cause failure of the component when the sliding is maintained for one million cycles or more.

Contacts which seem to be devoid of relative movement such as interference fits do in fact allow sliding on the scale of 1 [μm] when alternating and oscillating loads are carried. It is very difficult to eliminate such movements and the resultant fretting. Fretting wear and fretting fatigue are present in almost all machinery and are the cause of total failure of some otherwise robust components.

Surveys reveal that, unlike other forms of wear, the incidence of fretting problems in machinery has not declined over the past decades [1]. Fretting fatigue remains an important but largely unknown factor in the fracture of load-bearing components at very low levels of stress. A knowledge of fretting is therefore essential for any engineer or technologist concerned with the reliability of mechanical equipment which almost always contains a large number of small amplitude sliding contacts.

The most common wear mechanisms which can be classified as minor wear mechanisms are melting wear, wear due to electric discharges, diffusive wear and impact wear. In present day technology, these mechanisms occur rather rarely or in a few limited instances so that not much is known about them. But with changes in technology, these wear mechanisms may assume greater importance. For example, melting wear is a direct result of frictional temperature rises and careful investigations reveal that it is much more common than widely believed. This is surprisingly not a particularly destructive form of wear and is associated with low to moderate friction coefficients. Wear due to electric discharges occurs in all electric motors and pantograph-cable contacts. It is a specialized but important form of wear. Diffusive wear occurs when two dissimilar materials are in high temperature frictional contact and material diffuses from one body to the other. A classic example of this type of wear occurs in cutting tools where there is a high enough temperature to facilitate diffusion. Impact wear is found when one component impacts or hammers against another, in rotary percussive drills which cut rock by high frequency hammering or even in electrical contacts such as relays where impact between components occurs.

The questions of practical importance to engineers are: how can fretting be detected and recognized? Which interfaces are likely to suffer fretting? How does fretting combine with fatigue? Can lubrication prevent or suppress fretting? What influence does temperature have on fretting? What is melting wear and when is it likely to occur? What are the characteristics of diffusive wear and wear due to electrical discharges? In what way does impact wear differ from erosive wear? This chapter addresses these questions and others.

15.2 FRETTING WEAR

The fundamental characteristic of fretting is the very small amplitude of sliding which dictates the unique features of this wear mechanism. Under certain conditions of normal and tangential load applied to the contact a microscopic movement within the contact takes place even without gross sliding. The centre of the contact may remain stationary while the edges reciprocate with an amplitude of the order of 1 [μm] to cause fretting damage. Therefore from a practical standpoint, there is no lower limit to the tangential force required for fretting damage and this fact must be allowed for in the design of mechanical components. One of the characteristic features of fretting is that the wear debris produced are often retained within the contact due to small amplitude sliding. The accumulating wear debris gradually separates both surfaces and, in some cases, may contribute to the acceleration of the wear process by abrasion. The process of fretting wear can be further accelerated by corrosion, temperature and other effects.

Microscopic Movements within the Contact under Applied Loads

When two solids are pressed together and then subjected to a tangential force of increasing magnitude, there is a certain value of tangential force at which macroscopic sliding occurs. Although this is a well known experimental fact, it is less widely realised that at levels of tangential force below this limiting value tangential micromovements also occur in response to the applied force. It has been discovered that these micromovements are a fundamental feature of any Hertzian contact subjected to a tangential force [2,3]. There are two models which describe the behaviour of such contacts: an earlier 'elastic' model and a recently developed 'elasto-plastic' model.

· Elastic Model for Fretting Contacts

Normal stress '**p**' in a stationary Hertzian contact rises smoothly from zero at the edge of the contact to its maximum value at the centre of the contact as shown in Figure 15.1a. Assuming that the coefficient of static friction 'μ' across the contact is constant, the frictional stress 'μ**p**', resulting from the normal stress '**p**', also rises smoothly from zero at the edge of the contact to a maximum value at the centre as shown in Figure 15.1b. If an external tangential force **Q** < μ**W** is subsequently applied to the contact and no slip occurs, then the resulting tangential stress '**q**' rises from some finite value at the middle of the contact to an infinite value at the edges as shown in Figure 15.1b. The distribution of tangential stress '**q**' across, e.g., a circular contact, can be described by the following expression [4]:

$$q_x = \frac{Q}{2\pi a(a^2 - x^2)^{0.5}} \qquad (15.1)$$

where:

q_x is the calculated tangential stress along the '**x**' axis [Pa];

a   is the radius of the contact area [m];

Q   is the superimposed tangential force [N].

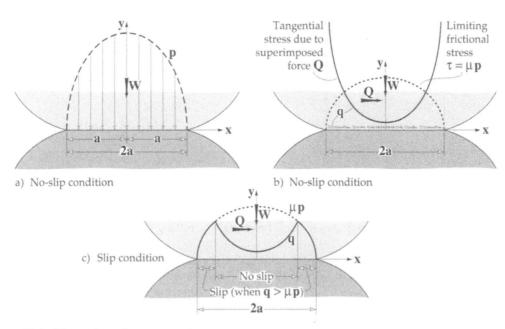

FIGURE 15.1 Normal and tangential stress fields for Hertzian contact with and without slip (adapted from [5]).

Cattaneo [2] and independently Mindlin [3] realized that the no-slip model could not correspond to real contacts and proposed that slip would occur wherever the calculated tangential stress 'q_x' exceeded the product of normal stress and the coefficient of friction 'μp' as shown in Figure 15.1c. In the region of slip, the real value of tangential stress was reasoned to be no greater than the product of local contact stress and the coefficient of friction. Therefore assuming, for example, that the normal load 'W' is constant and the tangential load 'Q' increases gradually from zero, then micro-slip occurs immediately at the edges of the contact area and spreads inwards until 'Q' approaches 'μW' and the 'stick' region reduces to a line for the line contacts, a point for the point contacts, etc. If 'Q' is increased further and exceeds 'μW', the contact starts to slide.

The contact is characterized by a central no slip region surrounded by an annular slip region. Slip, which reciprocates along with the tangential force, is the source of fretting damage and the edges of the contact are most vulnerable. The existence of a slip and a no-slip region in Hertzian contacts subjected to a tangential load has been confirmed experimentally [6,7]. In one of these studies a steel ball 5 [mm] in diameter was pressed against a glass plate under a load of 9.8 [N] and subjected to reciprocating sliding of varying amplitude [7]. At 1.25 [μm] amplitude, a thin ring of damaged glass surface on the edges of the Hertzian contact was evident. At 2.5 [μm] amplitude, the annular damage zone was much larger, leaving only a small circular unslipped region. A further increase in amplitude, above 3 [μm], resulted in gross sliding with no central unslipped zone. An example of the effect of increased amplitude of fretting between a hard steel ball and steel surface is shown in Figure 15.2 [6].

According to the model proposed by Mindlin, the ratio of the radius of the central unslipped region to the radius of the contact area is given by [3]:

$$a'/a = (1 - Q/\mu W)^{1/3}$$

(15.2)

where:

a'	is the radius of the central unslipped region [m];
a	is the contact radius [m];
Q	is the superimposed tangential force [N];
μ	is the coefficient of static friction;
W	is the normal load acting on the contact [N].

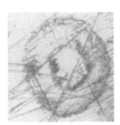 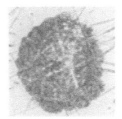

FIGURE 15.2 Effect of increased amplitude of fretting on fretting damage [6].

The relationship (15.2) between the ratio **a'/a** and the ratio of tangential force 'Q' to limiting frictional force 'μW' is shown in Figure 15.3a [7]. In experiments conducted with a steel ball oscillating on a glass surface, a reasonably good agreement between theoretical and experimental results was found [7]. It was assumed in the experiments that the diameter of the central unslipped region was identical to the area of contact remaining unobscured by wear debris. A typical fretting map showing all three regions of stick, partial slip and gross sliding as a function of contact load and slip amplitude is shown in Figure 15.3b [50,79]. Critical loads and amplitudes at the boundaries of the three major regions as well as boundary slopes depend on material properties, contact geometry and number of cycles.

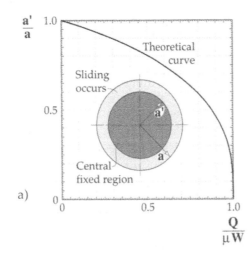 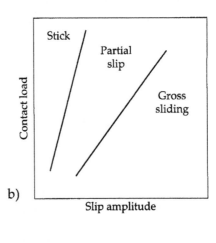

FIGURE 15.3 Relationship between the radius of central stationary zone and oscillating tangential load (a) [7] and schematic fretting regime map (b).

· *Elasto-Plastic Model for Fretting Contacts*

In the 'elastic' model of fretting it is assumed that relative displacement is accommodated by microslip between the surfaces in contact and elastic deformation of the contacting solids. However, it has been found from fretting experiments conducted on metals of varying hardness and under sufficiently high load to cause plastic yield that the maximum displacement amplitude which could be sustained without incipient gross slip was higher than predicted by elastic theory [8].

These discrepancies have been explained in terms of elasto-plastic behaviour of the material in the contact zone [9]. The 'elastic' model of fretting is based on the classical theory of friction which assumes that contact between the solids is achieved through contacts between the individual asperities. The theory assumes that the junctions between the asperities are rigid under load, and that when the surface shear stress exceeds a critical value slip will occur. Slip results from the sudden fracture of the asperity junction which takes place without any previous elastic or plastic deformation. This simplified assumption might be the reason for the discrepancy between the 'elastic' model and the experimental results [6].

The 'elasto-plastic' model of fretting contact has therefore been suggested [9]. According to this model, the asperities under the influence of a superimposed tangential force deform elastically in a central stick zone. This zone is surrounded by a zone in which the asperities have just yielded plastically but not fractured. The plastic deformation zone is in turn surrounded by a slip zone, where the asperities are subjected to fracture in a similar manner as in the 'elastic' model. This is illustrated in Figure 15.4. It can be seen from Figure 15.4 that the transition in surface stresses between the stick and slip regions is rounded as opposed to the sharp transition in Figure 15.1c.

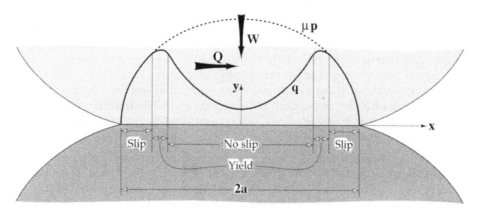

FIGURE 15.4 Surface stress distribution in an elasto-plastic fretting contact [9].

An analysis of a stress field during fretting revealed that plastic deformation occurs even for modest normal loads in most metals [10].

Fretting Regimes

As shown in Figure 15.3b there are three fretting regimes, i.e., stick, partial slip and gross slip. There is no damage observed when the fretting contact is operating under stick regime or conditions. The dominant damage under the partial slip regime is crack development, leading to, discussed later in this chapter, fretting fatigue. In the gross slip (or gross sliding) regime, damage by wear dominates. Often an additional regime, called mixed fretting, is included in this category. In this regime there is a competition between crack and wear-induced fretting damage.

Under the gross slip conditions two further wear regimes are often identified, i.e., a low wear regime when friction is low and a high wear regime when friction is high [81]. The low wear regime corresponds to mainly elastic or elastic shakedown situation, while plastic deformation dominates in the high wear regimes. The energy approach has been used to describe how wear debris is generated in the gross slip regime [81]. Firstly, a sufficient cumulated dissipated energy is required to plastically deform the material, changing the subsurface layer into a Tribological Transformed Structure (TTS). This layer then progressively disintegrates, releasing wear debris. Without a TTS, only minimal fretting wear occurs due to asperity deformation [81]. The concept of tribologically transformed structure in fretting is shown in Figure 15.5.

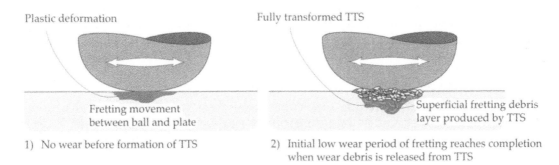

Figure 15.5   Model of a 'Tribological Transformed Structure' in fretting.

Effect of Amplitude and Debris Retention on Fretting Wear

Although it has been known for a considerable period of time that micromovements within the contact are the inevitable consequence of tangentially applied force and that this is the cause of fretting, a mechanistic definition of this wear process is still lacking. However, it is generally accepted that fretting wear increases with increasing amplitude of slip [11]. Plots for specific wear rates versus slip amplitude are often sigmoidal (resembling an elongated 'S') in shape [80]. The wear rates are low and often constant at low amplitudes (below 10-25 [μm]), increase linearly at moderate amplitudes (20-100 [μm]) and again tend to become constant at amplitudes above 100 [μm].

One of the special characteristics of fretting is a result of the prolonged retention of wear debris between the sliding surfaces when amplitude of sliding is minute. Debris retention can be explained in terms of the concept of the 'Mutual Overlap Coefficient' ('MOC') defined as the ratio of the contact area of the smaller of the sliding members to the wear track area [12]. In true sliding wear, the value of 'MOC' is very small, less than 0.1 as, for example, in the case of a piston ring sliding against a cylinder. On the other hand, in fretting wear the 'MOC' value approaches 1. The concept of 'MOC' is illustrated schematically in Figure 15.6 [13].

Experiments have revealed that at low 'MOC' values, wear debris left on exposed surfaces is quickly swept away by the leading edge of the smaller wearing body [12]. At high 'MOC' values, the majority of the worn surface is never exposed so there is hardly any expulsion of debris by the leading edge of the smaller wearing body. Wear debris accumulates between the sliding surfaces until it is eventually forced out by newly produced debris. It has been observed in experiments conducted with a chalk pin sliding against a glass surface that the wear debris produced could effectively support the load and separate the two surfaces in contact [12]. It has also been found in wear studies conducted with polymer pins sliding against steel surfaces that the 'MOC' can influence the fundamental characteristics of wear. For example, the effect of the 'MOC' on wear/temperature characteristics is illustrated in Figure 15.7 [13].

It can be seen from Figure 15.7 that the wear behaviour of polymers depends on the 'MOC'. The effect of the 'MOC' on wear rates is often radical in scale and the simple principles of practical aspects of wear control such as materials selection differ between ordinary sliding wear and fretting.

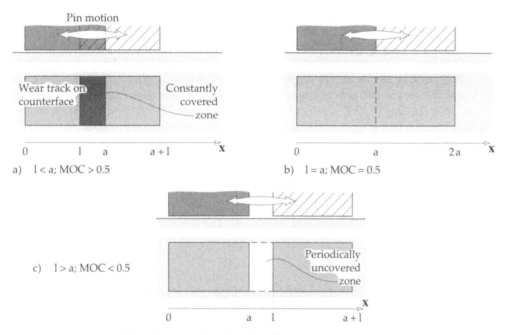

FIGURE 15.6   Concept of the 'Mutual Overlap Coefficient' [13].

A tentative definition of fretting as a wear process which occurs at amplitudes small enough to result in high 'MOC' values with debris entrapment taking place leads to wear characteristics which are completely different from normal sliding wear. It is generally accepted that debris retention is confined to amplitudes below 25 [μm] for most cases [5]. The process of debris entrapment is illustrated schematically in Figure 15.8.

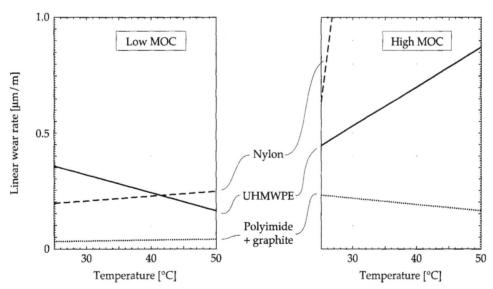

FIGURE 15.7   Wear rates of selected polymers versus temperature at low and high values of the 'Mutual Overlap Coefficient' [13].

There appears to be, however, no lower limit to the amplitude of sliding that can cause fretting wear [14]. Accumulation of wear debris at small amplitudes of fretting, e.g., 2.5 [μm], may result in apparently negative wear rate readings when wear volume is calculated from the depth of the wear scar [7]. Direct observation of a fretted contact by x-ray microscopy imaging revealed that debris accumulates in specific locations forming 'island layers' which may separate the fretting surfaces. Debris continues to accumulate in the worn contact until the wear scar becomes saturated with debris, i.e., after at least 10^5 cycles have elapsed. Beyond 10^5 cycles, the fretting surfaces are separated by a layer of debris which is progressively expelled from the wear scar as fretting wear particles [77]. In general, debris entrapment is harmless as long as the sliding surfaces are free to separate. For annular interference fits, however, debris entrapment may result in large increases in contact stresses.

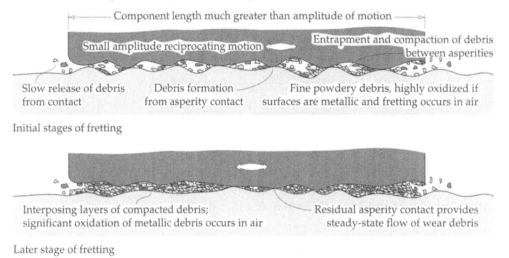

FIGURE 15.8 Mechanism of debris entrapment in a fretting contact.

Environmental Effects on Fretting Wear

Atmospheric oxygen and water have a strong effect on the fretting process, especially in metals [5]. It is generally found that oxygen and water accentuate fretting wear and surface damage [15] and that an inert atmosphere such as argon or nitrogen suppresses fretting of metals [15,16]. Debris from metallic contacts under fretting is high in oxide content [e.g., 5,17] and consists of a fine powder. When the fretted metal is iron or steel, the debris often shows a reddish-brown colour characteristic of iron oxide, for aluminium the debris is black.

Conductivity tests between metallic surfaces during fretting reveal a rise in electrical resistance consistent with a build up of oxide layer between the two surfaces [5]. It was found that when steel surfaces are fretted together the contact resistance increases from 10 [kΩ] to 1 [MΩ] or more in very dry air [18]. In nitrogen atmosphere, on the other hand, contact resistance remains very low until oxygen is admitted whereupon the resistance rises to very high levels as shown in Figure 15.9 [18].

In corrosive environments, e.g., saltwater, transient elevations in corrosion current tend to coincide with transient declines in the friction coefficient [82]. Experiments have shown that during fretting between titanium-aluminium-vanadium alloy and an alumina ball in salt water, where corrosion effects are expected to be significant, the corrosion current was controlled by the oxidation of titanium and subsequently a surface film of titanium oxide, and that imposed electrical potentials significantly influenced the overall volume of fretting wear [82].

In the absence of oxygen the mechanism of fretting wear is quite different. The small oscillating movements characteristic of fretting are ideal for causing plastic deformation and adhesion of the surfaces in contact. The limited amplitude of movement allows asperities to remain in contact or at least closely adjacent. While the fretting wear rate in nitrogen is about **3** times lower than in air, the coefficients of friction are in general much higher in nitrogen than in air [5]. For example, the coefficient of friction for aluminium and copper fretted in nitrogen is about **3**, whereas for steel this effect is not so pronounced and coefficients of friction remain below **1** [5]. Unless there is an intervening layer of oxide, fretted asperities tend to adhere and lock together, which causes the high friction coefficient.

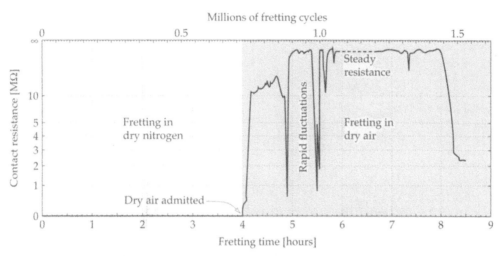

FIGURE 15.9   Variation of contact resistance for steel surfaces fretted first in nitrogen and then in air [18].

Fretting wear of non-noble metals is substantially lower in an inert atmosphere, i.e., in argon or nitrogen, than in air [15,16,19]. These findings seem to be consistent with the effects of the frequency of fretting motion on wear rates. A decrease in the wear rates of mild steels was reported as the frequency was increased up to about 1000 cycles per minute [16]. Beyond 1000 [cpm] wear was found to be almost independent of frequency. These frequency effects can be explained in terms of surface oxidation rates and the time available for the oxidation to occur between fretting cycles. With the increased frequency of sliding there is less time for a mechanically weak surface film to form in between asperity contacts, so the wear rate is lower. It is suggested that perhaps corrosion cracking, which seems to be very sensitive to frequency of sliding, contributes to the observed frequency effects in fretting wear [11].

Atmospheric pressure of air has a strong influence on fretting wear of steels due to the tribochemical metal oxidization. It has been found that the friction coefficient in fretting declines from values of 2 or more at pressures of 10^{-3} [Pa] to less than unity when the atmospheric pressure reaches 0.1 [Pa]. Reducing the amplitude of slip from 100 [μm] to 50 [μm] increases the maximum friction coefficient from 2 to 4 at low pressures of about 10^{-4}-10^{-3} [Pa]. However, if some oxidization is allowed, by the temporary admission of air to the test chamber, the friction coefficient rapidly declines. This can be explained by the effective retention of oxidized wear debris in a fretting contact. For reasons still not yet fully understood, fretting wear appears to reach a maximum at pressures of 1000 [Pa] (0.01 atmospheres) [83].

In general in the presence of oxygen the contact resistance between fretting surfaces tends to increase with the number of fretting cycles. This gradual rise in contact resistance, which is observed even with noble metals such as gold and platinum, can cause many problems with electrical contacts [20]. It has been found that with noble metals a friction polymer, which accumulates in the fretting contact instead of oxidized debris, is responsible for the increase in contact resistance [20]. The friction polymer is derived from organic contaminants in air and its polymerization is facilitated by the catalytic nature of many noble metal surfaces, e.g., platinum [20]. The process of friction polymer formation is illustrated schematically in Figure 15.10.

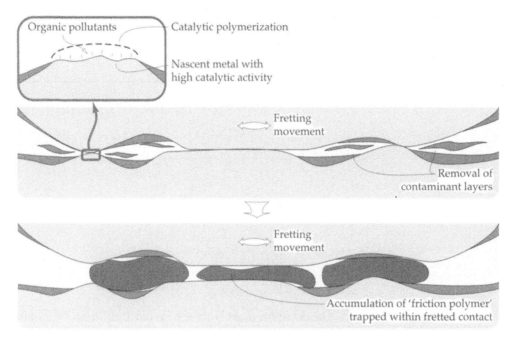

FIGURE 15.10 Mechanism of friction polymer formation in fretting electrical contacts.

This problem can easily be avoided by depositing on the electrical contact a thin film of liquid which dilutes the atmospheric contaminants and prevents adhesion of the friction polymer to the fretted surface. The polymer is then rapidly wiped off by movement of the fretting surfaces [20].

Current models of fretting wear of metals in air are usually based on the formation of a layer of oxide debris between the fretting surfaces. Oxide debris is assumed to originate from layers of oxide directly removed from the surface and from metallic wear debris which is subsequently broken up and re-oxidized many times [17]. Oxide debris formed during fretting in air has a structure that is either amorphous or nano-crystalline (crystals of only a few nanometres size). A mixture of amorphous and nano-crystalline debris can be generated where the composition is controlled by the rate of formation of the various metal oxides. For example, it was found that in the case of tin both SnO and SnO_2 are formed as a mixture of nano-crystalline grains [76]. The formation of oxidized debris proceeds by the same mechanism which was described for slow speed oxidative wear in Chapter 13. It has been suggested that a secondary wear mechanism of fretting is abrasion of the substrate by hard oxidized debris [17].

It has been shown that debris accumulations can deform the wearing substrate to produce depressions in the worn surface [21]. The formation of such pits in the fretted wear scar seems to be due more to the large contact stresses induced by debris entrapment than to true abrasion. An example of a fretted contact on steel is shown in Figure 15.11a. Characteristic features such as grooving by direct contact between asperities, accumulation of debris layers and the formation of depressions in the wear scar are clearly visible. Under similar fretting conditions ceramics are more resistant to fretting than steels [78]. Polishing, microfracture and tribolayer delamination are the dominating wear modes in ceramic fretting. An example of a fretted contact on ceramic is shown in Figure 15.11b.

FIGURE 15.11 SEM micrograph of the fretting contacts: a) between two crossed steel wires after 10^6 cycles, 5 [N] load and 25 [μm] fretting amplitude and b) between two ceramic surfaces after 10^6 cycles, 5 [N] load and 25 [μm] fretting amplitude.

Polymers, although not prone to oxidative corrosion, exhibit fretting behaviour which strongly depends on the presence of water and oxygen [22]. For example, polycarbonate in a fretting contact with a steel ball shows almost no wear in dry argon, nitrogen and oxygen. The wear rate is increased by a factor of 10 when 50% relative humidity air is introduced. Wet (85% relative humidity) nitrogen and oxygen cause a 30- and 50-fold increase in wear rate, respectively [22].

Although it is commonly assumed that frictional temperature rises are negligible in fretting, this may not always be so. The combination of high frequency of fretting movement, such as 100 [Hz] or more, friction coefficients reaching 1.0 and fretting amplitudes of at least 50 [μm], enables frictional temperature rises of several hundred degrees Celsius to be attained [84]. These high temperature rises may lead to significant tribochemical reactions for material combinations such as steel fretting against silicon nitride [84]. The additional effects of elevated ambient temperatures on fretting are described in the following section.

Effects of Temperature and Lubricants on Fretting

Temperature may affect the process of fretting in two ways. Firstly, the corrosion and oxidation rates usually increase with temperature, and secondly, the mechanical properties of materials change with temperature. In metals the temperature effects on fretting are best understood in terms of surface oxidation kinetics. Fretting wear rates usually decrease with increasing temperature if a stable and adherent oxide film is formed on the surface. On first consideration the decrease in fretting with temperature increase may be quite surprising since the oxidation rates of steel increase with temperature. However, thick, stable and mechanically strong oxide films forming on the surface act as a solid lubricant preventing metal-to-metal contact and hence reducing friction and surface damage [11]. The effectiveness of the oxide films formed at high temperatures depends on their mechanical properties and severity of fretting. If damage of these films occurs then fretting wear rates will most likely be much greater than at lower temperature because of the increased oxidation rate [5].

The formation of protective films has been observed in high temperature fretting of, for example, carbon steel [23], stainless steel [24], titanium alloys [25] and nickel alloys [26,27]. It has been found that after the transition temperature of about 200°C is reached the fretting wear rate of mild steel in air falls to a very low value which is maintained up to 500°C [23,28,29]. The oxide films generated vary in thickness and morphology and are formed at different temperature ranges for different materials. The 'glaze' type compacted oxide films were found on steel and nickel alloys [24,26] while titanium alloys were found to be often protected by thin oxide films [25].

Substitution of air by an inert gas such as argon resulted in a less dramatic reduction in fretting wear of mild steel after a transition temperature of about 200°C was reached. A further temperature increase, above 300°C, caused a sharp increase in fretting wear. This effect is illustrated in Figure 15.12 which shows fretting damage expressed as area of fretting scar times maximum depth of scar versus temperature at two levels of fretting cycles [28]. The decline in fretting damage around 200°C for tests in argon is attributed mainly to strain ageing of steel at this temperature and the rapid increase in fretting wear rate above 300°C is due to catastrophic surface failure by contact fatigue [5].

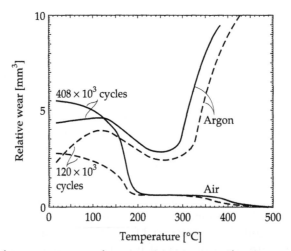

FIGURE 15.12 Effect of temperature and gas environment on fretting wear of mild steel [28].

Low temperatures, on the contrary, cause an increase in fretting damage which generally increases with the number of cycles [16]. An example of this trend is shown in Figure 15.13 [16].

It is generally recognized that because of the very low sliding speeds occurring in fretting, damage cannot be prevented by liquid lubricants [30]. However, they can provide a useful attenuation of fretting. The main purpose of a lubricant is to fill up the contact space and prevent the access of oxygen [5]. Comparative tests of fretting between dry steel surfaces and steel surfaces lubricated by various commercial oils demonstrated large reductions in fretting wear volume when lubrication was applied [31]. It was found that a simple lubricant such as a base mineral oil is effective in reducing fretting wear and friction [32]. In general, the effect of the lubricant is to suppress adhesive and corrosive wear occurring in the contact and allow wear to proceed by milder fatigue-based delamination wear mechanisms [31,33].

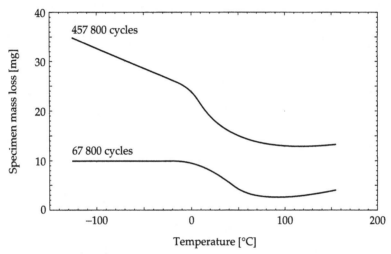

FIGURE 15.13 Effect of low temperatures on fretting wear of mild steel [16].

The literature available on the effects of lubricant additives on fretting is limited. Early work showed that phosphorous compounds, e.g., tricresylphosphate (TCP), diethyl hydrogen phosphite, etc., are effective in reducing fretting damage [34]. More recent studies have confirmed the effectiveness of commonly used compounds such ZnDDP [31,35] and sulphur based antiwear additives [32]. The reduction in fretting wear and friction is achieved through the formation of reaction films by lubricant additives on the contacting surfaces [e.g., 32,34]. Solid lubricants such as graphite and MoS_2 have also been shown to be effective in reducing fretting [36]. On the other hand, EP additives are less effective since the temperatures attained in fretting contacts are often not sufficiently high for the chemical reactions to take place and protective films to form [33,37].

Effect of Materials Properties and Surface Finish on Fretting

Manipulating hardness is found to be an unreliable means of improving fretting resistance [5]. Fretting studies of alloyed steel surfaces revealed that hardness has no direct relation to the level of fretting wear [38]. Instead microstructural factors, such as whether the steel is martensitic or austenitic, have a strong controlling influence on wear rate [39]. Microstructural phases also affect fretting wear of composite materials. For example, aluminium-silicon alloy matrix composites containing fibres of alumina as well as silica and graphite particles display moderate coefficients of friction when fretted against hardened steel. However, when there is a high concentration of hard silica particles or alumina fibres, rapid wear of the steel occurs. A mitigating factor is that below 20 [μm] of sliding amplitude, the wear of the steel is small [86].

In some cases, a harder material of a fretting pair can be worn as well as the softer one. For example, aluminium alloy, A357, is a non-ferrous alloy that is softer than 52100 bearing steel. When this alloy is fretted against a steel ball, there is an initial adhesion of aluminium alloy to the steel surface. If the fretting between these materials occurs in open air then the aluminium wear debris eventually oxidizes to form aluminium oxide. Since aluminium oxide is harder than steel, the oxidized debris is able to abrade the steel ball. As a result, the steel ball begins to wear after an initial period of protection by a film transfer. Increased amplitude of fretting, from e.g., 50 [μm] to 200 [μm], reduces the number of fretting cycles required to initiate wear of the steel ball [87].

It is commonly known that a high degree of surface finish accentuates damage due to fretting and to minimize the damage rough surfaces are preferred. At elevated temperatures, however, the converse is true, i.e., surfaces with better surface finish suffer less damage than rough surfaces [28].

Fretting Fatigue

Fretting fatigue is a phenomenon where the surface of a component subjected to alternating bulk stresses is also fretted, resulting in a severe reduction in fatigue life. For steels, there is often no endurance limit when fretting is combined with fatigue. A comparison between the number of cycles to fracture for pure fatigue and fretting fatigue for austenitic steel is shown in Figure 15.14 [40].

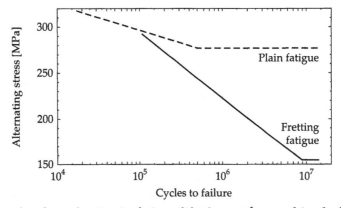

FIGURE 15.14 Example of a reduction in fatigue life due to the combined effect of fatigue and fretting [40].

Although fretting causes an alternating frictional surface stress to be superimposed on the pre-existing bulk cyclic stresses, of much greater importance is the generation of surface microcracks by asperity contact. Surface microcracks can be initiated due to repetitive fretting contacts as shown schematically in Figure 15.15. The number of surface microcracks increases with the increased amplitude. This is because the initiation of the microcrack relaxes the surface tensile stresses adjacent to it and to initiate another crack the contacting asperity must move a distance that is large compared to instantaneous crack length [11]. It has been found that fretting fatigue life rapidly decreases with increasing amplitude up to about 8 [μm] and after that is relatively insensitive to further increases in amplitude [11,75].

The surface microcracks generated due to fretting accelerate the initial stages of fatigue. An example of this effect is illustrated in Figure 15.16 which shows the relationship between the rate of crack propagation '**da/dN**' and the crack length '**a**' ('**N**' is the number of cycles).

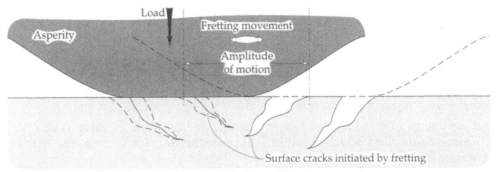

FIGURE 15.15 Mechanism of surface microcrack initiation in fretting contacts (adapted from [11]).

Although there are differences in the course of crack propagation depending on the loading conditions, the overall trend of higher initial crack growth rates for fretting fatigue is quite distinct. Adhesion, oscillating movement and tensile loads between fretting asperities are the basic cause of acceleration of early crack growth. A strong adhesive bond can form between a pair of fretted asperities, and instead of producing a wear particle, the bond may result in crack formation since the asperities never move very far from each other [5].

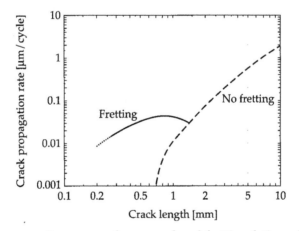

FIGURE 15.16 Crack propagation curves for normal and fretting fatigue (adapted from [41]).

Fretting fatigue cracks tend to occur at the boundary between fretted and non-fretted regions where shear stresses are sufficiently high [5]. A reduction in fatigue strength due to fretting can be estimated from the following expression [42]:

$$S_{fr} = S_o - 2\mu p_o[1 - e^{(-l/k)}] \tag{15.3}$$

where:

S_{fr} is the fretting fatigue strength [MPa];

S_o is the fatigue strength in the absence of fretting [MPa];

μ is the coefficient of friction;

p_o is the contact pressure [MPa];

l is the fretting amplitude [μm];

    k   is a constant. Typically **k = 3.8** [μm] [19], which renders the exponential term negligible for amplitudes of slip greater that **25** [μm].

For example, if the fatigue strength of the material in the absence of fretting is 800 [MPa], the coefficient of friction is 0.5, the contact pressure is 500 [MPa] and the amplitude of slip is 20 [μm] then the fretting fatigue strength according to equation (15.3) is about **303** [MPa]. It can be seen from equation (15.3) that a large reduction in fatigue strength can result if the coefficient of friction between the fretting surfaces is high.

It has been found that at elevated temperatures, providing that an alloy is able to produce thick surface oxide films which result in reduced adhesion and friction, the fretting fatigue life generally improves [11,43].

The combined action of corrosion and fretting can result in further reductions in the fatigue strength of some materials. For example, the fretting fatigue of steel is intensified in the presence of air compared to argon atmosphere [44,45]. A similar effect was observed when fretting fatigue tests were conducted in a sodium chloride (NaCl) solution [46]. Fretting only accelerates corrosion-fatigue when surface oxide films which can effectively prevent corrosion are disrupted. If a protective oxide film is broken, the nascent surface exposed is so reactive that the lack of corrosivity of air compared to saltwater makes little difference. In both cases, fretting fatigue in air and fretting fatigue in a corrosive medium, i.e., NaCl or saltwater, are accelerated by intensified localized corrosion at oxide film cracks, producing pits which initiate cracks and promote fatigue. Therefore fretting usually accelerates fatigue in corrosive environments in cases of alloys which depend on protective surface oxide films for corrosion resistance. The effect of combined corrosion and fretting on the fatigue life of corrodible and corrosion resistant metals is shown in Figure 15.17.

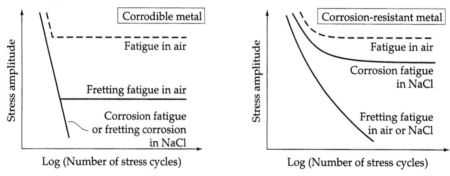

FIGURE 15.17 Effect of combined corrosion and fretting on fatigue life of corrodible and corrosion resistant metals [40].

The corrosive contribution to fretting wear and fretting fatigue is in fact quite large compared to the mechanical contribution alone. For example, it was found that the fretting fatigue of an aluminium alloy in a vacuum is about **10 - 15** times that in air [47]. Viewing fretting as a process dominated by corrosion, it would be expected that the wear rates would decrease with increased frequency of sliding since there is less time for a corrosion film to form. More information of the effect of combined corrosion and fretting on fatigue life can be found in Waterhouse and Dutta[40].

Practical Examples of Fretting

Many contacts that are nominally fixed in practice suffer fretting. These include most interference fits and devices subjected to vibration. The suppression of vibration is most important in the prevention of fretting wear and fretting fatigue. For example, in most railway wagon designs, steel wheels are press-fitted onto a steel axle. The rotation of the wheel and axle causes fretting and, more importantly, fretting fatigue. Interference fits of rotating assemblies therefore need to be carefully designed [5]. Heat exchangers provide a classical instance of a nominally static assembly which in fact suffers wear. Turbulent flow around the heat-exchange pipes causes them to vibrate against the baffle plates. Fretting wear and leakage may result. Assemblies of plates held together by rivets, i.e., air-frames, are also prone to fretting when vibration occurs. Typical examples of fretting occurring between the bearing outer ring and housing and in a riveted joint are shown in Figure 15.18.

Other common examples of fretting contacts can be found in flexible couplings, rolling bearings used for small oscillatory movements, wire ropes, electrical switch gears, etc. Some types of flexible couplings consist of interlocking gear teeth or a toothed connection. Since the teeth slide relative to each other by a few micrometres in a reciprocating mode during rotation of the coupling, fretting results. The debris from fretting wear can be sufficient to jam and seize the coupling or the teeth can wear out [5]. Rolling bearings are in principle not intended for small oscillatory movements but are often used due to the lack of a suitable alternative. Elastohydrodynamic films are unlikely to form when the rolling speed is extremely low and wear results. A wear scar forms under each roller or ball and this is known as 'false brinelling'. The individual wires of a wire rope must slide between each other for the rope to flex. The contacts between wires are therefore subjected to fretting and fretting fatigue [48,49]. It is extremely difficult to monitor the extent of fretting between wires without dismantling the rope. Fretting proceeds unnoticed until a wear scar has cut through half the thickness of a wire and that wire fractures. The performance of electrical contacts, e.g., relays, switches and selectors, is extremely sensitive to fretting. Fretting of electrical contacts in the open air causes the formation of insulating films between the contacting surfaces and an increase in contact resistance to unacceptably high levels. The insulating film is composed of oxidized fretting wear debris. The most effective way to prevent or control contact resistance is to ensure that only partial slip occurs in the fretting contact [88]. When gross slip (movement over the entire contact area) occurs, the accumulation of oxidized wear debris is inevitable followed by an increase in contact resistance. Protective coatings of noble metals offer protection but only until they are worn away [88]. More examples of case histories of fretting are described elsewhere [5,40].

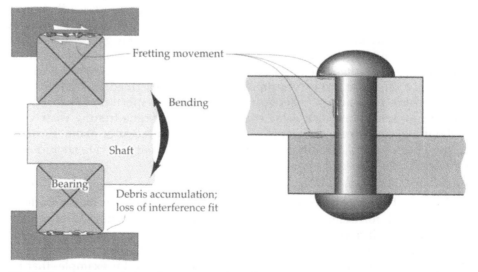

FIGURE 15.18 Examples of fretting occurring between two stationary surfaces due to oscillatory motion in a bearing assembly and in a riveted joint.

Means of Controlling Fretting

Fretting can be effectively controlled through design optimization and through the applications of surface treatments such as coatings and shot peening.

The suppression of fretting by design optimization usually involves geometric modifications of components aimed at eliminating excessive shear stress concentrations at the interface. An example of design optimization to minimize fretting is illustrated in Figure 15.19 where a press-fit shaft assembly is shown.

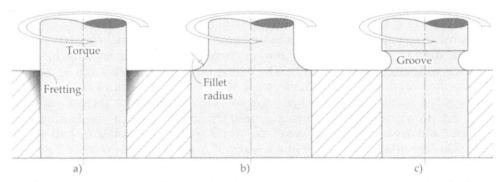

FIGURE 15.19 Suppression of fretting by design optimization in a press-fit shaft assembly (adapted from [5]).

Fretting occurs at the edge of the press-fit where slip is most likely to be present (Figure 15.19a). One of the solutions is to enlarge the press-fit diameter compared to the shaft diameter as shown in Figure 15.19b. Although increased rigidity of the press-fit suppresses slip and the resulting fretting, attention should be paid to the fillet radius on the shoulder between the two diameters. If it is impossible to allow an increased press-fit diameter then a groove could be machined close to the edge of the press-fit as shown in Figure 15.19c. This will also result in suppressing fretting. However, in this case of design optimization there is a compromise between reduction of strength caused by the groove and relief from fretting [5].

Fretting can also be suppressed by the application of surface coatings [5]. The basic principle is to cover the fretting surfaces with a non-metallic layer which suppresses adhesion and stops the oxidation caused by fretting of plain metal surfaces. The surfaces can be covered with a polymer which is intended to wear sacrificially [11]. An inorganic lubricant such as molybdenum disulphide or an anti-wear compound such as titanium carbide or chromium nitride can also be deposited on the surfaces [11,40,85]. The efficiency of these coatings depends strongly on the coating technique applied as described in Chapter 9. A major problem is to ensure good adhesion between the coating and substrate to prevent spalling of the coating. It has been shown that some coatings can reduce the fretting wear volume by a factor of about **100** or increase the fretting fatigue life by a factor of about **10** or more [11]. However, due to the great variability in the performance of coatings, care should be exercised in coating selection for a specific application.

One way of controlling fretting fatigue is perhaps best illustrated by equation (15.3) which shows the beneficial effect of reducing the coefficient of friction between the fretted surfaces. This can be achieved, for example, by introducing low friction coatings. Fretting fatigue can also be reduced by the introduction of compressive stresses in the surface layers of the fretted metal which suppress the growth of cracks quite effectively [40]. The compressive stresses can be introduced by the process of shot peening. It has been found, for example, that in rotating bending tests on an austenitic stainless steel shot peening almost cancels out the loss in fatigue strength due to fretting [40]. Tests with other metals and alloys showed a similar beneficial effect obtained from shot peening [40]. Plasma nitriding appears to be a useful treatment in elevating the fretting fatigue resistance of stainless steels. The nitriding temperature of 520°C was found to be most effective in achieving a coating with good fretting fatigue resistance. Coatings produced at lower temperature of 400°C were less effective. Following plasma nitriding at 520°C a hard, rich in chromium nitrides, surface layer forms, resulting in a fretting fatigue limit even higher than the plain fatigue limit.

15.3   MELTING WEAR

This is a universal form of wear since any material which melts without decomposing will show melting wear under appropriate conditions. The present concept of melting wear is based on the phenomenon that when sliding speed and load are increased, the frictional surface temperature also rises until the melting temperature is reached and layers of molten surface material begin to influence friction and wear [51]. This mechanism is illustrated schematically in Figure 15.20.

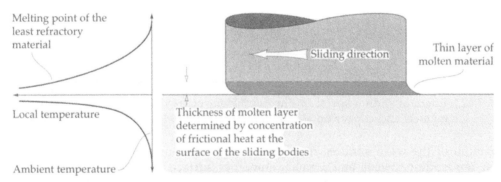

FIGURE 15.20 Schematic illustration of the formation of molten layer.

The generation of the layer of molten material usually dramatically reduces friction and this is the reason why snow skis and skates slide easily and why friction is reduced to dangerously low levels during the skidding of tyres at high speeds. For example, during skiing there is sufficient frictional heat to melt a layer of ice producing a film of water at the points where the asperities touch the ski. The skier simply hydroplanes on the layer of water generated by the friction. When the temperature drops below a critical level, about -10°C, the heat is conducted away too quickly to allow for this melting to occur and ice asperities adhere to the ski and are sheared during the sliding. At these conditions the coefficient of friction rises dramatically from about **0.02** to **0.4**. The increase in coefficient of friction with a progressive decrease in temperature caused considerable hardship to early polar explorers, since with the temperature drop pulling the sleds was increasingly harder.

One important practical use of the melting wear phenomenon is found in guns, at the interface between the gun barrel and the shell. Close contact between the shell and barrel is needed for accurate shooting while a low friction and wear coefficient is desirable to provide a long life for the barrel. The sealing rings on each shell are coated with a low-melting point metal alloy to facilitate melting wear. It has been calculated that when gilding metal (90% Cu and 10% Zn) is used in shell sealing rings, the friction coefficient rapidly falls to about 0.02 [52]. An example of the reduction in coefficient of friction with increased sliding speed is shown in Figure 15.21 [51].

It can be seen from Figure 15.21 that a continuous decline in the friction coefficient occurs for both bismuth and copper. Bismuth shows a decline in friction at lower sliding speeds because its melting point at 271°C is far lower than that of copper at 1083°C. The friction coefficients of 1.5 or more measured at moderate sliding speeds for copper are characteristic of severe adhesive wear, whereas the friction coefficient of 0.2 measured at 600 [m/s] indicates that a much milder wear process is taking place. The tests were run unlubricated in a vacuum to minimize drag forces on the ball. The wear rate measured for bismuth on steel increased dramatically from 5×10^{-13} [m³/Nm] at 100 [m/s] sliding speed to 2.5×10^{-11} [m³/Nm] at 400 [m/s]. This indicates that when there is sufficient frictional heating to effectively melt the surface material, rapid wear results, especially in low melting point metals.

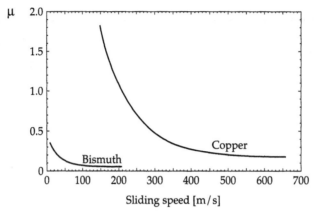

FIGURE 15.21 Effect of sliding speed on the coefficient of friction measured between bismuth on steel and copper on steel combinations [51].

Examination of the worn surfaces revealed a rough torn surface characteristic of adhesive wear at low sliding speeds and a much smoother surface with considerable smearing of surface material at high sliding speeds. With a low melting point metal such as bismuth, droplets of metal were also found to adhere to the surface just outside the contact. It has been speculated that films of molten material can even effectively separate two sliding surfaces, and hydrodynamic lubrication models have been developed to support the hypothesis that there is sufficient frictional heat to form the lubricating films of molten material [53,54].

A similar decline in friction was found for high melting point metals such as tungsten (melting point 3410°C). A coefficient of friction of 0.3 was measured at 200 [m/s] sliding speed, and a coefficient of friction of 0.1 at 700 [m/s], but clear signs of surface melting were not found. Instead, mutual diffusion of 'W' and 'Fe', and steel film transfer were observed [51]. The non-metals, glass, rubber, PTFE and nylon also showed a decline in coefficient of friction with rising sliding speed and surface melting at high sliding speeds [51,55]. Evidence of melting wear was found when magnesium alloys were slid against hardened steel counter-faces. The estimated depth of the melted layer was between 10 and 20 [μm] and the surface melting was associated with moderate levels of friction and wear [90]. The mechanism of melting wear is illustrated schematically in Figure 15.22.

Although it is usually implied that extremely high sliding speeds are a prerequisite for melting wear of metals [e.g., 55-57], it has been found that melting wear can occur also at moderate sliding speeds [58]. Therefore this form of wear could be of much greater practical significance than is generally supposed. Tests with cast iron sliding on cast iron in a vacuum at speeds of 5 [m/s] revealed the presence of ledeburite, which is a form of re-solidified cast iron on the worn surface [58].

Melting wear can also occur during momentary contacts, such as between a projectile and a target. During these contacts a very high contact stresses, between 1 and 2 [GPa], can be reached. Tests with aluminium alloys revealed that at these high contact pressures, the molten metal film can form. These films can maintain a shear resistance as high as 100 [MPa] [91].

Severe sliding conditions often result in the formation of hard white layers on the worn surface. These layers do not etch to reveal a microstructure and simply appear as white areas under the metallurgical microscope, hence the name. In some cases these layers form after rapid solidification of the molten top surface [e.g., 58-60]. For example, in gun barrels, where there is a combination of high stress, high temperature and a hostile environment, the white layers are formed by rapid solidification of molten material [60]. Investigation of the nature and formation of white layers on steels revealed that they consist of re-solidified metal with large amounts of iron carbides and oxides mixed in [59]. Wear particles were formed not by the escape of molten metal but by the fracture of the re-solidified layer and detachment from the surface. The grain structure of the white layers was found to be extremely fine because of the very rapid cooling associated with transient frictional temperatures.

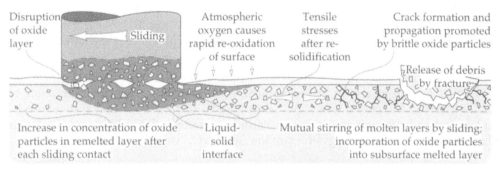

FIGURE 15.22 Schematic illustration of the mechanism of melting wear.

It is apparent that melting wear involves many complex phenomena, most of which are still poorly understood. For example, it is still uncertain whether the films of molten material support all the load or there is additional solid contact. However, it is found empirically that the friction characteristic of many materials in high speed sliding can be summarized by a conveniently simple formula [96-98]. This formula states that beyond a critical minimum value of sliding speed, the coefficient of friction is inversely proportional to the square root of the sliding speed, i.e.:

$$\mu = \mu_{threshold} \left(\frac{U_{threshold}}{U_{sliding}} \right)^{0.5}$$

(15.4)

where:

$U_{threshold}$ is the threshold speed [m/s], i.e., critical minimum value of sliding speed below which melting does not occur;

$U_{sliding}$ is the sliding speed [m/s];

$\mu_{threshold}$ is the friction coefficient at or below threshold speed.

In the model a constant coefficient of friction is assumed for speeds less than '$U_{threshold}$'. For $U_{sliding} > U_{threshold}$ coefficient of friction decreases according to equation (15.4). However, in real materials this decrease might be affected by the visco-elastic behaviour of the molten layer.

Apart from the early pioneering work of Bowden and Tabor [51], research on melting wear has unfortunately been very limited and it is still a matter for thorough experimental investigation to decide the conditions under which melting wear occurs. Polymers are quite susceptible to melting wear and more information on this topic can be found in Chapter 16.

15.4 WEAR DUE TO ELECTRICAL DISCHARGES AND PASSAGE OF ELECTRIC CURRENT ACROSS A CONTACT

Accelerated material loss caused by electric arcing between sliding surfaces occurs, for example, in pantograph-cable systems and between the slip ring and commutators of an electric motor. It is believed that arcing between the two surfaces when they are momentarily separated is the cause of accelerated wear. The average duration of such an arc is a few milliseconds and a close correlation has been found between the duration of arcing periods and the periods of separation between sliding surfaces [61]. Material from the contacting surface, which is usually metal, is thought to be vaporized or oxidized by the electrical energy discharge to leave a small pit. The mechanism of wear due to electrical discharge is illustrated schematically in Figure 15.23.

Wear rate due to electrical discharge has been found to be proportional to the amount of arcing, with almost no wear occurring in the absence of arcing [62]. Wear rate tends also to increase with speed decrease, e.g., the wires of a pantograph-cable system suffer very rapid wear below 14 [m/s] (about 50 [km/h]) [63].

The passage of electric current across a contact may also cause wear rates to vary due to the effect of electric potentials on tribochemical reactions. For example, tungsten carbide composites showed increased wear and friction rates during a scratch test where a direct current was passed through the tungsten carbide composite [92]. This was attributed to a growth of thick and rougher tungsten tri-oxide film (WO_3) on the scratched surface.

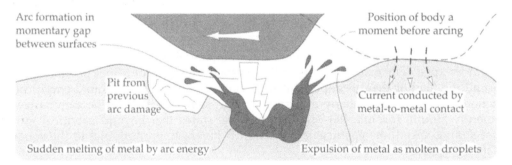

Arc formation in momentary gap between surfaces

Position of body a moment before arcing

Pit from previous arc damage

Current conducted by metal-to-metal contact

Sudden melting of metal by arc energy

Expulsion of metal as molten droplets

FIGURE 15.23 Schematic illustration of the wear mechanism due to electrical arcing between sliding surfaces.

The wear of electric commutator materials, which are usually made of carbon graphite, is also strongly affected by electric current. Without any electric current, micro-cutting dominates the wear of carbon-graphite against hardened steel. When an electric current is passed across this contact, the wear rate increases in proportion to the current but the friction coefficient is not significantly affected. However, for any given current there is a maximum wear rate for a specific contact load. The wear debris generated is finer, showing signs of very high temperatures at asperity contacts [93].

The electric current can also significantly affect wear and friction of steels under lubricated conditions [64]. It seems that the current direction is more important than the current intensity, e.g., cathodic surfaces were found to wear more than anodic surfaces. Cathodic surfaces suffer more wear because the development of protective films is suppressed by the imposed voltage. On the other hand, because of the favourable corrosion potential, anodic surfaces develop a thicker film of oxide. In terms of the wear mechanism, the imposed voltage causes a nascent metallic surface to last longer without oxide films on the cathodic surface than on the anodic surface. The greater proportion of nascent surface on the cathodic surface causes an increase in adhesive wear. It should be mentioned, however, that current traversing the contact is very small, e.g., about 1 [mA] for 0.2 [mm^2] of wear scar area. This current level is very small compared to that occurring in electric current collectors which suffer from entirely different wear mechanism, e.g., from spark erosion and surface melting.

As mentioned already lubricated contacts are not exempt from electric current effects. When there is sufficient voltage difference across a hydrodynamic oil film, electrical arc discharge occurs, causing melting of tin in white metal alloys (babbitts) and the formation of pits [94]. The low melting point of tin facilitates damage by electrical arcing. The limiting electrical voltage gradient or dielectric strength of the oil is approximately 15 [kV/mm] [95]. Thus an oil film of average thickness of 10 [μm] can withstand about 150 [V] potential difference, which is sufficiently small to become a problem in the operation of rotating machinery.

It was also found that when sulphur based additives are present, anodic surfaces suffer greater wear than cathodic surfaces. The reason for this reversal of wear bias is that sulphur based additives introduce more rapid corrosion into the wearing contacts. With rapid corrosion taking place, the anodic surface suffers corrosive wear while on the cathodic surface, corrosion is suppressed to a level closer to the optimum balance between corrosive and adhesive wear [65]. In contrast, phosphorus based additives, which appear to function by decomposition on the worn surface as opposed to corrosion, suppress the wear on both cathodic and anodic surfaces [65]. A study of the wear between two bodies of dissimilar size under boundary lubricated conditions also revealed that self-generated voltages induced by wear cause the smaller of two sliding bodies to gain a positive charge (become anodic) and suffer accelerated wear [66].

15.5 DIFFUSIVE WEAR

When there is true contact between the atoms of opposing surfaces and a high interface temperature, significant diffusion of chemical elements from one body to another can occur. The most widespread example of such a contact is the rake face of a cutting tool close to the cutting edge in high-speed machining. In this situation, there is almost the perfect contact between the tool and the metal chip due to the extreme contact stresses and very high temperatures, reaching 700°C or more [67]. The metal chip represents a continually refreshed supply of relatively pure metal whilst the tool is a high concentration mixture of some radically different elements, e.g., tungsten and carbon. Therefore, there is a tendency for some of the elements in the tool to diffuse into the chip where solubility conditions are more favourable. When the surface material of the tool loses a vital alloying element it becomes soft and is very soon worn away by the chip [67]. The mechanism of diffusive wear is illustrated schematically in Figure 15.24 with a tungsten carbide tool as an example.

In the early days of the introduction of carbide tools, tungsten carbide was widely used. Problems arose in machining of steel since the tungsten was rapidly lost to the chip. Slightly changing the tool composition by adding titanium carbide or tantalum carbide was found to remedy the problem. A similar mechanism is believed to be responsible for the excessive wear of silicon based ceramic cutting tools in the machining of steel. This time it is silicon which diffuses through grain boundaries to the workpiece [68]. The diffusive wear rate of cutting tools depends on the tool material solubility limits in the workpiece. The mechanism of diffusive wear has also been modelled mathematically [69].

15.6 IMPACT WEAR

Impact wear is caused by repetitive collision between opposing surfaces. A classic example of this form of wear is found on the heads of hammers. This form of wear involves flat surfaces or nearly flat surfaces with a large radius of curvature compared to the size of the wear scar. This feature distinguishes impact wear from erosive wear where a sharp particle indents a flat surface. In impact wear the surface is subjected to repetitive impact by a series of pulses of high contact stress combined with some energy dissipation in each impact as shown schematically in Figure 15.25.

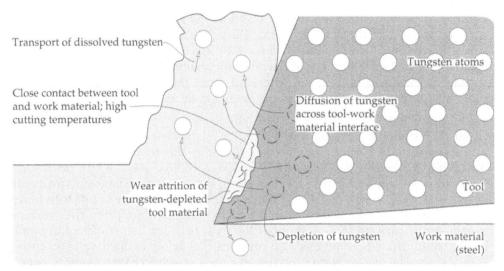

FIGURE 15.24 Schematic illustration of mechanism of diffusive wear.

FIGURE 15.25 Repetitive stress pulses under impact wear.

The mechanism of impact wear involves elastic and plastic deformation when impact energy is high and/or fatigue accompanied by wear debris release due to crack formation [70,71]. If oxygen is present and the wearing material can be oxidized then a corrosive or oxidative wear mechanism can also take place. Iron and steel components are susceptible to impact wear by tribo-oxidation, especially at elevated temperatures at which rapid oxidation occurs [70]. The mechanisms of impact wear are illustrated schematically in Figure 15.26.

In general, impact wear is dependent on the formation of deformed layers, particularly when wear by fatigue or crack formation is predominant [72]. In such cases, subsurface cracks extend parallel to the surface in a manner very similar to 'delamination' wear. The material through which the cracks propagate is very often plastically deformed and work-hardened as a result of contact stresses during impact [72]. Spallation and wear by crack formation can also occur in relatively brittle materials [73] so it seems that the presence of surface plastic deformation, in some cases, is not essential to this form of wear.

The type of material sustaining impact wear has a strong effect on the wear mechanism. Metals are prone to an oxidative form of impact wear while ceramics wear by cracking and spalling. Polymers tend to wear by plastic deformation, fatigue cracking or by chemical attack from hot compressed layers of oxygen and pollutants at the moment of impact.

The process of repeated impacts between mechanical components is often accompanied by small sliding movements and these can affect the wear mechanism occurring. Movement tangential to the contacting surfaces during impact is usually caused by elastic deformation of the supporting structures. For example, if a wearing surface is supported by a cantilever then the deflection of the cantilever will cause sliding movement during impact. Superimposed sliding during impact caused by elastic deflection is illustrated schematically in Figure 15.27.

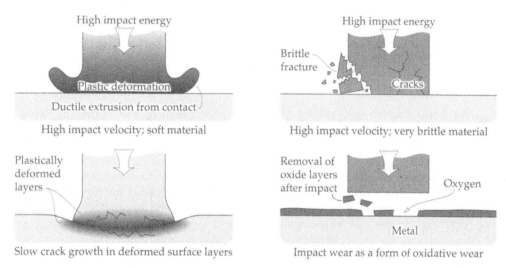

FIGURE 15.26 Schematic illustration of the mechanisms of impact wear.

Superimposed sliding can cause an acceleration in wear, for example, by fretting and is usually undesirable [70]. The effect of sliding on the impact wear of steel is to cause plastic deformation of the steel in the direction of sliding and the formation of thicker oxide layers on the worn faces [74]. This lateral movement of the impacted metal is a form of wear since material is removed from the wearing contact even though it may remain attached to the wearing component. With more brittle materials, such as the tungsten carbide composites used in rock drills, superimposed sliding results in more intense spalling of the impacting surfaces [73].

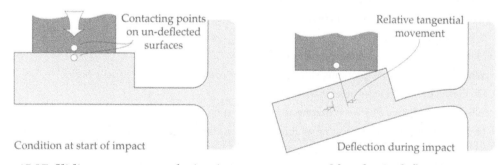

FIGURE 15.27 Sliding movements during impact wear caused by elastic deflection.

As may be expected, sufficient hardness of the impacted component is necessary to prevent rapid wear or extrusion of material from the contact by plastic deformation. In most situations this condition can be fulfilled by assuring an adequate hardness and then wear is controlled by other material characteristics. For example, wear by spalling or crack formation is controlled by material characteristics such as brittleness and microstructure. The use of materials with low concentrations of inclusions and material flaws would suppress impact wear by crack formation. Brittleness favours rapid crack growth and the formation of very large spalls or even macroscopic fracture of the component [73] while crack initiation is facilitated by inclusions [72].

Lubricants are useful in controlling impact wear in applications where they can be reasonably applied and providing that the lubricant does not cause chemical attack on the contacting surfaces. Hydrodynamic squeeze films can separate contacting surfaces for the duration of impact and the absence of sliding ensures that contact temperature rises are relatively small so that lubrication by adsorbed films is feasible. The basic limitation of lubricants, however, is their inability to significantly reduce the contact stresses during impact and wear is therefore only partially suppressed.

15.7   SUMMARY

Fretting has been known for a long time but only recently the forms of wear which distinguish it from other wear processes have been identified. The special feature of fretting is debris entrapment in the contact due to the restricted amplitude of sliding and the debris layer structures formed between fretting surfaces. The nature of the debris layers controls the fretting characteristics of a material. Fretting is very sensitive to local conditions and small changes in system design may solve a fretting problem and avoid an expensive equipment failure. Where such remedies are not available, a lubricant or surface coating may be effective. Fretting fatigue can cause quite unexpected reductions in the fracture stress of components, particularly when combined with corrosion. Nominally static contacts such as interference fits are prone to this form of damage.

Knowledge of wear mechanisms dominated by extremes of energy, i.e., frictional, electrical and impact, transmitted across the sliding interface is limited because under normal operation they do not occur or are limited to a restricted range of applications. From a conceptual point of view these wear mechanisms are important in defining the limits to wear under extreme conditions. It seems that almost any process of material transformation, mechanical, chemical or physical, may form the basis of a wear mechanism.

REVISION QUESTIONS

15.1   Why should fretting be a particular concern to tribologists?

15.2   Under what conditions does fretting wear occur?

15.3   Give examples of fretting. Try to think of an example not given in the text.

15.4   Why does debris accumulate in a fretting contact?

15.5   How is oxygen involved in the mechanism of fretting?

15.6   What is the range of reciprocating sliding amplitudes over which fretting usually occurs?

15.7   Do elevated temperatures accelerate fretting?

15.8   What is the difference between fretting wear and fretting fatigue?

15.9   How many cycles of reciprocating sliding must occur before the weakening effect of fretting fatigue is developed?

15.10   When does rapid wear and production of large amounts of oxide debris occur during the fretting of steel?

15.11   Indicate the reduction in strength of aluminium alloys and steels due to fretting fatigue.

15.12   Is fretting fatigue a well understood subject?

15.13   What effect on fretting does exclusion of oxygen have?

15.14   Give an example of surface coating to resist a) fretting wear and b) fretting fatigue.

15.15   Suggest a reason why coatings that reduce fretting fatigue do not reduce fretting wear.

15.16   By what process can just one alloying element of a metal be worn away?

15.17   In cutting tool catalogues different cutting tool materials are recommended for different workpiece materials. What could be the main reason for this?

15.18   Cemented carbide and high-speed steel (HSS) cutting tools suffer excessive wear in many cutting operations. What is the recent improvement in cutting tools?

15.19   What happens to contacting asperities at very high sliding speeds?

15.20   How is the frictional transient temperature affected when melting wear is established?

15.21   Name a significant physical (thermodynamic) parameter controlling melting wear. Gives some reasons why it should be significant.

15.22   During firing of a gun the projectile is lubricated by melting metal on the bands around the shell. Calculate the coefficient of friction at a projectile velocity of **1000** [m/s] when the threshold speed for melting wear is **50** [m/s] and the sliding coefficient of friction is **0.3**. (Ans. $\mu = 0.067$)

15.23   Can wear occur without sliding motion?

15.24   Give examples of impact wear.

15.25   What is the basic mechanism of wear for an impacting contact under dry, cool solutions?

REFERENCES

1   J. Sato, Recent Trend in Studies of Fretting Wear, *Transactions JSLE*, Vol. 30, 1985, pp. 853-858.

2   C. Cattaneo, Sul Contatto di Due Corpi Elastici: Distribuzione Locale Degli Sforzi, *Rendiconti dell'Accademia Nazionalle dei Lincei*, Vol. 27, Ser. 6, 1938, pp. 342-348, 434-436, 474-478.

3   R.D. Mindlin, Compliance of Elastic Bodies in Contact, *Journal of Applied Mechanics*, Vol. 71, 1949, pp. 259-268.

4   K.L. Johnson, Contact Mechanics, Cambridge University Press, 1985.

5   R.B. Waterhouse, Fretting Corrosion, Pergamon Press, Oxford, 1972.

6   K.L. Johnson, Surface Interaction between Elastically Loaded Bodies under Tangential Forces, *Proc. Roy. Soc., London*, Series A, Vol. 230, 1955, pp. 531-548.

7   J. Sato, M. Shima and T. Sugawara, A Fundamental Study of Fretting Damage to Glass Using an Improved Apparatus, *Wear*, Vol. 106, 1985, pp. 53-61.

8   U. Bryggman and S. Soderberg, Contact Conditions in Fretting, *Wear*, Vol. 110, 1986, pp. 1-17.

9   M. Odfalk and O. Vingsbo, An Elastic-Plastic Model for Fretting Contact, *Wear*, Vol. 157, 1992, pp. 435-444.

10   O. Vingsbo and M. Odfalk, Conditions for Elastic Contact in Fretting, Proc. Japan Int. Tribology Conf, Nagoya, October, 1990, Japanese Society of Tribologists, Tokyo, 1990, pp. 833-838.

11   R.C. Bill, Fretting Wear and Fretting Fatigue - How are They Related? *Transactions ASME, Journal of Lubrication Technology*, Vol. 105, 1983, pp. 230-238.

12   D. Play, Mutual Overlap Coefficient and Wear Debris Motion in Dry Oscillating Friction and Wear Tests, *ASLE Transactions*, Vol. 28, 1985, pp. 527-535.

13   S. Abarou, D. Play and F.E. Kennedy, Wear Transition of Self-Lubricating Composites Used in Dry Oscillating Applications, *ASLE Transactions*, Vol. 30, 1987, pp. 269-281.

14   R.C. Bill, Fretting of AISI 9310 Steel and Selected Fretting Resistant Surface Treatments, *ASLE Transactions*, Vol. 21, 1978, pp. 236-242.

15   R.C. Bill, The Role of Oxidation in the Fretting Wear Process, NASA TM-81570, AVRADCOM-TR-80-C-15, 1980.

16   I.M. Feng and H.H. Uhlig, Fretting Corrosion of Mild Steel in Air and in Nitrogen, *Transactions ASME, Journal of Applied Mechanics*, Vol. 21, 1954, pp. 395-400.

17   I.M. Feng and B.G. Rightmire, Experimental Study of Fretting, *Proc. Inst. Mech. Engrs.*, London, Vol. 170, 1956, pp. 1055-1064.

18   A.J. Fenner, K.H.R. Wright and J.Y. Mann, Fretting Corrosion and Its Influence on Fatigue Failure, Proc. Int. Conf. on Fatigue of Metals, 1956, pp. 386- 393.

19   R.C. Bill, Study of Fretting Wear in Titanium, Monel-400 and Cobalt-25 Percent Molybdenum Using Scanning Electron Microscopy, *ASLE Transactions*, Vol. 16, 1974, pp. 286-290.

20   M. Antler, Effect of Lubricants on Frictional Polymerization of Palladium Electrical Contacts, *ASLE Transactions*, Vol. 26, 1983, pp. 376-380.

21 Ch. Colombie, Y. Berthier, A. Floquet, L. Vincent, M. Godet, Fretting: Load Carrying Capacity of Wear Debris, *Transactions ASME, Journal of Tribology*, Vol. 106, 1984, pp. 194-201.

22 J. Sato, Recent Studies on Fretting Wear of Polymeric Materials, *Transactions JSLE*, Vol. 33, 1988, pp. 26-32.

23 P.L. Hurricks, The Fretting Wear of Mild Steel from 200°C to 500°C, *Wear*, Vol. 30, 1974, pp. 189-212.

24 T. Kayaba and A. Iwabuchi, The Fretting Wear of 0.45%C Steel and Austenitic Stainless Steel from 20 to 650°C in Air, *Wear*, Vol. 74, 1981, pp. 229-245.

25 R.B. Waterhouse and A. Iwabuchi, High Temperature Fretting of Four Titanium Alloys, *Wear*, Vol. 106, 1985, pp. 303-313.

26 M.M. Hamdy and R.B. Waterhouse, The Fretting Wear of Ti-6Al-4V and Aged Inconel 718 at Elevated Temperatures, *Wear*, Vol. 71, 1981, pp. 237-248.

27 R.C. Bill, Fretting of Nickel-Chromium-Aluminium Alloys at Temperatures to 816°C, NASA TN D-7570, 1974.

28 P.L. Hurricks and K.S. Ashford, The Effect of Temperature on the Fretting Wear of Mild Steel, *Proc. Inst. Mech. Engrs.*, London, Vol. 184, Pt. 3L, 1969-70, pp. 165-175.

29 P.L. Hurricks, The Fretting Wear of Mild Steel from Room Temperature to 200°C, *Wear*, Vol. 19, 1972, pp. 207-229.

30 R.B. Waterhouse, Introduction, *Wear*, Vol. 106, 1985, pp. 1-4.

31 A. Neyman, The Influence of Oil Properties on the Fretting Wear of Mild Steel, *Wear*, Vol. 152, 1992, pp. 171-181.

32 Y. Qiu and B.J. Roylance, The Effect of Lubricant Additives on Fretting Wear, *Lubrication Engineering*, Vol. 48, 1992, pp. 801-808.

33 J.R. McDowell, Fretting of Hardened Steel in Oil, *ASLE Transactions*, Vol. 1, 1958, pp. 287-295.

34 D. Godfrey, A Study of Fretting Wear in Mineral Oil, *Lubrication Engineering*, Vol. 12, 1956, pp. 37-42.

35 J. Sato, M. Shima, T. Sugawara and A. Tahara, Effect of Lubricants on Fretting Wear of Steel, *Wear*, Vol. 125, 1988, pp. 83-95.

36 E.E. Weismantel, Friction and Fretting with Solid Film Lubricants, *Lubrication Engineering*, Vol. 11, 1955, pp. 97-100.

37 D.D. Fuller, Theory and Practice of Lubrication for Engineers, John Willey and Sons Inc., New York, 1966.

38 A.W. Batchelor, G.W. Stachowiak, G.B. Stachowiak, P.W. Leech and O. Reinhold, Control of Fretting Friction and Wear of Roping Wire by Laser Surface Alloying and Physical Vapour Deposition, *Wear*, Vol. 152, 1992, pp. 127-150.

39 P.W. Leech, A.W. Batchelor and G.W. Stachowiak, Laser Surface Alloying of Steel Wire With Chromium and Zirconium, *Journal of Materials Science Letters*, Vol. 11, 1992, pp. 1121-1123.

40 R.B. Waterhouse, Fretting Fatigue, Applied Science Publishers Ltd., London, 1981.

41 K. Sato and H. Fuji, Crack Propagation Behaviour in Fretting Fatigue. *Wear*, Vol. 107, 1986, pp. 245-262.

42 K. Nishioka and K. Hirakawa, Fundamental Investigations of Fretting Fatigue, *Bulletin JSME*, Vol. 12, 1969, pp. 692-697.

43 M.M. Hamdy and R.B. Waterhouse, The Fretting Fatigue Behaviour of a Nickel Based Alloy (Inconel 718) at Elevated Temperatures, Proc. Int. Conf. on Wear of Materials, Proc. Int. Conf. on Wear of Materials, Dearborn, Michigan, 16-18 April 1979, editors: K.C. Ludema, W.A. Glaeser and S.K. Rhee, Publ. American Society of Mechanical Engineers, New York, 1979, pp. 351-355.

44 K. Endo and H. Goto, Initiation and Propagation of Fretting Fatigue Cracks, *Wear*, Vol. 38, 1976, pp. 311-324.

45 K. Endo and H. Goto, Effects of Environment on Fretting Fatigue, *Wear*, Vol. 48, 1978, pp. 347-367.

46 R.B. Waterhouse and M.K. Dutta, The Fretting Fatigue of Titanium and Some Titanium Alloys in Corrosive Environment, *Wear*, Vol. 25, 1973, pp. 171-175.

47 C. Poon and D. Hoeppner, The Effect of Environment on the Mechanism of Fretting Fatigue, *Wear*, Vol. 52, 1979, pp. 175-191.

48 B.R. Pearson, P.A. Brook and R.B. Waterhouse, Fretting in Aqueous Media, Particularly of Roping Steels in Seawater, *Wear*, Vol. 106, 1985, pp. 225-260.

49 G. Lofficial and Y. Berthier, L'Usure dans les Cables et Conduits Flexibles, une Etude de cas en Tribologie, Eurotrib, September 1985, Ecole Centrale de Lyon, Vol. 3, 1985, pp. 2.2.1-2.2.5.

50   L. Vincent, Y. Berthier, M.C. Dubourg and M. Godet, Mechanics and Materials in Fretting, *Wear*, Vol. 153, 1992, pp. 135-148.

51   F.P. Bowden and D. Tabor, The Friction and Lubrication of Solids, Part II, Clarendon Press, Oxford, 1964.

52   R.S. Montgomery, Surface Melting of Rotating Bands, *Wear*, Vol. 38, 1976, pp. 235- 243.

53   W.R.D. Wilson, Lubrication by a Melting Solid, *Transactions ASME, Journal of Lubrication Technology*, Vol. 98, 1976, pp. 22-26.

54   A.K. Stiffler, Friction and Wear with a Fully Melting Surface, *Transactions ASME, Journal of Lubrication Technology*, Vol. 106, 1984, pp. 416-419.

55   F.P. Bowden and P.A. Persson, Deformation, Heating and Melting of Solids in High-Speed Friction, *Proc. Roy. Soc., London*, Series A, Vol. 260, 1960, pp. 433- 458.

56   F.P. Bowden and E.H. Freitag, The Friction of Solids at Very High Speeds, I. Metal on Metal, II Metal on Diamond, *Proc. Roy. Soc., London*, Series A, Vol. 248, 1958, pp. 350-367.

57   R.S. Montgomery, Friction and Wear at High Sliding Speeds, *Wear*, Vol. 36, 1976, pp. 275- 298.

58   M. Kawamoto and K. Okabayashi, Wear of Cast Iron in Vacuum and the Frictional Hardened Layer, *Wear*, Vol. 17, 1971, pp. 123-138.

59   T.S. Hong, S. Maj and D.W. Borland, The Formation of White Layers During Sliding Wear, Proc. Int. Tribology Conference, Melbourne, The Institution of Engineers, Australia, National Conference Publication No. 87/18, December, 1987, pp. 193-197.

60   O. Botstein and R. Arone, The Microstructural Changes in the Surface Layer of Gun Barrels, *Wear*, Vol. 142, 1991, pp. 87-95.

61   A. Kohno, M. Itoh and N. Soda, Effect of Contact Arc on Wear of Materials for Current Collection, Part 3, *Transactions JSLE*, Vol. 29, 1984, pp. 458-462.

62   O. Oda, Y. Fuji and T. Kohida, Considerations on the Arc Vanishing Time and Contactstrip Wear of the Pantograph Connected by Means of Bus Conductor, *Transactions JSLE*, Vol. 29, 1984, pp. 463-468.

63   T. Teraoka and K. Fukuhara, Studies on Wear in Sliding Power Collection (1), *Transactions JSLE*, Vol. 30, 1985, pp. 878-882.

64   T. Katafuchi, Effects of Electric Current on Wear under Lubricated Conditions; I - Electric Current - Wear Characteristics, *Transactions JSLE*, Vol. 30, 1985, pp. 883-886.

65   T. Katafuchi, Effects of Electric Current on Wear under Lubricated Conditions; II - Effect of Addition of Additives on Current-Wear Characteristics, *Transactions JSLE*, Vol. 30, 1985, pp. 887-893.

66   I.L. Goldblatt, Self-Generated Voltages and Their Relationship to Wear Under Boundary Lubricated Conditions, AD-A-058611, Publ. NTIS, USA, July 1978.

67   E.M. Trent, Metal Cutting, Butterworths, London, 1977.

68   J.A. Yeomans and T.F. Page, The Chemical Stability of Ceramic Cutting Tool Materials Exposed to Liquid Metals, *Wear*, Vol. 131, 1989, pp. 163-175.

69   P.A. Dearnley, Rake and Flank Wear Mechanism of Coated Cemented Carbides, *Surface Engineering*, Vol. 1, 1985, pp. 43-58.

70   P.A. Engel, Impact Wear of Materials, Elsevier, Amsterdam, 1976.

71   P.A. Engel, Percussive Impact Wear, *Tribology International*, Vol. 11, 1978, pp. 169-176.

72   S.L. Rice, The Role of Microstucture in the Impact Wear of Two Aluminium Alloys, *Wear*, Vol. 54, 1979, pp. 291-301.

73   K.J. Swick, G.W. Stachowiak and A.W. Batchelor, Mechanism of Wear of Rotary-Percussive Drilling Bits and the Effect of Rock Type on Wear, *Tribology International*, Vol. 25, 1992, pp. 83-88.

74   N. Yahata and M. Tsuchida, Impact Wear Characteristics of Annealed Carbon Steels, *Tribologist*, Vol. 35, 1990, pp. 144-150.

75   A.J. Fenner and J.E. Field, Fatigue of an Aluminium Alloy Under Conditions of Friction, *Rev. de Metallurgie*, Vol. 55, 1958, pp. 475-485.

76   E. de Wit, L. Froyen and J-P. Celis, The Oxidation Reaction during Sliding Wear Influencing the Formation of Either Amorphous or Nanocrystalline Debris, *Wear*, Vol. 231, 1999, pp. 116-123.

77   Y. Fu, A.W. Batchelor and N.L. Loh, Study on Fretting Wear Behavior of Laser Treated Coatings by X-ray Imaging, *Wear*, Vol. 218, 1998, pp. 250-260.

78 G.B. Stachowiak and G.W. Stachowiak, Fretting Wear and Friction Behaviour of Engineering Ceramics, *Wear*, Vol. 190, 1995, pp. 212-218.

79 O. Vingsbo and D. Soderberg, On Fretting Maps, *Wear*, Vol. 126, 1988, pp. 131-147.

80 R.B. Waterhouse, Fretting Wear, ASM Handbook, Vol. 18, Publ. ASM International, 1992, p. 245.

81 S. Fouvry, Ph. Kapsa and L. Vincent, An Elastic-Plastic Shakedown Analysis of Fretting Wear, *Wear*, Vol. 247, 2001, pp. 41-54.

82 S. Barril, N. Debaud, S. Mischler and D. Landolt, A Tribo-Electrochemical Apparatus for in Vitro Investigation of Fretting-Corrosion of Metallic Implant Materials, *Wear*, Vol. 252, 2002, pp. 744-754.

83 R. Chen, A. Iwabuchi and T. Shimizu, Effects of Ambient Pressure on Fretting Friction and Wear Behaviour between SUS 304 Steels, *Wear*, Vol. 249, 2001, pp. 379-388.

84 M. Kalin and J. Vizintin, High Temperature Phase Transformations under Fretting Conditions, *Wear*, Vol. 249, 2001, pp. 172-181.

85 H. Chen, P.Q. Wu, C. Quaeyhaegens, K.W. Xu, L.H. Stals, J.W. He and J.-P. Celis, Comparison of Fretting Wear of Cr-Rich CrN and TiN Coatings in Air of Different Relative Humidities, *Wear*, Vol. 253, 2002, pp. 527-532.

86 H. Goto and K. Uchijo, Fretting Wear of Al-Si Alloy Matrix Composites, *Wear*, Vol. 256, 2004, pp. 630-638.

87 K. Elleuch and S. Fouvry, Wear Analysis of A357 Aluminium Alloy Under Fretting, *Wear*, Vol. 253, 2002, pp. 662-672.

88 S. Hannel, S. Fouvry, Ph. Kapsa and L. Vincent, The Fretting Sliding Transition as a Criterion for Electrical Contact Performance, *Wear*, Vol. 249, 2001, pp. 761-770.

89 C. Allen, C.X. Li, T. Bell and Y. Sun, The Effect of Fretting on the Fatigue Behaviour of Plasma Nitrided Stainless Steels, *Wear*, Vol. 254, 2003, pp. 1106-1112.

90 H. Chen and A.T. Alpas, Sliding Wear Map of Magnesium Alloy Mg-9Al-0.9Zn (AZ91), *Wear*, Vol. 246, 2000, pp. 106-116.

91 M. Okada, N-S. Liou, V. Prakash and K. Miyoshi, Tribology of High-Speed Metal-On-Metal Sliding at Near-Melt and Fully-Melt Interfacial Temperatures, *Wear*, Vol. 249, 2001, pp. 672-686.

92 L. Rapoport, N. Parkansky, I. Lapsker, A. Rayhel, B. Alterkop, R.L. Boxman, S. Goldsmith and L. Burstain, Effect of Transverse Current Injection on the Tribological Properties of WC Cemented Carbide, *Wear*, Vol. 249, 2001, pp. 1-5.

93 H. Zhao, G.C. Barber and J. Liu, Friction and Wear in High Speed Sliding with and without Electrical Current, *Wear*, Vol. 249, 2001, pp. 409-414.

94 C.M. Lin, Y.C. Chiou and R.T. Lee, Pitting Mechanism on Lubricated Surface of Babbitt Alloy/Bearing Steel Pair Under AC Electric Field, *Wear*, Vol. 249, 2001, pp. 133-142.

95 D. Busse, J. Erdman, R.J. Kerkman, D. Schlegel and G. Skibinski, System Electrical Parameters and Their Effects on Bearing Currents, *IEEE Trans. Ind. Application*, Vol. 33, No. 2, 1997, pp. 577-584.

96 R.S. Montgomery, Friction and Wear at High Sliding Velocities, *Wear*, Vol. 36, 1976, pp. 275-298.

97 C.M.McC. Ettles, Polymer and Elastomer Friction in the Thermal Control Regime, *ASLE Transactions*, Vol. 30, 1987, pp. 149-159.

98 S. Philippon, G. Sutter and A. Molinari, An Experimental Study of Friction at High Sliding Velocities, *Wear*, Vol. 257, 2004, pp. 777-784.

16 WEAR OF NON-METALLIC MATERIALS

16.1 INTRODUCTION

Although the wear and friction of non-metallic solids have some fundamental similarities to that of metals, there are also significant differences in the wear mechanisms involved and the level of friction or wear which occurs. These differences can be exploited to produce valuable new bearing materials which can change commonly accepted expectations of tribological performance. An example of this is a common polymer called Polytetrafluoroethylene (PTFE) which can provide a coefficient of friction as low as 0.05 in the complete absence of any lubricant. However, not all of the differences in tribological characteristics of metals and non-metals are in favour of the non-metals. For example, PTFE is soft and has a very high wear rate so that for most applications it must be blended carefully with other materials before a useful bearing material is created. A careful study of the tribology of non-metallic materials is a pre-requisite to their successful adaptation as bearing or wear resistant materials. Questions such as: in what applications should polymers be applied as bearing materials? Are the polymers superior to metals as bearing materials? Will ceramic bearing wear out a metal shaft? These and many others are of considerable interest to engineers.

The development of new technologies, often motivated by global issues such as environmental pollution, creates new requirements for bearings and wear resistant materials that cannot be satisfied by traditional metallic materials. A prime example of this research trend is the adiabatic combustion engine where ceramic materials are being developed as high temperature cylinder and piston materials. The adiabatic engine is potentially more efficient than a cooled engine since there is no need for lubricant cooling systems, i.e., the radiator and water cooled cylinder block. The development of such engines poses new challenges; the fundamental limitations of most ceramics as bearing materials have to be solved by a painstaking combination of fundamental studies into the wear and friction of ceramics and in addition new lubrication technologies have to be found. In this chapter the fundamental wear mechanisms operating in non-metallic materials together with some prognoses concerning the future developments of these materials are described.

16.2 TRIBOLOGY OF POLYMERS

The term 'polymers', which is associated with materials formed by the polymerization of hydrocarbons, is used to describe an enormous range of different substances. Most of these materials were never intended to be utilized as bearing or wear-resistant materials and, in fact, are usually unsuitable for this purpose. A few polymers, however, do have valuable tribological properties and most research is directed towards this relatively limited number of polymers. Common polymers, with actual or potential tribological function together with their basic tribological characteristics, are listed in Table 16.1.

TABLE 16.1 Tribological characteristics of typical polymers.

Polymer	Tribological characteristics
Polytetrafluoroethylene (PTFE)	Low friction but high wear rate; usually blended with other polymers or reinforced as a composite material. High operating temperature limit.
Nylons	Moderate coefficient of friction and low wear rate. Medium performance bearing material. Wear accelerated by water. Relatively low temperature limit.
Polyacetals	Performance similar to nylon. Durable in rolling contacts.
Polyetheretherketone (PEEK)	High operating temperature limit. Resistant to most chemical reagents. Suitable for high contact stress. High coefficient of friction in pure form.
Ultra high molecular weight polyethylene (UHMWPE)	Very high wear resistance even when water is present. Moderate coefficient of friction. Good abrasive wear resistance. Relatively low temperature limit.
Polyurethanes	Good resistance to abrasive wear and to wear under rolling conditions. Relatively high coefficient of friction in sliding.
Polyimides	High performance polymers, suitable for high contact stresses and high operating temperatures.
Epoxies and phenolics	Used as binders in composite materials.

The 'high operating temperature limit' in polymers refers to temperatures in excess of 150°C. Table 16.1 lists only the basic polymer types. Since most polymers used in engineering applications are blends of different polymers and additives, e.g., nylon containing PTFE, there is an enormous range of polymer materials that can be used for tribological applications. The use of composites, e.g., glass reinforced PTFE, extends the range of materials even further and this aspect of polymers is discussed later in this chapter. The list in Table 16.1 is not exhaustive and there are also other polymers under study for bearing materials. The physical properties of selected polymers are shown in Table 16.2.

TABLE 16.2 Physical properties of polymers used as bearing materials.

Polymer	Upper service temperature [°C]	Thermal conductivity [W/mK]	Thermal expansivity [K^{-1}]	Tensile Modulus [GPa]	Tensile strength [MPa]
Polycarbonate	125	0.2	70×10^{-6}	2.4	65
Polyamide (nylon 6)	110 - 180	0.25	90×10^{-6}	3.3	82
Polyetheretherketone	250	0.25	60×10^{-6}	2.2	85*
Polyethylene (UHMWPE)	95	0.45	170×10^{-6}	0.2 - 1.2	20*
Polyimide (kapton)	250 - 320	0.2	50×10^{-6}	2.5	70*
Polytetrafluoroethylene (PTFE)	260	0.25	130×10^{-6}	0.5	10*

* Significantly higher values reported under specific conditions.

Sliding Wear of Polymers, Transfer Layers on a Harder Counterface

Most polymer surfaces, when used as bearing materials, are worn by a harder counterface. The application of the hard metallic counterface rubbing against the polymer surface is dictated by mechanical design requirements and also by the fact that polymers are more effective against a metallic counterface than when sliding against themselves. A basic feature of almost all polymers is that a transfer film is formed when sliding against a harder counterface which has a strong influence on the tribology of polymers.

The polymer which provides a classic example of transfer film formation is PTFE, whose molecular and crystalline structure is shown in Figure 16.1. It has been demonstrated in experiments conducted with PTFE in vacuum that a strong adhesion occurs between PTFE and a metallic surface [1]. The cause of the adhesion is believed to be an interfacial chemical reaction between the fluorine and carbon in PTFE and the opposing metallic surface [1,2]. Although there is probably a strong adhesion between a metallic surface and any other polymer, the special molecular structure of PTFE causes a mechanism of film transfer which is particular to PTFE.

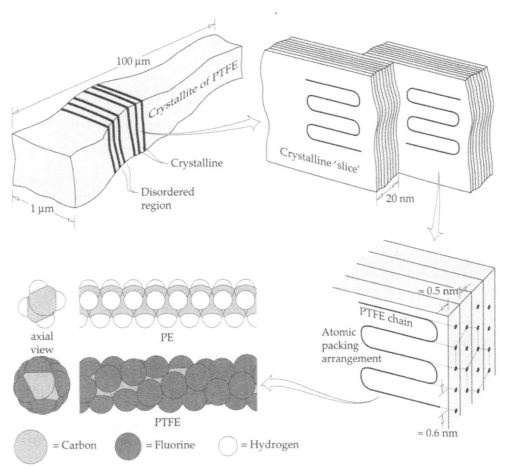

FIGURE 16.1   Crystalline structure of PTFE and molecular structures of PTFE and PE (adapted from [3] and [6]).

The lack of side groups and almost cylindrical form of the PTFE molecule ensures relatively easy movement between molecules under applied stress. The crystalline structure consists of layers of crystalline material between relatively weak layers of amorphous material and this favours deformation of the PTFE in a series of discrete laminae [4]. A block of PTFE in contact with a harder counterface loses material as a series of laminae resulting in low friction but a high wear rate [5,6]. The mechanism of sliding wear of PTFE is schematically illustrated in Figure 16.2.

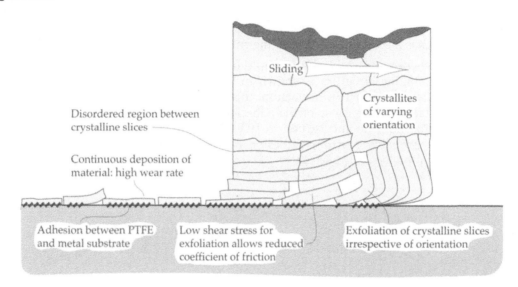

FIGURE 16.2 Wear and film transfer mechanism of PTFE.

A film transfer mechanism similar to that of PTFE has not been observed in any other polymer synthesized so far. The vast majority of polymers and polymer composites sustain a process of '**lumpy transfer**' when sliding against a hard counterface [7]. A few important exceptions to this rule are High Density Polyethylene (HDPE) and Ultra-High Molecular Weight Polyethylene (UHMWPE) [4]. Their similarity in behaviour to PTFE is believed to be caused by the common characteristic of a 'smooth molecular profile' or the absence of side groups and kinks in the polymer chain [8]. The initial friction of PTFE, UHMWPE and HDPE, which is effectively the static coefficient of friction, is approximately 50% higher than their kinetic coefficient of friction [9]. The cause for this difference is probably due to the extra force required to initiate the formation of a transfer film. The mechanism of lumpy transfer is shown in Figure 16.3 where lumps of polymer are shown as being removed from asperity peaks and left adhering on the counterface.

The size of these lumps is about 1 [μm] in average diameter [9]. Because the contact diameter is very small compared to the planar PTFE transfer film, their load capacity as a transfer film is also small. This form of transfer film does not contribute to better wear and friction characteristics of the sliding contacts and in fact most polymers which exhibit this lumpy film transfer are not very effective as bearing materials.

When a polymer slides against another polymer, the cohesively weaker polymer is worn preferentially to form a transfer film on the cohesively stronger polymer. The wear mechanism operating is essentially the same as observed in non-polymer counterfaces [10].

Influence of Counterface Roughness, Hardness and Material Type on Transfer Films and Associated Wear and Friction of Polymers

The counterface has considerable influence on the wear of any polymer and a good bearing design includes careful specification of the counterface. The counterface affects the wear of a polymer according to its hardness, roughness and 'surface energy'. The latter quantity is a

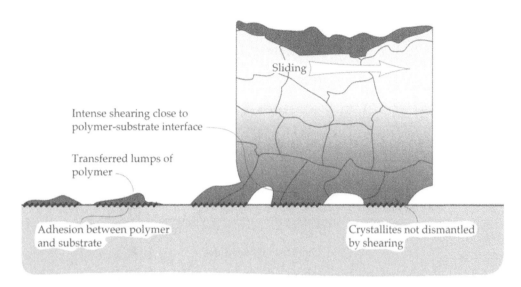

FIGURE 16.3 'Lumpy transfer' mechanism of most polymers.

poorly defined and difficult to measure parameter which is intended to define the difference between two surfaces, e.g., the difference between a gold and a calcium surface of equal hardness and roughness. In approximate terms, the surface energy is equivalent to the chemical potential of the counterface material.

· *Counterface Hardness*

It is generally accepted that the counterface should be much harder than the polymer and hardened steel is often recommended. This common engineering practice is so well accepted that nobody has tested the concept experimentally for many years. The prevailing view is that the counterface should be hard enough so that abrasion by stray contaminants, e.g., sand, will not cause it to become rough and abrade the polymer. A counterface hardness of 700 Vickers has been used in some studies of polymer wear and was recommended as sufficiently hard for most applications [11].

· *Counterface Roughness*

The counterface roughness has a rather more complex effect on polymer wear. While it has often been suggested that the roughness should be as low as possible to reduce abrasion of the polymer [11], more detailed research has demonstrated that for certain polymers an optimum roughness exists. Wear reaches a minimum for a finite level of roughness which is within practicable limits of manufacture [12]. The effects of surface roughness on the wear rate of UHMWPE sliding against a stainless steel counterface is shown in Figure 16.4 [13]. There seems to be an optimal surface roughness for low to moderate sliding speeds of about 1 to 5 [m/s] whereas at higher speeds of about 10 [m/s] the wear rate appears to be relatively insensitive to the counterface roughness. The loss of roughness dependence is due to a different wear mechanism prevailing at high sliding speeds.

Wear rates at extremely high levels of smoothness are comparable to wear rates on relatively rough surfaces. The reason for this is believed to be the lack of sharp edged grooves on a smooth surface which would otherwise abrade fragments of polymer to form a rudimentary transfer film. This model of controlled abrasion by sharp asperities which cease to abrade after being covered by the fragments of polymer is shown in Figure 16.5.

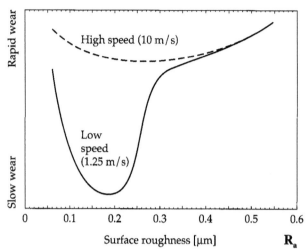

FIGURE 16.4   Effect of counterface roughness on the wear of UHMWPE [13].

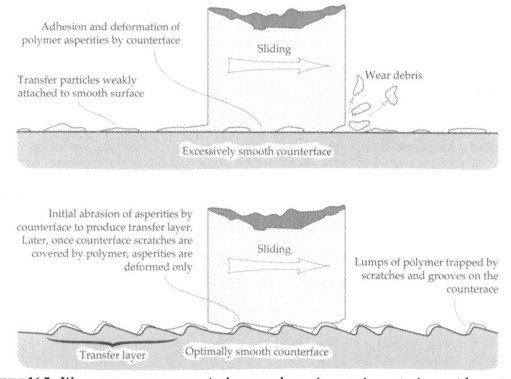

FIGURE 16.5   Wear process on excessively smooth surface and on surface with optimum roughness.

Abraded polymer debris is either trapped on the sharp edged grooves found on most rough surfaces or else is expelled as wear debris [13,14]. It appears that the strong adhesion between the polymer and the counterface is not always effective under typical engineering conditions where the polymer debris needs to be physically anchored. In these cases, it is thought that the loose interposed debris act as a 'transfer film' to reduce friction and wear.

When the surface becomes excessively rough wear is accelerated. It has been shown that the rough surface of a hard metal will abrade a polymer [15-17]. An example of polymer abrasion by a steel surface is shown in Figure 16.6.

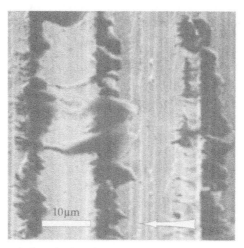

FIGURE 16.6   Accumulation of abraded polymer on a metallic counterface [15].

The wear of polymers against very rough surfaces can be modelled in terms of simple abrasion [15-17]. The wear model is based on the elementary idea of penetration depth of metallic asperities into a polymer as illustrated in Figure 16.7.

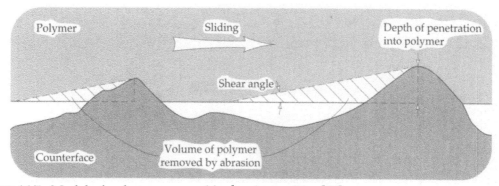

FIGURE 16.7   Model of polymer removal by hard asperities [15].

As shown in Figure 16.7 the wear rate of the polymer is determined by the penetration depth of metallic asperities, the shear angle of the polymer and the sliding distance. In practical situations wear does not proceed at the same rate with time, but because the asperities become covered with polymer, wear rate declines after an initial rapid period. The model, in general, works well but with some of the polymers, e.g., LDPE (Low Density Polyethylene) and PVC (Polyvinyl Chloride), the predicted wear rates are inaccurate for reasons not yet fully discovered.

The magnitude of counterface surface roughness is not the only factor affecting the wear and friction characteristics of the polymer. It was also found that the effect of counterface asperity height distribution on wear rate is significant. Significant differences in wear rate between surfaces with a Gaussian asperity height distribution and surfaces with a non-Gaussian distribution were recorded [18].

· *Counterface Surface Energy*

It has been observed that the surface energy of a counterface affects the wear of PTFE and the formation of PTFE transfer films [19]. A surface with relatively low energy, e.g., a noble or semi-noble metal such as copper, tends to generate thinner transfer films compared to that of a more chemically active metal such as zinc. With a less reactive metal, the wear debris produced also tends to be finer and the transfer film formed does not cover the surface uniformly, leaving gaps of exposed metal. With the more active metal, a thick multi-lamina transfer film is formed and PTFE is only removed as wear debris in large lumps which are physically ploughed away by the polymer surface. These two mechanisms of wear are illustrated schematically in Figure 16.8.

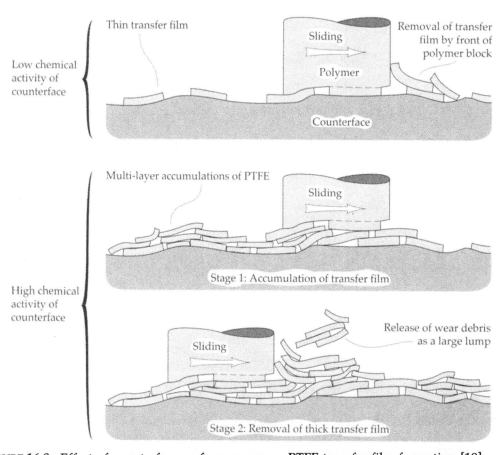

FIGURE 16.8   Effect of counterface surface energy on PTFE transfer film formation [19].

Unfortunately no similar studies for other polymers were reported in the literature but it may be assumed that the counterface surface energy does have some effect on wear for all polymers.

PV Limit

A useful parameter in polymer selection for a particular application is the 'PV' limit. The 'PV' limit is an empirical concept defining the conditions where rapid wear or overheating of any polymer will occur. 'P' is defined as the nominal contact pressure [Pa] and 'V' is the sliding speed [m/s]; the units of the 'PV' limit are therefore [Pam/s]. Each polymer has its own 'PV' limit and manufacturers of engineering polymers often quote 'PV' values as evidence of the quality of their product. The concept of 'PV' is related to frictional heating and implies that when the 'PV' limit is exceeded, then the polymer starts to melt or rapidly creep due to overheating. Below the 'PV' limit, wear is envisioned as a much slower, fracture process involving crack growth and the release of small wear particles.

Influence of Temperature on Polymer Wear and Friction

Most polymers melt at relatively low temperatures. This characteristic combined with the low thermal conductivity of polymers ensures that frictional contact temperatures can easily reach the melting point of a polymer and cause its surface to melt. When the polymer melts its friction and wear coefficients are markedly altered. This characteristic can be illustrated by considering a pad of butter in a heated saucepan. When the melting point of the butter is reached, the friction dramatically declines to allow the butter to slide across the pan. The 'wear rate' of the butter pad, however, tends to rise with temperature, particularly when melting of the butter occurs. Similar trends in polymer friction and wear can be found and the prevailing mechanism can be classified as a form of '**melting wear**'. The concept of melting wear of polymers is schematically illustrated in Figure 16.9.

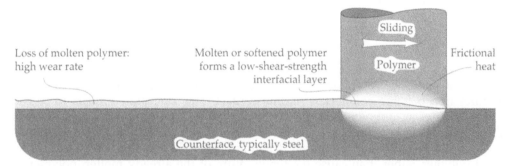

FIGURE 16.9 Melting wear of polymers caused by frictional contact temperatures.

A thin layer of molten polymer forms at the interface between the counterface and the polymer. In frictional heating the heat is confined to a very thin surface layer so the layer of molten polymer is also thin and is not 'squeezed out' of the contact. Since the counterface, e.g., steel, usually has a much higher melting point it is unaffected by the frictional heat.

Microscopic evidence of polymer surface melting was found when pins of the polymers PTFE, HDPE and polyoxymethylene were slid against a steel disc at speeds of up to 4 [m/s] [20,25]. Smooth tongues of re-solidified material were found on the surfaces of HDPE and polyoxymethylene but not on PTFE. This anomaly was rationalized by the known absence of liquid flow of PTFE when its melting point is reached [4]. Further evidence of polymer frictional melting was obtained by detecting the polymer vapour and decomposition products during a high speed sliding test in a vacuum [21]. Tests were performed on a wear test rig inside a vacuum chamber connected to a mass spectrometer. A schematic diagram of this apparatus is shown in Figure 16.10 [21].

Thermal decomposition tests conducted outside the apparatus also showed that these vapours were released only when melting of the polymer occurred [21].

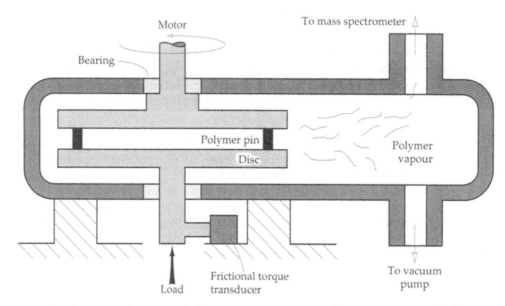

FIGURE 16.10 Schematic diagram of the wear apparatus to detect vapours and decomposition products emitted during high speed polymer sliding [21].

· *Limit on Frictional Temperature Rise Imposed by Surface Melting*

In common with any other form of melting wear, e.g., metals in high speed sliding, the latent heat of melting is believed to impose a temperature limit on the frictional temperatures in a polymer-counterface sliding contact. This condition appears to be contrary to the prediction of an unlimited rise in frictional temperature with sliding speed or load as described in Chapter 7 in the section on flash temperature. A temperature limit can be explained by the fact that any additional frictional heat released in a contact tends to melt additional polymer rather than cause the temperature of the already molten polymer to rise. This concept of limiting frictional temperature for polymers is schematically illustrated in Figure 16.11.

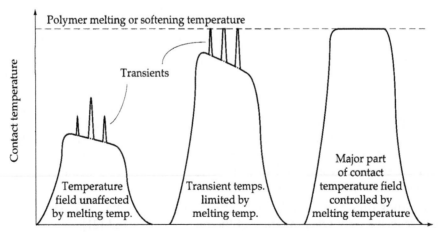

FIGURE 16.11 Limiting frictional temperature rise in the contact as dictated by melting or softening point of a polymer.

The basic concept of the limiting frictional temperature is the 'thermal control of friction' [22], which is defined in the following manner. When the melting temperature of the polymer is reached, the friction coefficient varies with sliding speed or load so that the temperature within the contact remains constant at the melting point [22]. The formula for average transient surface temperature rise for circular contacts due to friction (Table 7.5) is given in the following form:

$$\mathbf{T_{f_a}} = \mathbf{0.308} \frac{\mu \mathbf{W} |\mathbf{U_A} - \mathbf{U_B}|}{\mathbf{Ka}} \left(\frac{\chi}{\mathbf{Ua}} \right)^{0.5} \tag{16.1}$$

where:

$\mathbf{T_{f_a}}$	is the average flash temperature [°C];
μ	is the coefficient of friction;
$\mathbf{W}$	is the normal load [N];
$\mathbf{U_A}, \mathbf{U_B}$	are the surface velocities of solid '**A**' and solid '**B**', respectively [m/s];
$\mathbf{U}$	is the velocity of solid '**A**' or '**B**';
$\mathbf{a}$	is the radius of the contact circle [m];
χ	is the thermal diffusivity, $\chi = \mathbf{K}/\rho\sigma$, [m²/s];
$\mathbf{K}$	is the thermal conductivity [W/mK];
ρ	is the density [kg/m³];
σ	is the specific heat [J/kgK].

When the average temperature rise remains constant at the difference between external temperature and the melting point or the softening point of the polymer, then the coefficient of friction is inversely proportional to the square root of sliding velocity provided that material properties remain constant [22]:

$$\mu = \kappa/\mathbf{U}^{0.5} \tag{16.2}$$

where:

κ	is a constant [m⁰·⁵/s⁰·⁵].

The friction coefficients of polypropylene, nylon and LDPE sliding against steel showed good agreement with the model of limiting frictional temperature [22]. The relationship between the friction coefficient and the sliding speed for LDPE, polypropylene and nylon is shown in Figure 16.12.

It can be seen from Figure 16.12 that friction coefficients rise until a maximum value is achieved. At this point the friction coefficient determined by the 'thermal control model' equals the friction coefficient dictated by 'solid state friction'. The rise in friction coefficient up to this point can be viewed as analogous to the butter becoming sticky when in a heated saucepan. After reaching its maximum value the friction coefficient declines to a usefully low level which may be of benefit in bearing design.

An elevated external temperature has a strong effect on the melting wear mechanism. The prime effect is to shift the temperature of transition to thermal control of friction, i.e., transition occurring at reduced sliding speeds. A second effect is to increase the coefficient of friction in the lower range of sliding speeds that are insufficient to reach thermal control. This effect is demonstrated in Figure 16.13 which shows the relationship between the friction coefficient and the sliding velocity for steel sliding against unlubricated plexiglass at various external temperatures.

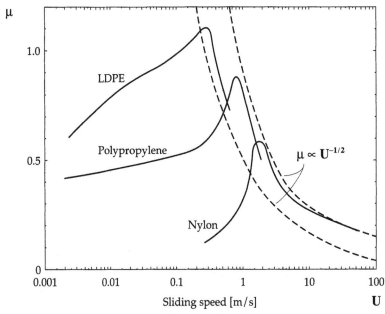

FIGURE 16.12 Relationship between the friction coefficient and the sliding speed for LDPE, polypropylene and nylon [22].

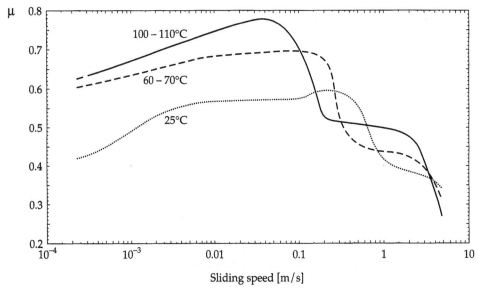

FIGURE 16.13 Effect of external temperature on friction coefficients of steel sliding against unlubricated plexiglass for varying sliding speed [23].

For harder polymers, e.g., phenolic resins, there may also be a gradual increase in friction coefficient with sliding distance when melting wear prevails. This friction rise is due to the counterface wear and subsequent material transfer from the counterface to the polymer [24]. In effect, the polymer and counterface contact degrades to a contact consisting mostly of counterface material. This trend is most pronounced for soft counterface materials such as aluminium but is hardly noticeable in steel and is unlikely to occur in ceramics [24]. It should be realized that any counterface is unlikely to remain completely intact in a high speed sliding contact against a polymer and that the extent of the counterface damage will influence friction and wear characteristics.

· *Effect of High Frictional Temperatures and Sliding Speeds on Wear*

The promising characteristic of a friction decline with increasing sliding speed is not matched by the wear rates at high sliding speeds or contact temperatures. A relatively large amount of polymer is lost at the sliding interface when melting occurs and this causes a high wear rate. This is illustrated in Figure 16.14 where the friction and wear characteristics of nylon 6 sliding against glass and steel are shown.

Melting wear was detected on nylon specimens sliding against glass by the presence of a thick recrystallized layer about 50 [μm] in depth beneath the worn specimen surfaces. The same recrystallized layer was not detected on the nylon specimens worn against steel. Steel has a

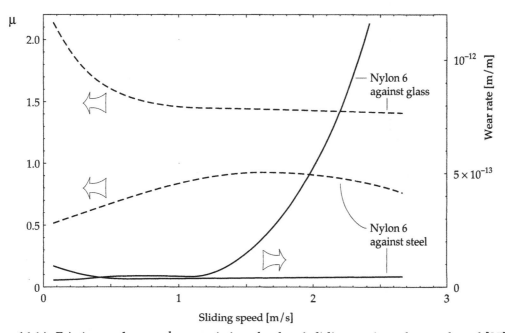

FIGURE 16.14 Friction and wear characteristics of nylon 6 sliding against glass and steel [25].

higher thermal conductivity than glass and so, for the same sliding speeds and loads, melting did not occur. For the glass, the friction coefficient showed a declining trend with sliding speed which is characteristic of melting wear, but at the same time the wear rate rose to very high levels. For the steel, contact temperatures were just sufficient to heat the nylon till it became 'sticky' resulting in a high coefficient of friction of about **1**. The wear rates, however, remained low.

There is consistent experimental evidence of a 'critical temperature' which initiates rapid wear in a polymer. Although it may be tempting to conclude that this critical temperature is equal to the melting or softening point of the polymer, more detailed experiments have shown this hypothesis to be false. At relatively high contact stresses and low sliding speeds, the critical temperature may be lower than the melting or softening point of a polymer [26].

Wear data collated from various tests on nylon with varying temperature, sliding speed and load are shown in Figure 16.15 as a function of frictional contact temperature.

It can be seen from Figure 16.15 that the wear rate of the nylon decreases initially with temperature to reach a minimum at approximately 125°C. The 'critical temperature' at which the wear rate rises rapidly to extremely high values is about 140°C. Since the melting point of crystalline nylon is about 220°C, it is clear that contact temperatures are too low for melting of the nylon to occur.

The conditions at which melting wear might occur for a particular polymer have not yet been established and await future research.

· *Combined Effect of High Surface Roughness and Elevated Contact Temperature on Wear*

The combination of a rough counterface and a high frictional contact temperature generally leads to rapid wear of a polymer even if continuous surface melting does not occur [13]. The wear process involved in such cases is mainly severe abrasion of the softened polymer surface.

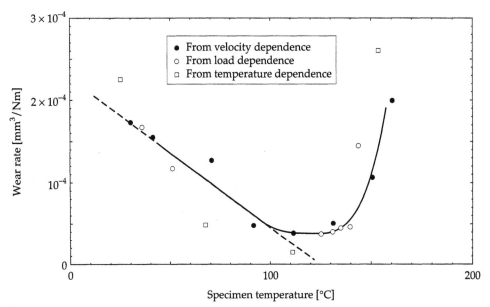

FIGURE 16.15 Wear rate of nylon in slow speed sliding contact as a function of frictional contact temperature [26].

With increased contact temperature there is a change in the wear kinetics from a linear constant rate process to a series of discrete rapid wear periods separated by longer periods of essentially negligible wear. When the counterface is rough this step form of wear can proceed effectively independent from an abrasive type of wear. In this case, the wear kinetics consists of small step jumps in wear added on to a slower linear wear. These two modes of wear observed for UHMWPE sliding against a stainless steel counterface are illustrated in Figure 16.16.

The mechanism of wear behind the step form of the wear kinetics is believed to result from the periodic release of molten polymer from the wearing contact. When the temperature is too low to sustain continuous melting, melting proceeds by a cycle of gradual formation of molten polymer. This is followed by the sudden release of the molten material when the entire wear surface is covered by molten polymer. Melting is initiated from the hottest point of the contact and spreads progressively over the entire wear surface. This model of polymer wear is illustrated schematically in Figure 16.17.

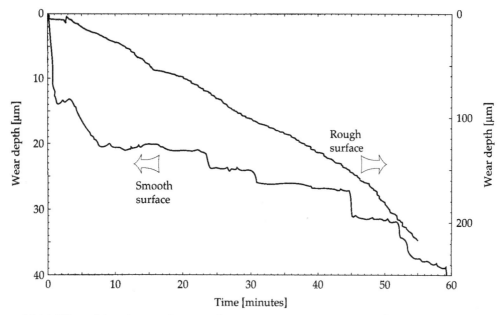

FIGURE 16.16 Wear kinetics at frictional contact temperatures slightly below transition to melting wear for UHMWPE [13].

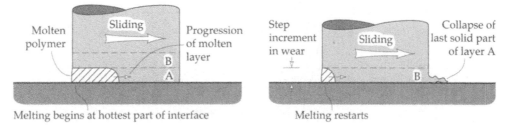

FIGURE 16.17 Mechanism of polymer wear at temperatures slightly lower than the transition to continuous melting wear [13].

Fatigue Wear of Polymers and Long Term Wear Kinetics

In polymers subjected to a large number of stress cycles from repeated sliding contact, a form of fatigue wear may occur. It was observed that when UHMWPE was slid against very smooth steel surfaces, an increase in wear rate occurred after several hundred kilometres of sliding distance [27] as shown in Figure 16.18.

This rise in wear rate coincided with the development of cracks and spalling on the worn polymer surface. It was found that after long sliding distances the fatigue wear, which is predominantly based on cracking and spalling of the surface, is superimposed on a pre-existing transfer film/adhesive wear process [27].

The transition to fatigue wear is controlled by the contact stress. For example, at low contact stresses less than 1 [MPa], a nearly infinite sliding distance is required before fatigue wear begins. Fatigue wear is therefore more likely to occur on heavily loaded, very smooth sliding surfaces after a long period of sliding. The experimental relationship between the onset of fatigue wear for UHMWPE and apparent contact stress is shown in Figure 16.19.

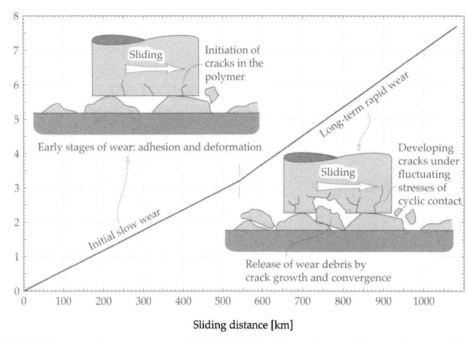

Sliding distance [km]

FIGURE 16.18 Increase in wear rate of UHMWPE after a long sliding distance due to the initiation of fatigue wear [27].

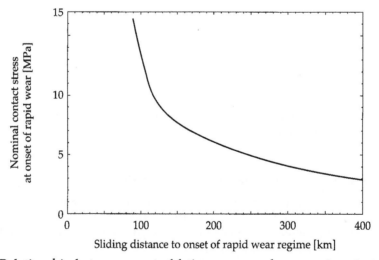

Sliding distance to onset of rapid wear regime [km]

FIGURE 16.19 Relationship between onset of fatigue wear and apparent contact stress [27].

Visco-Elasticity and the Rubbery State

Almost all polymers manifest visco-elasticity under certain conditions of load and strain rate, i.e., the stress exerted on a polymer is a function of the strain rate as well as the strain itself. Many polymers can be classified as rubbers and visco-elasticity is of profound importance to the tribology of rubbers. The basic concept of visco-elasticity is illustrated in Figure 16.20 where the asperities of two surfaces in dry contact are represented by a series of equivalent springs and dampers. This analogy is not entirely perfect because the amplitude of movement is limited whereas in the real case sliding can continue indefinitely. The model does, however, illustrate the nature of forces acting between contacting asperities.

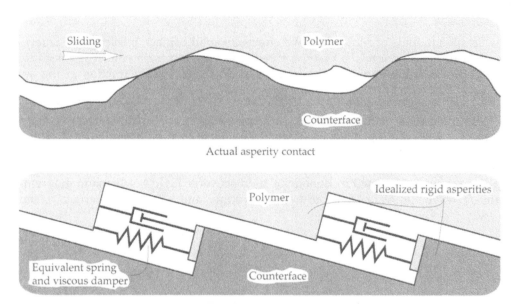

FIGURE 16.20 Mechanical analogy of visco-elastic asperity contact.

As can be seen from Figure 16.20, the forces, in particular the tangential or frictional forces, between asperities become a direct function of sliding speed. It should be noted, however, that the sliding speed also affects the coefficient of friction by virtue of its effect on frictional temperature. In any tribological experiment it can be difficult to distinguish between these two effects and, particularly in the older literature, this distinction was not fully observed. Visco-elasticity effects appear to be most clearly observable under lubricated or irrigated conditions where either a liquid cooling medium can effectively cool the frictional interface or the friction coefficient has been reduced to a low value. If liquids are present, the effect of speed dependent visco-elasticity must also be clearly separated from hydrodynamic effects [28].

Visco-elastic effects on friction were demonstrated for plexiglass sliding against steel when lubricated by sodium stearate (i.e., soap) [23]. It was found that the coefficient of friction measured as a function of external or counterface temperature correlated well with the vibration dissipation factor of plexiglass.

Friction and Wear in the Rubbery State

Rubbers and rubbery materials can provide a unique combination of low wear and high friction coefficients and are widely used for pneumatic tyres and pipe linings because of these characteristics. Rubbers or polymers in the rubbery state have a molecular structure which allows extremely large strains to be imposed before fracture occurs. The rubber or elastomer consists of long linear molecules which are coiled and tangled together to form an amorphous solid. When the material is strained, the molecules untangle and align in the direction of strain. Visco-elasticity is caused by relative movement between the polymer molecules and the mechanism of strain accommodation results in a very low tensile modulus while maintaining a comparatively high tensile strength [29]. Because of these mechanical properties the mechanics of asperity contact for rubbers are radically different from the prevailing patterns of behaviour for most other materials.

· *Schallamach Waves*

The low tensile modulus of rubber has two effects on solid contact, particularly contact with a harder surface. The first effect is that the true area of contact with rubber is relatively large compared to most other engineering materials and is a significant fraction of the apparent contact area. The second effect is that considerable tangential movement of the rubber parallel to the direction of sliding is possible without causing fracture and releasing wear debris. With most rigid materials, e.g., metals, asperity contacts break or form wear particles within a very short sliding distance. A further aspect of rubber tribology is that rubber adheres strongly by van der Waals bonding to many counterface materials, e.g., glass [30]. All of these factors acting together allow an abnormal form of sliding to occur which is known as a '**Schallamach wave**', named in honour of its discoverer [31]. A schematic diagram of the Schallamach wave sliding mechanism between rubber and a hard counterface is shown in Figure 16.21.

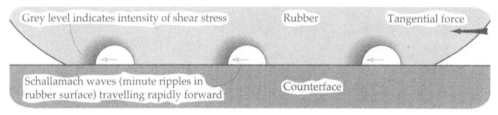

FIGURE 16.21 Schallamach wave mechanism of sliding between rubber and a hard counterface.

The Schallamach wave sliding mechanism works on the principle that a large proportion of the rubber contact area is strongly bonded to the opposing surface and cannot slide without a very substantial tangential force. However, at a much lower level of tangential force it is possible that a small area of rubber detaches from the counterface to form a 'ripple'. The 'ripple' or Schallamach wave then moves across the surface in the direction of applied tangential force to cause a smaller macroscopic tangential movement of the entire rubber body. The process is analogous to the movement of dislocations in a metal under shear. The Schallamach waves generated between a rubber sphere and a perspex plate during sliding are shown in Figure 16.22.

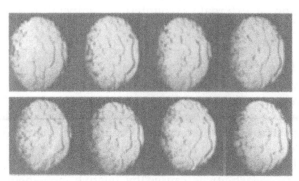

FIGURE 16.22 Schallamach waves generated between a rubber sphere and a perspex plate at a sliding speed of 0.43 [mm/s] [31].

The Schallamach waves move much faster than the two bodies in sliding. A wave velocity about 35 times faster than the sliding speed was observed at a sliding speed of 0.2 [mm/s], and about 15 times faster at a sliding speed of 0.9 [mm/s] [31]. There is evidently a limiting speed where the wave and the sliding speeds are equal and the mechanism fails. At this point it is probable that frictional heating and melting of the rubber cause a form of melting wear similar to other polymers [32].

· *Visco-Elasticity and Friction of Rubbers*

The strong influence of visco-elasticity on rubbers causes the dry friction coefficient to initially rise with sliding velocity before declining. This is in contrast to most other materials where the static coefficient of friction, i.e., the coefficient of friction in 'micro-sliding', is always greater than the kinetic coefficient of friction. This effect is demonstrated in Figure 16.23 where the relationship between the coefficient of friction and the product of sliding velocity and temperature visco-elasticity dependence parameter is shown. The temperature visco-elasticity dependence parameter is obtained from the following expression which is based on the Williams, Landel and Ferry equation [33]:

$$\log_{10}a_T = -\,8.86\,(T - T_s)/(101.5 + T - T_s) \tag{16.3}$$

where:

a_T is the temperature visco-elasticity dependence parameter [dimensionless];

T is the contact temperature [K];

T_s is the controlling standard temperature defined as: $T_s \approx T_g + 50K$ (T_g is the glass transition temperature of the polymer [K]).

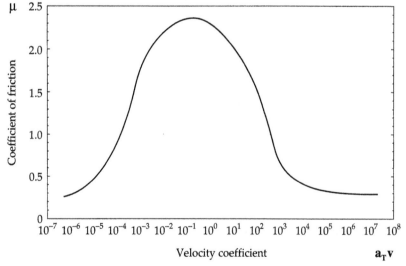

FIGURE 16.23 Relationship between the coefficient of friction and the product of sliding velocity and temperature visco-elasticity dependence parameter for acrylonitrile-butadiene rubber on glass [33].

It can be seen that the friction coefficient at negligible sliding speeds is quite small and increases to a very large maximum of about 2.4 at sliding speeds of the order of a few millimetres per second depending on the temperature. This maximum is determined by the balance between the increasing visco-elastic reaction force with increasing sliding speed and the decline in visco-elastic reaction force with increasing frictional temperature. It was found that the highest coefficient of friction was achieved at a temperature around 20°C. This implies that the friction of, for example, rubber tyres is greatest under moderate climatic conditions than under extremely cold or hot conditions.

It was also found that at high sliding speeds, the coefficient of friction conforms to the model of 'thermal control of friction' [22]. It declines from 0.4 at 15 [m/s] to less than 0.1 at 30 [m/s] (108 [km/h]). This clearly shows that when the rubber tyres of a car skid at freeway speeds there is an almost total loss of frictional grip.

· *Wear Mechanisms Particular to Rubbery Solids*

Rubbers are subject to abrasive, adhesive, fatigue and corrosive forms of wear as well as synergistic wear and thermal decomposition or pyrolysis at extreme sliding speeds [29]. These forms of wear are similar to wear mechanisms occurring with other materials and are not discussed here. A form of wear which is characteristic of rubbery materials is '**roll formation**'. Roll formation is a result of the large strain to fracture of a rubber and its mechanism is schematically illustrated in Figure 16.24.

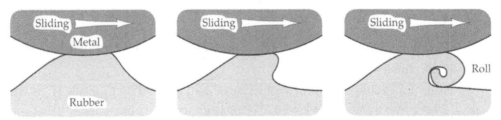

FIGURE 16.24 Mechanism of 'roll formation' on rubber surfaces.

The ability of rubber to withstand high strains without fracture ensures that adhesive asperity contact with a counterface causes tangential movement of the rubber rather than the formation of wear debris. At the interface between rubber and the counterface, rubber is pulled forward by adhesion to form a tongue and then rolled in on itself. Roll formation can occur even when abrasive wear, i.e., by a rough counterface, is present [29]. A major feature of this mechanism is that a far greater amount of frictional work is needed to form a wear particle than with any other mechanism of particle formation.

Effect of Lubricant, Corrosive Agents and Microstructure on Wear and Friction of Polymers

The wear of polymers is influenced by lubricants and chemical or corrosive agents in a manner similar to any other material. The microstructure of a polymer is also a significant factor affecting wear and frictional characteristics of polymers.

· *Effects of Lubricants*

Lubricants (including water), in general, reduce the friction of polymers to varying degrees. The reduction obtained depends on the type of polymer and the lubricant used. The addition of fatty acids, such as caproic, palmitic and stearic acids, causes a dramatic decline in friction from 0.4 to 0.09 in nylon [34]. The strong effect of lubricants on nylon is believed to be caused by the polar nature of the polyamide which constitutes nylon. The lubrication effect of fatty acids was also confirmed with nylon 11 and polyacetal [35]. Lubrication by non-polar organic substances such as hexane and benzene caused only a very marginal drop in the friction coefficient. Polyethylene, which is a less polar polymer than nylon, showed a slightly different trend in frictional characteristics when lubricated with the same lubricants. Also, no lubricating effect by fatty acids was found for polyetheretherketone (PEEK) [35].

Since water does not wet polyethylene it should not, in principle, cause any dramatic reduction in friction, except perhaps with a water adsorbing counterface such as glass. However, in some cases friction reduction in the presence of water is observed for polyethylene. In one study, a reduction in friction in the presence of water was found for many polymers, i.e., nylon 6, HDPE, LDPE, PTFE, polyacetal and polyimide. The most significant reduction in friction was found for nylon 6, where the friction coefficient dropped from 0.35 to 0.15 at low sliding speeds [28]. In another study simple addition of water to a dry nylon/glass contact resulted in a drop in the coefficient of friction to a value of about 0.12 [34]. It is possible that the water forms a softened layer on the polymer, which provides some form of a sacrificial lubrication. The effect of water on rubber appears to be essentially hydrodynamic [34] and the friction coefficient at very small sliding speeds is close to its dry friction value, i.e., $\mu = 1$.

Water has a significant influence on the wear of polymers, in many cases accelerating wear. Although the wear rates are generally increased this increase may vary depending on the type of polymer. For example, the wear rates of polyimide increase by a factor of 8 and nylon 6 by a factor of 3, while for LDPE and HDPE the wear rate increases by only 30% [28]. Polyetheretherketone sustains a significant increase in wear probably caused by the softening of the PEEK surface in the presence of water [122]. Polyphenyl sulphide (PPS) also shows an increase in wear upon exposure to water, although to a lesser extent than PEEK [122].

Increasing velocity in the presence of water causes the wear rate to decrease. For sufficiently high velocities, i.e., 1 [m/s], the wear rate decreases to a level lower than that obtained under dry conditions. The wear reduction appears to be caused by the hydrodynamic lubricating effect of water since it is accompanied by a decline in the friction coefficients to extremely low values of the order of 0.05 or less.

Formation of transfer films is usually suppressed by the presence of fluids, particularly water [36]. However, there are some exceptions. For example, polyphenyl sulphide, which contains sulphur, reacts with a steel counterface forming iron sulphide, which in turn promotes the transfer of PPS to the steel surface and the formation of a protective transfer film [122]. In general, however, the formation of transfer films is suppressed in the presence of fluids and the extent to which the lack of these films causes unsatisfactory levels of friction and wear is difficult to assess, but from the published data it appears that transfer films are fundamental to the maintenance of low wear and friction by polymers.

· *Effects of Corrosive Agents*

Polymer tribology is affected by a whole range of corrosive substances or chemical reagents. This aspect of polymer tribology, although important, is relatively neglected and existing research does not accurately portray the full range of substances that could affect polymer wear and friction.

An isolated example of how sensitive the friction and wear of polymers can be to external chemicals is provided by a study of the wear of HDPE in aqueous solutions of various metal chlorides against a steel counterface [37]. It was found that only ferric chloride caused an increase in wear rate as shown in Figure 16.25.

It can be seen from Figure 16.25 that the peak wear rate occurs at some critical concentration of ferric chloride. The critical concentration is about 0.01 wt% of ferric chloride (hydrated) and the corresponding wear rate is approximately equivalent to 1.7×10^{-13} [m³/Nm]. This value indicates that the wear is at least 100 times more rapid than what is considered to be an acceptable value for a bearing. The cause for this sudden rise in wear rate is not yet fully understood and it is suggested that the HDPE transfer film is disrupted at this particular concentration of ferric chloride. If the ferric chloride acted by simply corroding the steel counterface to produce abrasive pits with corrosion product debris, then there would be a monotonic increase in wear with ferric chloride concentration.

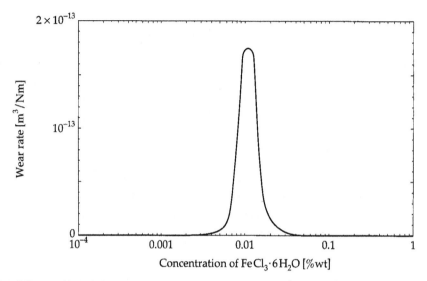

FIGURE 16.25 Wear of HDPE versus concentration of ferric chloride [37].

Polymers are soluble in many organic fluids and there can be a synergistic effect between an aggressive solvent and the polymer resulting in significant wear. If the solvent can penetrate the surface of the polymer it will have a detrimental effect on its wear behaviour. The rapid wear which results is believed to occur by aggravated cracking of the solvent weakened polymer during contact with the counterface [36]. This is schematically illustrated in Figure 16.26.

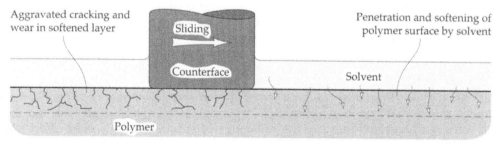

FIGURE 16.26 Synergism between wear of polymer and damage by a solvent.

Wear is most rapid in the presence of a solvent when the solubility parameters of the polymer and the solvent are the same [36]. The solubility parameter is defined as:

$$\partial = (\Delta E/V)^{0.5} \qquad (16.4)$$

where:

    ∂ is the solubility parameter [Pa$^{0.5}$]; (i.e., *n*-hexane $\partial = 15$ [MPa$^{0.5}$], methanol $\partial = 30$ [MPa$^{0.5}$] and water $\partial = 48$ [MPa$^{0.5}$] [36]);

    ΔE   is the energy of vaporization at zero pressure [J/mol];

    V   is the molar volume of fluid [m^3/mol].

Common solvents such as acetone, benzene, tetrachloromethane and toluene have solubility parameters close to the values of several bearing polymers and can cause accelerated wear when in contact with these polymers. PTFE has a relatively low solubility parameter and appears to be unaffected even by liquids with matching solubility parameters. Therefore PTFE is a suitable polymer when contact with organic liquids is inevitable. Other solvents such as methanol and water have solubility parameters higher than many polymers and some (except water) are more likely to reduce wear [36].

· Effect of Oxidizing and Biochemical Reagents

When oxidizing agents are present, common engineering polymers such as nylon and polyethylene display a form of wear that shares similarities with the corrosive wear of metals in features such as a reduced friction coefficient and elevated wear rates. It was found that the friction coefficient of nylon 6 and UHMWPE declines with increasing concentration of hydrogen peroxide while the wear rate conversely increases with concentration of hydrogen peroxide [112,113]. The scale of this effect is proportional to the level of surface damage by corrosive agents, e.g., nylon 6 showing more sensitivity to hydrogen peroxide than UHMWPE. This agrees with the widely observed high resistance of UHMWPE to most chemical reagents. Experiments have revealed that corrosive wear of polymers is controlled by the formation of a cracked and degraded layer of polymer on the worn surface. In mild cases of degradation, such as for UHMWPE, surface damage is limited to a network of cracks while with more severe degradation, such as for nylon 6, a layer of randomly shaped particles accumulates on the worn surface. This layer of particles is able to reduce the friction coefficient in perhaps the same way as sand particles can cause sliding between the shoe and the ground during walking. This is further evidence that friction and wear rates are not material properties but instead are controlled by the tribological system since neither the sand nor the shoe nor the ground can be considered as lubricants or self-lubricating.

In biological environments, where polymers, especially UHMWPE, are routinely used, e.g., in artificial joint implants, even seemingly benign substances, such as proteins, can cause the corrosive wear of polymers and also of metals. Pin-on-disc sliding wear tests conducted with UHMWPE pins slid against martensitic stainless steel discs revealed an acceleration of wear when human and animal proteins were supplied to the sliding contact. It appears that the effect of frictional heat and shearing causes the chemical modification of both the protein and the wearing material. The modified protein becomes sufficiently chemically active to degrade the UHMWPE and accelerate wear by the formation of a weakened surface layer on the UHMWPE. There is an increase in coefficient of friction since the protein forms a sticky layer on the surface of stainless steel. The effect depends on the protein type, e.g., gamma-globulin (a protein that is involved in the immune system) is particularly active while albumin (a very common protein) is only moderately active. This protein layer is not only sticky but can also became lumpy and cause abrasion of the UHMWPE which is usually a very tough and wear-resistant polymer. Wear also occurred on the martensitic stainless steel counter-sample which became pitted within the wear track. This is an example of corrosive wear caused by localised fractures occurring on the oxide film covering the metallic surface and sustaining intense localised anodic corrosion resulting in pits [114].

· *Effects of Polymer Microstructure*

The microstructure of polymers affects their wear and frictional characteristics. Two basic topics have appeared in studies of polymer microstructure:

- the relative merits of amorphous and crystalline polymers as bearing materials and
- the size of spherulites in crystalline polymers.

It is postulated that amorphous polymers can only provide low friction and wear rates close to the glass transition temperature [32]. At temperatures higher or lower than the glass transition, a high wear rate is expected. Crystalline polymers, on the other hand, offer a much wider temperature range for low wear and friction. The useful temperatures range from below room temperature (but above the brittleness temperature) to close to the melting point of the polymer [32]. Crystalline polymers are therefore commonly used as bearing materials given the variation in surface temperature with sliding speed and load. The amorphous polymer found usually on the surfaces of moulded items and formed as a result of rapid surface cooling is not recommended for use as a bearing material [38].

It has been theorized for some time that reducing the spherulite size of crystalline polymers should improve their wear resistance [38]. The basic argument for this is that the size of wear particles produced is proportional to the spherulite size as shown in Figure 16.27.

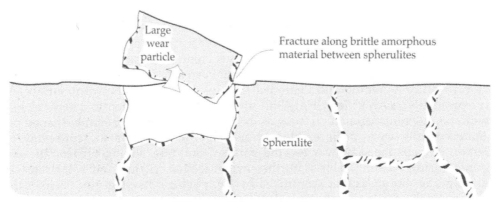

Polymer with large spherulites

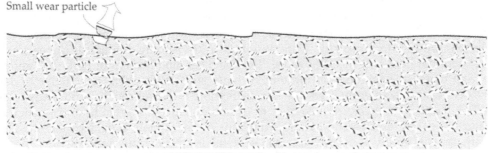

Polymer with small spherulites

FIGURE 16.27 Model of effect of the spherulite size on wear rate.

Spherulites of crystalline material in a polymer are separated by layers of more brittle amorphous material. According to the model shown in Figure 16.27, wear particles form by crack development between spherulites and their size is similar to that of the spherulites.

Any reduction in the spherulite size would therefore result in a reduction of the wear particle size and consequently the wear rate.

Tests at low sliding speeds and low contact stress conducted with polypropylene pins sliding against a relatively smooth (R_a = 0.02 [μm]) steel counterface confirmed the influence of spherulite size on wear rate [40], although the trend in wear rate is not exactly the same as predicted by the model. An optimal spherulite size was found which minimized both the friction and the wear coefficients. Experimental data of friction and wear coefficients as a function of spherulite diameter is shown in Figure 16.28.

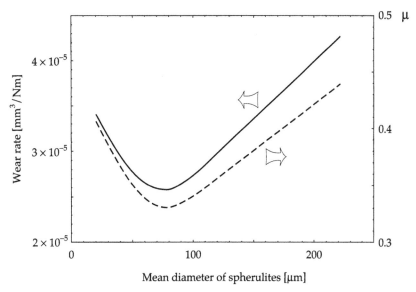

FIGURE 16.28 Variation of friction and wear coefficients versus spherulite diameter [40].

It can be seen from Figure 16.28 that the variation in wear rate versus spherulite diameter is quite small. A reduction of 35% in wear, compared to very small or very large spherulite sizes, occurs at 85 [μm] spherulite diameter.

In some applications, e.g., orthopaedic implants, UHMWPE components are sterilized by γ-ray irradiation which often causes a reduction in the wear resistance of this material. Despite the apparent disadvantages this procedure is generally accepted as the need for sterilization outweighs any consideration of reduced wear resistance. However, it has been found recently that γ-ray irradiation does not cause much destructive scission of UHMWPE molecular chains when applied in combination with heat treatment. Moderate elevations in temperature during or after γ-ray irradiation encourage cross-linking of the UHMWPE molecular chains rather than chain scission and this cross-linking is found to greatly improve sliding wear resistance [115].

16.3   TRIBOLOGY OF POLYMER COMPOSITES

Polymers are very rarely used as bearing materials in their pure form. Even a nominally pure polymer, e.g., nylon, contains plasticizers and colouring agents. As indicated in previous chapters, the wear of polymers is strongly influenced by these adjuvants. It has been shown, for example, that plasticizers cause a reduction in friction for polyethylene by diffusing to the polymer surface to form a lubricating layer [34].

The particular importance of polymer composites is that improvements in polymer tribology can be achieved by the deliberately engineered addition of strengthening and lubricating agents. This methodology is fundamentally different from the ad hoc exploitation of existing material characteristics. Composites are developed for superior mechanical strength and this objective often conflicts with the simultaneous achievement of superior wear resistance. World wide there is a common aim, however, of developing and producing high quality composite materials with special combinations of mechanical and tribological properties.

There is a wide variety of composite materials available, but the basic types of polymer composites are:

- · bulk polymer containing a lubricating polymer, e.g., nylon + PTFE,
- · polymer containing a metal or an inorganic powder, e.g., PTFE + lead powder,
- · metal wire or inorganic fibre reinforced polymer, e.g., glass reinforced PTFE,
- · reinforced polymer containing a second lubricating polymer, e.g., glass reinforced nylon + PTFE.

Polymer Blends

It is a common practice to add a polymer, usually PTFE, to another polymer in order to reduce the coefficient of sliding friction while maintaining a low wear rate. Examples of these materials are nylon or polyacetal with added PTFE. The tribological characteristics of these polymer composites appear to depend on the PTFE fraction depositing a thin transfer film on the counterface. This form of composite polymer, however, does not have the same potential for improvement of its tribological characteristics as other more complex composites such as fibre reinforced polymers.

Fibre Reinforced Polymers

Fibre reinforced polymers are a very important type of composite and a wide range of materials belong to this category. There are two basic forms of fibre-reinforced polymer:

- · polymers reinforced by randomly oriented chopped fibres,
- · polymers reinforced by unidirectional or woven fibres.

Glass, polymer, graphite and metals are usually used for fibres, although some metals, e.g., stainless steel, can be unsatisfactory [41].

· Chopped Fibre Reinforced Polymers

Chopped fibre reinforcements are effective in reducing wear [42] provided that there is a strong adhesion between the fibres and the matrix [43]. It was found that the limits of sliding speed and contact stress can be raised and the wear rate lowered by the incorporation of chopped fibres [42]. The major problem with fibre reinforcement, in particular with chopped fibres, is that the wear resistance depends more on fracture occurring between the fibres and the matrix than on the bulk mechanical properties of the material [43]. This is probably because wear occurs on a micro-scale where there is no mechanism of material strengthening comparable to bulk fracture which involves cracks passing through many fibre-matrix interfaces. For this reason the optimum volume concentration of chopped fibres for maximum wear resistance is close to 10% which is less than the concentration required for optimum toughness of the composite [43].

The type of reinforcement fibres and fillers used affects the tribological performance of composites to some degree. The inherent brittleness of most of the reinforcement fibres, e.g., glass, results in rapid fibre damage under abrasive wear [44] or erosive wear [45]. It was found that abrasive wear is more severe when the short chopped fibres filler is used rather than small spheres and that the improvement in adhesion between the filler particles and the matrix polymer results in the increase of abrasive wear resistance [46].

Polymer composites, although often exhibiting good wear resistance when in contact with a smooth counterface where adhesive or fatigue wear would prevail, often show inferior wear resistance compared to the unimproved polymer under conditions of abrasive or erosive wear [44]. This indicates that the composite materials for a particular application should be chosen very carefully.

· *Unidirectional and Woven Fibre Reinforcements*

The wear mechanisms in unidirectional and woven fibre polymer composites are the subject of quite intense research. Fibre orientation is critical to the tribology of the polymer composite. There are three principal fibre orientations relative to the sliding interface: parallel, anti-parallel and normal. These are illustrated in Figure 16.29.

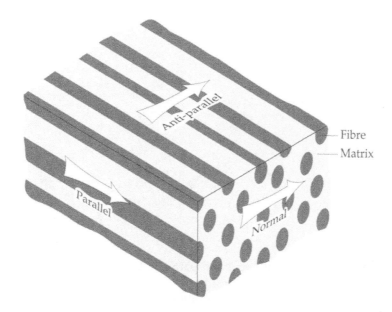

FIGURE 16.29 Orientation of reinforcement fibres to the sliding counterface [41].

Although the wear mechanisms involved in polymer composites with these three different fibre orientations are similar, they are not identical. The wear process of the parallel and anti-parallel orientations is illustrated in Figure 16.30. Wear of the matrix and fibre proceeds at the same rate until the depth of about half of the fibre diameter is worn away and the fibres start to detach in short segments from the matrix.

It can be seen from Figure 16.30 that wear debris originating from the fibres range from fine powder to complete segments of fibre as the wear proceeds. In contrast, wear debris from the matrix tend to be uniformly fine. It is possible that a fine transfer film of the matrix polymer may cover the exposed fibres and reduce the overall coefficient of friction [47]. This effect is especially marked when PTFE is used as the matrix polymer [47].

The wear mechanism of normally oriented fibres is different since partially worn fibres remain firmly attached in the matrix. During the process of wear the fibres are subjected to repeated bending which causes them to gradually debond from the matrix [48]. A simultaneous process of cracking and fragmentation at the fibre ends allows material to be eventually released as wear debris. The mechanism of wear during normal fibre orientation is schematically illustrated in Figure 16.31.

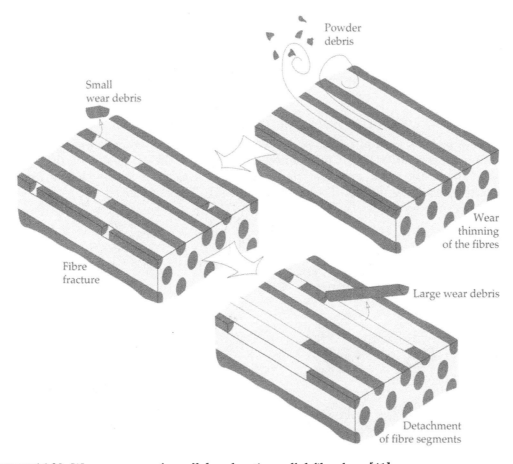

FIGURE 16.30 Wear process of parallel and anti-parallel fibre lays [41].

Polymer composites with parallel fibre orientation are the most preferable followed by the anti-parallel types [41]. Polymer composites with the normal fibre orientation give a low wear rate but at the risk of sudden seizure. The reason for this is that the exposed normal fibres tend to gouge into the counterface and initiate severe wear or seizure [41].

Unidirectional and woven reinforcements do not offer dramatic improvements over chopped fibre reinforcements for wear against smooth steel counterfaces. Wear rates under these conditions are usually controlled by crack propagation between fibres and matrix. The woven or unidirectional reinforcements offer far more favourable crack propagation conditions than short chopped fibres where many cracks are formed for each fibre segment [41,42]. This results in rapid wear by crack propagation to release wear particles. Woven fibre reinforcements, particularly made of tough materials such as Aramid, are useful in controlling abrasive wear [42]. Matrix-less weaves of PTFE fibres and glass fibres provide good wear resistance up to contact pressures of about a few [MPa], when slid against smooth steel surfaces. Lubricants or low friction polymers, e.g., PTFE, can also be added to the composite in order to lower the coefficient of friction while maintaining mechanical strength [49]. As mentioned already, brittle fibres cause rapid abrasive wear so the selection of fibre material is crucial to the characteristics of the composite.

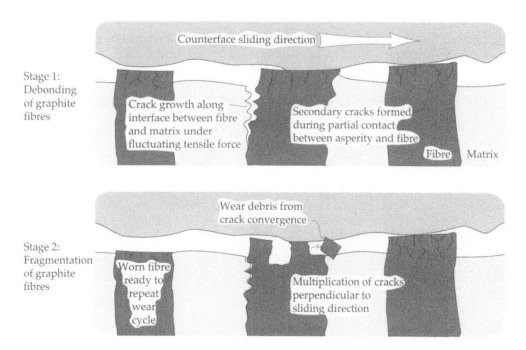

FIGURE 16.31 Mechanism of wear when reinforcement fibres are oriented normal to the counterface [48].

A significant restriction on the usage of PTFE is its poor wear resistance and one way to improve it is to reinforce PTFE with stronger fibres. Blending PTFE with reinforcement fibres such as carbon fibres and high strength polymer fibres, for example, poly-p-phenyleneterephtalamide, and solid lubricants such as molybdenum disulphide and graphite results in a large reduction in wear coefficients without significantly altering the already low coefficient of friction [123].

Fibre reinforced polymers are vulnerable to corrosive or chemical attack by many substances including lubricating oils and fuels [49]. Chemical attack of the composite usually causes fibre debonding which results in rapid wear [36]. Accelerated fibre debonding by solvents and the subsequent rapid wear are schematically illustrated in Figure 16.32.

Contact temperature can also have serious influence on the performance of composites. For example, carbon-carbon composites used for disc brakes display rapid wear, which is known as 'dusting', when the surface temperature reaches a critical value. At this critical temperature, often around 300°C, rapid oxidization of the composite occurs with disruption of protective film formation mechanisms [124].

· *Modelling of Wear of Fibre Reinforced Polymers*

A limited number of semi-empirical formulae have been developed to summarise the wear and friction behaviour of reinforced polymers [41,42]. For example, it was found that the coefficient of friction of reinforced polymers obeys a reciprocal law of mixtures [41], i.e.:

$$1/\mu = V_f/\mu_f + V_m/\mu_m \tag{16.5}$$

where:

μ is the coefficient of friction;

V is the volume fraction;

f, m   are the subscripts referring to fibre and matrix polymer, respectively.

The role of fibre 'lubricity' is described by this equation. For example, a high lubricity fibre which exhibits a low coefficient of friction against most counterfaces, e.g., graphite, allows the composite to have a lower coefficient of friction than it would have with a low lubricity fibre, e.g., glass. Equation (16.5) implies that transfer films of polymer smeared over exposed fibres do not have much effect on friction. This assumption, however, may be invalid in certain cases.

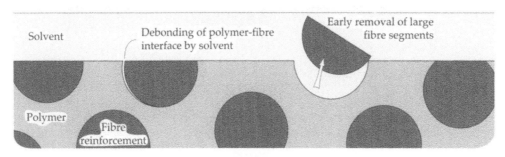

FIGURE 16.32 Wear mechanism of fibre composite in the presence of an aggressive solvent.

Powder Composites

A wide range of powdered metals and inorganic solids are often added to polymers. Commonly used powder reinforcements in polymers include copper, lead, zinc, non-ferrous alloys such as bronze and solid lubricants, such as graphite and molybdenum disulphide. Powdered glass has also been used as a reinforcement material [42]. As described already in Chapter 9, in general, the addition of solid lubricants to polymers results in some improvements in their wear and frictional characteristics. The degree of these improvements varies greatly depending on the type of polymer and lubricant used. For example, polymers of moderate lubricity such as nylon and polyimide exhibit the greatest improvements with the addition of solid lubricants, while the improvements to PTFE are very limited. Graphite usually shows superior performance to molybdenum disulphide. On the other hand, it has also been shown that, in some cases, molybdenum disulphide added to nylon oxidized during wear and failed to develop an effective transfer film and the friction performance of the composite was inferior to that of plain nylon [50].

Powdered metal reinforcements are particularly effective with PTFE, since the wear resistance is significantly increased while a low friction coefficient is maintained [4]. The reasons for this improvement are complex and involve many factors such as improved thermal conductivity to dissipate frictional heat. A recently developed model suggests that metal powders, in particular, function by bonding the crystalline lamellae of PTFE to the counterface and also by supporting part of the load. Soft plastic metals with some chemical activity, e.g., copper, exhibit these characteristics and are therefore useful fillers for PTFE [19]. The model of the function of powdered metal fillers in PTFE is schematically illustrated in Figure 16.33. As shown in Figure 16.33 a complex transfer layer of sheared PTFE and highly deformed powdered metal forms on the counterface. The properties of this layer determine friction and wear characteristics by an as yet unknown mechanism.

Recent development in powder composites are the nanoparticle reinforced polymers. These particles with diameters measured in nanometres, as opposed to much larger particles of several micrometres diameter, are found to produce valuable increases in wear resistance and reductions in friction coefficients when blended with polymers or polymer composites. Typically the particles are made of ceramics such as silicon dioxide (silica), silicon nitride or alumina. Unidirectional sliding tests on epoxy and epoxy-polyacrylamide composites showed that 2% blends of 9 [nm] silica particles had about one-third the wear rate and significantly lower friction coefficient than the same polymer with no added silica particles [125]. When blended with silica and silicon nitride nanoparticles, PEEK also showed similar improvements [126,127]. Surface treatment of the silica particles designed to improve matrix bonding may further enhance the tribological performance, i.e., wear resistance and friction coefficient, of these materials [125].

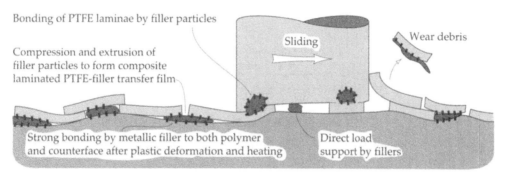

FIGURE 16.33 Influence of metal powder fillers on wear reduction of PTFE.

Blending alumina nanoparticles, of 38 [nm] average size, with PTFE resulted in a very large reduction in the wear rate, of about 600 times, and only a minor increase in friction coefficient [128]. The wear resistance was observed to increase up to the highest concentration studied, which was 20% by weight of alumina nanoparticles in PTFE.

Fullerenes are also a potentially useful additive for polyimides. When blended in concentrations of a few % by weight they result in significant reduction in wear rates of polyimides in dry sliding against smooth steel [129]. However, they show surprisingly little effect on the friction coefficient.

16.4 WEAR AND FRICTION OF CERAMICS

Ceramics are a special class of materials that include a wide range of hard refractory inorganic compounds, which are formed by heating the base material in powder form to a high temperature where sintering or solid state reaction occurs. The result of this process is a material which possesses superior hardness, good chemical resistance and, on occasions, much greater wear resistance than most metals.

An example of the superior wear resistance of ceramics over more traditional materials can be illustrated by a rocker and cam assembly of an internal combustion engine [51]. This particular assembly is a major cause of engine wear problems because of the combination of elevated temperature, sliding and high contact stresses. Experimental measurements of the wear depth on sintered ferrous and silicon nitride rocker arm pads are shown in Figure 16.34.

It is evident from Figure 16.34 that the replacement of the metallic pads by silicon nitride pads has caused wear to virtually cease after an initial wearing-in period.

Apart from the improvement in tribological performance, ceramics maintain their physical properties (hardness, strength) at elevated temperatures. Therefore they are being advocated as the new bearing materials, despite the fact that the manufacturing process required is much more difficult than that of metals. Ceramics used for bearings or wear resistant components usually consist of oxides, nitrides or carbides of aluminium, silicon and other metals. A list of typical ceramics with their physical properties is shown in Table 16.3.

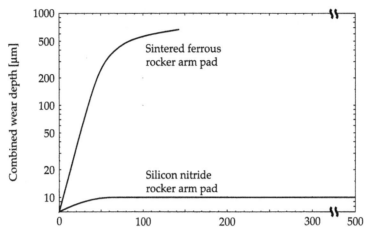

FIGURE 16.34 Effect of substitution of ferrous rocker arm pads by ceramic rocker arm pads on the wear resistance [51].

TABLE 16.3 Physical properties of typical engineering ceramics.

Ceramic		Density [kg/m³]	Hardness [Vickers]	Max. service temperature [°C]	Specific heat [J/kgK]	Thermal conductivity [W/mK]	Tensile Modulus [GPa]
Alumina		3900	1500 - 1650	1800	920	28 - 35	330 - 400
Sapphire (hard form of alumina)		3985	2500 - 3000	1800 - 1950	753	41.9	-
Aluminium nitride		3250	1200	1200	800	165	-
Boron carbide (hot pressed)		2450	3200	700 - 800	950	27 - 36	440 - 470
Boron nitride (hot pressed)		1800	Very soft	950 - 1200	780	15 - 33	20 - 100
Silicon carbide	(hot pressed)	3150	2400 - 2800	1500 - 1650	670 - 710	90 - 160	350 - 440
	(reaction bonded)	3100	2500 - 3500	1400 - 1650	670 - 710	200	410
Silicon nitride	(hot pressed)	3110	1700 - 2200	1100 - 1650	680 - 800	15 - 43	280 - 310
	(reaction bonded)	2400	800 - 1000	1200 - 1500	690	10 - 16	170 - 220
Sialon		3240	1650 - 1800	1500	620 - 710	20	280 - 300
Zirconia stab. with MgO		5740	1200	2200	470	1.3	200

The tribological characteristics of ceramics are complex and depend on the following factors: material composition and properties, sliding conditions (speed, load and temperature), the surrounding environment and the type of counterface [53].

Ceramics are often classified as 'oxides', 'carbides' or 'nitrides' and this classification is reflected by significant differences in friction and wear mechanisms. Some ceramics, e.g., aluminium oxide, consist mostly of a pure material to which small amounts of additives are added to promote sintering. Other ceramics are composite ceramics with mechanical and tribological characteristics that cannot be explained in terms of a single material. For example, sialon is a solid solution of aluminium oxide and silicon nitride; partially stabilized zirconia (PSZ) is zirconia containing a small amount of a stabilizer such as magnesium or yttrium oxides. The term 'ferrite' is also used but this is non-standard nomenclature for the magnetic iron oxides which are used in the electronics industry for data recording. The wear characteristic of ferrites determines the quality of data transmission from recording discs and has been intensively studied by researchers in the electronics industry [52].

Physical properties of ceramics depend, to a larger degree, on the manufacturing process, and for the same material different values of, for example, hardness and strength are obtained when different manufacturing processes are employed (Table 16.3). Other ceramic characteristics such as porosity and grain size also depend on the manufacturing process. Since all these properties influence wear and friction of ceramics, the method of manufacturing (e.g., hot pressing or reaction bonding) should always be included in material specifications [54,55]. For example, the effect of porosity on ceramic wear is schematically illustrated in Figure 16.35.

Each of the ceramics has particular merits and disadvantages. For example, silicon carbide and silicon nitride have good mechanical properties but require very high temperatures for processing. Aluminium oxide is hard but brittle, while tough partially stabilized zirconia loses its toughness at relatively low temperatures of around 500°C. Oxide ceramics are more chemically stable than nitride or carbide ceramics which can be oxidized, but some of the oxides, in particular zirconia ceramics, are susceptible to stress cracking in the presence of moisture. Variations in chemical reactivity between ceramics can affect their performance under conditions of corrosive wear. Despite the high hardness, ceramics often suffer severe wear, especially in dry conditions, and therefore very careful selection of these materials for a particular application is necessary.

It should, however, be mentioned that not all ceramics of interest to tribologists are hard. For example, hydroxyapatite (a form of calcium phosphate), which is the inorganic component of bone and teeth, has a hardness of approximately 50 [HV] [130]. In these ceramics, wear is accelerated by even increasing the roughness of the counterface, if it is made of a harder material [130]. Tribological behaviour of hydroxyapatite is an important area of study due to its relevance in joint replacement technology and dental restorations.

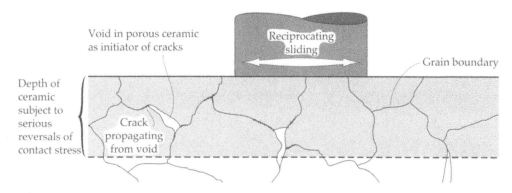

FIGURE 16.35 Effect of porosity on ceramic wear.

Unlubricated Wear and Friction of Ceramic-Ceramic Contacts

The unlubricated wear and friction of ceramics are strongly influenced by sliding conditions, temperature and the presence of moisture. Different wear mechanisms take place under dry (vacuum, dry gases) and moist (air, water) contacts and at elevated temperatures.

The deformation processes taking place in a dry ceramic contact can be loosely classified as either '**ductile**' or '**brittle**' and depend mainly on the speed/load conditions [53]. In ductile deformation, observed usually under moderate sliding conditions, an asperity contact causes plastic flow and displacement of material rather than its removal. Consequently, sliding results in low friction and little wear. In contrast, brittle deformation is characterized by extensive fracture along the grain boundaries during an asperity contact. This type of deformation dominates at high contact stresses and/or in systems where one counterface is much harder than the other. Entire grains of a ceramic can be detached by brittle fracture and debris is formed by the subsequent fragmentation of these grains. Severe wear usually accompanied by high friction is observed. The mechanisms of ductile and brittle deformations of ceramics are schematically illustrated in Figure 16.36.

Several wear mechanisms, such as abrasion, adhesion, micro-fracture and delamination, separate or combined, contribute to the wear damage in ceramic-ceramic sliding and rolling contacts. A fine powdery debris released during the asperity contact often accumulates to form debris layers on the worn surface. The formation of top layers, observed on both polished and ground surfaces, modifies surface topography and in some cases is responsible for lowering friction. The debris layers are further subject to smearing and, at sufficient stresses, are gradually worn by microfracture and/or delamination.

· *Dry Friction and Wear of Ceramics at Room Temperature*

In vacuum and in dry gases friction coefficients between self mated polycrystalline ceramics are usually high, in the range of **0.5** to **0.9** [56-58]. However, these coefficients are much lower than those of metals in a vacuum, which indicates the lower susceptibility of ceramics to seizure. In air a wider range of coefficients of friction, from **0.3** to **1.0**, is observed [56,57,59].

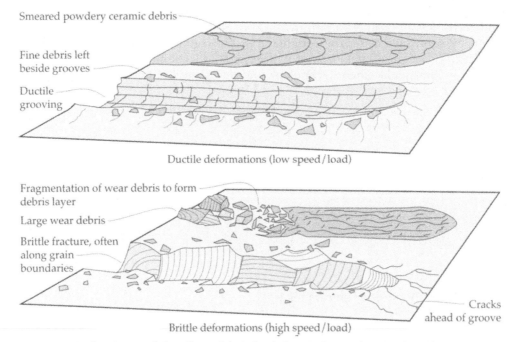

FIGURE 16.36 Mechanisms of ductile and brittle deformations of ceramics.

In air at low velocities and low contact pressures moderate friction and wear of ceramics are often observed [59,60]. A transition to severe wear occurs when speed and/or load is significantly increased [59,61]. High transient temperatures, particularly when caused by high sliding speed, can also result in a severe increase in the wear of ceramics. Self-mated ceramics with low thermal conductivity (e.g., PSZ) often suffer a steeper increase in wear rates at high speed than ceramics with higher thermal conductivity (e.g., SiC) [59]. Also, due to the brittleness of ceramics, cracking from thermal stresses after rapid cooling can occur [59]. Wear debris are then released as a result of crack growth and convergence, as illustrated schematically in Figure 16.37.

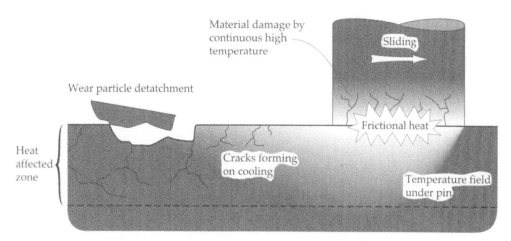

FIGURE 16.37 Wear mechanisms of ceramics by thermal stress.

When a sliding ceramic couple consists of two dissimilar materials, the wear depends on the materials' properties (i.e., hardness, thermal conductivity) and their configuration [58,62]. For example, harder pins cause higher wear on softer discs/rings which are used as a counterface in common tribometers, i.e., pin-on-disc machines, and lower thermal conductivity ceramics often suffer higher wear [58,62]. The tribological behaviour also changes when the pin and disc materials are reversed [63] but no clear explanation of this phenomenon is available.

· *Dry Friction and Wear of Ceramics at Elevated Temperatures*

The wear and friction of ceramics are usually increased at elevated temperatures [59,64,65], although in certain temperature ranges wear reduction has been recorded for silicon based [59,64], alumina [58] and PSZ ceramics [60,66,67]. Alumina ceramics suffer increased wear at high temperatures in air which is usually due to abrasion [59], however, the opposite trend has been observed in nitrogen [58]. Both an increased and a decreased wear of PSZ ceramics at elevated temperatures have been reported and the behaviour has been explained in terms of different phase transformations taking place [59,66]. The wear increase was associated with the presence of a cubic phase [59] while better wear resistance was observed when a tetragonal to monoclinic transformation occurred [66]. The examples of the wear tracks on PSZ at 25°C and 400°C after self-mated unlubricated sliding in air are shown in Figure 16.38.

Non-oxide ceramics such as silicon nitride and silicon carbide suffer tribo-oxidative wear (a combination of abrasive and oxidative wear) at high temperatures if air or oxygen is present [59,64]. A film of silicon dioxide formed on the worn ceramic surfaces is preferentially worn away because of its inferior mechanical properties. The mechanism of tribo-oxidative wear of non-oxide ceramics is shown schematically in Figure 16.39.

· *Friction and Wear of Ceramics in the Presence of Water or Humid Air*

Water and/or atmospheric moisture can affect the wear of ceramics in both positive and negative ways. The most beneficial effect of moisture is the formation of a thin soft hydrated layer on the ceramic surface which acts as a lubricant [68]. The lubricating layer can be formed on both alumina [56,69] and silicon-based ceramics [56,70]. However, if the depth of the hydrated layer becomes excessively large then a form of corrosive wear occurs in the presence of water [68]. The contrast between the beneficial lubricating effect of a thin hydrated layer and the effect of accelerated wear by a thick hydrated layer is schematically illustrated in Figure 16.40.

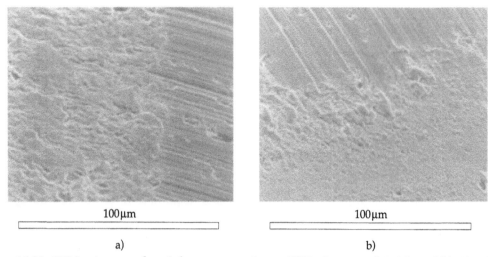

100μm   100μm

a)   b)

FIGURE 16.38 SEM micrographs of the wear tracks on PSZ plates at a) 25°C and b) 400°C after self-mated unlubricated sliding in air.

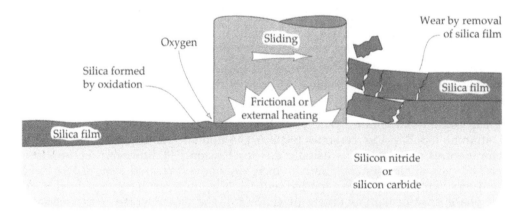

FIGURE 16.39 Mechanism of tribo-oxidative wear of non-oxide silicon ceramics.

Another effect of water is to introduce stress corrosion cracking by hydrolysis at the crack tip [56]. The stressed crystalline lattice at the apex of a crack provides a favourable reaction site for the dissociation of water and the production of hydroxide [56]. The hydroxide is much weaker than the corresponding oxide of the ceramic substrate and rapidly fails under tension, extending the crack and exposing fresh oxide. This effect is particularly noticeable in zirconia-based ceramics such as tetragonal zirconia polycrystals (Y-TZP), where accompanying tetragonal to monoclinic transformation at the crack tip introduces a net of tiny cracks [56,57]. Hydrolytic stress corrosion cracking and the corresponding wear mechanism are illustrated schematically in Figure 16.41.

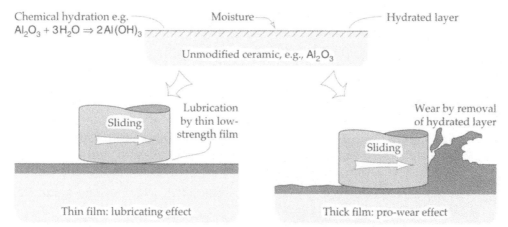

FIGURE 16.40 Lubrication and corrosive wear by hydrated layers on ceramics.

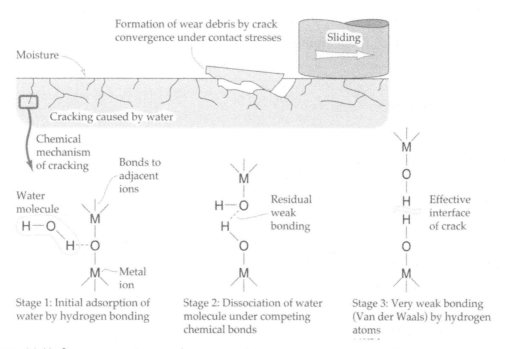

FIGURE 16.41 Stress corrosion cracking caused by water in oxide ceramics.

· *Wear Modelling of Ceramics*

Attempts have been made to model the wear of ceramics, especially to predict the transition from mild to severe wear regimes, and also to describe the wear loss analytically. One study of the dry, wet and lubricated sliding wear of ceramics at ambient temperatures revealed a close relationship between the maximum tensile stress in the wearing contact and the critical fracture stress of the ceramic [117]. Wear volume can be calculated from the following empirical formula [117]:

$$V = \frac{CWl\sigma_{max}}{H\sigma_D}$$

(16.6)

where:

V	is the wear volume [m³];
C	is the proportionality constant [non-dimensional];
W	is the normal load [N];
l	is the sliding distance [m];
H	is the Vickers hardness [Pa];
σ_D	is the critical fracture stress for a ceramic grain size of '**d**' [Pa]. This stress can be deduced from a modified form of the Griffith theory of brittle fracture where defect size is assumed to be directly proportional to grain size of the ceramic;
σ_{max}	is the maximum tensile stress in the contact [Pa], which can be found from the Hertzian contact theory, i.e.:

$$\sigma_{max} = \frac{3W}{2\pi a^2} \left[\frac{1 - 2\upsilon}{3} + \frac{\pi\mu(4 + \upsilon)}{8} \right]$$

where:

a	is the Hertzian contact radius [m];
υ	is Poisson's ratio;
μ	is the friction coefficient.

Rearranging equation (16.6) yields:

$$\boxed{\frac{V\,H}{W\,l} = C\frac{\sigma_{max}}{\sigma_D}} \tag{16.7}$$

The left-hand side of this equation is the Archards wear coefficient '**K**' for materials (see Chapter 10) [116] while the right-hand side of the equation contains a factor, 'σ_{max}/σ_D', which describes, at any time, the ratio of the applied stress to the critical damage stress for brittle materials under contact [117]. The 'σ_{max}/σ_D' ratio depends on material properties and operating conditions. An analysis of equation (16.7) indicates that its left-hand side may not be constant. This implies that during the wear of ceramics the wear mechanism may change, resulting in wear transition [117]. It has been found that during the sliding wear of ceramics a mild to severe wear transition with a step change, or precipitate increase, in the value of the wear coefficient '**C**' occurs. In the mild wear regime, '**C**' approximates to 10^{-6} (1/million) while in the severe wear regime, '**C**' approximates to 10^{-4} (1/10,000). The transition occurs when 'σ_{max}/σ_D' ratio exceeds unity. This means that severe wear is initiated by extensive cracking on the worn surface.

In another study of dry uni-directional sliding of ceramics at ambient temperatures, physical parameters appear to control the transition from mild to severe wear. In cases of alumina, toughened zirconia, silicon carbide and silicon nitride, mild wear only occurs when the values of both the 'mechanical severity' and 'thermal severity' parameters are less than the corresponding critical values [131]. The mechanical severity parameter is controlled by the coefficient of friction, Hertzian contact stress, the fracture toughness and the length of pre-existing cracks. The thermal severity parameter is defined as the product of the coefficient of friction, load and sliding speed divided by the product of frictional temperature rise, the nominal contact radius and a parameter related to the thermal conductivity. The thermal severity parameter indicates the likelihood of thermally induced cracking caused by the sudden increase in temperature as the sliding counter-face passes over the worn surface [131].

In mild wear the wear debris is very fine and a smooth worn surface is formed, while under severe wear the surface roughness is comparable to the grain size and the wear rate is about 1000 times higher than in case of mild wear [132]. Tribochemical reactions with atmospheric oxygen and lubricants, for example, in case of silicon nitride or carbide, contribute to the smoothing of the worn surface.

Closer inspection of the described models reveals their limitation, i.e., they apply only to a specific material pair working under usually narrow operating conditions. To allow for ceramic wear prediction across a wide range of materials and operating conditions a new modelling methodology has been proposed [133]. Firstly, the ceramics wear data is separated according to different wear regimes using a 'severity parameter'. Then, material normalization is carried out for each of the wear regimes. This methodology is capable of predicting wear of ceramics to ±1 order of magnitude. Given the obscure nature of wear and the difficulties in modelling complex chaotic processes, this represents a major advance in modelling of ceramic wear. More information on ceramic wear models can be found elsewhere [133].

Since wear of ceramics is affected by so many variables it is often very difficult for an engineer to assess a true benefit of using ceramics in a particular tribological application. In order to resolve this problem various user-friendly graphical wear maps, similar to those available for steels, have recently been developed [e.g., 119-121]. Different wear regimes, such as severe, mild and negligible wear, and the critical operating limits of load, speed or temperature for a particular ceramic can easily be determined from such graphs.

· *Dry Wear and Friction Characteristics of Individual Ceramics*

There are significant differences in wear and friction characteristics of individual ceramics and the tribology of the most wear resistant engineering ceramics is summarized in Table 16.4.

A common feature, however, is a similar response of all the ceramics to tribological stress. At low to moderate contact stress, wear of ceramics is controlled by ductile deformations resulting in mild wear regime. At large contact stresses a transition from mild to severe wear regime occurs and wear is controlled by cracks and fracture. Despite the relatively inert nature of ceramics in bulk, e.g., aluminium oxide, friction and wear characteristics may vary significantly with differing combinations of sliding ceramic pairs. For example, zirconia-on-zirconia in dry unidirectional sliding manifests higher wear rates and more overt signs of adhesive wear than zirconia-on-alumina. It appears that zirconia-on-zirconia sliding combination causes adhesive wear while alumina-on-alumina is self-lubricating [134]. Additional effects play an important role in environments where tribochemical reactions take place, e.g., humid air, water, high temperature [118]. In alumina ceramics, there is a strong effect of alkalinity and acidity on the friction coefficient in wet sliding conditions. These ceramics exhibit high friction coefficient in water for all but very low and very high levels of pH where the friction coefficient may decline to 0.2 from a mid-pH value of 0.6 [135]. Interestingly, the wear coefficients show the almost opposite trend to frictional behaviour. The wear rates are high at both extremely high and low pH and also at moderate pH of 8.5 (weakly alkaline). High wear in very acidic and very alkaline solutions is explained by alumina dissolution and absence of tribolayers while high wear at pH = 8.5 is due to the inferior properties of a tribolayer, related to the net surface charge [135].

Lubricated Wear and Friction of Ceramic-Ceramic Contacts

It is known that ceramics can experience severe wear damage and high friction during unlubricated sliding, especially under conditions of high loads, speeds and temperatures. Extensive research efforts have been directed towards incorporating liquid and solid lubrication into ceramic systems in an attempt to reduce friction and wear.

TABLE 16.4   Tribological characteristics of selected wear resistant ceramics in unlubricated self-mated contacts.

Ceramic and properties	Environment	Tribological characteristics
Alumina · High hardness · Low fracture toughness	Vacuum or dry gas	High friction and wear; susceptible to wear by brittle microfracture and grain pull-out [56,58] Evidence of plastic deformation at elevated temperatures with decreasing wear [58]
	Air	Lower wear and friction [58]
	Moist or wet	Can be lubricated in water by the hydrated top layers [56,69]
Transformation toughened zirconia (Mg-PSZ, Y-PSZ, Ce-PSZ) · High fracture toughness · Moderate hardness · Low heat conductivity	Vacuum or dry gas	High wear and friction [56]
	Air	Low wear and friction at very low speeds [60] High wear and friction at higher loads/speeds [59,61,67] Increase in wear and friction at elevated temperatures [58,59,67] with few exceptions [66,67] Evidence of plastic deformation and delamination [67]
	Moist or wet	Stress corrosion cracking in the presence of water molecules [56,57]
Silicon based ceramics (Silicon carbide, silicon nitride, sialon) · High hardness · Low to moderate fracture toughness	Vacuum or dry gas	High wear and friction [56]
	Air	Tribochemical reaction with the formation of silicon dioxide in the presence of air/oxygen [56,70] Wear by cracking and spalling of SiO_2 [56]
	Moist or wet	Tribochemical reaction in water and humid air; decrease in wear and friction [56,70]

· *Liquid Lubrication*

It is possible to achieve significant reductions in the friction and wear of ceramics by the application of oil lubricants which were originally formulated for metallic sliding surfaces. However, liquid lubrication of ceramics has two major drawbacks:

· limited reactivity of lubricant additives with ceramic surfaces, and

· a temperature barrier at which liquid lubricants start to decompose.

As discussed already in Chapter 8, there are two basic types of friction-reducing additives present in lubricating oils: surfactants, which produce adsorbate films, and reactive additives (EP additives), which form corrosion product films. With ceramics the former type is most effective [71,72] as they are almost inert to reactive additives. This distinction is illustrated in Figure 16.42.

Mineral and synthetic oils might provide effective lubrication only up to a certain temperature, which is dictated by the thermal stability of the lubricant. These critical temperatures, discussed already in Chapter 3, range from 150°C to 200°C for mineral oils and from 250°C to about 300°C for synthetic oils. Current developments in synthetic lubricants predict their use at temperatures up to 450°C in the near future [73]. Since the ceramics can successfully be used in systems operating even at higher temperatures, another form of lubrication must be found.

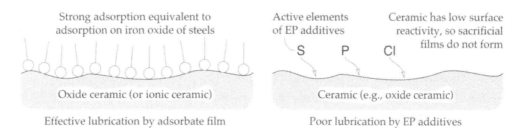

FIGURE 16.42 Effect of lubricant additives on ceramics.

Experimental results show that the wear and friction of ceramics are usually reduced in the presence of simple hydrocarbons and mineral oils compared to unlubricated contacts [57,60,74]. Synthetic liquid oils, such as polyol esters, polyphenyl ethers and perfluoropolyalkylethers, which can operate at higher temperatures than mineral oils, can also lubricate ceramics [75,76]. The coefficient of friction is usually reduced to about 0.1 but the wear performance of ceramic-ceramic pairs in oil is often worse than that of metallic or metal-ceramic pairs operating under the same conditions [62,75]. This indicates that there is a different mechanism of interaction between oil additives and ceramic and metallic surfaces [77].

The performance of selected ceramics in the presence of liquid lubricants is summarized in Table 16.5. Most of the tests were conducted at room temperature, and only limited data are available for elevated temperatures.

TABLE 16.5 Performance of selected ceramics in liquid lubricants.

Lubricant	Test conditions	Ceramic	Wear and friction	Ref.
Hexadecane	Low speed, moderate load	PSZ	Friction and wear reduction compared to air and water	[57]
Hexadecane + stearic acid				
n-Tridecane + stearic acid	Very low speed	PSZ	Friction reduction but wear increase compared to air	[60]
Mineral oil	Boundary conditions	PSZ	Wear higher than steel	[74]
		Al_2O_3	Wear comparable to steel	
Squalane	Concentrated contact	PSZ	Wear similar to steel	[75]
		SiC	Wear higher than steel	
		Si_3N_4		
Base oil, 90°C	4-ball test	PSZ, SiC	Less wear than steel for lower speeds and loads	[74]
		Al_2O_3	Wear higher than steel	
		Si_3N_4	Very low wear and friction	
Perfluoro-polyalkyl-ether	Concentrated contact	PSZ	Wear lower than steel	[75]
		SiC	Wear higher than steel	
		Si_3N_4		

The most durable and therefore effective adsorbate films are formed on ionic ceramics which are usually oxide ceramics, where a strong bond is developed between a hydrogen atom from a polar hydrocarbon and an oxygen from a ceramic [71,78]. It was found that at low loads and sliding speeds alumina is effectively lubricated by fatty acids with six or more carbon atoms

[71,78]. In comparison, longer chain fatty acids are needed for steel lubrication, as discussed in Chapter 8. The lubricating adsorption layers on alumina are also formed by ZnDDP additives with long n-alkyl chains [71,78]. Covalently bonded ceramics, e.g., silicon carbide, do not respond significantly to lubrication by fatty acids and ZnDDPs, since the adsorption force is governed by a weaker van der Waals bonding [71].

In boundary lubrication of metallic surfaces, the chemical reactions between oil additives and metallic surfaces generate reaction products which separate surfaces and reduce friction as described in Chapter 8. The chemical inertness of most ceramics minimizes the possibility of similar reactions occurring in ceramic-ceramic lubricated systems [71,77]. One view is that the effective lubrication of these systems is achieved by chemical reactions in the heated lubricant without any direct involvement of the ceramic surface [77]. The reaction products from the lubricants are deposited on the ceramic surface and a reduction in friction and wear usually results [77]. Small amounts of ZnDDP additives were found to reduce the wear of silicon carbide, silicon nitride and PSZ [74].

· *Solid Lubricants*

One of the main advantages of ceramics over metals is that they maintain their mechanical properties at high temperatures, i.e., they do not suffer from rapid softening as is the case with most metals. The prime objective therefore is to provide lubrication at high temperatures, above the levels set by mineral and synthetic lubricants. The other objective is to provide lubrication in a vacuum where ordinary oils quickly evaporate. The task can be accomplished by the use of solid lubricants in the form of powders, sticks or deposits produced by vapour or gas decomposition at elevated temperatures [65,73,79]. Alternatively, the application of thin surface coatings of a lubricious nature might also provide a solution to this problem.

In general, solid lubricants consist of metal oxides, fluorides or compounds based on sulphur and phosphorus. One proposed solution to minimize friction and wear in ceramic bearings at high temperatures is the use of powder lubricants, based on, for example, TiO_2 or MoS_2, provided via a gaseous carrier [79,80]. Another concept, 'Vapour Phase' (VP) lubrication, involves the deposition of lubricating films directly from an atmosphere of vapourized lubricant. This lubrication is particularly attractive for use in internal combustion engines. A vapourized lubricant is delivered in a carrier gas forming a thin lubricating film on the hot bearing surface [81]. A similar concept involves the deposition of carbon lubricants achieved by decomposition of gaseous hydrocarbons on hot substrate surfaces [82].

VP lubrication of ceramics by the heated vapour of tricresylphosphate (TCP) carried in a nitrogen/oxygen mixture gas has been demonstrated [83]. However, pre-treatment of the ceramic surface to produce ferric oxide was necessary in order to obtain good adsorption of TCP oxidation products. In some cases a coefficient of friction as low as 0.05 was achieved at a temperature of 280°C [83]. On the untreated ceramic surfaces the lubricating films were deposited by the mechanism of physisorption and thus failed to be effective in reducing friction and wear [81].

Spontaneous decomposition of ethylene to form a graphitic layer on a ceramic surface at high temperatures has been demonstrated for sapphire pins sliding on silicon nitride and silicon carbide discs [82]. It was found that the process of dry sliding induced sufficient catalytic activity on the ceramic surface for the ethylene gas to decompose and form a lubricating film, resulting in a low coefficient friction of about 0.1 at 550°C [82]. The concepts of high temperature lubrication of ceramic by gas and vapour are schematically illustrated in Figure 16.43.

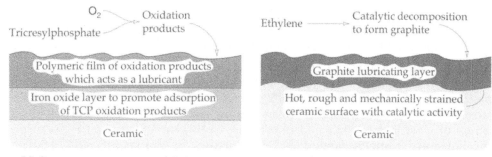

FIGURE 16.43 Concepts of high temperature lubrication of ceramic by gas.

As there is no underlying metal with its Kramer electrons to be exposed, ceramics do not display a high degree of reactivity in the 'nascent' state. However, they exhibit some chemical reactivity in sliding contacts. In some cases the resulting chemical compounds may affect wear and friction of ceramics or even act as solid lubricants. During sliding of a silicon carbide disk against an alumina ball in a chamber containing a variable pressure of HFC-134a gas (a hydro-fluorocarbon gas), friction and wear coefficients were found to be inversely related to the gas pressure and the low friction correlated well with the formation of siliconfluoride compounds on the worn surfaces [136]. A similar dependence on gas pressure was also found during sliding of an alumina ball against a zirconia disk in the presence of HFC-134a gas [137]. It was observed that wear and friction coefficients declined from their values in vacuum until a pressure of 0.1 atmosphere was reached. It seems that at higher gas pressures, fully fluorinated compounds of ZrF_4 and AlF_3 are forming instead of more lubricious non-stoichiometric ZrO_xF_y and AlO_mF_n compounds [137].

Another method of solid lubrication of ceramics involves the use of thin lubricious surface coatings. Coatings produced by various techniques such as plasma spraying [84,85], ion-beam-assisted deposition (IBAD) [86-89] and ion-implantation [90-92] have been investigated and a reduction in friction and wear was often found. Solid lubrication can be provided by the application of soft metallic coatings such as silver [87] and MoS_2 [90] or the generation of lubricious metallic oxides formed at high temperatures [86]. The application of modern coating techniques eliminates most of the problems associated with providing a good bond between a solid lubricant and a substrate surface.

Since the effectiveness of solid lubricant films is often limited by their short lifetime the idea of directly incorporating a solid lubricant into a ceramic matrix has been explored. There are two main problems associated with this concept: (i) finding a way to introduce a solid lubricant into the ceramic structure without sacrificing high ceramic strength and hardness, and (ii) the provision of a sufficient amount of lubricant release during sliding. The concept has been tested on silicon nitride and alumina ceramics with drilled holes filled with graphite [93]. The coefficient of friction was reduced for Si_3N_4-graphite composite sliding on steel but the wear was comparable to that of plain Si_3N_4.

Wear and Friction of Ceramics Against Metallic Materials

Metallic alloys, in particular steel [62,94-96] and cast iron [96,97], have been widely studied as sliding counterfaces to ceramics. The number of tested combinations of materials as sliding couples is quite large and both the metal and the ceramic contribute to the variation in friction and wear characteristics. The coefficient of friction in dry ceramic-metal contacts depends on the type of metallic counterface and the load/speed conditions which directly influence the transient interface temperature and the extent of metallic surface oxidation. Broad ranges of coefficients of friction for various metallic counterfaces have been reported:

0.2 - 0.8 for steels and cast irons [62,96,98], **0.2 - 0.5** for softer materials such as brass, bronze, aluminium and copper [60,61,96,98] and **0.3 - 0.4** for cobalt-chromium alloys [99]. These coefficients usually increase at elevated temperatures [67,98].

The common feature of almost all ceramic-metal interactions is that the metal adheres to the ceramic to form a transfer film [53,62,96,98]. This should always be taken into account in the interpretation of wear and friction results. The formation of a transfer film is the result of two factors: strong adhesion between clean ceramic and metal surfaces and the lower plastic flow stress of most metals compared to ceramics. Adhesion and friction between metal and ceramic surfaces strongly depend on the ductility of the metals [100]. Soft metallic counterfaces such as brass and bronze usually generate thicker transfer films [61,96] while steel and cast iron transfer films are fragmentaric [96]. The adhesion forces can be suppressed by the presence of contaminants and/or hard surface coatings such as borides [98,99]. In such cases negligible film transfer and lower friction are observed. At elevated temperatures the thickness and surface coverage of the metallic transfer film increase due to the removal of contaminants and softening of the metallic materials [67,98]. The mechanism of metal adhesion to a ceramic surface and the formation of a metallic transfer layer are illustrated in Figure 16.44. SEM micrographs of metallic transfer film onto ceramic surfaces at room and at an elevated temperature are shown in Figures 16.45 and 16.46, respectively.

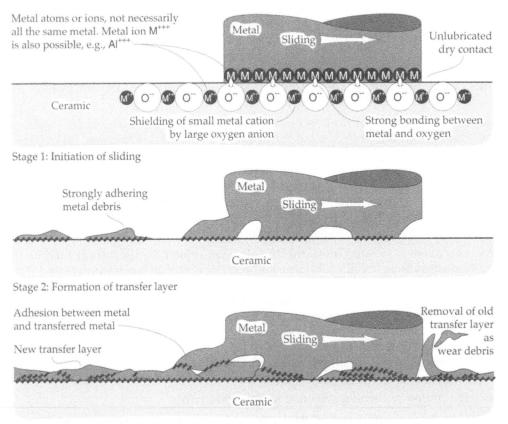

FIGURE 16.44 Mechanism of metal adhesion to a ceramic surface and the formation of a metallic transfer layer.

The damage sustained by a metallic counterface depends on the type of ceramic used and is usually lower with softer ceramics such as PSZ and higher with harder ceramics such as alumina [99]. In certain combinations of materials, e.g., when silicon-based ceramics are matched with ferrous alloys, the chemical reactions at the interface result in increased wear of the ceramic. This behaviour is often observed in metal cutting when silicon-based cutting tools are used to machine steel and cast iron [101]. At the extremely high contact stresses and temperatures present at the interface between a ceramic cutting tool and ferrous metal, diffusive wear of the ceramic can occur.

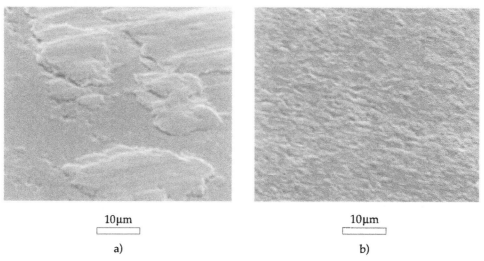

10μm

a)

10μm

b)

FIGURE 16.45 SEM micrographs of the metallic transfer films onto ceramic surfaces at room temperature; a) soft metal (brass) and b) hard metal (cast iron).

100μm

FIGURE 16.46 Effect of elevated temperature on the morphology of the transfer film (cast iron 400°C).

The removal of metal to form a transfer film involves fine metal particles which are susceptible to oxidation if air is present. High frictional temperatures promote rapid oxidation of the metallic layers and the transfer film can be converted from a layer of pure metal to metal covered by a layer of superficial oxide [67,94]. Tribo-oxidation of the metallic transfer layer is illustrated in Figure 16.47. The frictional heat also causes cracks to form in the ceramic and combined wear of the ceramic and metal may occur despite the large difference in the hardness of the two materials [94,98].

Secondary deformation of the partially oxidized metallic transfer layer may also occur to produce a mechanically alloyed layer of metal and metal oxide on the surface of the ceramic. This layer will eventually detach from the surface, either by adhesive failure at the ceramic surface or by the release of underlying ceramic debris.

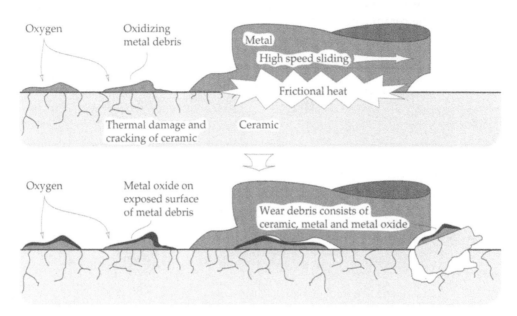

FIGURE 16.47 Tribo-oxidation of metallic transfer layers and thermal damage of ceramics as a wear mechanism for both metal and ceramic.

The effect of water and atmospheric humidity on metal-ceramic friction was found to depend on the chemical activity of the metals [102]. Easily oxidized metals showed an enhanced adhesion to oxide ceramics while lower friction was observed when low activity metals such as silver were used [102].

Wear and Friction of Ceramics Against Polymers

The pairing of polymers and ceramics leads to very useful sliding combinations which provide low friction in the absence of traditional lubricants. The best known applications of polymer-ceramic systems are orthopaedic endoprostheses where alumina has replaced metallic alloys in some of the prosthetic designs to become a counterface to ultra high molecular weight polyethylene (UHMWPE). Most of the published studies on the wear of ceramics against polymers have been concerned with this application [e.g., 103-106]. The experimental results have confirmed the advantage of an alumina-polyethylene configuration over a metal-polyethylene pair in reducing friction and wear of polyethylene [103,104,106]. This behaviour has been attributed to the chemical inertness and good wettability of alumina, as well as to the high resistance of alumina to scratching. The initial mirror surface finish of alumina is preserved during the lifetime of the implant and this minimizes the abrasion of polyethylene. Recently, new ceramics with increased toughness have been proposed as replacements for the brittle alumina in orthopaedic applications [105,106]. The wear of UHMWPE against toughened alumina and PSZ in various environments was found to be lower than that against traditional alumina [105,106].

As may be expected from the difference in hardness between polymers and ceramics, the wear damage is confined to the polymer, while a thin polymeric transfer film is observed on the ceramic surface, mainly in dry sliding [104,105]. The wear of polymers is strongly influenced by the surface finish and porosity of the ceramic and increases when the ceramic surface roughness is increased [105,107]. The surface irregularities on the original ceramic surface are preferentially filled by polymer particles and the ceramic surface is smoothed. Therefore the long time wear rates of polymers are usually lower than the initial rates which are dominated by the abrasive wear.

Wear and Friction of Ceramic Matrix Composites

The most promising ceramic composites consist of a ceramic matrix of alumina, silicon carbide, silicone nitride, carbon or glass, with ceramic, metallic fibres, or particulates incorporated. The use of whiskers as a strengthening agent is now radically diminished due to the health hazard. The inclusion of secondary phases into ceramic matrices can result in a higher flexural strength and fracture toughness and better reliability compared to unmodified ceramics. At the same time the tribological performance may be improved. However, the whole area of ceramic composite development and testing of their tribological characteristics is relatively new and the available literature scarce.

An example of a ceramic composite designed to improve the tribological characteristics of the matrix material is a graphite-fibre reinforced glass. The graphite fibres reduce the coefficient of friction of the composite to levels comparable with those of resins while the wear resistance is similar to that of glasses and ceramics [108].

Alumina, which is known for its brittle behaviour, is also a prime candidate for reinforcement. It has been shown that silicon carbide whiskers incorporated in an alumina matrix serve as a mechanical barrier to the passage of cracks and lower the brittle wear rate of the composite compared to pure alumina [109]. An additional tribochemical reaction at 800°C is responsible for low wear of this composite at high temperature [110]. However, whiskers of silicon carbide are very powerful carcinogens and therefore are gradually being withdrawn from the production of composite ceramics.

Abrasive wear resistance of ceramic matrix composites depends to a greater degree on the microstructure of the matrix (e.g., porosity), the type of inclusions and the bonding strength between the matrix and the second phase. For example, wear resistance of a silicon nitride matrix with SiC platelets was lower, slightly increased for alumina with SiC platelets and significantly increased for mullite matrix-SiC whiskers composite as compared to the unmodified matrix [111].

16.5   SUMMARY

The current state of development, potential benefits and limitations of non-metallic solids as bearing or wear resistant materials have been described in this chapter. Two classes of materials, polymers and ceramics, with entirely different characteristics have been discussed. Polymers can provide low friction and wear coefficients but their use is limited to lower temperatures and consequently low speeds and loads. Ceramics are resistant to high temperatures and often have a good wear resistance but their applications are limited by poor friction coefficients, especially in unlublicated applications. Ceramics and polymers are surprisingly vulnerable to accelerated wear in the presence of corrosive reagents and care should be taken in the selection of materials that are appropriate for particular operating conditions. Neither of these materials meets current or future needs, and much effort is being expended to develop new materials and improve the properties of existing materials for new and future applications. The development of polymer and ceramic matrix composites reinforced by fibres, platelets and particulates serves as an example of these efforts. Despite the restrictions on their usage, non-metallic materials provide a useful alternative for metals in many tribological applications and therefore are becoming more widely used.

REVISION QUESTIONS

16.1   Name an important low-wear high-friction polymer.

16.2   Polytetrafluoroethylene (PTFE) is well known as being difficult to stick to other materials (using glues or other conventional techniques). For this reason, does it show poor adhesion to metals under wearing contact, and does it therefore have low friction properties?

16.3   What happens when the transfer film is disrupted by some means, mechanical or chemical?

16.4   What is the most insidious aspect of transfer film disruption by chemicals such as ferric chloride?

16.5   Why are polymer bearings so vulnerable to overheating?

16.6   What is the 'PV' limit?

16.7   Name the two principal limitations of pure PTFE as a bearing material.

16.8   In general, which of the two kinds of polymers are the more useful tribologically: amorphous or crystalline? Indicate reasons for your answer.

16.9   How is the contact area between rubber and a contacting surface affected by adhesion forces?

16.10   What is the mechanism of low speed sliding that appears to be unique to rubber?

16.11   Apart from the high friction coefficient and resistance to wear, what other tribological property of rubber makes it especially suitable as a material for vehicle tyres?

16.12   Why should rubber be suitable as a material for bicycle brakes but not for car brakes?

16.13   Why are bicycle brakes totally ineffective when rain falls even though the tyres give reasonable friction coefficients?

16.14   Describe the relationship between the coefficient of friction for rubber sliding on a hard non-melting surface and the sliding speed.

16.15   Why should friction and wear between the sliding surfaces of an implanted orthopaedic prosthesis be different from typically lubricated surfaces? Discuss the role of the biochemicals involved.

16.16   Suppose that for a skidding rubber tyre, the friction coefficient is inversely proportional to the square root of the sliding (skidding speed) for speeds greater than **5** [m/s]. For speeds of **5** [m/s] and less, there is no melting wear occurring and the friction coefficient is $\mu = 1.0$. Calculate the friction coefficient at **50** [m/s] (**180** [km/h]) and **20** [m/s] (**72** [km/h]) skidding (sliding) speeds. Comment on the role of friction coefficients in skidding at high speeds. (Ans. $\mu_{50} = 0.316$, $\mu_{20} = 0.5$)

16.17   Suppose that for the same sliding system as in question 16.16, melting wear is controlled by a critical temperature of **150°C** (softening, flow temperature of the rubber). The data in question 16.16 was based on a road temperature of **30°C**. What would be the friction coefficients when the road temperature reaches **110°C**? Is the black colour of tarmac a desirable safety feature for any road anywhere in the world? (Ans. $\mu_{50} = 0.105$, $\mu_{20} = 0.167$)

16.18   A vehicle is travelling with the velocity of **180** [km/h]. Calculate the skidding distance for this vehicle decelerating to **72** [km/h] with the wheels locked. Assume that the softening temperature of the rubber is **125°C**, ambient (road) temperature is **25°C** and that the threshold velocity, i.e., the velocity at which the coefficient of friction reverts to its constant value of $\mu_{th} = 1$ is **5** [m/s] (i.e., **18** [km/h]). Comment on the difference between braking distance on Earth and on Mars.

Note that when a rubber tyre skids on a road, the frictional force of the tyre is controlled by the limiting frictional temperature which is the softening or melting point of rubber. A controlling equation for limiting friction temperature can be summarized as:

$$T_{limiting} = T_{ambient} + \Delta T_{frictional}$$

where $\Delta T_{frictional}$ is given by the following expression: $\Delta T_{frictional} = C\mu U^{0.5}$

where $T_{limiting}$ is temperature inside the road-tyre contact [°C], **C** is a constant [$Ks^{0.5}m^{-0.5}$], μ is the coefficient of friction and U is the skidding speed [m/s]. (Ans. x = 289.72 [m])

16.19   For the vehicle in problem 16.18 travelling at the same velocity, what would be the skidding distance during an abnormally hot day for the same speed reduction from **50** [m/s] to **20** [m/s], when the road surface temperature reaches **100°C** (the proverbial fried egg on the road)? Assume that the heat does not affect the road. (Ans. x = 1158.9 [m])

16.20   For the vehicle in question 16.18 travelling at the same velocity of **180** [km/h], what would be the stopping distance? Assume that the velocity at which the coefficient of friction reverts to its constant threshold value of μ_{th} = **1** is **5** [m/s] (i.e., **18** [km/h]). (Ans. x = 322.59 [m])

16.21   The '**PV**' value of a commercial polymer is quoted as **5** [MPam/s], meaning that the material would start to rapidly wear out at a sliding speed of **1** [m/s] and **5** [MPa] apparent contact stress or at **5** [m/s] sliding speed and **1** [MPa] contact stress. For the problem currently under study by an engineer, the prevailing operating temperature is **95**°C (i.e., simmering or boiling water at elevated altitudes). Most polymer wear tests are performed at ambient temperature, unless specified otherwise. In the commercial literature, no ambient test temperature is specified. In colder countries, ambient temperature is typically **20**°C while in warmer countries this temperature is about **25**°C. An independent check of the literature reveals that the softening temperature of the polymer is close to **170**°C. Provide a conservative but not timid estimate of the '**PV**' value at **100**°C. What other considerations should be pondered when deciding to use this polymer? (Ans. PV = 2.5 [MPam/s])

16.22   A wear coefficient of 10^{-14} [m³/Nm] is quoted in some sales literature for an engineering polymer. Does this value represent an interestingly high resistance to wear for a polymer expected to operate under **10** [N] load and **1000** [km] sliding distance? Discuss with an example of wear volume over a useful sliding distance and realistic load.

16.23   Tribological tests were conducted on a pin-on-disc machine using a partially stabilized zirconia (PSZ) pin sliding on a PSZ disc over a distance of **1000** [m] under a constant load of **3.4** [N]. The PSZ pin had a truncated conical shape, with a cone angle of **120°** and the initial diameter of the flat end was **0.5** [mm]. After **1000** [m] of sliding the diameter of the flat end increased to **1.26** [mm]. Calculate the specific wear coefficient of the PSZ pin and comment on its magnitude. (Ans. k = 4.2×10^{-14} [m³/Nm])

16.24   Why should electron discharge (sparking) be caused by crack formation at the surface of a worn ceramic?

REFERENCES

1   D.H. Buckley, Surface Effects in Adhesion, Friction, Wear and Lubrication, Elsevier, Amsterdam, 1981.

2   D. Gong, Q. Xue and H. Wang, ESCA Study on Tribochemical Characteristics of Filled PTFE, *Wear*, Vol. 148, 1991, pp. 161-169.

3   C.W. Bunn and E.R. Howells, Structures of Molecules and Crystals of Fluorocarbons, *Nature*, Vol. 174, 1954, pp. 549-551.

4   C.M. Pooley and D. Tabor, Friction and Molecular Structure: the Behaviour of Some Thermoplastics, *Proc. Roy. Soc., London*, Series A, Vol. 329, 1972, pp. 251-274.

5   K. Tanaka, Effects of Various Fillers on the Friction and Wear of PTFE-Based Composites, in Composite Materials Science, editor: K. Friedrich, Elsevier, Amsterdam, 1986, pp. 137-174.

6   K.R. Makinson and D. Tabor, The Friction and Transfer of Polytetrafluoroethylene, *Proc. Roy. Soc., London*, Series A, Vol. 281, 1964, pp. 49-61.

7   J.M. Thorpe, Tribological Properties of Selected Polymer Matrix Composites Against Steel Surfaces, in Composite Materials Science, editor: K. Friedrich, Elsevier, Amsterdam, 1986, pp. 89-135.

8   B. Briscoe, Wear of Polymers: an Essay on Fundamental Aspects, *Tribology International*, Vol. 14, 1981, pp. 231-243.

9   K. Tanaka and T. Miyata, Studies on the Friction and Transfer of Semi-Crystalline Polymers, *Wear*, Vol. 41, 1977, pp. 383-398.

10   V.K. Jain and S. Bahadur, Material Transfer in Polymer-Polymer Sliding, *Wear*, Vol. 46, 1978, pp. 177-198.

11   A. Birkett and J.K. Lancaster, Counterface Effects on the Wear of a Composite Dry-Bearing Liner, Proc. JSLE Int. Tribology Conference, 8-10 July 1985, Tokyo, Japan, Elsevier, pp. 465-470.

12   D. Dowson, J.M. Challen and J.R. Atkinson, The Influence of Counterface Roughness on the Wear Rate of Polyethylene, Proc. 3rd Leeds-Lyon Symposium on Tribology, Wear of Non-Metallic Materials, Sept. 1976, editors: D. Dowson, M. Godet and C.M. Taylor, Inst. Mech. Engrs. Publ., London, 1978, pp. 99-102.

13   T.S. Barrett, G.W. Stachowiak and A.W. Batchelor, Effect of Roughness and Sliding Speed on the Wear and Friction of Ultra-High Molecular Weight Polyethylene, *Wear*, Vol. 153, 1992, pp. 331-350.

14   T.A. Blanchett and F.E. Kennedy, The Development of Transfer Films in Ultra-High Molecular Weight Polyethylene/Stainless Steel Oscillatory Sliding, *Tribology Transactions,* Vol. 32, 1989, pp. 371-379.

15   N.S. Eiss, K.C. Wood, J.A. Herold and K.A. Smyth, Model for the Transfer of Polymer to Rough, Hard Surfaces, *Transactions ASME, Journal of Lubrication Technology*, Vol. 101, 1979, pp. 212-219.

16   J.H. Warren and N.S. Eiss, Depth of Penetration as a Predictor of the Wear of Polymers on Hard, Rough Surfaces, *Transactions ASME, Journal of Lubrication Technology*, Vol. 100, 1978, pp. 92-97.

17   N.S. Eiss and K.A. Smyth, The Wear of Polymers Sliding on Polymeric Films Deposited on Rough Surfaces, *Transactions ASME, Journal of Lubrication Technology*, Vol. 103, 1981, pp. 266-273.

18   D.F. Play, Counterface Roughness Effect on the Dry Steady State Wear of Self-Lubricating Polyimide Composites, *Transactions ASME, Journal of Lubrication Technology*, Vol. 106, 1984, pp. 177-184.

19   D. Gong, Q. Xue and H. Wang, Physical Models of Adhesive Wear of Polytetrafluoroethylene and Its Composites, *Wear*, Vol. 140, 1991, pp. 9-24.

20   M.K. Kar and S. Bahadur, Micromechanism of Wear at Polymer-Metal Sliding Interface, *Wear*, Vol. 46, 1978, pp. 189-202.

21   S.H. Rhee and K.C. Ludema, Mechanisms of Formation of Polymeric Transfer Films, *Wear*, Vol. 46, 1978, pp. 231-240.

22   C.M.McC. Ettles, Polymer and Elastomer Friction in the Thermal Control Regime, *ASLE Transactions*, Vol. 30, 1987, pp. 149-159.

23   A.M. Bueche and D.G. Flom, Surface Friction and Dynamic Mechanical Properties of Polymers, *Wear*, Vol. 2, 1958/1959, pp. 168-182.

24   Y. Mizutani, K. Kato and Y. Shimura, Friction and Wear of Phenolic Resin up to 200°C, Proc. JSLE International Tribology Conference, 8-10 July 1985, Tokyo, Japan, Elsevier, pp. 489-494.

25   K. Tanaka and Y. Uchiyama, Friction, Wear and Surface Melting of Crystalline Polymers, in Advances in Polymer Friction and Wear, editor: Lieng-Huang Lee, Plenum Press, New York, 1974, pp. 499-531.

26   M. Watanabe and H. Yamaguchi, The Friction and Wear Properties of Nylon, Proc. JSLE Int. Tribology Conf., 8-10 July 1985, Tokyo, Japan, Elsevier, pp. 483-488.

27   J.R. Atkinson, K.J. Brown and D. Dowson, The Wear of High Molecular Weight Polyethylene, Part 1 : The Wear of Isotropic Polyethylene against Dry Steel in Unidirectional Motion, *Transactions ASME, Journal of Lubrication Technology*, Vol. 100, 1978, pp. 208-218.

28   K. Tanaka, Friction and Wear of Semi-Crystalline Polymers Sliding against Steel under Water Lubrication, Proc. Int. Conf. on Wear of Materials, Dearborn, Michigan, 16-18 April 1979, editors: K.C. Ludema, W.A. Glaeser and S.K. Rhee, Publ. American Society of Mechanical Engineers, New York, 1979, pp. 563-572.

29   J.A. Schweitz and L. Ahman, Mild Wear of Rubber-Based Compounds, in Friction and Wear of Polymer Composites, editor: K. Friedrich, Elsevier, Amsterdam, 1986, pp. 289-327.

30   D. Tabor, Interaction between Surfaces: Friction and Adhesion, in Surface Physics of Materials, Vol. 2, Academic Press, New York, 1975, pp. 475-529.

31   A. Schallamach, How Does Rubber Slide?, *Wear*, Vol. 17, 1971, pp. 301-312.

32   G.M. Bartenev and V.V. Lavrentev, Friction and Wear of Polymers, editors: L.H. Lee and K.C. Ludema, Elsevier, 1981.

33   K.A. Grosch, The Relation between the Friction and Visco-Elastic Properties of Rubber, *Proc. Roy. Soc., London*, Series A, Vol. 274, 1963, pp. 21-39.

34   S.C. Cohen and D. Tabor, The Friction and Lubrication of Polymers, *Proc. Roy. Soc., London*, Series A, Vol. 291, 1966, pp. 186-207.

35   B.J. Briscoe, T.A. Stolarski and G.J. Davies, Boundary Lubrication of Thermoplastic Polymers in Model Fluids, *Tribology International*, Vol. 17, 1984, pp. 129-137.

36   D.C. Evans, Polymer-Fluid Interactions in Relation to Wear, Proc. 3rd Leeds-Lyon Symposium on Tribology, Wear of Non-Metallic Materials, Sept. 1976, editors: D. Dowson, M. Godet and C.M. Taylor, Inst. Mech. Engrs. Publ., London, 1978, pp. 47-71.

37   M. Watanabe, The Friction and Wear of High Density Polyethylene in Aqueous Solutions, Proc. Int. Conf. on Wear of Materials, Dearborn, Michigan, 16-18 April 1979, editors: K.C. Ludema, W.A. Glaeser and S.K. Rhee, Publ. American Society of Mechanical Engineers, New York, 1979, pp. 573-580.

38   F.J. Clauss, Solid Lubricants and Self-Lubricating Solids, Academic Press, 1972, New York.

39   V.A. Bely, A. I. Sviridenok, M.I. Petrokovets and V.G. Savkin, Friction and Wear in Polymer-Based Materials, Pergamon Press, 1982.

40   K. Tanaka and S. Ueda, Effect of Spherulite Size on the Friction and Wear of Semicrystalline Polymers, Proc. JSLE Int. Tribology Conf., 8-10 July 1985, Tokyo, Japan, Elsevier, pp. 459-464.

41 T. Tsukizoe and N. Ohmae, Friction and Wear Performance of Unidirectionally Oriented Glass, Carbon, Aramid and Stainless Steel Fibre-Reinforced Plastics, in Friction and Wear of Polymer Composites, editor: K. Friedrich, Amsterdam, Elsevier, 1986, pp. 205-231.

42 K. Friedrich, Wear of Reinforced Polymers by Different Abrasive Counterparts, in Friction and Wear of Polymer Composites, editor: K. Friedrich, Amsterdam, Elsevier, 1986, pp. 233-287.

43 S. Bahadur, Mechanical and Tribological Behavior of Polyester Reinforced with Short Fibres of Carbon and Aramid, *Lubrication Engineering*, Vol. 47, 1991, pp. 661-667.

44 J. Bijwe, C.M. Logani and U.S. Tewari, Influence of Fillers and Fibre Reinforcement on Abrasive Wear Resistance of Some Polymeric Composites, *Wear*, Vol. 138, 1990, pp. 77-92.

45 P.J. Mathias, W. Wu, K.C. Goretta, J.L. Routbort, D.P. Groppi and K.R. Karasek, Solid Particle Erosion of a Graphite-Fibre-Reinforced Bismaleimide Polymer Composite, *Wear*, Vol. 135, 1989, pp. 161-169.

46 A.C.M. Yang, J.E. Ayala, A. Bell and J.C. Scott, Effects of Filler Particles on Abrasive Wear of Elastomer-Based Composites, *Wear*, Vol. 146, 1991, pp. 349-366.

47 D. Gong, Q. Xue and H. Wang, Study of the Wear of Filled Polytetrafluoroethylene, *Wear*, Vol. 134, 1989, pp. 283-295.

48 O. Jacobs, Scanning Electron Microscopy Observation of the Mechanical Decomposition of Carbon Fibres under Wear Loading, *Journal of Materials Science Letters*, Vol. 10, 1991, pp. 838-839.

49 J.K. Lancaster, Composites for Aerospace Dry Bearing Applications, in Friction and Wear of Polymer Composites, editor: K. Friedrich, Amsterdam, Elsevier, 1986, pp. 363-397.

50 W. Liu, C. Huang, L. Gao, J. Wang and H. Dang, Study of the Friction and Wear Properties of MoS_2-Filled Nylon 6, *Wear*, Vol. 151, 1991, pp. 111-118.

51 M. Kano and I. Tanimoto, Wear Resistance Properties of Ceramic Rocker Arm Pads, *Wear*, Vol. 145, 1991, pp. 153-165.

52 B. Bhushan, Tribology Mechanics of Magnetic Storage Devices, Springer Verlag, 1990, Berlin.

53 D.H. Buckley and K. Miyoshi, Friction and Wear of Ceramics, *Wear*, Vol. 100, 1984, pp. 333-353.

54 W. Bundschuh and K.-H. Zum Gahr, Influence of Porosity on Friction and Sliding Wear of Tetragonal Zirconia Polycrystal, *Wear*, Vol. 151, 1991, pp. 175-191.

55 D.C. Cranmer, Ceramic Tribology - Needs and Opportunities, *Tribology Transactions*, Vol. 31, 1988, pp. 164-173.

56 S. Sasaki, The Effects of the Surrounding Atmosphere on the Friction and Wear of Alumina, Zirconia, Silicon Carbide and Silicon Nitride, *Wear*, Vol. 134, 1989, pp. 185-200.

57 T.E. Fischer, M.P. Anderson, S. Jahanmir and R. Salher, Friction and Wear of Tough and Brittle Zirconia in Nitrogen, Air, Water, Hexadecane and Hexadecane Containing Stearic Acid, *Wear*, Vol. 124, 1988, pp. 133-148.

58 C.S. Yust and F.J. Carignan, Observation on the Sliding Wear of Ceramics, *ASLE Transactions*, Vol. 28, 1985, pp. 245-253.

59 M. Woydt and K.-H. Habig, High Temperature Tribology of Ceramics, *Tribology International*, Vol. 22, 1989, pp. 75-87.

60 R.H.J. Hannink, M.J. Murray and H.G. Scott, Friction and Wear of Partially Stabilized Zirconia: Basic Science and Practical Applications, *Wear*, Vol. 100, 1984, pp. 355-356.

61 N. Gane and R. Breadsley, Measurement of the Friction and Wear of PSZ and Other Hard Materials Using a Pin on Disc Machine, Proc. Int. Tribology Conference, Melbourne, The Institution of Engineers, Australia, National Conference Publication No. 87/18, December, 1987, pp. 187-192.

62 K.H. Zum Gahr, Sliding Wear of Ceramic-Ceramic, Ceramic-Steel and Steel-Steel Pairs in Lubricated and Unlubricated Contact, *Wear*, Vol. 133, 1989, pp. 1-22.

63 P.J. Blau, An Observation of the Role Reversal Effects in Unlubricated Sliding Friction and Wear Tests of Alumina and Silicon Carbide, *Wear*, Vol. 151, 1991, pp. 193-197.

64 M.G. Gee, C.S. Matharu, E.A. Almond and T.S. Eyre, The Measurement of Sliding Friction and Wear of Ceramics at High Temperature, *Wear*, Vol. 138, 1990, pp. 169-187.

65 S. Gray, Friction and Wear of Ceramic Pairs Under High Temperature Conditions Representative of Advanced Engine Components, *Cer. Eng. Sci. Proc.*, Vol. 6, 1985, pp. 965-975.

66 V. Aronov, Friction Induced Strengthening Mechanisms of Magnesia Partially Stabilized Zirconia, *Transactions ASME, Journal of Tribology*, Vol. 109, 1987, pp. 531-536.

67   G.W. Stachowiak and G.B. Stachowiak, Unlubricated Wear and Friction of Toughened Zirconia Ceramics at Elevated Temperatures, *Wear*, Vol. 143, 1991, pp. 277-295.

68   Y. Tsuya, Tribology of Ceramics, Proc. JSLE Int. Tribology Conf., 8-10 July 1985, Tokyo, Japan, Elsevier, pp. 641-646.

69   R.S. Gates, S.M. Hsu and E.E. Klaus, Tribochemical Mechanisms of Alumina with Water, *Tribology Transactions*, Vol. 32, 1989, pp. 357-363.

70   H. Tomizawa and T.E. Fischer, Friction and Wear of Silicon Nitride and Silicon Carbide in Water: Hydrodynamic Lubrication at Low Sliding Speed Obtained by Tribochemical Wear, *ASLE Transactions*, Vol. 30, 1987, pp. 41-46.

71   P. Studt, Influence of Lubricating Oil Additives on Friction of Ceramics under Conditions of Boundary Lubrication, *Wear*, Vol. 115, 1987, pp. 185-191.

72   Y. Tsuya, Y. Sakuta, M. Akanuma, T. Murakami, T. Shimauchi, K. Chikugo, Y. Kiuchi, S. Takatsu, Y. Katsumura, M. Fukuhara, K. Umeda, G. Yaguchi, Y. Enomoto, K. Yamanaka, Compatibility of Ceramics with Oils, Proc. JSLE Int. Tribology Conf., 8-10 July 1985, Tokyo, Japan, Elsevier, pp. 167-172.

73   P. Sutor and W. Bryzik, Tribological Systems for High Temperature Diesel Engines, SAE 870157, SP-700.

74   J.J. Habeeb, A.G. Blahey and W.N. Rogers, Wear and Lubrication of Ceramics, Proc. Int. Trib. Conf. on Friction, Lubrication and Wear, Fifty Years On, July 1-3, 1987, Proc. Inst. Mech. Engrs., London, 1987, pp. 555-564.

75   W. Morales and D.H. Buckley, Concentrated Contact Sliding Friction and Wear Behaviour of Several Ceramics Lubricated with a Perfluoropolyalkylether, *Wear*, Vol. 123, 1988, pp. 345-354.

76   A. Erdemir, O.O. Ajayi, G.R. Fenske, R.A. Erck and J.H. Hsieh, The Synergistic Effects of Solid and Liquid Lubrication on the Tribological Behaviour of Transformation-Toughened ZrO_2 Ceramics, *Tribology Transactions*, Vol. 35, 1992, pp. 287-297.

77   E.E. Klaus, J.L. Duda and W.-T. Wu, Lubricated Wear of Silicon Nitride, *Lubrication Engineering*, Vol. 47, 1991, pp. 679-684.

78   P. Studt, Boundary Lubrication: Adsorption of Oil Additives on Steel and Ceramic Surfaces and Its Influence on Friction and Wear, *Tribology International*, Vol. 22, 1989, pp. 111-119.

79   B.G. Bunting, Wear in Dry-Lubricated, Silicon Nitride, Angular-Contact Ball Bearings, *Lubrication Engineering*, Vol. 46, 1990, pp. 745-751.

80   H. Heshmat and J.F. Dill, Traction Characteristics of High-Temperature Powder-Lubricated Ceramics (Si_3N_4/αSiC), *Tribology Transactions*, Vol. 35, 1992, pp. 360-366.

81   J.F. Makki and E.E. Graham, Vapor Phase Deposition on High Temperature Surfaces, *Tribology Transactions*, Vol. 33, 1990, pp. 595-603.

82   J.L. Lauer and S.R. Dwyer, Continuous High Temperature Lubrication of Ceramics by Carbon Generated Catalytically from Hydrocarbon Gases, *Tribology Transactions*, Vol. 33, 1990, pp. 529-534.

83   B. Hanyaloglu and E.E. Graham, Effect of Surface Condition on the Formation of Solid Lubricating Films at High Temperatures, *Tribology Transactions*, Vol. 35, 1992, pp. 77-82.

84   C. DellaCorte and H.E. Sliney, Composition Optimization of Self-Lubricating Chromium-Carbide-Based Composite Coatings for Use to 760°C, *ASLE Transactions*, Vol. 30, 1987, pp. 77-83.

85   W. Yinglong, J. Yuansheng and W. Shizhu, The Friction and Wear Performance of Plasma-Sprayed Ceramic Coatings at High Temperature, *Wear*, Vol. 129, 1989, pp. 223-234.

86   J. Lankford, W. Wei and R. Kossowsky, Friction and Wear Behaviour of Ion Beam Modified Ceramics, *Journal of Mat. Sci.*, Vol. 22, 1987, pp. 2069-2078.

87   A. Erdemir, D.E. Busch, R.A. Erck, G.R. Fenske and R. Lee, Ion-Beam-Assisted Deposition of Silver Films on Zirconia Ceramics for Improved Tribological Behavior, *Lubrication Engineering*, Vol. 47, 1991, pp. 863-872.

88   M. Kohzaki, S. Noda, H. Doi and O. Kamigaito, Tribology of Niobium-Coated SiC Ceramics and the Effects of High Energy Ion Irradiation, *Wear*, Vol. 131, 1989, pp. 341-351.

89   G.R. Fenske, A. Erdemir, R.A. Erck, C.C. Cheng, D.E. Busch, R.H. Lee and F.A. Nichols, Ion-Assisted Deposition of High-Temperature Lubricous Surfaces, *Lubrication Engineering*, Vol. 47, 1991, pp. 104-111.

90   R.S. Bhattacharya, A.K. Rai and V. Aronov, Co-Implantation of Mo and S in Al_2O_3 and ZrO_2 and their Tribological Properties, *Tribology Transactions*, Vol. 34, 1991, pp. 472-477.

91   C.S. Yust, C.J. McHargue and L.A. Harris, Friction and Wear of Ion-Implanted TiB_2, *Mat. Sci. Eng.*, Vol. A105/106, 1988, pp. 489-496.

92   W. Kowbel and A. Sathe, Effect of Boron Ion Implantation on Tribological Properties of CVD Si_3N_4, *Lubrication Engineering*, Vol. 46, 1990, pp. 645-650.

93   A. Gangopadhyay and S. Jahanmir, Friction and Wear Characteristics of Silicon Nitride-Graphite and Alumina-Graphite Composites, *Tribology Transactions*, Vol. 34, 1991, pp. 257-265.

94   T.A. Libsch, P.C. Becker and S.K. Rhee, Friction and Wear of Toughened Ceramics Against Steel, Proc. JSLE Int. Tribology Conf., 8-10 July 1985, Tokyo, Japan, Elsevier, pp. 185-190.

95   P.C. Becker, T.A. Libsch and S.K. Rhee, Wear Mechanisms of Toughened Zirconias, *Cer. Eng. Sci. Proc.*, Vol. 6, 1985, pp. 1040-1058.

96   G.W. Stachowiak, G.B. Stachowiak and A.W. Batchelor, Metallic Film Transfer during Metal-Ceramic Unlubricated Sliding, *Wear*, Vol. 132, 1989, pp. 361-381.

97   Y. Nakamura and S. Hirayama, Wear Tests of Grey Cast Iron Against Ceramics, *Wear*, Vol. 132, 1989, pp. 337-345.

98   G.M. Carter, R.M. Hooper, J.L. Henshall and M.O. Guillou, Friction of Metal Sliders on Toughened Zirconia Ceramic Between 298 and 973 K, *Wear*, Vol. 148, 1991, pp. 147-160.

99   C.V. Cooper, C.L. Rollend and D.H. Krouse, The Unlubricated Sliding Wear Behavior of a Wrought Cobalt-Chromium Alloy Against Monolithic Ceramic Counterfaces, *Transactions ASME, Journal of Tribology*, Vol. 111, 1989, pp. 668-674.

100   K. Miyoshi, Fundamental Considerations in Adhesion, Friction and Wear for Ceramic-Metal Contacts, *Wear*, Vol. 141, 1990, pp. 35-44.

101   S.K. Bhattacharya, E.O. Ezugwu and A. Jawaid, The Performance of Ceramic Tool Materials for the Machining of Cast Iron, *Wear*, Vol. 135, 1989, pp. 147-159.

102   K. Demizu, R. Wadabayashi and H. Ishigaki, Dry Friction of Oxide Ceramics against Metals: the Effect of Humidity, *Tribology Transactions*, Vol. 33, 1990, pp. 505-510.

103   M. Semlitsch, M. Lehmann, H. Webber, E. Doerre and H.G. Willert, New Prospects for a Prolonged Functional Life-Span of Artificial Hip Joints by Using the Material Combination Polyethylene/Aluminium Oxide Ceramic/Metal, *Journal of Biomedical Materials Research*, Vol. 11, 1977, pp. 537-552.

104   D. Dowson and P.T. Harding, The Wear Characteristics of UHMWPE Against a High Density Alumina Ceramic under Wet and Dry Conditions, *Wear*, Vol. 75, 1982, pp. 313-331.

105   A. Ben Abdallah and D. Treheux, Friction and Wear of Ultrahigh Molecular Weight Polyethylene Against Various New Ceramics, *Wear*, Vol. 142, 1991, pp. 43-56.

106   P. Kumar, M. Oka, K. Ikeuchi, K. Shimizu, T. Yamamuro, H. Okumura and Y. Kotoura, Low Wear Rate of UHMWPE Against Zirconia Ceramic (Y-TZP) in Comparison to Alumina Ceramic and SUS 316L Alloy, *Journal of Biomedical Materials Research*, Vol. 25, 1991, 813-828.

107   H. McKellop, I. Clarke, K. Markolf and H. Amstutz, Friction and Wear Properties of Polymer, Metal and Ceramic Prosthetic Joint Materials Evaluated on a Multichannel Screening Device, *Journal of Biomedical Materials Research*, Vol. 15, 1981, pp. 616-653.

108   E. Minford and K. Prewo, Friction and Wear of Graphite-Fiber-Reinforced Glass Matrix Composites, *Wear*, Vol. 102, 1985, pp. 253-264.

109   C.S. Yust, J.M. Leitnaker and C.E. Devore, Wear of an Alumina-Silicon Carbide Whisker Composite, *Wear*, Vol. 122, 1988, pp. 151-164.

110   C.S. Yust and L.F. Allard, Wear Characteristics of an Alumina-Silicon Carbide Whisker Composite at Temperatures to 800°C in Air, *Tribology Transactions*, Vol. 32, 1989, pp. 331-338.

111   D. Holz, R. Janssen, K. Friedrich and N. Claussen, Abrasive Wear of Ceramic-Matrix Composites, *J. Eur. Ceram. Soc.*, Vol. 5, 1989, pp. 229-232.

112   A.W. Batchelor and B.P. Tan, Effect of an Oxidizing Agent on the Friction and Wear of Nylon 6 Against a Steel Counterface, Proceedings of the 4th International Tribology Conference, AUSTRIB '94, 'Frontiers in Tribology', 5-8th December 1994, Volume I, (editor: G.W. Stachowiak), publ. Uniprint UWA, 1994, pp. 175-180.

113   N.W. Scott and G.W. Stachowiak, Long-Term Behaviour of UHMWPE in Hydrogen Peroxide Solutions, Proceedings of the 4th International Tribology Conference, AUSTRIB '94, 'Frontiers in Tribology', 5-8th December 1994, Volume I, (editor: G.W. Stachowiak), publ. Uniprint UWA, 1994, pp. 169-174.

114   M. Chandrasekaran, L.Y. Wei, K.K. Venkateshwaran, A.W. Batchelor and N.L. Loh, Tribology of UHMWPE Tested Against a Stainless Steel Counterface in Unidirectional Sliding in Presence of Model Synovial Fluid: Part 1, *Wear*, Vol. 223, 1998, pp. 13-21.

115 J. DeGaspari, Standing up to the Test, *Mechanical Engineering Magazine (ASME)*, Vol. 121, No. 8, 1999, pp. 69-70.

116 J.F. Archard, Wear Theory and Mechanisms, Wear Control Handbook, ASME, editors: M.B. Peterson and W.O. Winer, New York, 1980.

117 Y. Wang and S.M. Hsu, Wear and Wear Transition Modeling of Ceramics, *Wear*, Vol. 195, 1996, pp. 35-46.

118 Y. Wang and S.M. Hsu, Wear and Wear Transition Mechanisms of Ceramics, *Wear*, Vol. 195, 1996, pp. 112-122.

119 Y.S. Wang, S.M. Hsu and R.G. Munro, Ceramics Wear Maps: Alumina, *Lubrication Engineering*, Vol. 47, 1991, pp. 63-69.

120 X. Dong and S. Jahanmir, Wear Transition Diagram for Silicon Nitride, *Wear*, Vol. 165, 1993, pp. 169-180.

121 A. Blomberg, M. Olson and S. Hogmark, Wear Mechanisms and Tribo Mapping of Al_2O_3 and SiC in Dry Sliding, *Wear*, Vol. 171, 1994, pp. 77-89.

122 Y. Yamamoto and T. Takashima, Friction and Wear of Water Lubricated PEEK and PPS Sliding Contacts, *Wear*, Vol. 253, 2002, pp. 820-826.

123 J. Khedkar, I. Negulescu and E.I. Meletis, Sliding Wear of PTFE Composites, *Wear*, Vol. 252, 2002, pp. 361-369.

124 M. Gouider, Y. Berthier, P. Jaquemard, B. Rousseau, S. Bonnamy, H. Estrade-Szwarckopf, Mass Spectrometry During C/C Composite Friction: Carbon Oxidation Associated with High Friction Coefficient and High Wear Rate, *Wear*, Vol. 256, 2004, pp. 1082-1087.

125 M.Q. Zhang, M.Z. Rong, S.L. Yu, B. Wetzel and K. Friedrich, Effect of Particle Surface Treatment on the Tribological Performance of Epoxy Based Nanocomposites, *Wear*, Vol. 253, 2002, pp. 1086-1093.

126 Q. Wang, J. Xu, W. Shen and W. Liu, An Investigation of the Friction and Wear Properties of Nanometer Si_3N_4 Filled PEEK, *Wear*, Vol. 196, 1996, pp. 82-86.

127 Q. Wang, J. Xu and W. Shen, The Friction and Wear Properties of Nanometer SiO_2 Filled Polyetheretherketone, *Tribology International*, Vol. 30, No. 3, 1997, pp. 193-197.

128 W.G. Sawyer, K.D. Freudenberg, P. Bhimaraj and L.S. Schadler, A Study on the Friction and Wear of PTFE Filled with Alumina Nanoparticles, *Wear*, Vol. 254, 2003, pp. 573-580.

129 A.O. Pozdnyakov, V.V. Kudryavtsev and K. Friedrich, Sliding Wear of Polyimide-C_{60} Composite Coatings, *Wear*, Vol. 254, 2003, pp. 501-513.

130 M. Kalin, S. Jahanmir and L.K. Ives, Effect of Counterface Roughness on Abrasive Wear of Hydroxyapatite, *Wear*, Vol. 252, 2002, pp. 679-885.

131 H.S.C. Metselaar, B. Kerkwijk, E.J. Mulder, H. Verweij and D.J. Schipper, Wear of Ceramics Due to Thermal Stress: A Thermal Severity Parameter, *Wear*, Vol. 249, 2002, pp. 962-970.

132 K. Kato and K. Adachi, Wear of Advanced Ceramics, *Wear*, Vol. 253, 2002, pp. 1097-1104.

133 S.M. Hsu and M. Shen, Wear Prediction of Ceramics, *Wear*, Vol. 256, 2004, pp. 867-878.

134 Y. Morita, K. Nakata and K. Ikeuchi, Wear Properties of Zirconia/Alumina Combination for Joint Prostheses, *Wear*, Vol. 254, 2003, pp. 147-153.

135 M. Kalin, S. Novak and J. Vizintin, Wear and Friction Behavior of Alumina Ceramics in Aqueous Solutions with Different pH, *Wear*, Vol. 254, 2003, pp. 1141-1146.

136 P. Cong, K. Kobayashi, T. Li and S. Mori, The Role of Products of Tribochemical Reactions on the Friction and Wear Properties of Alumina/Silicon Carbide Couple Sliding in HFC-134a Gas, *Wear*, Vol. 252, 2002, pp. 467-474.

137 P. Cong, J. Imai and S. Mori, Effect of Gas Pressure on Tribological Properties and Tribochemical Reactions of Alumina Sliding Against Zirconia in HFC-134a, *Wear*, Vol. 249, 2001, pp. 143-149.

FUTURE DIRECTIONS IN TRIBOLOGY

17.1 INTRODUCTION

Tribology has been described as a young science with fundamentally new forms of tribology yet to be devised [1]. Since its birth in the 1960s the tribology has primarily been focused on solving the immediate industrial problems with wear and friction through the application of better materials, novel surface technologies, improved lubricants and lubrication methods. Once these immediate problems have been solved the research focus has shifted to the newly emerging areas like nanotechnology and biotribology. A new branch of nanotechnology, nanotribology, is now rapidly developing [2]. Nanotribology, as its name suggests, is the study of tribology in minute contacts such as those used in nanotechnology. The increasing pace of research and development generates a need to almost predict in advance what these developments will be. Will there be major advances in our fundamental knowledge of tribology? Will the range of application of tribology be significantly extended? In this chapter, some clues to future developments, based on currently available information, are briefly presented.

17.2 BIOTRIBOLOGY

The term biotribology is widely used to refer to tribological phenomena occurring in either the human body or in animals and possibly plants. There are two distinct themes in biotribology:

- tribological processes naturally occurring in or on the tissues and organs of animals, and
- tribological processes that may occur after implantation of an artificial device in the human body.

An example of the former is the wear of skin and its replenishment by new skin cells or the lubricated sliding of eyelids over the eye. An example of the latter is the wear of orthopaedic implants (artificial hips and knees), which releases alien debris into the body.

Biotribology of Living Tissues and Organisms

Animals, including humans, possess a wide variety of sliding and frictional interfaces. One of the most important interfaces is the contact between skin and external objects where the body secrets an oily substance, known as sebum, on the skin. The sebum reduces friction and wear of the skin when, for example, a hard rough object is gripped by a hand or during locomotion (to protect the feet). Studies have shown that when sweat inhibiting drugs are administered to rabbits, the rabbits lose traction on their feet because of reduced friction. It is believed that sweat washes away some of the sebum. Conversely if the sebum is effectively removed with organic solvents, the act of gripping causes a painful sticking of the hands to the gripped object.

Inside the body, lubrication is essential in the lungs so that the bronchioles can freely dilate and contract during respiration (breathing). A lubricating film present on the serous membrane found on the exterior of the lungs is vital to allow the lungs to move within the ribcage. Disruption of the serous membrane by pleuritis results in considerable pain for the sufferer. Other organs found within the visceral cavity such as the liver, intestines, etc., are also separated by lubricating films. During delivery of a baby, a white flaky layer, known as vernix caseosa, covers the baby to help reduce the friction between the skin of the baby and the birth canal. During the pregnancy, the vernix caseosa also helps protect the foetus from accumulated toxins in the amniotic fluid.

Human movements such as walking, running, jumping, flexing of the limbs and back, gripping by the hands all require articulating joints. Articulating joints consist of a synovial joint where closely conformal cartilaginous surfaces slide past each other. The cartilage is immersed in synovial fluid and a synovial membrane encloses the articulating joint. Close to the sliding interface, the cellular content of the tissue is very low and most of the cartilage consists of extra-cellular matrix. This is to prevent damage to living cells by sliding contact. Most synovial joints enjoy very low levels of friction and wear, unless illness or injury occurs. The most common problem with synovial joints is arthritis, which curtails movement by swelling of the joints and acute pain. The principal forms of arthritis are osteoarthritis and rheumatoid arthritis. Osteoarthritis is closely related to mechanical damage to the joint, by overload (excessive loading of the articulating surfaces), by injury (e.g., sport injury is a common precursor of osteoarthritis) or by lubrication failure. Rheumatoid arthritis occurs when the body's immune system is induced to attack the synovial joints, in particular the articular cartilage. Once the cells of the immune system have begun degradation of the synovial joints, additional mechanical damage may occur during even normal levels of human movement and a self-perpetuating damage process then ensues.

A key feature of tissues is their very low elastic modulus compared to most artificial materials. Even stiff tissues like bone have elastic moduli that are significantly lower than those of ferrous and non-ferrous alloys. The elastic modulus of low stiffness tissues such as membranes and cartilage is measured in [MPa]. The classical model of elastohydrodynamic lubrication is unlikely to be applicable to natural sliding contacts and instead low contact stresses with a high level of elastic or visco-elastic compliance would be more probable. Sliding speeds are also much lower, typically 0.1 [m/s] or less. This minimizes frictional heating which would denature proteins in the body fluids.

It is reasonable to surmise that any living tissue needs protection of high levels of friction and wear by an evolved lubricating film or system. In healthy synovial joints, a range of lubrication mechanisms are in operation; some of these correspond to the classic lubrication mechanisms found in machines, while other lubrication mechanisms are specific to living organisms. The mechanisms of synovial lubrication are not yet fully understood. Hydrodynamic lubrication is effective in healthy synovial joints where the synovial fluid has sufficient viscosity and the contact speeds are sufficient for the hydrodynamic film to form. In rheumatoid joints, the viscosity of synovial fluid is greatly reduced. The cartilage is porous and an ionic mechanism inside the cartilage prevents leakage of synovial fluid into the cartilage. The cartilage or cells beneath the cartilage secrete phospholipids to cover the surface with a lubricating film. These films are thought to act as boundary lubricating films in moving joints, providing a low coefficient of friction and protecting the articular cartilage against damage at low velocities. Experiments conducted with sheep synovial joints have shown that once these phospholipid films are removed from the articulating surfaces the wear of the synovial joints is rapidly accelerated [3]. The subsequent changes to the cartilage morphology and morphologies of wear particles are similar, if not the same, to those occurring in the osteoarthritic human joints [4-6].

In common with other body fluids, the synovial fluid contains proteins. Proteins are high molecular weight substances composed of amino acids. It is known that synovial proteins are effective in controlling friction and wear, yet the tribology is poorly understood. Proteins are entirely different from the much smaller molecules typical of lubricant additives. Classical theory of adsorption films, described in Chapter 8, is based on fatty acids with chain lengths of 16 to 20 carbon atoms with a well-defined atomic structure. Proteins by contrast have molecular weights reaching a million or more and have a molecular chain long enough to become tangled. Proteins attach to a surface by multi-site adsorption at many different positions along the molecular chain, as schematically illustrated in Figure 17.1. However, how such adsorption facilitates lubrication still remains unclear.

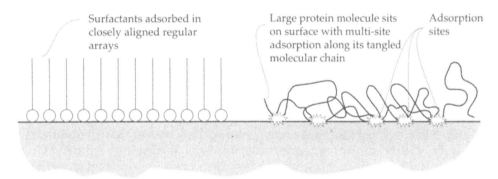

Figure 17.1   Schematic illustration of the difference between adsorption by conventional additives such as fatty acids and proteins.

It can be seen from Figure 17.1 that whereas additives are modelled as forming arrays of closely packed molecules on a surface, the much larger protein molecule sits on a surface with adsorption occurring at various positions along its molecular chain. It is also believed that non-proteins such as hyaluronic acid interact with synovial proteins, in particular, to reinforce the lubricating film. There is also evidence that material-specific chemical bonding between the hyaluronic acid and the substrate is a pre-requisite for lubrication [7]. The synovial proteins also influence the rheology of the synovial fluid to possibly enhance the hydrodynamic lubrication or EHL. A mysterious protein termed 'lubricin' has been cited as the main agent of synovial lubrication but its lubricating mechanism remains unclear [8]. When the various lubrication mechanisms of synovial fluid and synovial cartilage become impaired or fail, it is found that synovial cartilage cannot slide on itself without extremely high levels of friction and wear [3,4,9,10]. Another specialized series of proteins known as mucins provide not only lubrication but also exhibit stickiness, a useful property to trap bacteria and dust in the respiratory tract. The lubrication mechanism of mucins, like other human proteins, is still not fully understood. A detailed discussion of models of lubrication by proteins adsorbed on metallic surfaces can be found elsewhere [11].

Current findings from the research on polymers and metals in wet sliding tests suggest that adsorbed films of proteins tend to raise friction coefficients while lowering wear rates. This is probably because the thick adsorbed films of tangled protein molecules shield the surface while the protein molecules from opposing surfaces become entangled. This concept is shown schematically in Figure 17.2.

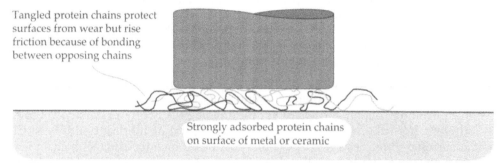

Tangled protein chains protect surfaces from wear but rise friction because of bonding between opposing chains

Strongly adsorbed protein chains on surface of metal or ceramic

Figure 17.2   Schematic illustration of the effect of adsorbed films of proteins on friction and wear.

Biotribology of Artificial Materials in Close Contact with Living Tissues

The widespread prevalence of osteoarthritis has led to the development of orthopaedic prostheses, e.g., the 'artificial hip' and 'artificial knee'. These devices provide pain relief and improved mobility to a huge number of arthritis sufferers. The 'artificial hip' does, however, present some unique tribological problems. Dental restorative materials, which are fitted to an even greater number of patients, have also their own wear problems.

Implantation of, for example, an 'artificial hip' places human or animal tissue with entirely alien materials. The environment of the body is typically saline (aqueous) solution with proteins and other biochemical substances present. This means that the aqueous form of corrosive wear will readily occur unless the material is carefully chosen. Orthopaedic prostheses are typically made of strong, non-toxic, corrosion resistant materials such as titanium and cobalt alloys or stainless steel for bone plates. Engineering ceramics, e.g., alumina, toughened zirconia, zirconia toughened alumina (ZTA) and wear resistant polymers, in most cases ultra-high molecular weight polyethylene (UHMWPE), are widely used in artificial implants [12]. In the hip replacement a common combination is a metal stem and ball sliding against a cup of UHMWPE. However, metal-on-metal joints are also used because of the reduced volume of wear debris. Ceramic materials, such as alumina (aluminium oxide), display very little wear as an orthopaedic implant [13]. These materials, due to their high hardness, are very brittle thus even a single scratch on the implant's surface can significantly affect its performance. Micro-separation of orthopaedic prostheses during normal motion is a concern for ceramic implants. Micro-separation is the detachment of the ball from the socket of the prostheses, which occurs when the patient's ligaments are not strong enough to main a positive closing force on the prosthesis. The impact of renewed contact between the ball and the socket may cause fracture or micro-fracture of the ceramic components [13]. The synovial fluid, especially its constituent proteins, influences the wear and friction of the implant materials significantly. In approximate terms (since the tribology is still poorly understood), the synovial proteins rapidly adsorb on the metal (or metal oxide) surfaces, forming a tenacious but 'sticky' film since there is no film of the neatly formed 'carpet' of aligned molecules found in classical boundary lubrication. These protein films formed may protect against wear but usually raise the friction coefficient considerably. A more detailed discussion on this topic can be found elsewhere [14].

Failure of orthopaedic prostheses is unlikely to be caused by gross wear of the implanted components. Apart from the ever-present risk of infection during implantation surgery, a common problem is the bodily reaction to wear debris. The wear debris, which becomes trapped in adjacent tissue, is found to be typical of low speed sliding between the prosthesis components with little modification in size or shape by chemical reagents in the body fluids and tissues [15]. New designs of orthopaedic implants now incorporate features such as a mobile bearing. In this new design a high density polyethylene insert is placed between the convex articulating surface and plate that are fitted to opposing bones, femoral (thigh) and tibial (shin bone), respectively, of the knee joint. The articulation takes place between the convex surfaces of the femoral component and the concave surface of the polyethylene insert. The polyethylene insert either can be rigidly fixed to the tibial component (fixed bearing prosthesis) or can be attached to the tibial component in a way that allows the polyethylene spacer to rotate together with the femoral portion of the prosthesis. This mobile bearing is believed to improve the conformity of the contacting surfaces, thereby reducing contact stress and minimizing the release of wear debris [16]. The sensitivity of the body to the minute wear debris produced by implants is a common cause of implant failure. A significant fraction of the released wear debris is of the right size to be ingested by cells from the immune system. These immune system cells gather around the implanted prosthesis and while attempting to ingest and destroy the wear debris they inflame the tissue around the joint. This leads to problems such as pain and destruction of bone (osteolysis). The finely divided metallic wear debris is also a source of metal ions of chromium and aluminium (both toxic) and titanium, vanadium and cobalt, which are only present in minute quantities in healthy tissues. The implications of metal ion release towards long-term health remain an issue of concern to health workers. In particular the ionic solubilized form of the metal is of concern, as these ions could modify the composition of the body fluids. For example, as long as the chromium metal remains in the solid state, insoluble to water, it is virtually harmless. With stainless steel cutlery and plates, almost all of the chromium remains in the solid state, either as metal or as oxide, so it is safe to use cutlery and plates made of stainless steel.

Dental restorative materials and tooth crowns also contain wear interfaces. Moderately high stress abrasive wear occurs during mastication of food or grinding of teeth and saliva not only lubricates the teeth but also contains calcium and other mineral ions that continuously restore the teeth. Dental materials for restoration and crowns should be sufficiently hard and robust to withstand repeated mastication and possibly bruxism (grinding of teeth during sleep) but should not be excessively hard. Very hard materials are found to cause excessive wear of the opposing teeth. This last condition greatly restricts the range of materials that can be used. A more detailed discussion on this topic is provided elsewhere [14].

Vascular prostheses such as artificial heart valves pose a rather unusual wear requirement. The critical requirement is that the size of the wear particles produced during valve operation should be less than the internal diameter of the smallest capillary. If the wear particle blocks a capillary then the adjacent tissue would be deprived of blood and may die. Experimental testing reveals that the most wear-resistant polymer, UHMWPE, produces excessively large wear particles. Thus faster-wearing materials, which produce smaller wear particles, for example, pyrolytic carbon, are chosen instead for heart valves.

17.3 ENVIRONMENTAL IMPLICATIONS OF TRIBOLOGY

There are two basic themes in environmental aspects of tribology:

· application of tribology to minimize wear of materials and energy consumption, and

· side effects or undesirable consequences of tribological practices.

The first theme was discussed in detail in the preceding chapters. The second theme has become a topical issue as concerns about environmental pollution become more acute.

A major issue is the disposal of used lubricating oils, since these are usually toxic and present in large volumes. There are about 40 millions tonnes of oils being produced annually worldwide. This is equivalent to a lake 4 [km] long, 500 [m] wide and 22 [m] deep. Some proportion of the used oil is reprocessed but most of it is disposed back into the environment. An accepted practice in arid, remote areas is to use the waste oil as a binder to form dirt roads. In populated high rain-fall areas, this practice obviously has its limitations. The problems associated with the used lubricants disposal or re-cycling are serious and hence there are now major research efforts focused on developing biodegradable lubricating oils, which can be harmlessly decomposed by bacteria and fungi after use. This research inevitably leads to natural oils such as vegetable oils, e.g., canola, sunflower, soybean, rapeseed, palm oil and others, since these oils are already present or can be grown in large quantities without ill effects. While natural oils can offer good lubricity and wear control and were once used as lubricants, their durability or resistance to oxidation, i.e., oxidation stability, remains unsatisfactory. This is the main reason for their still limited use. For example, rapeseed and olive oils used as lubricants in the past were replaced by mineral oils offering better and longer performance at elevated temperatures.

The use of lubricants in combustion engines and industrial machinery and their storage in tanks cause major pollution problems. The burning of lubricants, laced with additives containing sulphur, phosphorus and many other often toxic compounds, together with fuel in internal combustion engines is the direct cause of the acid rain and pollution in most of the cities across the globe. The storage tanks containing often more than 50,000 [m³] of oil can corrode at the base and leak the oil to the ground, thus polluting the ground water, as illustrated in Figure 17.3.

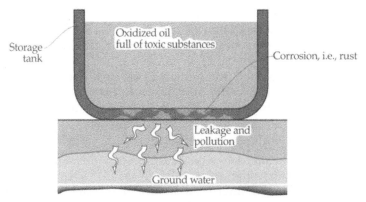

Figure 17.3   Pollution of ground waters by rusting oil storage tanks.

The oil additives in mineral oils, e.g., zinc dialkyldithiophosphate (ZnDDP), are the major contributors to the pollution problems caused by engine emission. They are also a source of toxicity. It might be possible to replace those additives with biodegradable compounds and possible replacements are being sought. Recently a possible replacement candidate for ZnDDP was found, i.e., organic borates. These compounds display good anti-oxidant and wear protection characteristics. However, these new additives still have some limitations which need to be resolved first before their wider applications [17].

Pollution caused by mobile earthmoving equipment such as bulldozers is of considerable concern. Bulldozers and excavators are often used in natural, possibly pristine environments, yet they continuously leak hydraulic fluid during operation [18]. It is thus vitally important that the hydraulic fluid is biodegradable and non-toxic or else very widespread pollution may occur, destroying the fragile environment. Thus international standards on hydraulic fluids are currently being discussed [18].

Research has also been conducted on the effectiveness of conventionally used additives such as ZnDDP when blended with the vegetable oils. The response of natural oils to anti-wear additives such as ZnDDP varies significantly with species of oil-plant, e.g., sunflower, soybean or rapeseed. It has been found that sunflower oil shows more improvement in wear protection with either zinc bis(O,O-diisopropyldithiophosphate) (ZnDTP) or dibutyl phosphonate (DBPo) than other oils such as soybean. As mentioned already natural oils are also vulnerable to oxidation and as oxidation progresses, the effectiveness of ZnDDP as an anti-wear additive declines. Such oxidation also occurs during storage and sunflower oil appears to show the greatest resistance to oxidation during storage [19].

Biodegradable lubricants from biological sources can have dual purpose, i.e., apart from performing as an ordinary lubricant they can also be used as a food supplement. The idea is not new. The railways in 19th century Russia used lard as a lubricant to lubricate the axle bearings. The lard was blended eventually with inedible soot in order to prevent starving peasants from eating the lard [20]. The soot was accidentally found to act as a solid lubricant and further reduced wear. While this practice of eating the lubricant may seem rather extreme, it may be feasible to provide edible lubricants as a source of emergency food, for example, for astronauts on future long missions to the planet Mars and beyond. Some natural oils, especially palm oil, are rich in vitamin E. An edible lubricant could also help prevent illnesses arising from vitamin E deficiency, such as lowered resistance to common infections, if the supply of dark leafy vegetables or nuts on the spaceship becomes depleted.

Owners and operators of ships face a perennial problem of oil leakage into the sea. With the rapid increase in world trade and a corresponding rise in the number of cargo ships, lubricant release at sea has to be carefully managed. Biodegradable lubricants are now available for stern-tube lubricants [21]. The stern-tube is the assembly of seals and bearings where the propeller shaft emerges from the hull.

17.4 NANOTRIBOLOGY - BASIC CONCEPTS

In current technical literature, nanotechnology and its specialist offshoot nanotribology have become topical subjects of discussion. Currently a major technological challenge is to produce machines that are minute as opposed to building enormous machines such as a two-deck airliner. 'Nano' is an adaptation of the SI system unit, 'nanometre', used to denote 10^{-9} [m]. As it happens the nanometre is comparable in size to an individual atom and hence the nanotechnology increasingly relates to devices and machines not much larger than a few hundred atom diameters in length, i.e., it often directly involves atoms and molecules. Such machines necessarily contain sliding contacts and these contacts are only of the size of a few atom diameters or possibly less. The direct application of atoms and molecules is a radically different concept from traditional technology which is based on a continuum model of materials. Under these conditions therefore it is no longer possible to apply continuum mechanics to solid contact, instead the contacting solids must be modelled as what they really are, a matrix of bonded atoms.

There is a scientific awareness of atoms through concepts such as the crystalline lattice and dislocations, but the individual atoms within the lattice are even more anonymous than a single house in a great metropolis. In nanotechnology, it may be pertinent to give individual atoms names or grid references to facilitate machine function. The properties of individual atoms are usually quite different from the bulk material, with the transition between bulk properties and nanoscale properties located around 10 [nm] film thickness [22], thus providing many interesting avenues for the developments in nanotechnology.

A major application of nanotechnology is in the development of computer disk drives with higher recording densities. Recording density is inversely proportional to film thickness between the recording head and the memory disk. In computer jargon, the film thickness between head and disk is known as the 'flying height' and is traditionally achieved by aerodynamic lubrication. Nanotribology is now being investigated and applied to allow film thicknesses as small as a few nanometres instead of the conventional 0.1 [μm] or 100 [nm] of air films [23] as this dramatically increases the recording densities. It would appear that such thin film thicknesses can only be achieved with monomolecular films.

After more than two thousand years, during which time most engineers have vigorously pursued the continuum theories of matter, the theories of Democritus about the fundamental particles of matter, i.e., atoms, are now beginning to be fully exploited in engineering. Nanotribology has its roots in earlier work, for example, the studies performed by Buckley of contact adhesion under high vacuum where a very fine stylus was used [24]. In retrospect, this can be regarded as an early form of nanotribology.

Relevance to Tribology

In common with most other disciplines within engineering science, tribology has traditionally invoked a continuum model of materials for all but a few exceptions such as stacking fault energy and void coalescence in delamination wear. This has been a reasonable assumption given the large size of most engineering contacts, when compared to atomic dimensions. Now, when the contact size is dramatically reduced, basic concepts in tribology are in need of revision. For example, how can there be wear particles when the size of the contact is much less than the average diameter of most known wear particles? Is plastic deformation possible when the contact diameter is less than the minimum spacing of dislocations? In 'nanotribology', as this subject area is now called, sliding ceases to be an example of perfectly linear motion, instead climb and descent over individual atoms become important. Nanotribology is still at an experimental stage, but it is already apparent that relative motion between surfaces is only accomplished by cyclic approach and separation between individual atoms of opposing surfaces. The scanning force microscope (SFM) and the scanning tunnelling microscope (STM) have become vital experimental tools in nanotribology. The wide range of experimental tools used in nanotribology is comprehensively reviewed elsewhere [2,25,26].

When using either the SFM or the STM, a very fine sharp stylus is drawn over a surface. As the tip of the stylus passes over the surface it registers contact with specific atoms revealing periodic variations in friction force and strong directional effects in friction. A schematic model of contact between a hypothetical sliding atom and an opposing surface, also made of atoms, is shown in Figure 17.4. The sliding atom could be the apex of a perfectly sharp stylus. As the sliding atom passes over the peak of each opposing atom and then dips into the space between atoms, the resultant friction force will also change. Thus the friction force will vary periodically as the sliding atom moves.

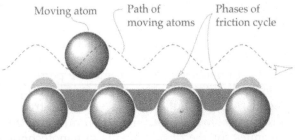

Figure 17.4 Simplified model of an atom sliding over another atomic surface.

Although the research on 'atomistic' solid contact is still in its infancy it is apparent that friction is controlled by the step-like motion of atoms from opposing surfaces as they slide past each other. Application of external exciting vibrations is found to reduce stick-slip friction to marginal levels, presumably when the frequency of vibrations coincides with the frequency of the atom-jumps. Conversely when the frequency of the external vibrations does not coincide with the atom-jump frequency then friction force increases [27].

Atomistic (i.e., involving explicit models of atoms) molecular dynamics simulations have been used to model the sliding contact. However, with current limitations of computing power, the sliding distance and sliding time are confined to a fraction of a micrometre and a microsecond, respectively. This is insufficient for anything more than a rudimentary analysis of the phenomena [27]. It is anticipated, however, that with increased computer power these studies, at a larger scale, will become possible in the near future.

So far most of the work conducted on nanoscale has been related to friction while the wear studies remain largely for the future. It has been stated that only zero or negligible wear is permissible in a micro-machine or else the sliding components would seize [28]. However, the answer to the question of wear particles and their morphology in minute contacts still remains largely unknown.

Fundamental principles of tribology such as Amonton's law are now being analysed in detail by computational models, i.e., the forces and energy flows on each atom within a sliding contact are being computed. A basic finding of this research is that due to the very limited range of interatomic forces, what may be called a 'fuzzy' or intermediate zone, where the opposing surfaces progressively develop contact forces to the maximum value, is extremely narrow. This means that even a surface protruberance, for example, a roughness feature of perhaps one or two atoms height, is sufficiently high to disrupt the 'fuzzy zone' for the rest of an otherwise smooth surface. The protruberance mechanically prevents the opposing surfaces from bringing their fuzzy zones into contact. This means that for practical purposes, all surfaces are functionally rough [29].

A limited amount of studies on nanoscale wear have also been performed. In one of the earlier, pioneering works a model of phonon exchange has revealed a strong velocity dependence on wear between sliding dry metals [30]. Detailed simulations, where a large number of atoms in and around the sliding contact were modelled, revealed that a sliding contact not only disturbs atoms close to the contact area but also those remote from the contact area [31].

Nanolubrication and Specialized Materials for Nanotribology

Nanotribology necessarily entails nanolubrication, since the reliable operation of minute contacts requires friction control just like larger contacts. A major difficulty in nanotribology is determining the level of contact force, since small scale effects such as electrostatic attraction can easily increase the contact force to many times the nominal contact load. Surface coatings and surface coatings technology is responding to this challenge but traditional lubrication is being re-fashioned to the needs of nanotechnology. Two basic methods of lubrication are now being developed:

·   monomolecular films of lubricants with very low vapour pressure (non-volatile), and

·   vapour lubrication by, e.g., tricresylphosphate (TCP).

This latter form of lubrication, also mentioned in Chapter 16, is called 'vapour phase' lubrication and involves:

·   volatilization of a thermally stable lubricant,

·   its diffusion or convected flow to the wearing contact and

·   its condensation to form a useful lubricant film [32].

Vapour phase lubrication appears mostly suited for a slightly larger scale of machine, i.e., micro-electromechanical systems (MEMS), which are typically sized in micrometres. In contrast, for durable monomolecular films, volatility has to be minimized and the remarkably non-volatile perfluoropolyethers are used as lubricants [23]. The monomolecular films are used in systems where maintenance or lubricant replenishment is highly undesirable such as computer disk drives and vapour lubrication is probably more suited to heavily loaded contacts requiring frequent lubricant replenishment.

Specialized materials, for example, buckminster fullerene, have been developed for nanotribological applications. Buckminster fullerene is a recently discovered molecular form of carbon. The unique feature of fullerene is that the carbon molecules can form extremely small spheres (buckyballs) and shafts (nanotubes). Other more common forms of carbon, such as graphite and diamond, can only form lamellae and tetrahedra, respectively. Although it was thought that it might be possible to develop nanoscopic bearings and rotating shafts with fullerene [33] the profitable realization of this concept still remains largely elusive. In theory, atomic repulsion between the opposing carbon atoms should lead to negligible or zero friction between the sliding surfaces. However, like many theories, hitherto neglected secondary effects obstruct the practical development of such bearings [34].

Nanolubrication represents a break from the tradition of accepting the lubrication processes facilitated by the use of natural or processed mineral lubricants. In nanolubrication, the adsorbed film is deliberately engineered to fulfil its function by varying parameters such as chain length and cross-linking [35]. This analytical approach to tribology can perhaps be said to best represent the new tribology, which might form a topic for future editions of this book.

Nanotribology is inevitably linked to the device miniaturization, i.e., one of the frontier technologies of the 21st century. Introduction of these technologies may change the ways in which people and machines interact with the physical world. Device miniaturization involves the manufacturing of the micro scale mechanical components, i.e., of the size ranging from a few to a few hundred micrometres, with high tolerances using a broad range of engineering materials. These devices find applications in medicine, biotechnology, optics, electronics, aviation and many others. Mechanisms of material removal at nano/micro scale, especially in ductile materials, are becoming vital in the development of nano/micro grinding technologies. These technologies are needed in the production of, for example, micro lenses with good surface finish for miniaturized endoscopes. An example of the aspherical tungsten carbide mould insert manufactured using nano/micro grinding is shown in Figure 17.5 [36].

17.5 SUMMARY

The content and scope of tribology are rapidly changing. It is no longer sufficient to minimize friction and wear in isolation, instead it is necessary to consider the wider implications of tribology as it is practised. Environmental tribology is now developing in response to this external view of tribology and it is suggested that some reductions in, for example, pollution may occur as a result. After a period of obsolescence, natural oils are now being studied as environmentally sensitive replacements for mineral oils.

Tribology has always closely followed developments in physics and the new move towards microscopic or even 'nanoscopic' machinery is now leading to fundamental changes in how we view friction, wear and sliding motion. Experiments conducted at a nanolevel have already led to a better understanding of the origins and validity of Amonton's law, which was never proven theoretically [29,37]. At the nanoscale, surface tension effects may cause significant deviations from a simple proportionality between load and friction force. A fundamental question is the definition of the load, i.e., the contact force or the externally imposed force, as this would affect the validity and definition of Amonton's law.

Despite several decades of intense research our understanding of tribology is still far from complete, with most knowledge directed towards practical systems such as steel lubricated by oil. New materials and novel operating conditions of emerging nano-technologies would require a much better understanding of the fundamentals of tribology.

Beyond artificial materials, many interesting and practically important problems have been found in the sliding interfaces of living organisms. These problems range from excessively slippery feet to minimizing the escape of wear debris from a medical implant, which may compromise the health of a patient. As it is impossible to discuss all emerging new trends in tribology, in sufficient detail, in the context of this book for more information on the topics discussed in this chapter the readers are referred to more specialized literature.

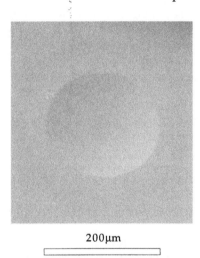

200μm

Figure 17.5   Example of an aspherical tungsten carbide mould insert manufactured using nano/micro grinding. Note that the diameter of the insert is around 200 [μm], i.e., comparable to the thickness of human hair [36].

REVISION QUESTIONS

17.1   How significant is biotribology to the normal functioning of a human body?

17.2   Why do proteins function so differently from conventional lubricant additives when used to control wear and friction.

17.3   Why has environmental tribology become so important? Suggest a major reason.

17.4   Suggest some special problems likely to affect nano-technology (e.g., nano-machines)?

17.5   What material properties are important in materials selection for artificial joints (for example, femoral head rotating against acetabular cup)?

17.6   Give examples of materials used in the human body where the tribological properties are taken into consideration.

17.7   Vegetable oils are biodegradable and therefore environmentally-friendly. Can vegetable oils replace mineral oils in engineering applications?

REFERENCES

1   S.M. Hsu, Oral Presentation at the International Tribology Conference, Yokohama, Japan, November, 1995.

2   S.M. Hsu and Z.C. Ying (editors), Nanotribology, Critical Assessment and Research Needs, Kluwer Academic Publishers, 2003, Dordrecht, The Netherlands.

3   C. Jones, K. Stoffel, H. Ozturk and G.W. Stachowiak, The Effects of Surface Active Phospholipids on Wear and Lubrication of Osteoarthritic Sheep Joints: Wear, *Tribology Letters*, Vol. 16, No. 4, 2004, pp. 291-296.

4   G.C. Ballantine and G.W. Stachowiak, The Effects of Lipid Depletion on Osteoarthritic Wear, *Wear*, Vol. 253, 2002, pp. 385-393.

5   P. Podsiadlo, M. Kuster and G.W. Stachowiak, Numerical Analysis of Wear Particles from Non-Arthritic and Osteoarthritic Human Knee Joints, *Wear*, Vol. 210, 1997, pp. 318-325.

6   M. Kuster, P. Podsiadlo and G.W. Stachowiak, Shape of the Particles Found in Human Knee Joints and Their Relationship to Osteoarthritis, *British Journal of Rheumatology*, Vol. 3, 1998, pp. 978-984.

7   M. Benz, N. Chen and J. Israelachvili, Lubrication and Wear Properties of Grafted Polyelectrolytes, Hyaluronan and Hylan, Measured in the Surface Forces Apparatus, *Journal of Biomedical Materials Research*, Vol. 71A, 2004, pp. 6-15.

8   G.D. Jay, D.A. Harris and C.J. Cha, Boundary Lubrication by Lubricin is Mediated by O-linked beta (1-3) GalNAc Oligosaccharides, *Glycoconjugates Journal*, Vol. 18, 2001, pp. 807-815.

9   S.L. Graindorge and G.W. Stachowiak, Changes Occurring in the Surface Morphology of Articular Cartilage During Wear, *Wear*, Vol. 241, 2000, pp. 143-151.

10   H. Ozturk, K. Stoffel, C. Jones and G.W. Stachowiak, The Effects of Surface Active Phospholipids on Wear and Lubrication of Osteoarthritic Sheep Joints: Friction, *Tribology Letters*, Vol. 16, 2004, 283-289.

11   C. Sittig, M. Textor, N.D. Spencer, M. Wieland, P.-H. Vallotton, Surface Characterization of Implant Materials c.p. Ti, Ti-6Al-7Nb and Ti-6Al-4V With Different Pretreatments, *Journal of Materials Science: Materials in Medicine*, Vol. 10, 1999, pp. 35-46.

12   G.W. Stachowiak, Friction and Wear of Polymers, Ceramics and Composites in Biomedical Applications, Chapter 14 in Advances in Composites Tribology, Vol. 8, Composite Materials Series, editor: Klaus Friedrich, Elsevier, 1993, pp. 509-557.

13   H. Oonishi, I.C. Clarke, V. Good, H. Amino and M. Ueno, Alumina Hip Joints Characterized by Run-In Wear and Steady-State Wear to 14 Million Cycles in Hip-Simulator Model, *Journal of Biomedical Materials Research*, Vol. 70A, 2004, pp. 523-532.

14   A.W. Batchelor and M. Chandrasekaran, Service Characteristics of Biomedical Materials, Imperial College Press, 2004.

15   I. Catelas, J.B. Medley, P.A. Campbell, O.L. Huk and J.D. Bobyn, Comparison of in-Vitro with in-Vivo Characteristics of Wear Particles from Metal-Metal Hip Implants, *Journal of Biomedical Materials Research*, Part B, Vol. 70B, 2004, pp. 167-178.

16   Y. Minoda, A. Kobayashi, H. Iwaki, M. Miyaguchi, Y. Kadoya, H. Ohashi and K. Takaoka, Characteristics of Polyethylene Wear Particles Isolated from Synovial Fluid after Mobile-Bearing and Posterior-Stabilized Total Knee Arthroplasties, *Journal of Biomedical Materials Research*, Part B, Vol. 71B, 2004, pp. 1-6.

17   K. Varlot, M. Kasrai, G.M. Bancroft, E.S. Yamaguchi, P.R. Ryason and J. Igarashi, X-ray Adsorption Study of Antiwear Films Generated from ZDDP and Borate Micelles, *Wear*, Vol. 249, 2001, pp. 1029-1035.

18   K. Carnes, Offroad Hydraulic Fluids: Beyond Biodegradability, *Tribology and Lubrication Technology*, Vol. 60, 2004, pp. 32-40.

19   I. Minami and S. Mitsumune, Anti-Wear Properties of Phosphorous-Containing Compounds in Vegetable Oils, *Tribology Letters*, Vol. 13, No. 2, 2002, pp. 95-101.

20   D. Tabor, Discussion on D. Dowson paper Friction and Traction in Lubricated Contacts, I.L. Singer and H.M. Polock (editors), Fundamentals of Friction: Macroscpopic and Microscopic Processes, Kluwer Academic Publishers, 1992.

21   Mer (Official Journal of IMAREST), SMM 2000 Preview, September 2004, pp. 43-58.

22   G. He and M.O. Robbins, Scale Effects and the Molecular Origins of Tribological Behaviour, pp. 29-44, S.M. Hsu and Z.C. Ying (editors), Nanotribology, Critical Assessment and Research Needs, Kluwer Academic Publishers, 2003, Dordrecht, The Netherlands.

23   T.E. Karis, Nanotribology of Thin Film Magnetic Recording Media, pp. 291-325, S.M. Hsu and Z.C. Ying (editors), Nanotribology, Critical Assessment and Research Needs, Kluwer Academic Publishers, 2003, Dordrecht, The Netherlands.

24   D.H. Buckley, Surface Effects in Adhesion, Friction, Wear and Lubrication, Elsevier, Amsterdam, 1981.

25   B. Bhushan (editor), Handbook of Micro/Nano Tribology, CRC Series Mechanics and Materials Science, CRC Press Inc., 1995.

26   G.W. Stachowiak, A.W. Batchelor and G.B. Stachowiak, Experimental Methods in Tribology, Elsevier, 2004.

27   M. Urbakh, J. Klafter, D. Gourdon and J. Israelachvili, The Nonlinear Nature of Friction, *Nature*, Vol. 430, 2004, pp. 525-528.

28   S.M. Hsu, Nanotribology: the Link to Macrotribolgy, Reply to Questions, Proc. Intern. Tribology Conf., AUSTRIB '02, editor: G.W. Stachowiak, 2-5 December 2002, Perth, Australia, pp. 9-15.

29   J. Gao, W.D. Luedtke, D. Gourdon, M. Ruths, J.N. Israelachvili and U. Landman, Frictional Forces and Amonton's Law: From the Molecular to the Macroscopic Scale, *Journal of Physical Chemistry, Part B*, Vol. 108(11), 2004, pp. 3410-3425.

30   A. Hayd and M. Maurer, Solid State Physics of Friction, *Wear*, Vol. 113, 1986, pp. 87-101.

31   J.A. Harrison, P.T. Mikulski, S.J. Stuart and A.B. Tutein, Dependence of Frictional Properties of Hydrocarbon Chains on Tip Contact Area, pp. 55-62, S.M. Hsu and Z.C. Ying (editors), Nanotribology, Critical Assessment and Research Needs, Kluwer Academic Publishers, 2003, Dordrecht, The Netherlands.

32   A.J. Gellman, Vapor Lubricant Transport in MEMS Devices, *Tribology Letters*, Vol. 17, No. 3, 2004, pp. 455-461.

33   M.F. Yu, M.J. Dyer and R.S. Ruoff, Carbon Nanotubes: Objects of Well-Defined Geometry for new Studies in Nanotribology, pp. 109-113, S.M. Hsu and Z.C. Ying (editors), Nanotribology, Critical Assessment and Research Needs, Kluwer Academic Publishers, 2003, Dordrecht, The Netherlands.

34   M.N. Gardos, Self-Lubricating Buckyballs and Buckytubes for Nanobearings and Gears- Science or Science f(r)iction?, pp. 95-108, S.M. Hsu and Z.C. Ying (editors), Nanotribology, Critical Assessment and Research Needs, Kluwer Academic Publishers, 2003, Dordrecht, The Netherlands.

35   S.M. Hsu, Nanolubrication: Concept and Design, pp. 327-346, S.M. Hsu and Z.C. Ying (editors), Nanotribology, Critical Assessment and Research Needs, Kluwer Academic Publishers, 2003, Dordrecht, The Netherlands.

36   H. Huang, Private Communication, University of Western Australia, 2005.

37   W. Tysoe and N. Spencer, Why Does Amontons' Law Work so Well?, *Tribology and Lubrication Technology*, Vol. 60, 2004, p. 64.

INTRODUCTION

The following programs, written in Matlab version 5.2, are provided to help the reader find the solutions to tribological problems:

'**VISCOSITY**'   provides the solution to the 'Vogel' equation and calculates viscosity at any given temperature.

'**SIMPLE**'   calculates the effective viscosity, bearing load capacity, bearing effective temperature and eccentricity ratio in a journal bearing.

'**PARTIAL**'   calculates the dimensionless load, attitude angle, Petroff multiplier and dimensionless (normalized) friction coefficient in a journal bearing. The solution is based on an isoviscous model of hydrodynamic lubrication with no elastic deflection of the bearing.

'**THERMAL**'   calculates the temperature field and load capacity of a thermohydrodynamic, perfectly rigid pad bearing where frictional heating of the lubricant causes variation in viscosity.

'**DEFLECTION**'   calculates the load capacity and pivot point of a Michell pad that is subjected to elastic deflection for an isoviscous model of hydrodynamic lubrication.

'**GROOVE**'   calculates the dimensionless load and lubricant flows for a journal bearing with two axial grooves positioned at 90° to the load line. It also provides the side flow from the bearing and the distribution of lubricant consumption between the grooves. An isoviscous model of lubrication and a perfectly rigid bearing are assumed.

'**STABILITY**'   calculates the load capacity and non-dimensional critical angular velocity of a partial perfectly rigid arc bearing. Vibration stability as a function of eccentricity, 'L/D' ratio, angle of partial arc bearing and misalignment ratio are also calculated together with the stiffness and damping coefficients in directions parallel and normal to the load line.

A.1   USER-FRIENDLY INTERFACE

This interface provides a pop-up menu to operate all the programs.

```
function tribology(action)
munlock; global tbv tbs tbp tbt tbd tbg tba tbinfo tbclose;
if nargin<1, action='initialize'; end;
if strcmp(action,'initialize'),
  oldFigNumber=watchon;
```

```
figNumber=figure('Name','Engineering Tribology','NumberTitle','off', 'Visible','off','Color',[0.85 0.6 0],'ColorMap',[0 0 1]);
figDefaultAxesPos=[0.10 0.10 0.63 0.80]; set(figNumber,'DefaultAxesPosition',figDefaultAxesPos);
 axes('Units','normalized','Position',figDefaultAxesPos);
%===================================
% Set up the program window
top=0.35;left=0.05;right=0.75;bottom=0.05; labelHt=0.05;spacing=0.005;
%===================================
% Information for all buttons
labelColor=[0.2 0.2 0.2]; top=0.95; left=0.78; btnWid=0.18; btnHt=0.06; spacing=0.03;
%===================================
% The CONSOLE frame
frmBorder=0.02; yPos=0.05-frmBorder; frmPos=[left-frmBorder yPos btnWid+2*frmBorder 0.9+2*frmBorder];
uicontrol('Style','frame','Units','normalized','Position',frmPos,'BackgroundColor', [0 0.5 1]);
%===================================
% The VISCOSITY button
btnNumber=1; yPos=top-(btnNumber-1)*(btnHt+spacing); labelStr='VISCOSITY'; callbackStr='tribology("tribobutton")';
btnPos=[left yPos-btnHt btnWid btnHt];
 tbv = uicontrol('Style','radiobutton', ...
                'Units','normalized', ...
                'Position',btnPos, ...
                'BackgroundColor',[0.9 0.5 0], ...
                'ForegroundColor',[1 1 1], ...
                'FontWeight', 'bold', ...
                'String',labelStr, ...
                'Callback',str2mat('viscosity'));
%===================================
% The SIMPLE button
btnNumber=2; yPos=top-(btnNumber-1)*(btnHt+spacing); labelStr='SIMPLE'; callbackStr='tribology("tribobutton")';
btnPos=[left yPos-btnHt btnWid btnHt];
 tbs = uicontrol('Style','radiobutton', ...
                'Units','normalized', ...
                'Position',btnPos, ...
                'BackgroundColor',[0.9 0.5 0], ...
                'ForegroundColor',[1 1 1], ...
                'FontWeight', 'bold', ...
                'String',labelStr, ...
                'Callback',str2mat('simple'));
%===================================
% The PARTIAL button
btnNumber=3; yPos=top-(btnNumber-1)*(btnHt+spacing); labelStr='PARTIAL'; callbackStr='tribology("tribobutton")';
btnPos=[left yPos-btnHt btnWid btnHt];
 tbp = uicontrol('Style','radiobutton', ...
                'Units','normalized', ...
                'Position',btnPos, ...
                'BackgroundColor',[0.9 0.5 0], ...
                'ForegroundColor',[1 1 1], ...
                'FontWeight', 'bold', ...
                'String',labelStr, ...
                'Callback',str2mat('partial'));
%===================================
% The THERMAL button
btnNumber=4; yPos=top-(btnNumber-1)*(btnHt+spacing); labelStr='THERMAL'; callbackStr='tribology("tribobutton")';
btnPos=[left yPos-btnHt btnWid btnHt];
 tbt = uicontrol('Style','radiobutton', ...
                'Units','normalized', ...
                'Position',btnPos, ...
                'BackgroundColor',[0.9 0.5 0], ...
                'ForegroundColor',[1 1 1], ...
                'FontWeight', 'bold', ...
                'String',labelStr, ...
                'Callback',str2mat('thermal'));
%===================================
% The DEFLECTION button
btnNumber=5; yPos=top-(btnNumber-1)*(btnHt+spacing); labelStr='DEFLECTION'; callbackStr='tribology("tribobutton")';
btnPos=[left yPos-btnHt btnWid btnHt];
 tbd = uicontrol('Style','radiobutton', ...
                'Units','normalized', ...
                'Position',btnPos, ...
                'BackgroundColor',[0.9 0.5 0], ...
                'ForegroundColor',[1 1 1], ...
                'FontWeight', 'bold', ...
                'String',labelStr, ...
```

```
                            'Callback',str2mat('deflection'));
   %====================================
   % The GROOVE button
   btnNumber=6; yPos=top-(btnNumber-1)*(btnHt+spacing); labelStr='GROOVE'; callbackStr='tribology("tribobutton")';
   btnPos=[left yPos-btnHt btnWid btnHt];
   tbg = uicontrol('Style','radiobutton', ...
                'Units','normalized', ...
                'Position',btnPos, ...
                'BackgroundColor',[0.9 0.5 0], ...
                'ForegroundColor',[1 1 1], ...
                'FontWeight', 'bold', ...
                'String',labelStr, ...
                'Callback',str2mat('groove'));
   %====================================
   % The STABILITY button
   btnNumber=7; yPos=top-(btnNumber-1)*(btnHt+spacing); labelStr='STABILITY'; callbackStr='tribology("tribobutton")';
   btnPos=[left yPos-btnHt btnWid btnHt];
   tba = uicontrol('Style','radiobutton', ...
                'Units','normalized', ...
                'Position',btnPos, ...
                'BackgroundColor',[0.9 0.5 0], ...
                'ForegroundColor',[1 1 1], ...
                'FontWeight', 'bold', ...
                'String',labelStr, ...
                'Callback',str2mat('stability'));
   %====================================
   % The INFO button
   tbinfo = uicontrol('Style','pushbutton', ...
                'Units','normalized', ...
                'Position',[left bottom+1.5*btnHt+spacing btnWid 1.5*btnHt], ...
                'BackgroundColor',[0.9 0 0.2], ...
                'ForegroundColor',[1 1 1], ...
                'FontWeight', 'bold', ...
                'String','Info', ...
                'Callback','tribology("info")');
   %====================================
   % The CLOSE button
   tbclose = uicontrol('Style','pushbutton', ...
                'Units','normalized', ...
                'Position',[left bottom btnWid 1.5*btnHt], ...
                'BackgroundColor',[0.9 0 0.2], ...
                'ForegroundColor',[1 1 1], ...
                'FontWeight', 'bold', ...
                'String','Close', ...
                'Callback','close(gcf)');
   % Uncover the figure
   set(figNumber,'Visible','on');
   watchoff(oldFigNumber); figure(figNumber);
elseif strcmp(action,'tribobutton'),
   cmdStr=get(gco,'UserData'); mcwHndl=findobj(gcf,'Tag','mcw'); set(mcwHndl,'String',cmdStr); evalmcw(mcwHndl);
elseif strcmp(action,'eval'),
   mcwHndl = findobj(gcf,'Tag','mcw'); cmdStr = get(mcwHndl,'String'); evalmcw(mcwHndl);
elseif strcmp(action,'info'),
   strhlp = {'Tribology'       help('introduction');
            'Viscosity'      help('viscosity');
            'Simple'         help('simple');
            'Partial'        help('partial');
            'Thermal'        help('thermal');
            'Deflection'     help('deflection');
            'Groove'         help('groove');
            'Stability'      help('stability');};
   helpwin(strhlp,'Tribology','Engineering Tribology - Info');
end;
```

A.2   PROGRAM 'VISCOSITY'

This program provides solutions to the Vogel viscosity equation shown in Table 2.1, i.e.:

$$\eta = ae^{b/(T - c)}$$

where:

η is the dynamic viscosity of the lubricant at temperature 'T' [Pas];

T is the temperature [°C];

a, b, c are constants.

The program also includes further options providing either viscosity at a given temperature or the temperature for a particular viscosity.

```
clc; cla reset; echo off;
global tbv tbs tbp tbt tbd tbg tba tbinfo tbclose;
set([tbs tbp tbt tbd tbg tba tbinfo tbclose],'Enable','off');
% BEGIN OF INPUT DATA
% --------------------
prompt = {'Maximum temperature [°C]:', 'Middle temperature [°C]:', 'Minimum temperature [°C]:'};
title = 'INPUT DATA (VISCOSITY)'; lineno = 1; def = {'100', '60', '20'}; answer = inputdlg(prompt,title,lineno,def);
if size(answer) == 0, % PROGRAM IS TERMINATED
    set(tbv, 'Value', get(tbv, 'Min')); set([tbs tbp tbt tbd tbg tba tbinfo tbclose],'Enable','on'); break; end;
[t1, t2, t3] = deal(answer{:}); TEMP = [str2num(t1) str2num(t2) str2num(t3)]; TEMP = TEMP + 273.15;
prompt = {'Viscosity at maximum temperature [Pas]:', 'Viscosity at middle temperature [Pas]:',
          'Viscosity at minimum temperature [Pas]:'};
title='INPUT DATA (VISCOSITY)'; lineno=1; def={'0.0099', '0.0351', '0.387'}; answer= inputdlg(prompt,title,lineno,def);
if size(answer) == 0, % PROGRAM IS TERMINATED
    set(tbv, 'Value', get(tbv, 'Min')); set([tbs tbp tbt tbd tbg tba tbinfo tbclose],'Enable','on'); break; end;
[v1, v2, v3] = deal(answer{:}); VISC(1) = str2num(v1); VISC(2) = str2num(v2); VISC(3) = str2num(v3);
VISC = [str2num(v1) str2num(v2) str2num(v3)]; VISC = VISC;
% END OF INPUT DATA
% --------------------
% FUNCTION TO FIND CONSTANTS OF VOGEL VISCOSITY EQUATION
[a1,b1,c1,a2,b2,c2] = vogel(TEMP,VISC);
prompt = {'Choose solution: 1- First solution; 2 - Second solution', 'First solution (a1, b1, c1)',
          'Second solution (a2, b2, c2)'};
title='SOLUTIONS OF VOGEL EQUATION'; lineno=1;
def={'1',num2str([a1 b1-273.15 c1-273.15]),num2str([a2 b2-273.15 c2-273.15])}; answer= inputdlg(prompt,title,lineno,def);
if size(answer) == 0, % PROGRAM IS TERMINATED
    set(tbv, 'Value', get(tbv, 'Min')); set([tbs tbp tbt tbd tbg tba tbinfo tbclose],'Enable','on'); break; end;
choice = str2num(deal(answer{1}));
if choice == 1, a = a1; b = b1; c = c1; end;
if choice == 2, a = a2; b = b2; c = c2; end;
% PLOT OF KINEMATIC VISCOSITY VS TEMPERATURE
clear VISCPLOT TEMPPLOT;
iv = 1; it = TEMP(3);
  while (it >= TEMP(3)) & (it <= TEMP(1))
    VISCPLOT(iv) = (a*exp(b/(it-c))); TEMPPLOT(iv) = it-273.15; iv = iv + 1; it = it + 1;
  end;
subplot(1,1,1); plot(TEMPPLOT,VISCPLOT);
axis([fix(TEMP(3)-273.15) ceil(TEMP(1)-273.15) fix(0.9*min(VISCPLOT)) ceil(1.1*max(VISCPLOT))]);
xlabel('Temperature [°C]'); ylabel('Kinematic Viscosity [Pas]');
text('units','normalized','position',[0.1 1.05],'string', 'VISCOSITY VS TEMPERATURE (no ASTM plot)');
buttonname = questdlg('Do you want to find viscosity at a specific temperature?',...
                      'VISCOSITY AND TEMPERATURE','No','Yes','Yes');
helpdlg('Now please activate the command window');
switch buttonname,
  case 'No',
    choice = 0;
  case 'Yes',
    choice = 1;
end;
if choice == 1,
    morepoint = 1; hold on; ipoint = 1; clear TEMP_VAL VISC_VAL;
    disp(' '); disp('FIND VISCOSITY AT GIVEN TEMPERATURE');
    disp('Please read carefully all the steps listed below before selecting viscosity or temperature.');
    disp('1. Select zoom key from tools menu and then use mouse to select area of interest on plot');
    disp('2. In command window hit any key to activate crosshairs');
    disp('3. On plot window select exact point of interest and then press enter key');
    disp('4. Return to command window to get temperature and viscosity required'); disp(' ');
    while morepoint == 1,
      zoom on; pause;
      [tempa,visca] = ginput; hold on; plot(tempa,visca,'go'); text(tempa,visca,[' ' int2str(ipoint)],'Erase','back');
```

```
      TEMP_VAL(ipoint) = tempa; VISC_VAL(ipoint) = visca;
      fprintf('Temperature = %0.5g [°C]\t',tempa); fprintf('Viscosity = %0.5g [Pas]\n',visca); ipoint = ipoint + 1;
      buttonname = questdlg('Do you want to enter another point?', 'NEXT POINT','No','Yes','Yes');
      switch buttonname,
        case 'No',
          morepoint = 0;
        case 'Yes',
          morepoint = 1;
      end;
  end;
  zoom off;
end;
clc;
% PRINT OUT VALUES OF INPUT AND OUTPUT DATA
% ----------------------------------------
fprintf(' \n'); fprintf(' INPUT AND OUTPUT DATA FOR PROGRAM VISCOSITY\n'); fprintf(' \n');
fprintf(' INPUT DATA:\n');
fprintf('  Maximum temperature = %0.5g°C\n',TEMP(1)-273.15);
fprintf('  Middle temperature = %0.5g°C\n',TEMP(2)-273.15);
fprintf('  Minimum temperature = %0.5g°C\n',TEMP(3)-273.15);
fprintf('  Viscosity at maximum temperature = %0.5g [Pas]\n',VISC(1));
fprintf('  Viscosity at middle temperature = %0.5g [Pas]\n',VISC(2));
fprintf('  Viscosity at minimum temperature = %0.5g [Pas]\n',VISC(3));
fprintf(' \n'); fprintf(' OUTPUT DATA:\n');
fprintf('  Vogel parameter a = %0.5g [Pas]\n', a);
fprintf('  Vogel parameter b = %0.5g°C\n', b-273.15);
fprintf('  Vogel parameter c = %0.5g°C\n', c-273.15);
if choice == 1, disp('  Viscosity [Pas] against temperature [°C]'); disp([VISC_VAL'   TEMP_VAL']); end;
fprintf(' \n'); fprintf(' PROGRAM VISCOSITY HAS BEEN COMPLETED\n');
set(tbv, 'Value', get(tbv, 'Min')); set([tbs tbp tbt tbd tbg tba tbinfo tbclose],'Enable','on');

% CALCULATION OF COEFFICIENTS IN VOGEL EQUATION:
% VOGEL EQUATION VISCOSITY = a*exp(b/(t-b))
coeffc2 = (log(VISC(1)) - log(VISC(2)))*(TEMP(3) - TEMP(1)) - (log(VISC(1)) - log(VISC(3)))*(TEMP(2) - TEMP(1));
coeffc1 = (log(VISC(1)) - log(VISC(2)))*(TEMP(1) - TEMP(3))*(TEMP(2) + TEMP(1)) + (log(VISC(1)) - log(VISC(3)))* ...
          (TEMP(2) - TEMP(1))*(TEMP(1) + TEMP(3));
coeffc0 = (log(VISC(1)) - log(VISC(2)))*(TEMP(3) - TEMP(1))*TEMP(1)*TEMP(2) - (log(VISC(1)) - ...
          log(VISC(3)))*(TEMP(2) - TEMP(1))*TEMP(1)*TEMP(3);
% APPLY SOLUTION OF QUADRATIC EQUATIONS
term1 = coeffc1*coeffc1 - 4*coeffc2*coeffc0;
if (term1 < 0), error('NEGATIVE ROOT OF SOLUTION ONLY, PROGRAM TERMINATED.'); end;
c1 = (-coeffc1 + sqrt(term1))/(2*coeffc2);
b1 = (log(VISC(1)) - log(VISC(2)))*(TEMP(1)*TEMP(2) - c1*(TEMP(1) + TEMP(2)) + c1*c1)/(TEMP(2) - TEMP(1));
a1 = VISC(1)*exp(-b1/(TEMP(1) - c1)); c2 = (-coeffc1 - sqrt(term1))/(2*coeffc2);
b2 = (log(VISC(1)) - og(VISC(2)))*(TEMP(1)*TEMP(2) - c2*(TEMP(1) + TEMP(2)) + c2*c2)/(TEMP(2) - TEMP(1));
a2 = VISC(1)*exp(-b2/(TEMP(1) - c2));
```

Program Description

The Vogel viscosity equation is solved by substitution of **3** temperatures and **3** corresponding viscosities into **3** equations defining intermediate variables which are '**coeffc1**', '**coeffc2**' and '**coeffc0**'. The function called '**vogel.m**' is used to solve this equation. The constant '**c**' of the Vogel equation is found from the following quadratic equation:

$$coeffc2*c\text{^}2 + coeffc1*c + coeffc0 = 0$$

Before the quadratic solution is applied, the values of '**coeffc1**', '**coeffc2**' and '**coeffc0**' are tested for the existence of real solutions. If there is no real solution, the program is terminated. When a real solution exists, the '**c**' values are calculated. Solving a quadratic equation provides two solutions and this is allowed for by two variables '**c1**' and '**c2**'. Once '**c**' is calculated, the corresponding values of '**a**' and '**b**' can be determined. Each variable has two values hence the terms '**a1**' and '**a2**', '**b1**' and '**b2**'. Since two solutions are obtained it is necessary to decide which of them is appropriate. The rest of the program is devoted to predicting values of viscosity as a function of temperature or vice versa. The program is designed to allow this function to be repeated indefinitely. When this final stage of the program is exited, a warning is given that the program is completed and execution stops.

List of Variables

TEMP	Array of three temperature values for Vogel equation;
VISC	Array of three viscosity values for Vogel equation;
choice	Control variable to determine program options;
coeffc0, coeffc1,	
coeffc2	Coefficients of inverse equation to find exponents in Vogel equation;
term1	Solution to quadratic equation in terms of '**coeffc1**', '**coeffc2**' and '**coeffc0**';
a1, a2, b1,	
b2, c1, c2	Upper and lower values of quadratic solution to Vogel equation;
tempq	Test temperature for applying Vogel equation;
visca	Viscosity calculated for the test temperature;
viscq	Test viscosity for applying Vogel equation;
tempa	Temperature calculated for the test viscosity.

EXAMPLE

<u>Input data:</u> TEMP(1) = 100°C, TEMP(2) = 60°C, TEMP(3) = 20°C, VISC(1) = 0.0099 [Pas], VISC(2) = 0.0351 [Pas], VISC(3) = 0.387 [Pas].

<u>Output data:</u> First solution: a1 = 0.165E-3 [Pas], b1 = 418.98°C, c1 = -69.24°C. Second solution: a2 = 188.309E-3 [Pas], b2 = -273.15°C, c2 = 100°C.

A.3 PROGRAM 'SIMPLE'

This program calculates the effective viscosity, bearing load capacity, bearing effective temperature and eccentricity ratio in a journal bearing. It begins with data acquisition which offers an option of data units for the rotational speed of the bearing, i.e., speed can be expressed in radians per second [rad/s], revolutions per second [rps] or revolutions per minute [rpm], while the load must be expressed in Newtons [N]. Other data such as the lubricant density and specific heat are only accepted in SI units. The constants '**a**' and '**c**' of the ASTM viscosity equation (2.7) are also required as data. The program then proceeds to offer a choice between accepting standard iteration parameters or selecting different values.

```
clc; cla reset; echo off;
global tbv tbs tbp tbt tbd tbg tba tbinfo tbclose;
set([tbv tbp tbt tbd tbg tba tbinfo tbclose],'Enable','off');
% BEGIN OF INPUT DATA
% -------------------
prompt = {'Bearing length [m]:', 'Radial clearance [m]:', 'Bearing radius [m]:'};
title = 'INPUT DATA (SIMPLE)'; lineno = 1; def = {'0.1','0.0001','0.1'}; answer = inputdlg(prompt,title,lineno,def);
if size(answer) == 0, % PROGRAM IS TERMINATED
   set(tbs, 'Value', get(tbs, 'Min')); set([tbv tbp tbt tbd tbg tba tbinfo tbclose],'Enable','on'); break; end;
[length0, clearnce, radius] = deal(answer{:}); length0 = str2num(length0); clearnce = str2num(clearnce);
radius = str2num(radius);
prompt =  {'Bearing load [N]:', 'Bearing velocity units: 1 - rad/s; 2 - R.P.S.; 3 - R.P.M', 'Bearing velocity:'};
```

```
def = {'100000','1','314'}; title = 'INPUT DATA (SIMPLE)'; lineno = 1; answer = inputdlg(prompt,title,lineno,def);
if size(answer) == 0, % PROGRAM IS TERMINATED
   set(tbs, 'Value', get(tbs, 'Min')); set([tbv tbp tbt tbd tbg tba tbinfo tbclose],'Enable','on'); break; end;
[load, choice2, speed] = deal(answer{:}); load = str2num(load); choice2 = str2num(choice2); speed = str2num(speed);
if choice2 == 2
   speed = speed*2*pi;
elseif choice2 == 3
   speed = speed*2*pi/60;
end;
u0 = speed*radius;
prompt = {'Inlet lubricant temperature [°C]:',
          'Lubricant density [kg/m^3]:',
          'Lubricant specific heat [J/kgK]:',
          'Constant a of Walthers equation:'
          'Constant c of Walthers equation:'};
def = {'50','800','2000','0.1','0.05'}; title='INPUT DATA (SIMPLE)'; lineno=1; answer= inputdlg(prompt,title,lineno,def);
if size(answer) == 0, % PROGRAM IS TERMINATED
   set(tbs, 'Value', get(tbs, 'Min')); set([tbv tbp tbt tbd tbg tba tbinfo tbclose],'Enable','on'); break; end;
[tinlet, density, cp, adash, cdash] = deal(answer{:}); tinlet = str2num(tinlet); density = str2num(density);
cp = str2num(cp); adash = str2num(adash); cdash = str2num(cdash);
prompt = {'Terminating residual of narrow bearing eccentricity iteration:',
          'Max number of cycles for narrow bearing eccentricity iteration:',
          'Relaxation factor for narrow bearing eccentricity iteration:',
          'Terminating residual of effective temperature iteration:',
          'Max number of cycles for effective temperature iteration:',
          'Relaxation factor for effective temperature iteration:',
          'Terminating residual of effective viscosity iteration:',
          'Max number of cycles for effective viscosity iteration:',
          'Relaxation factor for effective viscosity iteration:'};
def={'0.000001', '1000', '0.01', '0.000001', '400', '0.5', '0.0001', '100','1'}; title='INPUT DATA (SIMPLE)'; lineno=1;
answer= inputdlg(prompt,title,lineno,def);
if size(answer) == 0, % PROGRAM IS TERMINATED
   set(tbs, 'Value', get(tbs, 'Min')); set([tbv tbp tbt tbd tbg tba tbinfo tbclose],'Enable','on'); break; end;
[reslim1,nlim1,factor1,reslim2,nlim2,factor2,reslim3,nlim3,factor3] = deal(answer{:});
reslim1 = str2num(reslim1); reslim2 = str2num(reslim2); reslim3 = str2num(reslim3);
nlim1 = str2num(nlim1); nlim2 = str2num(nlim2);nlim3 = str2num(nlim3);
factor1 = str2num(factor1); factor2 = str2num(factor2); factor3 = str2num(factor3);
% END OF INPUT DATA
% ------------------
subplot(1,1,1);
text('units','normalized','position',[0.2 0.55], 'FontWeight', 'bold','color',[1 0 0], 'string', 'CALCULATIONS IN PROGRESS');
figure(1);
% CALCULATE INLET VISCOSITY (visc0)
logtemp = log10(tinlet + 273.15); value = adash - cdash*logtemp; term = log(10)*value; value1 = exp(term);
term = log(10)*value1; visc0 = 0.001*density*(exp(term) - 0.6); epsilon0 = 0.5; diametr = 2*radius;
% ITERATION FOR ECCENTRICITY RESIDUAL
visc2 = 0; n3 = 0;  resid3 = reslim3 + 10;
while (resid3 > reslim3) & (n3 < nlim3),
   % FACTOR OF 0.001 INCLUDED BELOW TO CONVERT FROM MPAS TO PAS
   if n3 < 0.01, visc1 = visc0*0.001; end;
   if n3 > 0.01, visc1 = visc2*0.001; end;
   truedelta = (load*clearnce^2)/(u0*visc1*length0*radius^2); epsilon = epsilon0;
   % INVERSE EQUATION ECCENTRICITY
   sum = 0; n1 = 0; residl = reslim1 + 10;
   while (residl > reslim1) & (n1 < nlim1),
      delta = (length0^2/diametr^2)*(pi*epsilon*sqrt(0.62*epsilon^2 + 1))/((1-epsilon^2)^2);
      diff = (truedelta - delta)/delta; epsilon = epsilon + (1-epsilon)*diff*factor1;
      if epsilon < 0, epsilon = abs(epsilon); end;
      sum1 = abs(diff); residl = abs((sum1-sum)/sum1); sum = sum1; n1 = n1+1;
   end;
   % ITERATION FOR EFFECTIVE VISCOSITY
   petroff = sqrt(1-epsilon^2); coeff = 2*pi*diametr*u0/(density*cp*clearnce^2*epsilon*petroff);
   if n3 == 0, teff = tinlet; end;
   % START ITERATION
   n2 = 0; resid = reslim2 + 10;
   while (resid > reslim2) & (n2 < nlim2),
      logoldt = log10(teff + 273.15); value = adash - cdash*logoldt; term = log(10)*value; value1 = exp(term);
      term = log(10)*value1; visc1 = 0.001*density*(exp(term) - 0.6); deltat = coeff*visc1/1000;
      tstore = teff;teff1 = 0.8*deltat + tinlet; teff = (teff1-tstore)*factor2 + tstore; resid = abs((teff-tstore)/teff);
      n2 = n2+1;
   end;
   % MODIFY OVERALL VISCOSITY VALUE IN FINAL CYCLE OF ITERATION
```

```
      visc3 = visc2; visc2 = visc3 + factor3*(visc1 - visc3); resid3 = abs((visc2-visc3)/visc2); n3 = n3 + 1;
end;
attang = atan2(pi*sqrt(1-epsilon^2),4*epsilon)*180/pi;
plot(0,0);
text('units','normalized','position',[0.2,0.55],'string', 'NO PLOT FOR PROGRAM SIMPLE');
% PRINT OUT VALUES OF INPUT AND OUTPUT DATA
% -----------------------------------------
fprintf(' \n'); fprintf(' INPUT AND OUTPUT DATA FOR PROGRAM SIMPLE\n'); fprintf(' \n');
fprintf(' INPUT DATA:\n');
fprintf('   Bearing length = %0.5g [m]\n',length0);
fprintf('   Radial clearance = %0.5g [m]\n',clearnce);
fprintf('   Bearing radius = %0.5g [m]\n',radius);
fprintf('   Bearing load = %0.5g [N]\n',load);
if choice2 == 1
   fprintf('   Bearing velocity = %0.5g [rad/s]\n',speed);
elseif choice2 == 2
   fprintf('   Bearing velocity = %0.5g [R.P.S]\n',speed/(2*pi));
elseif choice2 == 3
   fprintf('   Bearing velocity = %8.5g [R.P.M]\n',speed*60/(2*pi));
end;
fprintf('   Inlet lubricant temperature = %0.5g°C\n',tinlet);
fprintf('   Lubricant density = %0.5g [kg/m^3]\n',density);
fprintf('   Specific heat of lubricant = %0.5g [J/kgK]\n',cp);
fprintf('   Constant a of ASTM viscosity-temperature relation = %0.5g\n',adash);
fprintf('   Constant c of ASTM viscosity-temperature relation = %0.5g\n',cdash);
fprintf(' \n');
fprintf(' OUTPUT DATA:\n');
fprintf('   Eccentricity ratio = %0.5g\n',epsilon);
fprintf('   Attitude angle = %0.5g°\n',attang);
fprintf('   Sliding speed = %0.5g [m/s]\n',u0);
fprintf('   Effective temperature = %0.5g°C\n',teff);
fprintf('   Effective viscosity = %0.5g [Pas]\n',visc2/1000);
fprintf(' \n'); fprintf(' PROGRAM SIMPLE HAS BEEN COMPLETED\n');
set(tbs, 'Value', get(tbs, 'Min')); set([tbv tbp tbt tbd tbg tba tbinfo tbclose],'Enable','on');
```

Program Description

An initial value of eccentricity equal to **0.5** is assigned for the first step of the iteration. The bearing inlet viscosity is then calculated from the ASTM equation and this is the final step before starting the iteration cycle. Calculations during the first cycle of iteration are based on the inlet viscosity and the conditional equation in '**while**' statement is applicable. For all subsequent calculations, the effective viscosity is applied. It is also necessary to express the viscosity in [Pas] because the result from the ASTM viscosity calculation is in [mPas]. When the immediate iterative value of viscosity has been decided, the program proceeds to the inner cycle of iteration where an iteration for eccentricity in terms of load, rotational speed, bearing geometry and effective viscosity is performed. Next, the frictional power dissipation (i.e., '**petroff**') and rise in lubricant temperature relative to viscosity (i.e., '**coeff**') are calculated.

Once the inner iteration for eccentricity is completed the new value of effective temperature is found by iteration. The revised or new estimate of effective viscosity is calculated from the existing value of effective temperature or the inlet temperature for the first cycle of iteration. An increment of temperature change is then found from the product of the rise in lubricant temperature relative to viscosity and the new value of viscosity. The existing value of effective temperature is then saved in a variable called '**tstore**' and a provisional value of the revised effective temperature is found by applying equation (4.127), i.e.:

$$\mathbf{T_{eff}} = \mathbf{T_{inlet}} + 0.8\Delta\mathbf{T}$$

If no relaxation was used then this value would be applied to the iteration, but instead a relaxation factor is applied in the subsequent statement to determine the revised value of '$\mathbf{T_{eff}}$'. A residual for temperature is found next and the program proceeds to a conditional statement on whether to terminate or continue the iteration for effective temperature.

Once the local or inner iterations for eccentricity and effective temperature are completed, the revised value of viscosity is found based on a relaxation formula. A similar conditional statement to the inner iterations is applied to determine whether the outer iteration in viscosity should proceed or be terminated. If the iteration is finished then the attitude angle is calculated and this is printed together with the eccentricity and data input parameters such as the sliding speed. Execution of the program is then terminated.

List of Variables

length0	Axial length of bearing [m];
clearnce	Radial clearance between shaft and journal [m];
radius	Bearing radius [m];
visc0	Lubricant inlet viscosity [Pas];
choice	Control variable to give option of data input units;
load	Bearing load [N];
speed	Angular velocity between shaft and journal, [radians/s];
u0	Sliding speed between shaft and journal [m/s];
truedelta	Dimensionless load, equivalent to 'Δ';
epsilon0	Datum value of eccentricity for start of iteration;
diametr	Diameter of bearing [m];
sum1	Modulus of relative difference between iterated value of dimensionless load and exact value of dimensionless load;
sum	Equivalent of '**sum1**' for previous cycle of iteration;
delta	Value of dimensionless load calculated by iteration equation;
diff	Relative difference between iterated and exact value of dimensionless load;
epsilon	Iterated value of eccentricity;
residl	Residual of iteration to find eccentricity, expressed in terms of error in dimensionless load;
term	Parameter to estimate attitude angle;
reslim1	Terminating residual of narrow bearing eccentricity iteration;
nlim1	Maximum number of cycles for narrow bearing eccentricity iteration;
factor1	Relaxation factor for narrow bearing eccentricity iteration;
reslim2	Terminating residual of effective temperature iteration;
nlim2	Maximum number of cycles for effective temperature iteration;
factor2	Relaxation factor for effective temperature iteration;
reslim3	Terminating residual of effective viscosity iteration;
nlim3	Maximum number of cycles for effective viscosity iteration;
factor3	Relaxation factor for effective viscosity iteration;
n1	Cycle number of narrow bearing eccentricity iteration;
n2	Cycle number of effective temperature iteration;
n3	Cycle number of effective viscosity iteration;
tinlet	Bearing inlet temperature [°C];

density	Lubricant density [kg/m^3];
cp	Lubricant specific heat [J/kgK];
cdash	Constant 'c' of ASTM viscosity-temperature relation;
adash	Constant 'a' of ASTM viscosity-temperature relation;
teff	Effective temperature;
logoldt	Logarithm of effective temperature estimated from previous cycle of iteration;
value1	Single exponentiation of core term in ASTM temperature-viscosity equation;
visc1	Value of viscosity calculated using the ASTM equation from the preceding estimate of effective temperature;
deltat	Temperature rise due to viscous heating;
tstore	Stored estimate of effective temperature from previous iteration cycle;
teff1	Direct iterated value of effective temperature;
petroff	Eccentricity dependent factor to calculate bearing friction from Petroff formula;
coeff	Factor to calculate rise in lubricant temperature during passage through bearing;
visc2, visc3	Storage variables for effective viscosity, i.e., 'visc2' holds the value of effective viscosity from the preceding cycle and 'visc3' gives temporary storage during calculation of a revised value of viscosity.

The viscosity temperature relationship is approximated by the ASTM (Walther's) equation (2.7):

$$\log_{10}\log_{10}(\upsilon_{cS} + 0.6) = a' - c\log_{10}T$$

where:

υ_{cS} is the kinematic viscosity of the lubricant at temperature 'T' [cS];

a', c are constants.

EXAMPLE

Input data: Bearing length = 0.1 [m], radial clearance = 0.0001 [m], bearing radius = 0.1 [m], bearing load = 100,000 [N], bearing velocity = 314 [rad/s] (approximately 3000 [rpm]), inlet lubricant temperature = 50°C, lubricant density = 800 [kg/m^3] specific heat of lubricant = 2000 [J/kgK], Walther's equation constants **a = 0.1** and **c = 0.05**.

Output data: Eccentricity ratio = 0.7849, attitude angle = 31.799°, sliding speed = 31.4 [m/s], effective temperature = 76.277°C, effective viscosity = 0.00647 [Pas].

A.4   PROGRAM 'PARTIAL'

This program calculates the dimensionless load, attitude angle, Petroff multiplier and dimensionless (normalized) friction coefficient for a specified eccentricity, angle of partial arc bearing, 'L/D' ratio and misalignment ratio. The solution is based on an isoviscous model of hydrodynamic lubrication with no elastic deflection of the bearing.

```
clc; cla reset; echo off;
global tbv tbs tbp tbt tbd tbg tba tbinfo tbclose;
set([tbv tbs tbt tbd tbg tba tbinfo tbclose],'Enable','off');
% BEGIN OF INPUT DATA
% -------------------
prompt = {'Eccentricity ratio:', 'L/D ratio:', 'Arc bearing angle [°]:', 'Misaligment parameter from interval [0,0.5]:'};
title='INPUT DATA (PARTIAL)'; lineno=1; def={'0.8','1', '120', '0'}; answer= inputdlg(prompt,title,lineno,def);
if size(answer) == 0, % PROGRAM IS TERMINATED
   set(tbp, 'Value', get(tbp, 'Min')); set([tbv tbs tbt tbd tbg tba tbinfo tbclose],'Enable','on'); break; end;
[epsilon,loverd,alpha,t] = deal(answer{:}); epsilon = str2num(epsilon); loverd = str2num(loverd);
alpha = str2num(alpha); t = str2num(t); slender = 0.5/loverd; alpha = alpha*pi/180;
if t < 0, t = 0; end; if t > 0.5, t = 0.5; end;
% SET MESH CONSTANTS
prompt = {'Number of nodes in the i or x direction:',
          'Number of nodes in the j or y direction:',
          'Terminating value of residual for iter. to solve Vogelpohl equation:',
          'Terminating value of residual for iter. to find attitude angle:',
          'Relaxation factor of iter. to solve Vogelpohl equation:',
          'Relaxation factor of iter. to find attitude angle:',
          'Max number of cycles during iter. to solve Vogelpohl equation:',
          'Max number of cycles during iter. to find attitude angle:'};
title='INPUT DATA (PARTIAL)'; lineno=1; def={'11','11','0.000001', '0.0001', '1.2', '1', '100','30'};
answer= inputdlg(prompt,title,lineno,def);
if size(answer) == 0, % PROGRAM IS TERMINATED
   set(tbp, 'Value', get(tbp, 'Min')); set([tbv tbs tbt tbd tbg tba tbinfo tbclose],'Enable','on'); break; end;
[inode,jnode,reslim1,reslim2,factor1,factor2,nlim1,nlim2] = deal(answer{:});
inode = str2num(inode); jnode = str2num(jnode); reslim1 = str2num(reslim1); reslim2 = str2num(reslim2);
factor1 = str2num(factor1); factor2 = str2num(factor2); nlim1 = str2num(nlim1); nlim2 = str2num(nlim2);
% END OF INPUT DATA
% -----------------
subplot(1,1,1);
text('units','normalized','position',[0.2 0.55], 'FontWeight', 'bold','color',[1 0 0], 'string', 'CALCULATIONS IN PROGRESS');
figure(1); slender = 0.5/loverd; deltax = alpha/(inode-1); deltay = 1/(jnode-1);
% DIFFERENTIAL QUANTITIES FOR STABILITy CALCULATIONS
% INITIALIZE VALUES OF M(I,J), SWITCH(I,J) AND P(I,J)
M = zeros(inode,jnode); P = zeros(inode,jnode);
% SET INITIAL VALUE OF OFFSET ANGLE
beta = 0;
% ENTER ATTITUDE ANGLE ITERATION CYCLE, CALCULATE H, F AND G VALUES
n2 = 0; betas = 0; residb = reslim2 + 10;
while (residb > reslim2) & (n2 < nlim2),
   n2 = n2 + 1;
   for i = 1:inode,
      xaux = (i-1)*deltax + pi - 0.5*alpha; theta = xaux - beta;
      for j = 1:jnode,
         y = (j-1)*deltay - 0.5; h0 = y*t*cos(xaux) + epsilon*cos(theta) + 1;
         dhdx0 = -y*t*sin(xaux) - epsilon*sin(theta); d2hdx20 = -y*t*cos(xaux) - epsilon*cos(theta);
         dhdy0 = t*cos(xaux); d2hdy20 = 0; H(i,j) = h0; G(i,j) = dhdx0/h0^1.5;
         F(i,j) = 0.75*(dhdx0^2 + (slender*dhdy0)^2)/h0^2 +  1.5*(d2hdx20 + d2hdy20*slender^2)/h0;
      end;
   end;
   coeff1 = 1/deltax^2; coeff2 = (slender/deltay)^2;
   % ------------------------------------------
   % SUBROUTINE TO SOLVE THE VOGELPOHL EQUATION
   sum2 = 0; n1 = 0; residp = reslim1 + 10;
   while (residp > reslim1) & (n1 < nlim1),
      n1 = n1 + 1; sum = 0;
      for i = 2:inode-1,
         for j = 2:jnode-1,
            store = ((M(i+1,j) + M(i-1,j))*coeff1 + (M(i,j+1) + M(i,j-1))*coeff2 -G(i,j))/(2*coeff1 + 2*coeff2 + F(i,j));
            M(i,j) = M(i,j) + factor1*(store-M(i,j));
            if M(i,j) < 0,  M(i,j) = 0; end;
            sum = sum + M(i,j);
         end;
      end;
      residp = abs((sum-sum2)/sum); sum2 = sum;
   end;
   % ------------------------------------------
   % FIND PRESSURE FIELD FROM VOGELPOHL PARAMETER
   for i = 2:inode-1,
      for j = 2:jnode-1,
         P(i,j) = M(i,j)/H(i,j)^1.5;
```

```
        end;
      end;
    % ITERATION RESIDUAL ON ATTITUDE ANGLE ITERATION
    % CALCULATE TRANSVERSE AND AXIAL LOADS
    % ------------------------------------
    % SUBROUTINE TO INTERGRATE FOR FORCES
    for i = 1:inode,
      SUMY(i) = 0;
      for j = 2:jnode, SUMY(i) = SUMY(i) + P(i,j) + P(i,j-1); end;
      SUMY(i) = SUMY(i)*0.5*deltay;
    end;
    axialw = 0; transw = 0;
    for i = 2:inode,
      x = (i-1)*deltax + pi - 0.5*alpha; x2 = (i-2)*deltax + pi - 0.5*alpha;
      axialw = axialw - cos(x)*SUMY(i) - cos(x2)*SUMY(i-1); transw = transw + sin(x)*SUMY(i) + sin(x2)*SUMY(i-1);
    end;
    axialw = axialw*deltax*0.5; transw = transw*deltax*0.5;
    % ------------------------
    loadw = sqrt(axialw^2 + transw^2); attang = atan(transw/axialw);
    if axialw > 0, attang1 = attang; end; if axialw < 0, attang1 = -attang; end;
    beta = beta + factor2*attang1; residb = abs((beta-betas)/beta); betas = beta;
  end;
  % -------------------------------------------
  % SUBROUTINE TO CALCULATE PETROFF MULTIPLIER
  for j = 1:jnode, ICAV(j) = 1000; end;
  for j = 2:jnode-1,
    for i = 2:inode,
      if (M(i,j) == 0) & (ICAV(j) == 1000), ICAV(j) = i; end;
    end;
  end;
  % EXTRAPOLATED VALUES OF ICAV(J) AT EDGES OF BEARING
  ICAV(1) = 2*ICAV(2) - ICAV(3); if ICAV(1) < 1, ICAV(1) = 1; end;
  if ICAV(1) > inode, ICAV(1) = inode; end; ICAV(jnode) = 2*ICAV(jnode-1) - ICAV(jnode-2);
  if ICAV(jnode) < 1, ICAV(jnode) = 1; end; if ICAV(jnode) > inode, ICAV(jnode) = inode; end;
  % CALCULATE FRICTION COEFFICIENT
  % FIND VALUES OF DIMENSIONLESS SHEAR STRESS
  for i = 1:inode,
    for j = 1:jnode,
      % CALCULATE dpdx FROM DOWNSTREAM VALUES
      if i > 1, dpdx = (P(i,j) - P(i-1,j))/deltax; end;
      % VALUE OF dpdx FOR i = 1
      if i == 1, dpdx = P(2,j)/deltax; end; if i < ICAV(j), TORR(i,j) = 1/H(i,j) + 3*dpdx*H(i,j); end;
      if i == ICAV(j), TORR(i,j) = 1/H(i,j); end; i10 = ICAV(j);
      if i > ICAV(j), TORR(i,j) = H(i10,j)/H(i,j)^2; end;
    end;
  end;
  % INTEGRATE FOR TORR(i,j) OVER X AND Y
  for i = 1:inode,
    % LINE INTEGRAL IN Y-SENSE
    SUMY(i) = 0;
    for j = 2:jnode, SUMY(i) = SUMY(i) + TORR(i,j) + TORR(i,j-1); end;
    SUMY(i) = SUMY(i)*0.5*deltay;
  end;
  friction = 0;
  for i = 2:inode, friction = friction + SUMY(i) + SUMY(i-1); end;
  friction = friction*0.5*deltax;
  % --------------------------
  % CALCULATE DIMENSIONLESS FRICTION COEFFICIENT
  myu = friction/loadw;
  % SEARCH FOR MAXIMUM PRESSURE
  pmax = 0;
  for i = 2:inode-1,
    for j = 2:jnode-1,
      if P(i,j) > pmax, pmax = P(i,j); end;
    end;
  end;
  % EXPRESS ALL PRESSURES AS PERCENTAGE OF MAXIMUM PRESSURE
  for i = 1:inode,
    for j = 1:jnode,
      P(i,j) = P(i,j)*100/pmax;
    end;
  end;
```

```
% PRESSURE FIELD PLOT
xi = 0:inode-1; yj = 0:jnode-1; xi = (xi*alpha*180/pi)/(inode-1); yj = yj/(jnode-1);
colormap([0.5 0.5 0.5]); subplot(2,1,1); surfl(xi,yj,P'); axis([0 max(xi) 0 1 0 100]);
zlabel('Dimensionless pressure [%]'); subplot(2,1,2); surfl(xi,yj,TORR');
axis([0 max(xi) 0 1 0 ceil(max(max(TORR')))]); zlabel('Dimensionless friction force [%]');
xlabel('Degrees'); ylabel('Inlet');
text('units','normalized','position',[0.1 2.5],'string',...
     'PRESSURE AND FRICTION FORCE FIELDS FOR PARTIAL BEARING');
% PRINT OUT VALUES OF INPUT AND OUTPUT DATA
% ---------------------------------------
fprintf(' \n'); fprintf(' INPUT AND OUTPUT DATA FOR PROGRAM PARTIAL\n'); fprintf(' \n');
fprintf(' INPUT DATA:\n');
fprintf('  Eccentricity ratio = %0.5g\n',epsilon);
fprintf('  L/D ratio = %0.5g\n',loverd);
fprintf('  Bearing arc angle = %0.5g°\n',alpha*180/pi);
fprintf('  Misaligment parameter = %0.5g\n',t);
fprintf(' \n');
fprintf(' OUTPUT DATA:\n');
fprintf('  Dimensionless load = %0.5g\n', loadw);
fprintf('  Attitude angle = %0.5g°\n',beta*180/pi);
fprintf('  Petroff multiplier = %0.5g\n',friction);
fprintf('  Dimensionless friction coefficient = %0.5g\n', myu)
fprintf('  Maximum dimensionless pressure = %0.5g\n',pmax);
fprintf(' \n'); fprintf(' PROGRAM PARTIAL HAS BEEN COMPLETED\n');
set(tbp, 'Value', get(tbp, 'Min')); set([tbv tbs tbt tbd tbg tba tbinfo tbclose],'Enable','on');
```

Program Description

Program '**PARTIAL**' begins with a request for data input from the operator. The controlling variables are eccentricity 'ε', 'L/D' ratio, angle of partial arc and misalignment ratio. This last parameter is defined as the gradient of 'h^*' with respect to 'y^*' along a line immediately above the load vector, i.e., at '$x^* = 0$'. If the eccentricity is zero (i.e., shaft and bush are concentric) and, if for example, the misalignment parameter '$t = 0.5$', then 'h^*' varies along a line directly above the load vector from **0.75** to **1.25** for 'y^*' varying from **-0.5** to **+0.5** (i.e., y = ± **L/2**). The values of relaxation factor, limiting residual, limiting number of iterations and node numbers in the '**i**' and '**j**' directions can be changed, as required, in the input data window.

With all data recorded, the program proceeds to execution and in the first step a finite difference mesh is established. Values of 'δx^*' are generated by dividing the range of 'x^*' or 'y^*' by the number of steps between nodes, i.e.:

deltax = alpha/(inode-1); deltay = 1/(jnode-1);

where:

alpha	is the subtended angle of the partial arc bearing in radians;
inode, jnode	are the number of nodes in the 'x^*' and 'y^*' directions, respectively;
deltax, deltay	correspond to 'δx^*' and 'δy^*'.

Zero values are then assigned to the values of the Vogelpohl parameter at each node, (in all programs the Vogelpohl parameter is written as '**M**' or '**M(i,j)**'). A zero value is also assigned to the term '**beta**' which is the angle between the line of centres and the load line (i.e., the attitude angle). The solution for journal or partial arc bearings requires two levels of iteration since the pressure or Vogelpohl parameter must be solved and the attitude angle is unknown. The attitude angle '**beta**' is thus iterated for as an outer level of iteration until the load calculated from the pressure field bisects the bearing.

Once the zero value of '**beta**' has been assigned, the program enters the outer level of iteration for '**beta**'. It is now necessary to calculate values of '$F(i,j)$' and '$G(i,j)$' (which are the equivalent of parameters '$F_{i,j}$' and '$G_{i,j}$'). Values of 'h^*', '$\partial h^*/\partial x^*$', '$\partial h^*/\partial y^*$' and '$\partial^2 h^*/\partial x^{*2}$' are required and are calculated from equations (5.24), (5.25), (5.26) and (5.27). The term '$\partial^2 h^*/\partial y^{*2}$' is zero and is not included in the calculations. The terms listed in equations (5.25), (5.26) and (5.27) are incorporated into the program with '**dhdx0**' representing '$\partial h^*/\partial x^*$', '**dhdy0**' equivalent to '$\partial h^*/\partial y^*$', '**d2hdx20**' as '$\partial^2 h^*/\partial x^{*2}$' and '**d2hdy20**' as '$\partial^2 h^*/\partial y^{*2}$'. Values of '$F(i,j)$' and '$G(i,j)$' are then found from expressions based on equations (5.5) and (5.6). Expressions for 'C_1' and 'C_2' are included later in the program although they could be performed earlier, and are represented by the parameters '**coeff1**' and '**coeff2**', respectively.

The program now proceeds to the iteration for '$M(i,j)$'. A parameter '**sum**' is arranged to collect the total of values of '$M(i,j)$' generated during one round of iteration. An initial value of zero is assigned to '**sum**' and then a pair of nested '**for**' loops commence the iteration for '$M(i,j)$' within the range of '**i**' from **2** to '**inode-1**' and '**j**' from **2** to '**jnode-1**'. This range of '**i**' and '**j**' covers all the nodes except those on the edge of the bearing as illustrated in Figure 5.2. The misaligned bearing is also analyzed in this program so that an entire bearing domain is covered by the iteration. The finite difference equation (5.9) is modified to enable 'over-relaxation'. Instead of '$M(i,j)$', the value of the right-hand side of (5.9) is assigned to a variable '**store**'. The final value of '$M(i,j)$' is then calculated from:

M(i,j) = M(i,j)+factor1*(store-M(i,j));

where the terms on the right hand side of the computing expression are the old values and the term on the left-hand side is the new value. The term '**factor1**' is the Gauss-Seidel relaxation factor whose value is typically **1.3** for this iteration. Where negative values of '$M(i,j)$' are generated they are immediately suppressed by the statement:

if M(i,j) < 0, M(i,j) = 0; end;

The final value of '$M(i,j)$' is then added to the collector term '**sum**' via the statement:

sum = sum+M(i,j);

which completes the statements within the '**for**' loops.

Outside of the '**for**' loops, a residual is compiled from the difference between the current value of '**sum**' and a stored value '**sum2**' from the previous round of iteration:

residp = abs((sum-sum2)/sum);

The value of '**sum**' is then transferred to '**sum2**' and '**residp**' is tested to determine if it is below the convergence limit. To test convergence an inequality is used with a limit on the number of sweeps or iteration rounds included:

while (residp > reslim1) & (n1 < nlim1),

where:

reslim1	is the prescribed value of residual to terminate the iteration;
n1	is a counter variable for the number of sweeps;
nlim1	the limiting number of sweeps.

When the iteration for the Vogelpohl parameter is completed, the program proceeds to the calculation for attitude angle. To calculate the attitude angle, values of the pressure integral parallel and normal to the load-line are required. Values of '$P(i,j)$' are found from the following expression:

P(i,j) = M(i,j)/H(i,j)^1.5;

Calculations of the integrals are based on the 'trapezium rule' with 'P(i,j)' multiplied by '-cos(x*)' for the pressure integral parallel to the load line and multiplied by 'sin(x*)' for the film force pressure integral normal to the load line. The trapezium rule is applied by a 'for' statement where, at each node, the value of the function to be integrated at this node and the preceding node are added to the integral. The 'for' statements run from the second to the final node in any given line of nodal values to be integrated. In this way, all nodes except the first contribute twice the nodal value to the integral sum as required by the trapezium rule. The sequence of steps for compiling a series of 'y*' line integrals over the complete solution domain is as follows:

```
for i = 1 : inode,
  SUMY(i) = 0;
  for j = 2 to jnode,
    SUMY(i) = SUMY(i)+P(i,j)+P(i,j-1);
  end;
  SUMY(i) = SUMY(i)*0.5*deltay;
end;
```

The film force term parallel to the load line is called 'axialw' and the normal film force is called 'transw'. Integration is first performed in the 'y*' direction to generate an array of integrals. The array is then multiplied by either 'sin(x*)' or 'cos(x*)' and integrated with respect to 'x*'. The sequence of steps is:

```
axialw = 0; transw = 0;
for i = 2 : inode;
  x = (i-1)*deltax+pi-0.5*alpha; x2 = (i-2)*deltax+pi-0.5*alpha;
  axialw = axialw-cos(x)*SUMY(i)-cos(x2)*SUMY(i-1);
  transw = transw+sin(x)*SUMY(i)+sin(x2)*SUMY(i-1);
end;
axialw = axialw*deltax*0.5; transw = transw*deltax*0.5;
```

The attitude angle can now be calculated from the arctangent of the ratio of 'transw' to 'axialw':

```
attang = atn(transw/axialw);
```

where 'attang' is the attitude angle. A conditional statement is also included to allow for the possibility of a negative axial film force where 'attang1' would be the negative of the above expression. The value of 'beta' is found from over-relaxation of the attitude angle 'attang1' with the refinement that the desired value of 'attang1' is already known, i.e., it is zero since the load vector should pass through 180°. This leads to the step

```
beta = beta+factor2*attang1;
```

A residual of the iteration in 'beta' is then calculated in a similar manner to the residual of the Vogelpohl parameter, i.e.:

```
residb = abs((beta-betas)/beta);
```

where 'betas' is the value of 'beta' from the previous iteration for 'beta'. A conditional return to calculations of 'F(i,j)' and 'G(i,j)' is then based on whether 'residb' has reached the termination value and that the number of iteration rounds has not exceeded a predetermined limit.

Finally the dimensionless friction coefficient is calculated. This begins with a search to find the cavitation front which is defined by an array 'ICAV(j)'. Values of 'ICAV(j)' are found by searching for zero 'M(i,j)' values in the down-stream direction and assigning to 'ICAV(j)' the first 'i' value where 'M(i,j)' is zero and 'ICAV(j)' still has a default value.

An array of dimensionless shear stress, 'TORR(i,j)', is then calculated with two expressions supplied for the cavitated and uncavitated regions of the bearing, i.e.:

if i < ICAV(j), TORR(i,j) = 1/H(i,j)+3*dpdx*H(i,j); end;

if i > ICAV(j), TORR(i,j) = H(i10,j)/H(i,j)^2; end;

where:

dpdx is the local value of 'dp^*/dx^*' calculated by linear interpolation;

i10 is a substitute for 'ICAV(j)' to simplify the array expression in the second statement.

The dimensionless friction force and the friction coefficient parameter defined in equations (5.17) and (5.21), respectively, are also calculated. These parameters are calculated by double summation with respect to the 'x' and 'y' axes of the dimensionless shear stress as defined in equation (5.16). The summation is performed inside nested 'for' loops for 'I' and 'J' with a logical statement to cover the possibility of cavitation. Where cavitation occurs it is necessary to allow for division of the film into streamers of lubricant and streamers of air or gas. In the uncavitated regions this is not necessary and so a conditional statement is inserted to exclude the uncavitated regions from calculations with streamers based on equation (5.23).

After summation, the dimensionless shear stresses and the dimensionless friction force are printed out. The friction coefficient parameter as defined equation (5.22) is then calculated and printed out.

When both levels of iteration for the Vogelpohl parameter and the attitude angle are completed and the friction coefficient determined, the final task remaining is to print out the values of load and pressure profile. A search is made for the maximum pressure and all pressure values are displayed as percentages of the maximum pressure. This practice avoids any difficulties with format when the range of values of dimensionless pressure is large. The absolute value of maximum pressure (dimensionless) is stored and printed out for reference purposes.

List of variables

M	Variable corresponding to 'M_v', the Vogelpohl parameter;
P	Dimensionless pressure 'p^*';
H	Dimensionless film thickness 'h^*';
F	Parameter 'F' of Vogelpohl equation;
G	Parameter 'G' of Vogelpohl equation;
SUMY	Storage parameter for integrating in 'y^*' direction;
ICAV	Integer variable to locate cavitation front;
TORR	Dimensionless shear stress 'τ^*' in friction calculations;
epsilon	Eccentricity ratio 'ε';
loverd	'L/D' ratio;
slender	'R/L' ratio;
alpha	Arc angle or subtended angle of partial arc bearing;

t	Misalignment parameter '**t**';
inode	Number of nodes in the '**i**' or '**x***' direction;
jnode	Number of nodes in the '**j**' or '**y***' direction;
reslim1	Terminating value of residual for iteration to solve Vogelpohl equation;
reslim2	Terminating value of residual for iteration to find attitude angle;
factor1	Relaxation factor of iteration to solve Vogelpohl equation;
factor2	Relaxation factor of iteration to find attitude angle;
nlim1	Maximum allowable number of cycles during iteration to solve Vogelpohl equation;
nlim2	Maximum allowable number of cycles during iteration to find attitude angle;
deltax	Mesh dimension in '**x***' direction;
deltay	Mesh dimension in '**y***' direction;
beta	Angle between line of centres and load line of partial arc bearing, i.e., attitude angle;
betas	Storage value of '**beta**' during successive iteration cycles;
n2	Number of cycles of iteration for attitude angle;
i	Number of nodes from '$x^* = 0$' position on finite difference mesh;
x	Equivalent to '**x***', dimensionless distance in direction of sliding, datum position located with respect to partial arc, not the line of shaft and journal centres;
theta	Equivalent to 'θ', the angle between specified position and line of shaft and journal centres;
j	Number of nodes from '$y^* = 0$' position on finite difference mesh;
y	Equivalent to '**y***';
h0	Equivalent to '**h***';
dhdx0	Equivalent to '$\partial h^*/\partial x^*$';
d2hdx20	Equivalent to '$\partial^2 h^*/\partial x^{*2}$';
dhdy0	Equivalent to '$\partial h^*/\partial y^*$';
d2hdy20	Equivalent to '$\partial^2 h^*/\partial y^{*2}$';
coeff1	Equivalent to 'C_1' in Vogelpohl equation;
coeff2	Equivalent to 'C_2' in Vogelpohl equation;
loadw	Equivalent to '**W***';
attang	Attitude angle in radians;
transw	Component of '**W***' normal to bisector of partial arc;
axialw	Component of '**W***' codirectional with bisector of partial arc;
friction	Dimensionless frictional force from integration of 'τ^*' over bearing area;
myu	Dimensionless friction coefficient, equivalent to 'μ^*';
pmax	Maximum nodal dimensionless pressure;

sum2   Storage variable of sum of 'M(i,j)' values to calculate residual during iteration to solve Vogelpohl equation;

sum   Similar function to 'sum2' except it relates to compilation of residual in the current cycle of iteration;

store   Temporary storage of calculated value of 'M(i,j)' from finite difference equation, used to facilitate over-relaxation;

residp   Residual of iteration to solve Vogelpohl equation;

xaux   Auxiliary variable used during integration by trapezium rule.

friccoeff   Dimensionless friction coefficient.

EXAMPLE

Input data: Eccentricity ratio = **0.8**, L/D ratio = **1**, bearing arc angle = **120°**, misalignment parameter = **0**, standard program settings for iteration (i.e., no additional input required).

Output data: Dimensionless load = **0.98048**, attitude angle = **27.52°**, Petroff multiplier = **7.5444**, dimensionless friction coefficient = **7.6946**, maximum dimensionless pressure = **1.7078**.

PRESSURE AND FRICTION FORCE FIELDS FOR PARTIAL BEARING

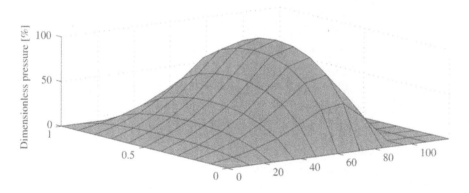

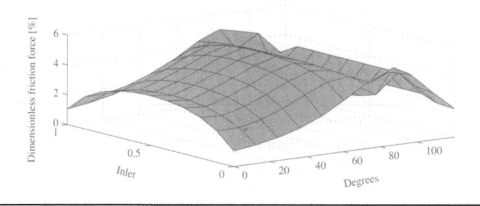

A.5   PROGRAM 'THERMAL'

This program calculates the temperature field and load capacity of a thermohydrodynamic pad bearing where frictional heating of the lubricant causes variation in viscosity. A perfectly rigid bearing is assumed. The pad geometry and film thickness are pre-determined in this program and a corresponding load capacity is calculated.

```
clc; cla reset; echo off; clear P U W T VISC;
global tbv tbs tbp tbt tbd tbg tba tbinfo tbclose;
set([tbv tbs tbp tbd tbg tba tbinfo tbclose],'Enable','off');
% BEGIN OF INPUT DATA
% --------------------
prompt = {'Select variable: 0 - isothermal; 1 - adiabatic bearing',
          'Maximum film thickness [mm]:',
          'Minimum film thickness [mm]:',
          'Sliding velocity [m/s]:',
          'Pad length [m]:'};
title='INPUT DATA (THERMAL)'; lineno=1; def={'1', '0.1', '0.05', '10', '0.1'}; answer= inputdlg(prompt,title,lineno,def);
if size(answer) == 0, % PROGRAM IS TERMINATED
   set(tbt, 'Value', get(tbt, 'Min')); set([tbv tbs tbp tbd tbg tba tbinfo tbclose],'Enable','on'); break; end;
[select,hmax,hmin,u0,length0] = deal(answer{:}); select = str2num(select); hmax = str2num(hmax)/1000;
hmin = str2num(hmin)/1000;u0 = str2num(u0); length0 = str2num(length0);
prompt = {'Viscosity at bearing inlet or datum temperature [Pas]:',
          'Exponential viscosity-temperature coefficient [1/K]:',
          'Specific heat of lubricant [J/kgK]:',
          'Density of lubricant [kg/m^3]:',
          'Thermal conductivity of lubricant [W/mK]:',
          'Bearing inlet temperature [°C]:'};
title='INPUT DATA (THERMAL)'; lineno=1; def={'0.5','0.05','2000','900','0.15','50'};
answer= inputdlg(prompt,title,lineno,def);
if size(answer) == 0, % PROGRAM IS TERMINATED
   set(tbt, 'Value', get(tbt, 'Min')); set([tbv tbs tbp tbd tbg tba tbinfo tbclose],'Enable','on'); break; end;
[visc0,gamma,cp1,rh0,condoil,tinlet] = deal(answer{:}); visc0= str2num(visc0); cp1= str2num(cp1);
rh0= str2num(rh0); tinlet= str2num(tinlet) + 273.15; condoil= str2num(condoil); gamma = str2num(gamma);
prompt = {'Number of nodes in the i or x direction:',
          'Number of nodes in the k or z direction:',
          'Terminating value of residual for pressure field iter.:',
          'Terminating value of residual for temperature field iter.:',
          'Terminating value of residual for temperature-pressure iter.:',
          'Terminating value of residual for iter. to find value of w:'};
title='INPUT DATA (THERMAL)'; lineno=1; def={'6','5','0.0000005','0.000001','0.0001','0.000001'};
answer= inputdlg(prompt,title,lineno,def);
if size(answer) == 0, % PROGRAM IS TERMINATED
   set(tbt, 'Value', get(tbt, 'Min')); set([tbv tbs tbp tbd tbg tba tbinfo tbclose],'Enable','on'); break; end;
[inode,knode,reslim1,reslim2,reslim3,reslim4] = deal(answer{:}); inode = str2num(inode); knode = str2num(knode);
reslim1 = str2num(reslim1); reslim2 = str2num(reslim2); reslim3 = str2num(reslim3); reslim4 = str2num(reslim4);
prompt = {'Relaxation factor for pressure field iter.:',
          'Relaxation factor for temperature field iter.:',
          'Relaxation factor for temperature-pressure iter.:',
          'Max number of cycles for pressure field iter.:',
          'Max number of cycles for temperature field iter.:',
          'Max number of cycles for pressure-temperature iter.:'};
title='INPUT DATA (THERMAL)'; lineno=1; def={'1','0.7','0.14','100','200','100'}; answer= inputdlg(prompt,title,lineno,def);
if size(answer) == 0, % PROGRAM IS TERMINATED
   set(tbt, 'Value', get(tbt, 'Min')); set([tbv tbs tbp tbd tbg tba tbinfo tbclose],'Enable','on'); break; end;
[factor1,factor2,factor3,nlim1,nlim2,nlim3] = deal(answer{:}); factor1 = str2num(factor1); factor2 = str2num(factor2);
factor3 = str2num(factor3); nlim1 = str2num(nlim1); nlim2 = str2num(nlim2); nlim3 = str2num(nlim3);
% END OF INPUT DATA
% -----------------
subplot(1,1,1);
text('units','normalized','position',[0.2 0.55], 'FontWeight', 'bold','color',[1 0 0], 'string', 'CALCULATIONS IN PROGRESS');
figure(1); if select == 0, klim = knode-1; end; if select == 1, klim = knode; end;
deltax = length0/(inode-1);
for i = 1:inode+1, DELTAZ(i) =(hmax - (i-1)*(hmax-hmin)/(inode-1))/(knode-1); end;
i = 0; deltazd = (hmax-(i-1)*(hmax-hmin)/(inode-1))/(knode-1);
% SET MESH CONSTANTS OF PAD
% INITIALIZE TEMPERATURE, VISCOSITY AND PRESSURE FIELDS
W = zeros(inode,knode+1); P = zeros(1,inode); T = ones(inode,knode)*tinlet; TSTORE = T;
VISC = ones(inode,knode)*visc0;
% OUTER LOOP OF ITERATION,OUTER LOOP OF RESIDUAL COUNTER
```

```
ncount3 = 0; psumold = 0; residp2 = reslim3 +10;
while (residp2 > reslim3) & (ncount3 < nlim3),
  psum = 0;
 % SOLUTION FOR PRESSURE
 for i = 1:inode,
    for k = 1:knode,
      F1(k) = 1/VISC(i,k); F2(k) = DELTAZ(i)*(k-1)/VISC(i,k);
    end;
    % FIRST ORDER INTERGRALS OF VISCOSITY
    for k = 1:knode,
      INT1(k) = 0; INT2(k) = 0;
      for m = 1:k-1,
        INT1(k) = INT1(k) + 0.5*F1(m) + 0.5*F1(m+1); INT2(k) = INT2(k) + 0.5*F2(m) + 0.5*F2(m+1);
      end;
      INT1(k) = INT1(k)*DELTAZ(i); INT2(k) = INT2(k)*DELTAZ(i);
    end;
    % COMPUTE SECONDARY INTERGRALS
    for k = 1:knode, MM(i,k) = INT2(k)-INT2(knode)*INT1(k)/INT1(knode); NN(i,k) = 1 - INT1(k)/INT1(knode); end;
    % SECOND ORDER INTERGRALS
    INTM(i) = 0; INTN(i) = 0;
    for k = 1:knode - 1,
      INTM(i) = INTM(i) + 0.5*MM(i,k)+0.5*MM(i,k+1); INTN(i) = INTN(i) + 0.5*NN(i,k)+0.5*NN(i,k+1);
    end;
    INTM(i) = INTM(i)*DELTAZ(i); INTN(i) = INTN(i)*DELTAZ(i);
 end;
 % PRESSURE, 1-D ITERATION OF SECOND DIFFERENTIAL EQUATION FOR PRESSURE
 ncount1 = 0; psum11 = 0; residp = reslim1 +10;
 while (residp > reslim1) & (ncount1 < nlim1),
    psum1 = 0;
    for i = 2:inode-1,
      store = P(i);
      store2 = 0.5*(P(i+1)+P(i-1)) + 0.125*(P(i+1)-P(i-1))*(INTM(i+1) - INTM(i-1))/INTM(i) + u0*(INTN(i+1)-INTN(i-1))*...
          deltax*0.25/INTM(i);
      P(i) = store + (store2-store)*factor1; psum1 = psum1 + abs(P(i));
    end;
    residp = abs((psum1-psum11)/psum1); psum11 = psum1; ncount1 = ncount1 + 1;
 end;
 % START TEMPERATURE SOLUTION
 % COMPUTE VELOCITY FIELDS
 for i = 2:inode-1, DPDX(i) = 0.5*(P(i+1)+P(i-1))/deltax; end;
 DPDX(1) = P(2)/deltax; DPDX(inode) = -P(inode-1)/deltax;
 for i = 1:inode,
    for k = 1:knode,
      U(i,k) = MM(i,k)*DPDX(i)+NN(i,k)*u0;
    end;
 end;
 for i = 1:inode, U(i,knode) = 0; end;
 % DUMMY ARRAY OF U AT knode+1
 for i = 1:inode, U(i,knode+1) = 2*U(i,knode)-U(i,knode-1); end;
 % FIND VALUES OF DUDZ
 for i = 1:inode,
    for k = 2:knode,
      if k < knode, DUDZ(i,k) = (U(i,k+1)-U(i,k-1))/(2*DELTAZ(i)); end;
      if k == knode,
        DUDZ(i,k) = (-2*U(i,k-1)+1.5*U(i,k)+0.5*U(i,k-2))/DELTAZ(i);
      end;
    end;
 end;
 % SUPPLY DUMMY ARRAY OF U at inlet
 for k = 1:knode,
    UD(k) = 2*U(1,k) - U(2,k); U(inode+1,k) = 2*U(inode,k) - U(inode-1,k);
 end;
 % FIND VELOCITY NORMAL TO FILM, COMPUTE WTERM
 for i = 2:inode,
    for k = 2:knode-1,
      use = U(i+1,k-1); usw = U(i-1,k-1); une = U(i+1,k+1); unw = U(i-1,k+1);
      WTERM(i,k) = 0.125*DELTAZ(i)*(une-unw-use+usw)/deltax;
    end;
 end;
 % ITERATE FOR VELOCITY NORMAL TO FILM
 for i = 2:inode,
    wsumold = 0; residw = reslim4 +10;
```

```
    while residw > reslim4,
       wsum = 0;
      for k = 2:knode-1,
         W(i,k) = WTERM(i,k) + 0.5*W(i,k+1) + 0.5*W(i,k-1); wsum = wsum + abs(W(i,k));
      end;
       residw = abs((wsum-wsumold)/wsum); wsumold = wsum;
    end;
  end;
 for k = 1:knode, W(1,k) = 2.5*W(2,k) - 2*W(3,k) + 0.5*W(4,k); end;
% COMPUTE COEFFICIENTS OF TEMPERATURE ITERATION
 for k = 2:klim,
    ae = -0.25*(DELTAZ(1) + DELTAZ(2))*(U(2,k) +  U(1,k))*0.5*cp1*rh0;
    aw = 0.25*(deltazd + DELTAZ(1))*(UD(k)+ U(1,k))*cp1*rh0*0.5;
    at = (condoil*deltax/DELTAZ(1)) - (deltax*(W(1,k+1) +  W(1,k))*cp1*0.25*rh0);
    ab = (condoil*deltax/DELTAZ(1))+(deltax*(W(1,k-1) +  W(1,k))*cp1*0.25*rh0);
    s = DUDZ(1,k)^2*VISC(1,k); sp = -gamma*s; sc = s*(1+gamma*T(1,k));
    b = sc*deltax*DELTAZ(1); ap = ae + aw + at + ab -sp*deltax*DELTAZ(1);
    e1 = abs(ae) + abs(aw) + abs(at) + abs(ab) - abs(ap);
    if e1 < 0,  e = 0; end; if e1 == 0, e = 0; end; if e1 > 0,  e = e1; end;
    CE(1,k) = ae/(ap+e); CW(1,k) = aw/(ap+e); CN(1,k) = at/(ap+e); CS(1,k) = ab/(ap+e);
    CB(1,k) = b/(ap+e); CP(1,k) = e/(ap+e);
 end;
 for i = 2:inode,
    for k = 2:klim,
       ae = -0.25*(DELTAZ(i) + DELTAZ(i+1))*(U(i+1,k) +  U(i,k))*0.5*cp1*rh0;
       aw = 0.25*(DELTAZ(i-1)+DELTAZ(i))*(U(i-1,k)+ U(i,k))*0.5*cp1*rh0;
       at = (condoil*deltax/DELTAZ(i)) - (deltax*(W(i,k+1) +  W(i,k))*0.25*cp1*rh0);
       ab = (condoil*deltax/DELTAZ(i)) + (deltax*(W(i,k-1) +  W(i,k))*cp1*.25*rh0);
       s = DUDZ(i,k)^2*VISC(i,k); sp = -gamma*s;
       sc = s*(1 + gamma*T(i,k)); ap = ae + aw + at + ab - sp*deltax*DELTAZ(i);
       b = sc*deltax*DELTAZ(i); e1 = abs(ae) + abs(aw) + abs(at) + abs(ab) - abs(ap);
       if e1 < 0, e = 0; end; if e1 == 0, e = 0; end; if e1 > 0, e = e1; end;
       CE(i,k) = ae/(ap+e); CW(i,k) = aw/(ap+e); CN(i,k) = at/(ap+e);
       CS(i,k) = ab/(ap+e); CB(i,k) = b/(ap+e); CP(i,k) = e/(ap+e);
    end;.
 end;
% TEMPERATURE ITERATION CYCLE BEGINS
ncount2 = 0; sumold=0; residt = reslim2 +10;
while (residt > reslim2) & (ncount2 < nlim2),
  sum = 0;
% SMOOTHING OF TEMPERATURES ON PAD SURFACE FOR ADIABATIC CASE
  for i = 2:inode,
    if T(i,knode) < T(i-1,knode), T(i,knode) = T(i-1,knode); end;
  end;
  for k = 1:knode,
     TD(k) = 2*T(1,k)-T(2,k); T(inode+1,k) = T(inode,k);
  end;
  T(1,knode) = tinlet;
  for i = 1:inode+1, T(i,knode+1) = T(i,knode); end;
  for i = 1:inode,
     for k = 2:klim,
        nostore = 1; nostore1 = 1;
        if U(1,k) > 0, T(1,k) = tinlet; end; if U(1,k) == 0, T(1,k) = tinlet; end; if (i == 1) & (U(1,k) > 0), nostore = 0; end;
        if nostore == 1,
           if (i == 1) & (U(1,k) == 0), nostore = 0; end;
        end;
        if nostore == 1,
          if (i == 1) & (U(1,k) < 0),
             store = CE(i,k)*T(i+1,k) + CW(i,k)*TD(k) +  CN(i,k)*T(i,k+1) + CS(i,k)*T(i,k-1) + CB(i,k) +  cp(i,k)*T(i,k);
             nostore1 = 0;
          end;
          if nostore1 == 1,
             store = CE(i,k)*T(i+1,k) + CW(i,k)*T(i-1,k) + CN(i,k)*T(i,k+1) + CS(i,k)*T(i,k-1) + CB(i,k) +  CP(i,k)*T(i,k);
          end;
          T(i,k) = T(i,k) + (store-T(i,k))*factor2; if T(i,k) < tinlet, T(i,k) = tinlet; end; sum = sum + abs(T(i,k));
        end;
     end;
  end;
% DUMMY ARRAY AT knode+1 FOR ADIABATIC CONDITION
% EVALUATE CONVERGENCE OF TEMPERATURE PROFILE
  residt = abs((sum-sumold)/sum); sumold = sum; sum = 0; ncount2 = ncount2 + 1;
end; % THE END OF while LOOP
```

```
% OUTER ITERATION LOOP
% CALCULATE VISCOSITY WITH NEW TEMPERATURE FIELD
for i = 1:inode,
   for k = 2:klim,
      T(i,k) = (T(i,k)-TSTORE(i,k))*factor3 + TSTORE(i,k); TSTORE(i,k) = T(i,k); term10 = gamma*(T(i,k)-tinlet);
      VISC(i,k) = visc0/exp(term10);
   end;
end;
% PRESSURE FIELD RESIDUALS OF OUTER ITERATION CYCLE
psum = 0;
for i = 1:inode, psum = psum+abs(P(i)); end;
residp2 = abs((psum-psumold)/psum); ncount3 = ncount3 + 1; psumold = psum;
end; % THE END OF while LOOP
% OUTPUT PRESSURE, TEMPERATURE, VELOCITY AND VISCOSITY DATA
% FIND MAXIMUM TEMPERATURE, PRESSURE, VELOCITIES AND VISCOSITY
umax = 0; wmax = 0;
for i = 1:inode,
   for k = 1:knode,
      if abs(U(i,k)) > umax, umax = abs(U(i,k)); end; if abs(W(i,k)) > wmax, wmax = abs(W(i,k)); end;
   end;
end;
pmax = max(abs(P)); tmax = max(max(abs(T))); viscmax= max(max(abs(VISC))); P = P/pmax;
% OUTPUT LOAD CAPACITY
sum = 0;
for i = 1:inode-1, sum = sum + 0.5*P(i) + 0.5*P(i+1); end;
sum = sum*deltax*pmax;
% DIVIDE PRESSURE, TEMPERATURE, VELOCITIES AND VISCOSITY BY
% CORRESPONDING MAXIMUM
for i = 1:inode,
   P(i) = P(i)/pmax;
   for k = 1:knode,
      U(i,k) = U(i,k)/umax; W(i,k) = W(i,k)/wmax; T(i,k) = T(i,k)/tmax; VISC(i,k) = VISC(i,k)/viscmax;
   end;
end;
% CALCULATE AVERAGE SHEAR STRESS
for i=1:inode
      storetorr = -VISC(i,1)*(U(i,2)-U(i,1))/DELTAZ(i)-...
      VISC(i,knode)*(U(i,knode)-U(i,knode-1))/DELTAZ(i);
   for k=2:knode-1
      torrlocal = -VISC(i,k)*(U(i,k+1)-U(i,k-1))/(2*DELTAZ(i));
      storetorr = storetorr + torrlocal;
   end;
   TORR(i)=storetorr/knode;
end;
% FRICTION FORCE
friction = TORR(1)*deltax*0.5;
for i=2:inode-1,
   friction = friction+0.5*deltax*(TORR(i)+TORR(i+1));
end;
% FRICTION COEFFICIENT
friccoef = friction/sum;
subplot(2,1,1); colormap([0 0 0]);
xaux = linspace(inode,0,inode); clear x y;
x(1:inode,1) = xaux';
y(1,1:knode) = linspace(1,knode,knode);
for k=2:knode, x(1:inode,k) = x(1:inode,k-1); end;
for i=2:inode, y(i,1:knode) = y(i-1,1:knode); end;
mesh(x,y,T(1:inode,1:knode)*100); view(-235,35);
zlabel('Dimensionless temperature [%]');xlabel('Length (nodes)');
ylabel('Inlet (nodes)');
subplot(2,1,2);
x = 0:inode-1; x = x/(inode-1);
plot(x,TORR);grid on;
ylabel('Frictional stress [Pa]');xlabel('Distance from bearing inlet');
if select == 0,
   text('units','normalized','position',[-0.1 2.55],'string',...
      'TEMPERATURE FIELD AND FRICTIONAL STRESS IN ISOTHERMAL PAD BEARING');
else
   text('units','normalized','position',[-0.1 2.55],'string',...
      'TEMPERATURE FIELD AND FRICTIONAL STRESS IN ADIABATIC PAD BEARING');
end;
clc;
```

```
% PRINT OUT VALUES OF INPUT AND OUTPUT DATA
% ------------------------------------------
fprintf(' \n'); fprintf(' INPUT AND OUTPUT DATA FOR PROGRAM THERMAL\n'); fprintf(' \n');
fprintf(' INPUT DATA:\n');
if  select == 0, fprintf(' Isothermal pad bearing.\n'); end;
if  select == 1, fprintf(' Adiabatic pad bearing.\n'); end;
fprintf(' Max. film thickness = %0.5g [mm]\n',hmax*1000);
fprintf(' Min. film thickness = %0.5g [mm]\n',hmin*1000);
fprintf(' Sliding velocity = %0.5g [m/s]\n',u0);
fprintf(' Pad width = %0.5g [m]\n',length0);
fprintf(' Viscosity of lubricant at bearing inlet = %0.5g [Pas]\n',visc0);
fprintf(' Exponential viscosity index = %0.5g [1/K]\n',gamma);
fprintf(' Specific heat of lubricant = %0.5g [J/kgK]\n',cp1);
fprintf(' Density of lubricant = %0.5g [kg/m^3]\n',rh0);
fprintf(' Bearing inlet temperature = %0.5g°C\n',tinlet-273.15);
fprintf(' Themal conductivity of lubricant = %0.5g [W/mK]\n',condoil);
fprintf(' \n');
fprintf(' OUTPUT DATA:\n');
fprintf(' Max. pressure  = %0.5g [MPa]\n',pmax/10^6);
fprintf(' Critical max. temperature = %0.5g°C\n',tmax-273.15);
fprintf(' Load capacity = %0.5g [MN/m]\n',sum/10^6);
fprintf(' Max. lubricant velocity in plane of oil film = %0.5g [m/s]\n',umax);
fprintf(' Max. lubricant velocity normal to oil film = %0.5g [mm/s]\n',wmax*1000);
fprintf(' Min. viscosity = %0.5g [Pas]\n',min(min(VISC)));
fprintf(' Friction force = %0.5g [N/m]\n',friction);
fprintf(' Friction coefficient = %0.5g [Dimensionless]\n',friccoef);
fprintf(' \n'); fprintf(' PROGRAM THERMAL HAS BEEN COMPLETED\n');
set(tbt, 'Value', get(tbt, 'Min')); set([tbv tbs tbp tbd tbg tba tbinfo tbclose],'Enable','on');
```

Program Description

Program 'THERMAL' begins with a series of prompts for data. Much more information about the physical properties of the lubricant is required for this program than in the isoviscous case. The data is required for the following parameters: the maximum and minimum film thicknesses 'hmax' and 'hmin', entraining velocity of the bearing 'u0', bearing width 'length0' ('length0' is the variable representing the length of the bearing in the direction of sliding, i.e., width), viscosity index of the lubricant 'gamma', datum viscosity of the lubricant 'visc0', specific heat of the oil 'CP', density of the lubricant 'rh0', the conductivity of the oil 'condoil' and the inlet temperature 'tinlet' which, in this program, is assumed to be identical to the sliding surface temperature. The data prompts are structured into two sections, the first as already described and the second relating to the solution parameters, such as iteration limits, which are similar to programs discussed previously. The program is written with values of the latter parameters built in, which are valid for most problems. If, however, extremes of the heating are to be studied a reduced iteration factor in the temperature iteration 'factor2' and a large cycle limit number 'nlim2' may be required to ensure convergence. The program is written to operate with a coarse mesh which is quite adequate for estimations of load capacity and basic bearing analysis. If a more detailed solution is required, a finer mesh can be set but the computing time increases sharply with the fineness of the mesh. An extra variable called 'select', which is also included in the data prompts, enables the choice of isothermal or adiabatic solution.

After inputting all required data, the program proceeds to mesh generation which is similar to the isoviscous programs, except that a dimension normal to the plane of the hydrodynamic film is introduced, i.e., the 'z' direction. The step length in the 'z' direction, 'DELTAZ' is not constant but is proportional to the local film thickness hence the following statement is required to facilitate generation of the mesh:

DELTAZ(i) = (hmax-(i-1)*(hmax-hmin)/(inode-1))/(knode-1);

where 'knode' is the number of nodal steps in the 'z' or 'k' direction. The notation 'i', 'j' and 'k' is based on unit directions in the cartesian coordinate system.

At this stage, the limits of iteration must also be defined. For example, for isothermal bearings, iteration proceeds from 'k = 2' to 'k = knode-1' and for adiabatic bearings the iteration proceeds from 'k = 2' to 'k = knode'. This is controlled by the statements:

if select = 0, klim = knode-1; end;

if select = 1, klim = knode; end;

where:

> **select**   is a control variable whose value is **0** for isothermal bearings or **1** for adiabatic bearings;
>
> **klim**   is the variable prescribing the limits of iteration.

Initialization of the pressure, velocity and temperature fields is performed next. The initial value of the pressure field is zero, the oil inlet temperature is assigned to the temperature field and the datum viscosity is assigned to the viscosity field. The fluid velocity normal to the lubricant film, 'W(i,k)' is also given zero values at this stage. On completion of these tasks, the program enters the first cycle of iteration.

The iteration cycle begins with computation of the viscosity integrals 'dz/η' and 'zdz/η' (equation (5.34)), based on an initially assumed constant viscosity. Two variables, 'F1(k)' and 'F2(k)', are introduced to provide an array of values corresponding to 'dz/η' and 'zdz/η', respectively. The position along the 'z' axis is found from the product of 'DELTAZ' and the number of mesh lengths, 'k-1'. The first order integrals of 'dz/η' and 'zdz/η' are then calculated using a trapezium rule to form the variables 'INT1(k)' and 'INT2(k)'. Three levels of '**for**' loops are used in this process, two outer loops in 'i' and 'k' to locate the calculations for a particular nodal point and an inner loop in 'm' which allows computation of the integrals of 'F1(m)' and 'F2(m)' from the nominal 'm = 1' case to 'm = k'. Variables 'MM(i,k)' and 'NN(i,k)' which correspond to 'M' and 'N' in equation (5.37) are then calculated. These variables are then integrated once again to form the variables 'INTM(i)' and 'INTN(i)' which form the coefficients of the finite difference equation for pressure.

The pressure field is then solved according to equation (5.41) based on the calculated values of 'm' and 'n'. Solution occurs within a '**for**' loop to cover values of 'P(i)' from 'i = 2' to 'inode-1'. Positions 'i = 1' and 'i = inode' which are the bearing inlet and outlet positions, respectively, have zero pressure values. The '**for**' loop is repeated until the residual of 'P(i)' converges to a limit '**reslim1**' or exceeds a critical number of iteration cycles, '**nlim1**'. This procedure is very similar to that already described for the isoviscous bearing. Derivatives of pressure are found from the central difference approximation, discussed in section 5.4, according to the following expressions:

for i = 2 : inode-1, DPDX(i) = 0.5*(P(i+1)+P(i-1))/deltax; end;

DPDX(1) = P(2)/deltax; DPDX(inode) = -P(inode-1)/deltax;

The central difference approximation is also used to find the derivatives with respect to 'x' of the integrals of 'Mdz' and 'Ndz', equations (5.38) and (5.39). When these expressions are substituted into equation (5.41) and the equation is rearranged so as to provide an expression for 'P(i)' the following statement results. A storage term '**store2**' is used to facilitate over-relaxation of this equation:

store2 = 0.5*(P(i+1)+P(i-1))+0.125*(P(i+1)-P(i-1))*(INTM(i+1)- ...

        INTM(i-1))/INTM(i)+u0*(INTN(i+1)-INTN(i-1))*deltax*0.25/INTM(i);

where:

> **u0**   is the bearing entraining velocity [m/s].

The relaxation factor is applied next in the consecutive statement:

P(i) = store+(store2-store)*factor1;

where:

    store   is the value of 'P(i)' from the preceding cycle of iteration;

    factor1   is the iteration factor for the pressure field.

Once the pressure field is known the values of 'u(x,z)' are found from equation (5.37), i.e.:

U(i,k) = MM(i,k)*DPDX(i)+NN(i,k)*u0;

where 'DPDX(i)' (which corresponds to 'dp/dx') was found a few steps before from the array of values of 'P(i)'. The derivative of 'u', ($\partial u/\partial z$), is found next from equation (5.61) as this is required to calculate the viscous dissipation term in the solution for temperature. Knowing 'u' allows the velocity 'w' to be found from equation (5.42). The velocity 'w' is found by iterating individual columns of nodes, i.e., nodes with a common value of 'i'. The iteration equation is a finite difference approximation to (5.42) and is performed in the following manner:

use = U(i+1,k-1); usw = U(i-1,k-1); une = U(i+1,k+1); unw = U(i-1,k+1);
WTERM(i,k) = 0.125*DELTAZ(i)*(une-unw-use+usw)/deltax;

where 'WTERM(i,k)' is the equivalent of '$0.5\partial z^2\left[\dfrac{\partial}{\partial z}\left(\dfrac{\partial u}{\partial x}\right)\right]$' and the terms 'use', 'usw', 'une' and 'unw' correspond to a classification of the various nodal values of 'u' adjacent to 'U(i,k)' according to the points on a compass. Nodal values of the velocity 'w', i.e., 'W(i,k)', are found by iteration of the finite difference expression for 'W(i,k)' which is based on (5.42), i.e.:

W(i,k) = WTERM(i,k)+0.5*W(i,k+1)+0.5*W(i,k-1);

These statements are enclosed in the '**for**' loops and conditional '**while**' statements required by a consecutive iteration of nodes.

With the velocity fields known, the program proceeds to a solution for temperature. A large section of the program therefore relates exclusively to the calculation of coefficients defined in equations (5.52) to (5.60). Two sets of coefficients, an extra set for a column of nodes immediately up-stream of the bearing inlet are also required. The lack of zero dimensions in the variable array requires the introduction of an additional set of dummy computer variables, e.g., 'UD(k)' which corresponds to 'U(0,k)'. In order to prevent excessive calculations within the iteration a final set of reduced coefficients are calculated, i.e.:

CE(i,k) = ae/(ap+e); CW(i,k) = aw/(ap+e); CN(i,k) = at/(ap+e);
CS(i,k) = ab/(ap+e); CB(i,k) = b/(ap+e); CP(i,k) = e/(ap+e);

where 'ap' corresponds to 'a_p', 'ae' to 'a_E', etc.

Before iterating for temperature it is necessary to maintain the boundary conditions. This is done by assigning new values to the arrays of 'T(i,k)' external to the true solution domain. A variable 'TD(k)' is created to provide values of temperature up-stream of the bearing. The value of 'TD(k)' is determined by linear extrapolation from adjacent nodes as described previously. Smoothing of the temperature values calculated at the pad-lubricant interface is necessary for the adiabatic bearing and this is controlled by the statement:

if T(i,knode) < T(i-1,knode) , T(i,knode) = T(i-1,knode); end;

After completing the described operations it is now possible to iterate the temperature field using a pair of nested '**for**' loops with limits '**i = 1**' to '**inode**' and '**k = 2**' to '**klim**'. The iteration at '**i =1**' is only provided to allow for reverse flow. Unless reverse flow occurs the nodes at '**i =1**' are excluded from iteration by a sequence of '**while**' statements positioned in front of the iteration equations for temperature. The controlling iteration statements are:

> **store = CE(i,k)*T(i+1,k)+CW(i,k)*T(i-1,k)+CN(i,k)*T(i,k+1)+ ...**
> **CS(i,k)*T(i,k-1)+CB(i,k)+CP(i,k)*T(i,k);**
> **T(i,k) = T(i,k)+(store-T(i,k))*factor2;**

where '**T(i,k)**' are the nodal values of temperature and '**factor2**' is the relaxation factor for the temperature field.

A similar statement is provided to allow for reverse flow at the inlet where it is necessary to use '**CW(i,k)*TD(k)**' rather than '**CW(i,k)*T(i-1,k)**', i.e.:

> **store = CE(i,k)*T(i+1,k)+CW(i,k)*TD(k)+CN(i,k)*T(i,k+1)+CS(i,k)*T(i,k-1)+CB(i,k)+CP(i,k)*T(i,k);**

The residual of temperature is then collected and iteration proceeds until convergence is reached or the cycle number limit is exceeded.

Once the temperature field has been found, the next task is to relate the calculated values of temperature to a revised viscosity field. Under-relaxation of the temperature change is required or the solution will oscillate between extremes of low and high viscosity without convergence. Under-relaxation is the application of a relaxation factor less than **1**. The reason for using it is that in the first cycle of iteration, the calculated temperatures are very high since the viscosity of unheated lubricant is used in the calculations. During the second cycle of iteration, without sufficient under-relaxation, the resulting low values of viscosity cause the temperature field to fall to unrealistically low values because there is no computed viscous dissipation of heat, leading to a repetition of the process in subsequent cycles. To avoid this the temperature field is therefore stabilized by the statements:

> **T(i,k) = (T(i,k)-TSTORE(i,k))*factor3+TSTORE(i,k);**
> **TSTORE(i,k) = T(i,k);**

where '**factor3**' is the relaxation factor for the overall pressure-temperature iteration.

This ensures that the difference in temperatures between successive cycles is reduced by a factor which can be as low as **0.05**. Once the temperature is known, the viscosity is calculated. To complete the combined two-level cycle of iteration in temperature and pressure described already, it is necessary to evaluate the residual of pressure between successive cycles of temperature iteration. This residual is quite different from the residual obtained at the end of an iteration for '**P(i)**' for a particular viscosity field. The latter residual is calculated by taking the sum of '**P(i)**' and comparing it with a similar sum from the previous iteration. Returning to the beginning of the program's iteration cycle, i.e., returning to the calculations of '**M**' and '**N**' from equations (5.38) and (5.39), depends on whether the residual exceeds the convergence criterion and whether the cycle numbers are less than the specified cycle limit.

As a compromise between physical accuracy and precision in calculating differential quantities from a series of discrete values describing a function, the average value of shear stress though the film thickness is calculated. For each nodal position of '**I**', a '**for**' loop in '**K**' is performed where successive values of '**TORR(I)**' are calculated based on the formula $\tau = \eta du/dx$. Values of '**du/dx**' are estimated from the difference in velocity values between adjacent nodes divided by '**DELTZ(I)**'. The values of '**du/dz**' at the pad and runner surfaces (z= 0 and z =h positions) are assumed to be the same as the values in adjacent nodes, respectively. This is an acceptable approximation for fine meshes (large values of '**K**'). Each nodal position in 'I' and 'K', a value of 'du/dz', is calculated, multiplied by the local viscosity, i.e., 'VISC(I,K)' and then assigned to a storage variable 'torrlocal'. Values of 'torrlocal' are then successively added to the 'sum', which is then divided by the number of nodes to obtain the average value of shear stress for the nodal position 'I'. In subsequent steps, this value of shear stress is summed to obtain the friction force and friction coefficient.

Once the iterations are completed, the calculations are performed to find the maximum values of the pressure, temperature, viscosity and velocity fields. The load per unit length is also found by integrating the pressure with respect to distance. When these quantities are found, the program is completed by printing the values of load, pressure, temperature, viscosity and fluid velocity as fractions of their respective peak values.

List of Variables

P	Hydrodynamic pressure 'p' [Pa];
T	Local film temperature 'T' [°C];
U	Equivalent to 'u' [m/s];
W	Equivalent to 'w' [m/s];
VISC	Local value of viscosity;
DELTAZ	Value of mesh dimension in 'z' direction for a specified distance from bearing inlet;
INT1	Integral of 'dz/η' through film thickness;
INT2	Integral of 'zdz/η' through film thickness;
F1	Local value of 'dz/η';
F2	Local value of 'zdz/η';
MM	Equivalent to 'M' in thermohydrodynamic equations;
NN	Equivalent to 'N' in thermohydrodynamic equations;
INTM	Integral of 'Mdz';
INTN	Integral of 'Ndz';
DPDX	Node value of 'dp/dx';
WTERM	Term used to iterate for 'W';
DUDZ	Equivalent to '$\partial u/\partial z$';
UD	Dummy array of 'u' at bearing inlet;
TD	Dummy array of 'T' at bearing inlet;
TSTORE	Storage of nodal temperature field for purposes of iteration;
TORR(I)	Array of shear stress values along the 'x' axis from pad inlet to exit;
CE, CW, CN, CS,	Reduced coefficients of the finite difference equation;
CP	Finite difference coefficients for the solution of temperature field;
select	Control variable for the selection of adiabatic or isothermal solution;
hmax	Maximum film thickness [m];
hmin	Minimum film thickness [m];
u0	Sliding velocity of the bearing [m/s];
length0	Length of pad in the direction of sliding (i.e., width 'B') [m];

gamma	Equivalent to 'γ' [K^{-1}];
visc0	Viscosity at bearing inlet [Pas];
cp1	Specific heat of lubricant [J/kgK];
rh0	Density of lubricant [kg/m^3];
tinlet	Bearing inlet temperature [°C];
condoil	Conductivity of lubricant [W/mK];
choice	Variable to control data input;
factor1	Relaxation factor for pressure field iteration;
factor2	Relaxation factor for temperature field iteration;
factor3	Relaxation factor for temperature-pressure iteration;
reslim1	Terminating value of residual for pressure field iteration;
reslim2	Terminating value of residual for temperature field iteration;
reslim3	Terminating value of residual for temperature-pressure iteration;
reslim4	Terminating value of residual for iterations to find local value of '**w**';
nlim1	Maximum number of cycles for pressure field iteration;
nlim2	Maximum number of cycles for temperature field iteration;
nlim3	Maximum number of cycles for pressure-temperature iteration;
inode	Number of nodes in the '**i**' or '**x**' direction;
knode	Number of nodes in the '**k**' or '**z**' direction;
klim	Limit of nodes in '**k**' direction for boundary of iteration;
deltax	Mesh dimension in '**i**' or '**x**' direction;
deltazd	Mesh dimension in '**k**' or '**z**' direction of dummy array at bearing inlet;
i	Position in mesh along '**x**' direction;
k	Position in mesh along '**z**' direction;
ncount3	Number of cycles in pressure-temperature iteration;
Psumold	Sum of '**P(i)**' values to compute residual in pressure-temperature iteration;
psum	Similar function to '**Psumold**' but relates to current cycle of iteration;
m	Auxiliary mesh position parameter in '**z**' direction;
ncount1	Number of cycles in iteration for pressure field;
psum11	Sum of '**P(i)**' values to compute residual in pressure field iteration;
psum1	Auxiliary variable to '**psum11**';
store	Storage variable to facilitate over-relaxation of '**P(i)**' during iteration. Also used for similar purpose in temperature iteration;
store2	Auxiliary variable to '**store**';
residp	Residual of '**P(i)**' after iteration;
use, usw	Nodal values of '**u**' adjacent to '**U(i,k)**';
une, unw	Nodal values of '**u**' adjacent to '**U(i,k)**';
wsumold	Sum of '**W(i,k)**' for computation of residual in '**W(i,k)**' during iteration;
wsum	Auxiliary variable to '**wsumold**';

residw	Residual in '**W(i,k)**';
ae	Equivalent to 'a_E' equation (5.53);
aw	Equivalent to 'a_W' equation (5.54);
at	Equivalent to 'a_N' equation (5.55);
ab	Equivalent to 'a_S' equation (5.56);
s	Equivalent to '**S**' equation (5.61);
sp	Equivalent to 'S_p' equation (5.63);
sc	Equivalent to 'S_c' equation (5.64);
b	Equivalent to '**b**' equation (5.58);
ap	Equivalent to 'a_p' equation (5.57);
e1	Equivalent to '**E1**' equation (5.60);
e	Equivalent to '**E**' equation (5.59);
ncount2	Number of cycles in temperature field iteration;
sum	Sum of '**T(i,k)**' to compute residual;
sumold	Sum of '**T(i,k)**' from previous iteration cycle;
residt	Residual of temperature field iteration;
term10	Storage variable to facilitate under-relaxation in pressure-temperature iteration;
residp2	Residual of pressure in pressure-temperature iteration;
tmax	Maximum nodal value of temperature;
umax	Maximum nodal value of '**u**';
wmax	Maximum nodal value of '**w**';
viscmax	Maximum nodal value of viscosity;
friction	Summation of shear stress values with respect to '**DELTAX**'. This becomes the friction force per unit width of the bearing in [N/m];
friccoef	Dimensionless friction coefficient;
storetorr	Temporary storage variable for summing up the local values of '**TORR**' prior to calculating the average;
torrlocal	Local value of '**TORR**', would correspond to '**TORR(I,K)**' if this were written into the program.

The viscosity temperature relationship is approximated by equation (5.65):

$$\eta = \eta_0 e^{-\gamma T}$$

where:

η	is the dynamic viscosity of the lubricant at temperature '**T**' [Pas];
η_0	is the input viscosity of the lubricant [Pas];
γ	is the exponential viscosity constant, typically $\gamma = 0.05$ [K^{-1}].

EXAMPLE

Input data: Adiabatic pad, maximum film thickness = **0.1** [mm], minimum film thickness = **0.05** [mm], sliding velocity = **10** [m/s], pad width (i.e., length in the direction of sliding) = **0.1** [m]. Viscosity of lubricant at bearing inlet = **0.5** [Pas], exponential viscosity index γ = **0.05** [K^{-1}], specific heat of lubricant = **2000** [J/kgK], density of lubricant = **900** [kg/m³], thermal conductivity of lubricant = **0.15** [W/mK] and bearing inlet temperature is **50°C**.

Output data: Maximum pressure = **19.8239** [MPa], critical maximum temperature **115.62°C**, load capacity = **1.2494** [MN/m], maximum lubricant velocity in plane of oil film = **10.328** [m/s], maximum velocity normal to oil film = **1.4133** [mm/s], minimum viscosity = **0.018897** [Pas], friction force = **546.73** [N/m], fiction coefficient = **0.0004376** [dimensionless].

TEMPERATURE FIELD AND FRICTIONAL STRESS IN ADIABATIC PAD BEARING

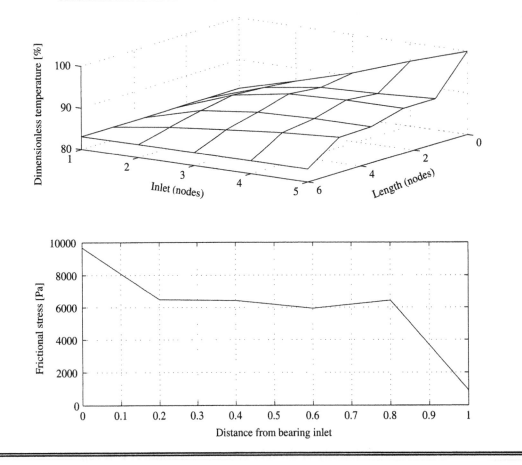

A.6 PROGRAM 'DEFLECTION'

This program calculates the load capacity and pivot point of a Michell pad that is subjected to elastic deflection. An initial film geometry is specified and the effect of elastic deflection on this film geometry is determined. An isoviscous model of hydrodynamic lubrication is assumed.

```
clc; cla reset; echo off;
clear H P D M DSTORE DHDX S INTM D1;
global tbv tbs tbp tbt tbd tbg tba tbinfo tbclose;
set([tbv tbs tbp tbt tbg tba tbinfo tbclose],'Enable','off');
% BEGIN OF INPUT DATA
% -------------------
prompt = {'Undeflected maximum film thickness [mm]:',
          'Undeflected minimum film thickness [mm]:',
          'Bearing length [m]:',
          'Lubricant viscosity [Pas]:',
          'Elastic modulus of pad material [GPa]:',
          'Pad thickness [mm]:',
          'Sliding velocity [m/s]:'};
title='INPUT DATA (DEFLECTION)'; lineno=1; def={'2', '1', '1', '1','207', '100',' 1'}; answer= inputdlg(prompt,title,lineno,def);
if size(answer) == 0, % PROGRAM IS TERMINATED
    set(tbd, 'Value', get(tbd, 'Min')); set([tbv tbs tbp tbt tbg tba tbinfo tbclose],'Enable','on'); break; end;
[hmax,hmin,length0,visc,emod,padth,u0] = deal(answer{:}); hmax = str2num(hmax); hmin = str2num(hmin);
emod = str2num(emod); length0 = str2num(length0); visc = str2num(visc); padth = str2num(padth); u0 = str2num(u0);
hmax = hmax/1000; hmin = hmin/1000; emod = emod*10^9; padth = padth/1000;
%ITERATION AND MESH PARAMETERS
prompt = {'Number of nodes in the i or x direction:',
          'Termination value of residual for pressure field iteration:',
          'Termination value of residual for pad bending iteration:',
          'Relaxation factor for pressure field iteration:',
          'Relaxation factor for pad bending iteration:',
          'Limiting number of cycles for pressure field iteration:',
          'Limiting number of cycles for pad bending iteration:'};
title='INPUT DATA (DEFLECTION)'; lineno=1; def={'11', '0.0000001', '0.000001', '1.2', '0.8','100','100'};
answer= inputdlg(prompt,title,lineno,def);
if size(answer) == 0, % PROGRAM IS TERMINATED
    set(tbd, 'Value', get(tbd, 'Min')); set([tbv tbs tbp tbt tbg tba tbinfo tbclose],'Enable','on'); break; end;
[inode,reslim1,reslim2,factor1,factor2,nlim1,nlim2] = deal(answer{:}); inode = str2num(inode); reslim1 = str2num(reslim1);
reslim2 = str2num(reslim2); factor1 = str2num(factor1); factor2 = str2num(factor2); nlim1 = str2num(nlim1);
nlim2 = str2num(nlim2);
% END OF INPUT DATA
% -----------------
subplot(1,1,1);
text('units','normalized','position',[0.2 0.55], 'FontWeight', 'bold','color',[1 0 0], 'string', 'CALCULATIONS IN PROGRESS');
figure(1);
% CALCULATE MESH PARAMETERS
deltax = length0/(inode-1);
% CALCULATE STIFFNESS OF PAD
ei = emod*padth^3/12;
% CALCULATE INITIAL FILM THICKNESS AND ASSIGN ZERO VALUES TO
% PRESSURES AND DEFLECTIONS
for i = 1:inode, H(i) = hmax-((hmax-hmin)*(i-1))/(inode-1); P(i) = 0; D(i) = 0; DSTORE(i) = 0; end;
sumd = 0;
% CALCULATE DH/DX
n2 = 0; residd = reslim2 + 10;
while (residd > reslim2) & (n2 < nlim2),
    for i = 2:inode-1, DHDX(i) = (H(i+1)-H(i-1))/(2*deltax); end;
    % ITERATE FOR PRESSURE
    n1 = 0; sum1 = 0; residp = reslim1 + 10;
    while (residp > reslim1) & (n1 < nlim1),
        sum = 0;
        for i = 2:inode-1,
            store = 0.5*(P(i+1)+P(i-1)) + ((P(i+1)-P(i-1))*deltax*0.75*DHDX(i)/H(i)) - ((3*visc*u0*DHDX(i)*deltax^2)/H(i)^3);
            P(i) = P(i) + (store-P(i))*factor1; if P(i) < 0, P(i) = 0; end; sum = sum+abs(P(i));
        end;
        n1 = n1 + 1; residp = abs((sum-sum1)/sum); sum1 = sum;
    end;
% SEARCH WITHIN THE PAD BOUNDARIES FOR A POSSIBLE CAVITATION FRONT
    icav = 0;
    for i=2:inode-1,
        if (P(i)==0 & icav==0), icav=1; end;
        if icav==1,
            hcav = H(i);
            icav=i;
        end;
    end;
    % CALCULATE TORR VALUES FROM VISCOUS REYNOLDS EQUATION
    for i=2:inode,
```

```
        TORR(i) = visc*u0/H(i) + ((P(i)-P(i-1))*H(i)*0.5)/deltax;
   end;
% SET EDGE VALUE OF TORR(i) EQUAL TO NEXT ADJACENT NODE VALUE (FAIR
% APPROXIMATION)
TORR(1) = TORR(2);
   % CALCULATE ARRAY OF SHEAR STRESSES, ALLOWING FOR A POSSIBLE CAVITATION
% FRONT (THIS IS VERY LIKELY FOR FLEXIBLE PADS)
friction = 0;
for i=2:inode,
    if (icav > 0 & i > icav),
        friction = friction + (hcav/H(i))*(TORR(i-1)+TORR(i));
    else
        friction = friction + TORR(i-1) + TORR(i);
    end;
end;
% CALCULATE DEFLECTIONS, START WITH CENTROID
int1 = 0; int2 = 0;
for i = 2:inode, int1 = int1 + (i-1)*deltax*P(i) + (i-2)*deltax*P(i-1); int2 = int2 + P(i) + P(i-1); end;
int1 = int1*deltax*0.5; int2 = int2*deltax*0.5; centroid = int1/int2; S(1) = 0; S(inode) = 0; M(1) = 0; M(inode) = 0;
for i = 2:inode,
    S(i) = 0;
    for i1 = 2:i, S(i) = S(i) +deltax*(P(i1)+P(i1-1))*0.5; end;
end;
for i = 2:inode,
    M(i) = 0;
    for i1 = 2:i, M(i) = M(i)+deltax*(S(i1)+S(i1-1))*0.5; end;
end;
% MODIFY M(i) TO ALLOW FOR STEP CHANGES AT PIVOT POINT
for i = 2:inode,
    x = (i-1)*deltax;
    if x >= centroid, term1 = (x-centroid)*int2; M(i) = M(i)-term1; end;
end;
% FIND i POSITION TO THE LEFT OF CENTROID
for i = 2:inode-1,
    x = (i-1)*deltax; if x < centroid, icentrd = i; end;
end;
% CALCULATE DEFLECTIONS ASSUMING FREE ENDS
% FIND INTEGRALS OF M(i)
INTM(1) = 0;
for i = 2:inode,
    INTM(i) = 0;
    for i1 = 2:i, INTM(i) = INTM(i) + M(i1) + M(i1-1); end;
    INTM(i) = INTM(i)*deltax*0.5;
end;
INTM2(1) = 0;
for i = 2:inode,
    INTM2(i) = 0;
    for i1 = 2:i, INTM2(i) = INTM2(i) + INTM(i1) + INTM(i1-1); end;
    INTM2(i) = INTM2(i)*deltax*0.5;
end;
% FIND INTEGRAL VALUES AT CENTROID BY LINEAR INTERPOLATION
% FIRST INTEGRAL
 deltax1 = centroid - ((icentrd-1)*deltax);
intmcen = INTM(icentrd) + (deltax1/deltax)*(INTM(icentrd+1)-INTM(icentrd));
intm2cn = INTM2(icentrd) + (deltax1/deltax)*(INTM2(icentrd+1) - INTM2(icentrd));
% BOUNDARY CONSTANTS
 c1 = (-intmcen)/ei; c2 = (-intm2cn)/ei + centroid*intmcen/ei;
% CALCULATE VALUES OF DEFLECTION, INITIAL VALUES
 for i = 1:inode, D1(i) = (INTM2(i)/ei) + c1*(i-1)*deltax + c2; end;
% CALCULATE VALUES OF DEFLECTION WITH RELAXATION
 sum = 0;
 for i = 1:inode, D(i) = DSTORE(i) + factor2*(D1(i)-DSTORE(i)); sum = sum + abs(D(i)); DSTORE(i) = D(i); end;
 n2 = n2 + 1; residd = abs((sum-sumd)/sum); sumd = sum;
% NEW VALUES OF FILM THICKNESS
 for i = 1:inode, H(i) = hmax - (hmax-hmin)*(i-1)/(inode-1) + D(i); end;
 % INEQUALITY TO REPEAT OR TERMINATE ITERATION
end;
% FIND MAXIMUM NODAL PRESSURE
pmax = 0; imax = 0;
for i=1:inode,
  if P(i) > pmax, pmax = P(i); imax = i; end;
end;
```

```
imax = (imax - 1)/(inode-1);
% PLOT PRESSURE [kPa]
x = 0:inode-1; x = x/(inode-1); subplot(3,1,1); plot(x,P/1000); grid on;
ylabel('Pressure [kPa]'); axis([0 1 0 (1.1*pmax)/1000]);
text('units','normalized','position',[0.35 1.05],'string','PRESSURE PLOT');
% HYDRODYMANIC FILM THICKNESS [mm]
subplot(3,1,2); plot(x,H*1000); grid on;
ylabel('Film thickness [mm]');
axis([0 1 0 ceil(1.1*max(H)*1000)]);
text('units','normalized','position',[0.2 1.05],'string','HYDRODYNAMIC FILM THICKNESS PLOT');
% SHEAR STRESS [Pa]
subplot(3,1,3); plot(x,TORR); grid on;
xlabel('Distance from bearing inlet'); ylabel('Shear stress [Pa]');
axis([0 1 0 ceil(1.1*max(TORR))]);
text('units','normalized','position',[0.35 1.05],'string',...'string',...
    'SHEAR STRESS PLOT');
clc;
% PRINT OUT VALUES OF INPUT AND OUTPUT DATA
% -------------------------------------
fprintf(' \n'); fprintf(' INPUT AND OUTPUT DATA FOR PROGRAM DEFLECTION\n'); fprintf(' \n');
fprintf(' INPUT DATA:\n');
fprintf(' Undeflected maximum film thickness = %0.5g [mm]\n',hmax*1000);
fprintf(' Undeflected minimum film thickness = %0.5g [mm]\n',hmin*1000);
fprintf(' Bearing length = %0.5g [m]\n',length0);
fprintf(' Lubricant viscosity = %0.5g [Pas]\n',visc);
fprintf(' Elastic modulus of pad material = %0.5g [GPa]\n',emod/10^9);
fprintf(' Pad thickness = %0.5g [mm]\n',padth*1000);
fprintf(' Sliding velocity = %0.5g [m/s]\n',u0);
fprintf(' \n');
fprintf(' OUTPUT DATA:\n');
fprintf(' Load = %0.5g [N/m]\n',int2);
fprintf(' Position of centroid of pressure field from bearing inlet = %8.5g [m]\n',centroid);
fprintf(' Max. nodal pressure = %0.5g[kPa] at %0.5g [m] from bearing inlet\n',pmax/1000,imax);
fprintf(' Film thickness at bearing inlet = %0.5g [mm]\n',H(1)*1000);
fprintf(' Film thickness at bearing outlet = %0.5g [mm]\n',H(inode)*1000);
fprintf(' Friction force = %0.5g [N/m]\n',friction *deltax*0.5);
fprintf(' Friction coefficient = %0.5g [Dimensionless]\n',(friction*deltax*0.5)/int2);
if (icav == 0),
    fprintf(' Cavitation is not present in this bearing\n');
else
    fprintf(' Cavitation is present in this bearing\n');
    fprintf(' Position of the cavitation front = %0.5g \n',icav/(inode-1));
end;
fprintf('\n'); fprintf(' PROGRAM DEFLECTION HAS BEEN COMPLETED\n');
set(tbd, 'Value', get(tbd, 'Min')); set([tbv tbs tbp tbt tbg tba tbinfo tbclose],'Enable','on');
```

Program Description

Program 'DEFLECTION' utilizes many features already described in the preceding programs, e.g., the method of iteration and data acquisition, and these are not discussed in this section. Program steps that are significant and particular to this problem relate to the specification of film thickness with bearing deformation and the calculation of pad deflection.

A data parameter that is specific to bearing deformation is the flexural stiffness of a pad. The stiffness is calculated from the standard bending moment formulae for a beam of rectangular cross section. The dimension of the pad perpendicular to the direction of bending, i.e., the length of the beam, is not included in the program. The reason for this is that since a bearing of nominally infinite length is being considered, quantities such as load, bending moment and stiffness are calculated relative to a unit length. The flexural stiffness is calculated in the following manner:

ei = emod*padth^3/12;

In the initial stage of the program, an undeformed pad is assumed, so that the film thickness at any position 'x' can be calculated from a linear variation of film thickness between the inlet and the outlet. The undeformed film thickness is given by:

```
H(i) = hmax-((hmax-hmin)*(i-1))/(inode-1);
```

Once the iteration for pressure is initiated, the controlling finite difference equation (5.72) is applied sequentially over all the nodal positions between inlet and outlet. The equation is applied with provision for a relaxation coefficient and calculated values are subject to a cavitation test (i.e., negative values are set to zero) in the statement following immediately after. This is performed by the following statements:

```
store = 0.5*(P(i+1)+P(i-1))+((P(i+1)-P(i-1))*deltax*0.75*DHDX(i)/ ...
        H(i))-((3*visc*u0*DHDX(i)*deltax^2)/H(i)^3);
P(i) = P(i)+(store-P(i))*factor1;
if P(i) < 0, P(i) = 0; end;
```

Once the iteration for pressure is completed, the position of the centroid of the pressure integral is computed. The centroid is designated as the pivot point and must be determined before bending moments and pad deflections can be calculated. The position of the centroid is found from the quotient of moment calculated from the bearing inlet and the pressure integral. The moment relative to the bearing inlet is the integral of '**pxdx**'. The program sequence to find the centroid consists of an initial '**for**' loop to find the two integrals using the trapezium rule followed by division of the integrals to find the centroid, i.e.:

```
for i = 2: inode,
   int1 = int1+(i-1)*deltax*P(i)+(i-2)*deltax*P(i-1); int2 = int2+P(i)+P(i-1);
end;
int1 = int1*deltax*0.5; int2 = int2*deltax*0.5; centroid = int1/int2;
```

The program then proceeds to calculate nodal values of shear forces and bending moments by applying the trapezium integration rule according to the same method as described previously. The step change in shear force at the pivot point necessitates some modification of nodal moment values derived. The shear force changes in value by an amount equal to the pivot load. For nodes lying between the centroid and the bearing outlet, the change in the computed value of moment equals the pivot load times the distance between the node and the centroid. This correction of nodal moment values is achieved in the following manner:

```
for i = 2: inode,
   x = (i-1)*deltax;
   if x >= centroid,
      term1 = (x-centroid)*int2; M(i) = M(i)-term1;
   end;
end;
```

Once the nodal bending moments have been determined, it is then necessary to find the location of the node immediately adjacent to the centroid but lying between the centroid and the bearing inlet. Knowledge of this node is necessary for calculating the pad deflection as will be shown later. The sequence of statements to find the node adjacent to the centroid is:

```
for i = 2: inode-1,
   x = (i-1)*deltax;
   if x < centroid, icentrd = i; end;
end;
```

The integral of bending moment with respect to 'x' and the double integral of moment with respect to 'x' are calculated by a sequence of statements different from those for the calculation of quantities 'M' and 'N' in the program '**THERMAL**'. While nodal values of bending moment integral can be calculated by this method, it is necessary to apply interpolation to find values of integrals at the centroid which are required for equations (5.78) and (5.79). The centroid values of bending moment integrals found by linear interpolation are calculated by applying the following steps:

> **deltax1 = centroid-((icentrd-1)*deltax);**
>
> **intmcen = INTM(icentrd)+(deltax1/deltax)*(iNTM(icentrd+1)-INTM(icentrd));**
>
> **intm2cn = INTM2(icentrd)+(deltax1/deltax)*(INTM2(icentrd+1)-INTM2(icentrd));**
>
> **BOUNDARY CONSTANTS**
>
> **c1 = -intmcen/ei; c2 = -intm2cn/ei+centroid*intmcen/ei;**

Once the constants expressed by equations (5.78) and (5.79) are known the program proceeds to the calculation of pad deflection. Nodal values of calculated pad deflections are initially assigned to a storage array '**D1(i)**'. Adopted values of pad deflection are determined from the difference between the calculated deflection and the previously established value of deflection times an under-relaxation coefficient. If this practice is not included and calculated values are used directly, numerical instability results. The cause of this is that the initial undeflected film thickness produces an excessively large hydrodynamic pressure field and consequently exaggerated values of pad deflection are calculated. If these deflection values are then added to the film thickness without under-relaxation, then the pressure field in the subsequent iteration becomes unrealistically small. A cyclic process results, with an oscillation between large and small pressure and deflection values. This process is called 'pad flapping'. This numerical instability is controlled by the following statements:

> **CALCULATE VALUES OF DEFLECTION, INITIAL VALUES**
>
> **for i = 1 : inode,**
>
>   **D1(i) = (INTM2(i)/ei)+c1*(i-1)*deltax+c2;**
>
> **end;**
>
> **CALCULATE VALUES OF DEFLECTION WITH RELAXATION**
>
> **sum = 0;**
>
> **for i = 1 : inode,**
>
>   **D(i) = DSTORE(i)+factor2*(d1(i)-DSTORE(i));**
>
>   **sum = sum+abs(D(i));**
>
>   **DSTORE(i) = D(i);**
>
> **end;**

In the last stage of deflection iteration the film thickness with the deformed pad is calculated in preparation for the next cycle of iteration. The '**for**' loop used to calculate the nodal values of film thickness is given by:

> **for i = 1 to inode,**
>
>   **H(i) = hmax-(hmax-hmin)*(i-1)/(inode-1)+D(i);**
>
> **end;**

Once these calculations are completed, a conditional '**while**' statement involving the deflection residual is used to decide whether to continue with the iteration or to proceed to output the film thickness and pressure field.

The isoviscous model, i.e., of constant viscosity, is used to find the friction coefficient. Equations 4.114 and its preceding expression, i.e.:

$$F = \int_0^L \int_0^B \left[\left(z - \frac{h}{2} \right) \frac{dp}{dx} + \frac{U\eta}{h} \right] dx\, dy$$

and

$$\frac{du}{dz} = \left(2z - h \right) \frac{1}{2\eta} \frac{dp}{dx} + \frac{U}{h}$$

are adapted to find the corresponding shear stress on the moving surface of the bearing. This means that the '$z = 0$' value of the 4.114 is selected and then multiplied by the viscosity on the basis that the shear stress equals velocity gradient multiplied by viscosity.

An array of shear stresses '**TORR(I)**' is thus compiled in a similar fashion to that performed in '**THERMAL**'. Important distinction from '**THERMAL**' is that the velocity gradient strictly applies to the surface and that the possibility of cavitation must be allowed for. With program '**DEFLECTION**', very flexible pads can be studied where a '**U**'-shaped film profile may be generated as schematically shown in Figure A.1.

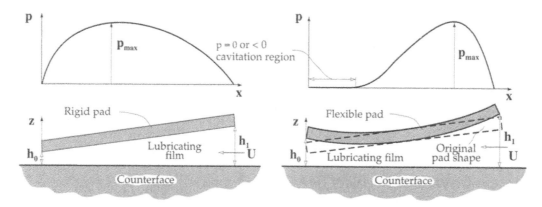

Figure A.1 Schematic illustration of the pressure distribution in a flexible pad.

If the '**U**'-shaped profile leads to cavitatation, the local friction force is reduced by a fraction equal to the film thickness where cavitation first occurs divided by the local film thickness. A switching variable '**icav**' is used to indicate the location of cavitation within the corresponding film thickness '**hcav**'. Shear stress, unlike friction force, remains unaffected by cavitation since the cavitated flow is divided between streamers of lubricant and cavitated air. From the point of view of materials failure, the shear stress generated by lubricant in the streamers is the controlling parameter.

It should be noted that this program is based on real quantities, not dimensionless quantities like those used in the programs '**PARTIAL**' and '**GROOVE**'.

List of Variables

P	Hydrodynamic pressure [Pa];
S	Shear force [N];
M	Bending moment [Nm];
H	Film thickness [m];
DHDX	Derivative of film thickness '**dh/dx**';

D	Deflection of pad under bending [m];
D1	Provisional value of displacement used during iteration;
DSTORE	Storage of displacement to facilitate relaxation;
INTM	Integral of '**M'dx**';
INTM2	Double integral of '**M'dx**';
TORR(I)	Array of shear stress values along the '**x**' axis from pad inlet to exit;
hmax	Specified maximum film thickness [m];
hmin	Specified minimum film thickness [m];
length0	Length of bearing in direction of sliding [m];
visc	Lubricant viscosity [Pas];
emod	Elastic modulus of pad material [Pa];
padth	Thickness of pad [m];
u0	Sliding velocity [m/s];
inode	Number of nodes in '**i**' or '**x**' direction;
factor1	Relaxation factor for pressure field iteration;
factor2	Relaxation factor for pad bending iteration;
nlim1	Limiting number of cycles for pressure field iteration;
nlim2	Limiting number of cycles for pad bending iteration;
reslim1	Termination value of residual for pressure field iteration;
reslim2	Termination value of residual for pad bending iteration;
deltax	Mesh dimension in '**x**' direction;
ei	Pad stiffness, equivalent to '**EI**', where '**E**' is the elastic modulus and '**I**' is the second moment of inertia;
i	Position along mesh in '**i**' or '**x**' direction;
inode	Number of nodes in '**i**' or '**x**' direction;
sumd	Sum of '**D(i)**' to compute residual of '**D(i)**';
n2	Number of cycles of iteration for '**D(i)**';
n1	Number of cycles of iteration for '**P(i)**';
sum1	Sum of '**P(i)**' to find residual of '**P(i)**';
sum	Auxiliary variable to '**SUM1**';
int1	Term to find integral of '**pxdx**';
int2	Term to find integral of '**pdx**';
centroid	Position of centroid of pressure field from bearing inlet;
i1	Auxiliary counter to '**i**';
x	Equivalent to '**x**' distance from bearing inlet;
term1	Term used to compensate for step change in shear force at bearing pivot point;
deltax1	Increment of '**x**' between precise position of centroid and adjacent node in the direction of bearing inlet;
intmcen	Interpolated value of first integral of '**M'dx**' at centroid;

intm2cn	Interpolated value of double integral of 'M'dx' at centroid;
c1, c2	Equivalent to 'C_1' and 'C_2' (equations 5.78 and 5.79);
sum	Sum of '$D(i)$', used to find residual of '$D(i)$';
residd	Residual of '$D(i)$';
icav	Marker variable to determine the location of the cavitation front;
hcav	Film thickness where the cavitation front is first detected;
friction	Friction force per unit width of the bearing [N/m].

EXAMPLE

Input data: Undeflected maximum film thickness = 2 [mm], undeflected minimum film thickness = 1 [mm], bearing width (i.e., length in the direction of sliding) = 1 [m], lubricant's viscosity = 1 [Pas], elastic modulus of pad material = 207 [GPa], pad thickness = 30 [mm], sliding velocity 1 [m/s].

Output data: Load = 54258 [N/m], position of centroid of pressure field from bearing inlet = 0.43901 [m], maximum nodal pressure = 100.74 [kPa] at 0.4 [m] from bearing inlet, film thickness at bearing inlet = 2.3679 [mm], film thickness at bearing outlet = 1.552 [mm], friction force = 652.11 [N/m], friction coefficient = 0.012019 [dimensionless], cavitation is not present in this bearing.

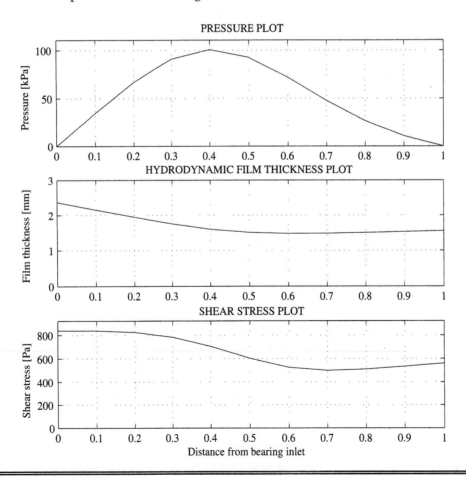

A.7 PROGRAM 'GROOVE'

This program calculates the dimensionless load and lubricant flows for a journal bearing with two axial grooves positioned at 90° to the load line. An isoviscous model of lubrication and a perfectly rigid bearing are assumed. The side flow from the bearing and the distribution of lubricant consumption between the grooves are calculated.

```
clc; cla reset; echo off;
global tbv tbs tbp tbt tbd tbg tba tbinfo tbclose;
set([tbv tbs tbp tbt tbd tba tbinfo tbclose],'Enable','off');
% BEGIN OF INPUT DATA
% --------------------
prompt = {'Eccentricity ratio:',
          'L/D ratio:',
          'Misaligment parameter from interval [0,0.5]:',
          'Relative groove width [m]:',
          'Groove subtended angle [°]:',
          'Dimensionless groove pressure:',};
title='INPUT DATA (GROOVE)'; lineno=1; def={'0.7', '1', '0', '0.4', '36', '0.05'}; answer= inputdlg(prompt,title,lineno,def);
if size(answer) == 0, % PROGRAM IS TERMINATED
    set(tbg, 'Value', get(tbg, 'Min')); figure(1); set([tbv tbs tbp tbt tbd tba tbinfo tbclose],'Enable','on'); break; end;
[epsilon,loverd,t,groovy,groovx,groovp] = deal(answer{:}); epsilon  = str2num(epsilon); loverd = str2num(loverd);
t = str2num(t); groovy = str2num(groovy); groovx = str2num(groovx); groovx = groovx*pi/180; groovp = str2num(groovp);
prompt = {'Number of nodes in the i or x* direction:',
          'Number of nodes in the j or y* direction:',
          'Terminating value of residual for iter. to solve Vogelpohl equation:',
          'Terminating value of residual for iter. to find attitude angle:',
          'Relaxation factor of iter. to solve Vogelpohl equation:',
          'Relaxation factor of iter. to find attitude angle:',
          'Max number of cycles during iter. to solve Vogelpohl equation:',
          'Max number of cycles during iter. to find attitude angle:'};
title='INPUT DATA (GROOVE)'; lineno=1; def={'21', '11', '0.000001', '0.0001', '1.2','0.8','200','100'};
answer= inputdlg(prompt,title,lineno,def);
if size(answer) == 0, % PROGRAM IS TERMINATED
    set(tbg, 'Value', get(tbg, 'Min')); figure(1); set([tbv tbs tbp tbt tbd tba tbinfo tbclose],'Enable','on'); break; end;
[inode,jnode,reslim1,reslim2,factor1,factor2,nlim1,nlim2] = deal(answer{:});
inode = str2num(inode); jnode = str2num(jnode); reslim1 = str2num(reslim1); reslim2 = str2num(reslim2);
factor1 = str2num(factor1); factor2 = str2num(factor2); nlim1 = str2num(nlim1); nlim2 = str2num(nlim2);
% END OF INPUT DATA
% -----------------
subplot(1,1,1);
text('units','normalized','position',[0.2 0.55], 'FontWeight', 'bold','color',[1 0 0], 'string', 'CALCULATIONS IN PROGRESS');
figure(1);
% SET MESH CONSTANTS
slender = 0.5/loverd; deltax = 2*pi/(inode-1); deltay = 1/(jnode-1); grad = deltax/deltay;
% VALUE OF GRAD WITH ROUND-OFF ALLOWANCE
grad1 = 0.999*grad;
% MIDDLE POSITION IN J-AXIS
jmid = round(0.5*(jnode-1)) + 1;
% INITIALIZE VALUES OF P AND SWITCH1 ARRAY
SWITCH1 = zeros(inode,jnode); P = zeros(inode,jnode);
% ESTABLISH GROOVE GEOMETRY
i1 = round(groovx/deltax); j1 = round(groovy/deltay); ig1 = ((inode - 1)/4) - i1 + 1; ig2 = ((inode - 1)/4) + i1 + 1;
ig3 = ((inode - 1)*3/4) - i1 + 1; ig4 = ((inode - 1)*3/4) + i1 + 1; jg1 = ((jnode - 1)/2) - j1 + 1; jg2 = ((jnode - 1)/2) + j1 + 1;
for j = jg1:jg2,
    for i = ig1:ig2,
        P(i,j) = groovp; SWITCH1(i,j) = 1;
    end;
end;
for j = jg1:jg2,
    for i = ig3:ig4,
        P(i,j) = groovp; SWITCH1(i,j) = 1;
    end;
end;
% INITIALIZE VALUES OF M(i,j)
M = zeros(inode+1,jnode);
% SET INITIAL VALUES OF OFFSET ANGLE AND ITERATION COUNTER
beta = 0; n2 = 0; betas = 0; residb = reslim2 + 1;
while (residb > reslim2) & (n2 < nlim2),
    n2 = n2 + 1;
```

```
% CALCULATE FILM THICKNESSES AND F AND G COEFFICIENTS
for i = 1:inode+1,
    x = (i-1)*deltax; theta = x - beta;
    for j = 1:jnode,
        y = (j-1)*deltay - 0.5; h0 = y*t*cos(x) + epsilon *cos(theta) + 1; dhdx = -y*t*sin(x) - epsilon*sin(theta);
        dhdy = t*cos(x); d2hdx2 = -y*t*cos(x) - epsilon*cos(theta); d2hdy2 = 0;
        H(i,j) = h0; G(i,j) = dhdx/h0^1.5;
        F(i,j) = 0.75*(dhdx^2 + (slender*dhdy)^2)/h0^2 + 1.5*(d2hdx2 + d2hdy2*slender^2)/h0;
    end;
end;
for i = 1:inode,
    for j = 1:jnode,
        if SWITCH1(i,j) == 1, M(i,j) = groovp*H(i,j)^1.5; end;
    end;
end;
% FINITE DIfFERENCE COEFFICIENTS
 coeff1 = 1/deltax^2; coeff2 = (slender/deltay)^2;
% INITIALIZE CAVITATION & REFORMATION FRONTS VARIABLES
for j = 1:jnode, ICAV(j) = inode; IREF(j) = 2; end;
n1 = 0; sum2 = 0; residp = reslim1 + 1;
while (residp > reslim1) & (n1 < nlim1)
if (n1 == 1) & (n2 == 1), MNEW = M; ICAVNEW = ICAV; end;
    sum = 0;
  % PERFORM MESH SWEEPS FROM DOWNSTREAM OF UPSTREAM GROOVE
    for l = 1:2,
        if l == 1,
            i20 = ig2 + 1; i21 = inode;
        elseif l == 2,
            i20 = 2; i21 = ig2;
        end;
        for i = i20:i21,
            for j100 = 1:jmid-1,
                for l100 = 1:2,
                    if l100 == 1,
                        j = jmid + j100 - 1;
                    elseif l100 == 2,
                        j = jmid - j100 + 1;
                    end;
                    if (j100 ~= 1) | (l ~= 2)
                        % REFORMATION FRONT GEOMETRY VARIABLE, ASSIGN
                        % NON-FUNCTIONING VALUE SWITCH1 VARIABLE TO REMOVE
                        % GROOVE AREAS FROM ITERATION
                        if SWITCH1(i,j) ~= 1,
                            % FINITE DIfFERENCE EQUATION
                            store = ((M(i+1,j)+M(i-1,j))*coeff1 + (M(i,j+1) + M(i,j-1))*coeff2 - G(i,j))/(2*coeff1 + 2*coeff2 + F(i,j));
                            % CALCULATED VALUE OF M(i,j) ALLOWING FOR RELAXATION
                            M(i,j) = M(i,j) + factor1*(store-M(i,j));
                            % ALLOW FOR CAVITATION
                            if (M(i-1,j) == 0) & (M(i,j) < 0), M(i,j) = 0; end;
                            if (M(i,j) < 0), M(i,j) = 0; end;
                            % DEFINE CAVITATION FRONT
                            if (M(i-1,j) > 0) & (M(i,j) < 0),
                                ICAV(j) = i; M(i,j) = 0;
                            end;
                            if M(i,j) ~= 0,
                              % POSITIVE VALUES OF M(i,j) BETWEEN REFORMATION AND
                              % CAVITATION FRONTS EXCLUDED FROM REFORMATION TEST
                                if (M(i,j) < 0) | (i > ICAV(j)) | (i < IREF(j)),
                                    if (ICAV(j) > ig1) | (i < ig4) | (i < IREF(j)) | (M(i,j) < 0),
                                        % APPLY REFORMATION CONDITION
                                        i10 = ICAV(j); hcav = H(i10,j);
                                        term1 = (H(i,j)-hcav)/H(i,j)^3;
                                            if term1 >= 0,
                                        % SEARCH FOR REFORMATION FRONT
                                        % FIND REFORMATION FRONT FROM j = 1 side
                                            jref1 = 0;
                                            for j2 = 2:jnode-1,
                                                if M(i-1,j2) ~= 0,
                                                    if (M(i-1,j2) > 0) & (jref1 == 0), jref1 = j2; end;
                                                end;
                                            end;
                                        % FIND REFORMATION FRONT FROM j = jnode SIDE
```

```
            jref2 = 0;
           for j2 = jnode-1:-1:2,
             if M(i-1,j2) ~= 0,
               if (M(i-1,j2) > 0) & (jref2 == 0), jref2 = j2; end;
             end;
           end;
         % CHOOSE VALUE OF JREF
         if jref1 > j,
           jref = jref1;
         elseif jref2 < j,
           jref = jref2;
         elseif jref1 == j,
           jref = jref1;
         elseif jref2 == j,
           jref = jref2;
         elseif (jref1 == 0) & (jref2 == 0),
           jref = 0;
         else
           jref = j;
         end;
     % DEFINE ANGLE OF REFORMATION FRONT
       % ALLOW FOR SIGMA = pi/2
       if jref ~= j,
         tansig = grad/(jref-j);
         % ALLOW FOR SIGMA = 0
         if jref == 0, tansig = 0; end;
       % STEEP CURVATURES OF REFORMATION FRONT
         tansig1 = abs(tansig);
         if tansig1 >= grad1,
           ilim = round(inode/5); isign = 1;
           for i100 = 1:ilim,
             i101 = i - i100;
             if isign ~= 0,
               if i101 < 1, i101 = i101 + inode - 1; end;
               if M(i101,j+1) > 0,
                 tansig = grad*i100;
               else
                 isign = 0;
               end;
             end;
           end;
           isign = 1;
           for i100 = 1:ilim,
             i101 = i - i100;
             if isign ~= 0,
               if i101 < 1 , i101 = i101 + inode - 1; end;
               if M(i101,j-1) > 0,
                 tansig = -grad*i100;
               else
                 isign = 0;
               end;
             end;
           end;
         end;
     % FIND PRESSURES for REFORMATION INEQUALITY
       p1 = M(i,j)/H(i,j)^1.5; p2 = M(i+1,j)/H(i+1,j)^1.5;
       p3 = M(i,j+1)/H(i,j+1)^1.5; p4 = M(i,j-1)/H(i,j-1)^1.5;
       if tansig < 0,
         term2 = ((p2-p1)/deltax) + ((p1-p4)/deltay)*tansig*slender^2;
       elseif tansig > 0,
         term2 = ((p2-p1)/deltax) + ((p3-p1)/deltay)* tansig*slender^2;
       elseif tansig == 0,
         term2 = ((p2-p1)/deltax);
       end;
       tansig = 0;
     % REFORMATION INEQUALITY
       if term2 < term1, M(i,j) = 0; end;
       if (M(i-1,j) == 0) & (M(i,j) > 0), IREF(j) = i; end;
     end;
   end;
 end;
end;
```

```
              end;
            end;
          sum = sum + abs(M(i,j));
          if (i == ig2) & (j > jg1) & (j < jg2),
             IREF(j) = ig1;
          elseif (i == ig4) & (j > jg1) & (j < jg2),
             IREF(j) = ig3;
          elseif (i == ig2) & (j == jg1),
             IREF(j) = ig1;
          elseif (i == ig2) & (j == jg2),
             IREF(j) = ig1;
          elseif (i == ig4) & (j == jg1),
             IREF(j) = ig3;
          elseif (i == ig4) & (j == jg2),
             IREF(j) = ig3;
          elseif (i == ig2) & (j > jg1) & (j < jg2),
             ICAV(j) = inode;
          elseif (i == ig4) & (j > jg1) & (j < jg2),
             ICAV(j) = inode;
          elseif (i == ig2) & (j == jg1),
             ICAV(j) = inode;
          elseif (i == ig2) & (j == jg2),
             ICAV(j) = inode;
          elseif (i == ig4) & (j == jg1),
             ICAV(j) = inode;
          elseif (i == ig4) & (j == jg2),
             ICAV(j) = inode;
          end;
        end;
      end;
    end;
  end;
end;
% CHECK FOR SERRATION ON REFORMATION FRONT OF UPSTREAM GROOVE
CHECK = zeros(1,inode);
for i = ig2:-1:1
  for j = 2:jnode-1
    if CHECK(i) ~= 2,
      if (M(i,j) <= 0) & (M(i,j-1) > 0), CHECK(i) = 1; end; if (CHECK(i) == 1) & (M(i,j) > 0), CHECK(i) = 2; end;
    end;
  end;
end;
% CHECK WHETHER SERRATION HAS OCCURRED
iserrate = 0;
for i = ig2:-1:1,
  if iserrate ~= 1,
    if CHECK(i) > 1,
      iserrate = 1;
    elseif CHECK(i) == 1,
      iserrate = 0;
    elseif CHECK(i) == 0,
      iserrate = 0;
    end;
  end;
end;
% NON-APPLICATION OF ANTI-SERRATION WHEN NO SERRATION DETECTED
if iserrate ~= 0,
  % REMOVE SERRATIONS
  for i = ig2:-1:1,
    if CHECK(i) ~= 0,
      for j = 2:jnode - 1
        % EXCLUDE GROOVES FROM NEUTRALIZATION
        if SWITCH1(i,j) ~= 1,
          imid =((inode-1)/2) + 1;
          if ICAV(j) >= imid - 1,
            if (i < ICAV(j)), M(i,j) = 0; end;
          else
            if (ICAV(j) < imid-1) & (i > ICAV(j)), M(i,j) = 0; end;
          end;
        end;
      end;
    end;
  end;
```

```
        end;
     end;
     if n1 == 0, MNEW1 = M; end;
     % END OF SERRATION NEUTRALIZATION
     % EQUALIZE M(i,j) AT x = 0 & x = 2 pi & AT i = 2 & inode+1
     for j = 1:jnode, M(1,j) = M(inode,j); M(inode+1,j) = M(2,j); end;
     % SUPPRESS ZERO PRESSURES AT CENTRE OF UPSTREAM GROOVE
     for i=2:ig1-1
        if (M(i,jmid) < M(i,jmid-1)) & (M(i,jmid) < M(i,jmid+1)), M(i,jmid) = 0.5*M(i,jmid-1) + 0.5*M(i,jmid+1); end;
     end;
     residp = abs((sum-sum2)/sum); sum2 = sum; n1 = n1 + 1;
  end;
  % COMPUTE CENTRE OF LOAD, FIND VALUES OF P(i,j)
  for i = 1:inode,
     for j = 1:jnode,
        if SWITCH1(i,j) == 0, P(i,j) = M(i,j)/H(i,j)^1.5; end;
     end;
  end;
  % DOUBLE INTEGRATION FIND Y INTEGRALS
  for i = 1:inode,
     INTY(i) = 0;
     for j = 2:jnode,
        INTY(i) = INTY(i) + P(i,j) + P(i,j-1);
     end;
     INTY(i) = INTY(i)*0.5*deltay;
  end;
  % INTEGRATE Y-INTEGRALS IN X-DIRECTION, ALLOWING FOR LOAD ANGLE
  axialw = 0; transw = 0;
  for i = 2:inode,
     x = (i-1)*deltax; x2 = (i-2)*deltax; axialw = axialw - cos(x)*INTY(i) - cos(x2)*INTY(i-1);
     transw = transw + sin(x)*INTY(i) + sin(x2)*INTY(i-1);
  end;
  axialw = axialw*deltax*0.5; transw = transw*deltax*0.5; loadw = (axialw^2 + transw^2)^0.5;
  % CALCULATE OFFSET ANGLE
  attang = atan(transw/axialw);
  if axialw > 0,
     attang1 = attang;
  elseif axialw < 0,
     attang1 = -attang;
  end;
  beta = beta + factor2*attang; residb = abs((beta-betas)/beta); betas = beta;
end;
% SEARCH FOR MAXIMUM PRESSURE
pmax = max(max(abs(P)));
% CALCULATE OIL FLOWS, SIDE FLOW
qside1 = 0; qside2 = 0;
for i = 2:inode,
   qside1 = qside1 + (slender^2*P(i,2)*H(i,1)^3/deltay) +  (slender^2*P(i-1,2)*H(i-1,1)^3/deltay);
   qside2 = qside2 + (slender^2*P(i,jnode-1)*H(i,jnode)^3/deltay) + (slender^2*P(i-1,jnode-1)*H(i-1,jnode)^3/deltay);
end;
qside1 = qside1*deltax*0.5; qside2 = qside2*deltax*0.5; qtotal = qside1 + qside2;
% CALCULATE FLOWS AROUND GROOVES
qgroov1 = 0; qgroov2 = 0;
for j = jg1+1:jg2,
   % FLOW AROUND UPSTREAM GroovE
   qgroov1 = qgroov1 + ((P(ig1,j)-P(ig1-1,j))*H(ig1,j)^3/deltax) - H(ig1,j) + ((P(ig1,j-1)-P(ig1-1,j-1))*H(ig1,j-1)^3/deltax) -...
        H(ig1,j-1);
   qgroov1 = qgroov1 + ((P(ig2,j)-P(ig2+1,j))*H(ig2,j)^3/deltax) + H(ig2,j) + ((P(ig2,j-1)-P(ig2+1,j-1))*H(ig2,j-1)^3/deltax) +...
        H(ig2,j-1);
   % FLOW AROUND DOWNSTREAM GroovE
   qgroov2 = qgroov2 +((P(ig3,j) - P(ig3-1,j))*H(ig3,j)^3/deltax) - H(ig3,j) + ((P(ig3,j-1) - P(ig3-1,j-1))*H(ig3,j-1)^3/deltax) -...
        H(ig3,j-1);
   qgroov2 = qgroov2 + ((P(ig4,j) - P(ig4+1,j))*H(ig4,j)^3/deltax) + H(ig4,j) +((P(ig4,j-1) - P(ig4+1,j-1))*H(ig4,j-1)^3/deltax)+...
        H(ig4,j-1);
end;
qgroov1 = qgroov1*0.5*deltay; qgroov2 = qgroov2*0.5*deltay;
% CALCULATE SIDE FLOW OF GROOVES
% SIDE FLOW OF UPSTREAM GROOVE
qgroovs1 = 0;
for i = ig1+1:ig2,
   qgroovs1 = qgroovs1 + slender^2*H(i,jg1)^3*(P(i,jg1)-P(i,jg1-1))/deltay + slender^2*H(i-1,jg1)^3*(P(i-1,jg1) - ...
        P(i-1,jg1-1))/deltay;
```

```
        qgroovs1 = qgroovs1 + slender^2*H(i,jg2)^3*(P(i,jg2)-P(i,jg2+1))/deltay + slender^2*H(i-1,jg2)^3*(P(i-1,jg2) - ...
                P(i-1,jg2+1))/deltay;
end;
qgroovs1 = qgroovs1*0.5*deltax;
% SIDE FLOW OF DOWNSTREAM GROOVE
qgroovs2 = 0;
for i = ig3+1:ig4,
        qgroovs2 = qgroovs2 + slender^2*H(i,jg1)^3*(P(i,jg1)-P(i,jg1-1))/deltay + slender^2*H(i-1,jg1)^3*(P(i-1,jg1) - ...
                P(i-1,jg1-1))/deltay;
        qgroovs2 = qgroovs2 + slender^2*H(i,jg2)^3*(P(i,jg2)-P(i,jg2+1))/deltay + slender^2*H(i-1,jg2)^3*(P(i-1,jg2) - ...
                P(i-1,jg2+1))/deltay;
end;
qgroovs2 = qgroovs2*0.5*deltax;
% ADJUSTMENT FOR TRUNCATION ERROR IN FLOW INTEGRALS
qstrunc = (ig2-ig1 + 1)/(ig2-ig1); qaxtrunc = (jg2-jg1 + 1)/(jg2-jg1); qgroovs1 = qgroovs1*qstrunc;
qgroovs2 = qgroovs2*qstrunc; qgroov1 = qgroov1*qaxtrunc; qgroov2 = qgroov2*qaxtrunc;
% CALCULATE SUMS OF GROOVE FLOWS
% UPSTREAM GROOVE
q1 = qgroov1 + qgroovs1;
% DOWNSTREAM GROOVE
q2 = qgroov2 + qgroovs2;
% CALCULATE PERCENTAGE ERROR IN GROOVE FLOW CALCULATIONS
qerror = (q1 + q2 - qtotal)*100/qtotal;
% PRESSURE FIELD [kPa]
x = -180:18:180; y = 0:jnode-1; y = y/(jnode-1); colormap([0.5 0.5 0.5]); subplot(1,1,1); surfl(x,y,P'*100/pmax); grid on;
axis([min(x) max(x) 0 1 0 100]); xlabel('Degrees to load line');
ylabel('Inlet'); zlabel('Dimensionless pressure [%]');
text('units','normalized','position',[0.1 1.05],'string', 'PRESSURE FIELD FOR GROOVED JOURNAL BEARING');
% PRINT OUT VALUES OF INPUT AND OUTPUT DATA
% ----------------------------------------
fprintf(' \n'); fprintf(' INPUT AND OUTPUT DATA FOR PROGRAM GROOVE\n'); fprintf(' \n');
fprintf(' INPUT DATA:\n');
fprintf('  Eccentricity ratio = %0.5g\n',epsilon);
fprintf('  L/D ratio = %0.5g\n',loverd);
fprintf('  Misaligment parameter = %0.5g\n',t);
fprintf('  Relative groove width = %0.5g [m]\n',groovy);
fprintf('  Groove length or subtended angle = %0.5g°\n',groovx*180/pi);
fprintf('  Dimensionless groove pressure = %0.5g\n',groovp);
fprintf(' \n');
fprintf(' OUTPUT DATA:\n');
fprintf('  Dimensionless load = %0.5g\n',loadw);
fprintf('  Attitude angle = %0.5g°\n',beta*180/pi);
fprintf('  Max. dimensionless pressure = %0.5g\n',pmax);
fprintf('  Dimensionless lubricant side flow = %0.5g\n',qtotal);
fprintf('  Dimensionless up-stream groove flow = %0.5g\n',q1);
fprintf('  Dimensionless down-stream groove flow = %0.5g\n',q2);
fprintf('  Discrepancy between total groove flow and size flow = %0.5g [%]\n',qerror);
fprintf(' \n'); fprintf(' PROGRAM GROOVE HAS BEEN COMPLETED\n');
set(tbg, 'Value', get(tbg, 'Min')); set([tbv tbs tbp tbt tbd tba tbinfo tbclose tbinfo tbclose],'Enable','on');
```

Program Description

Program **'GROOVE'** begins with the routine steps of data input. Mesh spacing, **'deltax'** and **'deltay'** are calculated by dividing the total lengths of the **'x*'** and **'y*'** domains by the number of mesh steps which is equal to the number of nodes minus one. Groove geometry is then established by defining the widths and lengths of the grooves in terms of node numbers, i.e.:

i1 = round(groovx/deltax); j1 = round(groovy/deltay);

The corners of each groove are then specified by using parameters **'ig1'** and **'ig2'** to denote the up-stream and down-stream faces of the leading groove, i.e.:

ig1 = ((inode-1)/4)-i1+1; ig2 = ((inode-1)/4)+i1+1;

where:

ig1   is the **'i'** node up-stream from the load-line of the groove;

ig2   is the down-stream 'i' node of the same groove;

inode   is the number of nodes in the 'i' direction.

A similar relationship applies to the parameters 'ig3' and 'ig4' concerning the down-stream groove. In the lateral direction, 'jg1' and 'jg2' specify the ends of the grooves. A variable 'SWITCH1(i,j)' is then arranged with values of **1** and **0** only. The default value is zero and this is initially assigned to all nodes. For nodes enclosed by grooves, the value **1** is assigned to exclude those nodes from iteration. The pressure in the grooves is also set at the value '**groovp**' which is the dimensionless pressure of the groove.

Once the definition of groove geometry is completed, the program proceeds with a solution of the Vogelpohl equation as already described. Refinements to allow for the reformation equation are incorporated into the iteration sweep after the finite difference equations. The sweep pattern is also modified to prevent an asymmetrical reformation front from forming. Instead of iterating from 'j = 1' to 'j = jnode', iteration begins at the middle value of 'j' and nodes on either side of the middle node are iterated simultaneously. In the 'i' direction, sweeping begins at 'ig2+1' or immediately down-stream of the leading groove. This position ensures positive values of 'M(i,j)' which simplifies the process of discriminating between true and false positive values of 'M(i,j)' during the first sweep of an iteration.

The steps needed to deduce the location of the reformation front are written after the finite difference equations but within the nested '**for**' loops of the iteration scheme. A method of successive elimination from the most common to the most particular case is applied. The most common case is false positive nodal pressure up-stream of the reformation front. The procedure shown in Figure 5.24 is applied. The majority of nodal values are sent directly to the end of the '**for**' loop by '**if**' conditional statements. The first task in this process is to eliminate cavitated nodes. The cavitation front is located by noting that at the front the calculated value of 'M(i,j)' is negative while the node immediately up-stream has a positive value. This is controlled by the statement:

   if (M(i−1,j) > 0) & (M(i,j) < 0), ICAV(j) = i; end;

where 'ICAV(j)' describes the position of the cavitation front in terms of 'i' and 'j' nodes where the transition to zero pressure occurs.

The negative values of 'M(i,j)' are then suppressed by setting the negative value to zero, i.e.:

   if M(i,j) < 0, M(i,j) = 0; end;

This is followed by the statement:

   if M(i,j) ~= 0,

which excludes all zero values of 'M(i,j)', i.e., excludes these nodes from further scrutiny. Nodal values that have not yet been affected by the conditional statements described, i.e., positive values, are now checked to determine whether the reformation condition should be applied. If a positive value of 'M(i,j)' is exempt from reformation tests, then the node position lies down-stream of a reformation front and up-stream of a cavitation front. This condition is controlled by the statement:

   if (M(i,j) < 0) | (i > ICAV(j)) | (i < IREF(j)),

The cavitation front may straddle the solution boundary of the unwrapped film at 'x* = 2π' which requires another escape condition, i.e.:

   if (ICAV(j) > ig1) | (i < ig4) | (i < IREF(j)) | (M(i,j) < 0),

Before the first sweep of the iteration, all values of '**ICAV(j)**' and '**IREF(j)**' are set to '**inode**' and **2**, respectively. This allows the first values of '**M(i,j)**' to escape elimination. Positive values of '**M(i,j)**' which have not been excluded from further examination are now subjected to the reformation condition. The first step of this process is to calculate the term '$(h^* - h^*_{cav})/h^{*3}$', which becomes:

> **i10 = ICAV(j);**
> **hcav = H(i10,j);**
> **term1 = (H(i,j)-hcav)/H(i,j)^3;**

where:

> **H(i,j)**   is the local film thickness;
>
> **hcav**   is the film thickness at the same '**j**' nodal position on the cavitation front immediately up-stream;
>
> **i10, term1**   are the storage variables.

The above statements are followed by a condition that if 'h^*' is less than 'h^*_{cav}', then the reformation condition is not applicable, i.e.:

> **if term1 >= 0,**

In the next step a value of '**tanϕ**' is deduced. The method adopted for this purpose involves searching along the '**i-1**' row for the first positive value of '**M**'. The search is performed from both ends of the row, i.e., from '**j = 1**' and from '**j = jnode**'. A subsidiary '**for**' loop in terms of '**j2**' as opposed to '**j**' is set up and the following condition applied:

> **if (M(i-1,j2) > 0) & (jref1 == 0), jref1 = j2; end;**

Two '**for**' loops are applied to perform these searches because the shape of the reformation front determines whether the first positive value of '**M(i-1,j)**' detected is or is not adjacent to the node under examination as illustrated in Figure A.2.

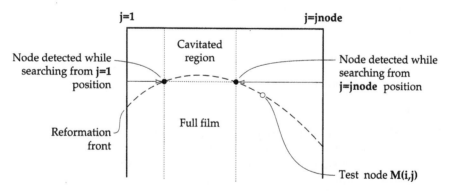

FIGURE A.2   Method for determining the shape of the reformation front.

As shown in Figure A.1, the position of the reformation front is searched for by sequentially examining whether nodal pressures are positive from '**j = 1**' to '**jnode**' for a constant '**k**' value. The next task is to select which value of '**jref**', '**jref1**' or '**jref2**' is appropriate. The deciding condition is that if '**tanϕ**' is positive then '**jref**' or the value of '**j**' on the reformation front at '**i-1**' will be greater than '**j**'. If '**tanϕ**' is positive then the value of '**jref**' found by searching from '**j = 1**' is valid, i.e., '**jref1**'. The converse applies if '**tanϕ**' is negative. This leads to the following conditional statements:

```
if jref1 > j,
    jref = jref1;
elseif jref2 < j,
    jref = jref2;
elseif jref1 == j,
    jref = jref1;
elseif jref2 == j,
    jref = jref2;
elseif (jref1 == 0) & (jref2 == 0),
    jref = 0;
else
    jref = j;
end;
```

The last conditional statement allows for the fact that the '(i,j)' node may lie at the apex of the reformation front. If 'jref = j' then the reformation front lies parallel to the direction of 'x*' and 'tanϕ' equals infinity. In this case the reformation condition is automatically satisfied and the program moves onto the next node, i.e.:

```
if jref ~= j,
```

The value of 'tanϕ' can now be calculated. A new term 'grad' defined as 'grad = deltax/deltay' enables 'tanϕ' to be expressed as:

```
tansig = grad/(jref-j);
```

Another possibility that cannot be excluded is of values of 'ϕ' greater than 'tan⁻¹grad' ('ϕ' is represented as 'grad' in the program). This condition is found when the reformation front is sharply curved. Large values of 'grad' are found by searching along a 'j' column until a positive value of 'M' is detected. The principle behind the search for large values of 'tanϕ' is shown in Figure A.3.

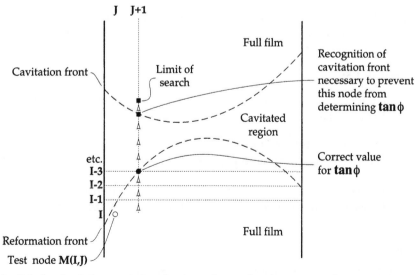

FIGURE A.3 Method of determining large values of reformation front curvature.

Both the 'j+1' column and the 'j-1' column are tested in this manner. A search is made in an up-stream direction for a distance no greater than one-fifth of the circumference of the bearing. This range easily exceeds the limit of values of 'tanφ' found in practice. Small values of 'tanφ' are excluded by the statements:

```
tansig1 = abs(tansig);
if tansig1 >= grad1;
```

For values of 'tanφ' not affected by the 'if' conditional statement, the search proceeds as follows:

```
ilim = round(inode/5);
isign = 1;
for i100 = 1 : ilim,
  i101 = i-i100;
  if isign ~= 0,
    if i101 < 1, i101 = i101+inode-1; end;
    if M(i101,j+1) > 0,
      tansig = grad*i100;
    else
      isign = 0;
    end;
  end;
end;
```

A similar search is made in the 'j-1' column with exactly the same statements except that 'j+1' is substituted by 'j-1'. The principle behind the search is the definition of 'tanφ' as a multiple of 'deltax/deltay' since the computed slope of the reformation front is a multiple of 'δx*' and 'δy*'. The process is repeated for increasing values of 'i100' until cavitation is detected by a zero value in 'M(i101,j+1)'. Once cavitation is found then any further modification of 'tanφ' is prevented by the control variable 'isign'.

On completion of the compilation of a value for 'tanφ', the values of 'P(i,j)', 'P(i+1,j)' and 'P(i,j+1)' or 'P(i,j-1)' can be calculated and used to test the reformation inequality. These values of 'P' are found from the equation:

```
P(i,j) = M(i,j)/H(i,j)^1.5;
```

In the program, the terms 'P(i,j)', etc., are not specifically used, instead local variables are used to simplify the statements, i.e.:

```
p1 = M(i,j)/H(i,j)^1.5; p2 = M(i+1,j)/H(i+1,j)^1.5;
p3 = M(i,j+1)/H(i,j+1)^1.5; p4 = M(i,j-1)/H(i,j-1)^1.5;
```

where the last term 'p4' is used when 'tanφ' is negative. The definition of the reformation condition then follows in three versions to allow for negative, zero or positive values of 'tanφ', i.e.:

```
if tansig < 0,
  term2 = ((p2-p1)/deltax)+((p1-p4)/deltay)*tansig*slender^2;
elseif tansig > 0,
```

```
    term2 = ((p2-p1)/deltax)+((p3-p1)/deltay)*tansig*slender^2;
elseif tansig == 0,
    term2 = ((p2-p1)/deltax);
end;
```

The reformation inequality can now be applied to determine the validity of the calculated positive value of 'M(i,j)':

```
if term2 < term1, M(i,j) = 0; end;
```

If a positive value of 'M' satisfies the reformation criterion then the position of the reformation node is found by applying the following condition:

```
if M(i-1,j) == 0 & M(i,j) > 0, IREF(j) = i; end;
```

The program now returns to the main iteration with the statement:

```
sum = sum+abs(M(i,j);
```

which allows collection of the iteration sweep residual. Before moving onto the end of the 'for' loop the values of 'ICAV(j)' and 'IREF(j)' on the down-stream faces of the grooves are reset. 'ICAV(j)' is reset to 'inode' and 'IREF(j)' is reset to 'ig1' when 'i = ig2' and reset to 'ig3' when 'i = ig4'. Resetting is confined to the values 'jg1 < j < jg2'. This practice ensures that positive values of 'M(i,j)' down-stream of a groove are not subjected to the reformation condition unless a cavitation front lies between the node and the groove immediately up-stream of the node.

An additional refinement, which is not necessary for the partial arc bearing program, is required after the calculation of 'M(i,j)'. It is often necessary to forestall the establishment of physically unrealistic pressures at the computed reformation front. The problem lies in the fact that a deeply indented or serrated front is also predicted by the reformation inequality in certain cases. To resolve this problem the serrations are detected by checking whether a sequence of zero and non-zero values of 'M(i,j)' occur in any 'i' row of nodes and then all non-zero 'M(i,j)' values up-stream of the deepest serration are eliminated ('M(i,j)' values in grooves are excluded from the procedure).

Outside of the iteration for 'M(i,j)', the program reverts to the form applied in the analysis of the partial arc bearing. Integration of pressure to find the attitude angle is performed and a revised value of attitude angle is calculated for a second level of iteration to ensure that the load-line bisects the bearing at '$x^* = \pi$'. With low values of groove pressure, the interaction between attitude angle and pressure field is weak so that rapid convergence of attitude angle is possible. When the lubricant supply pressure to the groove exceeds the maximum hydrodynamic pressure, convergence of attitude angle is poor or, in some cases, impossible.

After completion of both levels of iteration, lubricant flows are calculated around the groove and along the sides of the bearing. The terms in the integrals (5.85) and (5.86) are calculated based on the difference in pressure between the groove or bearing edge and an adjacent node. For example, to compute the side flow of one edge of the bearing the following statements, based on the trapezium integration rule of '$h^{*3} (\partial p^*/\partial x^*)$', are applied:

```
qside1 = 0;
for i = 2 : inode,
    qside1 = qside1+(slender^2*P(i,2)*H(l,1)^3/deltay+(slender^2*P(i-1,2)*H(i-1,1)^3/deltay);
end;
qside1 = qside1*deltax*0.5;
```

where:

qside1	is the dimensionless side flow on one side of the bearing;
slender	is the ratio '**R/L**' of bearing radius and length;
deltax, deltay	are node step lengths in the '**x***' and '**y***' directions, respectively;
P(i,2)	is the nodal pressure adjacent to the edge of the bearing.

For computing purposes, there should be at least **3** nodes enclosed by each groove in both the '**x***' and '**y***' directions. For particularly narrow grooves, the mesh should include a sufficient number of nodes to meet this condition. The trapezium form of numerical integration is used in all calculations. Flow quantities are divided into side flows from the bearing '**qside1**' and '**qside2**', net axial flows from the up-stream and down-stream grooves '**qgroov1**' and '**qgroov2**' and the side flows from the up-stream and down-stream grooves '**qgroovs1**' and '**qgroovs2**', respectively. To ensure reasonable agreement between the computed oil inflow through the grooves and outflow through the sides of the bearing, an allowance for truncation error in the finite difference expressions for integrals is made. It is difficult to calculate '$\partial p^*/\partial x^*$' or '$\partial p^*/\partial y^*$' at the edges of the grooves because there is a step change from varying pressure to uniform pressure. It is, however, possible to obtain a pressure gradient between the groove edge and the nearest adjacent pressure node. The pressure gradient then corresponds to the position '**0.5*deltax**' or '**0.5*deltay**' from the groove edge. The closest perimeter around a groove which allows computation of flow is a perimeter which is one step-length longer and wider than the groove.

To allow for this increased perimeter, the computed groove flows '**qgroov1**', '**qgroov2**', '**qgroovs1**' and '**qgroovs2**' are multiplied by truncation factors '**qaxtrunc**' and '**qstrunc**' which are the ratios of the corresponding perimeter and groove dimensions, i.e.:

qstrunc = (ig2-ig1+1)/(ig2-ig1); qaxtrunc = (jg2-jg1+1)/(jg2-jg1);

qgroovs1 = qgroovs1*qstrunc; qgroovs2 = qgroovs2*qstrunc;

qgroov1 = qgroov1*qaxtrunc; qgroov2 = qgroov2*qaxtrunc;

A similar reasoning applies to the side flow except that the perimeter has a length equal to the circumference of the bearing and therefore the truncation factor in this case is unity and can be omitted.

The program finishes by printing all the calculated data including the non-dimensional pressure '**p***' expressed as 'p^*/p^*_{max}'.

List of Variables

Since this program is an extension of the program '**PARTIAL**', the terms common to both programs are not listed here.

SWITCH1	Switching parameter to exclude grooves from iteration;
INTY	Storage parameter for '**y***' integrals;
ICAV	Parameter to locate cavitation front;
IREF	Parameter to locate reformation front;
CHECK	Parameter to indicate serration in reformation front;
groovx	Half of subtended angle of groove in radians;
groovy	Half of dimensionless axial length of groove;
groovp	Dimensionless groove pressure;
grad	Ratio of mesh dimensions in '**x***' and '**y***' directions;

grad1	Reduced value of '**grad**' to prevent errors in conditional statements;
jmid	Mid-position in '**y***' direction on mesh;
i1	Mesh dimension in step number of groove in '**x***' direction;
j1	Mesh dimension in step number of groove in '**y***' direction;
ig1	Up-stream edge '**i**' value of up-stream groove;
ig2	Down-stream edge '**i**' value of up-stream groove;
ig3	Up-stream edge '**i**' value of down-stream groove;
ig4	Down-stream edge '**i**' value of down-stream groove;
l	Auxiliary integer variable to '**i**', used to control sequence of iterated nodes;
i20	Auxiliary integer variable to '**i**';
i21	Auxiliary integer variable to '**i**';
j100	Auxiliary integer variable to '**j**';
i10	Local substitute for '**ICAV(j)**';
hcav	Local value of film thickness at cavitation front;
term1	Difference of flow between full and cavitated conditions as required by reformation condition;
jref1	Position of reformation front adjacent to node under scrutiny, found by searching from '**j = 1**' to '**j = jnode**';
jref2	Position of reformation front, searching in opposite direction to '**jref1**';
jref	Final value of either '**jref1**' or '**jref2**';
tansig	Gradient of reformation front;
tansig1	Modulus of '**tansig**';
ilim	Range of '**i**' node positions for search of steep angle of reformation front;
i100	Variable to allow search along '**i**' node rows within a simultaneous '**for**' loop in '**i**';
i101	Similar to '**i100**' but with allowance for film boundary at '**i = 1**' or '**i = inode**';
isign	Control variable to prevent '**M(i,j)**' values of the up-stream cavitation front being treated as part of the reformation front;
p1, p2, p3, p4	'**P(i,j)**' equivalents of '**M(i,j)**' as required for determination of the reformation front;
term2	Sum of pressure terms in the reformation condition;
iserrate	Control parameter to apply anti-serration process;
imid	Node position in '**i**', midway between **1** and '**inode**';
qside1	Dimensionless side flow along '**j = 1**' side of the bearing;
qside2	Dimensionless side flow along '**j = jnode**' side of the bearing;
qtotal	Total of dimensionless side flow from the bearing;
qgroov1	Dimensionless axial flow from up-stream groove;
qgroov2	Dimensionless axial flow from down-stream groove;
qgroovs1	Dimensionless side flow from up-stream groove;

qgroovs2	Dimensionless side flow from down-stream groove;
qstrunc	Factor to allow for truncation error in calculating the side flow from the grooves;
qaxtrunc	Factor to compensate for truncation error in calculating axial flow from the grooves;
q1	Total flow from up-stream groove;
q2	Total flow from down-stream groove;
qerror	Relative error between total groove flow and bearing side flow.

EXAMPLE

Input data: Eccentricity ratio = **0.7**, L/D ratio = **1**, misalignment parameter = **0**, half of relative groove length = **0.4** (groove length = **0.8** of bearing length), half of groove length or subtended angle = **36°**, dimensionless groove pressure = **0.05**.

Output data: Dimensionless load = **0.52462**, attitude angle = **29.626°**, maximum dimensionless pressure = **0.80176**, dimensionless lubricant side flow = **1.3614**, dimensionless up-stream groove flow = **0.57353**, dimensionless down-stream groove flow = **0.9591**, discrepancy between total groove flow and side flow = **12.58%**.

PRESSURE FIELD FOR GROOVED JOURNAL BEARING

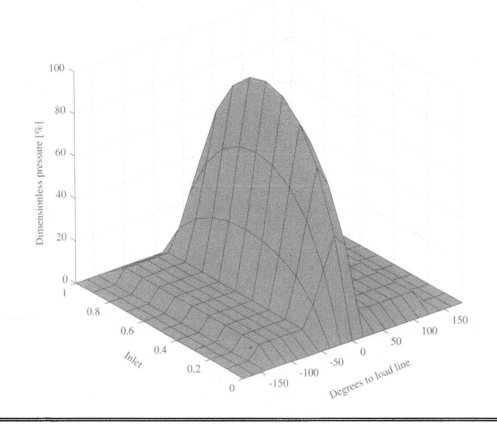

A.8   PROGRAM 'STABILITY'

This program calculates the load capacity and non-dimensional critical angular velocity of a partial arc bearing. The stiffness and damping coefficients in directions parallel and normal to the load line are also determined. Uniform viscosity and perfect rigidity of the bearing are assumed. The controlling parameters are the same as for the program for a non-vibrating partial arc bearing with the exception that an additional parameter, the dimensionless exciter mass, is also required. Vibration stability as a function of eccentricity, 'L/D' ratio, angle of partial arc bearing and misalignment ratio are calculated by this program.

```
clc; cla reset; echo off; warning off;
global reslim1 nlim1 inode jnode SWITCH1 M G F coeff1 coeff2 sum2;
global SUMY P deltay deltax axialw transw alpha factor1 residp;
clear global reslim1 nlim1 inode jnode SWITCH1 M G F coeff1 coeff2 sum2;
clear global SUMY P deltay deltax axialw transw alpha factor1 residp;
global tbv tbs tbp tbt tbd tbg tba tbinfo tbclose;
set([tbv tbs tbp tbt tbd tbg tbinfo tbclose],'Enable','off');
% BEGIN OF INPUT DATA
% ------------------
prompt = {'Eccentricity ratio:',
          'L/D ratio:',
          'Arc bearing angle [°]:',
          'Misaligment parameter from interval [0,0.5]:',
          'Dimensionless exciter mass:'};
title='INPUT DATA (STABILITY)'; lineno=1; def={'0.7', '1', '120', '0.3','0.1'}; answer= inputdlg(prompt,title,lineno,def);
if size(answer) == 0, % PROGRAM IS TERMINATED
   set(tba, 'Value', get(tba, 'Min')); set([tbv tbs tbp tbt tbd tbg tbinfo tbclose],'Enable','on'); break; end;
[epsilon,loverd,alpha,t,gamma] = deal(answer{:}); epsilon = str2num(epsilon); loverd = str2num(loverd);
alpha = str2num(alpha); t = str2num(t); gamma = str2num(gamma); alpha = alpha*pi/180;
prompt = {'Number of nodes in the i or x direction:',
          'Number of nodes in the j or y direction:',
          'Terminating value of residual for iter. to solve Vogelpohl equation:',
          'Terminating value of residual for iter. to find attitude angle:',
          'Relaxation factor of iter. to solve Vogelpohl equation:',
          'Relaxation factor of iter. to find attitude angle:'};
title='INPUT DATA (STABILITY)'; lineno=1; def={'11', '11', '0.0000001', '0.00001', '1.2','1'};
answer= inputdlg(prompt,title,lineno,def);
if size(answer) == 0, % PROGRAM IS TERMINATED
   set(tba, 'Value', get(tba, 'Min')); set([tbv tbs tbp tbt tbd tbg tbinfo tbclose],'Enable','on'); break; end;
[inode,jnode,reslim1,reslim2,factor1,factor2] = deal(answer{:}); inode = str2num(inode); jnode = str2num(jnode);
reslim1 = str2num(reslim1); reslim2 = str2num(reslim2); factor1 = str2num(factor1); factor2 = str2num(factor2);
prompt = {'Max number of cycles during iter. to solve Vogelpohl equation:',
          'Max number of cycles during iter. to find attitude angle:',
          'Incremental displacement in x direction:',
          'Incremental displacement in y direction:',
          'Incremental squeeze velocity in x direction of the above axes:',
          'Incremental squeeze velocity in y direction of the above axes:'};
title='INPUT DATA (STABILITY)'; lineno=1; def={'100','30','0.001','0.001','0.001','0.001'};
answer= inputdlg(prompt,title,lineno,def);
if size(answer) == 0, % PROGRAM IS TERMINATED
   set(tba, 'Value', get(tba, 'Min')); set([tbv tbs tbp tbt tbd tbg tbinfo tbclose],'Enable','on'); break; end;
[nlim1,nlim2,dx,dy,w0x,w0y] = deal(answer{:}); nlim1 = str2num(nlim1); nlim2 = str2num(nlim2);
dx = str2num(dx); dy = str2num(dy); w0x = str2num(w0x); w0y = str2num(w0y);
% END OF INPUT DATA
% -----------------
subplot(1,1,1);
text('units','normalized','position',[0.2 0.55], 'FontWeight', 'bold','color',[1 0 0], 'string', 'CALCULATIONS IN PROGRESS');
figure(1); deltax = alpha/(inode-1); deltay = 1/(jnode-1); slender = 0.5/loverd;
% INITIALIZE VALUES OF M(i,j), SWITCH1(i,j) & P(i,j)
M = zeros(inode,jnode); SWITCH1 = zeros(inode,jnode); P = zeros(inode,jnode);
% SET INITIAL VALUE OF OFFSET ANGLE
beta = 0;
% ENTER ATTITUDE ANGLE ITERATION CYCLE, CALCULATE H,F & G VALUES
n2 = 0; betas = 0; residb = reslim2 + 10;
while (residb > reslim2) & (n2 < nlim2),
   n2 = n2+1;
   for i = 1:inode,
      x = (i-1)*deltax + pi - 0.5*alpha; theta = x - beta;
      for j = 1:jnode,
```

```
            y = (j-1)*deltay - 0.5; h0 = y*t*cos(x) + epsilon *cos(theta) + 1; dhdx0 = -y*t*sin(x) - epsilon *sin(theta);
            d2hdx20 = -y*t*cos(x) - epsilon *cos(theta); dhdy0 = t*cos(x); d2hdy20 = 0; H(i,j) = h0; G(i,j) = dhdx0/h0^1.5;
            F(i,j) = 0.75*(dhdx0^2 + (slender*dhdy0)^2)/h0^2 + 1.5*(d2hdx20 + d2hdy20*slender^2)/h0;
            DHDX(i,j) = dhdx0; D2HDX2(i,j) = d2hdx20; DHDY(i,j) = dhdy0;
        end;
    end;
 coeff1 = 1/deltax^2; coeff2 = (slender/deltay)^2;
 % SUBROUTINE TO SOLVE THE VOGELPOHL EQUATION
 vogel_stability;
 % FIND PRESSURE FIELD FROM VOGELPOHL PARAMETER
 for i = 2:inode-1,
    for j = 2:jnode-1,
        P(i,j) = M(i,j)/H(i,j)^1.5;
    end;
 end;
 % ITERATION RESIDUAL ON ATTITUDE ANGLE ITERATION
 % SUBROUTINE TO CALCULATE TRANSVERSE AND AXIAL LOADS
 loads_stability; loadw = sqrt(axialw^2 + transw^2); attang = atan(transw/axialw);
 if axialw > 0, attang1 = attang; end; if axialw < 0, attang1 = -attang; end;
 beta = beta + factor2*attang1; residb = abs((beta-betas)/beta); betas = beta;
end;
% STABILIZATION OF M(i,j) FIELD BEFORE COMPUTATION OF STIFFNESS AND
% DAMPING COEFFICIENTS *** FIX LOCATION OF CAVITATION FRONT USING SWITCH1(i,j)
for i = 2:inode-1,
  for j = 2:jnode-1,
    if M(i,j) == 0, SWITCH1(i,j) = 1; end;
  end;
end;
% CALCULATE VALUES OF F(i,j) AND G(i,j) USING FINAL VALUE OF beta
for i = 1:inode,
  x = (i-1)*deltax + pi - 0.5*alpha; theta = x - beta;
    for j = 1:jnode,
            y = (j-1)*deltay - 0.5; h0 = y*t*cos(x) + epsilon *cos(theta) + 1; dhdx0 = -y*t*sin(x) - epsilon *sin(theta);
            d2hdx20 = -y*t*cos(x) - epsilon *cos(theta); dhdy0 = t*cos(x); d2hdy20 = 0; H(i,j) = h0; G(i,j) = dhdx0/h0^1.5;
            F(i,j) = 0.75*(dhdx0^2 + (slender*dhdy0)^2)/h0^2 + 1.5*(d2hdx20 + d2hdy20*slender^2)/h0;
            DHDX(i,j) = dhdx0; D2HDX2(i,j) = d2hdx20; DHDY(i,j) = dhdy0;
    end;
end;
% RE-ITERATE: REMOVE ANY CAVITATION INDUCED INSTABILITIES IN M FIELD
% SUBROUTINE TO SOLVE THE VOGELPOHL EQUATION
vogel_stability;
% SAVE VALUES OF M(i,j)
for i = 1:inode,
  for j = 1:jnode,
    MSAVE(i,j) = M(i,j);
  end;
end;
% SUBROUTINE TO CALCULATE TRANSVERSE AND AXIAL LOADS
loads_stability; loadw = sqrt(axialw^2 + transw^2); loadw1 = loadw;
% CALCULATE STIFFNESS COEFFICIENTS kxx AND kyx
for i = 1:inode,
  x = (i-1)*deltax + pi -0.5*alpha;
    for j = 1:jnode,
      h0 = H(i,j) + dx*cos(x); dhdx0 = DHDX(i,j) - dx*sin(x); d2hdx20 = D2HDX2(i,j) - dx*cos(x); G(i,j) = dhdx0/h0^1.5;
      F(i,j) = 0.75*(dhdx0^2 + (slender*DHDY(i,j))^2)/h0^2 + 1.5*d2hdx20/h0;
    end;
end;
% SUBROUTINE TO SOLVE THE VOGELPOHL EQUATION
vogel_stability;
% CALCULATE CHANGE IN FORCES
for i = 1:inode,
  x = (i-1)*deltax + pi - 0.5*alpha;
    for j = 1:jnode,
      P(i,j) = M(i,j)/(H(i,j)+dx*cos(x))^1.5- MSAVE(i,j)/H(i,j)^1.5;
    end;
end;
% SUBROUTINE TO CALCULATE TRANSVERSE AND AXIAL LOADS
loads_stability; kxx = axialw/(dx*loadw1); kyx = -transw/(dx*loadw1);
% CALCULATE STIFFNESS COEFFICIENTS kxx AND kyx
for i = 1:inode,
  x = (i-1)*deltax + pi - 0.5*alpha;
    for j = 1:jnode,
```

```
     h0 = H(i,j) + dy*sin(x); dhdx0 = DHDX(i,j) + dy*cos(x); d2hdx20 = D2HDX2(i,j) - dy*sin(x); G(i,j) = dhdx0/h0^1.5;
     F(i,j) = 0.75*(dhdx0^2 + (slender*DHDY(i,j))^2)/h0^2 + 1.5*d2hdx20/h0;
  end;
end;
% SUBROUTINE TO SOLVE THE VOGELPOHL EQUATION
vogel_stability;
% CALCULATE CHANGE IN FORCES
for i = 1:inode
   x = (i-1)*deltax + pi - 0.5*alpha;
   for j = 1:jnode, P(i,j) = M(i,j)/(H(i,j)+dy*sin(x))^1.5 - MSAVE(i,j)/H(i,j)^1.5; end;
end;
% SUBROUTINE TO CALCULATE TRANSVERSE AND AXIAL LOADS
loads_stability; kyy = -transw/(dy*loadw1); kxy = axialw/(dy*loadw1);
% CALCULATE DAMPING FORCES cxx AND cyx
for i = 1:inode,
   x = (i-1)*deltax + pi - 0.5*alpha; w = w0x*cos(x);
   for j = 1:jnode,
      G(i,j) = (DHDX(i,j) + 2*w)/H(i,j)^1.5;
      F(i,j) = 0.75*(DHDX(i,j)^2 + (slender*DHDY(i,j))^2)/H(i,j)^2 + 1.5*D2HDX2(i,j)/H(i,j);
   end;
end;
% SUBROUTINE TO SOLVE THE VOGELPOHL EQUATION
vogel_stability;
% CALCULATE CHANGE IN FORCES
for i = 1:inode,
   for j = 1:jnode, P(i,j) = M(i,j)/H(i,j)^1.5 - MSAVE(i,j)/H(i,j)^1.5; end;
end;
% SUBROUTINE TO CALCULATE TRANSVERSE AND AXIAL LOADS
loads_stability;
% CALCULATE DAMPING FORCES cxx AND cyx
cxx = axialw/(w0x*loadw1); cyx = -transw/(w0x*loadw1);
for i = 1:inode,
   x = (i-1)*deltax + pi - 0.5*alpha; w = w0y*sin(x);
   for j = 1:jnode,
      G(i,j) = (DHDX(i,j) + 2*w)/H(i,j)^1.5;
   end;
end;
% SUBROUTINE TO SOLVE THE VOGELPOHL EQUATION
vogel_stability;
% CALCULATE CHANGE IN FORCES
for i = 1:inode,
   for j = 1:jnode,
      P(i,j) = M(i,j)/H(i,j)^1.5 - MSAVE(i,j)/H(i,j)^1.5;
   end;
end;
% SUBROUTINE TO CALCULATE TRANSVERSE AND AXIAL LOADS
loads_stability;
% CALCULATE DAMPING FORCES cxx AND cyx
cyy = -transw/(w0y*loadw1); cxy = axialw/(w0y*loadw1);
% APPLY ROUTH-HURWITZ CRITERION
a1 = kxx*cyy - kxy*cyx - kyx*cxy + kyy*cxx; a2 = kxx*kyy - kxy*kyx;
a3 = cxx*cyy - cxy*cyx; a4 = kxx + kyy; a5 = cxx + cyy;
vc = a1*a3*a5^2/((a1^2 + a2*a5^2 - a1*a4*a5)*(a5 + gamma*a1)); vc = sqrt(abs(vc));
figure(1); plot(0,0); text('units','normalized','position',[0.2,0.55],'string', 'NO PLOT FOR PROGRAM STABILITY');
% PRINT OUT VALUES OF INPUT AND OUTPUT DATA
% ----------------------------------------
fprintf(' \n'); fprintf(' INPUT AND OUTPUT DATA FOR PROGRAM STABILITY\n'); fprintf(' \n');
fprintf(' INPUT DATA:\n');
fprintf('  Eccentricity ratio = %0.5g\n',epsilon);
fprintf('  L/D ratio = %0.5g\n',loverd);
fprintf('  Arc angle = %0.5g°\n',alpha*180/pi);
fprintf('  Misaligment parameter = %0.5g\n',t);
fprintf('  Dimensionless exciter mass = %0.5g\n',gamma);
fprintf(' \n');
fprintf(' OUTPUT DATA:\n');
fprintf('  Dimensionless load = %0.5g\n',loadw);
fprintf('  Attitude angle = %0.5g°\n',beta*180/pi);
fprintf('  Dimensionless critical frequency = %0.5g\n',vc);
fprintf('  Dimensionless stiffness and damping ceofficients:\n');
fprintf('   kxx = %0.5g\t kxy = %0.5g\t kyy = %0.5g\t kyx = %0.5g\n',kxx,kxy,kyy,kyx);
fprintf('   cxx = %0.5g\t cxy = %0.5g\t cyy = %0.5g\t cyx = %0.5g\n',cxx,cxy,cyy,cyx);
fprintf(' \n'); fprintf(' PROGRAM STABILITY HAS BEEN COMPLETED\n');
```

```
set(tba, 'Value', get(tba, 'Min')); set([tbv tbs tbp tbt tbd tbg tbinfo tbclose],'Enable','on');

% SUBROUTINE TO SOLVE THE VOGELPOHL EQUATION (VOGEL_STABILITY.M)
global residp reslim1 nlim1 inode jnode SWITCH1 M G F coeff1 coeff2 sum2;
sum2 = 0; n1 = 0; residp = reslim1 + 10;
while (residp > reslim1) & (n1 < nlim1),
   n1 = n1 + 1; sum = 0;
   for i = 2:inode-1,
      for j = 2:jnode-1,
         if SWITCH1(i,j) ~= 1,
            store = ((M(i+1,j) + M(i-1,j))*coeff1 + (M(i,j+1) + M(i,j-1))*coeff2 -G(i,j))/(2*coeff1 + 2*coeff2 + F(i,j));
            M(i,j) = M(i,j) + factor1*(store-M(i,j));
            if M(i,j) < 0, M(i,j) = 0; end; sum = sum + M(i,j);
         end;
         if SWITCH1(i,j) == 1,  sum = sum + M(i,j); end;
      end;
   end;
residp = abs((sum-sum2)/sum); sum2 = sum;
end;

% SUBROUTINE TO CALCULATE TRANSVERSE AND AXIAL LOADS (LOADS_STABILITY.M)
global SUMY P deltay deltax axialw transw alpha;
for i = 1:inode,
   SUMY(i) = 0;
   for j = 2:jnode,
      SUMY(i) = SUMY(i) + P(i,j) + P(i,j-1);
   end;
   SUMY(i) = SUMY(i)*0.5*deltay;
end;
axialw = 0; transw = 0;
for i = 2:inode,
   x = (i-1)*deltax + pi - 0.5*alpha; x2 = (i-2)*deltax + pi - 0.5*alpha;
   axialw = axialw - cos(x)*SUMY(i) - cos(x2)*SUMY(i-1); transw = transw + sin(x)*SUMY(i) + sin(x2)*SUMY(i-1);
end;
axialw = axialw*deltax*0.5; transw = transw*deltax*0.5;
```

Program Description

Program 'STABILITY' begins with the standard input requests for values of the eccentricity ratio, 'L/D' ratio, angle of partial arc and misalignment ratio. The dimensionless exciter mass is also required in this program. The dimensionless exciter mass refers to the rotating mass of the shaft or attached disc which provides the energy for vibration.

The program then proceeds to solve the Vogelpohl equation according to the steps adapted from the program 'PARTIAL' with a small modification by the subroutine called 'VOGEL_STABILITY.M'. This modification requires that another subroutine called 'LOADS_STABILITY.M' is needed to perform the integrations for load and the iteration procedure. Inside the iteration subroutine a variable 'SWITCH1(i,j)' is used to exempt cavitated nodes from the iteration.

This variable is only activated, i.e., non-zero values assigned to further iteration, after completion of the iteration for the static case. The purpose of 'SWITCH1(i,j)' is to prevent any oscillation of the cavitated region during calculation of stiffness and damping coefficients. Calculation of these coefficients is based on very small differences between the equilibrium pressure field and a perturbed pressure field resulting from small displacements and squeeze velocities. Accurate values of the coefficients are only obtained when the static or equilibrium pressure field is iterated to a high degree of accuracy. To meet this requirement, once the initial double iteration for pressure field and attitude angle is completed, cavitated nodes are isolated by applying the following steps:

```
for i = 2 : inode-1;
   for j = 2 : jnode-1;
      if M(i,j) == 0, SWITCH1(i,j) = 1; end;
```

```
    end;
  end;
```

Values of 'F' and 'G' in the Vogelpohl equation are then calculated with the final value of 'β' from the attitude angle iteration. The iteration for 'M(i,j)' is then repeated once more to eliminate any possible errors caused by a mobile cavitation front. When these steps are completed, it is found that accurate values of 'M(i,j)' are obtained. The final values of 'M(i,j)' are then stored in an array 'MSAVE(i,j)'.

The program then proceeds to the calculation of stiffness and damping coefficients via the introduction of infinitesimal displacements, '**dx**' and '**dy**', and squeeze velocities, '**w0x**' and '**w0y**'. The solution for 'M(i,j)' with perturbations begins with calculations of 'F(i,j)' and 'G(i,j)' with the perturbations included. In general, the perturbations refer to displacements or squeeze velocities. The displacements '**dx**', '**dy**' and squeeze velocities '**w0x**' and '**w0y**' are applied in order to find the stiffness '**kxx**', '**kyx**', '**kyy**', '**kxy**' and damping '**cxx**', '**cyx**', '**cyy**', '**cxy**' coefficients. For a given perturbation, once the values of 'F(i,j)' and 'G(i,j)' have been calculated then the subroutine for the iteration of 'M(i,j)' is applied. For example, the procedure for the calculation of the stiffness coefficients 'K_{xx}^{*}' and 'K_{yx}^{*}' is initiated by the following steps:

```
    for i = 1 : inode,
      x = (i-1)*deltax+pi-0.5*alpha;
      for j = 1 : jnode,
        P(i,j) = M(i,j)/(H(i,j)+dx*cos(x))^1.5-MSAVE(i,j)/H(i,j)^1.5;
      end;
    end;
```

The array '**P(i,j)**' is used to store the difference in pressures rather than the absolute value of pressures so that it can be read directly in the subroutine for load integration to find the stiffness or damping coefficients. The value of '**H(i,j)**' of the perturbed solution must also include the displacement, which in the example shown above is '**dy**'. If this were omitted inaccurate values of stiffness would be obtained. For squeeze perturbations, the film thickness is identical to the static case. The difference in load capacity is then obtained by applying the load integration subroutine to the values of '**P(i,j)**'. The values of stiffness coefficients, e.g., '**kxx**' and '**kxy**', are then found by applying the following steps:

```
    kxx = axialw/(dx*loadw); kyx = -transw/(dx*loadw);
```

An analogous set of statements is employed to calculate the damping coefficients. For example, the damping coefficients 'C_{xx}^{*}' and 'C_{yx}^{*}' are calculated according to the following steps:

```
    cxx = axialw/(w0x*loadw1); cyx = -transw/(w0x*loadw1);
```

Once the values of stiffness and damping coefficients are known, the program then proceeds to calculate the dimensionless critical frequency by applying equations (5.104-5.111). The program concludes with a print out of the values of dimensionless critical frequency, stiffness and damping coefficients.

List of Variables

Since this program is also an extension of program '**PARTIAL**', terms common to both programs are not listed here.

MSAVE	Storage of 'M(i,j)' obtained from static solution;
DHDX	Storage of node values of '$\partial h/\partial x$' under static conditions;
D2HDX2	Storage of node values of '$\partial^2 h/\partial x^2$' under static conditions;
DHDY	Storage of node values of '$\partial h/\partial y$' under static conditions;
SWITCH1	Control variable used to exclude cavitated nodes from iteration beyond initial static case;
gamma	Dimensionless exciter mass, equivalent to 'W/kc';
dx	Incremental displacement in 'X' direction;
dy	Incremental displacement in 'Y' direction;
w0x	Incremental squeeze velocity in 'X' direction of the above axes;
w0y	Incremental squeeze velocity in 'Y' direction of the above axes;
kxx	Equivalent to 'K^*_{xx}';
kyx	Equivalent to 'K^*_{yx}';
kyy	Equivalent to 'K^*_{yy}';
kxy	Equivalent to 'K^*_{xy}';
cxx	Equivalent to 'C^*_{xx}';
cxy	Equivalent to 'C^*_{xy}';
cyy	Equivalent to 'C^*_{yy}';
cyx	Equivalent to 'C^*_{yx}';
a1	Equivalent to 'A_1';
a2	Equivalent to 'A_2';
a3	Equivalent to 'A_3';
a4	Equivalent to 'A_4';
a5	Equivalent to 'A_5';
vc	Equivalent to 'ω^*_c'.

EXAMPLE

Input data: Eccentricity ratio = 0.7, L/D ratio = 1, arc angle = 120°, misalignment parameter = 0.3, dimensionless exciter mass = 0.1.

Output data: Dimensionless load = 0.56612, attitude angle = 31.843°, dimensionless critical frequency = 2.0589, dimensionless stiffness and damping coefficients, K^*_{xx} = 5.0482, K^*_{xy} = 3.696, K^*_{yy} = 1.1953, K^*_{yx} = -0.047477, C^*_{xx} = 7.8693, C^*_{xy} = 1.4988, C^*_{yy} = 0.92425 and C^*_{yx} = 1.4982.

INDEX

Printed and bound by CPI Group (UK) Ltd, Croydon, CR0 4YY

03/10/2024

01040339-0020